Birkhäuser

Monografie Matematyczne
Instytut Matematyczny Polskiej Akademii Nauk (IMPAN)

Volume 73
(New Series)

Founded in 1932 by
S. Banach, B. Knaster, K. Kuratowski,
S. Mazurkiewicz, W. Sierpinski, H. Steinhaus



Volumes 31–62 of the series
Monografie Matematyczne were published by
PWN – Polish Scientific Publishers, Warsaw

Pavel Plotnikov • Jan Sokołowski

# Compressible Navier–Stokes Equations

## Theory and Shape Optimization

Birkhäuser

Pavel Plotnikov
Siberian Branch of RAS
Lavrentyev Institute of Hydrodynamics
Novosibirsk
Russia

Jan Sokołowski
Institut Élie Cartan de Nancy, UMR 7502
Université de Lorraine, CNRS
Vandœuvre-lès-Nancy
France

ISBN 978-3-0348-0366-3 ISBN 978-3-0348-0367-0 (eBook)
DOI 10.1007/978-3-0348-0367-0
Springer Basel Heidelberg New York Dordrecht London

Library of Congress Control Number: 2012945409

Mathematics Subject Classification (2010): 35Q30, 35Q35, 35Q93, 49Q10, 49Q12, 49K20, 49K40, 76N10, 76N25



Printed on acid-free paper

Springer Basel AG is part of Springer Science+Business Media (www.birkhauser-science.com)

# Contents

# Preface

This book is a result of scientific collaboration, for more than ten years, between three countries, France, Russia and Poland. The main support for the joint research comes from Université de Lorraine, Institut Élie Cartan and from Centre Nationale de la Recherche Scientifique (CNRS) for the visits of Pavel I. Plotnikov in France. Also the support of the Systems Research Institute of the Polish Academy of Sciences for the visits in Poland is acknowledged. The research of Pavel I. Plotnikov was partially supported by Contract 02.740.11.0617 with the Ministry of Education and Science of the Russian Federation and by grant 10-01-00447-a from the Russian Foundation of Basic Research, while the research of Jan Sokołowski was partially supported by the project ANR-09-BLAN-0037 *Geometric analysis of optimal shapes (GAOS)* financed by the French Agence Nationale de la Recherche (ANR) and by grant N51402132/3135 of the Polish Ministerstwo Nauki i Szkolnictwa Wyższego "Optymalizacja z wykorzystaniem pochodnej topologicznej dla przepływów w ośrodkach ściśliwych" in the Systems Research Institute of the Polish Academy of Sciences. The monograph is published in the Polish mathematical series by Birkhäuser Basel. The authors are greatly indebted to Jerzy Trzeciak for his careful work on editing the monograph, and his patience.

Pavel Plotnikov
Lavrentyev Institute of Hydrodynamics and
Novosibirsk State University

Jan Sokołowski
Université de Lorraine, Institut Élie Cartan and
Systems Research Institute of the Polish Academy of Sciences

Nancy, Novosibirsk and Warsaw, March 2012

# Introduction

This monograph is devoted to the study of boundary value problems for equations of viscous gas dynamics, named compressible Navier-Stokes equations. The mathematical theory of Navier -Stokes equations is very interesting in its own right, but its principal significance lies in the central role Navier-Stokes equations now play in fluid dynamics. Most of this book concentrates on those aspects of the theory that have proven useful in applications.

The mathematical study of compressible Navier-Stokes equations dates back to the late 1950s. It seems that Serrin [121] and Nash [92] were the first to consider the mathematical questions of compressible viscous fluid dynamics. An intensive treatment of compressible Navier-Stokes equations starts with pioneering papers by Itaya [60], Matsumura & Nishida [87], Kazhikhov & Solonnikov [63], and Hoff [56] on the local theory for nonstationary problems, and by Beirão da Veiga [11, 12], Padula [103], and Novotný & Padula [98, 99] on the theory of stationary problems for small data. A global theory of weak solutions to compressible Navier-Stokes equations was developed by P.-L. Lions in 1998. These results were essentially improved, sharpened and generalized by E. Feireisl. We refer the reader to the books by Lions [80], Feireisl [34], Novotný & Straškraba [101], and Feireisl & Novotný [37] for the state of the art in the domain.

Although the theory is satisfactory in what concerns local time behavior and small data, many issues of global behavior of solutions for large data are far from being understood. There are a vast range of unsolved problems concerning questions such as regularity of solutions to compressible Navier-Stokes equations, the theory of weak solutions for small adiabatic exponents, existence theory for heat conducting fluids. However, these problems will not be our primary concern here. We are mainly interested in three problems that we describe briefly below.

*Existence theory.* This issue is important since no progress in the mathematical theory of Navier-Stokes equations can be made without answering the basic questions on their well-posedness. We focus on existence results for the inhomogeneous in/out flow problem, in particular the problem of the flow around a body placed in a finite domain. Notice that the majority of known results are related to viscous gas flows in domains bounded by impermeable walls. In/out flow problems are still poorly investigated. We refer to the paper by Novo [93], where an existence theorem was proved for constant boundary data, and recent work by Girinon [52], where the existence of a weak solution was established for convex flow domains with inlet independent of the time variable. We give an existence result in the general nonstationary case without imposing restrictions on the geometry of the flow domain and the behavior of boundary data. In contrast, the question of existence of global weak solutions to the stationary in/out flow problem remains essentially unsolved. Local strong solutions close to the uniform flow have been studied by Farwig [29] and Kweon & Kellogg [69, 72, 73]. With applications to shape optimization theory in mind, we consider the problem of the flow around a body placed in a bounded domain for small Mach and Reynolds numbers.

*Stability of solutions with respect to nonsmooth data and domain perturbations. Propagation of rapid oscillations in compressible fluids.* In compressible viscous flows, any irregularities in the initial and boundary data are transferred inside the flow domain along fluid particle trajectories. The transport of singularities in viscous compressible flows was studied by Hoff [56, 57]. In this book we discuss the propagation of rapid oscillations of the density, which can be regarded as acoustic waves. The main idea is that any rapidly oscillating sequence is associated with some stochastic field named the Young measure (see Tartar [128] and Perthame [106] for basic ideas). We establish that the distribution function of this stochastic field satisfies a kinetic equation of a special form, which leads to a rigorous model for propagation of nonlinear acoustic waves. Notice that oscillations can be induced not only by oscillations of initial and boundary data, but also by irregularities of the boundary of the flow domain.

*Domain dependence of solutions to compressible Navier-Stokes equations.* This issue is important because of applications to shape optimization theory. The latter is a branch of the general calculus of variations which deals with the shapes of geometric and physical objects instead of parameters and functions. The classic examples of shape optimization problems are the isoperimetric problem and Newton's problem of the body of minimal resistance. We refer to [126], [18], [21], [46], [54], [62], [91] for a general account of the theory and the relevant references. The first global result on domain dependence of solutions to compressible Navier-Stokes equations is due to Feireisl [33], who proved that the set of solutions to compressible Navier-Stokes equations is compact provided the set of flow domains is compact in the Kuratowski-Mosco topology and their boundaries have "uniformly small" volumes. We prove that the compactness result holds true if the set of flow domains is compact in the Kuratowski-Mosco topology, and also that some cost functionals, such as the drag and the work of hydrodynamical forces, are continuous in this topology. With applications to shape optimization in mind, we consider the shape differentiability of strong solutions and give formulae for the shape derivative of the drag functional. Let us also mention that in the incompressible case the shape differentiability of the drag functional was considered in [14], [15], [123]. Finally, we refer to Mohammadi & Pironneau [90] for the relevant references in applied shape optimization for fluids.

We also discuss the mathematical questions which are not related directly to Navier-Stokes equations. Among them are the theory of boundary value problems for transport equations and the problem of vanishing viscosity for diffusion equations with convective terms.

The basic idea of the book is to give as more details as possible and to avoid using complicated mathematical tools. In particular, we do not use compensated compactness results, the Bogovski lemma, or semigroup theory. Only the undegraduate background in mathematical analysis and elementary facts from functional analysis are assumed of the reader.

The material is organized in twelve chapters and an appendix. We start in Chapter 1 with a review of standard topics from real and functional analysis. The chapter includes, mainly without proofs, basic facts on measure and integral, functional analysis, elliptic and parabolic equations. We focus on measure theory, since the notion of Young measure is widely used throughout the book. Most of the material will be familiar to the reader and can be omitted. Possible exceptions are Section 1.3.1 containing a general formula for integration by parts in the Lebesgue-Stieltjes integral, Section 1.4 where the notion of Young measure is introduced, and Sections 1.1.2 and 1.5 devoted to interpolation theory and Sobolev spaces.

In Chapter 2 we collect the basic physical facts concerning compressible Navier-Stokes equations including the formulation of equations in a moving coordinate frame and formulae for the hydrodynamical forces and the work of these forces. In Chapter 3 we give the mathematical formulation of the main boundary value problem for compressible Navier-Stokes equations and discuss the notions of weak and renormalized solutions.

Chapters 4–11 can be considered as the core of the book. Our considerations are based on the approach developed by P.-L. Lions and E. Feireisl. The main ingredients of their method are Lions's result on weak continuity of the effective viscous flux, $L^p$-estimates of the pressure function, and the theory of the oscillation defect measure developed by E. Feireisl. Some of these results are of general character and hold true for any system of mass and impulse-momentum laws.

We assemble all such results in Chapter 4, where we consider the system of balance laws which is formulated as follows. Assume that a medium occupies a domain $\Omega \subset \mathbb{R}^d$, $d = 2, 3$. We want to find a velocity field $\mathbf{u} : \Omega \times (0, T) \to \mathbb{R}^d$ and a density function $\varrho : \Omega \times (0, T) \to \mathbb{R}^+$ satisfying the momentum and mass balance equations

$$\begin{aligned} &\partial_t(\varrho \mathbf{u}) + \operatorname{div}(\varrho\, \mathbf{u} \otimes \mathbf{u}) = \operatorname{div} \mathbb{T} + \varrho \mathbf{f} \quad \text{in } \Omega \times (0, T), \\ &\partial_t \varrho + \operatorname{div}(\varrho \mathbf{u} - \mathbf{g}) = 0 \quad \text{in } \Omega \times (0, T), \end{aligned} \tag{0.0.1}$$

where $\mathbf{f}$, $\mathbf{g}$ are given vector fields, and $\mathbb{T}$ is the stress tensor. The class of such systems includes compressible Navier-Stokes equations and their numerous modifications. At this stage we do not specify the form of the stress tensor and do not impose boundary and initial conditions on the density and the velocity. Instead we assume that they satisfy some integrability conditions. In Chapter 4 we consider the basic properties of solutions to equations (0.0.1). The results we present include elementary facts on integrability of functions with finite energy (Section 4.2), and the standard material concerning weak compactness properties of the impulse momentum $\varrho\mathbf{u}$ and the kinetic energy tensor $\varrho\mathbf{u}\otimes\mathbf{u}$ (Section 4.4). Section 4.6 is more important: we prove there the general form of P.-L. Lions's result on weak continuity of the viscous flux, which is the most important result of the mathematical theory of viscous compressible flows.

Chapter 5 deals with existence theory for the nonstationary in/out flow boundary value problem for compressible Navier-Stokes equations, which are a particular case of system (0.0.1) with the stress tensor of the form

$$\mathbb{T} = \nabla\mathbf{u} + (\nabla\mathbf{u})^\top + (\lambda - 1)\,\mathrm{div}\,\mathbf{u}\,\mathbb{I} - p(\varrho)\,\mathbb{I}, \tag{0.0.2}$$

where $\lambda$ is some constant, and $p(\varrho)$ is a monotone function such that $p(\varrho)\sim\varrho^\gamma$ at infinity. The in/out flow problem for equations (0.0.1)–(0.0.2) can be formulated as follows. Let a vector field $\mathbf{U}:\Omega\times(0,T)\to\mathbb{R}^d$ and a nonnegative function $\varrho_\infty:\Omega\times(0,T)\to\mathbb{R}$ be given. We want to find a solution of (0.0.1)–(0.0.2) satisfying the initial and boundary conditions

$$\begin{aligned} &\mathbf{u} = \mathbf{U}, \quad \varrho = \varrho_\infty \quad \text{on } \partial\Omega\times\{t=0\},\\ &\mathbf{u} = \mathbf{U} \quad \text{on } \partial\Omega\times(0,T), \quad \varrho = \varrho_\infty \quad \text{on } \Sigma_{\mathrm{in}}, \end{aligned} \tag{0.0.3}$$

where the inlet $\Sigma_{\mathrm{in}}$ is the open subset of $\partial\Omega\times(0,T)$ which consists of all points $(x,t)$ such that the vector $\mathbf{U}(x,t)$ points to the inside of $\Omega\times(0,T)$. The peculiarity of this problem is that we deal with the boundary value problem for the mass balance equations. In Chapter 5 we prove that for the adiabatic exponent $\gamma>2d$ and smooth initial and boundary data satisfying the compatibility conditions, the problem has a renormalized solution. We follow the multilevel regularization scheme proposed by E. Feireisl, but with a different regularization technique. The main ingredient of our method is the estimates of the normal derivatives of solutions to singularly perturbed transport equations (Section 5.3.8). These estimates are nontrivial and their derivation is based on Aronson-type inequalities for the heat kernels of diffusion equations with convective terms. Another essential ingredient of our method is the systematic use of Young measure theory.

Chapter 6 is of technical character. There we prove that for the solution to problem (0.0.1)–(0.0.3) constructed in Chapter 5, the pressure $p(\varrho)$ is locally integrable with some exponent greater than 1.

In Chapter 7 the results obtained are extended to the range of adiabatic exponents $(3/2,\infty)$ common for homogeneous boundary value problems. In this chapter we propose a new approach to the boundary value problems with fast oscillating boundary data and develop a theory of such problems based on the kinetic formulation of the governing equation. We deal with the sequence of solutions $(\mathbf{u}_\epsilon,\varrho_\epsilon)$, $\epsilon>0$, to problem (0.0.1)–(0.0.3) with regularized pressure functions of the form $p_\epsilon = p(\varrho)+\varepsilon\varrho^n$, and the initial and boundary data $\varrho_\infty^\epsilon$. We assume that the sequence $\varrho_\infty^\epsilon$ is only bounded, but need not converge to any limit in the strong sense. In particular this class of data includes rapidly oscillating functions of the form

$$\varrho_\infty^\epsilon = F\Big(x, t, \frac{x}{\epsilon}, \frac{t}{\epsilon}\Big),$$

where $F(x,t,y,\tau)$ is a bounded function, periodic in $y$ and $\tau$. Under these assumptions, the sequence $(\varrho_\epsilon, \mathbf{u}_\epsilon, p(\varrho_\epsilon))$ converges only weakly to some limit $(\overline{\varrho}, \mathbf{u}, \overline{p})$ as $\epsilon \to 0$. Following [128] we conclude that this limit admits a representation

$$\overline{\varrho}(x,t) = \int_{\mathbb{R}} s\, d\mu_{x,t}(s), \quad \overline{p}(x,t) = \int_{\mathbb{R}} p(s)\, d\mu_{x,t}(s), \tag{0.0.4}$$

where $\mu_{x,t}$ is a probability measure on the real line named the Young measure. It is completely characterized by the distribution function $f(x,t,s) = \mu_{x,t}(-\infty, s]$. Notice that the sequence $\varrho_\epsilon$ converges strongly if and only if the distribution function is deterministic, i.e., $f(1-f) = 0$. The basic idea underlying the method of kinetic equations (see [106]) is that the distribution function satisfies a differential relation named a kinetic equation. Usually the kinetic equation contains some undefined terms and cannot be considered as an equation in the strict sense of this word. A remarkable property of compressible Navier-Stokes equations is that in this particular case the kinetic equation can be written in closed form as

$$\partial_t f + \operatorname{div}(f\mathbf{u}) - \partial_s\Big(sf \operatorname{div}\mathbf{u} + \frac{s}{\lambda+1}\int_{(-\infty,s]} (p(\tau) - \overline{p})\, d_\tau f(x,t,\tau)\Big) = 0.$$

In Chapter 7 we derive the kinetic equation and show that, when combined with relations (0.0.4) and the momentum balance equations, it gives a closed system of integro-differential equations which describes the propagation of rapid oscillations in a compressible viscous flow. We also prove that if the data are deterministic and the function $f$ satisfies some integrability condition, then any solution to the kinetic equation satisfying some integrability conditions is deterministic. This fact is a general property of the kinetic equation and has no connection with the theory of Navier-Stokes equations. It follows that if $\varrho_\infty^\epsilon$ converges strongly, then so does $\varrho_\epsilon$.

In the next chapters we apply the kinetic equation method to the analysis of the domain dependence of solutions to compressible Navier-Stokes equations. We restrict our considerations to the problem of the flow around an obstacle placed in a fixed domain. In this problem $\Omega = B \setminus S$ is a condenser type domain, $B$ is a fixed *hold-all domain* and $S$ is a compact obstacle. It is assumed that $\mathbf{U}$ vanishes on $S \times (0,T)$.

In Chapter 8 we collect the basic facts concerning domain convergence and related questions from capacity theory. The most important is Hedberg's theorem (Theorem 8.2.22) on approximation of Sobolev functions. In this chapter we also introduce the notion of $\mathcal{S}$-convergence, which plays a key role in the next chapters. Denote by $C_S^\infty(B)$ the set of all smooth functions defined in $B$ and vanishing on $S \Subset B$. Let $W_S^{1,2}(B)$ be the closure of $C_S^\infty(B)$ in the $W^{1,2}(B)$-norm. It is clear that $W^{1,2}_S(B)$ is a closed subspace of $W^{1,2}(B)$. A sequence of compact sets $S_n \Subset B$ is said to $\mathcal{S}$-converge to $S$ if

- there is a compact set $B' \Subset B$ such that $S_n, S \subset B'$ and $S_n$ converges to $S$ in the standard Hausdorff metric;
- for any sequence $u_n \rightharpoonup u$ weakly convergent in $W^{1,2}(B)$ with $u_n \in W^{1,2}_{S_n}(B)$, the limit element $u$ belongs to $W^{1,2}_S(B)$;
- whenever $u \in W^{1,2}_S(B)$, there is a sequence $u_n \in W^{1,2}_{S_n}(B)$ with $u_n \to u$ strongly in $W^{1,2}(B)$.

We investigate in great detail the properties of $\mathcal{S}$-convergence and give examples of classes of obstacles which are compact with respect to this convergence.

In Chapter 9 we prove the central result on the domain stability of solutions to compressible Navier-Stokes equations. We show that if a sequence $S_n$ of compact obstacles $\mathcal{S}$-converges to a compact obstacle $S$ then the sequence of corresponding solutions to the in/out flow problem contains a subsequence which converges to a solution to the in/out flow problem in the limiting domain. In Chapter 10 we sharpen this result by proving that the typical cost functionals, such as the work of hydrodynamical forces, are continuous with respect to $\mathcal{S}$-convergence. As a conclusion we establish the solvability of the problem of minimization of the work of hydrodynamical forces in the class of obstacles with a given fixed volume.

Chapter 11 is devoted to the shape sensitivity analysis of the stationary boundary problem for compressible Navier-Stokes equations. Here we prove the local existence and uniqueness results for the in/out flow problem for compressible Navier-Stokes equations under the assumption that the Reynolds and Mach numbers are sufficiently small. We show the weak differentiability of solutions with respect to the shape of the flow domain and derive formulae for the derivatives and corresponding system of adjoint equations, which are of a practical interest. The results obtained are based on the theory of strong solutions to boundary value problems for transport equations which is presented in Chapter 12.

# Chapter 1

# Preliminaries

We use the following notation throughout the monograph.

A vector in $n$-dimensional Euclidean space is denoted by $\mathbf{v} = (v_1, \ldots, v_n)$, whether it is a column vector or a row vector. For a matrix $\mathbf{A} = (A_{ij})$, $i, j = 1, \ldots, n$, with real entries $A_{ij}$, $i$ is the row index and $j$ the column index, and $\mathbf{A}^\top = (A_{ji})$ stands for the transposed matrix. The product of a matrix by a column vector is denoted by $\mathbf{Av}$, and of a row vector by a matrix by $\mathbf{vA}$. The product of two matrices $\mathbf{C} = \mathbf{AB}$ is a matrix with the entries $C_{ij} = A_{ik}B_{kj}$ with the summation convention over repeated indices.

In particular, the scalar product of two vectors is $\mathbf{v} \cdot \mathbf{u} = v_i u_i$, and $(\mathbf{Av})_i = A_{ij} v_j$, while

$$\mathbf{vA} = \mathbf{A}^\top \mathbf{v}$$

with $(\mathbf{vA})_i = (\mathbf{A}^\top \mathbf{v})_i = A_{ji} v_j$.

The tensor product of two vectors $\mathbf{u}, \mathbf{v} \in \mathbb{R}^n$ is the matrix $\mathbf{A} := \mathbf{u} \otimes \mathbf{v}$ with the entries $A_{ij} := u_i v_j$, $i, j = 1, \ldots, n$. For the product of this matrix with a vector we have

$$(\mathbf{u} \otimes \mathbf{v})\mathbf{w} = \mathbf{u}(\mathbf{v} \cdot \mathbf{w}) \quad \text{and} \quad \mathbf{w}(\mathbf{u} \otimes \mathbf{v}) = (\mathbf{w} \cdot \mathbf{u})\mathbf{v}.$$

The derivatives of a scalar or a vector function with respect to the time variable are denoted by $\partial_t \mathbf{v} = \frac{\partial \mathbf{v}}{\partial t}$, and similarly for the spatial variables. There is a difference between the Jacobian of a vector function and its gradient: the Jacobian is denoted by

$$D\mathbf{v} = (\partial_{x_j} v_i) = \left[\frac{\partial \mathbf{v}}{\partial x_1}, \frac{\partial \mathbf{v}}{\partial x_2}, \frac{\partial \mathbf{v}}{\partial x_3}\right] = \begin{bmatrix} \dfrac{\partial v_1}{\partial x_1} & \dfrac{\partial v_1}{\partial x_2} & \dfrac{\partial v_1}{\partial x_3} \\ \dfrac{\partial v_2}{\partial x_1} & \dfrac{\partial v_2}{\partial x_2} & \dfrac{\partial v_2}{\partial x_3} \\ \dfrac{\partial v_3}{\partial x_1} & \dfrac{\partial v_3}{\partial x_2} & \dfrac{\partial v_3}{\partial x_3} \end{bmatrix},$$

and the gradient is its transpose

$$\nabla \mathbf{v} = D\mathbf{v}^\top = (\partial_{x_i} v_j) = [\nabla v_1, \nabla v_2, \nabla v_3] = \begin{bmatrix} \dfrac{\partial v_1}{\partial x_1} & \dfrac{\partial v_2}{\partial x_1} & \dfrac{\partial v_3}{\partial x_1} \\ \dfrac{\partial v_1}{\partial x_2} & \dfrac{\partial v_2}{\partial x_2} & \dfrac{\partial v_3}{\partial x_2} \\ \dfrac{\partial v_1}{\partial x_3} & \dfrac{\partial v_2}{\partial x_3} & \dfrac{\partial v_3}{\partial x_3} \end{bmatrix}.$$

Therefore, the nonlinear term in the Navier-Stokes equations is a vector denoted by $\varrho \mathbf{v}\nabla\mathbf{v} = (\varrho(v_j \partial_{x_j} v_i)) = [\varrho\mathbf{v}\cdot\nabla v_1, \varrho\mathbf{v}\cdot\nabla v_2, \varrho\mathbf{v}\cdot\nabla v_3]$, where we sum over the repeated indices $j = 1, 2, 3$, and $\varrho\mathbf{v}\cdot\nabla v_1$, $\varrho\mathbf{v}\cdot\nabla v_2$, $\varrho\mathbf{v}\cdot\nabla v_3$ stand for column vectors according to our convention. In general, for a function $u : \mathbb{R}^d \to \mathbb{R}$ we denote by

$$\partial^\alpha u = \frac{\partial^{|\alpha|} u}{\partial x_1^{\alpha_1} \dots \partial x_d^{\alpha_d}}$$

the partial derivative of order $|\alpha| = \alpha_1 + \dots + \alpha_d$ with the multiindex $\alpha = (\alpha_1, \dots, \alpha_d)$, for integers $\alpha_i$, $i = 1, \dots, d$.

We also use the simplified notation e.g., $\partial_x^2 \varrho$ for the collection of all the second order derivatives $\partial^2_{x_i x_j} \varrho = \dfrac{\partial^2 \varrho}{\partial x_i \partial x_j}$, $i, j = 1, \dots, d$, of a scalar function $\mathbb{R}^d \ni x \mapsto \varrho(x) \in \mathbb{R}$, and $\partial_x^k \mathbf{u}$, $k = 1, 2$, for the collections of the first order and of the second order derivatives of a vector function $\mathbb{R}^d \ni x \mapsto \mathbf{u}(x) \in \mathbb{R}^d$. For simplicity we write, e.g., $\mathbf{u} \in L^p(\Omega)$ to mean that all components $u_j$ of a vector function $\mathbf{u} = (u_1, \dots, u_d)$ belong to the space $L^p(\Omega)$, in other words $L^p(\Omega)$ stands here for $L^p(\Omega; \mathbb{R}^d)$. The same convention is used for other spaces, e.g., in our notation $(\mathbf{v}, \varphi, \zeta) \in C^{1+\gamma}(\Omega) \times C^\gamma(\Omega)^2$ means that all components of the vector function $\mathbf{v}$ belong to the space $C^{1+\gamma}(\Omega)$, and the scalar functions $\varphi, \zeta$ belong to $C^\gamma(\Omega)$.

The notation $\partial_x^2 \mathbf{u} \in L^p(\Omega)$ means that all the second order derivatives belong to $L^p(\Omega)$. In this way we avoid the notation with multiindices unless strictly necessary.

For a given symmetric tensor $\mathbb{S} = (S_{ij})$, its divergence is the vector denoted by $\operatorname{div}\mathbb{S}$ with the components $\operatorname{div}\mathbb{S}_i = \partial_{x_j} S_{ij}$, summed over $j = 1, 2, 3$. The product of two tensors is the scalar $\mathbb{A} : \mathbb{B} = A_{i,j} B_{i,j}$.

On the other hand, points in $\mathbb{R}^d$ are denoted by $x, y$ with coordinates $x = (x_i), y = (y_i)$; this is an exception from the vector notation.

## 1.1 Functional analysis

### 1.1.1 Banach spaces

We recall some well known facts; the main sources are [24] and [131]. A *normed space* $A$ is a linear space over the field of real numbers equipped with a norm

$\|\cdot\|_A : A \to \mathbb{R}$ such that

$$\|u\|_A \geq 0, \quad \|u\|_A = 0 \quad \text{if and only if} \quad u = 0,$$
$$\|\lambda u\|_A = |\lambda|\, \|u\|_A \quad \text{for all } \lambda \in \mathbb{R} \text{ and } u \in A,$$
$$\|u + v\|_A \leq \|u\|_A + \|v\|_A \quad \text{for all } u, v \in A.$$

A sequence $u_n \in A$ converges (converges strongly) to $u \in A$ if $\|u_n - u\|_A \to 0$ as $n \to \infty$. In this case we write $u_n \to u$ or $\lim_{n\to\infty} u_n = u$. A normed $A$ space is *complete* if $\|u_m - u_n\|_A \to 0$ as $m, n \to \infty$ implies the existence of $u \in A$ such that $u_n \to u$. Complete normed spaces are named *Banach spaces.*

A set $G \subset A$ is open if for any $a \in A$ there is $\varepsilon > 0$ such that the ball $\{x \in A : \|x - a\|_A < \varepsilon\}$ is contained in $A$. A set $F \subset A$ is closed if for any sequence $F \ni u_n \to u$ the limit $u$ belongs to $F$. Obviously $F$ is closed if and only if $A \setminus F$ is open. The closure of $D \subset A$ is denoted by $\operatorname{cl} D$ or $\overline{D}$. We say that a set $D \subset A$ is *dense* in a set $E \subset A$ if $E \subset \operatorname{cl} D$.

*Embedding.* We say that a Banach space $A$ is *continuously embedded* in a Banach space $B$ or that the *embedding* of $A$ in $B$ *is bounded* if $A \subset B$ and there exists $c > 0$ such that $\|u\|_B \leq c\|u\|_A$ for all $u \in A$. In this case we write $A \hookrightarrow B$.

*Product, sum and intersection.* The Cartesian product $A \times B$ of Banach spaces $A$, $B$ consists of all pairs $(u, v)$, where $u \in A$, $v \in B$, and is equipped with the norm $\|u\|_A + \|v\|_A$. Let $A$, $B$ be Banach spaces, both subsets of an ambient Banach space $Z$. Then the intersection $A \cap B$ equipped with the norm $\|u\|_{A\cap B} = \|u\|_A + \|u\|_B$ and the algebraic sum $A + B := \{w = u + v : u \in A,\ v \in B\}$ equipped with the norm

$$\|w\|_{A+B} = \inf\{\|u\|_A + \|v\|_B : u + v = w\}$$

are Banach spaces.

*Linear operators.* Let $A$, $B$ be Banach spaces. Linear mappings $T : A \to B$ are called *linear operators.* A linear operator is *bounded* if it has a finite norm

$$\|T\|_{\mathcal{L}(A,B)} := \sup_{\|u\|_A \leq 1} \|Tu\|_B = \inf\{c : \|Tu\|_B \leq c\|u\|_A \text{ for all } u \in A\}.$$

Equipped with this norm, the set $\mathcal{L}(A, B)$ of bounded linear operators becomes a Banach space.

*Duality.* The *dual space* $A'$ of a Banach space $A$ consists of all continuous linear functionals $u' : A \to \mathbb{R}$. The duality pairing between $A'$ and $A$ is defined by $\langle u', u\rangle := u'(u)$. Equipped with the norm $\|u'\|_{A'} := \sup_{\|u\|_A \leq 1} |\langle u', u\rangle|$, $A'$ becomes a Banach space. We have (see [49, Thm. 5.13])

**Theorem 1.1.1.** *Let Banach spaces $A$ and $B$ be subsets of a Banach space $Z$ and suppose $A \cap B$ is dense in $Z$. Then*

$$(A \cap B)' = A' + B', \quad (A + B)' = A' \cap B'.$$

The structure of the dual of a Cartesian product is simpler. For every element $w'$ of the dual space $(A\times B)'$, there are $u' \in A'$ and $v' \in B'$ such that $\langle w', (u,v)\rangle = \langle u', u\rangle + \langle v', v\rangle$ and $\|w'\|_{(A\times B)'} = \|u'\|_{A'} + \|v'\|_{B'}$. Hence $(A\times B)'$ can be identified with $A' \times B'$.

*Hahn-Banach theorem.* We will use the following result which is a particular case of the famous Hahn-Banach theorem ([24, Ch. II.3, Thm. 11] and [24, Ch. V.1, Thm. 12]).

**Theorem 1.1.2.** *(i) Let $B$ be a closed linear subspace of a Banach space $A$ and $v' \in B'$. Then there is $w' \in A'$ whose restriction to $B$ coincides with $v'$ and $\|w'\|_{A'} = \|v'\|_{B'}$.*

*(ii) Let $\mathcal{K}$ be a closed convex subset of a Banach space $A$ and $e \in A \setminus \mathcal{K}$. Then there is $u' \in A'$ such that $\langle u', e\rangle > \sup_{u\in\mathcal{K}}\langle u', u\rangle$.*

*Weak convergence. Weak topology.* Let $A$ be a Banach space. A sequence $u_n \in A$ is said to *converge weakly* to $u \in A$ (denoted $u_n \rightharpoonup u$) if $\langle u', u_n\rangle \to \langle u', u\rangle$ as $n \to \infty$ for all $u' \in A'$. A sequence $u'_n \in A'$ is said to *converge weakly*$^\star$ to $u' \in A'$ (denoted $u'_n \rightharpoonup u'$) if $\langle u'_n, u\rangle \to \langle u', u\rangle$ as $n \to \infty$ for all $u \in A$.

A set $G \subset A$ is said to be open in the *weak topology* if for every $u \in G$ there are finite collections of $u'_i \in A'$ and $\varepsilon_i > 0$, $i = 1, \dots, n$, such that

$$\{v \in A : |\langle u'_i, v - u\rangle| < \varepsilon_i,\ i = 1, \dots, n\} \subset G.$$

Weakly closed sets are complements of weakly open sets. For every $F, E \subset A$, a mapping $\varphi : F \to E$ is continuous in the weak topology (weakly continuous) if for every $v = \varphi(u)$ with $u \in F$, and every weakly open set $G \ni v$, there is a weakly open set $D \ni u$ such that $\varphi(w) \in G \cap E$ for all $w \in F \cap D$.

A mapping $\varphi : F \to E$ is *sequentially weakly continuous* if for any sequence $F \ni u_n \rightharpoonup u \in F$, we have $\varphi(u_n) \rightharpoonup \varphi(u)$.

A Banach space $A$ is *weakly complete* if for any sequence $u_n \in A$, $n \geq 1$, with the property

$$\langle u', u_m - u_n\rangle \to 0 \quad \text{as } m, n \to \infty \text{ for all } u' \in A',$$

there is $u \in A$ such that $u_n \rightharpoonup u$.

*Reflexivity.* Let $A$ be a Banach space. For every $u \in A$, the mapping $\varkappa_u : A' \to \mathbb{R}$ defined by $\varkappa_u(u') = \langle u', u\rangle$ is continuous and so can be regarded as an element of $A''$. Thus we get the so-called natural isometry $A \ni u \mapsto \varkappa_u \in A''$. The space $A$ is *reflexive* if the natural isometry takes $A$ onto $A''$, i.e., for every $u'' \in A''$ there exists a unique $u \in A$ such that $u'' = \varkappa_u$. In this case we can identify $A$ and $A''$.

The Eberlein-Šmulian theorem ([131, Ch. 5. 4]) shows that a reflexive Banach space is weakly complete:

**Theorem 1.1.3.** *A Banach space is weakly complete if and only if it is reflexive.*

*Weak compactness.* Let $A$ be a Banach space. A set $K \subset A$ is *weakly compact* if any family $\{G_\alpha\}$ of weakly open sets such that $K \subset \bigcup_\alpha G_\alpha$ contains a finite subfamily $G_{\alpha_i}$, $i = 1, \dots, n$, such that $K \subset \bigcup_i G_{\alpha_i}$.

A set $K$ is *sequentially weakly compact* if any sequence $\{u_n\} \subset K$ contains a subsequence $\{u_{n_k}\}$ which converges weakly to some element $u \in K$.

In Banach spaces bounded sets usually have some weak compactness properties. The following theorem ([67, Ch. 4.3, Thm. 4]) is a simplest result of this sort.

**Theorem 1.1.4.** *Let a Banach space $A$ be separable, i.e. $A$ contains a dense countable subset. Then every bounded sequence $\{u'_n\} \subset A'$ contains a subsequence $\{u'_{n_k}\}$ which converges weakly$^\star$ to some element $u' \in A'$.*

In the nonseparable case we have the following Alaoglu theorem [24, Ch. V.4, Thm. 2].

**Theorem 1.1.5.** *Let $A$ be a reflexive Banach space. Then every closed ball in $A$ is weakly compact.*

The connection between weak compactness and sequential weak compactness is established by the second Eberlein-Šmulian theorem ([24, Ch. V.6, Thm. 1]):

**Theorem 1.1.6.** *Let $A$ be a Banach space and $\mathcal{K} \subset A$. Then every sequence $u_n \in \mathcal{K}$ contains a subsequence $u_{n_k} \rightharpoonup u \in A$ if and only the weak closure of $\mathcal{K}$ is weakly compact.*

On the other hand, in view of [24, Ch. V.3, Thm. 13], every convex closed set of a Banach space is weakly closed. This result along with Theorems 1.1.5 and 1.1.6 leads to

**Theorem 1.1.7.** *Every closed ball in a reflexive Banach space $A$ is sequentially weakly compact. In particular, every bounded sequence $\{u_n\} \subset A$ contains a subsequence $\{u_{n_k}\}$ which converges weakly to some element $u \in A$.*

*Metrizability of the weak topology.* Let $A$ be a Banach space and $\mathcal{K}$ be a closed ball in $A$. The weak topology on $\mathcal{K}$ (induced weak topology) is said to be metrizable if there is a symmetric distance function $\rho : \mathcal{K} \times \mathcal{K} \to \mathbb{R}$ with the properties

- $\rho(u, v) \geq 0$, $\rho(u, v) = 0$ if and only if $u = v$,
- $\rho(u, w) \leq \rho(u, v) + \rho(v, w)$,
- a set $E \subset \mathcal{K}$ is the intersection $\mathcal{K} \cap G$, where $G$ is weakly open, if and only if for every $u \in E$, the set $E$ contains a ball $\{v \in \mathcal{K} : \rho(u, v) < \epsilon\}$, $\varepsilon > 0$.

The following theorem ([24, Ch. V.5, Thm. 2]) connects metrizability and separability.

**Theorem 1.1.8.** *If $A$ is a Banach space and $A'$ is separable, then the weak topology on every closed ball of $A$ is metrizable.*

As a consequence we get

**Lemma 1.1.9.** *Let $A$ be a Banach space such that $A'$ is separable and let $\mathcal{K}$ be a closed ball in $A$. If a mapping $\varphi : \mathcal{K} \to \mathcal{K}$ is sequentially weakly continuous, then it is weakly continuous.*

*Proof.* Let $\varphi : \mathcal{K} \to \mathcal{K}$ be sequentially weakly continuous. Assume, contrary to our claim, that it is not weakly continuous. Then there is $u \in \mathcal{K}$ and a weakly open set $V \ni \varphi(u)$ with the property that every weakly open set $G \ni u$ contains $w \in G \cap \mathcal{K}$ such that $\varphi(w)$ does not belong to $V$. Since $V$ is weakly open and $\varphi(u)$ is in $V$, there are $u_i' \in A'$ and $\varepsilon_i > 0$, $i = 1, \ldots, m$, such that

$$\{v : |\langle u_i', v - \varphi(u)\rangle| < \varepsilon_i,\, i = 1, \ldots, m\} \subset V,$$

hence

$$\max_i |\langle u_i', \varphi(w) - \varphi(u)\rangle| \geq \min \varepsilon_i =: \delta > 0.$$

Since the weak topology on $\mathcal{K}$ is metrizable, there are weakly open sets $G_n$ such that

$$G_n \cap \mathcal{K} = \{v : \rho(u, v) < 1/n\}.$$

Hence there are $w_n \in \mathcal{K}$, $n \geq 1$, such that

$$\rho(w_n, u) < 1/n \quad \text{and} \quad \max_{1 \leq i \leq m} |\langle u_i', \varphi(w_n) - \varphi(u)\rangle| \geq \delta > 0. \tag{1.1.1}$$

Let us prove that $w_n \rightharpoonup u$. Choose $u' \in A'$ and $\varepsilon > 0$. Then the set $U_\varepsilon = \{v : |\langle u_i', v - u\rangle| < \varepsilon\}$ is weakly open and contains $u$. Hence $U_\varepsilon \cap \mathcal{K}$ contains some ball $\{v \in \mathcal{K} : \rho(v, u) < r\}$. It follows that for $n \geq 1/r$, the elements $w_n$ belong to $U_\varepsilon$, which leads to $\limsup_{n\to\infty} |\langle u_i', w_n - u\rangle| \leq \varepsilon$. Since $\varepsilon$ is arbitrary, we conclude that $\langle u_i', w_n - u\rangle \to 0$ and hence $w_n \rightharpoonup u$. By the sequential continuity of $\varphi$ it follows that $\varphi(w_n) \rightharpoonup \varphi(u)$, which contradicts (1.1.1). □

### 1.1.2 Interpolation of Banach spaces

In this section we recall some results from interpolation theory (see [16] for the proofs). We only consider the so-called real interpolation method, which is different from the widely used complex interpolation method (see [16], [129]). Let $A_0$ and $A_1$ be Banach spaces, contained in some ambient normed space. For $t > 0$ introduce nonnegative functions $K : A_0 + A_1 \to \mathbb{R}$ and $J : A_0 \cap A_1 \to \mathbb{R}$ defined by

$$K(t, u, A_0, A_1) = \inf_{\substack{u = u_0 + u_1 \\ u_i \in A_i}} (\|u_0\|_{A_0} + t\|u_1\|_{A_1}),$$

$$J(t, u, A_0, A_1) = \max\{\|u\|_{A_0}, t\|u\|_{A_1}\}.$$

The function $K$ is positive, increasing and concave in $t$. For each $s \in (0, 1)$, $1 < r < \infty$, the *$K$-interpolation space* $[A_0, A_1]_{s,r,K}$ consists of all elements $u \in A_0 + A_1$

having the finite norm

$$\|u\|_{[A_0,A_1]_{s,r,K}} = \left( \int_0^\infty t^{-1-sr} K(t,u,A_0,A_1)^r \, dt \right)^{1/r}. \tag{1.1.2}$$

On the other hand, the *J-interpolation space* $[A_0, A_1]_{s,r,J}$ consists of all elements $u \in A_0 + A_1$ which admit a representation

$$u = \int_0^\infty \frac{v(t)}{t} \, dt, \quad v(t) \in A_1 \cap A_0 \quad \text{for } t \in (0,\infty), \tag{1.1.3}$$

where $v$ is Bochner integrable, and have the finite norm

$$\|u\|_{[A_0,A_1]_{s,r,J}} = \inf_{v(t)} \left( \int_0^\infty t^{-1-sr} J(t,v(t),A_0,A_1)^r \, dt \right)^{1/r} < \infty, \tag{1.1.4}$$

where the infimum is taken over the set of all $v(t)$ satisfying (1.1.3).

The first main result of interpolation theory states that for all $s \in (0,1)$ and $r \in (1,\infty)$ the Banach spaces $[A_0, A_1]_{s,r,K}$ and $[A_0, A_1]_{s,r,J}$ are isomorphic topologically and algebraically. Hence the two norms introduced are equivalent, and we can omit the indices $J$ and $K$. The following properties of interpolation spaces will be used throughout this book. The first concerns the interpolation of subspaces and directly follows from the definitions.

**Lemma 1.1.10.** *If $\tilde{A}_i$ are closed subspaces of $A_i$ for $i = 0, 1$, then $[\tilde{A}_0, \tilde{A}_1]_{s,r} \subset [A_0, A_1]_{s,r}$ and $\|u\|_{[A_0,A_1]_{s,r}} \le \|u\|_{[\tilde{A}_0,\tilde{A}_1]_{s,r}}$.*

The second lemma, which is a particular case of [16, Thm. 3.4.2], establishes the density properties of $A_1 \cap A_0$.

**Lemma 1.1.11.** *Let $0 < s < 1$ and $1 < r < \infty$. Then the linear space $A_1 \cap A_0$ is dense in $[A_0, A_1]_{s,r}$. In particular, if $A_1 \subset A_0$, then $A_1$ is dense in $[A_0, A_1]_{s,r}$ (even if $A_1$ is not dense in $A_0$).*

The following representation for the interpolation of dual spaces ([16, Thm. 3.7.1]) is important.

**Lemma 1.1.12.** *Let $A_i$, $i = 0, 1$, be Banach spaces such that $A_i$ are linear subspaces of some normed space $A$ and $A_1 \cap A_0$ is dense in $A_0$ and $A_1$ in the topology of $A$. Then the Banach spaces $[(A_0)', (A_1)']_{s,r'}$ and $([A_0, A_1]_{s,r})'$ are isomorphic topologically and algebraically. Hence the spaces can be identified with equivalent norms.*

It follows from this and Lemma 1.1.11 that if $A_1 \subset A_0$ is dense in $A_0$, then, since $A_0' \subset A_1'$, the Banach space $A_0'$ is dense in $([A_0, A_1]_{s,r})' = [(A_0)', (A_1)']_{s,r'}$ for $0 < s < 1 < r < \infty$.

The following lemma is a central result of interpolation theory.

**Lemma 1.1.13.** *Let $A_i$, $B_i$, $i = 0, 1$, be Banach spaces, such that $A_i$ are linear subspaces of some Banach space $A$, and $B_i$ are linear subspaces of some normed space $B$, and let $T : A_i \to B_i$ be a bounded linear operator. Then, for all $s \in (0,1)$ and $r \in (1,\infty)$, the operator $T : [A_0, A_1]_{s,r} \to [B_0, B_1]_{s,r}$ is bounded and*

$$\|T\|_{\mathcal{L}([A_0,A_1]_{s,r},[B_0,B_1]_{s,r})} \leq c \max\big\{\|T\|_{\mathcal{L}(A_0,B_0)}, \|T\|_{\mathcal{L}(A_1,B_1)}\big\}.$$

### 1.1.3 Invertibility of linear operators

**Lemma 1.1.14.** *Let $V$ be a reflexive Banach space, $V'$ its dual, and $\mathbb{A} : V \to V'$ a linear operator satisfying, for some $\sigma > 0$,*

$$\langle \mathbb{A}[\varphi], \varphi \rangle \geq \sigma \|\varphi\|_V^2 \quad \textit{for all } \varphi \in V.$$

*Then $\mathbb{A}$ has an inverse $\mathbb{A}^{-1} : V' \to V$, and*

$$\|\mathbb{A}^{-1}[\psi]\|_V \leq \sigma^{-1} \|\psi\|_{V'} \quad \textit{for all } \psi \in V'. \tag{1.1.5}$$

*Proof.* Clearly, there is an inverse $\mathbb{A}^{-1} : \operatorname{Im} \mathbb{A} \to V$. Moreover, for $\varphi = \mathbb{A}^{-1}[\psi]$,

$$\|\mathbb{A}^{-1}[\psi]\|_V \|\psi\|_{V'} \geq \langle \psi, \mathbb{A}^{-1}[\psi] \rangle = \langle \mathbb{A}[\varphi], \varphi \rangle \geq \sigma \|\varphi\|_V^2 = \sigma \|\mathbb{A}^{-1}[\psi]\|_V^2,$$

which gives (1.1.5). It remains to prove that $\operatorname{Im} \mathbb{A} = V'$. Since $A^{-1}$ is bounded, $\operatorname{Im} \mathbb{A}$ is a closed subspace of $V'$. If $\operatorname{Im} \mathbb{A} \neq V'$ then by the Hahn-Banach theorem 1.1.2, there is an element $0 \neq \omega \in V$ such that $\langle \psi, \omega \rangle = 0$ for all $\psi \in \operatorname{Im} \mathbb{A}$. This gives $\langle \mathbb{A}[\omega], \omega \rangle = 0$, a contradiction. □

### 1.1.4 Fixed point theorems

The following three famous theorems on solvability of nonlinear functional equations (see [68] for instance) are exploited in this book. The first is the Schauder theorem:

**Theorem 1.1.15.** *Let $K$ be a convex, bounded, closed subset of a Banach space $X$, and $\Xi : K \to K$ a continuous mapping such that*

$$\Xi(K) \textit{ is a relatively compact subset of } X.$$

*Then there is a point $x \in K$ such that $\Xi(x) = x$.*

The continuity and compactness conditions can be replaced by the following hypothesis: The mapping $\Xi$ takes each weakly convergent sequence $\{x_n\} \subset K$ to the strongly convergent sequence $\{\Xi(x_n)\} \subset K$.

The second result is the more general Leray-Schauder theorem:

**Theorem 1.1.16.** *Let $G$ be an open bounded subset of a Banach space $X$, and let $\Xi(\tau, \cdot) : \operatorname{cl} G \to X$, $\tau \in [0,1]$, be a one-parameter family of mappings that satisfies:*

- *The mapping* $\Xi : [0,1] \times \operatorname{cl} G \to X$ *is continuous.*
- *For* $\tau \in [0,1]$ *and any bounded set* $A \subset G$, *the image* $\Xi(\tau, A)$ *is a relatively compact subset of* $X$.
- $\Xi(\tau, x) \neq x$ *for all* $\tau \in [0,1]$ *and* $x \in \partial G$, *i.e., the mapping* $\Xi(\tau, \cdot)$ *has no fixed points at the boundary of* $G$.
- *The mapping* $\Xi(0, \cdot)$ *has a unique fixed point* $x_0 \in G$. *Moreover, the mapping* $I - \Xi(0, \cdot)$ *is a homeomorphism of some neighborhood of* $x_0$, *where* $I$ *stands for the identity mapping.*

*Then the mapping* $\Xi(1, \cdot)$ *has a fixed point* $x_1 \in G$, *that is,* $x_1 = \Xi(1, x_1)$.

In many applications it is useful to replace the Schauder fixed point theorem by the following Tikhonov fixed point theorem (see [24, Ch. V.10, Thm. 5]):

**Theorem 1.1.17.** *Let* $A$ *be a locally convex linear topological space and* $\mathcal{K}$ *be a convex compact set in* $A$. *Assume furthermore that a mapping* $\varphi : \mathcal{K} \to \mathcal{K}$ *is continuous in the topology of* $A$. *Then there exists* $w \in \mathcal{K}$ *such that* $w = \varphi(w)$.

For our purposes the Tikhonov theorem is applied in the following less abstract form.

**Theorem 1.1.18.** *Let* $A$ *be a reflexive Banach space such that* $A'$ *is separable. Let* $\mathcal{K} \subset A$ *be a closed ball and* $\varphi : \mathcal{K} \to \mathcal{K}$ *be a sequentially continuous mapping, i.e.,* $\varphi(u_n) \rightharpoonup \varphi(u)$ *for any* $\mathcal{K} \ni u_n \rightharpoonup u \in \mathcal{K}$. *Then there exists* $w \in \mathcal{K}$ *such that* $w = \varphi(w)$.

*Proof.* The space $A$ equipped with the weak topology (the system of weakly open sets) becomes a locally convex linear topological space. In view of Theorem 1.1.5, the closed ball $K$ is a compact convex subset of this space. It remains to note that, by Lemma 1.1.9, every sequentially continuous mapping $\varphi : \mathcal{K} \to \mathcal{K}$ is weakly continuous, and apply Theorem 1.1.17. □

## 1.2 Function spaces

### 1.2.1 Hölder spaces

**Definition 1.2.1.** Let $\Omega$ be a domain in $\mathbb{R}^d$. We denote by $C(\Omega)$ the Banach space of all bounded continuous functions $u : \Omega \to \mathbb{R}$ with the norm

$$\|u\|_{C(\Omega)} = \sup_{x \in \Omega} |u(x)|.$$

All such functions have unique continuous extensions to the closure $\operatorname{cl} \Omega$; the extensions will be denoted by the same letter,

We say that a function $u : \Omega \to \mathbb{R}$ is *compactly supported in* $\Omega$ if

$$\operatorname{supp} u = \operatorname{cl} \{x \in \Omega : u(x) \neq 0\} \Subset \Omega,$$

where $\operatorname{supp} u \Subset \Omega$ means that $\operatorname{supp} u$ is a compact subset of the open set $\Omega$. We denote by $C_c(\Omega)$ the linear space of all continuous functions $u : \Omega \to \mathbb{R}$ compactly supported in $\Omega$.

**Definition 1.2.2.** For every integer $m \geq 0$, we denote by $C^m(\Omega)$ the Banach space which consists of all functions $u$ on $\Omega$ such that

$$\partial^\alpha u \in C(\Omega) \quad \text{for all } |\alpha| \leq m \quad \text{and} \quad \|u\|_{C^m(\Omega)} = \max_{|\alpha|\leq m} \|\partial^\alpha u\|_{C(\Omega)} < \infty.$$

For every integer $m \geq 0$ and real $\beta \in (0,1)$, we denote by $C^{m+\beta}(\Omega)$ the Banach space which consists of all functions $u \in C^m(\Omega)$ with the finite norm

$$\|u\|_{C^{m+\beta}(\Omega)} = \|u\|_{C^m(\Omega)} + \sum_{|\alpha|=m} \sup_{x,y\in\Omega, x\neq y} \frac{|\partial^\alpha u(x) - \partial^\alpha u(y)|}{|x-y|^\beta}.$$

We will identify $C^0(\Omega)$ with $C(\Omega)$ and $C^{m+0}(\Omega)$ with $C^m(\Omega)$. We denote by $C^\infty(\Omega)$ the linear space of functions $u : \Omega \to \mathbb{R}$ such that all derivatives of $u$ belong to the class $C(\Omega)$. The symbol $C_0^\infty(\Omega)$ denotes the subspace of $C^\infty(\Omega)$ which consists of compactly supported functions.

### Domains

**Definition 1.2.3.** Let $\Omega \subset \mathbb{R}^d$ be a bounded domain, $m \geq 0$ be an integer and $\beta \in [0,1)$. We say that $\partial\Omega$ (or equivalently $\Omega$) belongs to the class $C^{m+\beta}$, denoted $\partial\Omega \in C^{m+\beta}$, if it has the following property. For every point $x_0 \in \partial\Omega$, there exist a neighborhood $U_{x_0}$ of $x_0$ in $\mathbb{R}^d$, a Cartesian coordinate system

$$y = (y_1, \dots, y_{d-1}, y_d) := (\overline{y}, y_d),$$

centered at $x_0$, and a function $\eta \in C^{m+\beta}(B_{d-1})$, $B_{d-1} = \{|\overline{y}| < 1\}$, such that in the new coordinates

$$U_{x_0} \cap \partial\Omega = \{y : y_d = \eta(\overline{y}),\ \overline{y} \in B_{d-1}\}. \tag{1.2.1}$$

In other words, $\partial\Omega$ is the graph of a $C^{m+\beta}$ function in a neighborhood of each of its points.

Let $\partial\Omega \in C^{m+\beta}$. Assume that an integer $l \geq 0$ and $\sigma \in [0,1)$ satisfy the inequality $l + \sigma \leq m + \beta$. We say that a function $u : \partial\Omega \to \mathbb{R}$ belongs to the class $C^{l+\sigma}(\partial\Omega)$ if for every $x_0 \in \partial\Omega$ there exists $\varphi \in C^{l+\sigma}(B_{d-1})$ such that

$$u|_{U_{x_0}\cap\partial\Omega} = \varphi(\overline{y}).$$

**Definition 1.2.4.** Let $\Omega \subset \mathbb{R}^d$ be a bounded domain. We say that $\partial\Omega$ (or equivalently $\Omega$) is *Lipschitz* if it admits a representation (1.2.1) with a Lipschitz function $\eta$ in a neighborhood of every point $x_0 \in \partial\Omega$.

### Hölder spaces of Banach space valued functions

**Definition 1.2.5.** Let $X$ be a Banach space and $T \in (0,\infty)$. We will denote by $C(0,T;X)$ the Banach space of all bounded continuous functions $u : [0,T] \to X$ with the norm

$$\|u\|_{C(0,T;X)} = \sup_{t\in[0,T]} \|u(t)\|_X.$$

We say that the mapping $u : [0,T] \to X$ is *differentiable* at $t \in (0,T)$ if there is $\partial_t u(t) \in X$ such that

$$\lim_{s\to t} \left\| \partial_t u(t) - \frac{u(s)-u(t)}{s-t} \right\|_X = 0.$$

**Definition 1.2.6.** Let $X$ be a Banach space, $T \in [0,\infty)$, and $m \geq 0$ be an integer. We denote by $C^m(0,T;X)$ the linear space of all mappings $u : [0,T] \to X$ such that for every integer $k \in [0,m]$, the mapping $\partial_t^k u : (0,T) \to X$ is continuous and has a continuous extension onto $[0,T]$. Endowed with the norm

$$\|u\|_{C^m(0,T;X)} = \max_{0\leq k\leq m} \|\partial_t^k u\|_{C(0,T;X)},$$

$C^m(0,T;X)$ becomes a Banach space. For $\beta \in [0,1)$, we denote by $C^{m+\beta}(0,T;X)$ the Banach space which consists of all functions $u \in C^m(0,T;X)$ with the finite norm

$$\|u\|_{C^{m+\beta}(0,T;X)} = \|u\|_{C^m(0,T;X)} + \sup_{s,t\in[0,T],\, t\neq s} \frac{\|\partial_t^m u(t) - \partial_s^m u(s)\|_X}{|t-s|^\beta}.$$

### Arzelà-Ascoli theorem

**Theorem 1.2.7.** *Let $\Omega \subset \mathbb{R}^d$ be a bounded domain. Let $u_n \in C(\Omega)$, $n \geq 1$, be a bounded and equicontinuous sequence, i.e., for every $\varepsilon > 0$ there is $\delta > 0$ such that*

$$|u_n(x) - u_n(y)| \leq \varepsilon \quad \text{for all } x,y \in \Omega \text{ with } |x-y| \leq \delta, \text{ and all } n \geq 1.$$

*Then there exists a subsequence $\{u_{n_k}\} \subset \{u_n\}$ and $u \in C(\Omega)$ such that $u_{n_k} \to u$ in $C(\Omega)$ as $k \to \infty$.*

For Banach space valued functions, the Arzelà-Ascoli theorem reads:

**Theorem 1.2.8.** *Let $X$ be a Banach space, $\mathcal{K} \subset X$ be compact and $T \in (0,\infty)$. Let $u_n : [0,T] \to \mathcal{K}$, $n \geq 1$, be an equicontinuous sequence, i.e., for every $\varepsilon > 0$ there is $\delta > 0$ such that*

$$\|u_n(t) - u_n(s)\|_X \leq \varepsilon \quad \text{for all } \ s,t \in [0,T] \text{ with } |s-t| \leq \delta \text{ and all } n \geq 1.$$

*Then there exists a subsequence $\{u_{n_k}\} \subset \{u_n\}$ and $u \in C(0,T;X)$ such that $u_{n_k} \to u$ in $C(0,T;X)$ as $k \to \infty$.*

**Corollary 1.2.9.** *Let $X$ and $Y$ be Banach spaces and $m+\beta>0$. Assume that the embedding $Y \hookrightarrow X$ is compact. Then every bounded sequence $u_n \in C^{m+\beta}(0,T;Y)$, $n \geq 1$, contains a subsequence which converges in $C(0,T;X)$.*

**Lemma 1.2.10.** *Let $\Omega$ be a bounded domain in $\mathbb{R}^d$ with $\partial\Omega \in C^k$. Assume that nonnegative integers $l,m$ and real $\mu,\nu \in [0,1)$ satisfy $l+\sigma < m+\beta \leq k$. Then for every $u \in C^{m+\beta}(\Omega)$,*

$$\|u\|_{C^{l+\sigma}(\Omega)} \leq c\|u\|_{C(\Omega)}^{1-\lambda}\|u\|_{C^{m+\beta}(\Omega)}^{\lambda}, \quad \lambda = (l+\sigma)/(m+\beta),$$

*where $c$ is independent of $u$.*

**Corollary 1.2.11.** *Under the assumptions of the previous lemma, the embedding $C^{m+\beta}(\Omega) \hookrightarrow C^{l+\sigma}(\Omega)$ is compact.*

### 1.2.2 Measure and integral

**General theory.** Let $X$ be a nonempty set. A collection $\mathcal{A}$ of subsets of $X$ is said to be a *$\sigma$-algebra* if the following conditions are fulfilled:

- $\emptyset \in \mathcal{A}$; if $E \in \mathcal{A}$ then $X \setminus E \in \mathcal{A}$.
- If $E_n \in \mathcal{A}$, $n \geq 1$, then $\bigcup_n E_n \in \mathcal{A}$.

Notice that if $E_n \in \mathcal{A}$, $n \geq 1$, then $\bigcap_n E_n \in \mathcal{A}$.

**Example 1.2.12.** The most important example of a $\sigma$-algebra is the algebra of Borel sets which is defined as follows. Let $\Omega$ be an arbitrary open subset of $\mathbb{R}^d$ (most often, $\Omega = \mathbb{R}^d$). The *Borel $\sigma$-algebra* or the $\sigma$-algebra of *Borel sets* $\mathcal{B}(\Omega)$ is the minimal $\sigma$-algebra in $\Omega$ containing all open subsets of $\Omega$.

**Remark 1.2.13.** For any countable collections of open sets $G_i \subset \mathbb{R}^d$ and closed sets $F_i \subset \mathbb{R}^d$, the sets

$$\bigcup_i G_i \cap \Omega, \quad \bigcap_i G_i \cap \Omega, \quad \bigcup_i F_i \cap \Omega, \quad \bigcap_i F_i \cap \Omega,$$

belong to $\mathcal{B}(\Omega)$. There is not any constructive way to recognize Borel sets among other subsets of $\Omega$.

**Measures.** A map $\mu : \mathcal{A} \to [0,\infty]$ is called a *measure* (or a nonnegative measure) if

$$\mu(\emptyset) = 0, \quad \mu\Big(\bigcup_n E_n\Big) = \sum_n \mu(E_n)$$

for every countable collection of pairwise disjoint sets $E_n \in \mathcal{A}$. A measure $\mu$ is said to be *$\sigma$-finite* if $X$ can be represented as a countable union of finite measure sets; $\mu$ is said to be *finite* if $\mu(X) < \infty$. A measure $\mu$ is a *probability measure* if $\mu(X) = 1$. The triple $(X, \mathcal{A}, \mu)$ is called a *measure space.*

**Integral.** A function $u : X \to \mathbb{R}$ is said to be *measurable with respect to* $\mathcal{A}$ if $u^{-1}(E) \in \mathcal{A}$ for every Borel set $E \subset \mathbb{R}$.

A *simple function* is a measurable function $u : X \to \mathbb{R}$ whose range consists of finitely many points, i.e.,

$$u(x) = \sum_k u_k \chi_k(x), \quad u_k \in \mathbb{R}, \quad \chi_k(x) = 1 \text{ for } x \in E_k, \quad \chi_k(x) = 1 \text{ otherwise},$$

and $\{E_k\} \subset \mathcal{A}$ is a finite collection of pairwise disjoint measurable sets covering $X$. Notice that for any measurable function $u : X \to \mathbb{R}$ there is an increasing sequence of simple functions $u^{(n)}$ such that

$$u^{(n)}(x) \le u^{(n+1)}(x), \quad \lim_{n\to\infty} u^{(n)}(x) = u(x) \quad \text{in } X. \tag{1.2.2}$$

A simple function $u \ge 0$ is *integrable* if

$$\int_X u \, d\mu := \sum_k u_k \mu(E_k) < \infty.$$

Here, we adopt the standard convention $0 \cdot \infty = 0$ for $\mu(E_k) = \infty$.

**Definition 1.2.14.** A nonnegative measurable function $u : X \to \mathbb{R}$ is *integrable* (or $\mu$-integrable) if there is a sequence of nonnegative simple functions $u^{(n)}$, $n \ge 1$, with the following properties:

$$u^{(n)}(x) \nearrow u(x) \quad \text{as } n \to \infty \quad \text{in } X,$$

the limit

$$\lim_{n\to\infty} \int_X u^{(n)} \, d\mu =: \int_X u \, d\mu$$

exists. An arbitrary measurable function $u : X \to \mathbb{R}$ is defined to be integrable if there are nonnegative integrable functions $u^{\pm}$ such that $u = u^+ - u^-$.

The following three theorems play an important role in integration theory. The first is the famous Lebesgue dominated convergence theorem.

**Theorem 1.2.15.** *Let $(X, \mathcal{A}, \mu)$ be a measure space, and let $u_n : X \to [-\infty, \infty]$ be a sequence of integrable functions such that the limit*

$$\lim_{n\to\infty} u_n(x) = u(x)$$

*exists for $\mu$-a.e. $x \in X$, i.e., for all $x$ except a set of zero measure. If there is an integrable function $v$ such that*

$$|u_n(x)| \le v(x) \quad \textit{for } \mu\textit{-a.e. } x \in X,$$

*then $u$ is integrable and*

$$\lim_{n\to\infty} \int_X u_n \, d\mu = \int_X u \, d\mu.$$

The second is the monotone convergence Fatou theorem [42]:

**Theorem 1.2.16.** *Let $(X, \mathcal{A}, \mu)$ be a measure space, and let $u_n : X \to [-\infty, \infty]$ be a sequence of integrable functions such that*

$$u_n \le u_{n+1} \quad \text{a.e. in } X \quad \text{and} \quad \sup_n \int_X u_n \, d\mu < \infty.$$

*Then $u(x) = \lim_{n\to\infty} u_n(x)$ is integrable and*

$$\lim_{n\to\infty} \int_X u_n \, d\mu = \int_X u \, d\mu.$$

The third is the Levi theorem:

**Theorem 1.2.17.** *Let $(X, \mathcal{A}, \mu)$ be a measure space, and let $u_n : X \to [0, \infty]$ be a sequence of integrable nonnegative functions such that the limit*

$$\lim_{n\to\infty} u_n(x) = u(x)$$

*exists for $\mu$-a.e. $x \in X$. Then $u$ is integrable and*

$$\int_X u \, d\mu \le \liminf_{n\to\infty} \int_X u_n \, d\mu.$$

**Radon measures.** Let $\Omega$ be an open subset of $\mathbb{R}^d$, and let $\mathcal{B}(\Omega)$ be the $\sigma$-algebra of Borel subsets of $\Omega$. A function $u : \Omega \to \mathbb{R}$ is called a *Borel function* if it is measurable with respect to $\mathcal{B}(\Omega)$. A measure $\mu : \mathcal{B}(\Omega) \to [0, \infty]$ is called a *Borel measure* (or a nonnegative Borel measure).

**Definition 1.2.18.** A Borel measure $\mu : \mathcal{B}(\Omega) \to [0, \infty]$ is a *Radon measure* if $\mu(K) < \infty$ for every compact set $K \subset \Omega$.

Since $\mathbb{R}^d$ is locally compact and every open set in $\mathbb{R}^d$ is a union of a countable collection of its compact subsets, every Radon measure in $\Omega$ has the following regularity properties:

- For every open set $D \subset \Omega$,

$$\mu(D) = \sup \{\mu(K) : K \subset D, \ K \text{ compact}\}.$$

- For every Borel $E \subset \Omega$,

$$\mu(E) = \inf \{\mu(A) : E \subset A, \ A \text{ open}\}.$$

**Remark 1.2.19.** Any bounded Borel function is integrable over a Borel set $E$ of finite measure. If $\mu(\Omega) < \infty$, then any bounded Borel function is integrable over $\Omega$.

**Remark 1.2.20.** If $E \subset \Omega$ is a Borel set and $u : \Omega \to \mathbb{R}$ is a Borel function, then

$$\mu(E) := \int_\Omega \chi_E(x)\, d\mu, \qquad \int_E u(x)\, dx := \int_\Omega \chi_E(x) u(x)\, d\mu,$$

provided that the integrals on the right hand sides exist. Here, $\chi_E$ is the characteristic function of the set $E$, i.e., $\chi_E(1 - \chi_E) = 0$ a.e.

**Signed measures.** Let $\Omega$ be an open set in $\mathbb{R}^d$. A *signed Radon measure* on $\Omega$ is a function $\mu : \mathcal{B}(\Omega) \to [-\infty, \infty]$ such that

- $\mu(\emptyset) = 0$ and $\mu$ takes at most one of the values $\pm\infty$, so that we have either $\mu : \mathcal{B}(\Omega) \to (-\infty, \infty]$ or $\mu : \mathcal{B}(\Omega) \to [-\infty, \infty)$.
- For every countable collection $E_n$, $n \geq 1$, of pairwise disjoint sets $E_n \in \mathcal{B}(\Omega)$ we have
$$\mu\Big(\bigcup_n E_n\Big) = \sum_n \mu(E_n).$$
- $|\mu(K)| < \infty$ for every compact $K \subset \Omega$.

Notice that the series on the right hand side of the last equality converges absolutely if the left hand side is finite. The following Jordan decomposition theorem (see [42]) shows that the theory of signed measures is the same as the theory of nonnegative measures.

**Theorem 1.2.21.** *Let $\mu : \mathcal{B}(\Omega) \to [-\infty, \infty]$ be a signed Borel measure. Then there is a unique pair of nonnegative mutually singular Borel measures $\mu^\pm$, one of which is finite, such that $\mu = \mu^+ - \mu^-$.*

Recall that measures $\mu^\pm$ are *mutually singular* if there exists a Borel set $F \subset \Omega$ such that for any Borel $E \subset \Omega$,

$$\mu^+(E) = \mu^+(F \cap E), \quad \mu^-(E) = \mu^-\big((\Omega \setminus F) \cap E\big).$$

The nonnegative Radon measure $|\mu| := \mu^+ + \mu^-$ is called the *absolute variation* of the signed measure $\mu$.

**Riesz theorem** (see [42]). Let $\Omega$ be an open subset of $\mathbb{R}^d$. Recall that $C_c(\Omega)$ denotes the linear space of continuous functions $u : \Omega \to \mathbb{R}$ compactly supported in $\Omega$.

**Definition 1.2.22.** A linear functional $\mathfrak{l} : C_c(\Omega) \to \mathbb{R}$ is *locally bounded* if for every compact set $K \subset \Omega$, there exists a constant $C_K$ such that

$$|\mathfrak{l}(u)| \leq C_K \|u\|_{C(\Omega)}$$

for all continuous functions $u$ with $\operatorname{supp} u \subset K$.

**Theorem 1.2.23.** *Let $\mathfrak{l} : C_c(\Omega) \to \mathbb{R}$ be a locally bounded linear functional. Then there exists a unique signed Radon measure $\mu : \mathcal{B}(\Omega) \to [-\infty, \infty]$ such that*

$$\mathfrak{l}(u) = \int_\Omega u\, d\mu \quad \text{for every } u \in C_c(\Omega). \tag{1.2.3}$$

*Moreover if $\mathfrak{l}(u) \geq 0$ for every nonnegative $u \in C_c(\Omega)$, then $\mu$ is a nonnegative Radon measure.*

**Banach space of Radon measures.** Denote by $C_0(\Omega)$ the closed subspace of $C(\Omega)$ which consists of all continuous functions $u \in C(\Omega)$ vanishing at $\partial\Omega$ such that $u(x) \to 0$ as $|x| \to \infty$ (if $\Omega$ is unbounded). It is clear that $C_c(\Omega) \subset C_0(\Omega)$, hence any continuous functional $\mathfrak{l} : C_0(\Omega) \to \mathbb{R}$ is locally bounded in $C_c(\Omega)$. In particular, $\mathfrak{l}$ meets all requirements of Theorem 1.2.23 and admits representation (1.2.3) with a Radon measure $\mu$. The following result is a consequence of the Riesz theorem.

**Theorem 1.2.24.** *Let $\mathfrak{l} : C_0(\Omega) \to \mathbb{R}$ be a bounded linear functional. Then there exists a unique signed Radon measure $\mu : \mathcal{B}(\Omega) \to (-\infty, \infty)$ such that*

$$\mathfrak{l}(u) = \int_\Omega u\, d\mu \quad \text{for every } u \in C_0(\Omega). \tag{1.2.4}$$

*Moreover,*

$$|\mu|(\Omega) = \mu^+(\Omega) + \mu^-(\Omega) = \|\mathfrak{l}\|_{C_0(\Omega)'} := \sup_{\{u \in C_0(\Omega) : \|u\|_{C(\Omega)} = 1\}} |\mathfrak{l}(u)|,$$

*where $\mu^\pm$ is the Jordan decomposition of $\mu$ defined by Theorem 1.2.21. Conversely, for any signed Radon measure $\mu$ in $\Omega$ with $|\mu(\Omega)| < \infty$, relation (1.2.4) defines a continuous functional $\mathfrak{l} \in C_0(\Omega)'$ such that $\|\mathfrak{l}\| = |\mu|(\Omega)$.*

The totality of all signed Radon measures $\mu$ with

$$\|\mu\|_{\mathcal{M}(\Omega)} := |\mu|(\Omega) < \infty$$

forms a Banach space $\mathcal{M}(\Omega)$ which is isomorphic to the dual space $C_0(\Omega)'$.

### 1.2.3 Lebesgue measure in $\mathbb{R}^d$. Lebesgue spaces

There is a unique Radon measure $\mu_L$ in $\mathbb{R}^d$ with the property

$$\mu_L(Q) = \prod_{i=1}^d (b_i - a_i) \quad \text{for any parallelepiped } Q = \prod_{i=1}^d (a_i,\, b_i).$$

The following construction is due to Lebesgue. We say that $E \subset \mathbb{R}^d$ has Lebesgue measure zero if for any $\epsilon > 0$, there is a Borel set $B$ such that

$$E \subset B \quad \text{and} \quad \mu_L(B) < \epsilon.$$

Next, we say that a set $A \subset \mathbb{R}^d$ is *Lebesgue measurable* if $A = B \cup E$, where $B$ is a Borel set and $E$ is a zero measure set. In this case we set

$$m(A) = \mu_L(B).$$

Denote by $\mathcal{L}_d$ the totality of all Lebesgue measurable sets. It is known that $\mathcal{L}_d$ is a $\sigma$-algebra in $\mathbb{R}^d$, and $m : \mathcal{L}_d \to \mathbb{R}$ is a nonnegative measure in $\mathbb{R}^d$. The measure $m$ is called the *Lebesgue measure* in $\mathbb{R}^d$ or the Lebesgue extension of the measure $\mu_L$. The integral with respect to the Lebesgue measure is named the *Lebesgue integral.*

**Remark 1.2.25.** Obviously this procedure may be applied to all Borel measures $\mu$ on $\mathbb{R}^d$. Thus we can get the Lebesgue extension denoted by $\overline{\mu}$ of any Borel measure $\mu$, but the classes of $\overline{\mu}$-measurable sets and $\overline{\mu}$-integrable functions strongly depend on $\mu$.

Further, when no measure is explicitly indicated, we say simply measure, measurable function, integrable function for Lebesgue measure and Lebesgue measurable or integrable function. We also use the standard notation $dx := dm$ and write meas $A$ for $m(A)$.

Let $\Omega \subset \mathbb{R}^d$ be a measurable set. We say that two measurable functions $u$ and $v$ are *equivalent* in $\Omega$ if $u(x) = v(x)$ for a.e. $x \in \Omega$ (i.e. for all $x \in \Omega$, possibly except a set of zero measure). From now on we identify a measurable function with its equivalence class.

**Definition 1.2.26.** Let $\Omega \subset \mathbb{R}^d$ be a measurable set and $1 \le p \le \infty$. Then we define the *Lebesgue space*

$$L^p(\Omega) := \{u : \Omega \to \mathbb{R} : u \text{ is measurable and } \|u\|_{L^p(\Omega)} < \infty\},$$

where for $1 \le p < \infty$,

$$\|u\|_{L^p(\Omega)} := \Big( \int_\Omega |u(x)|^p \, dx \Big)^{1/p},$$

and if $p = \infty$, then

$$\|u\|_{L^\infty(\Omega)} := \operatorname*{ess\,sup}_\Omega |u| = \inf\{k \in \mathbb{R} : |u(x)| \le k \text{ for a.e. } x \in \Omega\}.$$

It follows from the Minkowski inequality

$$\|u_1 + \cdots + u_n\|_{L^p(\Omega)} \le \|u_1\|_{L^p(\Omega)} + \cdots + \|u_n\|_{L^p(\Omega)}$$

that $L^p(\Omega)$ is a normed space. Moreover, it is a Banach space.

We say that $u \in L^p_{\text{loc}}(\Omega)$ if $u \in L^p(K)$ for every compact set $K \subset \Omega$. Note that $L^p_{\text{loc}}\Omega)$ is just a linear space.

**Hölder inequality. Duality.** The main tool for studying Lebesgue spaces is the Hölder inequality

$$\|\prod_{i=1}^{n} u_i\|_{L^1(\Omega)} \leq \prod_{i=1}^{n} \|u_i\|_{L^{p_i}(\Omega)}$$

which holds, provided that the right hand side is finite, for all exponents

$$p_i \in [1, \infty] \quad \text{with} \quad \frac{1}{p_1} + \cdots + \frac{1}{p_n} = 1.$$

If $\operatorname{meas} \Omega$ is finite and

$$p, p_i \in [1, \infty], \quad \frac{1}{p} \geq \frac{1}{p_1} + \cdots + \frac{1}{p_n},$$

then there is a constant $c$, depending on $\Omega$ and $p_i$, such that for all $u_i \in L^{p_i}(\Omega)$,

$$\|\prod_{1=1}^{n} u_i\|_{L^p(\Omega)} \leq c \prod_{1=1}^{n} \|u_i\|_{L^{p_i}(\Omega)}. \tag{1.2.5}$$

The Hölder inequality leads to the following description of the dual spaces.

**Theorem 1.2.27.** *Let $\Omega \subset \mathbb{R}^d$ be a measurable set and $1 \leq p < \infty$. Then for any continuous linear functional $\mathfrak{l} \in L^p(\Omega)'$ there is a unique function $v \in L^q(\Omega)$, with $q = p/(p-1)$ for $p > 1$, and $q = \infty$ for $p = 1$, such that for all $u \in L^p(\Omega)$,*

$$\mathfrak{l}(u) = \int_\Omega u(x)v(x)\, dx.$$

*Moreover*

$$\|\mathfrak{l}\|_{L^p(\Omega)'} := \sup_{\{u \in L^p(\Omega) : \|u\|_{L^p(\Omega)} = 1\}} |\mathfrak{l}(u)| = \|v\|_{L^q(\Omega)}.$$

*In other words, the dual space $L^p(\Omega)'$ is isomorphic to $L^q(\Omega)$.*

**Weak convergence in Lebesgue spaces.** Let $\Omega$ be a measurable subset of $\mathbb{R}^d$. Assume that $1 \leq q < \infty$, $q' = q/(q-1)$, i.e.,

$$\frac{1}{q} + \frac{1}{q'} = 1.$$

Recall that a sequence $(f_k)_{k \geq 1} \subset L^q(\Omega)$ *converges weakly to* $f \in L^q(\Omega)$ in $L^q(\Omega)$, written

$$f_k \rightharpoonup f \quad \text{weakly in } L^q(\Omega),$$

provided that

$$\int_\Omega f_k g\, dx \to \int_\Omega f g\, dx \quad \text{as } k \to \infty \tag{1.2.6}$$

for each $g \in L^{q'}(E)$. A sequence $(f_k)_{k\geq 1} \subset L^q(\Omega)$ is *weakly convergent* if the limit

$$\lim_{k\to\infty} \int_\Omega f_k g\, dx \tag{1.2.7}$$

exists for all $g \in L^{q'}(\Omega)$. If $1 < q < \infty$, then each weakly convergent sequence converges weakly to some element $f \in L^q(\Omega)$. This is not true for $q = 1$. The theory of weak convergence in $L^\infty(\Omega)$ is complicated. For this reason we replace it with the weak* convergence. Recall that a sequence $(f_k)_{k\geq 1} \subset L^\infty(\Omega)$ *converges weakly** to $f \in L^\infty(\Omega)$ in $L^\infty(\Omega)$ provided that

$$\int_\Omega f_k g\, dx \to \int_\Omega fg\, dx \quad \text{as } k \to \infty$$

for all $g \in L^1(\Omega)$.

Now we recall some boundedness properties of weakly convergent sequences.

**Lemma 1.2.28.** *Assume that $f_k \rightharpoonup f$ weakly in $L^q(\Omega)$. Then $(f_k)_{k\geq 1}$ is bounded in $L^q(\Omega)$ and $\|f\|_{L^q(\Omega)} \leq \liminf_{k\to\infty} \|f_k\|_{L^q(\Omega)}$. Moreover, if $1 < q < \infty$ and $\|f\|_{L^q(\Omega)} = \lim_{k\to\infty} \|f_k\|_{L^q(\Omega)}$ then $f_k \to f$ converges strongly (i.e., in norm) in $L^q(\Omega)$ as $k \to \infty$.*

For $1 < q < \infty$, the Banach space $L^q(\Omega)$ is reflexive, hence bounded sequences are relatively weakly compact:

**Lemma 1.2.29.** *Assume $1 < q < \infty$ and the sequence $(f_k)_{k\geq 1}$ is bounded in $L^q(\Omega)$. Then there is a subsequence $(f_{k_j})_{j\geq 1}$ and an element $f \in L^q(\Omega)$ such $f_{k_j} \rightharpoonup f$ weakly in $L^q(\Omega)$ as $j \to \infty$.*

For $q = 1$ the boundedness is necessary but not sufficient for the weak compactness. In order to formulate the corresponding result it is convenient to give the following definition.

**Definition 1.2.30.** We say that a set $\mathfrak{M} \subset L^1(\Omega)$ is *equi-integrable* if for any $\varepsilon > 0$ there exists $\delta(\varepsilon) > 0$, depending only on $\mathfrak{M}$ and $\varepsilon$, such that

$$\int_A |f|\, dx < \varepsilon$$

for each $A \subset \Omega$ with $\operatorname{meas} A < \delta(\varepsilon)$ and for all $f \in \mathfrak{M}$.

This definition is equivalent to the following property: $\mathfrak{M} \subset L^1(\Omega)$ is equi-integrable if and only if there is a continuous function $\Phi : [0,\infty) \to [0,\infty)$ with

$$t^{-1}\Phi(t) \to \infty \quad \text{as } t \to \infty$$

and

$$\sup_{f\in\mathfrak{M}} \int_\Omega \Phi(|f|)\, dx < \infty.$$

**Lemma 1.2.31.** *Assume the sequence $(f_k)_{k\geq 1}$ is bounded in $L^1(\Omega)$ and equi-integrable. Then there exists a subsequence $f_{k_j}$ and a function $f \in L^1(\Omega)$ with $f_{k_j} \rightharpoonup f$ weakly in $L^1(\Omega)$.*

## 1.3 Compact sets in $L^p$ spaces

Throughout this book we use the following simple compactness criterion in $L^p$ spaces (see [42]).

**Proposition 1.3.1.** *Let $\Omega \subset \mathbb{R}^d$ be a bounded Lebesgue measurable set. A set $\mathcal{K} \subset L^p(\Omega)$, $1 \le p < \infty$, is relatively compact in $L^p(\Omega)$ if and only if the following two conditions are satisfied:*

*(i) Any sequence of functions $f_n \in \mathcal{K}$ contains a subsequence $f_{n_m}$ which converges in measure, i.e.,*

$$\lim_{m,k\to\infty} \operatorname{meas}\{x \in \Omega : |f_{n_m}(x) - f_{n_k}(x)| \ge \epsilon\} = 0 \quad \text{for all } \epsilon > 0.$$

*(ii) For any $\epsilon > 0$ there is $\delta > 0$ such that*

$$\int_A |f|^p\,dx < \epsilon \quad \text{for all } f \in \mathcal{K} \quad \text{and} \quad A \subset \Omega \quad \text{with } \operatorname{meas} A < \delta.$$

The condition (ii) is equivalent to the following: There is a continuous function $\Phi : \mathbb{R}^+ \to \mathbb{R}^+$ with the properties

$$\lim_{t\to\infty} t^{-1}\Phi(t) = \infty, \quad \sup_{f\in\mathcal{K}} \int_\Omega \Phi(|f|^p)\,dx < \infty.$$

In particular, we have the following condition.

**Lemma 1.3.2.** *Let $\Omega \subset \mathbb{R}^d$ be a bounded Lebesgue measurable set. If a sequence $f_n \in L^r$, $1 < r \le \infty$, converges in measure and is bounded in $L^r(\Omega)$, then $f_n$ converges strongly in any $L^s(\Omega)$ with $1 \le s < r$.*

### 1.3.1 Functions of bounded variation

Let $I$ be an arbitrary interval (open, closed, half-open, or infinite). The *total variation* of a function $F : I \to \mathbb{R}$ over $I$ is defined by

$$\bigvee_I F := \sup \sum_{k=0}^{q} |F(t_{k+1}) - F(t_k)|, \tag{1.3.1}$$

where the supremum is taken over all $t_0 < \cdots < t_q$ such that $t_i \in I$. A function $F$ is said to be *of bounded variation* on $I$ if the total variation of $F$ over $I$ is finite. Every monotone bounded function is a function of bounded variation. Conversely, every function of bounded variation on $I$ can be represented as a difference of two monotone bounded functions:

$$F = F^+ - F^-, \quad F^+(t) = \bigvee_{(-\infty,t]\cap I} F, \quad F^- = F^+ - F.$$

In particular, every function $F : I \to \mathbb{R}$ of bounded variation is continuous everywhere on $I$ except on a countable set of jump points $t$ at which the limits

$$F(t+0) = \lim_{z \searrow t} F(z), \quad F(t-0) = \lim_{z \nearrow t} F(z)$$

exist. Further we use the following simple lemma.

**Lemma 1.3.3.** *If $F \in L^1(I)$ and $f \in L^1(I)$ satisfy*

$$\int_I \eta'(t)F(t)\,dt = -\int_I \eta(t)f(t)\,dt \quad \textit{for all } \eta \in C_0^\infty(I),$$

*then $F$ can be changed on a set of zero measure in such a way that $F \in C[0,T]$,*

$$\bigvee_I F \le \int_I |f(t)|dt, \quad \|F\|_{C[0,T]} \le c\|F\|_{L^1(I)} + \|f\|_{L^1(I)}.$$

The following theorem due to Helly (see [67]) is the most important result in the theory of bounded variation functions.

**Theorem 1.3.4.** *Let a sequence of functions $F_n : [0,T] \to \mathbb{R}$, $T < \infty$, satisfy the condition*

$$\sup_n \bigvee_{[0,T]} F_n < \infty.$$

*Then there is a subsequence, still denoted by $F_n$, and a function $F$ of bounded variation such that*

$$F_n(t) \to F(t) \quad \textit{as } n \to \infty \quad \textit{for all } t \in [0,T], \quad \bigvee_{[0,T]} F \le \liminf_n \bigvee_{[0,T]} F_n.$$

**Lebesgue-Stieltjes integral.** Let $I$ be an arbitrary interval and $U \supset I$ be an open interval. Let $F : U \to \mathbb{R}$ be a function of bounded variation. Then there exists a unique signed Borel measure $\mu_F$ on $U$ with the properties

$$\mu_F[a,b] = F(b+0) - F(a-0), \quad \mu_F(a,b) = F(b-0) - F(a+0),$$
$$\mu_F[a,b) = F(b-0) - F(a-0), \quad \mu_F(a,b] = F(b+0) - F(a+0).$$

The Borel measure $\mu_F$ is called the *Lebesgue-Stieltjes measure* associated with $F$. Every finite Borel measure $\mu$ on an open interval $U$ can be regarded as the Lebesgue-Stieltjes measure associated with the function $F(s) = \mu\big((-\infty, s] \cap U\big)$. If functions of bounded variation $F, \tilde{F} : U \to \mathbb{R}$ coincide everywhere except on a countable set, then $\mu_F = \mu_{\tilde{F}}$.

The integral with respect to the Borel measure $\mu_F$ is called the *Lebesgue-Stieltjes integral* and is denoted by

$$\int_I f(s)\,dF(s) := \int_I f(s)\,d\mu_F \quad \text{for all Borel functions } f.$$

**Example 1.3.5.** Let $g \in C[a,b]$ and suppose $f : [a,b] \to \mathbb{R}$ admits the representation

$$f(t) = f_{ac}(t) + f_j(t), \quad f_{ac}(t) = \int_0^t f'(s)\,ds, \quad f_j(t) = \sum_k c_k \theta(t - a_k),$$

where

$$\theta(t) = 0 \quad \text{for } t < 0 \quad \text{and} \quad \theta(t) = 1 \quad \text{for } t \geq 1,$$

$f' \in L^1(a,b)$ and $\{a_k\} \subset [a,b]$ is a finite or infinite sequence. Then

$$\bigvee_{[a,b]} f = \int_a^b |f'(s)|\,ds + \sum_{k:a_k > a} |c_k|$$

and

$$\int_{[a,b]} g(s)\,df(s) = \int_a^b g(s)f'(s)\,ds + \sum_k c_k g(a_k). \tag{1.3.2}$$

**Example 1.3.6.** Let $\mu \in \mathcal{M}(\mathbb{R}) = C_0(\mathbb{R})'$ be a nonnegative finite Radon measure on $\mathbb{R}$. Then the *distribution function* $f$ of the measure $\mu$,

$$f(s) = \mu(-\infty, s], \tag{1.3.3}$$

has bounded variation and $\mu_f = \mu$. The function $f$ is nondecreasing and satisfies

$$f(s) = \lim_{h \searrow 0} f(s+h), \qquad \lim_{s \to -\infty} f(s) = 0. \tag{1.3.4}$$

The most important fact in the theory of the Lebesgue-Stieltjes integral is the following integration by parts formula:

**Lemma 1.3.7.** *Let $I$ be an arbitrary interval and $U \supset I$ an open interval. Let $f, g : U \to \mathbb{R}$ be functions of bounded variation, and let $S$ denote the set of points $s \in I$ at which both $f$ and $g$ are discontinuous. Then*

$$\int_I f\,dg + \int_I g\,df = \mu_{fg}(I) + \sum_{s \in S} A(s),$$

*where*

$$\begin{aligned} A(s) = {} & [f(s) - \tfrac{1}{2}(f(s+0) + f(s-0))]\mu_g(\{s\}) \\ & + [g(s) - \tfrac{1}{2}(g(s+0) + g(s-0))]\mu_f(\{s\}), \end{aligned}$$

$$\begin{aligned} \mu_{fg}[\alpha,\beta] &= f(\beta+0)g(\beta+0) - f(\alpha-0)g(\alpha-0), \\ \mu_{fg}(\alpha,\beta) &= f(\beta-0)g(\beta-0) - f(\alpha+0)g(\alpha+0), \\ \mu_{fg}[\alpha,\beta) &= f(\beta-0)g(\beta-0) - f(\alpha-0)g(\alpha-0), \\ \mu_{fg}(\alpha,\beta] &= f(\beta+0)g(\beta+0) - f(\alpha+0)g(\alpha+0), \end{aligned}$$

$$f(\beta-0)g(\beta-0) = \lim_{s\to\infty} f(s)g(s) \quad \textit{for } \beta = \infty,$$

$$f(\alpha+0)g(\alpha+0) = \lim_{s\to-\infty} f(s)g(s) \quad \textit{for } \alpha = -\infty.$$

*In particular:*

*(i) If $S$ is empty or if $f(s) = \frac{1}{2}\big(f(s+0)+f(s-0)\big)$ and $g(s) = \frac{1}{2}\big(g(s+0)+g(s-0)\big)$ for all $s \in S$, then*

$$\int_I f\,dg + \int_I g\,df = \mu_{fg}(I).$$

*(ii) If $f$ and $g$ are continuous from the right at all points of $S$, then*

$$\int_I f\,dg + \int_I g\,df = \mu_{fg}(I) + \frac{1}{4}\sum_{s\in S}\mu_f(\{s\})\mu_g(\{s\}).$$

### 1.3.2 $L^p$ spaces of Banach space valued functions

**Definition 1.3.8.** Let $\Omega \subset \mathbb{R}^d$ be a (Lebesgue) measurable set, and let $Y$ be a Banach space. We say that a function $s : \Omega \to Y$ is *simple* if $s$ admits a representation

$$s(x) = \sum_{i=1}^{N} c_i \chi_{E_i}(x), \tag{1.3.5}$$

where the sets $E_i \subset \Omega$ are measurable and mutually disjoint, $c_i \in Y$ are distinct elements, and $\chi_{E_i}$ is the characteristic function of the set $E_i$.

**Definition 1.3.9.** (i) A function $u : \Omega \to Y$ is *strongly measurable* (or simply *measurable*) if there is a sequence $\{s_n\}$ of simple measurable functions such that

$$\lim_{n\to\infty} \|s_n(x) - u(x)\|_Y = 0 \quad \text{for a.e. } x \in \Omega.$$

(ii) A function $u : \Omega \to Y$ is *weakly measurable* if for any $u' \in Y'$ the function

$$x \in \Omega \mapsto \langle u', u(x)\rangle \in \mathbb{R} \quad \text{is measurable.}$$

(iii) A function $u : \Omega \to Y'$ is *weakly$^\star$ measurable* if for any $v \in Y$ the function

$$x \in \Omega \mapsto \langle u(x), v\rangle \in \mathbb{R} \quad \text{is measurable.}$$

If a function $u$ is strongly measurable, then the scalar function $\|u(x)\|_Y$ is measurable. It is difficult to check whether a given function is strongly measurable or not. The relation between strong measurability and the standard notion of measurability is given by the following theorem ([24, Ch. III.6, Lemma 9]).

**Theorem 1.3.10.** *Let $\Omega \subset \mathbb{R}^d$ be a measurable set and $Y$ be a Banach space. A function $u : \Omega \to Y$ is strongly measurable if and only if the following conditions hold:*

*(I) There is a set $E$ of zero Lebesgue measure such that the set $u(\Omega \setminus E)$ is separable, i.e., contained in the closure of some countable set.*

*(II) $u^{-1}(G)$ is measurable for every open set $G \subset Y$, or equivalently*

*(II') $u^{-1}(G)$ is measurable for every Borel set $G \subset Y$.*

Condition (I) is obviously fulfilled if $Y$ is separable. The following theorem due to Pettis shows that the separability of $Y$ guarantees the equivalence of strong and weak measurability (see [42, Thm. 2.104], [24, Ch. III.6, Thm. 11]).

**Theorem 1.3.11.** *Let $\Omega \subset \mathbb{R}^d$ be a measurable set of finite Lebesgue measure and let $Y$ be a Banach space. A function $u : \Omega \to Y$ is strongly measurable if and only if it is weakly measurable and there is a set $E \subset \Omega$ of zero measure such that the set $u(\Omega \setminus E) \subset Y$ is separable. In particular, if $Y$ is separable, then any weakly measurable function is strongly measurable.*

The Egoroff and Lusin theorems remain true for strongly measurable functions:

**Theorem 1.3.12.** *Let $\Omega \subset \mathbb{R}^d$ be a measurable set of finite Lebesgue measure and let $Y$ be a Banach space. Assume that strongly measurable functions $u_n$ converge a.e. in $\Omega$ to a measurable function $u$, i.e., for almost every $x \in \Omega$, the sequence $u_n(x)$ converges to $u(x)$ strongly in $Y$. Then for every $\varepsilon > 0$ there is a measurable set $E \subset \mathbb{R}^d$ with $\operatorname{meas} E < \varepsilon$ such that $\|u_n(x) - u(x)\|_Y \to 0$ uniformly on $\Omega \setminus E$.*

**Theorem 1.3.13.** *Let $\Omega \subset \mathbb{R}^d$ be a measurable set of finite Lebesgue measure, $Y$ be a Banach space, and $u : \Omega \to Y$ be a strongly measurable function. Then for every $\varepsilon > 0$ there is an open set $G \subset \mathbb{R}^d$ with $\operatorname{meas} G < \varepsilon$ such $u$ is continuous on $\Omega \setminus G$.*

**Bochner integral.** Let $\Omega \subset \mathbb{R}^d$ be a Lebesgue measurable set. A simple measurable function $s$ of the form (1.3.5) is integrable if $c_i = 0$ whenever $\operatorname{meas} E_i = \infty$. The *Bochner integral* of $s$ with respect to the Lebesgue measure is defined as follows:

$$\int_\Omega s(x)\,dx = \sum_{i=1}^N c_i \operatorname{meas} E_i \quad \text{and} \quad \int_E s(x)\,dx = \sum_{i=1}^N c_i \operatorname{meas}(E_i \cap E)$$

for any measurable set $E \subset \Omega$.

**Definition 1.3.14.** Let $\Omega \subset \mathbb{R}^d$ be a measurable set. A strongly measurable function $u : \Omega \to Y$ is *Bochner integrable* if there is a sequence $\{s_n\}$ of simple integrable functions such that

$$\lim_{n\to\infty} \|s_n(x) - u(x)\|_Y = 0 \quad \text{for a.e. } x \in \Omega,$$

$$\lim_{n\to\infty} \int_\Omega \|s_n(x) - u(x)\|_Y\,dx = 0.$$

The *Bochner integral* of $u$ over $\Omega$ is defined to be the limit

$$\int_\Omega u(x)\,dx = \lim_{n\to\infty} \int_\Omega s_n(x)\,dx.$$

This limit exists and it is independent of the choice of the sequence $s_n$ convergent to $u$.

The following important theorem due to Bochner shows that measurability and integrability in norm are equivalent to Bochner integrability ([42, Thm. 2.108]).

**Theorem 1.3.15.** *A strongly measurable function $u$ is Bochner integrable over $\Omega$ if and only if the norm $\|u(x)\|_Y$ is Lebesgue integrable over $\Omega$. Moreover,*

$$\left\| \int_\Omega u\, dx \right\|_Y \leq \int_\Omega \|u(x)\|_Y \, dx.$$

Many general results concerning the properties of real-valued integrable functions remain true for Bochner integrals. In particular we have the following version of the Lebesgue dominated convergence theorem ([24, Ch. III.6, Cor. 16]).

**Theorem 1.3.16.** *Let $\Omega \subset \mathbb{R}^d$ be a Lebesgue measurable set, $Y$ be a Banach space, and suppose strongly measurable functions $u_n : \Omega \to Y$ converge a.e. in $\Omega$ to a function $u$. Furthermore assume that there is a nonnegative function $g \in L^1(\Omega)$ such that $\|u_n(x)\|_Y \leq g(x)$ a.e. in $\Omega$. Then $u$ is Bochner integrable and*

$$\lim_{n\to\infty} \int_\Omega \|u_n(x) - u(x)\|_Y \, dx = 0.$$

The next theorem due to Bochner justifies the "differentiation" of the Bochner integral ([131, Ch. 5,5, Thm. 2]).

**Theorem 1.3.17.** *Let $\Omega \subset \mathbb{R}^d$ be an open set, $Y$ be a Banach space, and $u : \Omega \to Y$ be a Bochner integrable function. Then for $a \to 0$ and for a.e. $x_0 \in \Omega$,*

$$a^{-d} \int_{Q_a(x_0)} u(x)\, dx \to u(x_0) \quad \textit{strongly in } Y.$$

*Here $Q_a(x_0)$ denotes the cube centered at $x_0$ with side length $a$.*

**$L^p$ spaces.** We are now in a position to define $L^p$ spaces of functions which take values in Banach spaces. We distinguish between strongly and weakly measurable functions.

**Definition 1.3.18.** Let $\Omega$ be a measurable subset of $\mathbb{R}^d$ and $L^p(\Omega; Y)$ be the linear space of all strongly measurable functions $u : \Omega \to Y$ with the finite norm

$$\|u\|_{L^p(\Omega;Y)} = \left( \int_\Omega \|u(x)\|_Y^p \, dx \right)^{1/p},$$

with $1 \leq p < \infty$. If we identify all functions in the same equivalence class, then $L^p(\Omega; Y)$ becomes a Banach space.

$L^\infty(\Omega; Y)$ is the normed space of all (equivalence classes of) strongly measurable functions $u : \Omega \to Y$ with the finite norm

$$\|u\|_{L^\infty(\Omega;Y)} = \operatorname*{ess\,sup}_{x\in\Omega} \|u(x)\|_Y := \inf\{c : \|u(x)\| < c \text{ for a.e. } x \in \Omega\}.$$

**Remark 1.3.19.** It follows from Theorem 1.3.15 that all functions $u \in L^p(\Omega; Y)$ are Bochner integrable.

The following theorem describes the basic properties of the spaces $L^p(\Omega; Y)$ ([42, Thm. 2.110]).

**Theorem 1.3.20.** *Let $Y$ be a Banach space and $\Omega$ be a measurable subset of $\mathbb{R}^d$. Then*

*(i) $L^p(\Omega; Y)$ is a Banach space for $1 \leq p \leq \infty$;*

*(ii) the family of all integrable simple functions is dense in $L^p(\Omega; Y)$ for $1 \leq p < \infty$, i.e., for any $u \in L^p(\Omega; Y)$ and $\varepsilon > 0$ there is a simple function $s$ of the form (1.3.5) such that $\|s - u\|_Y \leq \varepsilon$.*

**Definition 1.3.21.** Let $\Omega$ be a measurable subset of $\mathbb{R}^d$, let $Y$ be a Banach space, $Y'$ the dual space to $Y$, and $1 \leq p < \infty$. Then $L^p_w(\Omega; Y')$ is the space of all (equivalence classes of) weakly$^\star$ measurable functions $u : \Omega \to Y'$ with the finite norm

$$\|u\|_{L^p_w(\Omega;Y')} = \Big( \int_\Omega \|u(x)\|_{Y'}\, dx \Big)^{1/p}.$$

If we identify functions with their equivalence classes, then $L^p_w(\Omega; Y')$ becomes a Banach space. $L^\infty_w(\Omega; Y')$ is the Banach space of all (equivalence classes of) weakly$^\star$ measurable functions $u : \Omega \to Y'$ with the finite norm

$$\|u\|_{L^\infty_w(\Omega;Y')} = \operatorname*{ess\,sup}_{x\in\Omega} \|u(x)\|_{Y'}.$$

The properties of the spaces $L^p(\Omega; Y)$ are similar to the properties of $L^p(\Omega)$; in particular we have the following extension of Theorem 1.3.16 ([24, Ch. III.6, Cor. 16]).

**Theorem 1.3.22.** *Let $\Omega \subset \mathbb{R}^d$ be a Lebesgue measurable set, $Y$ be a Banach space, $p \in [1, \infty)$, and suppose that strongly measurable functions $u_n : \Omega \to Y$ converge a.e. in $\Omega$ to a function $u$. Furthermore, assume that there is a nonnegative function $g \in L^1(\Omega)$ such that $\|u_n(x)\|^p_Y \leq g(x)$ a.e. in $\Omega$. Then $u \in L^p(\Omega; Y)$ and*

$$\lim_{n\to\infty} \|u_n - u\|_{L^p(\Omega;Y)} = 0.$$

**Subspaces of $L^p(\Omega; Y)$.** In applications, various subspaces $Z$ of some "hold-all" Banach spaces $Y$ are considered. A natural question is whether $L^p(\Omega; Z)$ is then a subspace of $L^p(\Omega; Y)$. The answer is given by the following simple lemma.

**Lemma 1.3.23.** *Let $Z$ be a closed subspace of a Banach space $Y$, and suppose a function $u : \Omega \to Z$ belongs to $L^p(\Omega; Y)$ for some $p \in [1, \infty]$. Then $u \in L^p(\Omega; Z)$.*

*Proof.* Since $\|u(x)\|_Z = \|u(x)\|_Y$, it suffices to show that $u : \Omega \to Z$ is strongly measurable. By Theorem 1.3.10 there are a set $E \subset \mathbb{R}^d$ of zero measure and a countable subset $\mathcal{Y} = \{y_i\}$ of $Y$ with the property that for every $x \in \Omega \setminus E$ and $\varepsilon > 0$ there is $y_i \in \mathcal{Y}$ such that $\|u(x) - y_i\| \le \varepsilon$.

Next for every $i$, choose $z_i \in Z$ such that

$$\|z_i - y_i\|_Y \le 2\,\mathrm{dist}(y_i, Z).$$

and set $\mathcal{Z} = \{z_i\} \subset Z$. Note that if $\|u(x) - y_i\|_Y \le \varepsilon$, then $\mathrm{dist}(y_i, Z) \le \varepsilon$ and so

$$\|u(x) - z_i\|_Y \le \|u(x) - y_i\|_Y + \|z_i - y_i\|_Y \le 3\varepsilon.$$

In other words, for every $x \in \Omega \setminus E$ and $\varepsilon > 0$, there exists $z_i \in \mathcal{Z}$ satisfying $\|u(x) - z_i\|_Y \le 3\varepsilon$. This means that

$$u(X \setminus E) \subset \mathrm{cl}\,\mathcal{Z} \subset Z. \tag{1.3.6}$$

Let us prove that for every open $G \subset Z$, the set $u^{-1}(G)$ is measurable. Notice that the set $F = Z \setminus G$ is closed in $Y$. Since the mapping $u : \Omega \to Y$ is measurable, it follows from assertion (II') in Theorem 1.3.10 that $u^{-1}(F)$ is measurable and hence the set $u^{-1}(G) = \Omega \setminus u^{-1}(F)$ is measurable. From this and (1.3.6) we conclude that $u : \Omega \to Z$ meets all requirements of assertions (I)–(II) in Theorem 1.3.10 and hence is strongly measurable. □

**Duality.** The following result (see [42, Thm. 2.112]) is important for us since it leads to the notion of Young measure which is widely used throughout this book.

**Theorem 1.3.24.** *Let $\Omega$ be a measurable subset of $\mathbb{R}^d$, let $Y$ be a Banach space and let $1 \le p < \infty$, $q = p/(p-1)$.*

- *Assume that $Y$ is separable. Then a functional $V : L^p(\Omega; Y) \to \mathbb{R}$ belongs to $(L^p(\Omega; Y))'$ if and only if there is $v \in L^q_w(\Omega; Y')$ such that*

$$\langle V, u\rangle = \int_\Omega \langle v(x), u(x)\rangle\, dx.$$

*Moreover $v$ is unique and the norm of the functional $V$ equals $\|v\|_{L^q_w(\Omega, Y')}$.*

- *Assume that $Y$ is reflexive. Then a functional $V : L^p(\Omega; Y) \to \mathbb{R}$ belongs to $(L^p(\Omega; Y))'$ if and only if there is $v \in L^q(\Omega; Y')$ such that*

$$\langle V, u\rangle = \int_\Omega \langle v(x), u(x)\rangle\, dx.$$

*Moreover, $v$ is unique and the norm of $V$ is equal to $\|v\|_{L^q(\Omega, Y')}$.*

**Corollary 1.3.25.** *If $Y$ is reflexive and $1 < p < \infty$, then $L^p(\Omega; Y)$ is reflexive.*

**Anisotropic spaces.** The anisotropic Lebesgue spaces are the most important class of $L^p$ spaces. In the analysis of nonstationary problems we often need functions with different properties with respect to time and the spatial variables. In this case it is convenient to replace the standard Lebesgue space by an anisotropic space which takes into account the features of the problem.

Let $\Omega \subset \mathbb{R}^d$ be a measurable set, $T > 0$ and let $Q$ be the cylinder $\Omega \times (0,T)$. For any measurable function $u : Q \to \mathbb{R}$ we write $u(t)$ instead $u(\cdot,t)$, i.e., $u(x,t)$ denotes the value of $u$ at the point $(x,t)$, but $u(t)$ denotes a function of the spatial variable $x$ which is defined in $\Omega$ and takes the value $u(x,t)$ at $x$.

**Definition 1.3.26.** Let $p, q \in [1,\infty]$ be given. Then

$$L^q(0,T;L^p(\Omega)) := \big\{u : Q \to \mathbb{R} : u \text{ measurable and } \|u\|_{L^q(0,T;L^p(\Omega))} < \infty\big\},$$

where for $1 \le q < \infty$,

$$\|u\|_{L^q(0,T;L^p(\Omega))} := \Big\{\int_0^T \|u(t)\|_{L^p(\Omega)}\,dt\Big\}^{1/q},$$

while if $q = \infty$, then

$$\|u\|_{L^\infty(0,T;L^p(\Omega))} := \operatorname*{ess\,sup}_{t\in(0,T)} \|u(t)\|_{L^p(\Omega)}.$$

The theory of anisotropic Lebesgue spaces is the same as the theory of standard Lebesgue spaces. In particular, we have the following generalization of the Hölder inequality.

Assume that the domain $\Omega$ is of finite measure, and the exponents $p, q$, $p_i$, $q_i$, $1 \le i \le n$, satisfy the conditions

$$p, q, p_i, q_i \in [1,\infty], \quad \frac{1}{p} \ge \sum_{i=1}^n \frac{1}{p_i}, \quad \frac{1}{q} \ge \sum_{i=1}^n \frac{1}{q_i}. \tag{1.3.7}$$

Then there is a constant $c$, depending on $\Omega$ and $p$, $q$, $p_i$, $q_i$, such that for all $u_i \in L^{q_i}(0,T;L^{p_i}(\Omega))$,

$$\Big\|\prod_{i=1}^n u_i\Big\|_{L^q(0,T;L^p(\Omega))} \le c \prod_{i=1}^n \|u_i\|_{L^{q_i}(0,T;L^{p_i}(\Omega))}. \tag{1.3.8}$$

**Remark 1.3.27.** If in (1.3.7) there are equalities instead of inequalities, then inequality (1.3.8) is fulfilled for all measurable sets $\Omega$. Moreover, in this case the constant $c$ is 1.

Next, we have the following representation for dual spaces. For any exponent $p \in [1,\infty]$ set

$$p' = p/(p-1) \quad \text{for } 1 < p < \infty, \quad p' = \infty \quad \text{for } p = 1, \quad p' = 1 \quad \text{for } p = \infty.$$

**Theorem 1.3.28.** *Let $\Omega \subset \mathbb{R}^d$ be a measurable set and $1 \leq p, q < \infty$. Then for any continuous functional $\mathfrak{l} \in L^q(0,T;L^p(\Omega))'$ there is a unique function $v \in L^{q'}(0,T;L^{p'}(\Omega))$ such that for all $u \in L^p(\Omega)$,*

$$\mathfrak{l}(u) = \int_Q u(x,t)v(x,t)\,dxdt.$$

*Moreover*

$$\|\mathfrak{l}\|_{L^q(0,T;L^p(\Omega))'} := \sup_{\|u\|_{L^q(0,T;L^p(\Omega))}=1} |\mathfrak{l}(u)| = \|v\|_{L^{q'}(0,T;L^{p'}(\Omega))}.$$

*In other words, $L^q(0,T;L^p(\Omega))'$ is isomorphic to $L^{q'}(0,T;L^{p'}(\Omega))$.*

In accordance with Theorem 1.3.28 we have the following definitions for weak convergence in $L^q(0,T;L^p(\Omega))$:

**Definition 1.3.29.** Let $1 < p, q \leq \infty$. A sequence $f_n \in L^q(0,T;L^p(\Omega))$, $n \geq 1$, *converges to $f$ weakly$^\star$ in $L^q(0,T;L^p(\Omega))$* if

$$\lim_{n\to\infty} \int_0^T \int_\Omega f_n\xi\,dxdt = \int_0^T \int_\Omega f\xi\,dxdt \quad \text{for any } \xi \in L^{q'}(0,T;L^{p'}(\Omega)). \qquad (1.3.9)$$

If (1.3.9) holds and $1 < p, q < \infty$ then the sequence $f_n \in L^q(0,T;L^p(\Omega))$, $n \geq 1$, *converges to $f$ weakly in $L^q(0,T;L^p(\Omega))$.*

The following result establishes the weak compactness of bounded sets.

**Theorem 1.3.30.** *Let $f_n \in L^q(0,T;L^p(\Omega))$, $n \geq 1$, be a bounded sequence in $L^q(0,T;L^p(\Omega))$, with $1 < p, q \leq \infty$. Then there are a subsequence, still denoted by $f_n$, and a function $f \in L^q(0,T;L^p(\Omega))$ such that $f_n$ converges to $f$ weakly$^\star$ (weakly for $p, q < \infty$) in $L^q(0,T;L^p(\Omega))$.*

## 1.4 Young measures

Let $\Omega \subset \mathbb{R}^d$ be a measurable set. A Young measure (sliced measure, local probability distribution) is a family of Borel probability measures $\mu_x$ in $\mathbb{R}$, depending on a parameter $x \in \Omega$, such that the mapping $x \mapsto \mu_x$ is weakly measurable with respect to the Lebesgue measure in $\mathbb{R}^d$.

Recall that $C_0(\mathbb{R}) \subset C(\mathbb{R})$ denotes the closed subspace of all continuous functions $u : \mathbb{R} \to \mathbb{R}$ such that $u(x) \to 0$ as $|x| \to \infty$, and $\mathcal{M}(\mathbb{R})$ denotes the dual space $C_0(\mathbb{R})'$ of Radon measures defined by Theorem 1.2.24. It is clear that the Banach space $Y := C_0(\mathbb{R})$ is separable and it meets all requirements of Theorem 1.3.24. Applying that theorem to the space $L^1(\Omega; C_0(\mathbb{R}))$ we conclude that

$$(L^1(\Omega; C_0(\mathbb{R})))' = L^\infty_w(\Omega; \mathcal{M}(\mathbb{R})).$$

Thus we introduce the following definition.

**Definition 1.4.1.** Let $\Omega \subset \mathbb{R}^d$ be a measurable set. A *Young measure* $\mu$ is an element of $L^\infty_w(\Omega; \mathcal{M}(\mathbb{R}))$ such that

$$\mu_x \geq 0, \quad \langle \mu_x, 1 \rangle = 1 \quad \text{for a.e. } x \in \Omega,$$

i.e. $\mu_x$ is a probability measure on $\mathbb{R}$ for a.e. $x \in \Omega$.

It is clear that the norm in $L^\infty_w(\Omega; \mathcal{M}(\mathbb{R}))$ of every Young measure is equal to 1. Hence a Young measure belongs to the unit sphere in this space and the set of Young measures is not compact in the weak* topology. It is of interest when the weak limit of Young measures is also a Young measure. The following lemma gives an answer to this question.

**Lemma 1.4.2.** *Let $\Omega \subset \mathbb{R}^d$ be a bounded measurable set and $\{\mu^n\}_{n\geq 1}$ be a sequence of Young measures. Furthermore, assume that for any $\varepsilon > 0$ there is $N$ such that for all $n$,*

$$\int_{|\lambda|\geq N} d\mu^n_x(\lambda) < \varepsilon.$$

*Then there is a subsequence, still denoted by $\{\mu^n\}$, and a Young measure $\mu \in L^\infty_w(\Omega; \mathcal{M}(\mathbb{R}))$ such that*

$$\mu^n \rightharpoonup \mu \quad \textit{weakly}^\star \textit{ in } L^\infty_w(\Omega; \mathcal{M}(\mathbb{R})).$$

*Proof.* Since $L^1(\Omega; C_0(\mathbb{R}))$ is separable, any bounded subset of the dual space is weak* sequentially compact. Hence the sequence $\{\mu^n\}$ contains a subsequence, still denoted by $\mu^n$, which converges weakly* in $L^\infty_w(\Omega; \mathcal{M}(\mathbb{R}))$ to an element $\mu$. By Theorem 1.3.24, $\mu$ can be represented as a weak* measurable family of Radon measures $\mu_x$. It remains to show that $\mu_x$ is a probability measure for a.e. $x \in \Omega$. To this end choose an arbitrary nonnegative function $\psi \in C_0(\Omega)$, and select $\eta \in C_0(\mathbb{R})$ with $0 \leq \eta \leq 1$, and $\eta(x) = 1$ for $|x| \leq N$. We have

$$\begin{aligned}(1-\varepsilon)\int_\Omega \psi(x)\,dx &\leq \int_\Omega \psi(x)\,dx - \int_\Omega \psi(x)\Big\{\int_\mathbb{R} (1-\eta(x/N))\,d\mu^n_x\Big\}\,dx \\ &= \int_\Omega \psi(x)\Big\{\int_\mathbb{R} \eta(x/N)\,d\mu^n_x\Big\}\,dx \leq \int_\Omega \psi(x)\,dx.\end{aligned}$$

Letting $n \to \infty$ we obtain

$$(1-\varepsilon)\int_\Omega \psi(x)\,dx \leq \int_\Omega \psi(x)\Big\{\int_\mathbb{R} \eta(x/N)\,d\mu_x\Big\}\,dx \leq \int_\Omega \psi(x)\,dx.$$

Letting $N \to \infty$ and next $\varepsilon \to 0$ we arrive at

$$\int_\Omega \psi(x)\Big\{\int_\mathbb{R} d\mu_x\Big\}\,dx = \int_\Omega \psi(x)\,dx \quad \text{for all nonnegative } \psi \in C_0(\Omega).$$

Hence $\mu_x$ is a probability measure for a.e. $x \in \Omega$. □

Further applications require two important definitions.

**Definition 1.4.3.** Let $\Omega \subset \mathbb{R}^d$ be a measurable set and $B \subset \mathbb{R}^m$ be a Borel set. A function $f : \Omega \times B \to \mathbb{R}$ is said to be a *normal integrand* if for a.e. $x \in \Omega$ the function $f(x, \cdot)$ is lower semicontinuous and there exists a Borel function $g : \Omega \times B \to \mathbb{R}$ such that $f(x, \cdot) = g(x, \cdot)$ for a.e. $x \in \Omega$.

**Definition 1.4.4.** Let $\Omega \subset \mathbb{R}^d$ be a measurable set and $B \subset \mathbb{R}^m$ be a Borel set. A function $f : \Omega \times B \to \mathbb{R}$ is said to be a *Carathéodory integrand* if for a.e. $x \in \Omega$ the function $f(x, \cdot)$ is continuous and for every $y \in B$ the function $f(\cdot, y)$ is measurable. Every Carathéodory integrand is a normal integrand.

The following fundamental theorem on Young measures is the most important result of the theory.

**Theorem 1.4.5.** *Let $\Omega$ be a bounded subset of $\mathbb{R}^d$ and let $\{v_n : \Omega \to \mathbb{R}\}$ be a sequence of measurable functions with the following property:*

$$\lim_{t\to\infty} \limsup_{n\to\infty} \operatorname{meas}\{x : |v_n(x)| > t\} = 0. \tag{1.4.1}$$

*Then there is a subsequence, still denoted by $v_n$, and a Young measure $\mu \in L^\infty_w(\Omega; \mathcal{M}(\mathbb{R}))$ such that for any $\varphi \in C_0(\mathbb{R})$:*

- *$\varphi(v_n) \rightharpoonup \overline{\varphi}$ weakly$^\star$ in $L^\infty(\Omega)$, where $\overline{\varphi}(x) = \langle \mu_x, \varphi \rangle$.*
- *For any normal integrand $F : \Omega \times \mathbb{R} \to \mathbb{R}$ bounded from below,*

$$\liminf_{n\to\infty} \int_\Omega F(x, v_n(x))\, dx \geq \int_\Omega \overline{F}(x)\, dx, \quad \textit{where } \overline{F}(x) = \langle \mu_x, F(x, \cdot) \rangle.$$

- *For any Carathéodory integrand $F : \Omega \times \mathbb{R} \to \mathbb{R}$ such that the functions $\{F(x, v_n(x))\}$ are equi-integrable,*

$$\lim_{n\to\infty} \int_\Omega F(x, v_n(x))\, dx = \int_\Omega \overline{F}(x)\, dx, \quad \textit{where } \overline{F}(x) = \langle \mu_x, F(x, \cdot) \rangle.$$

**Remark 1.4.6.** Condition (1.4.1) is equivalent to the following: There is a continuous function $g : \mathbb{R}^+ \to \mathbb{R}$ such that $\lim_{s\to\infty} g(s) = \infty$ and

$$\sup_n \int_\Omega g(|v_n|)\, dx < \infty.$$

As a corollary we obtain the following result on weak convergence in $L^r(\Omega)$.

**Theorem 1.4.7.** *Let $\Omega \subset \mathbb{R}^d$ be a bounded measurable set. Assume that a sequence $v_n \in L^r(\Omega)$, $1 < r < \infty$, converges weakly in $L^r(\Omega)$ to a function $v$. Then after passing to a subsequence we can assume that there is a Young measure $\mu \in L^\infty_w(\Omega; \mathcal{M}(\mathbb{R}))$ such that for all continuous functions $g : \mathbb{R} \to \mathbb{R}$ satisfying*

$$\lim g(s)/|s|^r \to 0 \quad \textit{as } |s| \to \infty, \tag{1.4.2}$$

*$g(v_n)$ converges weakly in $L^1(\Omega)$ to $\overline{g}(x) = \langle \mu_x, g \rangle$. The Young measure $\mu$ has the properties $\langle \mu_x, |s|^r \rangle \in L^1(\Omega)$,*

$$\overline{g}(x) = \langle\, \mu_x,\, g\, \rangle \quad a.e.\ in\ \Omega$$

*for all continuous $g$ satisfying (1.4.2), and*

$$|v(x)|^r = |\langle\, \mu_x, s\, \rangle|^r \leq \langle\, \mu_x, |s|^r\, \rangle \quad a.e.\ in\ \Omega.$$

*Moreover, $v_n$ converges to $v$ strongly in any $L^s(\Omega)$, $1 \leq s < r$, if and only if $\mu_x$ is the Dirac measure at $v(x)$, i.e. $\langle \mu_x, g \rangle = g(v(x))$.*

## 1.5 Sobolev spaces

First we recall some basic facts from the theory of Sobolev-Slobodetsky spaces which can be found in [2] and [129]. Let $\Omega$ be an open subset of $\mathbb{R}^n$. For every multi-index $\alpha = (\alpha_1, \dots, \alpha_d)$ with nonnegative integers $\alpha_i$ and $f \in L^1_{\text{loc}}(\Omega)$, a function $g \in L^1_{\text{loc}}(\Omega)$ is the $\alpha$-th *generalized* (or *distributional*) *derivative* of $f$ (denoted $g = \partial^\alpha f$) if

$$\int_\Omega g\varphi\, dx = (-1)^{\alpha_1 + \cdots + \alpha_d} \int_\Omega f \partial^\alpha \varphi\, dx$$

for all $\varphi \in C_0^\infty(\Omega)$. The number $|\alpha| = \alpha_1 + \cdots + \alpha_d$ is the *order* of the derivative $g$.

For an integer $l \geq 0$ and for an exponent $r \in [1, \infty)$, we denote by $W^{l,r}(\Omega)$ the *Sobolev space* of functions having all generalized derivatives up to order $l$ in $L^r(\Omega)$. Endowed with the norm $\|u\|_{W^{l,r}(\Omega)} = \sup_{|\alpha| \leq l} \|\partial^\alpha u\|_{L^r(\Omega)}$, it becomes a Banach space.

For real $0 < s < 1 < r < \infty$, the fractional Sobolev space $W^{s,r}(\Omega)$ is obtained by the real interpolation method between $L^r(\Omega)$ and $W^{1,r}(\Omega)$, i.e., $W^{s,r}(\Omega) = [L^r(\Omega), W^{1,r}(\Omega)]_{s,r}$, and consists of all measurable functions with the finite norm

$$\|u\|_{W^{s,r}(\Omega)} = \|u\|_{L^r(\Omega)} + |u|_{s,r,\Omega},$$

where

$$|u|^r_{s,r,\Omega} = \int_{\Omega \times \Omega} |x - y|^{-d-rs} |u(x) - u(y)|^r\, dx dy. \tag{1.5.1}$$

In view of Lemma 1.1.11, the space $W^{1,r}(\Omega)$ is dense in $W^{s,r}(\Omega)$. In particular, if $\Omega$ is a bounded domain of class $C^1$, then $C^\infty(\Omega)$ is dense in $W^{s,r}(\Omega)$.

In general, the Sobolev space $W^{l+s,r}(\Omega)$, $0 < s < 1 < r < \infty$, $l \geq 0$ an integer, is defined as the space of measurable functions with the finite norm $\|u\|_{W^{l+s,r}(\Omega)} = \|u\|_{W^{l,r}(\Omega)} + \sup_{|\alpha|=l} \|\partial^\alpha u\|_{W^{s,r}(\Omega)}$.

Furthermore, the notation $W_0^{s,r}(\Omega)$, $0 \leq s \leq 1$, stands for the closed subspace of $W^{s,r}(\mathbb{R}^d)$ which consists of all $u \in W^{s,r}(\mathbb{R}^d)$ vanishing outside of $\Omega$. We identify functions of $W_0^{s,r}(\Omega)$ with their restrictions to $\Omega$. We have $W^{s,r}(\Omega) = W_0^{s,r}(\Omega)$ for $sr < 1$.

For $1 < r < \infty$, set $\mathcal{W}_0^{0,r}(\Omega) = W_0^{0,r}(\Omega)$ and $\mathcal{W}_0^{1,r}(\Omega) = W_0^{1,r}(\Omega)$. For all $0 < s < 1$ and $1 < r < \infty$, we denote by $\mathcal{W}_0^{s,r}(\Omega)$ the interpolation space $[\mathcal{W}_0^{0,r}(\Omega), \mathcal{W}_0^{1,r}(\Omega)]_{s,r}$ endowed with one of the equivalent norms defined by the real interpolation method. In other words, $\mathcal{W}_0^{s,r}(\Omega)$ is obtained by real interpolation of the subspaces $\mathcal{W}_0^{1,r}(\Omega) = W_0^{1,r}(\Omega) \subset W^{1,r}(\Omega)$ and $\mathcal{W}_0^{0,r}(\Omega) = L^r(\Omega)$.

Notice that in view of Lemma 1.1.11, the space $W_0^{1,r}(\Omega)$ is dense in $\mathcal{W}_0^{s,r}(\Omega)$ for $0 < s < 1 < r < \infty$. It follows that $C_0^\infty(\Omega)$ is dense in $\mathcal{W}_0^{s,r}(\Omega)$ and hence $\mathcal{W}_0^{s,r}(\Omega)$ is the completion of $C_0^\infty(\Omega)$ in the $\mathcal{W}_0^{s,r}$-norm. But in general $\mathcal{W}_0^{s,r}(\Omega)$ does not coincide with the completion of $C_0^\infty(\Omega)$ in the $W^{s,r}$-norm.

Generally speaking, $\mathcal{W}_0^{s,r}(\Omega)$ is not a subspace of $W^{s,r}(\Omega)$, and its norm is stronger than the norm of $W^{s,r}(\Omega)$.

**Embedding theorems.** Let $\Omega$ be the whole $\mathbb{R}^n$ or a bounded domain with $C^1$ boundary. Let $sr > d$ and $0 \le \alpha < s - d/r$ be real numbers. Then the embedding $W^{s,r}(\Omega) \hookrightarrow C^\alpha(\Omega)$ is continuous. If $\Omega$ is a bounded domain of class $C^1$, then the embedding is compact, i.e., every bounded subset of $W^{s,r}(\Omega)$ is relatively compact in $C^\alpha(\Omega)$. Moreover, for $sr > d$, and all $u, v \in W^{s,r}(\Omega)$,

$$\|uv\|_{W^{s,r}(\Omega)} \le c(r,s)\|u\|_{W^{s,r}(\Omega)}\|v\|_{W^{s,r}(\Omega)}. \tag{1.5.2}$$

Let $\Omega$ be a bounded domain of class $C^1$. If $sr < d$ and $t^{-1} = r^{-1} - d^{-1}s$, then the embedding $W^{s,r}(\Omega) \hookrightarrow L^t(\Omega)$ is continuous. If $\Omega$ is an arbitrary domain of class $C^1$, then the embedding is continuous under the additional assumption $r < d$ [2, Thm. 7.57].

We also have (see [2, Thm. 7.58]), for $s > 0$, $1 < r < \beta < \infty$ and $\alpha = s - dr^{-1} + d\beta^{-1}$,

$$\|u\|_{W^{\alpha,\beta}(\Omega)} \le c(r,s,\alpha,\beta,\Omega)\|u\|_{W^{s,r}(\Omega)}. \tag{1.5.3}$$

**Duality.** Let $\Omega$ be the whole $\mathbb{R}^n$ or a bounded domain with $C^1$ boundary. We fix $s \in [0,1]$. We define

$$\langle u, v\rangle = \int_\Omega uv\,dx \tag{1.5.4}$$

for all functions such that the right hand side makes sense. For $r \in (1,\infty)$, each element $v \in L^r(\Omega)$ determines a functional $\mathcal{L}_v$ in $(W_0^{s,r'}(\Omega))'$, $r' = r/(r-1)$, $s \ge 0$, by the identity $\mathcal{L}_v(u) = \langle u, v\rangle$. We introduce the $(-s,r)$-norm of an element $v \in L^r(\Omega)$ to be, by definition, the norm of the functional $\mathcal{L}_v$, that is,

$$\|v\|_{W^{-s,r}(\Omega)} = \sup_{\substack{u\in W_0^{s,r'}(\Omega)\\ \|u\|_{W_0^{s,r'}(\Omega)}=1}} |\langle v, u\rangle|. \tag{1.5.5}$$

Let $W^{-s,r}(\Omega)$ be the completion of $L^r(\Omega)$ with respect to the $(-s,r)$-norm. For an integer $s$, the space $W^{-s,r}(\Omega)$ is topologically and algebraically isomorphic to

$(W_0^{s,r'}(\Omega))'$ and can be identified with it. Notice that for $0 < s < 1$, the Banach space $W^{-s,r}(\Omega)$ is not an interpolation space between $L^r(\Omega)$ and $W_0^{1,r}(\Omega)$.

Similarly, we can define $\mathcal{W}^{-s,r}(\Omega)$ to be the completion of $L^r(\Omega)$ in the norm

$$\|v\|_{\mathcal{W}^{-s,r}(\Omega)} = \sup_{\substack{u\in\mathcal{W}_0^{s,r'}(\Omega)\\ \|u\|_{\mathcal{W}_0^{s,r'}(\Omega)}=1}} |\langle v, u\rangle|. \tag{1.5.6}$$

We intensively exploit the spaces $\mathcal{W}^{-s,r}(\Omega)$ in Chapter 11. The main properties of these spaces are listed in the following lemma.

**Lemma 1.5.1.** *Let real $s, r$ satisfy $0 < s < 1 < r < \infty$. Then*

- *$\mathcal{W}^{-s,r}(\Omega)$ is topologically and algebraically isomorphic to the Banach space $(\mathcal{W}_0^{s,r'}(\Omega))'$ and can be identified with it;*
- *$\mathcal{W}^{-s,r}(\Omega)$ is topologically and algebraically isomorphic to the interpolation space $[L^r(\Omega), W_0^{-1,r}(\Omega)]_{s,r} = ([L^{r'}(\Omega), W_0^{1,r'}(\Omega)]_{s,r'})'$ and can be identified with it;*
- *$(L^{r'}(\Omega))'$, which can be identified with $L^r(\Omega)$, is dense in $\mathcal{W}^{-s,r}(\Omega)$ and hence $C_0^\infty(\Omega)$ is dense in $\mathcal{W}^{-s,r}(\Omega)$.*

*Proof.* Observe that for $v \in L^r(\Omega)$, $\mathcal{L}_v : u \mapsto \langle v, u\rangle$ is a continuous functional on $L^{r'}(\Omega)$, i.e., $\mathcal{L}_v \in (L^{r'}(\Omega))'$. Since $\mathcal{W}_0^{s,r'}(\Omega)$ is continuously embedded in $L^{r'}(\Omega)$, $\mathcal{L}_v$ is a continuous functional on $\mathcal{W}_0^{s,r'}(\Omega)$, i.e., $\mathcal{L}_v \in (\mathcal{W}_0^{s,r'}(\Omega))'$. In view of (1.5.6), the norm $\|v\|_{\mathcal{W}^{-s,r}(\Omega)}$ is exactly the norm $\|\mathcal{L}_v\|_{(\mathcal{W}_0^{s,r'}(\Omega))'}$. From this and the definition of $\mathcal{W}^{-s,r}(\Omega)$ we conclude that $\mathcal{W}^{-s,r}(\Omega)$ is the completion of $(L^{r'}(\Omega))'$ in the $(\mathcal{W}_0^{s,r'}(\Omega))'$-norm.

Notice that $W_0^{1,r'}(\Omega)$ is dense in $L^{r'}(\Omega)$. Setting $A = A_0 = L^{r'}(\Omega)$, $A_1 = W_0^{1,r'}(\Omega)$ and applying Lemma 1.1.12 we conclude that

$$([L^{r'}(\Omega), W_0^{1,r'}(\Omega)]_{s,r'})' = [(L^{r'}(\Omega))', (W_0^{1,r'}(\Omega))']_{s,r}$$

By the definition of $\mathcal{W}_0^{s,r'}(\Omega)$, we have

$$(\mathcal{W}_0^{s,r'}(\Omega))' \equiv ([L^{r'}(\Omega), W_0^{1,r'}(\Omega)]_{s,r'})'.$$

Thus we get

$$(\mathcal{W}_0^{s,r'}(\Omega))' = [(L^{r'}(\Omega))', (W_0^{1,r'}(\Omega))']_{s,r}. \tag{1.5.7}$$

Next, the inclusion $W_0^{1,r'}(\Omega) \subset L^{r'}(\Omega)$ implies $(L^{r'}(\Omega))' \subset (W_0^{1,r'}(\Omega))'$. Applying Lemma 1.1.11 we find that $(L^{r'}(\Omega))'$ is dense in $[L^{r'}(\Omega)', (W_0^{1,r'}(\Omega))']_{s,r}$. From this and (1.5.7) we conclude that $L^{r'}(\Omega)'$ is dense in $(\mathcal{W}_0^{s,r'}(\Omega))'$. Hence $(\mathcal{W}_0^{s,r'}(\Omega))'$

is the completion of $(L^{r'}(\Omega))'$ in the $(\mathcal{W}_0^{s,r'}(\Omega))'$-norm. But so is $\mathcal{W}^{-s,r}(\Omega)$. Thus we get

$$\mathcal{W}^{-s,r}(\Omega) = (\mathcal{W}_0^{s,r'}(\Omega))' = [L^r(\Omega), W_0^{-1,r}(\Omega)]_{s,r} = ([L^{r'}(\Omega), W_0^{1,r'}(\Omega)]_{s,r'})'.$$

It remains to note that $C_0^\infty(\Omega)$ is dense in $L^r(\Omega)$ which is identified with $(L^{r'}(\Omega))'$. Hence $C_0^\infty(\Omega)$ is dense in $\mathcal{W}^{-s,r}(\Omega)$. □

It is well known that the operator $\nabla : W^{s,r}(\Omega) \to W^{s-1,r}(\Omega)$ is continuous for $s = 0, 1$. It follows that $\nabla : W^{s,r}(\Omega) \to \mathcal{W}^{s-1,r}(\Omega)$ is continuous for all $s \in (0,1)$, and hence

$$\|\nabla u\|_{\mathcal{W}^{s-1,r}(\Omega)} \le c(s,r)\|u\|_{W^{s,r}(\Omega)} \quad \text{for all } s \in (0,1), \ \ 1 < r < \infty. \tag{1.5.8}$$

**Extension.** Let $k \ge 1$ be an integer and $\Omega \subset \mathbb{R}^d$ be a bounded domain of class $C^k$. Then for every $r \in [1,\infty]$ there exists a bounded extension operator $\Xi : W^{k,r}(\Omega) \to W^{k,r}(\mathbb{R}^d)$ such that $\Xi u = u$ in $\Omega$ and

$$\|\Xi u\|_{W^{k,r}(\mathbb{R}^d)} \le c\|u\|_{W^{k,r}(\Omega)},$$

where $c$ depends on $k$ and $\Omega$.

## 1.6 Mollifiers and DiPerna & Lions lemma

A nonnegative function $\theta : \mathbb{R} \to \mathbb{R}^+$ is said to be a *mollifying kernel* in $\mathbb{R}$ if

$$\theta \in C_0^\infty(\mathbb{R}), \quad \int_{\mathbb{R}} \theta(s)\, ds = 1, \quad \operatorname{supp} \theta \subset (-1,1), \quad \theta(-s) = \theta(s). \tag{1.6.1}$$

For any $f \in L^1_{\text{loc}}(\mathbb{R}^d \times \mathbb{R})$ define the mollifiers

$$[f]_{,k}(x,s) := k\int_{\mathbb{R}} \theta(k(\tau - s)) f(x,\tau)\, d\tau, \tag{1.6.2}$$

$$[f]_{m,}(x,s) := m^d \int_{\mathbb{R}^d} \Theta(m(y-x)) f(y,s)\, dy, \quad \Theta(x) = \prod_{i=1}^d \theta(x_i). \tag{1.6.3}$$

We write simply $[f]_{m,k}$ instead of $[[f]_{m,}]_{,k}$. Notice that the composition of mollifiers is also a mollifier. More precisely, we have

$$[[f]_{,k}]_{,k}(x,s) = k\int_{\mathbb{R}} \theta_2(k(\tau-s)) f(x,\tau)\, d\tau, \quad \text{where } \theta_2(s) = \int_{\mathbb{R}} \theta(s-\xi)\theta(\xi)\, d\xi. \tag{1.6.4}$$

Recall some simple properties of mollifying operators. For any function $f$ locally integrable in $\mathbb{R}^d$, the mollified function $[f]_{m,}$ has continuous derivatives of all orders in $\mathbb{R}^d$. Let $\Omega$ be a bounded domain in $\mathbb{R}^d$ and extend $f \in L^r(\Omega)$, $1 \le r < \infty$, by zero outside of $\Omega$ over the whole $\mathbb{R}^d$. Then

$$\|f - [f]_{m,}\|_{L^r(\Omega)} \to 0 \quad \text{as } m \to \infty.$$

If, in addition, $f$ depends on the time variable $t \in (0, T)$ and belongs to the space $L^q(0, T; L^r(\Omega))$, $1 \le q \le \infty$, $1 \le r < \infty$,

$$\|f - [f]_{m,}\|_{L^q(0,T;L^r(\Omega))} \to 0 \quad \text{as } m \to \infty \quad \text{when } q < \infty$$

and

$$\|f - [f]_{m,}\|_{L^s(0,T;L^r(\Omega))} \to 0 \quad \text{as } m \to \infty \quad \text{for all } s < \infty.$$

Next if $f \in L^q(0, T; W^{s,r}(\mathbb{R}^d))$, $s \in \mathbb{R}$, $1 < q, r < \infty$, then

$$\|f - [f]_{m,}\|_{L^q(0,T;W^{s,r}(\mathbb{R}^d))} \to 0 \quad \text{as } m \to \infty.$$

The commutator of a mollifying operator and the operator of multiplication by a function from a Sobolev space is bounded in a suitable Lebesgue space, and its norm vanishes as the parameter $m$ tends to zero. The following lemma is a particular case of a general result due to DiPerna and P.-L. Lions [23].

**Lemma 1.6.1.** *Suppose that $f \in L^\infty(0, T; L^r_{\text{loc}}(\mathbb{R}^d))$ and $\mathbf{u} \in L^1(0, T, W^{1,2}_{\text{loc}}(\mathbb{R}^d))$. This means that $f \in L^\infty(0, T; L^r(B))$ and $\mathbf{u} \in L^1(0, T, W^{1,2}(B))$ for every ball $B \subset \mathbb{R}^d$. Then for $q^{-1} = 2^{-1} + r^{-1}$ and a compact set $E \subset \mathbb{R}^d$,*

$$\operatorname{div}([f\mathbf{u}]_{m,} - [f]_{m,}\mathbf{u}) \to 0 \quad \text{in } L^1(0, T; L^q(E)) \quad \text{as } m \to \infty. \tag{1.6.5}$$

The main application of this lemma is to justify the renormalization procedure for weak solutions to the linear transport equation

$$\begin{aligned} &\partial_t \varrho + \operatorname{div}(\varrho \mathbf{u}) = F \quad \text{in } \mathbb{R}^d \times (0, T), \\ &\varrho(x, 0) = \varrho_0(x) \quad \text{in } \mathbb{R}^d. \end{aligned} \tag{1.6.6}$$

If $\varrho$ and $\mathbf{u}$ are sufficiently smooth, then it is easy to check that for any $C^1$ function $\varphi : \mathbb{R} \to \mathbb{R}$, the composite function $\varphi(\varrho)$ satisfies

$$\begin{aligned} &\partial_t \varphi(\varrho) + \operatorname{div}(\varphi(\varrho)\mathbf{u}) + (\varphi'(\varrho) - \varphi(\varrho)) \operatorname{div} \mathbf{u} = \varphi'(\varrho) F \quad \text{in } \mathbb{R}^d \times (0, T), \\ &\varphi(\varrho)(x, 0) = \varphi(\varrho_0)(x) \quad \text{in } \mathbb{R}^d. \end{aligned} \tag{1.6.7}$$

(1.6.7) is called the renormalized equation and derivation of (1.6.7) from the original equation (1.6.6) is called the renormalization procedure. The justification of the renormalization procedure for weak nonsmooth solutions is nontrivial. Throughout this book we use the following lemma which is a particular case of a general result established in [23]:

**Lemma 1.6.2.** *Assume that functions* $\varrho \in L^\infty(0,T;L^r(\mathbb{R}^d))$, $\varrho_0 \in L^r(\mathbb{R}^d)$ *with*

$$2 < r \le \infty, \quad r^{-1} \le 2^{-1} - d^{-1}$$

*vanish for* $|x| > R$, *and a function* $F \in L^1(0,T;L^1(\Omega))$ *also vanishes for* $|x| > R$, *with large* $R$. *Furthermore, assume that a vector field* $\mathbf{u} \in L^\alpha(0,T;W^{1,2}(\mathbb{R}^d))$, $\alpha > 1$, *and the functions* $\varrho$, $F$ *satisfy the integral identity*

$$\int_0^T \int_{\mathbb{R}^d} \varrho((\partial_t \psi + \nabla\psi \cdot \mathbf{u}) + \psi F)\, dxdt + \int_{\mathbb{R}^d} \psi(x,0)\varrho_0\, dx = 0 \tag{1.6.8}$$

*for all* $\psi \in C^\infty(\mathbb{R}^d \times (0,T))$ *vanishing for all sufficiently large* $|x|$ *and for* $t = T$. *Then the integral identity*

$$\int_0^T \int_{\mathbb{R}^d} \varphi(\varrho)(\partial_t\psi + \nabla\psi\cdot\mathbf{u})\, dxdt + \int_0^T \int_{\mathbb{R}^d} \varphi'(\varrho)\psi F\, dxdt$$
$$- \int_0^T \int_{\mathbb{R}^d} (\varphi'(\varrho)\varrho - \varphi'(\varrho)) \operatorname{div}\mathbf{u}\, dxdt + \int_{\mathbb{R}^d} \psi(x,0)\varrho_0\, dx = 0 \tag{1.6.9}$$

*holds for any* $C^1$ *function* $\varphi : \mathbb{R} \to \mathbb{R}$ *such that*

$$|\varphi(s)| + |(1+|s|)\varphi'(s)| \le M < \infty.$$

**Mollifiers and functions of bounded variation.** The following lemma is widely used when applying the kinetic equation method to Navier-Stokes equations.

**Lemma 1.6.3.** *Let* $f : \mathbb{R} \to \mathbb{R}$ *be a monotone function of bounded variation and let* $H$ *be a bounded function on* $\mathbb{R}$ *such that the limits* $\lim_{\tau\to s\pm 0} H(\tau)$ *exist at each* $s \in \mathbb{R}$. *Then for any* $\psi \in C(\mathbb{R})$,

$$\lim_{k\to\infty} [\psi[H]_{,k}]_{,k} = \lim_{k\to\infty} [\psi H]_{,k} = \psi\tilde{H}, \tag{1.6.10}$$

*where*

$$\tilde{H}(s) = \frac{1}{2}\lim_{h\searrow 0} H(s-h) + \frac{1}{2}\lim_{h\searrow 0} H(s+h).$$

*The function* $\tilde{H}$ *is Borel and*

$$\lim_{k\to\infty} \int_{\mathbb{R}} \psi[H]_{,k}\partial_s[f]_{,k}\, ds \equiv \lim_{k\to\infty} \int_{\mathbb{R}} \psi[H]_{,k}\, d[f]_{,k} = \int_{\mathbb{R}} \psi\tilde{H}\, df. \tag{1.6.11}$$

*In particular,*

$$2\int_{\mathbb{R}} \psi\tilde{f}\, d_s f(s) = \int_{\mathbb{R}} \psi\, d_s f^2(s). \tag{1.6.12}$$

*Proof.* We have

$$[\psi[H]_{,k}]_{,k}(s) = \psi[[H]_{,k}]_{,k}(s) + k\int \theta(ks-k\tau)(\psi(\tau)-\psi(s))[H]_{,k}(\tau)\,d\tau.$$

Since $\psi$ is continuous and $H$ is bounded, we have

$$\begin{aligned} k\Big|\int_{\mathbb{R}} \theta(k\tau-ks)(\psi(\tau)-\psi(s))[H]_{,k}(\tau)\,d\tau\Big| &\le ck\int_{\mathbb{R}} \theta(k\tau-ks)|\psi(\tau)-\psi(s)|\,d\tau \\ &= c\int_{\mathbb{R}} \theta(\tau)|\psi(s+\tau/k)-\psi(s)|\,d\tau \to 0 \quad \text{as } k\to\infty. \end{aligned} \tag{1.6.13}$$

Notice that the kernel $\theta_2$ is even, which yields

$$\tilde{H}(s) = H^-(s)k\int_{-\infty}^{s} \theta_2(k(\tau-s))\,d\tau + H^+(s)k\int_{s}^{\infty}\theta_2(k(\tau-s))\,d\tau,$$

where $H^\pm(s) = \lim_{\tau\to s\pm 0} H(\tau)$. From this and (1.6.4) we get

$$\begin{aligned} [[H]_{,k}]_{,k} - \tilde{H}(s) = &\int_{-\infty}^{s} \theta_2(\tau)(H(s+\tau/k)-H^-(s))\,d\tau \\ &+\int_{s}^{\infty} \theta_2(\tau)(H(s+\tau/k)-H^+(s))\,d\tau \to 0 \end{aligned} \tag{1.6.14}$$

as $k\to\infty$. Combining (1.6.13) and (1.6.14) we obtain $\lim_{k\to\infty}[\psi[H]_{,k}]_{,k} = \psi\tilde{H}$. Repeating these arguments we obtain $\lim_{k\to\infty}[\psi H]_{,k} = \psi\tilde{H}$. It follows that $\tilde{H}$ is a pointwise limit of continuous functions $[H]_{,k} = \psi\tilde{H}$ and hence it is Borel. Next,

$$\int_{\mathbb{R}} \psi[H]_{,k}\,d[f]_{,k} = \int_{\mathbb{R}} \psi[H]_{,k}(s)\partial_s\Big\{k\int_{\mathbb{R}}\theta(k(\tau-s))f(\tau)\,d\tau\Big\}\,ds. \tag{1.6.15}$$

Integrating by parts in the Stieltjes integral gives

$$\begin{aligned} \partial_s\Big\{k\int_{\mathbb{R}}\theta(k(\tau-s))f(\tau)\,d\tau\Big\} &= -k^2\int_{\mathbb{R}}\theta'(k(\tau-s))f(\tau)\,d\tau \\ &= k\int_{\mathbb{R}}\theta(k(\tau-s))\,d\tau f. \end{aligned}$$

Since $\psi[H]_{,k}$ is continuous and bounded, the Fubini theorem yields

$$\begin{aligned} &\int_{\mathbb{R}} \psi[H]_{,k}(s)\Big\{k\int_{\mathbb{R}}\theta(k(\tau-s))\,d\tau f\Big\} \\ &\quad = \int_{\mathbb{R}}\Big\{k\int_{\mathbb{R}}\psi[H]_{,k}(s)\theta(k(\tau-s))\,ds\Big\}\,d_\tau f(\tau) = \int_{\mathbb{R}}[\psi[H]_{,k}]_{,k}\,d_\tau f(\tau). \end{aligned}$$

Thus we get

$$\int_{\mathbb{R}} \psi[H]_{,k}\,d[f]_{,k} = \int_{\mathbb{R}}[\psi[H]_{,k}]_{,k}\,d_\tau f(\tau).$$

Letting $k \to \infty$, noting that the continuous functions $[\psi[H]_{,k}]_{,k}$ converge pointwise to $\psi\tilde{H}$ and applying the Lebesgue dominated convergence theorem we arrive at (1.6.11). It remains to prove (1.6.12). Assume for a moment that $\psi \in C^1(\mathbb{R})$. Notice that

$$2\int_{\mathbb{R}} \psi[f]_{,k}\, d[f]_{,k} = \int_{\mathbb{R}} \psi\, d[f]_{,k}^2 = -\int_{\mathbb{R}} \psi'[f]_{,k}^2\, ds \to -\int_{\mathbb{R}} \psi' \tilde{f^2}\, ds. \tag{1.6.16}$$

A function of bounded variation is continuous everywhere except on some countable set. Hence $\tilde{f^2} = f^2$ a.e. in $\mathbb{R}$. From this and the integration by parts formula in Lemma 1.3.7 we obtain

$$\int_{\mathbb{R}} \psi' \tilde{f^2}\, ds = \int_{\mathbb{R}} \psi' f^2\, ds = -\int_{\mathbb{R}} \psi\, d_s f^2. \tag{1.6.17}$$

On the other hand relation (1.6.11) with $H$ replaced by $f$ gives

$$2\lim_{k\to\infty} \int_{\mathbb{R}} \psi[f]_{,k}\, d[f]_{,k} = 2\int_{\mathbb{R}} \psi\tilde{f}\, df.$$

Combining this with (1.6.16) and (1.6.17) we obtain (1.6.12) for $\psi \in C^1(\mathbb{R})$. It remains to note that every function $\psi \in C(\mathbb{R})$ is a pointwise limit of a sequence of uniformly bounded functions $\psi_n \in C^1(\mathbb{R})$, say $\psi_n = [\psi]_{,n}$. □

## 1.7 Partial differential equations—selected facts

### 1.7.1 Elliptic equations

**Poisson equation.** The simplest elliptic equation is the Poisson equation in $\mathbb{R}^d$,

$$\Delta u = f \quad \text{in } \mathbb{R}^d, \quad d \geq 2 \tag{1.7.1}$$

which is understood in the sense of the theory of distributions. This means that the functions $u$ and $f$ are integrable on every compact subset of $\mathbb{R}^d$ and

$$\int_{\mathbb{R}^d} u\Delta\varphi\, dx = \int_{\mathbb{R}^d} f\varphi\, dx \quad \text{for all} \quad \varphi \in C_0^\infty(\mathbb{R}^d).$$

Its solution is given by explicit formulae, which along with the classical estimates of the Newton potential lead to the following result:

**Lemma 1.7.1.** *For any bounded set $\Omega$ and*

$$f \in W^{s-2,r}(\mathbb{R}^d), \quad s \in \mathbb{R}, \quad 1 < r < \infty,$$

*with* $\operatorname{supp} f \subset \Omega$, *equation* (1.7.1) *has a solution with the properties:*

- *This solution is analytic outside of* $\Omega$.
- *It satisfies*

$$\limsup_{|x|\to\infty} |x|^{d-2}|u(x)| < \infty \quad \textit{for } d > 2,$$
$$\limsup_{|x|\to\infty} (\log|x|)^{-1}|u(x)| < \infty \quad \textit{for } d = 2,$$
$$\|u\|_{W^{s,r}(B_R)} \le c(R, s, r, \Omega)\|f\|_{W^{s-2,r}(\mathbb{R}^d)}.$$

*Here,* $B_R$ *is the ball* $\{x \in \mathbb{R}^d : |x| < R\}$ *of an arbitrary radius* $R < \infty$*, and the constant* $c$ *depends only on* $R$*,* $s$*,* $r$*, and* $\Omega$*.*

The relation $f \to u$ determines a linear operator $u = \Delta^{-1}[f]$. In this framework we can define the linear operators

$$A_j = \partial_{x_j}\Delta^{-1}, \quad R_{kj} = \partial_{x_k}\partial_{x_j}\Delta^{-1}, \quad 1 \le k, j \le d.$$

The *Riesz operator* $R_{kj}$ is a singular integral operator and by the Zygmund-Calderón theorem (see [127]) it is bounded in any space $L^p(\mathbb{R}^d)$ with $1 < p < \infty$. The Zygmund-Calderón theorem and Lemma 1.7.1 imply

**Corollary 1.7.2.** *Under the assumptions of Lemma 1.7.1 for every* $s \in \mathbb{R}$*,* $r \in (1,\infty)$*, and* $f \in W^{s,r}(\mathbb{R}^d)$*,*

$$\|A_j[f]\|_{W^{s+1,r}(B_R)} \le c_0\|f\|_{W^{s,r}(\mathbb{R}^d)},$$
$$\|R_{kj}[f]\|_{W^{s,r}(\mathbb{R}^d)} \le c_1\|f\|_{W^{s,r}(\mathbb{R}^d)}.$$

*where* $c_0$ *depends only on* $R > 0$*,* $s$*,* $r$*, and* $c_1$ *depends only on* $s$ *and* $r$*.*

**Corollary 1.7.3.** *For every* $r \in (1,\infty)$ *and* $f \in L^r(\mathbb{R}^d)$*,* $g \in L^{r'}(\mathbb{R}^d)$*,* $r' = r/(r-1)$*,*

$$\int_{\mathbb{R}^d} gR_{ij}[f]\,dx = \int_{\mathbb{R}^d} fR_{ij}[g]\,dx.$$

**Boundary value problems for elliptic equations.** In Chapter 5 we use existence and uniqueness results for solutions to elliptic equations. Denote by $\alpha$ a vector with nonnegative integer components

$$\alpha = (\alpha_1, \dots, \alpha_d), \quad \alpha_i \in \mathbb{N} \cup \{0\}, \quad |\alpha| = \alpha_1 + \dots + \alpha_d,$$

and set

$$\xi^\alpha = \xi_1^{\alpha_1}\dots\xi_d^{\alpha_d} \quad \text{for } \xi \in \mathbb{R}^d, \quad \partial^\alpha = \partial_{x_1}^{\alpha_1}\cdots\partial_{x_d}^{\alpha_d}.$$

**Remark 1.7.4.** We also use the notation $\partial_x^k u$ for the totality of derivatives of order $k$ of a function $u$, and the notation $\partial_n^k u$ for the normal derivative of order $k$ of $u$ at the boundary of the domain of definition of $u$.

Let $\Omega$ be a bounded domain in $\mathbb{R}^d$ with $C^\infty$ boundary. Assume that functions $b_{\alpha\beta} \in C^\infty(\Omega)$, $|\alpha| = |\beta| = m$, satisfy the ellipticity condition

$$c_b^{-1}|\xi|^{2m} \le b_{\alpha\beta}\xi^\alpha\xi^\beta \le c_b|\xi|^{2m} \quad \text{for all } \xi \in \mathbb{R}^d, \quad c_b > 0$$

(recall the summation convention). Let us consider the boundary value problem

$$\sum_{|\alpha|=|\beta|=m} \partial^\beta(b_{\alpha\beta}\partial^\alpha v) = f \quad \text{in } \Omega, \quad \partial^\kappa v = 0 \quad \text{on } \partial\Omega, \quad 0 \le |\kappa| \le m-1. \quad (1.7.2)$$

The following theorem (see [3], [78]) is a particular case of the general theory of boundary value problems for elliptic equations.

**Theorem 1.7.5.** *Let $s \ge -m$ be an integer. Then for any $f \in W^{s,2}(\Omega)$ problem (1.7.2) has a unique solution*

$$v \in W^{2m+s,2}(\Omega), \quad \text{with} \quad v \in W_0^{m,2}(\Omega),$$

*which satisfies the estimate*

$$\|v\|_{W^{2m+s,2}(\Omega)} \le c\|f\|_{W^{s,2}(\Omega)}.$$

*Here, the constant $c$ depends only on $b_{\alpha\beta}$, $\Omega$ and $s$.*

**Remark 1.7.6.** For $s \in [-m, 0]$ equation (1.7.2) is understood in the sense of distributions, which means that we have the integral identity

$$(-1)^{|\beta|}\int_\Omega b_{\alpha\beta}\partial^\alpha v\partial^\beta\varphi\,dx = \langle f, \varphi\rangle \quad \text{for all } \varphi \in C_0^\infty(\Omega).$$

We also have the following result:

**Lemma 1.7.7.** *Let $\Omega \subset \mathbb{R}^d$, $d = 2, 3$, be a bounded domain with $C^\infty$ boundary. Let $b \in C^\infty(\Omega)$ with $b > \sigma > 0$. Furthermore, assume that $\varrho \in L^1(\Omega)$ is a nonnegative function and the operator $\mathbb{A} : W_0^{4,2}(\Omega) \to W^{-4,2}(\Omega)$ is defined by*

$$\langle\mathbb{A}[\varphi], \psi\rangle = \int_\Omega (b\Delta^2\varphi\Delta^2\psi + b\varphi\psi + \varrho\varphi\psi)\,dx \quad \textit{for all } \varphi, \psi \in W_0^{4,2}(\Omega). \quad (1.7.3)$$

*Then the inverse $\mathbb{A}^{-1} : W^{-4,2}(\Omega) \to W_0^{4,2}(\Omega)$ exists and satisfies*

$$\|\mathbb{A}^{-1}[\psi]\|_{W_0^{4,2}(\Omega)} \le \sigma^{-1}\|\psi\|_{W^{-4,2}(\Omega)} \quad \textit{for all } \psi \in W^{-4,2}(\Omega). \quad (1.7.4)$$

*Proof.* Note that $W_0^{4,2}(\Omega) \hookrightarrow C^2(\Omega)$ and

$$\|\varphi\|_{W_0^{4,2}(\Omega)} \sim \Big\{\int_\Omega (|\Delta\varphi|^2 + |\varphi|^2)\,dx\Big\}^{1/2}.$$

Hence the space $V = W_0^{4,2}(\Omega)$ and the operator $\mathbb{A}$ meet all requirements of Lemma 1.1.14, and the proof is complete. □

As a consequence of Theorem 1.7.5 and Corollary 1.7.7 we obtain the following lemma for systems of elliptic equations.

**Lemma 1.7.8.** *Let $d = 2, 3$, $\tau \in (0, 1)$, and*

$$0 \le \varrho \in L^2(\Omega), \quad b \in C^\infty(\Omega), \quad b > \sigma > 0.$$

*Then for all vector fields $\mathbf{f} \in L^2(\Omega)$, the boundary value problem*

$$\frac{1}{\tau}\Delta^2(b\Delta^2\mathbf{v}) + \frac{1}{\tau}b\mathbf{v} + \varrho\mathbf{v} = \mathbf{f} \quad \text{in } \Omega, \quad \mathbf{v} \in W_0^{4,2}(\Omega), \tag{1.7.5}$$

*has a unique solution; moreover, this solution satisfies*

$$\|\mathbf{v}\|_{W^{8,2}(\Omega)} \le c\tau\|\mathbf{f}\|_{L^2(\Omega)}.$$

*Here, the constant $c$ depends only on $b$, $\varrho$, and $\Omega$.*

*Proof.* Since all requirements of Lemma 1.7.7 are satisfied, it follows that problem (1.7.5) has a unique solution, and

$$\|\mathbf{v}\|_{W^{4,2}(\Omega)} \le c\tau\|\mathbf{f}\|_{L^2(\Omega)}.$$

Noting that the embedding $W^{4,2}(\Omega) \hookrightarrow C(\Omega)$ is compact we conclude that the vector field $\mathbf{F} = \tau\mathbf{f} - \tau\varrho\mathbf{v} - b\mathbf{v}$ satisfies the estimate

$$\|\mathbf{F}\|_{L^2(\Omega)} \le \tau\|\mathbf{f}\|_{L^2(\Omega)} + \tau\|\varrho\|_{L^2(\Omega)}\|\mathbf{v}\|_{W^{4,2}(\Omega)} + \|b\|_{C(\Omega)}\|v\|_{W^{4,2}(\Omega)} \le c\tau\|\mathbf{f}\|_{L^2(\Omega)}.$$

On the other hand, the elliptic equation (1.7.5) with the homogeneous boundary conditions can be rewritten as

$$\Delta^2(b\Delta^2\mathbf{v}) + bv = \mathbf{F} \quad \text{in } \Omega, \quad \mathbf{v} \in W_0^{4,2}(\Omega).$$

Since the operator $\mathbf{z} \mapsto \Delta^2(b\Delta^2\mathbf{z})$ satisfies all hypotheses of Theorem 1.7.5 we have

$$\|\mathbf{v}\|_{W^{8,2}(\Omega)} \le c\|\mathbf{F}\|_{L^2(\Omega)},$$

completing the proof. □

### 1.7.2 Stokes problem

We need an auxiliary result for the Stokes problem which we give with the proof.

**Stokes equations.** The following lemma is a straightforward consequence of classical results on solvability of the first boundary value problem for Stokes equations (see [50]) and interpolation theory. For $\Omega$ bounded, consider the projection of $L^1(\Omega)$ onto the subspace of codimension one which consists of the mean value zero functions,

$$\Pi q := q - \frac{1}{\operatorname{meas}\Omega}\int_\Omega q\,dx, \quad q \in L^1(\Omega). \tag{1.7.6}$$

**Lemma 1.7.9.** *Let $\Omega \subset \mathbb{R}^3$ be a bounded domain with $\partial\Omega \in C^2$ and $(F, G) \in \mathcal{W}^{s-1,r}(\Omega) \times W^{s,r}(\Omega)$ for $0 < s < 1, 1 < r < \infty$, $(F, G) \in W^{s-1,r}(\Omega) \times W^{s,r}(\Omega)$ for $s = 0, 1, 1 < r < \infty$. Then the boundary value problem*

$$\begin{aligned} &\Delta \mathbf{u} - \nabla \pi = F, \quad \operatorname{div} \mathbf{u} = \Pi G \quad \text{in } \Omega, \\ &\mathbf{u} = 0 \quad \text{on } \partial\Omega, \quad \Pi\pi = \pi, \end{aligned} \tag{1.7.7}$$

*has a unique solution $(\mathbf{u}, \pi) \in W^{s+1,r}(\Omega) \times W^{s,r}(\Omega)$ such that.*

$$\|\mathbf{u}\|_{W^{s+1,r}(\Omega)} + \|\pi\|_{W^{s,r}(\Omega)} \le c(\Omega, r, s)(\|F\|_{\mathcal{W}^{s-1,r}(\Omega)} + \|G\|_{W^{s,r}(\Omega)}) \tag{1.7.8}$$

*for $0 < s < 1 < r < \infty$ and*

$$\|\mathbf{u}\|_{W^{s+1,r}(\Omega)} + \|\pi\|_{W^{s,r}(\Omega)} \le c(\Omega, r, s)(\|F\|_{W^{s-1,r}(\Omega)} + \|G\|_{W^{s,r}(\Omega)}) \tag{1.7.9}$$

*for $s = 0, 1$ and $1 < r < \infty$.*

*Proof.* Note that, by Theorem 6.1 in [50], for any $F \in W^{s-1,r}(\Omega)$ and $G \in W^{s,r}(\Omega)$ with $s = 0, 1$, problem (1.7.7) has a unique solution $\mathbf{u}, \pi$ satisfying (1.7.9). Thus the relation $(F, G) \mapsto (\mathbf{u}, \pi)$ determines a linear operator $T : W^{s-1,r}(\Omega) \times W^{s,r}(\Omega) \to W^{s+1,r}(\Omega) \times W^{s,r}(\Omega)$. Therefore, Lemma 1.7.9 is a consequence of Lemma 1.1.13. □

**Remark 1.7.10.** We stress that for every $s \in [0, 1]$, the solution to problem (1.7.7) given by the lemma belongs to the class $W_0^{1,r}(\Omega) \times L^r(\Omega)$ and is a weak solution in the sense of the definition given in [50, Ch. 4]. Notice also that the statement of the lemma does not hold for $0 < s < 1$ if we replace $\mathcal{W}^{s-1,r}$ by $W^{s-1,r}$ since the latter is not an interpolation space for $s \in (0, 1)$.

### 1.7.3 Parabolic equations

Let $\Omega$ be a bounded domain in $\mathbb{R}^d$ with $C^{2+\alpha}$ boundary and $Q = \Omega \times (0, T)$. For any $\alpha \in (0, 1)$ denote by $C^{\alpha/2,\alpha}(Q)$ and $C^{1+\alpha/2,2+\alpha}(Q)$ the Banach spaces of all functions $u : \operatorname{cl} Q \to \mathbb{R}$ with the finite norms

$$\begin{aligned} \|u\|_{C^{\alpha/2,\alpha}(Q)} = \|u\|_{C(Q)} + &\sup_{(t,\tau,x)\in[0,T]^2\times\overline{\Omega}} \frac{|u(x,t) - u(x,\tau)|}{|t-\tau|^{\alpha/2}} \\ + &\sup_{(t,x,y)\in[0,T]\times\overline{\Omega}^2} \frac{|u(x,t) - u(y,t)|}{|x-y|^{\alpha}} \end{aligned}$$

(where $\overline{\Omega} = \operatorname{cl}\,\Omega$) and

$$\begin{aligned} \|u\|_{C^{1+\alpha/2,2+\alpha}(Q)} = \|u\|_{C^{1,2}(Q)} + &\sup_{(t,\tau,x)\in[0,T]^2\times\overline{\Omega}} \frac{|\partial_t u(x,t) - \partial_t u(x,\tau)|}{|t-\tau|^{\alpha/2}} \\ + \sum_{i,j} &\sup_{(t,x,y)\in[0,T]\times\overline{\Omega}^2} \frac{|\partial_{x_i}\partial_{x_j} u(x,t) - \partial_{x_i}\partial_{x_j} u(y,t)|}{|x-y|^{\alpha}}. \end{aligned}$$

**Remark 1.7.11.** By the Arzelà-Ascoli compactness theorem, the embedding

$$C^{\kappa}(0,T;C^1(\Omega)) \hookrightarrow C^{\alpha/2,\alpha}(Q)$$

is compact for all $0<\alpha<\kappa\le 1$.

The following results are due to Ladyzhenskaya, Solonnikov and Ural'tseva [74].

**Theorem 1.7.12.** *Let $\Omega\subset\mathbb{R}^d$ be a bounded domain with $C^{2+\alpha}$ boundary and let*

$$\mathbf{b}\in C^{\alpha/2,\alpha}(Q) \quad \text{and} \quad k\in C^{\alpha/2,\alpha}(Q).$$

*Furthermore, assume that $\varrho_\infty \in C^{1+\alpha/2,2+\alpha}(Q)$ and $f\in C^{\alpha/2,\alpha}(Q)$ satisfy the compatibility condition*

$$\partial_t\varrho_\infty - \varepsilon\Delta\varrho_\infty + \mathbf{b}\cdot\nabla\varrho_\infty + k\varrho_\infty = f \quad \text{for } (x,t)\in\partial\Omega\times\{t=0\}.$$

*Then the boundary value problem*

$$\partial_t\varrho - \varepsilon\Delta\varrho + \mathbf{b}\cdot\nabla\varrho + k\varrho = f \quad \text{in } Q, \tag{1.7.10a}$$

$$\varrho = \varrho_\infty \quad \text{on } \partial\Omega\times(0,T), \qquad \varrho=\varrho_\infty \quad \text{for } t=0, \tag{1.7.10b}$$

*has a unique solution $\varrho\in C^{1+\alpha/2,2+\alpha}(Q)$, and*

$$\|\varrho\|_{C^{1+\alpha/2,2+\alpha}(Q)} \le c(\|\varrho_\infty\|_{C^{1+\alpha/2,2+\alpha}(Q)} + \|f\|_{C^{\alpha/2,\alpha}(Q)}).$$

*Here, the constant $c$ depends only on $T$, $\Omega$, $\varepsilon$ and $\|\mathbf{b}\|_{C^{\alpha/2,\alpha}(Q)}$, $\|k\|_{C^{\alpha/2,\alpha}(Q)}$. Moreover, if $\varrho_\infty > c_\infty > 0$ and $f\ge 0$, then $\inf_Q \varrho(x,t) > c > 0$, where the constant $c$ depends only on $\Omega$, $T$ and $\|b\|_{C^{\alpha/2,\alpha}(Q)}$, $\|k\|_{C^{\alpha/2,\alpha}(Q)}$, $c_\infty$.*

**Theorem 1.7.13.** *Let $\Omega\subset\mathbb{R}^d$ be a bounded domain with $C^{2+\alpha}$ boundary and suppose that functions $\varrho_\infty$, $f$ satisfy the condition*

$$\|\partial_t\varrho_\infty\|_{L^2(Q)} + \|\varrho_\infty\|_{L^2(0,T;W^{2,2}(\Omega))} + \|\varrho_\infty\|_{C(0,T;W^{1,2}(\Omega))} + \|f\|_{L^2(Q)} < \infty.$$

*Then the boundary value problem*

$$\partial_t\varrho - \varepsilon\Delta\varrho = f \quad \text{in } Q, \tag{1.7.11a}$$

$$\varrho = \varrho_\infty \quad \text{on } \partial\Omega\times(0,T), \qquad \varrho=\varrho_\infty \quad \text{for } t=0, \tag{1.7.11b}$$

*has a unique solution $\varrho$, and the solution satisfies*

$$\begin{aligned}&\|\partial_t\varrho\|_{L^2(Q)} + \|\varrho\|_{L^2(0,T;W^{2,2}(\Omega))} + \|\varrho\|_{C(0,T;W^{1,2}(\Omega))}\\ &\quad \le c\big(\|\partial_t\varrho_\infty\|_{L^2(Q)} + \|\varrho_\infty\|_{L^2(0,T;W^{2,2}(\Omega))} + \|\varrho_\infty\|_{C(0,T;W^{1,2}(\Omega))} + \|f\|_{L^2(Q)}\big),\end{aligned}$$

*where the constant $c$ depends only on $T$, $\Omega$ and $\varepsilon$.*

# Chapter 2

# Physical background

## 2.1 Governing equations

Suppose a compressible fluid (which we may also call a gas) occupies a domain $\Omega \subset \mathbb{R}^3$ named the *flow domain*. The flow domain can vary in time and its position and even shape can depend on the time variable $t$. In this case we write $\Omega_t$ to stress the dependence on $t$. The state of the fluid is characterized completely by the macroscopic quantities: the *density* $\varrho(x,t)$, the *velocity* $\mathbf{u}(x,t)$, and the *temperature* $\vartheta(x,t)$. These quantities are called *state variables* in the following. The *governing equations* represent three basic principles of fluid mechanics: the *mass balance*

$$\partial_t \varrho + \operatorname{div}(\varrho \mathbf{u}) = 0 \quad \text{in } \Omega, \tag{2.1.1a}$$

the *balance of momentum*

$$\partial_t(\varrho \mathbf{u}) + \operatorname{div}(\varrho \mathbf{u} \otimes \mathbf{u}) + \nabla p = \varrho \mathbf{f} + \operatorname{div} \mathbb{S}(\mathbf{u}) \quad \text{in } \Omega, \tag{2.1.1b}$$

and the *energy conservation law*

$$\partial_t E + \operatorname{div}((E+p)\mathbf{u}) = \operatorname{div}(\mathbb{S}(\mathbf{u})\mathbf{u}) + \operatorname{div}(\kappa \nabla \vartheta) + (\varrho \mathbf{f}) \cdot \mathbf{u} + \varrho Q. \tag{2.1.1c}$$

Here, the vector field $\mathbf{f}$ denotes the density of external mass force, the *heat conduction coefficient* $\kappa$ is a positive constant, the given function $Q$ is the intensity of the external energy flux, the *viscous stress tensor* $\mathbb{S}$ has the form

$$\mathbb{S}(\mathbf{u}) = \nu_1 \left( \nabla \mathbf{u} + \nabla \mathbf{u}^\top - \frac{2}{3} \operatorname{div} \mathbf{u} \mathbb{I} \right) + \nu_2 \operatorname{div} \mathbf{u} \mathbb{I}, \tag{2.1.1d}$$

in which the *viscosity coefficients* $\nu_i$, $i = 1, 2$, satisfy the inequality $\frac{4}{3}\nu_1 + \nu_2 > 0$, the *energy density* $E$ is given by

$$E = \frac{1}{2} \varrho |\mathbf{u}|^2 + \varrho e,$$

where $e$ is the density of *internal energy*. The physical properties of the fluid are reflected through *constitutive equations* relating the state variables to the pressure $p$ and the internal energy density $e$. The common point of view is that $p$ and $e$ can be represented as functions of $\varrho$ and $\vartheta$. The functions $p(\varrho, \vartheta)$ and $e(\varrho, \vartheta)$ are not arbitrary but should satisfy the Gibbs equation

$$\frac{1}{\vartheta} de(\varrho, \vartheta) - \frac{p(\varrho, \vartheta)}{\vartheta \varrho^2} d\varrho = ds(\varrho, \vartheta),$$

which means that the left hand side is the exact differential of some function $s$ named *entropy density*. In the classical case of *perfect polytropic* gases, the pressure and the internal energy density are defined by the following formulae, called the *constitutive law*:

$$p = R_m \varrho \vartheta, \quad e = c_v \vartheta. \tag{2.1.1e}$$

Here, $R_m$ is a positive constant inversely proportional to the molecular weight of the gas that is

$$R_m = c_p - c_v, \quad \text{with } \gamma := c_p/c_v > 1,$$

where $c_v$ is the specific heat capacity at constant volume and $c_p$ is the specific heat capacity at constant pressure; $c_v, c_p$ are positive constants. In this notation the entropy density $s$ takes the form

$$s = \log e + (\gamma - 1) \log \varrho. \tag{2.1.2}$$

The system of differential equations (2.1.1) is called the *compressible Navier-Stokes-Fourier* equations.

It is useful to rewrite the governing equations in dimensionless form which is widely used in applications. To this end, we denote by $u_c$, $\varrho_c$, $p_c$, $\vartheta_c$ the typical values of velocity, density, pressure, and temperature, and by $l_c$ and $T_c$ the typical values of length scale and time intervals. For simplicity we assume that $u_c = l_c/T_c$. Under this assumption, the characteristic values form four dimensionless combinations which are named: the Reynolds number, the Pecle number, the Mach number, and the viscosity ratio (see [119]),

$$\mathsf{Re} = \frac{\varrho_c u_c l_c}{\nu_1}, \quad \mathsf{Pe} = \frac{p_c l_c u_c}{\kappa_c \vartheta_c}, \quad \mathsf{Ma}^2 = \frac{\varrho_c u_c^2}{p_c}, \quad \lambda = \frac{1}{3} + \frac{\nu_2}{\nu_1}.$$

Denote also by $f_c$ and $Q_c$ the characteristic values of mass force and heat influx. They form two dimensionless combinations

$$\mathsf{Fr}_m^2 = \frac{u_c^2}{f_c l_c}, \quad \Theta = \frac{\varrho_c Q_c l_c}{p_c u_c}.$$

Note that here the characteristic quantities $\varrho_c$, $\vartheta_c$, and $p_c$ should be compatible with the constitutive law. For instance if the pressure is defined by (2.1.1e), then $p_c = R_m \, \varrho_c \, \vartheta_c$. Observe that the specific values of the constants $\gamma$, $\lambda$, and

Pe depend only on the physical properties of the fluid. For example, for the air under standard conditions, we have $\gamma = 7/5$ and $\mathsf{Pe} = 7/10$. The passage to the dimensionless variables is defined as follows:

$$\begin{aligned} &x \to l_c x, \quad t \to T_c t, \quad \mathbf{u} \to u_c \mathbf{u}, \quad \varrho \to \varrho_c \varrho, \\ &\vartheta \to \vartheta_c \vartheta, \quad \varrho \mathbf{f} \to \varrho_c f_c \varrho \mathbf{f}, \quad \kappa \to \kappa_c \kappa, \end{aligned} \tag{2.1.3}$$

and when performed in (2.1.1), leads to the following system of differential equations for dimensionless quantities in the scaled domain $l_c^{-1}\Omega$, still denoted by $\Omega$:

$$\partial_t(\varrho \mathbf{u}) + \operatorname{div}(\varrho \mathbf{u} \otimes \mathbf{u}) + \frac{1}{\mathsf{Ma}^2} \nabla p = \frac{1}{\mathsf{Re}} \operatorname{div} \mathbb{S}(\mathbf{u}) + \frac{1}{\mathsf{Fr}_m^2} \varrho \mathbf{f} \quad \text{in } \Omega, \tag{2.1.4a}$$

$$\partial_t \varrho + \operatorname{div}(\varrho \mathbf{u}) = 0 \quad \text{in } \Omega, \tag{2.1.4b}$$

$$\begin{aligned} &\partial_t(\varrho s) + \operatorname{div}(\varrho s \mathbf{u}) - \frac{1}{\mathsf{Pe}} \operatorname{div}\left(\frac{\kappa}{\vartheta} \nabla \vartheta\right) \\ &\qquad = \frac{1}{\vartheta}\left(\frac{\mathsf{Ma}^2}{\mathsf{Re}} \mathbb{S}(\mathbf{u}) : \nabla \mathbf{u} + \frac{\kappa}{\mathsf{Pe}\vartheta} |\nabla \vartheta|^2\right) + \Theta \frac{\varrho Q}{\vartheta}, \end{aligned} \tag{2.1.4c}$$

where $\mathbf{u} \otimes \mathbf{u}$ stands for the tensor product of two vectors and the dimensionless viscous stress tensor is defined by

$$\mathbb{S}(\mathbf{u}) = \nabla \mathbf{u} + \nabla \mathbf{u}^\top + (\lambda - 1) \operatorname{div} \mathbf{u} \, \mathbb{I}, \quad \operatorname{div} \mathbb{S}(\mathbf{u}) = \Delta \mathbf{u} + \lambda \nabla \operatorname{div} \mathbf{u}. \tag{2.1.5}$$

If the flow domain $\Omega_t$ varies in time, the above equations should be considered in the moving scaled domain.

### 2.1.1 Isentropic flows. Compressible Navier-Stokes equations

The flow is *barotropic* if the pressure depends only on the density. The most important example of such flows are *isentropic flows*. In order to deduce the governing equations for isentropic flows we note that for a perfect fluid with $\nu_i = \kappa = 0$, $i = 1, 2$, the entropy takes a constant value at each material point. Hence in this case the governing equations have a family of explicit solutions with the entropy $s = \text{const}$. By (2.1.1e) and (2.1.2) in this case we have

$$p(\varrho) = (\gamma - 1) \exp(s_c) \varrho^\gamma,$$

where the positive constant $s_c$ is a characteristic value of the entropy (without loss of generality we can take $(\gamma - 1)\exp(s_c) = 1$). Assuming that this relation holds for $\nu_i \neq 0$, $i = 1, 2$, we arrive at the system of *compressible Navier-Stokes equations* for isentropic flows of a viscous compressible fluid in dimensionless form:

$$\partial_t(\varrho\mathbf{u}) + \operatorname{div}(\varrho\mathbf{u}\otimes\mathbf{u}) + \frac{1}{\mathsf{Ma}^2}\nabla p(\varrho) = \frac{1}{\mathsf{Re}}\operatorname{div}\,\mathbb{S}(\mathbf{u}) + \frac{1}{\mathsf{Fr}_m^2}\varrho\mathbf{f} \quad \text{in } \Omega, \tag{2.1.6a}$$

$$\partial_t\varrho + \operatorname{div}(\varrho\mathbf{u}) = 0 \quad \text{in } \Omega. \tag{2.1.6b}$$

Recall that the exponent $\gamma$ depends on the physical properties of the fluid. In particular [25], $\gamma = 5/3$ for mono-atomic, $\gamma = 7/5$ for diatomic and $\gamma = 4/3$ for polyatomic gases.

It is worth noting that the entropy production for a viscous gas is proportional to $\mathbb{S}(\mathbf{u}) : \nabla\mathbf{u}$. Generally speaking, this means that the total entropy of a viscous gas increases in time and in contrast to the nonviscous case the existence of isentropic solutions for viscous gas dynamics equations is unlikely. In this sense, the compressible Navier-Stokes equations are thermodynamically inconsistent. The equations can be considered as an approximation of the real physical problem. However, this situation is not unusual in mathematical physics. We recall that the standard heat equation is also thermodynamically inconsistent.

Nevertheless, the compressible Navier-Stokes equations play an important role in the theory of compressible fluid dynamics as the only example of physically relevant equations for which we have nonlocal existence results.

## 2.2 Boundary and initial conditions

**Boundary conditions.** The governing equations should be supplemented with boundary conditions. The typical boundary conditions for the velocity field are: the first boundary condition (Dirichlet-type condition)

$$\mathbf{u} = \mathbf{U} \quad \text{on } \partial\Omega, \tag{2.2.1}$$

the second boundary condition (Neumann-type condition)

$$(\mathbb{S}(\mathbf{u}) - p\mathbb{I})\mathbf{n} = \mathbf{S}_n \quad \text{on } \partial\Omega, \tag{2.2.2}$$

where $\mathbf{n}$ is the outward normal vector to $\partial\Omega$, and $\mathbf{U}$ and $\mathbf{S}_n$ are given vector fields. Important particular cases are the *no-slip boundary condition* with $\mathbf{U} = 0$, and the zero normal stress condition with $\mathbf{S}_n = 0$. A third type of physically and mathematically reasonable condition is the *no-stick boundary condition*

$$\mathbf{u}\cdot\mathbf{n} = 0, \quad ((\mathbb{S}(\mathbf{u}) - p\mathbb{I})\mathbf{n}) \times \mathbf{n} = 0 \quad \text{on } \partial\Omega,$$

which corresponds to the case of frictionless boundary. The typical boundary conditions for the temperature are the Dirichlet and Neumann boundary conditions.

The formulation of boundary conditions for the density is a more delicate task. Assume that the velocity $\mathbf{u}$ satisfies the first boundary condition (2.2.7), and split the boundary of the flow region into three disjoint sets called the *inlet* $\Sigma_{\text{in}}$, the *outgoing set* $\Sigma_{\text{out}}$, and the *characteristic set* $\Sigma_0$; these sets are defined by

$$\begin{aligned}\Sigma_{\text{in}} &= \{x \in \partial\Omega : \mathbf{U}\cdot\mathbf{n} < 0\}, \quad \Sigma_{\text{out}} = \{x \in \Sigma : \mathbf{U}\cdot\mathbf{n} > 0\},\\ \Sigma_0 &= \{x \in \partial\Omega : \mathbf{U}\cdot\mathbf{n} = 0\}.\end{aligned} \tag{2.2.3}$$

The density should be given on the inlet:

$$\varrho = \varrho_\infty \quad \text{on } \Sigma_{\text{in}}. \tag{2.2.4}$$

The boundary conditions for the density are not needed in the case of $\Sigma_{\text{in}} = \emptyset$. In particular, there are no boundary conditions for the density if the velocity satisfies the no-slip or no-stick conditions, i.e., whenever $\Sigma_{\text{in}} = \Sigma_{\text{out}} = \emptyset$.

If the flow domain depends on $t$, i.e. $\Omega = \Omega_t$, and material points of $\partial\Omega_t$ are moving with a velocity $\mathbf{V}(x,t)$ then the boundary conditions become

$$\mathbf{u} = \mathbf{U} \quad \text{on } \partial\Omega_t, \quad \varrho = \varrho_\infty \quad \text{on } \Sigma^t_{\text{in}}, \tag{2.2.5}$$

where the time dependent inlet is defined by

$$\Sigma^t_{\text{in}} = \{x \in \partial\Omega_t : (\mathbf{U}(x,t) - \mathbf{V}(x,t))\cdot\mathbf{n} < 0\}.$$

**Initial conditions.** At the initial time $t = 0$ the distributions of the velocity field, density and temperature should be prescribed in $\Omega := \Omega_0$ for solutions of the Navier-Stokes-Fourier equations. The velocity field and density should be prescribed for solutions of the compressible Navier-Stokes equations

$$\mathbf{u}(x,0) = \mathbf{u}_0(x), \quad \varrho(x,0) = \varrho_0(x) \quad \text{in } \Omega. \tag{2.2.6}$$

**Flows around moving bodies.** The problem of a gas flow around a moving body is of practical importance. It presents an example of an exterior boundary value problem and is formulated as follows:

Suppose that at time $t$ a body occupies a compact set $S_t \subset \mathbb{R}^d$. Its position and shape can vary in time, but we assume that there is no *mass flux* through its boundary. Let material points of $\partial S_t$ be moving with the velocity $\mathbf{V}(x,t)$. Then the problem is to find the velocity field $\mathbf{u}$ and the density $\varrho$ satisfying the compressible Navier-Stokes equations (2.1.6) along with the boundary and initial conditions

$$\begin{aligned}&\mathbf{u} = \mathbf{V} \quad \text{on } \partial S_t,\\ &\mathbf{u}(x,0) = \mathbf{u}_0(x), \quad \varrho(x,0) = \varrho_0(x) \quad \text{in } \Omega_0,\\ &\mathbf{u}(x,t) \to 0, \quad \varrho \to \varrho_\infty \quad \text{as } |x| \to \infty.\end{aligned} \tag{2.2.7}$$

## 2.3 Power and work of hydrodynamic forces

The *stress tensor* in a viscous compressible flow is defined by

$$\mathbb{T} = \mathbb{S}(\mathbf{u}) - p\,\mathbb{I}, \tag{2.3.1}$$

where the viscous stress tensor $\mathbb{S}(\mathbf{u})$ is given by (2.1.1d), and the force acting from the side of the flow at the boundary point $x \in \partial\Omega$ is equal to

$$\mathbf{R}_f = -\mathbb{T}\,\mathbf{n} = (-\mathbb{S}(\mathbf{u}) + p\,\mathbb{I})\,\mathbf{n}. \tag{2.3.2}$$

Recall that $\mathbf{n}$ is the outward normal vector to $\partial\Omega$. If the flow domain $\Omega = \Omega_t$ varies in time, and material points on its boundary are moving with a given velocity $\mathbf{V}_s(x,t)$, the surface density of the power developed by the hydrodynamic force $\mathbf{R}_f$ is given by

$$J_{\text{dens}} = -\mathbb{T}\,\mathbf{n}\cdot\mathbf{V}_s = (-\mathbb{S}(\mathbf{u}) + p\,\mathbb{I})\,\mathbf{n}\cdot\mathbf{V}_s. \tag{2.3.3}$$

The total hydrodynamic force $\mathbf{R}_\Omega$ and the power $J_\Omega$ developed by the hydrodynamic forces are equal to

$$\mathbf{R}_\Omega = \int_{\partial\Omega_t} (-\mathbb{S}(\mathbf{u}) + p\,\mathbb{I})\,\mathbf{n}\,dS, \quad J_\Omega = \int_{\partial\Omega_t} (-\mathbb{S}(\mathbf{u}) + p\,\mathbb{I})\,\mathbf{n}\cdot\mathbf{V}_s\,dS. \tag{2.3.4}$$

Therefore, the total work of the hydrodynamic forces over the time period $[0,T]$ is

$$W_\Omega = \int_0^T \int_{\partial\Omega_t} (-\mathbb{S}(\mathbf{u}) + p\,\mathbb{I})\,\mathbf{n}\cdot\mathbf{V}_s\,dS\,dt. \tag{2.3.5}$$

After scaling (2.1.3) of independent and state variables and the following scaling of the hydrodynamic force, power, and work:

$$\mathbf{R}_\Omega \to \varrho_c u_c^2 l_c^2\,\mathbf{R}_\Omega, \quad J_\Omega \to \varrho_c u_c^3 l_c^2\,J_\Omega, \quad W_\Omega \to T_c \varrho_c u_c^3 l_c^2\,W_\Omega,$$

expressions (2.3.4)–(2.3.5) can be rewritten in dimensionless form as

$$\begin{aligned} \mathbf{R}_\Omega &= -\frac{1}{\mathsf{Re}} \int_{\partial\Omega_t} (\nabla\mathbf{u} + (\nabla\mathbf{u})^\top + (\lambda - 1)\operatorname{div}\mathbf{u} - \sigma p\,\mathbb{I})\,\mathbf{n}\,dS, \\ J_\Omega &= -\frac{1}{\mathsf{Re}} \int_{\partial\Omega_t} (\nabla\mathbf{u} + (\nabla\mathbf{u})^\top + (\lambda - 1)\operatorname{div}\mathbf{u} - \sigma p\,\mathbb{I})\,\mathbf{n}\cdot\mathbf{V}_s\,dS, \end{aligned} \tag{2.3.6}$$

and

$$W_\Omega = -\frac{1}{\mathsf{Re}} \int_0^T \int_{\partial\Omega_t} (\nabla\mathbf{u} + (\nabla\mathbf{u})^\top + (\lambda - 1)\operatorname{div}\mathbf{u} - \sigma p\,\mathbb{I})\,\mathbf{n}\cdot\mathbf{V}_s\,dSdt. \tag{2.3.7}$$

Here, we use the notation

$$\lambda = \frac{1}{3} + \frac{\nu_2}{\nu_1}, \quad \sigma = \frac{\mathsf{Re}}{\mathsf{Ma}^2}. \tag{2.3.8}$$

## 2.4 Navier-Stokes equations in a moving frame

If the flow region varies in time, then it is convenient, for technical and practical reasons, to reduce the corresponding boundary value problem for fluid dynamics equations to a problem in a fixed domain by a change of independent variables. In this section we describe such a change of variables in the case when $\Omega_t$ evolves like an absolutely rigid body. Assume that $x$ and $t$ are dimensionless variables. Recall that a one-parameter family of mappings $y \mapsto x(y,t)$ represents a rigid body motion in Euclidean space $\mathbb{R}^d$ if and only if

$$x = \mathbb{U}(t)y + \mathbf{a}(t), \tag{2.4.1}$$

where $\mathbf{a}(t)$ is an arbitrary vector function and $\mathbb{U}(t)$ is an arbitrary one-parameter family of special orthogonal matrices, i.e., $\mathbb{U}\,\mathbb{U}^\top = \mathbb{I}$ and $\det \mathbb{U} = 1$. For simplicity, we assume that $\mathbf{a}$ and $\mathbb{U}$ are twice continuously differentiable in $t$ and

$$\mathbb{U}(0) = \mathbb{I}, \quad \mathbf{a}(0) = 0.$$

The velocity $\mathbf{V}$ of the rigid body motion (2.4.1) is defined by

$$\mathbf{V}(y,t) = \dot{\mathbb{U}}(t)y + \dot{\mathbf{a}}(t), \tag{2.4.2}$$

where $\dot{\mathbf{a}}(t) = \frac{d}{dt}\mathbf{a}(t)$. Introduce also the vector field

$$\mathbf{W}(y,t) = \mathbb{U}^\top(t)\,\mathbf{V}(y,t) = \mathbb{U}^\top(t)\dot{\mathbb{U}}(t)y + \mathbb{U}^\top(t)\dot{\mathbf{a}}(t). \tag{2.4.3}$$

Next set

$$\mathbf{v}(y,t) = \mathbb{U}^\top(t)\,\mathbf{u}(x(y,t),t) - \mathbf{W}(y,t), \quad \rho(y,t) = \varrho(x(y,t),t). \tag{2.4.4}$$

**Lemma 2.4.1.** *Let the flow domain be of the form $\Omega_t = \mathbb{U}(t)\,\Omega_0 + \mathbf{a}(t)$. Then smooth functions $(\mathbf{u},\varrho)$ satisfy equations (2.1.6) in $\Omega_t$ if and only if the functions $(\mathbf{v},\rho)$, defined by (2.4.4), satisfy the equations*

$$\partial_t(\rho\,\mathbf{v}) + \operatorname{div}(\rho\,\mathbf{v}\otimes\mathbf{v}) - \frac{1}{\mathsf{Re}}\operatorname{div}\mathbb{S}(\mathbf{v}) + \frac{1}{\mathsf{Ma}^2}\nabla p(\rho) + \mathbb{C}\,\mathbf{v} = \rho\,\mathbf{f}^* \quad \text{in } \Omega_0, \tag{2.4.5a}$$

$$\partial_t(\rho) + \operatorname{div}(\rho\mathbf{v}) = 0 \quad \text{in } \Omega_0, \tag{2.4.5b}$$

*where the viscous stress tensor takes the form*

$$\mathbb{S}(\mathbf{v}) = \nabla\mathbf{v} + \nabla\mathbf{v}^\top + (\lambda - 1)\operatorname{div}\mathbf{v},$$

*and the skew-symmetric matrix $\mathbb{C} = (C_{ij})_{d\times d}$ and the vector field $\mathbf{f}^*$ are defined by*

$$C_{ij} = \frac{\partial W_i}{\partial y_j} - \frac{\partial W_j}{\partial y_i},$$

*and*

$$\mathbf{f}^* = \frac{1}{\mathsf{Fr}_m^2}\mathbb{U}^\top \mathbf{f}(x(y,t),t) - \partial_t \mathbf{W} + \frac{1}{2}\nabla|\mathbf{W}|^2.$$

*Here, $W_i$ are the components of the vector field $\mathbf{W}$ given by* (2.4.3). *The expressions* (2.3.6) *and* (2.3.7) *for the power developed by the hydrodynamic forces and the work produced by these forces read in the new variables*

$$J_\Omega = -\frac{1}{\mathsf{Re}}\int_{\partial\Omega_0} (\nabla\mathbf{v} + (\nabla\mathbf{v})^\top + (\lambda - 1)\operatorname{div}\mathbf{v}\,\mathbb{I} - \sigma p(\rho)\,\mathbb{I})\,\mathbf{n}\cdot\mathbf{W}\,dS, \tag{2.4.6}$$

*and*

$$W_\Omega = -\frac{1}{\mathsf{Re}}\int_0^T\int_{\partial\Omega_0} \big(\nabla\mathbf{v} + (\nabla\mathbf{v})^\top + (\lambda - 1)\operatorname{div}\mathbf{v}\,\mathbb{I} - \sigma p(\rho)\,\mathbb{I}\big)\,\mathbf{n}\cdot\mathbf{W}\,dSdt. \tag{2.4.7}$$

It is worth noting that from the mathematical standpoint, the only difference between equations (2.1.6) and (2.4.5) is the presence of the *Coriolis force* $\mathbb{C}\mathbf{v}$ in (2.4.5).

**Remark 2.4.2.** In three dimensions the expression for the Coriolis force can be written in the traditional form of the cross product

$$\mathbb{C}\mathbf{v} = 2\boldsymbol{\omega}\times\mathbf{v}, \quad \text{where the angular velocity is} \quad 2\boldsymbol{\omega} = \operatorname{rot}\mathbf{W}. \tag{2.4.8}$$

A similar representation holds in the two-dimensional case if we assume that the angular velocity is orthogonal to the plane of motion.

**Stationary flows.** In general, the matrix $\mathbb{C}$ and the force $\mathbf{f}^*$ depend on the time variable $t$ and equations (2.4.5) are nonautonomous. The equations become autonomous and lead to stationary solutions only if

$$\partial_t\mathbf{W} = \partial_t\{\mathbb{U}^\top(t)\mathbf{f}(x(y,t),t)\} = 0. \tag{2.4.9}$$

By (2.4.3), the equality $\partial_t\mathbf{W} = 0$ is equivalent to the relations

$$\mathbb{U}(t) = e^{\mathbb{B}t}, \quad \dot{\mathbf{a}}(t) = e^{-\mathbb{B}t}\mathbf{b},$$

in which $\mathbb{B}$ is a skew-symmetric matrix and $\mathbf{b}$ is a constant vector. If this is the case, then the matrix $\mathbb{C} = \mathbb{B}$ and the corresponding angular velocity $\boldsymbol{\omega}$ are constants. In particular, condition (2.4.9) is fulfilled if the vector field $\mathbf{f}$ and the angular velocity $\boldsymbol{\omega}$ are constants such that

$$\mathbf{f}\times\boldsymbol{\omega} = 0.$$

**Example 2.4.3.** The most important case is the Galileo transformation such that

$$\mathbb{U} = \mathbb{I}, \quad \mathbf{a}(t) = \mathbf{v}_\infty t, \quad \mathbf{v}_\infty = \text{const}.$$

In this case

$$\mathbb{C} = 0, \quad \mathbf{f}^* = \frac{1}{\mathsf{Fr}_m^2}\mathbf{f}(y - \mathbf{v}_\infty t, t).$$

**Example 2.4.4.** Another important example is the motion of gas on rotating Earth. Assume that the origin of the coordinate system is at the center of Earth and the $x_1$ axis is directed to the north pole. In this case the dimensionless angular velocity is given by

$$\boldsymbol{\omega} = \omega \mathbf{e}_1, \quad \mathbf{e}_1 = (1,0,0), \quad \omega = \omega_e / T_c,$$

where $\omega_e$ is the angular velocity of Earth. If $(y,t)$ is a moving frame connected with Earth, then

$$x = e^{\mathbb{C}_\omega t} y, \quad \text{i.e. } \mathbb{U}(t) = e^{\mathbb{C}_\omega t}, \quad \mathbf{a}(t) = 0.$$

Here, the matrix $\mathbb{C}_\omega$ defined by $\mathbb{C}_\omega y = \boldsymbol{\omega} \times y$ has the form

$$\mathbb{C}_\omega = \omega \begin{pmatrix} 0 & 0 & 0 \\ 0 & 0 & -1 \\ 0 & 1 & 0 \end{pmatrix},$$

$$e^{\mathbb{C}_\omega t} = \cos \omega t \begin{pmatrix} \cos \omega t & 0 & 0 \\ 0 & 1 & 0 \\ 0 & 0 & 1 \end{pmatrix} + \sin \omega t \begin{pmatrix} \sin \omega t & 0 & 0 \\ 0 & 0 & -1 \\ 0 & 1 & 0 \end{pmatrix}.$$

The gravity force is determined from the Newton law and, after scaling, can be taken in the form

$$\mathbf{f} = -\frac{x}{|x|^3},$$

and $\mathbf{W} = \mathbb{C}_\omega y$. It follows from Lemma 2.4.1 that the functions $\mathbf{v}$ and $\rho$ satisfy (2.4.5) with

$$\mathbf{f}^* = -\frac{1}{\mathsf{Fr}_m^2} \frac{y}{|y|^3} + \frac{\omega^2}{2} \nabla (y_2^2 + y_3^2).$$

This means that in the rotating system of coordinates we have to take into account the Coriolis and centrifugal forces, and the system of governing equations is autonomous, i.e., no terms in the equations are explicitly time dependent, but the solutions are functions of the time and the spatial variables.

## 2.5 Flow around a moving body

Let us return to the flow of a viscous gas around a moving body. Assume that the body is rigid and its configuration $S_t$ at time $t$ is determined in dimensionless variables by the relations

$$S_t = \mathbb{U}(t) S + \mathbf{a}(t), \quad \mathbb{U}^\top \mathbb{U} = \mathbb{I}, \quad \mathbb{U}(0) = \mathbb{I}, \quad \mathbf{a}(0) = 0, \tag{2.5.1}$$

where the initial configuration $S$ is a fixed compact subset of $\mathbb{R}^d$. In particular, a point of $S$ evolves with the dimensionless velocity $\mathbf{V}(y,t) = \dot{\mathbb{U}}(t)\, y + \dot{\mathbf{a}}(t)$. Then the modified velocity field

$$\mathbf{v}(y,t) = \mathbb{U}^\top(t) \mathbf{u}(x(y,t),t) - \mathbf{W}(y,t) \quad \text{with } \mathbf{W} = \mathbb{U}^\top(t) \mathbf{V}(y,t)$$

and the density $\rho(y,t) = \varrho(x(y,t),t)$ satisfy equations (2.4.5) in the flow domain $\Omega_0 = \mathbb{R}^d \setminus S$. It follows from (2.2.7), (2.4.2), and (2.4.3) that in addition $\rho$ and $\mathbf{v}$ satisfy the following boundary and initial conditions:

$$\begin{aligned} &\mathbf{v} = 0 \quad \text{on } \partial S, \\ &\mathbf{v}(y,0) = \mathbf{v}_0(y), \quad \rho(y,0) = \varrho_0(y) \quad \text{in } \mathbb{R}^d \setminus S, \\ &\mathbf{v}(y,t) + \mathbf{W}(y,t) \to 0, \quad \rho \to \varrho_\infty \quad \text{as } |y| \to \infty, \end{aligned} \tag{2.5.2}$$

where the initial vector field is $\mathbf{v}_0(y) = \mathbf{u}_0(y) - (\dot{\mathbb{U}}(0)y - \dot{\mathbf{a}}(0))$.

For the purposes of numerical simulation, the boundary value problems in unbounded domains should be replaced by modified problems in bounded domains. To this end, we choose an arbitrary hold-all domain $B \subset \mathbb{R}^3$, for instance, a sufficiently large ball, such that $S \Subset B$. Next, we transfer the boundary conditions from infinity to $\partial B$ and arrive at the following boundary value problem for $\mathbf{v}$ and $\rho$, for a given $T > 0$.

**Problem 2.5.1.** *Find functions* $(\mathbf{v}, \rho)$ *satisfying*

$$\begin{aligned} \partial_t(\rho\mathbf{v}) + \operatorname{div}(\rho\mathbf{v}\otimes\mathbf{v}) - \frac{1}{\mathsf{Re}}\operatorname{div}\mathbb{S}(\mathbf{v}) \\ + \frac{1}{\mathsf{Ma}^2}\nabla p(\rho) + \mathbb{C}\,\mathbf{v} = \rho\,\mathbf{f}^* \quad \textit{in } \Omega\times(0,T), \end{aligned} \tag{2.5.3a}$$

$$\partial_t\rho + \operatorname{div}(\rho\mathbf{v}) = 0 \quad \textit{in } \Omega\times(0,T), \tag{2.5.3b}$$

$$\begin{aligned} &\mathbf{v} = 0 \quad \textit{on } \partial S\times(0,T), \quad \mathbf{v} = -\mathbf{W} \quad \textit{on } \partial B\times(0,T), \\ &\rho = \varrho_\infty \quad \textit{on } \Sigma_{\text{in}}, \\ &\mathbf{v}(y,0) = \mathbf{v}_0(y) \quad \textit{in } \Omega, \quad \rho(y,0) = \varrho_0(y) \quad \textit{in } \Omega, \end{aligned} \tag{2.5.3c}$$

*where*

$$\Omega = B\setminus S, \quad \Sigma_{\text{in}} = \{(y,t)\in\partial B\times(0,T) : \mathbf{W}(y,t)\cdot\mathbf{n}(y) > 0\}.$$

Now we recalculate the work $W_S$ of the hydrodynamic forces acting on the moving body. To this end we can use (2.4.7), which gives

$$W_S = -\frac{1}{\mathsf{Re}}\int_0^T\int_{\partial S}\big(\nabla\mathbf{v} + (\nabla\mathbf{v})^\top + (\lambda-1)\operatorname{div}\mathbf{v}\,\mathbb{I} - \sigma p(\rho)\,\mathbb{I}\big)\,\mathbf{n}\cdot\mathbf{W}\,dSdt. \tag{2.5.4}$$

But this formula is inconvenient since it involves the surface integral of $\nabla\mathbf{v}$ and $p$ which is not defined for weak solutions. We replace (2.5.4) by a more complicated but robust formula which only contains volume integrals.

**Lemma 2.5.2.** *Let* $(\mathbf{v}, \rho)$ *be a classical solution to problem* (2.5.3) *in a cylinder* $(B\setminus S)\times(0,T)$ *and suppose a function* $\eta \in C^\infty(B\setminus S)$ *satisfies the condition*

$$\eta(x) = 0 \textit{ in a neighborhood of } \partial B, \quad \eta(x) = 1 \textit{ in a neighborhood of } \partial S. \tag{2.5.5}$$

*Then*

$$W_S = -\int_{B\setminus S} \eta\{\rho(\cdot,T)\mathbf{v}(\cdot,T)\cdot\mathbf{W}(\cdot,T) - \rho_0\mathbf{v}_0\cdot\mathbf{W}(\cdot,0)\}\,dx +$$

$$\int_0^T\int_{B\setminus S}\Big\{\rho\eta\mathbf{v}\cdot\partial_t\mathbf{W} + \Big(\rho(\mathbf{v}\otimes\mathbf{v}) - \frac{1}{\mathsf{Re}}\mathbb{T}\Big) : \nabla(\eta\mathbf{W}) + \eta(\rho\mathbf{f}^* - \mathbb{C}\mathbf{v})\cdot\mathbf{W}\Big\}\,dxdt, \tag{2.5.6}$$

*where* $\mathbb{T} = \nabla\mathbf{v} + (\nabla\mathbf{v})^\top + (\lambda-1)\operatorname{div}\mathbf{v}\,\mathbb{I} - \sigma p(\rho)\,\mathbb{I}$.

*Proof.* Since $\eta$ vanishes on $\partial B$ and equals 1 on $\partial S$, we can integrate by parts to obtain

$$\begin{aligned} -\int_{\partial S}\mathbb{T}\mathbf{n}\cdot\mathbf{W}\,dS &= -\int_{\partial\Omega}\eta\mathbb{T}\mathbf{W}\cdot\mathbf{n}\,dS \\ &= -\int_{B\setminus S}\operatorname{div}\mathbb{T}\cdot(\eta\mathbf{W})\,dx - \int_{B\setminus S}\mathbb{T} : \nabla(\eta\mathbf{W})\,dx. \end{aligned}$$

From this, the identity

$$\frac{1}{\mathsf{Re}}\mathbb{S}(\mathbf{v}) - \frac{1}{\mathsf{Ma}^2}p(\rho)\mathbb{I} = \frac{1}{\mathsf{Re}}\mathbb{T},$$

and (2.5.3a) we conclude that

$$\begin{aligned} W_S = -\int_0^T\int_{B\setminus S}\Big\{&\eta\partial_t(\rho\mathbf{v})\cdot\mathbf{W} + \operatorname{div}(\rho\mathbf{v}\otimes\mathbf{v})\cdot(\eta\mathbf{W}) + \eta\mathbb{C}\mathbf{v}\cdot\mathbf{W} \\ &- \eta\rho\mathbf{f}^*\cdot\mathbf{W} + \frac{1}{\mathsf{Re}}\mathbb{T} : \nabla(\eta\nabla\mathbf{W})\Big\}\,dxdt. \end{aligned} \tag{2.5.7}$$

Since $\eta\mathbf{v}\otimes\mathbf{v} = 0$ on $\partial B\cup\partial S$, we can integrate by parts to obtain

$$\begin{aligned} -\int_0^T\int_{B\setminus S}&\{\eta\partial_t(\rho\mathbf{v})\cdot\mathbf{W} + \operatorname{div}(\rho\mathbf{v}\otimes\mathbf{v})\cdot(\eta\mathbf{W})\}\,dxdt \\ &= \int_0^T\int_{B\setminus S}\{\rho\eta\mathbf{v}\cdot\partial_t\mathbf{W} + \rho(\mathbf{v}\otimes\mathbf{v}) : \nabla(\eta\mathbf{W})\}dxdt \\ &\quad - \int_{B\setminus S}\eta\{\rho(\cdot,T)\mathbf{v}(\cdot,T)\cdot\mathbf{W}(\cdot,T) - \rho_0\mathbf{v}_0\cdot\mathbf{W}(\cdot,0)\}\,dx. \end{aligned}$$

Inserting this into (2.5.7) we arrive at (2.5.6). □

# Chapter 3

# Problem formulation

In this chapter we give the mathematical formulation of the problem which we intend to consider in this monograph. Recall that our main objective is to develop a mathematical theory for the drag optimization problem and investigation of the related mathematical questions in the theory of compressible Navier-Stokes equations. We focus on the problem of the compressible viscous gas flow around a moving rigid body (2.5.3) and on the question of optimal choice of the shape of this body in order to minimize the drag functional in the stationary case or the work in the nonstationary case.

By abuse of notation we return to the original notation and we will write $x$, $\mathbf{u}$, $\varrho$ instead of $y$, $\mathbf{v}$, $\rho$, and $\mathbf{f}$ instead of $\mathbf{f}^*$. Next notice that explicit values of the Reynolds and Mach numbers are not essential for the general mathematical theory of Navier-Stokes equations, provided that these numbers are separated from 0 and bounded from above. The Coriolis force is also immaterial for the mathematical theory. Hence without loss of generality we may assume that

$$\mathsf{Ma}^2 = \mathsf{Re} = 1 \quad \text{and} \quad l_c = u_c = T_c = 1, \quad \mathbb{C} = 0. \tag{3.0.1}$$

Next, in problem (2.5.3) of the flow around an obstacle, the velocity field vanishes on the obstacle $S$, equals $-\mathbf{W}$ on the boundary of the hold-all domain $B$, and coincides with a given vector field $\mathbf{v}_0$ at the initial time. It is convenient to introduce a vector field $\mathbf{U} : \mathbb{R}^d \times [0, T) \to \mathbb{R}^d$ such that

$$\mathbf{U} = -\mathbf{W} \text{ on } \partial B \times (0,T), \quad \mathbf{U} = 0 \text{ on } S \times (0,T), \quad \mathbf{U} = \mathbf{v}_0 \text{ on } (B \setminus S) \times \{t = 0\}. \tag{3.0.2}$$

When $\mathbf{v}_0$ is smooth and satisfies an obvious compatibility condition we can assume that $\mathbf{U}$ is also a smooth vector field. Thus we come to the following boundary value problem in the flow domain

$$\Omega = B \setminus S \subset \mathbb{R}^d, \quad d = 2, 3.$$

**Problem 3.0.3.** *For given $T > 0$ and for given functions*

$$\begin{aligned} \varrho_\infty &: \Omega \times [0,T) \to \mathbb{R}^+, \\ \mathbf{U} &: \Omega \times [0,T) \to \mathbb{R}^d, \\ \mathbf{f} &: \Omega \times (0,T) \to \mathbb{R}^d, \end{aligned} \tag{3.0.3}$$

*find a velocity field* $\mathbf{u} : \Omega \times [0,T) \to \mathbb{R}^d$ *and a density* $\varrho : \Omega \times [0,T) \to \mathbb{R}^+$ *satisfying*

$$\begin{aligned} \partial_t(\varrho\,\mathbf{u}) + \operatorname{div}(\varrho\,\mathbf{u} \otimes \mathbf{u}) + \nabla p(\rho) \\ = \operatorname{div} \mathbb{S}(\mathbf{u}) + \varrho\,\mathbf{f} \quad \text{in } \Omega \times (0,T), \end{aligned} \tag{3.0.4a}$$

$$\partial_t \varrho + \operatorname{div}(\varrho \mathbf{u}) = 0 \quad \text{in } \Omega \times (0,T), \tag{3.0.4b}$$

$$\begin{aligned} &\mathbf{u} = 0 \quad \text{on } \partial S \times (0,T), \\ &\mathbf{u} = \mathbf{U} \quad \text{on } \partial B \times (0,T), \\ &\varrho = \varrho_\infty \quad \text{on } \Sigma_{\text{in}}, \\ &\mathbf{u}(x,0) = \mathbf{U}(x,0) \quad \text{in } \Omega, \\ &\varrho(x,0) = \varrho_\infty(x,0) \quad \text{in } \Omega, \end{aligned} \tag{3.0.4c}$$

*where*

$$\mathbb{S}(\mathbf{u}) = \nabla \mathbf{u} + \nabla \mathbf{u}^\top + (\lambda - 1) \operatorname{div} \mathbf{u}, \quad \operatorname{div} \mathbb{S}(\mathbf{u}) = \Delta \mathbf{u} + \lambda \nabla \operatorname{div} \mathbf{u},$$

*and the inlet* $\Sigma_{\text{in}}$ *is defined by*

$$\Sigma_{\text{in}} = \{(x,t) \in \partial B \times (0,T) : \mathbf{U}(x,t) \cdot \mathbf{n}(x) < 0\}.$$

*Recall that* $\mathbf{n}$ *is the unit outward normal to the boundary* $\partial\Omega$.

The problem in the stationary case can be formulated as follows:

**Problem 3.0.4.** *For given*

$$\mathbf{U} : \partial\Omega \to \mathbb{R}^d, \quad \varrho_\infty : \partial\Omega \to \mathbb{R}^d, \quad \mathbf{f} : \Omega \to \mathbb{R}^d, \tag{3.0.5}$$

*find a velocity field* $\mathbf{u} : \Omega \to \mathbb{R}^d$ *and a density* $\varrho : \Omega \to \mathbb{R}^+$ *satisfying*

$$\operatorname{div}(\varrho\,\mathbf{u} \otimes \mathbf{u}) + \nabla p(\rho) = \operatorname{div} \mathbb{S}(\mathbf{u}) + \varrho\,\mathbf{f} \quad \text{in } \Omega, \tag{3.0.6a}$$

$$\operatorname{div}(\varrho \mathbf{u}) = 0 \quad \text{in } \Omega, \tag{3.0.6b}$$

$$\begin{aligned} &\mathbf{u} = 0 \quad \text{on } \partial S, \\ &\mathbf{u} = \mathbf{U} \quad \text{on } \partial B, \\ &\varrho = \varrho_\infty \quad \text{on } \Sigma_{\text{in}}, \end{aligned} \tag{3.0.6c}$$

*where the inlet* $\Sigma_{\text{in}}$ *is defined by*

$$\Sigma_{\text{in}} = \{x \in \partial B : \mathbf{U}(x) \cdot \mathbf{n}(x) < 0\}.$$

**Remark 3.0.5.** For stationary flows, if $\mathbf{U} \cdot \mathbf{n} = 0$ on $\partial\Omega$, the inlet is an empty set. In this particular case the total mass $M$ of the gas should be prescribed:

$$\int_\Omega \varrho(x)\,dx = M.$$

## 3.1 Weak solutions

All known results in the global existence theory for compressible Navier-Stokes equations concern the so-called weak solutions. The notion of a weak solution arises in a natural way from the formulation of the governing equations as a system of conservation laws. In our particular case weak solutions are defined as follows:

**Definition 3.1.1.** Let $\Omega \subset \mathbb{R}^d$ be a bounded domain with $C^3$ boundary $\partial\Omega$ and $Q$ be the cylinder $\Omega \times (0,T)$. A couple

$$(\varrho, \mathbf{u}) : Q \to \mathbb{R}^+ \times \mathbb{R}^d$$

is a *weak solution* to Problem (3.0.3) if the following conditions are satisfied:

- The state variables satisfy

$$\varrho, p, \varrho|\mathbf{u}|^2 \in L^\infty(0,T; L^1(\Omega)), \quad \mathbf{u} \in L^2(0,T; W^{1,2}(\Omega)), \tag{3.1.1}$$

$$\mathbf{u} = \mathbf{U} \quad \text{on } \partial\Omega \times [0,T]. \tag{3.1.2}$$

- The integral identity

$$\int_Q \big(\varrho\mathbf{u} \cdot \partial_t \boldsymbol{\psi} + \varrho\mathbf{u} \otimes \mathbf{u} : \nabla\boldsymbol{\psi} + p\,\mathrm{div}\,\boldsymbol{\psi} - \mathbb{S}(\mathbf{u}) : \nabla\boldsymbol{\psi}\big)\,dxdt$$
$$+ \int_Q \varrho\mathbf{f} \cdot \boldsymbol{\psi}\,dxdt + \int_\Omega \varrho_\infty \mathbf{U} \cdot \boldsymbol{\psi}(\cdot, 0)\,dx = 0 \tag{3.1.3}$$

  holds for all vector fields $\boldsymbol{\psi} \in C^\infty(Q)$ satisfying

$$\boldsymbol{\psi} = 0 \quad \text{in a neighborhood of } \{\partial\Omega \times [0,T]\} \cup \{\Omega \times \{T\}\},$$

  which means that the test vector fields vanish at the lateral side and the lid of the cylinder $Q$.

- The integral identity

$$\int_Q (\varrho\,\partial_t\zeta + \varrho\mathbf{u} \cdot \nabla\zeta)\,dxdt = \int_{\partial\Omega\times[0,T]} \zeta\varrho_\infty \mathbf{U} \cdot \mathbf{n}\,dSdt - \int_\Omega \varrho_\infty\zeta(\cdot, 0)\,dx \tag{3.1.4}$$

  holds for all functions $\zeta \in C^\infty(Q)$ satisfying

$$\zeta = 0 \quad \text{in a neighborhood of } \{(\partial\Omega \times [0,T]) \setminus \Sigma_{\text{in}}\} \cup \{\Omega \times \{T\}\},$$

  which means that the support of the test function $\zeta$ meets the boundary of the cylinder $Q$ at the inlet $\Sigma_{\text{in}}$ and the bottom $\Omega \times \{0\}$.

## 3.2 Renormalized solutions

The mass balance equation (3.0.4b) is a first order linear partial differential equation for the density $\varrho$. Given a smooth solution to this equation, the composite function $\varphi(\varrho)$ satisfies the first order differential equation

$$\partial_t \varphi(\varrho) + \operatorname{div}(\varphi(\varrho)\mathbf{u}) + (\varrho\varphi'(\varrho) - \varphi(\varrho)) \operatorname{div} \mathbf{u} = 0. \tag{3.2.1}$$

Thus, a smooth solution $\varrho$ to the mass balance equation generates a solution in the form $\varphi(\varrho)$ to the first order quasi-linear differential equation (3.2.1); the latter solution is associated with the prescribed function $\varphi$. Equation (3.2.1) derived for $\varphi(\varrho)$ is called a *renormalized equation*. The derivation of a renormalized equation is called *renormalization*. The renormalization procedure is simple for smooth solutions to the mass balance equation, but becomes nontrivial for weak solutions. It turns out that the renormalized equation is convenient for our purposes. A natural way to avoid the renormalization procedure for weak solutions to the mass balance equation is simple replacement of the mass balance equation by its renormalized version. This means that it is required that the density and the velocity satisfy (3.2.1) for a specific class of functions $\varphi$. Thus we come to the following definition:

**Definition 3.2.1.** Let

$$(\varrho, \mathbf{u}) : Q \to \mathbb{R}^+ \times \mathbb{R}^d$$

be a weak solution to Problem 3.0.3 satisfying all hypotheses of Definition 3.1.1. We say that $(\varrho, \mathbf{u})$ is a *renormalized solution* to Problem 3.0.3 if the integral identity

$$\int_Q \Big(\varphi(\varrho)\,\partial_t\zeta + \varphi(\varrho)\mathbf{u}\cdot\nabla\zeta - (\varrho\varphi'(\varrho) - \varphi(\varrho))\zeta \operatorname{div}\mathbf{u}\Big)\, dxdt$$
$$= \int_{\partial\Omega\times[0,T]} \zeta\varphi(\varrho_\infty)\mathbf{U}\cdot\mathbf{n}\, dSdt - \int_\Omega \varphi(\varrho_\infty)\zeta(\cdot,0)\, dx \tag{3.2.2}$$

holds for all functions $\zeta \in C^\infty(Q)$ satisfying

$$\zeta = 0 \quad \text{in a neighborhood of } \{(\partial\Omega)\setminus\Sigma_{\text{in}}\} \cup \{\Omega\times\{T\}\},$$

and for all functions $\varphi \in C_0^1(\mathbb{R})$.

**Remark 3.2.2.** Obviously, any $C^1$ function $\varphi : \mathbb{R} \to \mathbb{R}^+$ can be approximated by a sequence of compactly supported $C^1$ functions $\varphi_n$ such that

$$\varphi_n(s) \to \varphi(s), \quad \varphi_n'(s) \to \varphi'(s) \quad \text{as } n\to\infty \quad \text{uniformly on compact subsets of } \mathbb{R},$$

Hence the integral identity (3.2.2) is satisfied for all $C^1$ functions $\varphi$ such that $\varphi(\varrho), \varrho\varphi'(\varrho) \in L^2(Q)$.

## 3.3 Work functional. Optimization problem

If we adopt the conventions formulated at the beginning of this chapter, in particular if we assume that the dimensionless parameters satisfy condition (3.0.1), then Lemma 2.5.2 leads to the following expression for the work $W_S$ of the hydrodynamic forces acting on the moving obstacle $S$:

$$W_S = -\int_{B\setminus S} \eta\{\varrho(\cdot,T)\mathbf{u}(\cdot,T)\cdot\mathbf{W}(\cdot,T) - \varrho_0\mathbf{U}\cdot\mathbf{W}(\cdot,0)\}\,dx$$

$$+\int_0^T\int_{B\setminus S} \{\varrho\eta\mathbf{u}\cdot\partial_t\mathbf{W} + \big(\varrho(\mathbf{u}\otimes\mathbf{u}) - \mathbb{T}\big) : \nabla(\eta\mathbf{W}) + \eta(\varrho\mathbf{f})\cdot\mathbf{W}\}\,dxdt, \quad (3.3.1)$$

where

$$\mathbb{T} = \nabla\mathbf{u} + (\nabla\mathbf{u})^\top + (\lambda-1)\operatorname{div}\mathbf{u}\,\mathbb{I} - p(\varrho)\,\mathbb{I}.$$

Recall that if the obstacle $S$ moves in atmosphere like a solid body then its physical position $S_t$ at time $t$ is defined by (2.5.1) with $T_c = 1$ (because of (3.0.1)), i.e.,

$$S_t = \mathbb{U}(t)S + \mathbf{a}(t). \quad (3.3.2)$$

Here the unitary matrix $\mathbb{U}$ and the vector field $\mathbf{a}(t)$ can be defined by an appropriate flight planning scenario.

In this case the vector field $\mathbf{W}$ is given by (2.4.3), which in accordance with (3.0.1) reads

$$\mathbf{W}(x,t) = \mathbb{U}^\top(t)\dot{\mathbb{U}}(t)x + \mathbb{U}^\top(t)\dot{\mathbf{a}}(t). \quad (3.3.3)$$

**Remark 3.3.1.** Notice that both vector fields $\mathbf{U}$ and $\mathbf{W}$ are defined in the strip $\mathbb{R}^d\times[0,T]$, but the equality $\mathbf{U} = -\mathbf{W}$ holds only on the surface $\partial B\times(0,T)$. From the mathematical point of view this connection and representation (3.3.3) are not essential and we cam consider $\mathbf{U}$ and $\mathbf{W}$ as independent smooth vector fields.

When a solid body moves in atmosphere along the trajectory prescribed by (3.3.2), the corresponding value $W_S$ is nothing other than the work done by the engine which moves the body.

A natural question is to minimize this work by choosing an appropriate shape of $S$ from some suitable class $\mathfrak{S}$.

# Chapter 4

# Basic statements

## 4.1 Introduction

The mathematical theory of compressible Navier-Stokes equations is based on a number of statements which are not related directly to the specific structure of the Navier-Stokes equations themselves but are of a general character. In this chapter we assemble the basic facts of general character derived for solutions to the mass and momentum conservation equations

$$\begin{aligned} &\partial_t \varrho + \operatorname{div}(\varrho \mathbf{u} - \mathbf{g}) = 0, \\ &\partial_t(\varrho \mathbf{u}) + \operatorname{div}(\varrho \mathbf{u} \otimes \mathbf{u}) = \operatorname{div} \mathbb{T} + \rho \mathbf{f}. \end{aligned} \tag{4.1.1}$$

The majority of the statements formulated and proved here are independent of the form of the constitutive laws and pertain to both original and regularized equations of the motion of viscous fluids. We specify neither the form of the stress tensor $\mathbb{T}$ nor the mass source $\mathbf{g}$. In fact, system (4.1.1) is not complete and it can be regarded as a differential relation between the state variables $\varrho$, $\mathbf{u}$ and $\mathbb{T}$. Moreover, we do not impose boundary and initial conditions for the state variables at this stage. Instead we assume that they satisfy some integrability conditions. The results we present include elementary facts on integrability of functions with finite energy (Section 4.2), the standard material concerning weak compactness properties of solutions to the mass and momentum balance laws (Section 4.4), and the famous Lions result on the weak continuity of the so-called *viscous flux*, which is the most important result of the mathematical theory of viscous compressible flows (Section 4.6).

## 4.2 Bounded energy functions

In the mathematical theory of fluid mechanics, the stress tensor $\mathbb{T}$ is related to the density $\varrho$ and to the velocity vector field $\mathbf{u}$ by some constitutive relations. For example, for adiabatic flows of the viscous gas, the constitutive law reads

$$\mathbb{T} = \mathbb{S}(\mathbf{u}) - p(\varrho)\mathbb{I}, \quad p(\varrho) = c\varrho^\gamma,$$

where $\gamma$ is the adiabatic exponent, and $\mathbb{S}(\mathbf{u})$ is the viscous stress tensor. The constitutive relations should be in agreement with the second principle of thermodynamics, which yields the boundedness of the total energy and of the integral of energy dissipation rate. Thus

$$\begin{aligned} \|\varrho\|_{L^\infty(0,T;L^\gamma(\Omega))} + \|\varrho|\mathbf{u}|^2\|_{L^\infty(0,T;L^1(\Omega))} &\le E, \\ \|\mathbf{u}\|_{L^2(0,T;W^{1,2}(\Omega))} &\le E, \end{aligned} \tag{4.2.1}$$

where the constant $E$ depends on initial and boundary data. These inequalities are called the energy estimates or the first estimates for solutions of (4.1.1). Functions $(\varrho, \mathbf{u})$ satisfying (4.2.1) are said to be bounded energy functions. For the sake of generality we do not specify the constitutive relations and simply postulate estimates (4.2.1). In this section we derive the integrability properties of the density and of the velocity fields satisfying the energy estimates. This means that we obtain a collection of $L^p$-estimates for the momentum $\varrho\mathbf{u}$ and the kinetic energy tensor $\varrho\mathbf{u}\otimes\mathbf{u}$ in terms of the constant $E$ and the domain $\Omega$. The results are of general character and are used throughout the monograph.

Let $\Omega \subset \mathbb{R}^d$, $d = 2, 3$, be a bounded domain with Lipschitz boundary and $Q = \Omega \times (0, T)$. Assume that a density $\varrho : Q \to \mathbb{R}^+$ and a velocity field $\mathbf{u} : Q \to \mathbb{R}^d$ satisfy the energy inequalities (4.2.1). This raises the question on the existence of robust estimates for the momentum $\varrho\mathbf{u}$ and the kinetic energy tensor $\varrho\mathbf{u}\otimes\mathbf{u}$. The following proposition gives estimates which are sufficient for our purposes. We introduce the deviation parameter $\mathfrak{a}$ and the critical exponent $\mathfrak{b}$ by the equalities

$$\mathfrak{a} = \frac{2}{d} - \frac{1}{\gamma}, \quad \mathfrak{b} = \frac{2\gamma}{\gamma+1}. \tag{4.2.2}$$

**Proposition 4.2.1.** *Assume that $\varrho$ and $\mathbf{u}$ satisfy inequalities (4.2.1) with $\gamma > d/2$, and that $\alpha \in [2, \infty]$ and $\beta \in [1, \infty)$ satisfy the conditions*

$$\frac{\mathfrak{a}}{\alpha} + \frac{1}{\beta} \ge \frac{1}{\mathfrak{b}} \quad \textit{for } \alpha < \infty,\ d = 3 \quad \textit{and} \quad \alpha = \infty,\ d \ge 2, \tag{4.2.3}$$

$$\frac{\mathfrak{a}}{\alpha} + \frac{1}{\beta} > \frac{1}{\mathfrak{b}} \quad \textit{for } \alpha < \infty,\ d = 2. \tag{4.2.4}$$

*Then there is a constant $c_E$, depending only on $E$, $\Omega$, $\gamma$, $\alpha$, and $\beta$, such that*

$$\|\varrho\mathbf{u}\|_{L^\alpha(0,T;L^\beta(\Omega))} \le c_E. \tag{4.2.5}$$

*Proof.* First, we consider the most complicated case $\alpha < \infty$. Represent $\varrho^\beta|\mathbf{u}|^\beta$ in the form

$$\varrho^\beta|\mathbf{u}|^\beta = \varrho^\lambda(\sqrt{\varrho}|\mathbf{u}|)^\mu|u|^\omega$$

with nonnegative $\lambda, \mu, \omega$ satisfying

$$\lambda + \mu/2 = \mu + \omega = \beta. \tag{4.2.6}$$

Next, for all nonnegative exponents $m$, $n$, $l$ such that

$$\frac{1}{n} + \frac{1}{m} + \frac{1}{l} = 1, \quad \lambda n = \gamma, \quad \mu m = 2, \tag{4.2.7}$$

the Hölder inequality implies

$$\begin{aligned} \int_\Omega \varrho^\beta |\mathbf{u}|^\beta \, dx &\le \Big(\int_\Omega \varrho^{\lambda n} \, dx\Big)^{1/n} \Big(\int_\Omega \sqrt{\varrho}^{\mu m} |\mathbf{u}|^{\mu m} \, dx\Big)^{1/m} \Big(\int_\Omega |\mathbf{u}|^{\omega l} \, dx\Big)^{1/l} \\ &= \Big(\int_\Omega \varrho^\gamma \, dx\Big)^{1/n} \Big(\int_\Omega \varrho |\mathbf{u}|^2 \, dx\Big)^{1/m} \Big(\int_\Omega |\mathbf{u}|^{\omega l} \, dx\Big)^{1/l} \le c_E \Big(\int_\Omega |\mathbf{u}|^{\omega l} \, dx\Big)^{1/l}, \end{aligned}$$

which leads to

$$\|\varrho \mathbf{u}\|_{L^\beta(\Omega)} \le c_E \|\mathbf{u}\|^{\omega/\beta}_{L^{\omega l}(\Omega)}.$$

Now set

$$k = \omega l, \quad \alpha = 2\beta/\omega. \tag{4.2.8}$$

Thus we get

$$\begin{aligned} \|\varrho \mathbf{u}\|_{L^\alpha(0,T;L^\beta(\Omega))} &= \Big(\int_0^T \|\varrho(t)\mathbf{u}(t)\|^\alpha_{L^\beta(\Omega)} \, dt\Big)^{1/\alpha} \\ &\le c_E \Big(\int_0^T \|\mathbf{u}(t)\|^2_{L^k(\Omega)} \, dt\Big)^{1/\alpha} = c_E \|\mathbf{u}\|^{2/\alpha}_{L^2(0,T;L^k(\Omega))}. \end{aligned} \tag{4.2.9}$$

Notice that estimate (4.2.9) holds if and only if the exponents $m$, $n$, $l$ and $\lambda$, $\mu$, $\omega$ are nonnegative and satisfy relations (4.2.6) and (4.2.7). This leads to some restrictions on the range of the parameters $\beta$ and $\alpha = 2\beta/\omega$. To analyze such restrictions we notice that relations (4.2.6) and (4.2.7) give the following expressions for $n$ and $m$ in terms of $\omega$, $l$, and $\gamma$:

$$\frac{1}{n} = \frac{1}{1+\gamma}\Big(1 + \omega - \frac{1}{l}\Big), \quad \frac{1}{m} = \frac{1}{1+\gamma}\Big(\gamma - \omega - \frac{\gamma}{l}\Big). \tag{4.2.10}$$

It follows that $1/m \le \gamma/n$. Hence $m \ge 0$ yields $n \ge 0$. Moreover if $m$, $n$, $l$ and $\omega$ are nonnegative, then by (4.2.6) and (4.2.7), so are $\lambda$, $\mu$ and $\beta$. Hence the restrictions on the set of parameters are

$$m \ge 0, \quad l \ge 0, \quad \omega \ge 0. \tag{4.2.11}$$

Next, relations (4.2.6), (4.2.7), and (4.2.10) give

$$\beta = \frac{2}{1+\gamma}\Big(\gamma - \omega - \frac{\gamma}{l}\Big) + \omega,$$

which yields, in terms of $\gamma$ and $\beta$,

$$\frac{1}{l}=1+\frac{\gamma-1}{2\gamma}\,\omega-\frac{\gamma+1}{2\gamma}\,\beta=\beta\Big(\frac{1}{\beta}+\Big(1-\frac{1}{\gamma}\Big)\frac{1}{\alpha}-\frac{\gamma+1}{2\gamma}\Big).$$

Substituting this into (4.2.10) we can rewrite (4.2.11) in the form

$$\beta\geq 0,\quad \alpha\geq 2,\quad \frac{1}{\beta}+\frac{\gamma-1}{\gamma\alpha}\geq\frac{\gamma+1}{2\gamma}. \tag{4.2.12}$$

This system of inequalities is incomplete, because they do not guarantee the existence of the integral on the right hand side of (4.2.9). In order to cope with this difficulty notice that the embedding $W^{1,2}(\Omega)\hookrightarrow L^k(\Omega)$ is bounded for

$$1\leq k\leq 2d/(d-2)\quad \text{if } d\geq 3,\quad \text{and}\quad k<\infty\quad \text{if } d=2. \tag{4.2.13}$$

Recall that $k=2l\beta/\alpha$ and $\alpha<\infty$. It follows from this and (4.2.13) that the embedding $W^{1,2}(\Omega)\hookrightarrow L^k(\Omega)$ is bounded for

$$\frac{1}{l}\geq\frac{d-2}{d}\frac{\beta}{\alpha}\quad \text{and}\quad 1/l>0. \tag{4.2.14}$$

Combining this with (4.2.9) and (4.2.1) we deduce that $\|\varrho\mathbf{u}\|_{L^\alpha(0,T;L^\beta(\Omega))}\leq c_E$ for $\alpha,\beta$ satisfying (4.2.14). Let us consider inequalities (4.2.14) in more detail. By using the formula for $1/l$ we rewrite (4.2.14) as

$$\begin{aligned}1+\Big(\frac{2}{d}-\frac{1}{\gamma}\Big)\frac{\beta}{\alpha}-\frac{\gamma+1}{2\gamma}\beta\geq 0,\\ 1+\Big(1-\frac{1}{\gamma}\Big)\frac{\beta}{\alpha}-\frac{\gamma+1}{2\gamma}\beta>0.\end{aligned} \tag{4.2.15}$$

Since we are considering the case $\alpha<\infty$, these inequalities can be replaced by

$$1+\Big(\frac{2}{d}-\frac{1}{\gamma}\Big)\frac{\beta}{\alpha}-\beta\frac{\gamma+1}{2\gamma}\geq 0\quad \text{for } d\geq 3 \tag{4.2.16}$$

and

$$1+\Big(\frac{2}{d}-\frac{1}{\gamma}\Big)\frac{\beta}{\alpha}-\beta\frac{\gamma+1}{2\gamma}>0\quad \text{for } d=2. \tag{4.2.17}$$

Recalling the notations $\mathfrak{a}$ and $\mathfrak{b}$ we can rewrite (4.2.16)–(4.2.17) in the equivalent form

$$\frac{\mathfrak{a}}{\alpha}+\frac{1}{\beta}\geq\frac{1}{\mathfrak{b}}\quad \text{for } d\geq 3,\quad \frac{\mathfrak{a}}{\alpha}+\frac{1}{\beta}>\frac{1}{\mathfrak{b}}\quad \text{for } d=2. \tag{4.2.18}$$

This leads to the statement of Proposition 4.2.1 in the case of $\alpha<\infty$.

It remains to consider the case $\alpha = \infty$. By the Hölder inequality,

$$\int_\Omega (\varrho|\mathbf{u}|)^{2\gamma/(\gamma+1)}\,dx \le \Big(\int_\Omega \varrho^\gamma\,dx\Big)^{1/(\gamma+1)}\Big(\int_\Omega \varrho|\mathbf{u}|^2\,dx\Big)^{\gamma/(\gamma+1)},$$

which gives

$$\|\varrho\mathbf{u}\|_{L^\infty(0,T;L^{2\gamma/(\gamma+1)}(\Omega))} \le \|\varrho\|_{L^\infty(0,T;L^\gamma(\Omega))}\,\|\varrho|\mathbf{u}|^2\|_{L^\infty(0,T;L^1(\Omega))} \le c_E. \qquad \square$$

Later we use only the particular cases of Proposition 4.2.1 with $\alpha = \infty, 2$. The corresponding result is given by the following corollary:

**Corollary 4.2.2.** *Assume that $\varrho$ and $\mathbf{u}$ satisfy inequalities* (4.2.1) *with $\gamma > d/2$. Then there is a constant $c_E$, depending only on $E$, $\beta$, $\alpha$, and $\Omega$, such that*

$$\begin{aligned}
&\|\varrho\mathbf{u}\|_{L^\infty(0,T;L^{2\gamma/(\gamma+1)}(\Omega))} \le c_E,\\
&\|\varrho\mathbf{u}\|_{L^2(0,T;L^\beta(\Omega))} \le c_E \quad \text{for all } \beta \in [1,\infty] \quad \text{with} \quad \frac{1}{\beta} > \frac{1}{2} + \frac{1}{\gamma} - \frac{1}{d},\\
&\|\varrho\mathbf{u}\|_{L^\alpha(0,T;L^2(\Omega))} \le c_E \quad \text{for all } \alpha \in [1, 4\gamma/d - 2).
\end{aligned}$$

*Proof.* Take $\alpha = \infty, 2$ and next $\beta = 2$ in (4.2.3) and (4.2.4). $\square$

Proposition 4.2.1 leads to a useful estimate for the kinetic energy density.

**Corollary 4.2.3.** *Suppose that $\gamma > d/2$, the functions $\varrho$, $\mathbf{u}$ satisfy* (4.2.1), *and $a \ge 1$, $b \ge 1$ satisfy the inequalities*

$$1 \le a \le 2, \quad \mathfrak{a}\left(\frac{1}{a} - \frac{1}{2}\right) + \frac{1}{b} > \frac{1}{\mathfrak{b}} + \frac{d-2}{2d}. \tag{4.2.19}$$

*Then there is a constant $c_E$, depending only on $E$, $a$, $b$, $\gamma$ and $\Omega$, such that*

$$\|\varrho|\mathbf{u}|^2\|_{L^a(0,T;L^b(\Omega))} \le c_E. \tag{4.2.20}$$

*In particular*

$$\|\varrho|\mathbf{u}|^2\|_{L^2(0,T;L^\tau(\Omega))} \le c_E \quad \text{for all } \tau \in \big[1, (1 - 2^{-1}\mathfrak{a})^{-1}\big), \tag{4.2.21}$$

$$\|\varrho|\mathbf{u}|^2\|_{L^1(0,T;L^\tau(\Omega))} \le c_E \quad \text{for all } \tau \in \big[1, (1 - \mathfrak{a})^{-1}\big). \tag{4.2.22}$$

*Proof.* Proposition 4.2.1 and the Hölder inequality yield

$$\begin{aligned}
\|\varrho|\mathbf{u}|^2\|_{L^a(0,T;L^b(\Omega))} &\le \|\varrho\mathbf{u}\|_{L^\alpha(0,T;L^\beta(\Omega))}\,\|\mathbf{u}\|_{L^2(0,T;L^q(\Omega))}\\
&\le c\|\varrho\mathbf{u}\|_{L^\alpha(0,T;L^\beta(\Omega))}\,\|\mathbf{u}\|_{L^2(0,T;W^{1,2}(\Omega))} \le C(E)
\end{aligned}$$

for all nonnegative exponents $\alpha, \beta, q$ satisfying

$$\frac{1}{a} = \frac{1}{\alpha} + \frac{1}{2}, \quad \frac{1}{b} = \frac{1}{\beta} + \frac{1}{q},$$

and

$$\frac{\mathfrak{a}}{\alpha}+\frac{1}{\beta}>\frac{1}{\mathfrak{b}}, \quad q<\frac{2d}{d-2}.$$

It is easy to check that such exponents exist for all $a$ and $b$ satisfying (4.2.19). It remains to note that estimates (4.2.21) and (4.2.22) are particular cases of (4.2.19)–(4.2.20). □

## 4.3 Functions of bounded mass dissipation rate

Along with bounded energy functions $(\varrho, \mathbf{u})$ that satisfy the energy estimates (4.2.1), we consider functions $\varrho$ with the finite mass dissipation rate,

$$\varepsilon \int_Q (1+\varrho)^{\gamma-2}|\nabla \varrho(x,t)|^2\,dxdt \leq E < \infty, \tag{4.3.1}$$

where the mass diffusion coefficient $\varepsilon$ is a small parameter. Because such functions have some additional regularity compared to bounded energy functions, the results of Propositions 4.2.1 and Corollaries 4.2.2–4.2.3 can be improved. In this section we obtain an estimate for the convection term $\nabla\varrho\otimes\mathbf{u}$ and the "internal energy" $\varrho^\gamma$.

**Proposition 4.3.1.** *Let $\Omega$ be a bounded domain with Lipschitz boundary. Assume that $\varrho$ and $\mathbf{u}$ satisfy inequalities (4.2.1) and (4.3.1) with $\gamma > 2d$. Then there is a constant $c_E$, depending only on $E$, $\Omega$, $\gamma$, and $q$, such that*

$$\||\nabla\varrho|\,|\mathbf{u}|\|_{L^1(0,T;L^q(\Omega))} \leq \varepsilon^{-1/2}c_E \quad \text{for all } 1\leq q\leq \frac{d}{d-1}, \tag{4.3.2}$$

$$\|\varrho^{3\gamma/2}\|_{L^1(Q)} \leq c_E\varepsilon^{-1}. \tag{4.3.3}$$

*Proof.* Inequalities (4.2.1) and (4.3.1) yield

$$\varepsilon\|\varrho^{\gamma/2}\|_{L^2(0,T;W^{1,2}(\Omega))} = \varepsilon\Big(\int_Q\Big(\varrho^\gamma+\frac{\gamma^2}{4}\varrho^{\gamma-2}|\nabla\varrho|^2\Big)\,dxdt\Big)^{1/2} \leq \varepsilon^{1/2}c_E, \tag{4.3.4}$$

where the constant $c_E$ depends only on $\Omega$, $E$, and $\gamma$. The same arguments give

$$\varepsilon\|\varrho\|_{L^2(0,T;W^{1,2}(\Omega))} = \varepsilon\Big(\int_Q(\varrho^2+|\nabla\varrho|^2)\,dxdt\Big)^{1/2} \leq \varepsilon^{1/2}c_E. \tag{4.3.5}$$

On the other hand, the Hölder inequality implies the estimate

$$\|\varepsilon|\nabla\varrho(t)|\,|\mathbf{u}(t)|\|_{L^q(\Omega)} \leq c\|\varepsilon\nabla\varrho(t)\|_{L^2(\Omega)}\|\mathbf{u}(t)\|_{L^m(\Omega)}$$

whenever

$$\frac{1}{q}\geq\frac{1}{2}+\frac{1}{m}.$$

Since the embedding $L^m(\Omega) \hookrightarrow W^{1,2}(\Omega)$ is bounded for $m < 2d/(d-2)$, we conclude that

$$\varepsilon \|\nabla \varrho(t) \mathbf{u}(t)\|_{L^q(\Omega)} \le c\varepsilon^{1/2} \|\varepsilon^{1/2} \nabla \varrho(t)\|_{L^2(\Omega)} \|\mathbf{u}(t)\|_{W^{1,2}(\Omega)}$$

and hence

$$\varepsilon \|\nabla \varrho(t) \mathbf{u}(t)\|_{L^1(0,T;L^q(\Omega))} \le c\varepsilon^{1/2} c_E$$

whenever

$$\frac{1}{q} > \frac{1}{2} + \frac{d-2}{2d} \quad \Leftrightarrow \quad q < \frac{d}{d-1},$$

which yields (4.3.2). Next,

$$\int_\Omega \varrho^{3\gamma/2}\, dx \le \Big(\int_\Omega \varrho^\gamma\, dx\Big)^{1/2} \Big(\int_\Omega \varrho^{2\gamma}\, dx\Big)^{1/2} \le c_E \Big(\int_\Omega \varrho^{2\gamma}\, dx\Big)^{1/2} \le c_E \|\varrho^{\gamma/2}\|^2_{L^4(\Omega)}.$$

Since $4 < 2d/(d-2)$, we have

$$\|\varrho^{\gamma/2}(t)\|^2_{L^4(\Omega)} \le c \|\varrho^{\gamma/2}(t)\|^2_{W^{1,2}(\Omega)}.$$

Thus we get

$$\int_Q \varrho^{3\gamma/2}\, dxdt \le c_E \int_0^T \|\varrho^{\gamma/2}(t)\|^2_{W^{1,2}(\Omega)}\, dt \le c_E \varepsilon^{-1}. \qquad \square$$

## 4.4 Compactness properties

The compactness properties of solutions to the mass and momentum balance equations are crucial for the theory of weak solutions to Navier-Stokes equations. There are two reasons for this. The first is existence theory. Notice that the standard proof of existence of weak solutions to boundary value problems for Navier-Stokes equations consists of two steps. First, we construct a sequence of approximate solutions with uniformly bounded energy. Next, we prove that any weak limit of such a sequence is a solution to the governing equations. The latter step is almost trivial for incompressible fluids, but leads to difficult problems for compressible Navier-Stokes equations. The second reason for the necessity of a detailed study of weak compactness properties of the solution set to the Navier-Stokes equations is that the compactness of this set implies its stability with respect to perturbations of given data. In this framework, the term compactness is not appropriate and should be replaced by *weak compactness* or, more precisely, *weak closedness.* In this section we consider elementary weak compactness properties of bounded energy functions; integrability of such functions was studied in Section 4.2.

Assume that $\Omega$ is a bounded domain with Lipschitz boundary and set

$$Q = \Omega \times (0,T), \quad 0 < T < \infty.$$

Let us consider sequences of vector fields $\mathbf{u}_n : Q \to \mathbb{R}^d$ and of functions $\varrho_n, \varphi_n : Q \to \mathbb{R}$, $n \ge 1$, such that $\varrho_n \ge 0$. Assume that the following condition is satisfied:

**Condition 4.4.1.** • There are exponents $d/2 < \gamma < \infty$, $d/2 < s \le \infty$ and a constant $E$ independent of $n$ such that the functions $\varrho_n$, $\varphi_n$, and the vector fields $\mathbf{u}_n$ satisfy the "energy inequalities"

$$\|\varrho_n^\gamma\|_{L^1(Q)} + \|\mathbf{u}_n\|_{L^2(0,T;W^{1,2}(\Omega))} \le E, \tag{4.4.1}$$

$$\|\varrho_n|\mathbf{u}_n|^2\|_{L^\infty(0,T;L^1(\Omega))} \le E, \tag{4.4.2}$$

$$\|\varphi_n\|_{L^\infty(0,T;L^s(\Omega))} \le E. \tag{4.4.3}$$

• There are integrable vector fields $\mathfrak{u}_n, \mathfrak{v}_n, \mathfrak{w} : Q \to \mathbb{R}^d$, functions $\mathfrak{g}_n, \mathfrak{h}_n : Q \to \mathbb{R}$ and matrix-valued functions $\mathfrak{V}_n : Q \to \mathbb{R}^{d\times d}$ such that

$$\|\mathfrak{u}_n\|_{L^1(Q)} + \|\mathfrak{v}_n\|_{L^1(Q)} + \|\mathfrak{w}_n\|_{L^1(Q)} + \|\mathfrak{g}_n\|_{L^1(Q)} + \|\mathfrak{h}_n\|_{L^1(Q)} + \|\mathfrak{V}_n\|_{L^1(Q)} \le E \tag{4.4.4}$$

and

$$\begin{aligned} \partial_t\varrho_n &= \operatorname{div}\mathfrak{u}_n + \mathfrak{g}_n, \\ \partial_t\varphi_n &= \operatorname{div}\mathfrak{v}_n + \mathfrak{h}_n, \\ \partial_t(\varrho_n\mathbf{u}_n) &= \operatorname{div}\mathfrak{V}_n + \mathfrak{w}_n. \end{aligned} \tag{4.4.5}$$

The following theorem on weak compactness properties of the momentum $\varrho\mathbf{u}$ and of the kinetic energy tensor $\varrho\mathbf{u}\otimes\mathbf{u}$ is the main result of this section.

**Theorem 4.4.2.** *Let a sequence $(\mathbf{u}_n, \varrho_n, \varphi_n)$ satisfy Conditions 4.4.1. Then there is a subsequence, still denoted by $(\mathbf{u}_n, \varrho_n, \varphi_n)$, a vector field $\mathbf{u} \in L^2(0,T;W^{1,2}(\Omega))$, and functions $\varrho \in L^\infty(0,T;L^\gamma(\Omega))$, $\varphi \in L^\infty(0,T;L^s(\Omega))$ such that*

$$\begin{aligned} \varrho_n &\rightharpoonup \varrho \quad \text{weakly}^\star \text{ in } L^\infty(0,T;L^\gamma(\Omega)), \\ \varphi_n &\rightharpoonup \varphi \quad \text{weakly}^\star \text{ in } L^\infty(0,T;L^s(\Omega)), \\ \mathbf{u}_n &\rightharpoonup \mathbf{u} \quad \text{weakly in } L^2(0,T;W^{1,2}(\Omega)); \end{aligned} \tag{4.4.6}$$

*for $m^{-1} > 2^{-1} + \gamma^{-1} - d^{-1}$,*

$$\varrho_n\mathbf{u}_n \rightharpoonup \varrho\mathbf{u} \quad \text{weakly in } L^2(0,T;L^m(\Omega)), \tag{4.4.7}$$

$$\varphi_n\mathbf{u}_n \rightharpoonup \varphi\mathbf{u} \quad \text{weakly in } L^2(0,T;L^m(\Omega)); \tag{4.4.8}$$

*and for $1 < b < (1 - 2^{-1}\mathfrak{a})^{-1}$,*

$$\varrho_n\mathbf{u}_n\otimes\mathbf{u}_n \rightharpoonup \varrho\mathbf{u}\otimes\mathbf{u} \quad \text{weakly in } L^2(0,T;L^b(\Omega)). \tag{4.4.9}$$

*Moreover, for a.e. $t \in (0,T)$,*

$$\varrho_n(t) \rightharpoonup \varrho(t) \quad \text{weakly in } L^\gamma(\Omega), \tag{4.4.10a}$$

$$\varphi_n(t) \rightharpoonup \varphi(t) \quad \text{weakly (weakly}^\star \text{ for } s=\infty) \text{ in } L^s(\Omega), \tag{4.4.10b}$$

$$\varrho_n(t)\mathbf{u}_n(t) \rightharpoonup \varrho(t)\mathbf{u}(t) \quad \text{weakly in } L^{2\gamma/(\gamma+1)}(\Omega). \tag{4.4.10c}$$

The remaining part of this section is devoted to the proof of Theorem 4.4.2. It is based on two auxiliary compactness lemmas which describe the pointwise convergence of weakly convergent sequences and the weak continuity of products of two weakly convergent sequences. These results can be regarded as a simplest version, appropriate for our purposes, of the compensated compactness principle.

### 4.4.1 Two lemmas on compensated compactness

First, we investigate pointwise convergence of integrable functions depending on the time variable $t \in (0,T)$. Let us consider a bounded sequence of functions $\mathcal{H}_n \in L^\infty(0,T;L^r(\Omega))$, $n \ge 1$, $r \in (1,\infty]$. We prove a simple criterion for the weak convergence of the functions $\mathcal{H}_n(t,\cdot)$. For every $\xi \in C_0^\infty(\Omega)$ set

$$\mathcal{F}_n[\xi](t) = \int_\Omega \mathcal{H}_n(t)\xi\, dx.$$

Since the sequence $\mathcal{H}_n$ is bounded in $L^\infty(0,T;L^r(\Omega))$, the functions $\mathcal{F}_n[\xi]$ belong to $L^\infty(0,T)$.

**Lemma 4.4.3.** *Let $\mathcal{H}_n$ be a bounded sequence in $L^\infty(0,T;L^r(\Omega))$ for $r \in (1,\infty]$. Furthermore, assume that for every $\xi \in C_0^\infty(\Omega)$ and for $n \ge 1$, the function $\mathcal{F}_n[\xi]$ can be modified on a set of zero measure in such a way that*

$$\bigvee_{[0,T]} \mathcal{F}_n[\xi] \le C_\xi, \tag{4.4.11}$$

*where $C_\xi$ is independent of $n$. Then there is a function $\mathcal{H} \in L^\infty(0,T;L^r(\Omega))$ and a subsequence, still denoted by $\mathcal{H}_n$, such that for a.e. $t \in [0,T]$,*

$$\mathcal{H}_n(t) \rightharpoonup \mathcal{H}(t) \quad \textit{weakly (weakly}^\star \textit{ for } r=\infty\textit{) in } L^r(\Omega).$$

For the definition of the variation $\bigvee$ and elementary properties of functions of bounded variation we refer to Section 1.3.1.

*Proof.* For every $n$, there is a set $\mathcal{T}_n$ of full measure in $[0,T]$ such that

$$\|\mathcal{H}_n(t)\|_{L^r(\Omega)} \le \|\mathcal{H}_n\|_{L^\infty(0,T;L^r(\Omega))}, \quad t \in \mathcal{T}_n.$$

Set $\mathcal{T} = \bigcap_n \mathcal{T}_n$. Next, choose a countable set of functions $\xi_m \in C_0^\infty(\Omega)$, $m \ge 1$, which is dense in $L^{r'}(\Omega)$, where $r' = r/(r-1)$ for $r < \infty$ and $r' = 1$ for $r = \infty$. It suffices to prove that for all $m$ and $t \in \mathcal{T}$, the limit

$$\lim_{n\to\infty} \int_\Omega \mathcal{H}_n(t)\xi_m\, dx \tag{4.4.12}$$

exists. Indeed, fix an arbitrary $\xi \in L^{r'}(\Omega)$. For every $\delta > 0$, there is $m$ such that $\|\xi - \xi_m\|_{L^{r'}(\Omega)} < \delta$. In view of

$$\limsup_{p,q\to\infty} \left| \int_\Omega (\mathcal{H}_p(t) - \mathcal{H}_q(t))\xi_m\, dx \right| = 0,$$

we have for $t \in \mathcal{T}$,

$$\begin{aligned}\limsup_{p,q\to\infty}\left|\int_\Omega (\mathcal{H}_p(t)-\mathcal{H}_q(t))\xi\,dx\right| &\le \limsup_{p,q\to\infty}\left|\int_\Omega (\mathcal{H}_p(t)-\mathcal{H}_q(t))\xi_m\,dx\right| \\ &+ \limsup_{p,q\to\infty}\int_\Omega (|\mathcal{H}_p(t)|+|\mathcal{H}_q(t)|)|\xi_m-\xi|\,dx \\ &\le c\limsup_{p,q\to\infty}\big\||\mathcal{H}_p(t)|+|\mathcal{H}_q(t)|\big\|_{L^r(\Omega)}\|\xi-\xi_m\|_{L^{r'}(\Omega)} \le c\delta.\end{aligned}$$

Since $\delta > 0$ is arbitrary, we conclude that for all $t \in \mathcal{T}$ and $\xi \in L^{r'}(\Omega)$, the limit

$$\lim_{n\to\infty}\int_\Omega \mathcal{H}_n(t)\xi\,dx$$

exists. Hence the sequence $\mathcal{H}_n(t)$ converges weakly (weakly$^\star$ for $r = \infty$) in $L^r(\Omega)$. Recall that the Lebesgue space $L^r(\Omega)$ with $r \in (1,\infty)$ is weakly complete and $L^\infty(\Omega)$ is weak$^\star$ complete. Therefore, there exists $\mathcal{H}(t)$ such that $\mathcal{H}_n(t)$ converges to $\mathcal{H}(t)$ weakly in $L^r(\Omega)$ (weakly$^\star$ for $r = \infty$).

Let us prove the existence of the limit (4.4.12). It follows from (4.4.11) that for every fixed $m$ the functions

$$F_n^{(m)} = \mathcal{F}_n[\xi_m], \quad n \ge 1,$$

are uniformly bounded in $[0,T]$ and have uniformly bounded variations. Now we apply a diagonal process. By the Helly compactness theorem 1.3.4, there is a subsequence $\{\mathcal{H}_{1,n}\} \subset \{\mathcal{H}_n\}$ and a bounded function $F_{1,\infty}^{(1)} : [0,T] \to \mathbb{R}$ such that the functions

$$F_{1,n}^{(1)}(t) := \int_\Omega \mathcal{H}_{1,n}(t)\xi_1\,dx$$

converge to $F_{1,\infty}^{(1)}$ pointwise on $[0,T]$. In its turn, there is a subsequence $\{\mathcal{H}_{2,n}\} \subset \{\mathcal{H}_{1,n}\} \subset \{\mathcal{H}_n\}$ such that the functions

$$F_{2,n}^{(m)}(t) := \int_\Omega \mathcal{H}_{2,n}(t)\xi_m\,dx, \quad m = 1,2,$$

converge to some bounded functions $F_{2,\infty}^{(m)}$ pointwise on $[0,T]$ as $n \to \infty$. Repeating these arguments we conclude that there are sequences

$$\cdots \subset \{\mathcal{H}_{k,n}\} \subset \{\mathcal{H}_{k-1,n}\} \subset \cdots \subset \{\mathcal{H}_n\}$$

and bounded functions $F_{k,\infty}^{(m)}$, $m \ge 1$, such that for every $1 \le m \le k$, and for $n \to \infty$,

$$F_{k,n}^{(m)}(t) := \int_\Omega \mathcal{H}_{k,n}(t)\xi_m\,dx \to F_{k,\infty}^{(m)}(t) \quad \text{pointwise on } [0,T].$$

Hence for every $m \geq 1$, the diagonal sequence

$$\{F_{n,n}^{(m)}(t)\} := \int_\Omega \mathcal{H}_{n,n}(t)\xi_m \, dx$$

converges pointwise on $[0,T]$ to a function $F_\infty^{(m)} \in L^\infty(0,T)$. Notice that $\{\mathcal{H}_{n,n}\}$ is a subsequence of $\{\mathcal{H}_n\}$. By abuse of notation we write simply $\mathcal{H}_n$ instead of $\mathcal{H}_{n,n}$. In other words, there is a subsequence of $\mathcal{H}_n$, still denoted by $\mathcal{H}_n$, such that for every $m \geq 1$,

$$\int_\Omega \mathcal{H}_n(t)\xi_m \, dx \to F_\infty^{(m)}(t) \quad \text{pointwise on } [0,T], \tag{4.4.13}$$

which yields the existence of the limit (4.4.12). □

The second auxiliary lemma establishes additional sufficient conditions for pointwise weak convergence of sequences of time dependent integrable functions.

**Lemma 4.4.4.** *Let $r \in (1,\infty]$, $\{\mathcal{H}_n\}_{n\geq 1}$ a bounded sequence in $L^\infty(0,T;L^r(\Omega))$, and $\{\mathcal{S}_n\}_{n\geq 1}$ a bounded sequence in $L^\infty(0,T;W^{-k,h}(\Omega))$, with $k \geq 0$ and $1 <$ $h< \infty$, such that*

$$\|\mathcal{S}_n(t)\|_{W^{-k,h}(\Omega)} \to 0 \quad \text{as } n \to \infty \quad \text{for a.e. } t \in (0,T). \tag{4.4.14}$$

*Furthermore, assume that*

$$\partial_t(\mathcal{S}_n + \mathcal{H}_n) = \operatorname{div} \mathfrak{Q}_n + \mathfrak{q}_n, \tag{4.4.15}$$

*where the sequences $\{\mathfrak{Q}_n\}_{n\geq 1}$ and $\{\mathfrak{q}_n\}_{n\geq 1}$ are bounded in $L^1(Q)$. Then there is a function $\mathcal{H} \in L^\infty(0,T;L^r(\Omega))$ and a subsequence, still denoted by $\mathcal{H}_n$, such that*

$$\mathcal{H}_n(t) \rightharpoonup \mathcal{H}(t) \quad \text{weakly (weakly}^\star \text{ for } r=\infty) \text{ in } L^r(\Omega) \quad \text{for a.e. } t \in [0,T].$$

*Proof.* Choose a positive constant $M$ and a set $\mathcal{T}$ of full measure in $(0,T)$ such that

$$\|\mathcal{S}_n\|_{L^\infty(0,T;W^{-k,h}(\Omega))} + \|\mathcal{H}_n\|_{L^\infty(0,T;L^r(\Omega))} \leq M,$$
$$\|\mathcal{S}_n(t)\|_{W^{-k,h}(\Omega)} + \|\mathcal{H}_n(t)\|_{L^r(\Omega)} \leq M \quad \text{for all } t \in \mathcal{T}.$$

For any $\xi \in C_0^\infty(\Omega)$ set

$$\mathcal{F}_n[\xi](t) = \langle \mathcal{S}_n(t), \xi \rangle + \int_\Omega \mathcal{H}_n(x,t)\xi(x)\, dx,$$

where $\langle \cdot,\cdot \rangle$ stands for the duality pairing between $W^{-k,h}(\Omega)$ and $W_0^{k,h}(\Omega)$. We have

$$\int_0^T \eta'(t)\mathcal{F}_n[\xi](t)\, dt = \int_0^T \eta(t) N_n[\xi](t)\, dt \quad \text{for all } \eta \in C_0^\infty(0,T), \tag{4.4.16}$$

where

$$N_n[\xi](t) = \int_\Omega (\mathfrak{Q}_n \cdot \nabla\xi - \mathfrak{q}_n\xi)\,dx.$$

The linear mapping $\xi \mapsto N_n[\xi](t)$ admits the estimate

$$|N_n[\xi](t)| \le (\|\mathfrak{Q}_n(t)\|_{L^1(\Omega)} + \|\mathfrak{q}_n\|_{L^1(\Omega)})\|\xi\|_{C^1(\Omega)}.$$

Since the sequences $\{\mathfrak{Q}_n\}_{n\ge1}$ and $\{\mathfrak{q}_n\}_{n\ge1}$ are bounded in $L^1(Q)$, we have

$$\int_0^T |N_n[\xi](t)|\,dt \le c\|\xi\|_{C^1(\Omega)}.$$

Next we have, a.e. on $(0,T)$,

$$|\mathcal{F}_n[\xi](t)| \le \|\mathcal{S}_n(t)\|_{W^{-k,h}(\Omega)}\|\xi\|_{C^k(\Omega)} + \|\mathcal{H}_n\|_{L^r(\Omega)}\|\xi\|_{C(\Omega)} \le cM\|\xi\|_{C^k(\Omega)}.$$

It follows that the function $t \mapsto \mathcal{F}_n[\xi](t)$ meets all requirements of Lemma 1.3.3, so after a possible change of its values on a set of zero measure we obtain

$$\bigvee_{[0,T]} \mathcal{F}_n[\xi] \le c\|\xi\|_{C^k(\Omega)}.$$

It now follows from the Helly Theorem 1.3.4 that for every $\xi \in C_0^k(\Omega)$, there exists a subsequence, still denoted by $\mathcal{F}_n[\xi]$, which converges everywhere in $[0,T]$. Choosing a countable set $\{\xi_m\}$ dense in $C_0^k(\Omega)$, arguing as in the proof of Lemma 4.4.3, and applying the diagonal process, we conclude that there is a subsequence, still denoted by $\mathcal{F}_n$, such that for all $t \in (0,T)$ and $\xi \in C_0^k(\Omega)$, the limits $\lim_{n\to\infty} \mathcal{F}_n[\xi](t)$ exist. In particular, for all $t \in \mathcal{T}$ and $\xi \in C_0^k(\Omega)$, the limits

$$\lim_{n\to\infty} \left(\langle \mathcal{S}_n(t), \xi\rangle + \int_\Omega \mathcal{H}_n(x,t)\xi(x)\,dx\right)$$

exist. By (4.4.14) we can choose $\mathcal{T}$ in such a way that

$$\lim_{n\to\infty} \langle \mathcal{S}_n(t), \xi\rangle = 0 \quad \text{for all } t \in \mathcal{T} \text{ and } \xi \in C_0^k(\Omega).$$

Therefore, as $n \to \infty$,

$$\int_\Omega \mathcal{H}_n(x,t)\xi(x)\,dx \quad \text{converges for all } t \in \mathcal{T} \text{ and } \xi \in C_0^k(\Omega).$$

Since the sequence $\{\mathcal{H}_n(t)\}$ is bounded in $L^r(\Omega)$, it follows that it converges weakly in $L^r(\Omega)$ (weakly$^\star$ for $r = \infty$) to some $\mathcal{H}(t)$, obviously belonging to $L^\infty(0,T;L^r(\Omega))$. □

The next lemma concerns the continuity properties of a product of weakly convergent sequences.

**Lemma 4.4.5.** *Let $2d/(d+2) < r \le \infty$ and suppose sequences $\mathcal{H}_n$ and $\mathbf{u}_n$ satisfy the conditions*

$$\|\mathcal{H}_n\|_{L^\infty(0,T;L^r(\Omega))} + \|\mathbf{u}_n\|_{L^2(0,T;W^{1,2}(\Omega))} \le c,$$
$$\mathbf{u}_n \rightharpoonup \mathbf{u} \quad \text{weakly in } L^2(0,T;W^{1,2}(\Omega)),$$
$$\mathcal{H}_n(t) \rightharpoonup \mathcal{H}(t) \quad \text{weakly (weakly}^\star \text{ for } r=\infty) \text{ in } L^r(\Omega) \text{ for a.e. } t \in (0,T).$$

*Then $\mathcal{H}_n \mathbf{u}_n \rightharpoonup \mathcal{H}\mathbf{u}$ weakly in $L^2(0,T;L^q(\Omega))$ for all $q \in (1,\infty)$ with $q^{-1} > 2^{-1} + r^{-1} - d^{-1}$.*

*Proof.* The Hölder inequality implies

$$\|\mathcal{H}_n\mathbf{u}_n\|_{L^2(0,T;L^q(\Omega))} \le \|\mathcal{H}_n\|_{L^\infty(0,T;L^r(\Omega))}\|\mathbf{u}\|_{L^2(0,T;L^\sigma(\Omega))} \le c\|\mathbf{u}\|_{L^2(0,T;L^\sigma(\Omega))}$$

for all $\sigma \in [1,\infty]$ such that

$$q^{-1} \ge r^{-1} + \sigma^{-1} \quad \text{or equivalently} \quad \sigma^{-1} \le q^{-1} - r^{-1}. \tag{4.4.17}$$

It follows from the assumptions of the lemma that $q^{-1} - r^{-1} > (d-2)/(2d)$. Hence there is a $\sigma < 2d(d-2)$ satisfying (4.4.17). For such a $\sigma$ the embedding $W^{1,2}(\Omega) \hookrightarrow L^\sigma(\Omega)$ is bounded, which leads to the estimate

$$\|\mathcal{H}_n\mathbf{u}_n\|_{L^2(0,T;L^q(\Omega))} \le c\|\mathbf{u}\|_{L^2(0,T;L^\sigma(\Omega))} \le c\|\mathbf{u}\|_{L^2(0,T;W^{1,2}(\Omega))} \le c.$$

Hence, it suffices to show that for all vector fields $\boldsymbol{\omega} \in C_0^\infty(Q)^3$,

$$\lim_{n\to\infty} \int_Q (\mathcal{H}_n\mathbf{u}_n - \mathcal{H}\mathbf{u}) \cdot \boldsymbol{\omega}\, dxdt = 0. \tag{4.4.18}$$

Now set $r' = r/(r-1)$ for $1 < r < \infty$ and $r' = 1$ for $r = \infty$. Notice that $r' < 2d(d-2)$ for $r > 2d/(d+2)$ and

$$W_0^{1,2}(\Omega) \hookrightarrow L^{r'}(\Omega) \quad \text{is compact for } r' < 2d/(d-2). \tag{4.4.19}$$

The continuity of $L^2(0,T;W^{1,2}(\Omega)) \hookrightarrow L^2(0,T;L^{r'}(\Omega))$ implies that

$$\mathbf{u}_n - \mathbf{u} \rightharpoonup 0 \quad \text{weakly in } L^2(0,T;L^{r'}(\Omega)).$$

Since $\mathcal{H}\boldsymbol{\omega} \in L^2(0,T;L^r(\Omega))$ it follows that

$$\lim_{n\to\infty} \int_Q \mathcal{H}\boldsymbol{\omega} \cdot (\mathbf{u}_n - \mathbf{u})\, dxdt = 0.$$

Thus we get

$$\lim_{n\to\infty} \left| \int_Q (\mathcal{H}_n\mathbf{u}_n - \mathcal{H}\mathbf{u}) \cdot \boldsymbol{\omega}\, dxdt \right| \le \lim_{n\to\infty} \left| \int_Q (\mathcal{H}_n - \mathcal{H})\mathbf{u}_n \cdot \boldsymbol{\omega}\, dxdt \right|.$$

By (4.4.19) the embedding of the dual spaces $L^r(\Omega) \hookrightarrow W^{-1,2}(\Omega)$ is compact, which yields

$$\mathcal{H}_n(t) \to \mathcal{H}(t) \quad \text{strongly in } W^{-1,2}(\Omega) \quad \text{for a.e. } t \in (0,T). \tag{4.4.20}$$

Moreover, the sequence $\mathcal{H}_n$ is bounded in $L^\infty(0,T;W^{-1,2}(\Omega))$. This leads to

$$\begin{aligned}
&\lim_{n\to\infty}\Big|\int_Q (\mathcal{H}_n - \mathcal{H})\mathbf{u}_n \cdot \boldsymbol{\omega}\, dxdt\Big| \\
&\leq \lim_{n\to\infty}\int_0^T \|\mathcal{H}_n(t) - \mathcal{H}(t)\|_{W^{-1,2}(\Omega)}\|\boldsymbol{\omega}\cdot\mathbf{u}_n(t)\|_{W^{1,2}(\Omega)}\, dt \\
&\leq \lim_{n\to\infty}\Big(\int_0^T \|\boldsymbol{\omega}\cdot\mathbf{u}_n(t)\|^2_{W^{1,2}(\Omega)}\, dt\Big)^{1/2}\Big(\int_0^T \|\mathcal{H}_n(t) - \mathcal{H}(t)\|^2_{W^{-1,2}(\Omega)}\, dt\Big)^{1/2} \\
&\leq c\lim_{n\to\infty}\Big(\int_0^T \|\mathcal{H}_n(t) - \mathcal{H}(t)\|^2_{W^{-1,2}(\Omega)}\, dt\Big)^{1/2}.
\end{aligned}$$

It follows from (4.4.20) and the Lebesgue dominated convergence theorem that the last integral vanishes as $n \to \infty$. □

### 4.4.2 Proof of Theorem 4.4.2

Notice that by Theorem 1.3.30 on weak compactness of bounded sets there is $(\varrho, \varphi, \mathbf{u})$ such that, after passing to a subsequence of $(\varrho_n, \varphi_n, \mathbf{u}_n)$ if necessary, the convergences (4.4.6) hold. Next we derive the limit relations (4.4.10). We begin by proving (4.4.10a). Set $\mathcal{H}_n = \varrho_n$. It follows from (4.4.5) that for any $\xi \in C_0^\infty(\Omega)$ the functions of the time variable $t \in (0,T)$,

$$\mathcal{F}_n[\xi](t) := \int_\Omega \varrho_n(t)\xi\, dx,$$

satisfy the integral identity

$$\int_0^T \eta'(t)\mathcal{F}_n[\xi](t)\, dt = \int_0^T \eta(t)G_n[\xi](t)\, dt \quad \text{for all } \eta \in C_0^\infty(0,T), \tag{4.4.21}$$

where

$$G_n[\xi](t) = \int_\Omega (\mathfrak{u}_n \cdot \nabla\xi - \mathfrak{g}_n\xi)\, dx.$$

We have the inequality

$$|G_n[\xi](t)| \leq (\|\mathfrak{u}_n(t)\|_{L^1(\Omega)} + \|\mathfrak{g}_n\|_{L^1(\Omega)})\|\xi\|_{C^1(\Omega)},$$

which combined with condition (4.4.4) leads to

$$\int_0^T |G_n[\xi](t)|\, dt \leq E\|\xi\|_{C^1(\Omega)} =: C_\xi.$$

Applying Lemma 1.3.3 we conclude that, after possible changes of values on a set of zero measure,

$$\bigvee_{[0,T]} \mathcal{F}_n[\xi] \le C_\xi. \tag{4.4.22}$$

It follows that $\varrho_n$ meets all requirements of Lemma 4.4.3 with $\mathcal{H}_n$ and $r$ replaced by $\varrho_n$ and $\gamma$, respectively. This yields (4.4.10a). The same arguments give (4.4.10b).

To prove (4.4.10c), we observe that by (4.4.1)–(4.4.2), the sequences $\varrho_n$, $\mathbf{u}_n$ meet all requirements of Corollary 4.2.2. It follows that the functions $\varrho_n \mathbf{u}_n$ are uniformly bounded in $L^\infty(0,T;L^{2\gamma/(\gamma+1)}(\Omega))$. Next, set $\mathcal{H}_n(t) = \varrho_n(t)\mathbf{u}_n(t)$. It follows from (4.4.5) that the functions

$$\mathcal{F}_n[\boldsymbol{\xi}](t) = \int_\Omega \varrho_n(t)\mathbf{u}_n \cdot \boldsymbol{\xi}\, dx, \quad \boldsymbol{\xi} \in C_0^\infty(\Omega),$$

satisfy the integral identity

$$\int_0^T \eta'(t)\mathcal{F}_n[\boldsymbol{\xi}](t)\, dt = \int_0^T \eta(t)\mathcal{G}_n[\boldsymbol{\xi}](t)\, dt,$$

where the function

$$\mathcal{G}_n[\boldsymbol{\xi}](t) = \int_\Omega (\mathfrak{V}_n(t) : \nabla\boldsymbol{\xi} - \mathfrak{w}_n \cdot \boldsymbol{\xi})\, dx$$

satisfies

$$|\mathcal{G}_n[\boldsymbol{\xi}](t)| \le c(\|\mathfrak{V}_n(t)\|_{L^1(\Omega)} + \|\mathfrak{w}_n(t)\|_{L^1(\Omega)})\|\boldsymbol{\xi}\|_{C^1(\Omega)}.$$

From this and inequalities (4.4.4) in Condition 4.4.1 we obtain

$$\int_0^T |\mathcal{G}_n[\boldsymbol{\xi}](t)|\, dt \le C(E)\|\xi\|_{C^1(\Omega)} =: C_\xi.$$

Hence the sequence $\mathcal{F}_n[\xi]$ satisfies (4.4.22) and $\mathcal{H}_n = \varrho_n\mathbf{u}_n$ meets all requirements of Lemma 4.4.3, which yields (4.4.10c).

To prove (4.4.7), it suffices to notice that by (4.4.10a) and (4.4.6), the functions $\varrho_n$ and $\mathbf{u}_n$ meet all requirements of Lemma 4.4.5 with $\mathcal{H}_n$ replaced by $\varrho_n$ and $r$ replaced by $\gamma$. Repeating the same arguments we obtain (4.4.8).

It remains to prove (4.4.9). By (4.4.10c), the sequence $\mathcal{H}_n(t) = \varrho_n(t)\mathbf{u}_n(t)$ converges weakly in $L^{2\gamma/(\gamma+1)}(\Omega)$ to $\mathcal{H}(t) = \varrho(t)\mathbf{u}(t)$ for a.e. $t \in (0,T)$. Moreover, the functions $\mathcal{H}_n(t)$ are uniformly bounded in $L^\infty(0,T;L^{2\gamma/(\gamma+1)}(\Omega))$. Since $\gamma > d/2$, we have $2\gamma/(\gamma+1) > 2d/(d+2)$. Hence the functions $\mathcal{H}_n$ and $\mathbf{u}_n$ meet all requirements of Lemma 4.4.5 with $r = 2\gamma/(\gamma+1)$, so $\varrho_n\mathbf{u}_n \otimes \mathbf{u}_n$ converges to $\varrho\mathbf{u} \otimes \mathbf{u}$ weakly in $L^2(0,T;L^b(\Omega))$ for all $b \in (1,\infty)$ satisfying

$$b^{-1} > 2^{-1} - d^{-1} + r^{-1} = 2^{-1} - d^{-1} + (\gamma+1)(2\gamma)^{-1} = 1 - 2^{-1}\mathfrak{a}.$$

This completes the proof of Theorem 4.4.2.

## 4.5 Basic integral identity

The compactness results obtained in Section 4.4.2 follow from the general fact that the governing equations represent a system of conservation laws in divergence form. But the momentum $\varrho\mathbf{u}$ and the kinetic energy tensor $\varrho\mathbf{u}\otimes\mathbf{u}$ are not the only nonlinear terms in compressible Navier-Stokes equations. In particular, the stress tensor usually includes the pressure $p$, which is a nonlinear function of the density $\varrho$. The question of weak compactness properties for the composite functions $p(\varrho_n)$, $n \geq 1$, is difficult, and its resolution requires a sophisticated mathematical technique. The key observation is that the quantity $p-(\lambda+1)\operatorname{div}\mathbf{u}$, called the effective viscous flux or effective viscous pressure, enjoys better regularity properties compared to $p$ and $\operatorname{div}\mathbf{u}$. This results from a detailed examination of the projection of the momentum balance equation onto the space of potential vector fields, as is proved in [34], [80]. Fortunately, it is not necessary to investigate this projection in detail. All needed information can be obtained from a special integral identity which is derived in this section. Before we formulate the corresponding results let us introduce some notation.

Recall the operators (see Section 1.7.1 for details)

$$A_i = \partial_{x_i}\Delta^{-1}, \quad R_{ij} = \partial_{x_i}\partial_{x_j}\Delta^{-1}, \quad 1 \leq i,j \leq d.$$

Introduce also the vector-valued operator $\mathbf{A}$ and the matrix-valued operator $\mathbf{R}$ defined by

$$\mathbf{A} = (A_i)_{1\leq i\leq 3}, \quad \mathbf{R} = (R_{ij})_{3\times 3}.$$

The symbol $\mathbf{R}[\cdot]$ will stand for the action of the linear operator $\mathbf{R}$ on a function:

- for a scalar function $\eta$ it is a matrix with the entries $(\mathbf{R}[\eta])_{ij} := R_{ij}\eta$;
- for a vector function $\mathbf{v}$ it is a vector with the components $(\mathbf{R}[\mathbf{v}])_i := R_{ij}v_j$;
- for a tensor function $\mathbb{T}$ it is a matrix with the entries $(\mathbf{R}[\mathbb{T}])_{ij} := R_{ik}T_{kj}$.

Let $\Omega$ be a bounded domain in $\mathbb{R}^d$ with Lipschitz boundary, and denote $Q = \Omega\times(0,T)$. Suppose that the functions $(\varrho,\varphi,\varpi)$, the vector fields $(\mathbf{u},\mathbf{g},\mathbf{g}^\varphi)$, and the matrix-valued functions $(\mathbb{T},\mathbb{G})$ satisfy the following equations: The first is the generalized mass balance equation with a source

$$\partial_t\varrho + \operatorname{div}(\varrho\mathbf{u}-\mathbf{g}) = 0 \quad \text{in } Q, \tag{4.5.1}$$

the second is the generalized momentum balance equation

$$\partial_t(\varrho\mathbf{u}) + \operatorname{div}(\varrho\mathbf{u}\otimes\mathbf{u}) = \operatorname{div}(\mathbb{T}+\mathbb{G}) + \varrho\mathbf{f} \quad \text{in } Q, \tag{4.5.2}$$

and the third is the "renormalized" mass balance equation

$$\partial_t\varphi + \operatorname{div}(\varphi\mathbf{u}-\mathbf{g}^\varphi) + \varpi = 0 \quad \text{in } Q. \tag{4.5.3}$$

Furthermore, we assume that the following conditions are satisfied.

**Condition 4.5.1.** • There are exponents $\gamma > 1$, $1 < r < \infty$, $1 < s \le \infty$, and a constant $E$ such that

$$\|\varrho^\gamma\|_{L^1(Q)} + \|\mathbf{u}\|_{L^2(0,T;W^{1,2}(\Omega))} \le E, \tag{4.5.4}$$

$$\|\varrho|\mathbf{u}|^2\|_{L^\infty(0,T;L^1(\Omega))} \le E, \tag{4.5.5}$$

$$\|\varphi\|_{L^\infty(0,T;L^s(\Omega))} \le E. \tag{4.5.6}$$

- For $r' = r/(r-1)$,

$$\mathbb{T} \in L^1(Q),\ \mathbb{T} \in L^{r'}(Q'),\ \varphi \in L^r(Q') \text{ for every compact } Q' \Subset Q. \tag{4.5.7}$$

- We have

$$\mathbf{g}, \mathbf{g}^\varphi, \varpi \in L^2(Q), \quad \mathbb{G} \in L^1(0,T; L^q(\Omega)),$$

$$\text{where } q = s/(s-1) \quad \text{for } s < \infty, \quad \text{and} \quad 1 < q \le 2 \quad \text{for } s = \infty.$$

Throughout this section we assume that all functions considered are extended by zero over the whole space $\mathbb{R}^d$. The extended functions are denoted by the same symbol.

**Theorem 4.5.2.** *Under Condition 4.5.1, if the exponents $s$, $\gamma$ and functions $\mathbf{g}^\varphi$, $\mathbb{G}$ satisfy either*

$$2d < \gamma \le s < \infty, \tag{4.5.8}$$

*or*

$$\gamma > d/2, \quad s = \infty, \quad \mathbf{g}^\varphi = 0, \quad \mathbb{G} = 0, \tag{4.5.9}$$

*then the integral identity*

$$\int_Q \xi\varphi\mathbf{R} : [\zeta(\mathbb{T} + \mathbb{G})]\, dxdt = \int_Q \big(\mathbf{H}\cdot\mathbf{u} + \zeta\varrho(\mathbf{N} + \mathbf{P} + \mathbf{Q})\cdot\mathbf{u} + S\big)\, dxdt \tag{4.5.10}$$

*holds for all $\xi \in C_0^\infty(Q)$ and $\zeta \in C_0^\infty(\Omega)$. Here,*

$$\mathbf{H} = \mathbf{R}[\xi\varphi]\,(\zeta\varrho\mathbf{u}) - (\xi\varphi)\,\mathbf{R}[\zeta\varrho\mathbf{u}], \tag{4.5.11}$$

$$S = (\nabla\zeta \otimes \mathbf{A}[\xi\varphi]) : (\varrho\mathbf{u}\otimes\mathbf{u} - \mathbb{T} - \mathbb{G}) + \zeta\mathbf{A}[\xi\varphi]\cdot(\varrho\mathbf{f}), \tag{4.5.12}$$

$$\mathbf{N} = \mathbf{R}[\xi\,\mathbf{g}^\varphi], \quad \mathbf{P} = \mathbf{A}[\varphi\partial_t\xi + \varphi\mathbf{u}\cdot\nabla\xi], \quad \mathbf{Q} = -\mathbf{A}[\mathbf{g}^\varphi\cdot\nabla\xi + \varpi\xi]. \tag{4.5.13}$$

*Furthermore, all functions in (4.5.10) are integrable in $Q$ and*

$$\|\mathbf{H}\|_{L^\infty(0,T;L^\iota(\Omega))} + \|\mathbf{P}\|_{L^2(0,T;W^{1,2}(\Omega))} + \|\mathbf{A}[\xi\varphi]\|_{L^\infty(0,T;W^{1,2d}(\Omega))} \le c,$$

$$\|\mathbf{Q}\|_{L^2(0,T;W^{1,2}(\Omega))} + \|\mathbf{N}\|_{L^2(Q)} \le c(\|\mathbf{g}\|_{L^2(Q)} + \|\mathbf{g}^\varphi\|_{L^2(Q)} + \|\varpi\|_{L^2(Q)}). \tag{4.5.14}$$

*Here, the constant $c$ depends only on the functions $\xi$, $\zeta$, and on $E$, and the exponent $\iota \ge 1$ is arbitrary satisfying the inequalities*

$$\iota < \frac{2\gamma}{\gamma+1} \quad \textit{for } d/2 < \gamma \le 2d, \quad \iota < \frac{2\gamma}{\gamma+3} \quad \textit{for } \gamma > 2d.$$

The proof is based on three auxiliary lemmas which give the construction and describe the properties of a special vector field $\boldsymbol{\omega}$.

**Lemma 4.5.3.** *Assume that solutions to equations (4.5.1)–(4.5.3) meet all requirements of Theorem 4.5.2 and $\xi \in C_0^\infty(Q)$, $\zeta \in C_0^\infty(\Omega)$. Then the time derivative of the vector field $\boldsymbol{\omega} = \zeta \mathbf{A}[\xi\varphi]$ has the representation*

$$\partial_t \boldsymbol{\omega} = \zeta\big(\mathbf{N} + \mathbf{P} + \mathbf{Q} - \mathbf{R}[\xi\varphi\,\mathbf{u}]\big), \tag{4.5.15}$$

*where the remainders $\mathbf{N}$, $\mathbf{P}$, and $\mathbf{Q}$ are defined by (4.5.13) and satisfy estimates (4.5.14). The momentum $\varphi\mathbf{u}$ satisfies the estimate*

$$\|\varphi\mathbf{u}\|_{L^2(0,T;L^\rho(\Omega))} + \|\mathbf{R}[\xi\varphi\mathbf{u}\|_{L^2(0,T;L^\rho(\Omega))} \le c, \tag{4.5.16}$$

*where $\rho$ is an arbitrary number such that*

$$2 \le \rho < \infty, \quad \rho^{-1} > 2^{-1} + s^{-1} - d^{-1}, \tag{4.5.17}$$

*and the constant $c$ depends only on $\xi$, the constant $E$ in (4.5.4)–(4.5.7), the domain $\Omega$, and the exponents $\gamma$, $s$, $\rho$.*

*Proof.* Observe that (4.5.3) is equivalent to the integral identity

$$\int_Q \big(\partial_t \varsigma\, \varphi + (\varphi\mathbf{u} - \mathbf{g}^\varphi)\cdot\nabla\varsigma - \varpi\varsigma\big)\, dxdt = 0 \tag{4.5.18}$$

for every $\varsigma \in C_0^\infty(Q)$. Choose any $\eta \in C_0^\infty(Q)$ and set

$$\varsigma = \xi A_i[\eta\zeta] \in C_0^\infty(Q).$$

Substituting $\varsigma$ into (4.5.18) and noting that all functions in (4.5.18) are compactly supported in $Q$ we arrive at

$$\begin{aligned}
&\int_{\mathbb{R}^d\times[0,T]} \xi\varphi A_i[\zeta\partial_t\eta]\, dxdt + \int_{\mathbb{R}^d\times[0,T]} \partial_t\xi\varphi\, A_i[\eta\zeta]\, dxdt\\
&\quad + \int_{\mathbb{R}^d\times[0,T]} \xi\, A_i[\eta\nabla\zeta]\cdot(\varphi\mathbf{u} - \mathbf{g}^\varphi)\, dxdt + \int_{\mathbb{R}^d\times[0,T]} \xi\, A_i[\zeta\nabla\eta]\cdot(\varphi\mathbf{u} - \mathbf{g}^\varphi)\, dxdt\\
&\qquad + \int_{\mathbb{R}^d\times[0,T]} A_i[\eta\zeta]\nabla\xi\cdot(\varphi\mathbf{u} - \mathbf{g}^\varphi)\, dxdt - \int_{\mathbb{R}^d\times[0,T]} \xi A_i[\eta\zeta]\varpi\, dxdt = 0.
\end{aligned}$$

Using the skew-symmetry of $A_i$ we can rewrite this identity in the form

$$\begin{aligned}
&\int_{\mathbb{R}^d\times[0,T]} \partial_t\eta\{\zeta A_i[\xi\varphi]\}\, dxdt + \int_{\mathbb{R}^d\times[0,T]} \zeta A_i[\xi(\varphi\mathbf{u} - \mathbf{g}^\varphi)]\cdot\nabla\eta\, dxdt\\
&\quad + \int_{\mathbb{R}^3\times[0,T]} \eta\,\{\zeta A_i[\partial_t\xi\varphi]\}\, dxdt + \int_{\mathbb{R}^d\times[0,T]} \eta\{\nabla\zeta\cdot A_i[\xi(\varphi\mathbf{u} - \mathbf{g}^\varphi)]\}\, dxdt\\
&\quad + \int_{\mathbb{R}^d\times[0,T]} \eta\{\zeta\, A_i[\nabla\xi\cdot(\varphi\mathbf{u} - \mathbf{g}^\varphi)]\}\, dxdt - \int_{\mathbb{R}^d\times[0,T]} \eta\{\zeta A_i[\xi\varpi]\}\, dxdt = 0.
\end{aligned} \tag{4.5.19}$$

In order to derive expressions for partial derivatives of $\boldsymbol{\omega}$, we have to integrate by parts in the integrals of this identity. The justification of this procedure requires estimates of the integrands. To derive such estimates fix $\rho$ satisfying (4.5.17). By the Hölder inequality, we have

$$\|\varphi(t)\mathbf{u}(t)\|_{L^\rho(\Omega)} \le \|\varphi(t)\|_{L^s(\Omega)}\|\mathbf{u}(t)\|_{L^\upsilon(\Omega)},$$

where $\upsilon^{-1} = \rho^{-1} - s^{-1}$. By (4.5.17) we have $\upsilon < 2d/(d-2)$, hence the embedding $W^{1,2}(\Omega) \hookrightarrow L^\upsilon(\Omega)$ is bounded. Thus we get

$$\|\varphi(t)\mathbf{u}(t)\|_{L^\rho(\Omega)} \le c\|\varphi(t)\|_{L^s(\Omega)}\|\mathbf{u}(t)\|_{W^{1,2}(\Omega)}$$

and

$$\|\varphi\mathbf{u}\|_{L^2(0,T;L^\rho(\Omega))} \le c\|\varphi\|_{L^\infty(0,T;L^s(\Omega))}\|\mathbf{u}\|_{L^2(0,T;W^{1,2}(\Omega))} \le c(E). \tag{4.5.20}$$

Since $\zeta \in C_0^\infty(\mathbb{R}^d)$ is compactly supported in $\Omega$ and $\xi \in C_0^\infty(\mathbb{R}^{d+1})$ is compactly supported in $Q$, the vector field $\zeta A_i[\xi\varphi\mathbf{u}]$ is compactly supported in $Q$. From this and Corollary 1.7.2 we obtain

$$\begin{aligned}\|A_i[\xi\varphi\mathbf{u}]\,\|_{L^2(0,T;W^{1,2}(\Omega))} &\le \|A_i[\xi\varphi\mathbf{u}]\|_{L^2(0,T;W^{1,\rho}(\Omega))}\\ &\le \|\varphi\mathbf{u}\|_{L^2(0,T;L^\rho(\Omega))} \le c(\xi, E).\end{aligned} \tag{4.5.21}$$

The same arguments give

$$\begin{aligned}&\|A_i[\xi\mathbf{g}^\varphi]\|_{L^2(0,T;W^{1,2}(\Omega))} + \|A_i[\xi\varpi]\|_{L^2(0,T;W^{1,2}(\Omega))}\\ &\qquad\qquad \le c(\xi)(\|\mathbf{g}^\varphi\|_{L^2(Q)} + \|\varpi\|_{L^2(Q)}).\end{aligned} \tag{4.5.22}$$

Since $s \ge 2$, we have

$$\|A_i[\partial_t\xi\varphi]\|_{L^2(0,T;W^{1,2}(\Omega))} \le c(\xi)\|\varphi\|_{L^2(0,T;L^2(\Omega))} \le c(\xi, E). \tag{4.5.23}$$

We can now integrate by parts to obtain

$$\begin{aligned}\int_{\mathbb{R}^d\times[0,T]} \zeta A_i[\xi(\varphi\mathbf{u} - \mathbf{g}^\varphi)]\cdot\nabla\eta\,dxdt = &-\int_{\mathbb{R}^d\times[0,T]} \eta\zeta\,\mathrm{div}\,A_i[\xi(\varphi\mathbf{u} - \mathbf{g}^\varphi)]\,dxdt\\ &-\int_{\mathbb{R}^d\times[0,T]} \eta\nabla\zeta\cdot A_i[\xi(\varphi\mathbf{u} - \mathbf{g}^\varphi)]\,dxdt.\end{aligned}$$

Substituting this result into (4.5.19) we arrive at

$$\begin{aligned}&\int_{\mathbb{R}^d\times[0,T]} \partial_t\eta\{\zeta A_i[\xi\varphi]\}\,dxdt - \int_{\mathbb{R}^d\times[0,T]} \eta\zeta\,\mathrm{div}\,A_i[\xi(\varphi\mathbf{u} - \mathbf{g}^\varphi)]\,dxdt\\ &\qquad + \int_{\mathbb{R}^d\times[0,T]} \eta\zeta\,A_i[\partial_t\xi\varphi + \nabla\xi\cdot(\varphi\mathbf{u} - \mathbf{g}^\varphi) - \xi\varpi]\,dxdt = 0\end{aligned}$$

for all $i = 1, \dots, d$. We can rewrite this equality in vector form

$$\int_{\mathbb{R}^d \times [0,T]} \partial_t \eta \, \boldsymbol{\omega} \, dxdt - \int_{\mathbb{R}^d \times [0,T]} \eta\zeta \mathbf{R} \left[\xi(\varphi \mathbf{u} - \mathbf{g}^\varphi)\right] dxdt + \int_{\mathbb{R}^d \times [0,T]} \eta\zeta \, \mathbf{A} \left[\partial_t \xi \varphi + \nabla \xi \cdot (\varphi \mathbf{u} - \mathbf{g}^\varphi) - \xi \varpi\right] dxdt = 0,$$

or equivalently

$$\int_Q \partial_t \eta \, \boldsymbol{\omega} \, dxdt + \int_Q \eta\zeta(\mathbf{N} + \mathbf{P} + \mathbf{Q} - \mathbf{R}[\xi\varphi\mathbf{u}]) \, dxdt = 0.$$

Since $\eta \in C_0^\infty(Q)$ is arbitrary, this yields (4.5.15).

It remains to note that inequality (4.5.14) for $\mathbf{N}$ follows from the boundedness of the operator $\mathbf{R}$ in $L^2(\mathbb{R}^d)$. Inequalities (4.5.14) for $\mathbf{P}$, $\mathbf{Q}$ easily follow from (4.5.21)–(4.5.23). Finally, inequality (4.5.20) and the boundedness of $\mathbf{R}$ in Lebesgue spaces imply (4.5.16). □

General properties of $\boldsymbol{\omega}$ are listed in the second lemma.

**Lemma 4.5.4.** *Under the assumptions of Theorem 4.5.2 and Lemma 4.5.3, if the exponents $s$ and $\gamma$ satisfy condition* (4.5.8), *then*

$$\partial_t \boldsymbol{\omega} \in L^2(Q), \quad \nabla \boldsymbol{\omega} \in L^\infty(0,T; L^s(\Omega)) \cap L^r(Q). \tag{4.5.24}$$

*If $s$, $\gamma$, $\mathbb{G}_n$ and $\mathbf{g}_n^\varphi$ satisfy* (4.5.9), *then*

$$\partial_t \boldsymbol{\omega} \in L^2(0,T; L^\tau(\Omega)), \quad \nabla \boldsymbol{\omega} \in L^\infty(0,T; L^\sigma(\Omega)) \cap L^r(Q) \tag{4.5.25}$$

*for all $\tau$, $\sigma$ such that $1 < \tau < 2d/(d-2)$ and $1 < \sigma < \infty$.*

*Proof.* If $s$ and $\gamma$ satisfy (4.5.8), then (4.5.24) for $\partial_t \boldsymbol{\omega}$ follows from estimates (4.5.14), (4.5.16), and equality (4.5.15) (since $\rho > 2$ in (4.5.16)). Recall that estimates (4.5.14), (4.5.16), and equality (4.5.15) are established by Lemma 4.5.3. It remains to consider the case of $\gamma$, $s$, $\mathbb{G}$, and $\mathbf{g}^\varphi$ which satisfy (4.5.9). Then $\mathbf{N} = 0$ and

$$\partial_t \boldsymbol{\omega} = \zeta(\mathbf{P} + \mathbf{Q} - \mathbf{R}[\xi\varphi\mathbf{u}]).$$

It follows from estimates (4.5.14) and the Sobolev embedding theorem that for all $\tau < 2d/(d-2)$,

$$\|\mathbf{P}\|_{L^2(0,T;L^\tau(\Omega))} + \|\mathbf{Q}\|_{L^2(0,T;L^\tau(\Omega))} \le c\|\mathbf{P}\|_{L^2(0,T;W^{1,2}(\Omega))} + c\|\mathbf{Q}\|_{L^2(0,T;W^{1,2}(\Omega))} < \infty.$$

On the other hand, since $s = \infty$ by (4.5.9), estimate (4.5.16) implies

$$\|\mathbf{R}[\xi\varphi\mathbf{u}]\|_{L^2(0,T;L^\tau(\Omega))} \le c \quad \text{for all } \tau < 2d/(d-2).$$

Hence $\partial_t\boldsymbol{\omega}$ satisfies (4.5.25). To estimate $\nabla\boldsymbol{\omega}$ notice that

$$\nabla\boldsymbol{\omega} = \zeta\mathbf{R}[\xi\varphi] + \nabla\zeta \otimes \mathbf{A}[\xi\varphi].$$

Since $\varphi \in L^\infty(0,T;L^s(\Omega)) \cap L^r_{\text{loc}}(Q)$ and $\xi$ is compactly supported in $Q$, we have $\xi\varphi \in L^\infty(0,T;L^s(\Omega)) \cap L^r(Q)$. Hence inclusions (4.5.24)–(4.5.25) for $\nabla\boldsymbol{\omega}$ follow from the estimates of the operators $\mathbf{R}$ and $\mathbf{A}$ given by Corollary 1.7.2. □

**Lemma 4.5.5.** *The integral identity*

$$\int_Q \big(\partial_t\boldsymbol{\omega}\cdot(\varrho\mathbf{u}) + \nabla\boldsymbol{\omega} : (\varrho\mathbf{u}\otimes\mathbf{u} - \mathbb{T} - \mathbb{G}) + \boldsymbol{\omega}\cdot(\varrho\mathbf{f})\big)\, dxdt = 0 \tag{4.5.26}$$

*holds for all vector fields $\boldsymbol{\omega}$ compactly supported in $Q$ with the properties*

$$\begin{aligned} &\boldsymbol{\omega} \in L^1(0,T;L^{\gamma/(\gamma-1)}(\Omega)), \quad \partial_t\boldsymbol{\omega} \in L^2(0,T;L^{\beta'}(\Omega)),\\ &\nabla\boldsymbol{\omega} \in L^\infty(0,T;L^{b'}(\Omega)) \cap L^\infty(0,T;L^{q'}(Q)) \cap L^r(Q), \end{aligned} \tag{4.5.27}$$

*where $q' = q/(q-1)$ and $\beta'$, $b'$ are some numbers satisfying*

$$\kappa' = \kappa/(\kappa-1) < \beta' < \infty, \quad \big((1-2^{-1}\mathfrak{a})^{-1}\big)' = 2\mathfrak{a}^{-1} < b' < \infty, \tag{4.5.28}$$

*where $\kappa^{-1} = 2^{-1} + \gamma^{-1} - d^{-1}$.*

*Proof.* We first observe that by (4.5.2), the integral identity (4.5.26) holds for all $\boldsymbol{\omega} \in C_0^\infty(Q)$. Let us consider in detail all the terms involved in this identity. It follows from (4.5.4)–(4.5.5) that $(\varrho, \mathbf{u})$ are functions of bounded energy and meet all requirements of Corollaries 4.2.2 and 4.2.3. Applying Corollary 4.2.2 we obtain

$$\|\varrho\mathbf{u}\|_{L^2(0,T;L^\beta(\Omega))} \le c(E) \quad \text{for all } \beta \in [1,\kappa). \tag{4.5.29}$$

On the other hand, Corollary 4.2.3 implies

$$\varrho\mathbf{u}\otimes\mathbf{u} \in L^2(0,T;L^b(\Omega)) \quad \text{for all } b \in [\,1,(1-2^{-1}\mathfrak{a})^{-1}\,). \tag{4.5.30}$$

By Condition 4.5.1, we have

$$\varrho \in L^\infty(0,T;L^\gamma(\Omega)), \quad \mathbb{T} \in L^{r'}_{\text{loc}}(Q), \quad \mathbb{G} \in L^1(0,T;L^q(\Omega)). \tag{4.5.31}$$

Let now $\boldsymbol{\omega}$ be an arbitrary vector field satisfying (4.5.27) and compactly supported in $Q$. To give a rigorous proof of (4.5.26) for $\boldsymbol{\omega}$ we consider the mollification $[\boldsymbol{\omega}]_{m,m}$ of $\boldsymbol{\omega}$, defined by (1.6.1), (1.6.3). Since $\boldsymbol{\omega}$ is compactly supported in $Q$ and satisfies (4.5.27), the smooth function $[\boldsymbol{\omega}]_{m,m}$ is compactly supported in $Q$ for sufficiently large $m$. Inserting $[\boldsymbol{\omega}]_{m,m}$ in (4.5.26) we obtain

$$\int_Q \big(\partial_t[\boldsymbol{\omega}]_{m,m}\cdot(\varrho\mathbf{u}) + \nabla[\boldsymbol{\omega}]_{m,m} : (\varrho\mathbf{u}\otimes\mathbf{u} - \mathbb{T} - \mathbb{G}) + [\boldsymbol{\omega}]_{m,m}\cdot(\varrho\mathbf{f})\big)\, dxdt = 0. \tag{4.5.32}$$

By general properties of mollifiers we have

$$\begin{aligned}
\partial_t[\boldsymbol{\omega}]_{m,m} = [\partial_t\boldsymbol{\omega}]_{m,m} &\to \partial_t\boldsymbol{\omega} \quad \text{in } L^2(0,T;L^{\beta'}(\Omega)),\\
\nabla[\boldsymbol{\omega}]_{m,m} = [\nabla\boldsymbol{\omega}]_{m,m} &\rightharpoonup \nabla\boldsymbol{\omega} \quad \text{weakly}^\star \text{ in } L^\infty(0,T;L^{q'}(\Omega)),\\
\nabla[\boldsymbol{\omega}]_{m,m} = [\nabla\boldsymbol{\omega}]_{m,m} &\to \nabla\boldsymbol{\omega} \quad \text{in } L^2(0,T;L^{b'}(\Omega)) \cap L^r(Q),\\
[\boldsymbol{\omega}]_{m,m} &\to \boldsymbol{\omega} \quad \text{in } L^1(0,T;L^{\gamma/(\gamma-1)}(\Omega)).
\end{aligned}$$

Letting $m \to \infty$ in (4.5.32) and recalling (4.5.29)–(4.5.31) we conclude that $\boldsymbol{\omega}$ satisfies the integral identity (4.5.26). □

We are now in a position to complete the proof of Theorem 4.5.2. Our first task is to check that the special vector field $\boldsymbol{\omega} = \zeta\mathbf{A}[\xi\varphi]$, defined by Lemma 4.5.3, satisfies the assumptions of Lemma 4.5.5. This follows directly from (4.5.25) if $\gamma$ and $s$ satisfy (4.5.9). Otherwise if $\gamma$ and $s$ satisfy (4.5.8), then

$$2d < \gamma \le s < \infty \quad \text{and} \quad s = q', \tag{4.5.33}$$

which along with (4.5.24) implies that

$$\boldsymbol{\omega} \in L^\infty(0,T;W^{1,2d}(\Omega)) \subset L^\infty(Q),$$

thus $\boldsymbol{\omega}$ obviously satisfies (4.5.27). Next, for $\gamma > d/2$, the quantity $\kappa$ satisfies the inequalities

$$\kappa > 2 \Rightarrow \kappa/(\kappa-1) < 2.$$

On the other hand, inclusion (4.5.24) implies $\partial_t\boldsymbol{\omega} \in L^2(0,T;L^2(\Omega))$. Combining these results we conclude that $\partial_t\boldsymbol{\omega}$ also satisfies (4.5.27) with an arbitrary exponent $\beta' \in (\kappa/(\kappa-1), 2]$. Let us prove that $\nabla\boldsymbol{\omega}$ satisfies (4.5.27). Inequalities (4.5.33) imply

$$\mathfrak{a} = \frac{2}{d} - \frac{1}{\gamma} > 3/(2d), \quad 2\mathfrak{a}^{-1} < 4d/3.$$

In particular, we have $s > 2\mathfrak{a}^{-1}$ and we can take $b' = s \equiv q'$. On the other hand, inclusion (4.5.24) implies $\nabla\boldsymbol{\omega} \in L^\infty(0,T;L^s(\Omega)) \cap L^r(Q)$. Hence, for this choice of $q'$ and $b'$ we conclude that

$$\nabla\boldsymbol{\omega} \in L^\infty(0,T;L^{b'}(\Omega)) \cap L^\infty(0,T;L^{q'}(\Omega)) \cap L^r(Q). \tag{4.5.34}$$

Thus $\nabla\boldsymbol{\omega}$ also satisfies condition (4.5.27). Hence the vector field $\boldsymbol{\omega}$ satisfies (4.5.27), and we can substitute $\boldsymbol{\omega}$ into the integral identity (4.5.26) and next apply Lemma 4.5.3 to obtain

$$\int_Q \big(\zeta\varrho\,(\mathbf{N}+\mathbf{P}+\mathbf{Q}-\mathbf{R}[\xi\varphi\,\mathbf{u}])\cdot\mathbf{u} + \nabla\boldsymbol{\omega} : (\varrho\mathbf{u}\otimes\mathbf{u} - \mathbb{T} - \mathbb{G}) + \boldsymbol{\omega}\cdot(\varrho\mathbf{f})\big)\,dxdt = 0.$$

Recalling the relation

$$\nabla\boldsymbol{\omega} = \zeta\mathbf{R}[\xi\varphi] + \nabla\zeta \otimes \mathbf{A}[\xi\varphi],$$

and formulae (4.5.11)–(4.5.13) we can rewrite the above identity in the form

$$\int_Q \big(\zeta\mathbf{R}[\xi\varphi] : (\varrho\mathbf{u}\otimes\mathbf{u} - \mathbb{T} - \mathbb{G}) \\ + \zeta\varrho\,(\mathbf{N}+\mathbf{P}+\mathbf{Q}-\mathbf{R}[\xi\varphi\,\mathbf{u}])\cdot\mathbf{u} + S\big)\,dxdt = 0. \quad (4.5.35)$$

Since $\xi$ is compactly supported in $Q$ and $\zeta$ is compactly supported in $\Omega$, it follows from Corollary 4.2.2 and Lemma 4.5.3 that

$$\zeta\varrho\mathbf{u} \in L^2(0,T;L^\beta(\mathbb{R}^d)), \quad \xi\varphi\mathbf{u} \in L^2(0,T;L^\rho(\mathbb{R}^d))$$

for all $\beta$ and $\rho$ satisfying

$$1 > \beta^{-1} > 2^{-1} + \gamma^{-1} - d^{-1}, \quad 2^{-1} > \rho^{-1} > 2^{-1} + s^{-1} - d^{-1}. \quad (4.5.36)$$

On the other hand, inequalities (4.5.8) and (4.5.9) yield

$$2^{-1} + \gamma^{-1} - d^{-1} + 2^{-1} + s^{-1} - d^{-1} = 1 + \gamma^{-1} + s^{-1} - 2d^{-1} < 1.$$

Hence there exist $\beta$ and $\rho$ satisfying (4.5.36) such that $\beta^{-1} + \rho^{-1} = 1$. Applying Corollary 1.7.3 we obtain

$$\int_Q \mathbf{R}[\xi\varphi\mathbf{u}]\cdot(\zeta\varrho\mathbf{u})\,dxdt = \int_{\mathbb{R}^d\times(0,T)} \mathbf{R}[\xi\varphi\mathbf{u}]\cdot(\zeta\varrho\mathbf{u})\,dxdt \\ = \int_{\mathbb{R}^d\times(0,T)} \xi\varphi\mathbf{R}[\zeta\varrho\mathbf{u}]\cdot\mathbf{u}\,dxdt = \int_Q \xi\varphi\mathbf{R}[\zeta\varrho\mathbf{u}]\cdot\mathbf{u}\,dxdt. \quad (4.5.37)$$

By Condition 4.5.1 and (4.5.8), (4.5.9), we have $\varphi \in L^\infty(0,T;L^s(\Omega))$ and $s \geq q'$. Since $\xi$ is compactly supported in $Q$ we infer that $\xi\varphi \in L^\infty(0,T;L^{q'}(\mathbb{R}^d))$. Since $\zeta$ is compactly supported in $\Omega$, Condition 4.5.1 yields $\zeta\mathbb{G} \in L^1(0,T;L^q(\mathbb{R}^d))$. From this and the symmetry properties of the operator $\mathbf{R}$ given by Corollary 1.7.3 we conclude that

$$\int_Q \mathbf{R}[\xi\varphi] : \zeta\mathbb{G}\,dxdt = \int_{\mathbb{R}^d\times(0,T)} \mathbf{R}[\xi\varphi] : \zeta\mathbb{G}\,dxdt \\ = \int_{\mathbb{R}^d\times(0,T)} \xi\varphi\mathbf{R} : [\zeta\mathbb{G}]\,dxdt = \int_Q \xi\varphi\mathbf{R} : [\zeta\mathbb{G}]\,dxdt. \quad (4.5.38)$$

By Condition 4.5.1, we have $\varphi \in L^r_{\mathrm{loc}}(Q)$. Since $\xi$ is compactly supported in $Q$ we infer that $\xi\varphi \in \cap L^r(\mathbb{R}^d\times(0,T))$. Moreover, there is $h > 0$ such that $\xi\varphi$ and $\mathbf{R}[\xi\varphi]$ are supported in $\mathbb{R}^d \times [h, T-h]$. On the other hand, since $\mathbb{T} \in L^{r'}_{\mathrm{loc}}(Q)$ and $\zeta$ is compactly supported in $\Omega$ we have $\zeta\mathbb{T} \in L^{r'}(\mathbb{R}^d \times (h, T-h))$. Applying Corollary 1.7.3 we obtain

$$\int_Q \mathbf{R}[\xi\varphi] : \zeta\mathbb{T}\,dxdt = \int_{\mathbb{R}^d\times[h,T-h]} \mathbf{R}[\xi\varphi] : \zeta\mathbb{T}\,dxdt \\ = \int_{\mathbb{R}^d\times[h,T-h]} \xi\varphi\mathbf{R} : [\zeta\mathbb{T}]\,dxdt = \int_Q \xi\varphi\mathbf{R} : [\zeta\mathbb{T}]\,dxdt. \quad (4.5.39)$$

Inserting (4.5.37)–(4.5.39) into (4.5.35) we arrive at

$$\int_Q \xi\varphi\,\mathbf{R} : [\zeta(\mathbb{T}+\mathbb{G})]\,dxdt = \int_Q \Big(\zeta\mathbf{R}[\xi\varphi] : (\varrho\mathbf{u}\otimes\mathbf{u}) - \xi\varphi\,\mathbf{R}[\zeta\varrho\mathbf{u}]\cdot\mathbf{u} + \zeta\varrho\,(\mathbf{N}+\mathbf{P}+\mathbf{Q})\cdot\mathbf{u} + S\Big)\,dxdt, \quad (4.5.40)$$

which combined with the obvious identity

$$\zeta\mathbf{R}[\xi\varphi] : (\varrho\mathbf{u}\otimes\mathbf{u}) - \xi\varphi\mathbf{R}[\zeta\varrho\mathbf{u}]\cdot\mathbf{u} \equiv \mathbf{H}\cdot\mathbf{u}$$

leads to (4.5.10).

It remains to prove inequalities (4.5.14). For $\mathbf{N}$, $\mathbf{P}$, and $\mathbf{Q}$, they follow directly from Lemma 4.5.3. Next, the hypotheses of Theorem 4.5.2 imply $s > 2d$, hence

$$\|\varphi\|_{L^\infty(0,T;L^{2d}(\Omega))} \le c\|\varphi\|_{L^\infty(0,T;L^s(\Omega))} \le c,$$

which along with Corollary 1.7.2 yields (4.5.14) for $\mathbf{A}[\xi\varphi]$. It remains to estimate $\mathbf{H}$. The Hölder inequality and boundedness of $\mathbf{R}$ in Lebesgue spaces (see Corollary 1.7.2) gives

$$\|\mathbf{H}\|_{L^\infty(0,T;L^\iota(\Omega))} \le c\|\varphi\|_{L^\infty(0,T;L^\tau(\Omega))}\|\varrho\mathbf{u}\|_{L^\infty(0,T;L^\sigma(\Omega))}$$

for all $\tau$, $\sigma$ such that

$$1 < \sigma,\tau < \infty, \quad \frac{1}{\tau}+\frac{1}{\sigma} \le \frac{1}{\iota} < 1.$$

By (4.5.4)–(4.5.5) the functions $(\varrho,\mathbf{u})$ have bounded energy and meet all requirements of Proposition 4.2.1. Hence we can apply Corollary 4.2.2 of that proposition to obtain

$$\|\varrho\mathbf{u}\|_{L^\infty(0,T;L^\sigma(\Omega))} \le c(E) \quad \text{for all } \sigma\in(1,2\gamma/(\gamma+1)].$$

By (4.5.6) we have

$$\|\varphi\|_{L^\infty(0,T;L^\tau(\Omega))} \le c(E) \quad \text{for all } \tau\in(1,s].$$

Hence

$$\|\mathbf{H}\|_{L^\infty(0,T;L^\iota(\Omega))} \le c(\zeta,E) \quad (4.5.41)$$

for all $\iota$ satisfying

$$1/s + (\gamma+1)/(2\gamma) < 1/\iota < 1.$$

If $\gamma$ and $s$ satisfy (4.5.8), then $\infty > s \ge \gamma > 2d$ and

$$1/s + (\gamma+1)/(2\gamma) \le (\gamma+3)/(2\gamma).$$

Hence inequality (4.5.41) holds for all $\iota\in(1,2\gamma/(\gamma+3))$. If $\gamma$ and $s$ satisfy condition (4.5.9), then $s=\infty$ and $\gamma > d/2$. In this case $(\gamma+1)/(2\gamma) < 1/\iota < 1$, and inequality (4.5.41) holds for all $\iota\in(1,2\gamma/(\gamma+1))$. This leads to estimate (4.5.14) for $\mathbf{H}$. □

## 4.6 Proof of continuity of viscous flux for general stress tensors

This section is devoted to the famous P.-L. Lions result on weak continuity of "viscous flux". This result occupies an important place in the theory of compressible flows and plays a crucial role in the proof of existence of solutions to compressible Navier-Stokes equations. The result is formulated in a most general setting for the dynamics of a compressible medium with a general stress tensor $\mathbb{T}$. Therefore, a general system of mass balance (4.6.1) and moment balance equations (4.6.2) is considered. Such a general result can be applied to different modifications of compressible Navier-Stokes equations including different regularization schemes defined for these equations.

First we formulate the explicit assumptions on classes of functions under consideration and on the governing equations. Let $\Omega$ be a bounded domain in $\mathbb{R}^d$, $d = 2, 3$, with a Lipschitz boundary, and $Q = \Omega \times [0, T]$. Furthermore, assume that functions $\varrho_n : Q \to \mathbb{R}^+$, vector fields $\mathbf{u}_n : Q \to \mathbb{R}^d$, and matrix-valued functions $\mathbb{T}_n : Q \to \mathbb{R}^{d\times d}$, $n \geq 1$, satisfy the differential equations

$$\partial_t \varrho_n + \operatorname{div}(\varrho_n \mathbf{u}_n - \mathbf{g}_n) = 0 \quad \text{in } Q, \tag{4.6.1}$$

$$\partial_t(\varrho_n \mathbf{u}_n) + \operatorname{div}(\varrho_n \mathbf{u}_n \otimes \mathbf{u}_n) = \operatorname{div}(\mathbb{T}_n + \mathbb{G}_n) + \rho_n \mathbf{f} \quad \text{in } Q, \tag{4.6.2}$$

which are understood in the sense of distributions. Here, $\mathbf{f} \in C(Q)$ and $\mathbf{g}_n \in L^2(Q)$ are given vector fields. Along with the "density" and "velocity" field $(\varrho_n, \mathbf{u}_n)$ we consider the "renormalized density" $\varphi_n$, $n \geq 1$, satisfying

$$\partial_t \varphi_n + \operatorname{div}(\varphi_n \mathbf{u}_n - \mathbf{g}_n^{\varphi}) + \varpi_n = 0 \quad \text{in } Q, \tag{4.6.3}$$

for a given vector field $\mathbf{g}_n^{\varphi} \in L^2(Q)$ and function $\varpi_n \in L^2(Q)$. We impose neither boundary nor initial conditions for solutions to (4.6.1)–(4.6.3). Moreover, no relation is imposed between the "stress tensor" $\mathbb{T}_n$ and the "state variables" $(\varrho_n, \mathbf{u}_n)$. Instead it is assumed that the state variables satisfy the following conditions:

**Condition 4.6.1.** • There are exponents $1 < \gamma < \infty$, $1 < s \leq \infty$, $1 < r < \infty$ and a constant $E$ independent of $n$ such that

$$\|\varrho_n^{\gamma}\|_{L^1(Q)} + \|\mathbf{u}_n\|_{L^2(0,T;W^{1,2}(\Omega))} \leq E, \tag{4.6.4}$$

$$\|\varrho_n |\mathbf{u}_n|^2\|_{L^\infty(0,T;L^1(\Omega))} \leq E, \tag{4.6.5}$$

$$\|\varphi_n\|_{L^\infty(0,T;L^s(\Omega)} \leq E. \tag{4.6.6}$$

The stress tensors $\mathbb{T}_n$ are uniformly integrable,

$$\|\mathbb{T}_n\|_{L^1(Q)} \leq E. \tag{4.6.7}$$

Moreover, for every compact subset $Q' \Subset Q$ there is a constant $C(Q')$ such that

$$\|\varphi_n\|_{L^r(Q')} + \|\mathbb{T}_n\|_{L^{r'}(Q')} \leq C(Q'). \tag{4.6.8}$$

- The given elements in (4.6.1)–(4.6.3) satisfy

$$\|\mathbb{G}_n\|_{L^1(0,T;L^q(\Omega))} \to 0 \quad \text{as } n \to \infty, \tag{4.6.9}$$

$$\|\mathbf{g}_n\|_{L^2(Q)} + \|\mathbf{g}_n^{\varphi}\|_{L^2(Q)} \to 0 \quad \text{as } n \to \infty, \tag{4.6.10}$$

$$\|\varpi_n\|_{L^2(Q)} \le E, \tag{4.6.11}$$

  where $q = s/(s-1)$ for $s < \infty$ and $1 < q \le 2$ for $s = \infty$.

- There are functions

$$\varrho \in L^\infty(0,T;L^\gamma(\Omega)), \quad \mathbf{u} \in L^2(0,T;W^{1,2}(\Omega)),$$
$$\overline{\mathbb{T}} \in L^1(Q) \cap L^{r'}_{\text{loc}}(Q) \quad \text{and} \quad \varphi \in L^\infty(0,T;L^s(\Omega)) \cap L^r_{\text{loc}}(Q),$$

  such that the following convergences hold:

$$\begin{aligned} &\varrho_n \rightharpoonup \varrho \quad \text{weakly}^\star \text{ in } L^\infty(0,T;L^\gamma(\Omega)),\\ &\varphi_n \rightharpoonup \varphi \quad \text{weakly}^\star \text{ in } L^\infty(0,T;L^s(\Omega)),\\ &\mathbf{u}_n \rightharpoonup \mathbf{u} \quad \text{weakly in } L^2(0,T;W^{1,2}(\Omega)),\\ &\varphi_n \rightharpoonup \varphi \quad \text{weakly in } L^r(Q'), \quad \mathbb{T}_n \rightharpoonup \overline{\mathbb{T}} \quad \text{weakly in } L^{r'}(Q') \end{aligned} \tag{4.6.12}$$

  for any $Q' \Subset Q$. Here $r' = r/(r-1)$.

**Definition 4.6.2.** The *generalized viscous flux* $\mathcal{V}_n$, associated with the system of conservation laws (4.6.1)–(4.6.2), is defined as follows:

$$\mathcal{V}_n = \mathbf{R} : [\zeta \mathbb{T}_n] := R_{ij}[\zeta T_{n,ij}], \tag{4.6.13}$$

where $\zeta \in C_0^\infty(\Omega)$ is an arbitrary function of the spatial variable $x$.

In our framework, the expression for the viscous flux involves a nonlocal operator and an arbitrary cut-off function. The problem of localization of the viscous flux is discussed in Section 4.7. Notice that $\zeta$ is compactly supported in $\Omega$ and $\mathbb{T}$ belongs to $L^r_{\text{loc}}(\Omega \times (0,T))$. It follows that the matrix-valued functions $\zeta\mathbb{T}$ and $\mathbf{R}[\zeta\mathbb{T}]$ belong to $L^r(\mathbb{R}^d \times [h, T-h])$ for every $h \in (0,T)$. From this and condition (4.6.12) we conclude that

$$\mathcal{V}_n \rightharpoonup \overline{\mathcal{V}} = \mathbf{R} : [\zeta\mathbb{T}] \quad \text{weakly in } L^q(\mathbb{R}^d \times [h, T-h])) \text{ for } h \in (0,T). \tag{4.6.14}$$

The following theorem, which is the main result of this section, gives the weak continuity of the product $\mathcal{V}_n\varphi_n$.

**Theorem 4.6.3.** *Assume that $\varrho_n$, $\mathbf{u}_n$, $\mathbb{T}_n$ are solutions to equations* (4.6.1)–(4.6.3) *satisfying Condition 4.6.1. Furthermore, assume that the exponents $s$, $\gamma$, $q$ and the elements $\mathbf{g}_n^{\varphi}$, $\mathbb{G}_n$ satisfy either*

$$2d < \gamma \le s < \infty, \tag{4.6.15}$$

*or*

$$\gamma > d/2, \quad s = \infty, \quad \mathbf{g}_n^{\varphi} = 0, \quad \mathbb{G}_n = 0. \tag{4.6.16}$$

*Then for all $\xi \in C_0^\infty(Q)$ and $\zeta \in C_0^\infty(\Omega)$,*

$$\lim_{n\to\infty} \int_Q \xi\, \mathcal{V}_n \varphi_n \, dxdt = \int_Q \xi\, \overline{\mathcal{V}}\, \varphi \, dxdt. \tag{4.6.17}$$

*Since $\xi\varphi_n, \xi\varphi \in L^r(Q)$ are supported in $\Omega \times (h, T-h)$ for some $h > 0$, the integrals in* (4.6.17) *are well defined.*

*Proof.* Fix $\xi \in C_0^\infty(Q)$ and $\zeta \in C_0^\infty(\Omega)$. If $\varrho_n$, $\mathbf{u}_n$, $\mathbb{T}_n$ are solutions to equations (4.6.1)–(4.6.3) satisfying Condition 4.6.1 and either (4.6.15) or (4.6.16), then they obviously satisfy (4.5.1)–(4.5.7) and meet all requirements of Theorem 4.5.2. Hence they satisfy the integral identity (4.5.10), which in this case can be rewritten in the form

$$\begin{aligned}\int_Q \xi\, \varphi_n\, \mathcal{V}_n \, dxdt &= -\int_Q \xi\varphi_n \mathbf{R} : [\zeta \mathbb{G}_n]\, dxdt \\ &+ \int_Q \big(\mathbf{H}_n \cdot \mathbf{u_n} + \zeta\varrho_n(\mathbf{N}_n + \mathbf{P}_n + \mathbf{Q}_n)\cdot \mathbf{u}_n + S_n\big)\, dxdt. \end{aligned} \tag{4.6.18}$$

Here, we denote

$$\mathbf{H}_n = \mathbf{R}[\xi\varphi_n](\zeta\varrho_n\mathbf{u}_n) - (\xi\varphi_n)\,\mathbf{R}[\zeta\varrho_n\mathbf{u}_n], \tag{4.6.19}$$

$$S_n = (\nabla\zeta \otimes \mathbf{A}[\xi\varphi_n]) : (\varrho_n\mathbf{u}_n \otimes \mathbf{u}_n - \mathbb{T}_n - \mathbb{G}_n) + \zeta\mathbf{A}[\xi\varphi_n]\cdot(\varrho_n\mathbf{f}), \tag{4.6.20}$$

$$\mathbf{N}_n = \mathbf{R}[\xi\, \mathbf{g}_n^{\varphi}], \quad \mathbf{P}_n = \mathbf{A}[\varphi_n\partial_t\xi + \varphi_n\mathbf{u}_n\cdot\nabla\xi], \tag{4.6.21}$$

$$\mathbf{Q}_n = -\mathbf{A}[\mathbf{g}_n^{\varphi}\cdot\nabla\xi + \varpi_n\xi].$$

Our task is to pass to the limit on the right hand side of identity (4.6.18) as $n \to \infty$. The considerations are based on the following

**Lemma 4.6.4.** *Under the assumptions of Theorem 4.6.3, the sequence $(\varrho_n, \varphi_n, \mathbf{u}_n)$ meets all requirements of Theorem 4.4.2 and satisfies* (4.4.6)–(4.4.10c).

*Proof.* It suffices to show that $(\varrho_n, \varphi_n, \mathbf{u}_n)$ and $\gamma$, $s$ satisfy Condition 4.4.1. Observe that $s, \gamma > d/2$ by (4.6.15)–(4.6.16). Hence $s, \gamma$ satisfy Condition 4.4.1. Next, estimates (4.4.6) in this condition coincide with (4.6.4)–(4.6.6) in Condition 4.6.1 of Theorem 4.6.3. It follows from (4.6.1)–(4.6.3) that $(\varrho_n, \varphi_n, \mathbf{u}_n)$ satisfy equations (4.4.5) in Condition 4.4.1 with

$$\mathfrak{u}_n = -\varrho_n\mathbf{u}_n + \mathbf{g}_n, \quad \mathfrak{g}_n = 0,$$
$$\mathfrak{V}_n = \mathbb{T}_n + \mathbb{G}_n - \varrho_n\mathbf{u}_n\otimes\mathbf{u}_n, \quad \mathfrak{w}_n = \varrho_n\mathbf{f},$$
$$\mathfrak{v}_n = -\varphi_n\mathbf{u}_n + \mathbf{g}_n^{\varphi}, \quad \mathfrak{h}_n = -\varpi_n.$$

Estimates (4.6.4)–(4.6.7) and relations (4.6.10) in Condition 4.6.1 obviously imply the boundedness of the sequences $\mathfrak{u}_n$, $\mathfrak{v}_n$, $\mathfrak{h}_n$, $\mathfrak{w}_n$, $\mathfrak{g}_n$ and $\mathfrak{V}_n$ in $L^1(Q)$. Hence $(\varrho_n, \varphi_n, \mathbf{u}_n)$ satisfy Condition 4.4.1. □

The remaining part of the proof of Theorem 4.6.3 naturally falls into three steps.

**Step 1. Weak continuity of remainders.** First of all we prove that under the assumptions of the theorem,

$$\mathbf{P}_n \rightharpoonup \mathbf{P} = \mathbf{A}[\varphi\partial_t\xi + \varphi\mathbf{u}\cdot\nabla\xi], \quad \mathbf{Q}_n \rightharpoonup \mathbf{Q} = -\mathbf{A}[\varpi\xi] \tag{4.6.22}$$

weakly in $L^2(0,T;W^{1,2}(\Omega))$. Since $\mathbf{P}_n$ and $\mathbf{Q}_n$ are bounded in $L^2(0,T;W^{1,2}(\Omega))$, it suffices to show that for all $\eta \in C_0^\infty(Q)$,

$$\lim_{n\to\infty}\int_Q (\mathbf{P}_n - \mathbf{P})\eta\,dxdt = 0, \quad \lim_{n\to\infty}\int_Q (\mathbf{Q}_n - \mathbf{Q})\eta\,dxdt = 0. \tag{4.6.23}$$

As $\eta$ and $\xi$ are compactly supported in $Q$, and the operator $\mathbf{A}$ is skew-symmetric on the space of compactly supported square integrable functions, we have

$$\int_Q (\mathbf{P}_n - \mathbf{P})\eta\,dxdt = \int_Q (\varphi - \varphi_n)\xi_t\mathbf{A}[\eta]\,dxdt + \int_Q (\varphi\mathbf{u} - \varphi_n\mathbf{u}_n)\cdot\nabla\xi\mathbf{A}[\eta]\,dxdt.$$

Notice that $\nabla\xi\mathbf{A}[\eta] \in C_0^\infty(Q)$. By Lemma 4.6.4 and Theorem 4.4.2 the sequence $\varphi_n(t)$ converges weakly (weakly$^\star$ for $s=\infty$) in $L^s(\Omega)$ to $\varphi(t)$ for almost every $t \in [0,T]$. Moreover, $\mathbf{u}_n$ converges to $\mathbf{u}$ weakly in $L^2(0,T;W^{1,2}(\Omega))$. Since $s > 2d$, we can apply Lemma 4.4.5 with $\mathcal{H}_n = \varphi_n$ and $r = s$ to infer that $\varphi_n\mathbf{u}_n$ converges to $\varphi\mathbf{u}$ weakly in $L^2(0,T;L^m(\Omega))$ for some $m \in (1,\infty)$. This yields the first equality in (4.6.23). Next, we have

$$\int_Q (\mathbf{Q}_n - \mathbf{Q})\eta\,dxdt = \int_Q (\varpi_n - \varpi)\xi_t\mathbf{A}[\eta]\,dxdt + \int_Q \mathbf{g}_n^\varphi\cdot\nabla\xi\mathbf{A}[\eta]\,dxdt.$$

Notice that $\mathbf{g}_n^\varphi \to 0$ in $L^2(0,T;L^2(\Omega))$ and $\varpi_n \rightharpoonup \varpi$ weakly in $L^2(0,T;L^2(\Omega))$ which leads to the second equality in (4.6.23).

Next, by Lemma 4.6.4 and Theorem 4.4.2, we have $\varrho_n(t)\mathbf{u}_n(t) \rightharpoonup \varrho(t)\mathbf{u}(t)$ weakly in $L^r(\Omega)$, $r = 2\gamma/(\gamma+1)$, for almost every $t \in [0,T]$. Moreover, $\varrho_n\mathbf{u}_n$ is uniformly bounded in $L^\infty(0,T;L^r(\Omega))$. By the hypotheses of Theorem 4.6.3 we have

$$\gamma > d/2, \quad \text{hence} \quad r = 2\gamma/(\gamma+1) > 2d/(d+2).$$

Therefore, the functions $\mathcal{H}_n = \varrho_n\mathbf{u}_n$ meet all requirements of Lemma 4.4.5. Applying that lemma to $\mathcal{H}_n\cdot\mathbf{P}_n$ and $\mathcal{H}_n\cdot\mathbf{Q}_n$ and using (4.6.22) we conclude that

$$\varrho_n\mathbf{u}_n\cdot\mathbf{P}_n \rightharpoonup \varrho\mathbf{u}\cdot\mathbf{P}, \quad \varrho_n\mathbf{u}_n\cdot\mathbf{Q}_n \rightharpoonup \varrho\mathbf{u}\cdot\mathbf{Q} \quad \text{weakly in } L^2(0,T;L^m(\Omega)) \tag{4.6.24}$$

for some $m \in (1,\infty)$. Now our task is to prove that $\varrho_n\mathbf{u}_n\cdot\mathbf{N}_n$ vanishes as $n\to\infty$. If condition (4.6.16) is fulfilled, then $\mathbf{N}_n \equiv 0$. If (4.6.15) holds, then $\gamma > 2d$. By

Condition 4.6.1, the bounded energy functions $(\varrho_n, \mathbf{u}_n)$ meet all requirements of Proposition 4.2.1. Applying Corollary 4.2.2 of that proposition we conclude that

$$\|\varrho_n \mathbf{u}_n\|_{L^2(0,T;L^\beta(\Omega))} \le c \quad \text{whenever} \quad \beta^{-1} > 2^{-1} + \gamma^{-1} - d^{-1}, \quad \text{hence for } \beta = 2.$$

On the other hand, since the operator $\mathbf{R}$ is bounded in $L^2(\mathbb{R}^d)$, we have

$$\|\mathbf{N}_n\|_{L^2(Q)} \le c\|\mathbf{g}_n^\varphi\|_{L^2(Q)},$$

which along with (4.6.10) gives

$$\|\varrho_n \mathbf{u}_n \cdot \mathbf{N}_n\|_{L^1(Q)} \le c\|\varrho_n \mathbf{u}_n\|_{L^2(Q)}\|\mathbf{g}_n^\varphi\|_{L^2(Q)} \le c\|\mathbf{g}_n^\varphi\|_{L^2(Q)} \to 0. \tag{4.6.25}$$

Combining (4.6.22)–(4.6.25) we finally obtain

$$\int_Q \zeta\varrho_n\big(\mathbf{N}_n + \mathbf{P}_n + \mathbf{Q}_n\big)\cdot \mathbf{u}_n \, dxdt \to \int_Q \zeta\varrho\big(\mathbf{P} + \mathbf{Q}\big)\cdot \mathbf{u}\, dxdt \quad \text{as } n \to \infty. \tag{4.6.26}$$

**Step 2. Weak continuity of $S_n$.** Now our task is to prove that the functions $S_n$ defined by (4.6.20) satisfy the limit relation

$$\lim_{n\to\infty} \int_Q S_n \, dxdt = \int_Q S \, dxdt, \tag{4.6.27}$$

where

$$S = (\nabla\zeta \otimes \mathbf{A}[\xi\varphi]) : (\varrho\mathbf{u}\otimes\mathbf{u} - \overline{\mathbb{T}}) \; + \; \zeta\mathbf{A}[\xi\varphi]\cdot(\varrho\mathbf{f}).$$

Observe that, since $s \ge 2d$, Condition 4.6.1 implies

$$\|\varphi_n\|_{L^\infty(0,T;L^{2d}(\Omega))} \le \|\varphi_n\|_{L^\infty(0,T;L^s(\Omega))} \le c.$$

Recall that by Lemma 4.6.4 the functions $\varphi_n$ meet all requirements of Theorem 4.4.2. It now follows from (4.4.10b) that there is a set $\mathcal{T}$ of full measure in $[0,T]$ with the properties

$$\|\varphi_n(t)\|_{L^{2d}(\Omega)} \le c \quad \text{for all } t \in \mathcal{T}$$

and

$$\varphi_n(t) \rightharpoonup \varphi(t) \quad \text{weakly in } L^{2d}(\Omega) \quad \text{for all } t\in\mathcal{T}.$$

Since $\xi$ is compactly supported in $\Omega$, it follows that

$$\|\mathbf{A}[\xi\varphi_n](t)\|_{W^{1,2d}(\Omega)} \le c$$

and

$$\mathbf{A}[\xi\varphi_n](t) \rightharpoonup \mathbf{A}[\xi\varphi](t) \quad \text{weakly in } W^{1,2d}(\Omega) \quad \text{as } n\to\infty$$

for all $t \in \mathcal{T}$. Recalling the compactness of the embedding $W^{1,2d}(\Omega) \hookrightarrow C(\Omega)$ we conclude that for all $t\in\mathcal{T}$,

$$\|\mathbf{A}[\xi\varphi_n](t)\|_{C(\Omega)} \le c$$

and

$$\mathbf{A}[\xi\varphi_n](t) \to \mathbf{A}[\xi\varphi](t) \quad \text{in } C(\Omega) \quad \text{as } n \to \infty.$$

Therefore, the functions $\mathbf{A}[\xi\varphi_n]$ are uniformly bounded in $Q$ and the sequence $\mathbf{A}[\xi\varphi_n]$ converges a.e. to $\mathbf{A}[\xi\varphi]$. In particular $\mathbf{A}[\xi\varphi_n]$ converges strongly in any $L^m(Q)$ with $1 \le m < \infty$. On the other hand, $(\varrho_n, \mathbf{u}_n)$ satisfy all hypotheses of Theorem 4.4.2. It follows that $\varrho_n$ and the matrix-valued functions $\varrho_n\mathbf{u}_n \otimes \mathbf{u}_n$ converge weakly to $\varrho$ and $\varrho\mathbf{u}\otimes\mathbf{u}$, respectively, in $L^p(Q)$ for some $p > 1$. This leads to the following convergence:

$$\int_\Omega \{(\nabla\zeta \otimes \mathbf{A}[\xi\varphi_n]) : (\varrho_n\mathbf{u}_n \otimes \mathbf{u}_n) + \zeta\mathbf{A}[\xi\varphi_n] \cdot (\varrho_n\mathbf{f})\}\, dxdt$$
$$\to \int_\Omega \{(\nabla\zeta \otimes \mathbf{A}[\xi\varphi]) : (\varrho\mathbf{u} \otimes \mathbf{u}) + \zeta\mathbf{A}[\xi\varphi] \cdot (\varrho\mathbf{f})\}\, dxdt. \quad (4.6.28)$$

Next, note that $\xi \in C_0^\infty(Q)$ and hence there is $h > 0$ such that $\xi(x,t) = 0$ for $t \in (0,h) \cup (T-h,T)$. Hence there is a function $\eta \in C_0^\infty(0,T)$ such that $\xi\eta = \xi$ in $Q$. We have

$$\mathbf{A}[\xi\varphi_n] = \eta\mathbf{A}[\xi\varphi_n] \quad \text{and} \quad \nabla\zeta \otimes \mathbf{A}[\xi\varphi_n] = \eta\nabla\zeta \otimes \mathbf{A}[\xi\varphi_n].$$

Hence the function on the left hand side of the latter identity is compactly supported on the set

$$Q' = \operatorname{supp} \eta\zeta \Subset Q.$$

Recall that $\mathbf{A}[\xi\varphi_n]$ converges in $L^m(Q)$ for all $m > 1$ and, in particular, it converges in $L^r(Q)$. Since $\mathbb{T}_n$ converges to $\overline{\mathbb{T}}$ weakly in $L^{r'}(Q')$, we have

$$\int_Q (\nabla\zeta \otimes \mathbf{A}[\xi\varphi_n]) : \mathbb{T}_n\, dxdt = \int_{Q'} (\nabla\zeta \otimes \mathbf{A}[\xi\varphi_n]) : \mathbb{T}_n\, dxdt$$
$$\to \int_{Q'} (\nabla\zeta \otimes \mathbf{A}[\xi\varphi]) : \overline{\mathbb{T}}\, dxdt = \int_Q (\nabla\zeta \otimes \mathbf{A}[\xi\varphi]) : \overline{\mathbb{T}}\, dxdt. \quad (4.6.29)$$

On the other hand, relation (4.6.9) yields

$$\Big| \int_Q (\nabla\zeta \otimes \mathbf{A}[\xi\varphi_n]) : \mathbb{G}_n\, dxdt \Big| \le c \int_Q |\mathbb{G}_n|\, dxdt \to 0 \quad \text{as } n \to \infty. \quad (4.6.30)$$

Combining (4.6.28)–(4.6.30) we arrive at the desired relation (4.6.27).

**Step 3. Weak continuity of $\mathbf{H}_n$.** This is the most important step. We prove that the sequence of vector functions $\mathbf{H}_n(t)$ converges weakly for almost every $t \in [0,T]$. This results from the following lemma on the commutator of the operator $\mathbf{R}$ and multiplication by a smooth function.

**Lemma 4.6.5.** *Let $\eta \in C_0^\infty(\Omega)$ and define a matrix-valued operator $\mathbf{C}_\eta : L^p(\mathbb{R}^d) \to L^p(\mathbb{R}^d)$ with entries $C_{\eta,ij}$ by*

$$\mathbf{C}_\eta[\cdot] = \eta\mathbf{R}[\cdot] - \mathbf{R}[\eta\,\cdot].$$

*Then for any $p \in (1,\infty)$, there is a constant $c$ depending only on $p$, $d$ and $\eta$ such that*

$$\|\mathbf{C}_\eta v\|_{W^{1,p}(\Omega)} \le c\|v\|_{L^p(\mathbb{R}^d)} \quad \textit{for all } v \in L^p(\mathbb{R}^d).$$

*Proof.* It suffices to assume that $v \in C_0^\infty(\Omega)$. Then $\mathbf{C}_\eta v \in C^\infty(\mathbb{R}^d)$ and we can apply the Laplace operator to $\mathbf{C}_\eta$ to obtain

$$\Delta C_{\eta,ij}v \equiv \partial_{x_i}\partial_{x_j}\eta\, v - \Delta\eta R_{ij}v + 2\partial_{x_k}(\partial_{x_k}\eta\, R_{ij}v) - \partial_{x_i}(\partial_{x_j}\eta\, v) - \partial_{x_j}(\partial_{x_i}\eta\, v),$$

which leads to the identity

$$C_{\eta,ij}v \equiv \Delta^{-1}\big(\partial_{x_i}\partial_{x_j}\eta\, v - \Delta\eta R_{ij}v\big) + 2A_k(\partial_{x_k}\eta\, R_{ij}v) - A_i(\partial_{x_j}\eta\, v) - A_j(\partial_{x_i}\eta\, v).$$

The desired estimate obviously follows from the boundedness of $\mathbf{R}$ in Lebesgue spaces and the smoothing properties of the operators $\mathbf{A}$ and $\Delta^{-1}$ established in Corollary 1.7.2. □

Let us show that this lemma implies the weak continuity of $\mathbf{H}_n(t)$. Recall that the functions $(\varrho_n, \varphi_n, \mathbf{u}_n)$ satisfy all conditions of Theorem 4.5.2. By estimate (4.5.14) in that theorem, there is a full measure set $\mathcal{T}$ in $[0,T]$ such that

$$\|\mathbf{H}_n(t)\|_{L^\iota(\Omega)} \le c \quad \text{for all } t \in \mathcal{T} \text{ and } n \ge 1, \tag{4.6.31}$$

where the exponent $\iota$ is greater than $2d/(d+2)$. Hence it suffices to prove that for a suitable choice of $\mathcal{T}$, the equality

$$\lim_{n\to\infty} \int_\Omega \eta\mathbf{H}_n(t)\,dx = \int_\Omega \eta\mathbf{H}(t)\,dx \tag{4.6.32}$$

holds for all $\eta \in C_0^\infty(\Omega)$. Notice that (4.6.19) combined with the symmetry properties of $\mathbf{R}$ yields the identity

$$\begin{aligned}\int_\Omega \eta\mathbf{H}_n(t)\,dx &= \int_\Omega \big(\mathbf{R}[\xi\varphi_n](\zeta\varrho_n\mathbf{u}_n) - (\xi\varphi_n)\mathbf{R}[\zeta\varrho_n\mathbf{u}_n]\big)\eta\,dx\\ &= \int_\Omega \mathbf{C}_\eta[\xi(t)\varphi_n(t)]\,(\zeta\varrho_n(t)\mathbf{u}_n(t))\,dx.\end{aligned}$$

By Lemma 4.6.4, the functions $(\varrho_n, \varphi_n, \mathbf{u}_n)$ meet all requirements of Theorem 4.4.2. It follows from relation (4.4.10b) in that theorem and the inequality $s > 2d$ that for a suitable choice of $\mathcal{T}$, $\varphi_n(t) \rightharpoonup \varphi(t)$ weakly in $L^{2d}(\Omega)$ for all $t \in \mathcal{T}$. Now Lemma 4.6.5 implies that $\mathbf{C}_\eta[\xi\varphi_n(t)] \rightharpoonup \mathbf{C}_\eta[\xi\varphi(t)]$ weakly in $W^{1,2d}(\Omega)$ for all $t \in \mathcal{T}$. Since $W^{1,2d}(\Omega)$ is compactly embedded into $C(\Omega)$, we finally infer that for all $t \in \mathcal{T}$,

$$\mathbf{C}_\eta[\xi\varphi_n(t)] \to \mathbf{C}_\eta[\xi\varphi(t)] \quad \text{uniformly in } \Omega \quad \text{as } n \to \infty.$$

On the other hand, by (4.4.10c), we have

$$\varrho_n(t)\mathbf{u}_n(t) \rightharpoonup \varrho(t)\mathbf{u}(t) \quad \text{weakly in } L^\beta(\Omega),\ \beta = 2\gamma/(\gamma+1),$$

for a.e. $t \in [0,T]$. Hence there is a set $\mathcal{T}$ of full measure in $[0,T]$ such that

$$\varrho_n\mathbf{C}_\eta[\xi\varphi_n(t)]\mathbf{u}_n \rightharpoonup \varrho\mathbf{C}_\eta[\xi\varphi(t)]\mathbf{u}(t) \quad \text{weakly in } L^\beta(\Omega)$$

for all $t \in \mathcal{T}$. Thus we get

$$\begin{aligned}\lim_{n\to\infty}\int_\Omega \eta\mathbf{H}_n(t)\,dx &= \int_\Omega\int_\Omega \mathbf{C}_\eta[\xi(t)\varphi_n(t)]\,(\zeta\varrho_n(t)\mathbf{u}_n(t))\,dx \\ &= \int_\Omega \mathbf{C}_\eta[\xi(t)\varphi(t)]\,(\zeta\varrho(t)\mathbf{u}(t))\,dx = \int_\Omega \eta\mathbf{H}(t)\,dx\end{aligned}$$

for all $t \in \mathcal{T}$. This leads to (4.6.32). Thus, the sequence $\mathbf{H}_n$ is bounded in $L^\infty(0,T;L^\iota(\Omega))$ and satisfies (4.6.31)–(4.6.32). Since $\iota > 2d/(d+2)$, the sequences $\mathcal{H}_n = \mathbf{H}_n$ and $\mathbf{u}_n$ meet all requirements of Lemma 4.4.5, hence there is $m \in (1,\infty)$ such that

$$\mathbf{H}_n\cdot\mathbf{u}_n \rightharpoonup \mathbf{H}\cdot\mathbf{u} \quad \text{weakly in } L^2(0,T;L^m(\Omega)) \quad \text{as } n\to\infty.$$

In particular,

$$\lim_{n\to\infty}\int_Q \mathbf{H}_n\cdot\mathbf{u}_n\,dxdt = \int_Q \mathbf{H}\cdot\mathbf{u}\,dxdt. \tag{4.6.33}$$

**Step 4. Proof of Theorem 4.6.3.** We are now in a position to complete the proof of the theorem. Combining (4.6.26), (4.6.27), (4.6.33) and recalling the integral identity (4.6.18) we obtain

$$\begin{aligned}\lim_{n\to\infty}\int_Q \xi\varphi_n\mathcal{V}_n\,dxdt &= \lim_{n\to\infty}\int_Q \xi\varphi_n\mathbf{R} : [\zeta\mathbb{T}_n]\,dxdt \\ &= \lim_{n\to\infty}\int_Q \big(\mathbf{H}_n\cdot\mathbf{u}_n + \zeta\varrho_n(\mathbf{N}_n+\mathbf{P}_n+\mathbf{Q}_n)\cdot\mathbf{u}_n + S_n\big)\,dxdt \\ &\quad - \lim_{n\to\infty}\int_Q \xi\varphi_n\mathbf{R} : [\zeta\mathbb{G}_n]\,dxdt \\ &= \int_Q \big(\mathbf{H}\cdot\mathbf{u} + \zeta\varrho(\mathbf{P}+\mathbf{Q})\cdot\mathbf{u} + S\big)\,dxdt - \lim_{n\to\infty}\int_Q \xi\varphi_n\mathbf{R} : [\zeta\mathbb{G}_n]\,dxdt.\end{aligned}$$

Since $s \geq \gamma$ and $q \geq \gamma/(\gamma-1)$, it follows from (4.6.9) that

$$\begin{aligned}\Big|\int_Q \xi\varphi_n \mathbf{R} : [\zeta\mathbb{G}_n]\,dxdt\Big| &\leq c\|\varphi_n\|_{L^\infty(0,T;L^\gamma(\Omega))}\|\mathbf{R} : [\zeta\mathbb{G}_n]\|_{L^1(0,T;L^{\gamma/(\gamma-1)}(\Omega))} \\ &\leq c\|\zeta\mathbb{G}_n\|_{L^1(0,T;L^q(\Omega))} \to 0 \quad \text{as } n\to\infty,\end{aligned}$$

which leads to the convergence

$$\lim_{n\to\infty}\int_Q \xi\varphi_n \mathcal{V}_n\,dxdt = \int_Q \big(\mathbf{H}\cdot\mathbf{u} + \zeta\varrho(\mathbf{P}+\mathbf{Q})\,\cdot\mathbf{u} + S\big)\,dxdt. \tag{4.6.34}$$

Recall that by Lemma 4.6.4 the functions $(\varrho_n, \varphi_n, \mathbf{u}_n)$ satisfy the hypotheses of Theorem 4.4.2. It follows from (4.4.6)–(4.4.9) and relations (4.6.9)–(4.6.12) that the limit functions $(\varrho, \varphi, \mathbf{u})$ satisfy equations (4.6.1)–(4.6.3) with $\mathfrak{g}_n = \mathfrak{g}_n^\varphi = 0$, $\mathbb{G}_n = 0$, $\varpi_n = \varpi$, and $\mathbb{T}_n = \overline{\mathbb{T}}$. Obviously the limit functions satisfy estimates (4.6.4)–(4.6.6). Therefore, $(\varrho, \varphi, \mathbf{u})$ meet all requirements of Theorem 4.5.2, so

$$\int_Q \xi\varphi\mathbf{R} : [\zeta\overline{\mathbb{T}}]\,dxdt = \int_Q \big(\mathbf{H}\cdot\mathbf{u} + \zeta\varrho(\mathbf{P}+\mathbf{Q})\,\cdot\mathbf{u}_n + S\big)\,dxdt.$$

Since $\mathbf{R} : [\zeta\overline{\mathbb{T}}] = \overline{\mathcal{V}}$, this identity along with (4.6.34) implies (4.6.17), which completes the proof of Theorem 4.6.3. □

## 4.7 Viscous flux. Localization

Theorem 4.6.3 has a serious drawback that expression (4.6.13) for the viscous flux involves a nonlocal singular operator and an arbitrary cut-off function. In this section we show that in the practically important case of compressible Navier-Stokes equations this result can be formulated in a local form without singular integrals. Recall that in the specific case of Navier-Stokes equations, the stress tensor has the representation

$$\mathbb{T}_n = \mathbb{S}(\mathbf{u}_n) - p_n\,\mathbb{I}, \tag{4.7.1}$$

where the viscous stress tensor $\mathbb{S}$ is given by

$$\mathbb{S}(\mathbf{u}) = \nabla\mathbf{u} + \nabla\mathbf{u}^\top + (\lambda-1)\,\mathrm{div}\,\mathbf{u}\,\mathbb{I}.$$

Define the "genuine viscous flux" $V_n$ by

$$V_n = p_n - (1+\lambda)\,\mathrm{div}\,\mathbf{u}_n. \tag{4.7.2}$$

The following theorem, a straightforward consequence of Theorem 4.6.3, is the main result of this section. This is a generalization of known results, in the sense that the stress tensor $\mathbb{T}$ plays no special role, and the result is given for an arbitrary solution $\varphi$ of the mass balance transport equation instead of the density $\varrho$.

**Theorem 4.7.1.** *Suppose that, under the assumptions of Theorem 4.6.3, the sequence* $\mathbb{T}_n$, $n \geq 1$, *admits representation* (4.7.1) *and* $r \geq 2$. *Then there is a subsequence of* $V_n$, *still denoted by* $V_n$, *such that*

$$V_n \rightharpoonup \overline{V} = \overline{p} - (1+\lambda)\operatorname{div}\mathbf{u} \quad \text{weakly in } L^{r'}(Q') \quad \text{for every } Q' \Subset Q. \tag{4.7.3}$$

*Moreover, for every* $\eta \in C_0^\infty(Q)$, *we have*

$$\lim_{n\to\infty} \int_Q \eta\, V_n\, \varphi_n\, dxdt = \int_Q \eta\, \overline{V}\, \varphi\, dxdt. \tag{4.7.4}$$

*Proof.* By (4.6.4), the sequence $\mathbb{S}(\mathbf{u}_n)$ is bounded in $L^2(Q)$. Moreover, estimates (4.6.7)–(4.6.8) imply the boundedness of $\mathbb{T}_n$ in $L^1(Q)$ and $L^{r'}_{\text{loc}}(Q)$. Notice that $r' = r/(r-1) \leq 2$. Now formula (4.7.1) implies that $p_n$ is bounded in $L^1(Q)$ and $L^{r'}_{\text{loc}}(Q)$. Passing to a subsequence we can assume that $p_n$ converges weakly to some $\overline{p} \in L^1(Q) \cap L^{r'}_{\text{loc}}(Q)$ in $L^{r'}(Q')$ for every $Q' \Subset Q$. Furthermore, $\operatorname{div}\mathbf{u}_n \to \operatorname{div}\mathbf{u}$ weakly in $L^2(Q)$. These results along with (4.7.2) imply (4.7.3).

Now choose $\zeta \in C_0^\infty(\Omega)$ such that $\eta(x,t)\zeta(x) = \eta(x,t)$ for all $(x,t) \in \mathbb{R}^{d+1}$. It is easy to see that

$$\zeta\mathbb{T}_n = \nabla(\zeta\mathbf{u}_n) + \nabla(\zeta\mathbf{u}_n)^\top + \zeta\big((\lambda-1)\operatorname{div}\mathbf{u}_n - p_n\big)\,\mathbb{I} - \nabla\zeta\otimes\mathbf{u}_n - \mathbf{u}_n\otimes\nabla\zeta. \tag{4.7.5}$$

Notice that the operator $\mathbf{R} : W^{1,2}(\mathbb{R}^d) \to W^{1,2}(\mathbb{R}^d)$ is bounded and commutes with differentiation. In particular,

$$\begin{aligned}\mathbf{R} : [\nabla\mathbf{w} + \nabla(\mathbf{w})^\top] &= R_{ij}[\,\partial_{x_i}(w_j) + \partial_{x_j}(w_i)\,]\\ &= \partial_{x_i}\partial_{x_j}\Delta^{-1}[\,\partial_{x_i}(w_j) + \partial_{x_j}(w_i)\,] = \partial_{x_j}\partial^2_{x_i}\Delta^{-1}w_j + \partial_{x_i}\partial^2_{x_j}\Delta^{-1}w_i\\ &= \partial_{x_j}w_j + \partial_{x_i}w_i = 2\operatorname{div}\mathbf{w}\end{aligned}$$

for all $\mathbf{w} \in W^{1,2}(\mathbb{R}^d)$. Thus we get

$$\mathbf{R} : [\nabla(\zeta\mathbf{u}_n(t)) + \nabla(\zeta\mathbf{u}_n(t))^\top] = 2\operatorname{div}(\zeta\mathbf{u}_n(t)). \tag{4.7.6}$$

Next, for every $v \in L^r(\Omega)$ with $1 < r < \infty$, we have

$$\mathbf{R} : [v\,\mathbb{I}] = R_{ij}[v\delta_{ij}] = R_{ii}[v] = \sum_{i=1}^d \partial^2_{x_i}\Delta^{-1}v = v,$$

hence for a.e. $t \in [0,T]$,

$$\mathbf{R} : [\zeta((\lambda-1)\operatorname{div}\mathbf{u}_n(t) - p_n(t))\,\mathbb{I}] = \zeta(\lambda-1)\operatorname{div}\mathbf{u}_n(t) - \zeta p_n(t). \tag{4.7.7}$$

Combining (4.7.5)–(4.7.7) we obtain

$$\begin{aligned}\mathbf{R} : [\zeta\mathbb{T}_n] = 2\operatorname{div}(\zeta\mathbf{u}_n(t)) + \zeta(\lambda-1)\operatorname{div}\mathbf{u}_n(t)\\ - \zeta p_n(t) - \mathbf{R} : [\nabla\zeta\otimes\mathbf{u}_n + \mathbf{u}_n\otimes\nabla\zeta],\end{aligned}$$

which leads to

$$\mathbf{R} : [\zeta \mathbb{T}_n(t)] = \zeta\big((\lambda+1)\operatorname{div}\mathbf{u}_n(t) - p_n(t)\big) - W_n(t),$$

where

$$W_n(t) = \mathbf{R} : [\nabla\zeta \otimes \mathbf{u}_n(t) + \mathbf{u}_n(t) \otimes \nabla\zeta] - 2\nabla\zeta \cdot \mathbf{u}_n(t).$$

Recalling the definition (4.7.2) of $V_n$ we can rewrite this identity in the form

$$\zeta V_n(t) = -\mathbf{R} : [\zeta \mathbb{T}_n(t)] - W_n(t) = -\mathcal{V}_n(t) - W_n(t). \tag{4.7.8}$$

Repeating the same arguments we arrive at the following representation of the limit $\overline{V}$:

$$\zeta \overline{V}(t) = -\mathbf{R} : [\zeta \mathbb{T}(t)] - W(t) = -\overline{\mathcal{V}} - W(t), \tag{4.7.9}$$

where

$$W(t) = \mathbf{R} : [\nabla\zeta \otimes \mathbf{u}(t) + \mathbf{u}(t) \otimes \nabla\zeta] - 2\nabla\zeta \cdot \mathbf{u}(t). \tag{4.7.10}$$

It follows from (4.7.8) and the equality $\eta\zeta = \eta$ that

$$\lim_{n\to\infty} \int_Q \eta\, V_n\, \varphi_n\, dxdt = -\lim_{n\to\infty} \int_Q \eta\, \mathcal{V}_n\, \varphi_n\, dxdt - \lim_{n\to\infty} \int_Q \eta\, W_n \varphi_n\, dxdt. \tag{4.7.11}$$

Since $\mathbb{T}_n$ and $\varphi_n$ meet all requirements of Theorem 4.6.3, we have

$$\lim_{n\to\infty} \int_Q \eta\, \mathcal{V}_n\, \varphi_n\, dxdt = \int_Q \eta\, \overline{\mathcal{V}}\, \varphi\, dxdt.$$

Combining this with (4.7.9) and (4.7.11) we obtain

$$\lim_{n\to\infty} \int_Q \eta\, V_n\, \varphi_n\, dxdt = \int_Q \eta\, \overline{V} \varphi\, dxdt - \lim_{n\to\infty} \int_Q \eta\big(W_n \varphi_n - W\varphi\big)\, dxdt.$$

It remains to prove that the limit on the right hand side is simply zero. Since the operator $\mathbf{R} : W^{1,2}(\mathbb{R}^d) \to W^{1,2}(\mathbb{R}^d)$ is bounded and $\mathbf{u}_n \rightharpoonup \mathbf{u}$ weakly in $L^2(0,T; W^{1,2}(\Omega))$, we have

$$W_n \rightharpoonup W \quad \text{weakly in } L^2(0,T; W^{1,2}(\Omega)).$$

On the other hand, by Lemma 4.6.4, the sequence $\varphi_n$ meets all requirements of Theorem 4.4.2, so $\varphi_n(t) \rightharpoonup \varphi(t)$ weakly (weakly$^\star$ for $s=\infty$) in $L^s(\Omega)$. Moreover, $s > 2d$. Hence the couple $(\varphi_n, W_n)$ meets all requirements of Lemma 4.4.5, so $W_n\varphi_n \rightharpoonup W\varphi$ weakly in $L^2(0,T; L^m(\Omega))$ for all $m^{-1} > s^{-1} + (d-2)/(2d)$. This completes the proof of Theorem 4.7.1. □

# Chapter 5

# Nonstationary case. Existence theory

## 5.1 Problem formulation. Results

In this chapter we prove the existence of weak solutions to the general initial-boundary value problem for compressible Navier-Stokes equations, which includes problem (3.0.3) as a particular case. The results obtained are only of preliminary character: we assume that the pressure $p(\varrho)$ is a fast growing function of the density, and the initial and boundary data are sufficiently smooth and satisfy some compatibility conditions. These restrictions are removed in Section 10.1. We use a three-level regularization method similar to the method used in monographs [34] and [37], but our regularization scheme is different because of difficulties caused by inhomogeneous boundary conditions.

In order to make the presentation clearer and avoid unnecessary technical difficulties, we assume that the flow domain $\Omega$, the function $p$ in the constitutive law $p = p(\varrho)$, and the given functions $\mathbf{U}$, $\varrho_\infty$, and $\mathbf{f}$ satisfy the following hypotheses:

**Condition 5.1.1.** • The flow domain $\Omega \subset \mathbb{R}^d$, $d = 2, 3$, is a bounded domain with $C^\infty$ boundary. For given $T > 0$ we denote by $Q$ the cylinder with lateral surface $S_T$ and parabolic boundary $\sqcup_T$ defined by

$$Q = \Omega \times (0, T), \quad S_T = \partial\Omega \times (0, T), \quad \sqcup_T = S_T \cup (\mathrm{cl}\,\Omega \times \{t = 0\}). \tag{5.1.1}$$

- We consider the general constitutive law $p = p(\varrho)$ and the internal energy density

$$P(\varrho) = \varrho \int_0^\varrho s^{-2} p(s)\, ds \tag{5.1.2}$$

with the properties

$$\varrho\, P'(\varrho) - P(\varrho) = p(\varrho), \quad P''(\varrho) = \varrho^{-1} p'(\varrho). \tag{5.1.3}$$

- The function $p : [0,\infty) \to \mathbb{R}^+$ has continuous first derivative with

$$p'(\varrho) \geq 0, \quad p(0) = p'(0) = 0. \tag{5.1.4}$$

There are an exponent $\gamma > 1$ and a constant $c_E > 0$ such that

$$\varrho^\gamma \leq c_E(P(\varrho)+1), \quad \varrho^\gamma \leq c_E(p(\varrho)+1), \quad p(\varrho) \leq c_E(P(\varrho)+1). \tag{5.1.5}$$

The function $P$ is twice continuously differentiable on $\mathbb{R}^+$ and for all $\varrho \in [1,\infty)$,

$$c_E^{-1}\varrho^\gamma \leq p(\varrho) \leq c_E\varrho^\gamma, \quad c_E^{-1}\varrho^{\gamma-1} \leq p'(\varrho) \leq c_E\varrho^{\gamma-1}. \tag{5.1.6}$$

Moreover,

$$P''(\varrho) \geq c_E^{-1} > 0 \quad \text{for all } \varrho \in [0,\infty), \quad |\nabla P'(\varrho_\infty)| \leq c_E. \tag{5.1.7}$$

- There is an exponent $\upsilon \in (1,\gamma]$ and a constant $c_p > 0$ such that

$$\varrho^\upsilon \leq c_p(P(\varrho)+1), \quad \varrho^\upsilon \leq c_p(p(\varrho)+1), \quad p(\varrho) \leq c_p(P(\varrho)+1). \tag{5.1.8}$$

The constant $c_E$ in (5.1.5)-(5.1.7) is greater than $c_p$.

- Let $\Gamma = (S_T \setminus \Sigma_{\text{in}}) \cap \text{cl}\, \Sigma_{\text{in}}$ be the interface between the inlet and the outgoing subset of $S_T$. We assume that

$$\lim_{\sigma \to 0} \sigma^{-d} \operatorname{meas} \mathcal{O}_\sigma < \infty, \tag{5.1.9}$$

where $\mathcal{O}_\sigma = \{(x,t) \in \mathbb{R}^{d+1} : \text{dist}((x,t),\Gamma) \leq \sigma\}$ is a tubular neighborhood of $\Gamma$.

- The initial and boundary data satisfy $\varrho_\infty, \mathbf{U} \in C^\infty(Q)$, $\mathbf{f} \in C(Q)$, and

$$\|\mathbf{U}\|_{C^1(\mathbb{R}^d\times(0,T))} + \|\varrho_\infty\|_{L^\infty(Q)} + \|\mathbf{f}\|_{C(Q)} \leq c_e, \quad \|\varrho_\infty\|_{C^1(\mathbb{R}^d\times(0,T))} \leq c_E \tag{5.1.10}$$

where $c_e$ is some positive constant. Moreover, the data satisfy the compatibility condition

$$\varrho_\infty = 0 \quad \text{in the vicinity of the edge } \partial\Omega \times \{t=0\}. \tag{5.1.11}$$

**Remark 5.1.2.** Throughout this chapter we consider various modifications of the governing equations and their regularizations depending on some parameters. The main ingredients of the theory are a priori estimates of solutions to the governing equations. The a priori estimates are bounds for norms of the solutions in appropriate Banach spaces. These bounds depend on the flow domain, initial and boundary data, the constitutive law, and parameters in the equations. With an application to the shape optimization problem in mind, we distinguish a priori estimates which depend only on the diameter of the flow domain and on "rough" characteristics of the data, and more precise estimates, depending on fine geometric properties of the flow domain and on the smoothness properties of the functions involved in the governing equations. *Further, we denote by $c_e$ generic constants depending only on*

$$\lambda, \quad \gamma, \quad \operatorname{diam}\Omega, \quad T, \quad \|\varrho_\infty\|_{L^\infty(\sqcup_T)}, \quad \|\mathbf{U}\|_{C^1(Q)}, \quad \|\mathbf{f}\|_{C(Q)}, \quad c_p.$$

*The notation $c_{E,p}$ stands for a generic constant depending on $c_e$ and on $\Omega$, $c_E$, $\gamma$ and the norms $\|\varrho_\infty\|_{C^3(Q)}$, $\|\mathbf{U}\|_{C^4(Q)}$.* In other words, $c_e$ depends only on rough characteristics of the data and on the constitutive law, while $c_{E,p}$ depends on the fine structure of the flow domain, the pressure function, and the given data.

Now we formulate the evolution problem under consideration.

**Problem 5.1.3.** *Find a velocity field* $\mathbf{u}$ *and a density function* $\varrho$ *satisfying*

$$\partial_t(\varrho\mathbf{u}) + \operatorname{div}(\varrho\,\mathbf{u}\otimes\mathbf{u}) + \nabla p(\rho) = \operatorname{div}\mathbb{S}(\mathbf{u}) + \varrho\mathbf{f} \quad \text{in } Q, \tag{5.1.12a}$$

$$\partial_t\varrho + \operatorname{div}(\varrho\mathbf{u}) = 0 \quad \text{in } Q, \tag{5.1.12b}$$

$$\begin{aligned} &\mathbf{u} = \mathbf{U} \quad \text{on } \sqcup_T = S_T \cup (\operatorname{cl}\Omega\times\{t=0\}),\\ &\varrho = \varrho_\infty \quad \text{on } \Sigma_{\text{in}}, \quad \varrho|_{t=0} = \varrho_\infty \quad \text{in } \Omega, \end{aligned} \tag{5.1.12c}$$

*where the inlet is of the form* $\Sigma_{\text{in}} = \{(x,t)\in S_T : \mathbf{U}(x,t)\cdot\mathbf{n}(x) < 0\}$, *and*

$$\mathbb{S}(\mathbf{u}) = \nabla\mathbf{u} + \nabla\mathbf{u}^\top + (\lambda-1)\operatorname{div}\mathbf{u}, \quad \operatorname{div}\mathbb{S} = \Delta\mathbf{u} + \lambda\nabla\operatorname{div}\mathbf{u}.$$

We claim that system (5.1.12) admits a weak renormalized solution which is defined as follows:

**Definition 5.1.4.** A couple

$$\mathbf{u}\in L^2(0,T;W^{1,2}(\Omega)), \quad \varrho\in L^\infty(0,T;L^\gamma(\Omega))$$

is said to be a *weak renormalized solution* to problem (5.1.12) if $(\mathbf{u},\varrho)$ satisfies

- The kinetic energy is bounded, i.e., $\varrho|\mathbf{u}|^2\in L^\infty(0,T;L^1(\Omega))$.
- The velocity satisfies the nonhomogeneous Dirichlet boundary condition $\mathbf{u} = \mathbf{U}$ on $S_T$. This condition makes sense since $\mathbf{u}\in L^2(0,T;W^{1,2}(\Omega))$ has a trace on $S_T = \partial\Omega\times(0,T)$.
- The integral identity

$$\int_Q \big(\varrho\mathbf{u}\cdot\partial_t\boldsymbol{\xi} + \varrho\mathbf{u}\otimes\mathbf{u} : \nabla\boldsymbol{\xi} + p\operatorname{div}\boldsymbol{\xi} - \mathbb{S}(\mathbf{u}) : \nabla\boldsymbol{\xi}\big)\,dxdt$$
$$+ \int_Q \varrho\mathbf{f}\cdot\boldsymbol{\xi}\,dxdt + \int_\Omega (\varrho_\infty\mathbf{U}\cdot\boldsymbol{\xi})(x,0)\,dx = 0 \tag{5.1.13}$$

  holds for all vector fields $\boldsymbol{\xi}\in C^\infty(Q)$ equal to 0 in a neighborhood of the lateral side $S_T$ and of the top $\Omega\times\{t=T\}$.

- The integral identity

$$\int_Q \big(\varphi(\varrho)\partial_t\psi + \varphi(\varrho)\mathbf{u}\cdot\nabla\psi + \psi(\varphi(\varrho) - \varphi'(\varrho)\varrho)\operatorname{div}\mathbf{u}\big)\,dxdt$$
$$= \int_{\Sigma_{\rm in}} \psi\varphi(\varrho_\infty)\mathbf{U}\cdot\mathbf{n}\,d\Sigma - \int_\Omega (\varphi(\varrho_\infty)\psi)(x,0)\,dx \quad (5.1.14)$$

holds for all $\psi \in C^\infty(Q)$ vanishing in a neighborhood of the surface $S_T \setminus \Sigma_{\rm in}$ and in a neighborhood of the top $\Omega \times \{t = T\}$, and for all smooth functions $\varphi : [0,\infty) \to \mathbb{R}$ such that

$$\limsup_{\varrho\to\infty}(|\varphi(\varrho)| + |\varrho\varphi'(\varrho)|) < \infty. \quad (5.1.15)$$

This means that $\varphi$ has minimal admissible smoothness and $\varphi$ is bounded at infinity.

**Remark 5.1.5.** Since $\mathbf{u} \in L^2(0,T;W^{1,2}(\Omega))$, the integral identity (5.1.14) also holds for functions $\varphi$ satisfying the growth conditions

$$\limsup_{\varrho\to\infty}(\varrho^{-\gamma/2}|\varphi(\varrho)| + \varrho^{-\gamma/2+1}|\varphi'(\varrho)|) < \infty. \quad (5.1.16)$$

The existence of weak renormalized solutions to problem (5.1.12) is the main result of this chapter.

**Theorem 5.1.6.** *Assume that Condition 5.1.1 is fulfilled, and that $\gamma > 2d$. Then problem* (5.1.12) *has a weak renormalized solution which meets all requirements of Definition 5.1.4 and satisfies the estimate*

$$\|\mathbf{u}\|_{L^2(0,T;W^{1,2}(\Omega))} + \|\varrho|\mathbf{u}_\epsilon|^2\|_{L^\infty(0,T;L^1(\Omega))} + \|P(\varrho)\|_{L^\infty(0,T;L^1(\Omega))} \le c_e, \quad (5.1.17)$$

*where the constant $c_e$ is as in Remark 5.1.2. Estimate* (5.1.17) *depends only on the pressure estimate* (5.1.8) *from below.*

The remaining part of this chapter is devoted to the proof of this theorem. Proceeding from a recent construction [34, 37], it contains three contributions to the existence theory for problem (5.1.12). The first is solvability of regularized equations. The second states a priori estimates for solutions of the regularized equations; the estimates are independent of the regularization parameters. The third establishes compactness properties for these solutions.

We begin with the construction of regularized equations.

There are several ways of regularizing the governing equations and defining approximate solutions: the fictitious domain method with mollifying the velocity vector field in the mass balance equation; penalization of governing equations (see [80]); time discretization combined with existence results for the stationary problem (see [80]); parabolic or elliptic regularization of the mass balance equations combined with correction of the momentum balance equations (see [80], [34]). Notice that justification of the penalization method is known at present only for a time independent convex surface $\Sigma_{\text{in}}$ (see [93], [52]). We cannot use the time discretization method since the stationary nonhomogeneous boundary value problem is more difficult to analyze compared to nonstationary problems. In fact, the stationary problem remains essentially unsolved.

We are going to use a three-level approximation scheme to regularize the governing equations. The basic element of the scheme is the standard *parabolic regularization* proposed by P.-L. Lions and E. Feireisl (see [80], [34]),

$$\partial_t(\varrho_\varepsilon \mathbf{u}_\varepsilon) + \operatorname{div}((\varrho_\varepsilon \mathbf{u}_\varepsilon - \varepsilon\nabla\varrho_\varepsilon)\otimes \mathbf{u}_\varepsilon) + \nabla p_\varepsilon = \operatorname{div}\mathbb{S} + \varrho_\varepsilon \mathbf{f}, \tag{5.1.18}$$

$$\partial_t\varrho_\varepsilon + \operatorname{div}(\varrho_\varepsilon \mathbf{u}) = \varepsilon\Delta\varrho_\varepsilon, \tag{5.1.19}$$

with the boundary and initial conditions

$$\mathbf{u}_\varepsilon = \mathbf{U}, \quad \varrho_\varepsilon = \varrho_\infty \quad \text{on } \sqcup_T.$$

We emphasize that, in contrast to [34, 37], the density $\varrho_\varepsilon$ should satisfy the Dirichlet condition at least on the inlet $\Sigma_{\text{in}}$ in view of the boundary conditions (5.1.12c) for the density. Equations (5.1.18)–(5.1.19) can be considered as a mathematical model of compressible flows with mass diffusion and such equations are of independent interest. The peculiarity of this model is that the momentum balance equation degenerates for $\varrho_\varepsilon = 0$, therefore, the model needs additional corrections. To make such a correction we notice that the main difference between the homogeneous and nonhomogeneous boundary value problems is that the test function $\psi$ in the integral identity (5.1.14) is nonzero on $\Sigma_{\text{in}}$. The corresponding integral identity for the regularized mass balance equation reads

$$\int_Q \big(\varrho_\varepsilon\partial_t\psi + (\varrho_\varepsilon\mathbf{u}_\varepsilon - \varepsilon\nabla\varrho_\varepsilon)\cdot\nabla\psi\big)\,dxdt$$
$$= \int_{\Sigma_{\text{in}}} \psi(\varrho_\infty\mathbf{U} + \varepsilon\nabla\varrho_\varepsilon)\cdot\mathbf{n}\,d\Sigma - \int_\Omega \varrho_\infty\psi(x,0)\,dx.$$

In order to obtain, in the limit $\varepsilon \to 0$, the integral identity (5.1.14) for $\varrho$, i.e., with $\varphi(\varrho) = \varrho$, we have to show that solutions to the regularized mass balance equations (5.1.19) satisfy

$$\varepsilon\int_Q \nabla\varrho_\varepsilon\nabla\psi\,dxdt \to 0, \quad \varepsilon\int_{\Sigma_{\text{in}}} \psi\,\partial_n\varrho_\varepsilon\,d\Sigma \to 0 \quad \text{as } \varepsilon\to 0.$$

The first relation easily follows from the energy estimate, but the second is less trivial to justify. Here, we are dealing with the vanishing viscosity limit for the normal derivative of solutions to an equation (5.1.19) with a convective term. This question arises in a natural way in the theory of elliptic-parabolic equations with nonnegative characteristic form. In the elliptic case it was shown in [102] that

$\varepsilon\partial_n\varrho_\varepsilon$ tends to 0 on $\Sigma_{\rm in}$ provided that the vector fields $\mathbf{u}_\varepsilon$ have uniformly bounded derivatives in a neighborhood of $\Sigma_{\rm in}$ and the functions $\varrho_\varepsilon$ are uniformly bounded in this neighborhood. In Section 13.4 we extend this result to the parabolic case with integrable $\varrho_\varepsilon$, but the restriction on the smoothness of $\mathbf{u}_\varepsilon$ cannot be weakened. To make solutions smooth near the boundary as $\varepsilon \to 0$, we take the regularized equations in the form

$$\begin{aligned}\Delta^2(b\Delta^2\partial_t(\mathbf{u}-\mathbf{U})) + b\partial_t(\mathbf{u}-\mathbf{U}) + \partial_t(\varrho\mathbf{u})&\\ + \operatorname{div}((\varrho\mathbf{u}-\varepsilon\nabla\varrho)\otimes\mathbf{u}) + \nabla p(\rho) = \operatorname{div}\mathbb{S}(\mathbf{u}) + \varrho\mathbf{f} \quad \text{in } Q,& \qquad (5.1.20a)\end{aligned}$$

$$\partial_t\varrho + \operatorname{div}(\varrho\mathbf{u}) = \varepsilon\Delta\varrho \quad \text{in } Q, \qquad (5.1.20b)$$

along with the boundary and initial conditions

$$\begin{aligned}&\varrho = \varrho_\infty \quad \text{on } \sqcup_T,\\ &\mathbf{u}-\mathbf{U} \in C(0,T; W^{4,2}_0(\Omega)), \quad \mathbf{u}|_{t=0} = \mathbf{U} \quad \text{in } \Omega,\end{aligned} \qquad (5.1.20c)$$

where $b \in C^\infty(\Omega)$ is an arbitrary positive function. In Section 5.2 we apply the Leray-Schauder Theorem 1.1.16 to show that problem (5.1.20) has a strong solution and derive estimates for this solution. The results obtained are collected in Theorem 5.2.1.

Then we start a long way from the regularized equations to compressible Navier-Stokes equations. First, we take $b(x) = a(x) + \sigma$, $\sigma > 0$, where $a$ is a smooth cut-off function such that $a = 1$ in the $\delta/2$-neighborhood of $\partial\Omega$ and $a = 0$ outside the $\delta$-neighborhood of $\partial\Omega$. We consider $\sigma$, $\delta$ and $\varepsilon$ as the parameters of regularization. Thus we have a three-level regularization, and the proof of Theorem 5.1.6 involves three successive passages to the limit: $\sigma \to 0$, then $\varepsilon \to 0$, and finally $\delta \to 0$. The first step is made in Section 5.3. Actually, we show that there is a sequence $\sigma_n \to 0$ such that the corresponding solutions to problem (5.1.20) converge weakly in $Q$ to some functions $(\varrho_\varepsilon, \mathbf{u}_\varepsilon)$. We also prove that $\varrho_\varepsilon$ satisfies the mass diffusion equation (5.1.19) in $Q$, but the momentum balance equation (5.1.18) holds true only in a smaller cylinder. However, the functions $(\varrho_\varepsilon, \mathbf{u}_\varepsilon)$ remain smooth in a neighborhood of $\Sigma_{\rm in}$. The results obtained in Section 5.3 are collected in Theorem 5.3.1.

By Theorem 5.3.1 the velocity field $\mathbf{u}_\epsilon$ and the density $\varrho_\varepsilon$ still remain smooth in a neighborhood of the lateral surface $S_T$. We use this fact in Section 5.4, where we exploit Proposition 5.3.11 to prove that $\varepsilon\partial_n\varrho_\varepsilon \to 0$ on the inlet as $\varepsilon \to 0$. Using that proposition we show that the weak limit points of $\varrho_\varepsilon$ serve as renormalized solutions to the mass balance equation and satisfy the integral identity (5.1.14). However, we are unable to prove the strong convergence of $\varrho_\varepsilon$ at this stage. Instead we obtain a representation of a weak limit $\overline{p}$ of $p(\varrho_\varepsilon)$ in terms of a Young measure associated with the sequence $\varrho_\varepsilon$. Letting $\delta \to 0$ in Section 5.5 completes the proof of the three-level convergence of the regularizing scheme.

## 5.2 Regularized equations

This section is devoted to existence theory for the regularized problem (5.1.20). Here we prove that for the adiabatic exponent $\gamma > 2d$, the regularized problem has a solution in $C^{1,2}(Q)$, and derive estimates for this solution independent of the regularization parameter $b$. The result is given by the following theorem.

**Theorem 5.2.1.** *Let the data* $\mathbf{U}$, $\varrho_\infty$, $\mathbf{f}$ *and the pressure function* $p(\varrho)$ *satisfy Conditions 5.1.1 with* $\gamma > 2d$. *Let* $b \in C^\infty(\Omega)$ *satisfy* $b > \sigma > 0$. *Then there exists* $\varepsilon_0 > 0$, *depending only on* $c_{E,p}$, *such that for all* $\varepsilon \in (0, \varepsilon_0)$, *problem* (5.1.20) *has a solution*

$$\mathbf{u} \in C^1(0,T; W^{8,2}(\Omega)), \quad \varrho \in C^{1+\alpha/2, 2+\alpha}(Q), \quad 0 < \alpha < 1, \quad \varrho \geq 0. \tag{5.2.1}$$

*The couple* $(\varrho, \mathbf{u})$ *satisfies the energy estimates*

$$\sup_{t\in(0,T)} \int_\Omega \big(b|\Delta^2 \mathbf{u}|^2 + b|\mathbf{u}|^2 + \varrho|\mathbf{u}|^2 + P\big)(x,t)\,dx + \int_Q \big(|\nabla \mathbf{u}|^2 + \varepsilon P''(\varrho)|\nabla\varrho|^2\big)\,dxdt \leq c_{E,p}, \tag{5.2.2}$$

$$\|\mathbf{u}\|_{L^2(0,T;W^{1,2}(\Omega))} \leq c_{E,p}, \quad \|\varrho|\mathbf{u}|^2\|_{L^\infty(0,T;L^1(\Omega))} \leq c_{E,p}, \tag{5.2.3}$$

$$\|\varrho^\gamma\|_{L^\infty(0,T;L^1(\Omega))} + \varepsilon\|(1+\varrho)^{\gamma-2}|\nabla\varrho|^2\|_{L^1(Q)} \leq c_{E,p}, \tag{5.2.4}$$

$$\|\varrho\mathbf{u}\|_{L^2(Q)} + \|\varrho\mathbf{u}\|_{L^\infty(0,T;L^\beta(\Omega))} \leq c_{E,p}, \tag{5.2.5}$$

$$\|\varrho|\mathbf{u}|^2\|_{L^2(0,T;L^z(\Omega))} \leq c_{E,p}, \tag{5.2.6}$$

$$\varepsilon^{1/2}\||\nabla\varrho|\,|\mathbf{u}|\|_{L^1(0,T;L^s(\Omega))} \leq c_{E,p}, \tag{5.2.7}$$

$$\varepsilon\|\varrho^{3\gamma/2}\|_{L^1(Q)} \leq c_{E,p}, \tag{5.2.8}$$

*where* $\beta = 2\gamma/(\gamma+1)$, $s = d/(d-1)$, *and* $z$ *is an arbitrary number satisfying*

$$1 \leq z < (1 - 2^{-1}\mathfrak{a})^{-1}, \quad \mathfrak{a} = 2d^{-1} - \gamma^{-1}. \tag{5.2.9}$$

*For every* $t \in [0,T]$, $\varrho$ *and* $\mathbf{u}$ *satisfy*

$$\int_\Omega \big(b|\Delta^2\mathbf{u}|^2 + |b\mathbf{u}|^2 + \varrho|\mathbf{u}|^2 + P\big)(x,t)dx + \int_0^t\int_\Omega \big(|\nabla\mathbf{u}|^2 + \varepsilon(P''(\varrho) - \varepsilon^{1/3})|\nabla\varrho|^2\big)dxdt \leq c_e + M(t) + c_e\int_0^t e^{c_e(t-s)}M(s)\,ds, \tag{5.2.10}$$

*where the constant* $c_e$ *is specified by Condition 5.1.1 and*

$$M(t) = 2\int_\Omega b|\Delta^2\mathbf{U}|^2(x,t)\,dx + 8\varepsilon^{2/3}\int_0^t\int_\Omega |\nabla P(\varrho_\infty)|^2\,dxdt - 4\int_0^t\int_\Omega \varrho\big(\partial_t P'(\varrho_\infty) + \nabla P'(\varrho_\infty)\cdot\mathbf{u}\big)\,dxdt. \tag{5.2.11}$$

In order to apply a fixed point argument we reduce problem (5.1.20) to an operator equation in a space of smooth functions. Observe that the governing equations can be rewritten in an equivalent form

$$\Delta^2(b\Delta^2\mathbf{v}) + b\mathbf{v} + \varrho\mathbf{v} = \Psi[\mathbf{u}, \varrho] \quad \text{in } Q, \quad \mathbf{v} \in C(0,T; W_0^{4,2}(\Omega)), \tag{5.2.12a}$$

$$\mathbf{u}(x,t) = \mathbf{U}(x,t) + \int_0^t \mathbf{v}(x,s)\,ds \quad \text{in } Q, \tag{5.2.12b}$$

$$\partial_t\varrho + \operatorname{div}(\varrho\mathbf{u}) = \varepsilon\Delta\varrho \quad \text{in } Q, \tag{5.2.12c}$$

$$\varrho = \varrho_\infty \quad \text{on } \sqcup_T = S_T \cup (\operatorname{cl}\Omega \times \{t=0\}), \tag{5.2.12d}$$

where the nonlinear operator $\Psi$ is defined by

$$\Psi[\mathbf{u}, \varrho] = \operatorname{div}\mathbb{S}(\mathbf{u}) + \varrho\mathbf{f} - (\varrho\mathbf{u} - \varepsilon\nabla\varrho)\nabla\mathbf{u} - \nabla p(\rho) - \varrho\partial_t\mathbf{U}. \tag{5.2.13}$$

Let us consider the family of mappings $\Xi_\tau : (\tilde{\mathbf{u}}, \tilde{\varrho}) \mapsto (\mathbf{u}, \varrho)$ defined by

$$\Delta^2(b\Delta^2\mathbf{v}) + b\mathbf{v} + \tau\tilde{\varrho}\mathbf{v} = \tau\Psi[\tilde{\mathbf{u}}, \tilde{\varrho}] \quad \text{in } Q, \quad \mathbf{v} \in C(0,T; W_0^{4,2}(\Omega)), \tag{5.2.14a}$$

$$\mathbf{u}(x,t) = \mathbf{U}(x,t) + \int_0^t \mathbf{v}(x,s)\,ds \quad \text{in } Q, \tag{5.2.14b}$$

$$\partial_t\varrho + \operatorname{div}(\varrho\tilde{\mathbf{u}}) - \varepsilon\Delta\varrho = 0 \quad \text{in } Q, \tag{5.2.14c}$$

$$\varrho = \varrho_\infty \quad \text{on } \sqcup_T. \tag{5.2.14d}$$

Problem (5.2.12) can be written as a fixed point problem

$$(\mathbf{u}, \varrho) = \Xi_1(\mathbf{u}, \varrho), \quad (\mathbf{u}, \varrho) \in X, \tag{5.2.15}$$

in an appropriate Banach space $X$. We set

$$X = C^\kappa(0,T; C_0^2(\Omega)) \times C^{1+\alpha/2, 2+\alpha}(Q) \quad \text{with } 0 < \alpha < \kappa < 1. \tag{5.2.16}$$

Theorem 5.2.1 will be proved if we show that the one-parameter family of operators $\Xi_\tau$ meets all requirements of the Leray-Schauder theorem 1.1.16 in $X$. We check these in a sequence of lemmas. The first yields the continuity properties of $\Xi_\tau$.

**Lemma 5.2.2.** *The mapping $X \times [0,1] \ni (\mathbf{u}, \varrho, \tau) \mapsto \Xi_\tau(\mathbf{u}, \varrho) \in X$ is continuous. Moreover, for any $\tau \in [0,1]$ the mapping $\Xi_\tau : X \to X$ is compact, i.e., it takes bounded subsets of $X$ into relatively compact sets.*

*Proof.* First we show that $\Xi_\tau : X \to C^1(0,T; W^{8,2}(\Omega)) \times C^{1+\iota/2, 2+\iota}(Q)$ is continuous for all $\tau \in [0,1]$ and any $\iota \in (0, \kappa]$. Let $(\tilde{\mathbf{u}}_n, \tilde{\varrho}_n) \to (\tilde{\mathbf{u}}, \tilde{\varrho})$ strongly in $X$. Since the embedding $X \hookrightarrow C(0,T; C^2(\Omega)) \times C^1(Q)$ is bounded, we have

$$\tilde{\mathbf{u}}_n \to \tilde{\mathbf{u}} \quad \text{in } C(0,T; C^2(\Omega)), \quad \tilde{\varrho}_n \to \tilde{\varrho} \quad \text{in } C^1(Q) \quad \text{as } n \to \infty.$$

It now follows from (5.2.13) that $\Psi(\tilde{\mathbf{u}}_n, \tilde{\varrho}_n) \to \Psi(\tilde{\mathbf{u}}, \tilde{\varrho})$ in $C(0,T; C(\Omega))$. By Lemma 1.7.8, the unique solution $\mathbf{v}_n(t)$ to the equation

$$\Delta^2(b\Delta^2\mathbf{v}_n(t)) + b\mathbf{v}_n(t) + \tau\tilde{\varrho}_n(t)\mathbf{v}_n(t) = \tau\Psi(\tilde{\mathbf{u}}_n, \tilde{\varrho}_n)(t), \quad \mathbf{v}_n(t) \in W_0^{4,2}(\Omega),$$

satisfies the inequality

$$\|\mathbf{v}_n(t)\|_{W^{8,2}(\Omega)} \le c\tau \|\Psi(\tilde{\mathbf{u}}_n(t), \tilde{\varrho}_n(t))\|_{L^2(\Omega)} \le c\tau,$$

where $c$ is independent of $n$. Thus, the sequence $\mathbf{v}_n$ is bounded in $C(0,T;W^{8,2}(\Omega))$. Next, we have

$$\begin{aligned}\Delta^2(b\Delta^2(\mathbf{v}-\mathbf{v}_n)) + b(\mathbf{v}-\mathbf{v}_n) + \tau\tilde{\varrho}(\mathbf{v}-\mathbf{v}_n)&\\ = \tau\big(\Psi(\tilde{\mathbf{u}},\tilde{\varrho}) - \Psi(\tilde{\mathbf{u}}_n,\tilde{\varrho}_n)\big) + \tau(\tilde{\varrho}_n - \tilde{\varrho})\mathbf{v}_n,&\end{aligned}$$

which along with Lemma 1.7.8 leads to

$$\begin{aligned}\|\mathbf{v}_n - \mathbf{v}\|_{C(0,T;W^{8,2}(\Omega))} \le{}& c\max_{(0,T)} \|\Psi(\tilde{\mathbf{u}}(t),\tilde{\varrho}(t)) - \Psi(\tilde{\mathbf{u}}_n(t),\tilde{\varrho}_n(t))\|_{C(\Omega)}\\ &+ c\max_{(0,T)}\|\tilde{\varrho}(t) - \tilde{\varrho}_n(t)\|_{C(\Omega)} \max_{(0,T)}\|\tilde{\mathbf{v}}_n(t)\|_{C(\Omega)} \to 0\end{aligned}$$

as $n\to\infty$. Hence $\|\mathbf{u}_n - \mathbf{u}\|_{C^1(0,T;W^{8,2}(\Omega))} \to 0$ as $n\to\infty$. Therefore, equations (5.2.14a) and (5.2.14b) determine a continuous mapping

$$\{(\tilde{\mathbf{u}},\tilde{\varrho}) \mapsto \mathbf{u}\} : X \to C^1(0,T;W^{8,2}(\Omega)).$$

Let us now return to the parabolic boundary value problem (5.2.14c)-(5.2.14d). We rewrite (5.2.14c) in the form

$$\partial_t\varrho - \varepsilon\Delta\varrho + \tilde{\mathbf{u}}\cdot\nabla\varrho + (\operatorname{div}\tilde{\mathbf{u}})\varrho = 0. \tag{5.2.17}$$

Notice that for any $(\tilde{\mathbf{u}},\tilde{\varrho}) \in X$, the coefficients $\tilde{\mathbf{u}}$ and $\operatorname{div}\tilde{\mathbf{u}}$ of equation (5.2.17) belong to $C^{\kappa/2,\kappa}(Q)$. Therefore, (5.2.14c) meets all requirements of Theorem 1.7.12 and problem (5.2.14c)–(5.2.14d) has a unique solution $\varrho \in C^{1+\kappa/2,2+\kappa}(Q)$. Thus equations (5.2.14c)–(5.2.14d) define a mapping

$$\{(\tilde{\mathbf{u}},\tilde{\varrho}) \mapsto \varrho\} : X \to C^{1+\kappa/2,1+\kappa}(Q).$$

Let us prove that the mapping is continuous. Assume that $(\tilde{\mathbf{u}}_n,\tilde{\varrho}_n)$ converges to $(\tilde{\mathbf{u}},\tilde{\varrho})$ in $X$. We have

$$\tilde{\mathbf{u}}_n \to \tilde{\mathbf{u}}, \quad \operatorname{div}\tilde{\mathbf{u}}_n \to \operatorname{div}\tilde{\mathbf{u}} \quad \text{in } C^{\kappa/2,\kappa}(Q).$$

By Theorem 1.7.12, the corresponding solutions $\varrho_n$ to problem (5.2.14c)–(5.2.14d) are uniformly bounded in $C^{1+\kappa/2,2+\kappa}(Q)$. The difference $\varrho - \varrho_n$ satisfies

$$\begin{aligned}\partial_t(\varrho-\varrho_n) - \varepsilon\Delta(\varrho-\varrho_n) + \tilde{\mathbf{u}}\cdot\nabla(\varrho-\varrho_n) + (\operatorname{div}\tilde{\mathbf{u}})(\varrho-\varrho_n)&\\ = (\tilde{\mathbf{u}}_n - \tilde{\mathbf{u}})\nabla\varrho_n + \operatorname{div}(\tilde{\mathbf{u}}_n-\tilde{\mathbf{u}})\varrho_n,& \qquad (5.2.18)\\ \varrho - \varrho_n = 0 \quad \text{on } \sqcup_T S_T \cup (\operatorname{cl}\Omega\times\{t=0\}).& \qquad (5.2.19)\end{aligned}$$

Since $\varrho = \varrho_n = \varrho_\infty$ on $\sqcup_T$ and $\varrho_\infty$ vanishes near the edge $\partial\Omega \times \{t = 0\}$, the initial and boundary data in equation (5.2.18) satisfy the compatibility conditions of Theorem 1.7.12. It follows that

$$\begin{aligned}
&\|\varrho - \varrho_n\|_{C^{1+\kappa/2,2+\kappa}(Q)} \\
&\qquad \le c(\|\mathbf{u}_n - \mathbf{u}\|_{C^{\kappa/2,\kappa}(Q)} + \|\operatorname{div}(\mathbf{u}_n - \mathbf{u})\|_{C^{\kappa/2,\kappa}(Q)})\|\varrho_n\|_{C^{1+\kappa/2,2+\kappa}(Q)} \\
&\qquad \le c\|\mathbf{u}_n - \mathbf{u}\|_{C^\kappa(0,T;C^2(\Omega))} \to 0 \quad \text{as } n \to \infty,
\end{aligned}$$

and the mapping $\{(\tilde{\mathbf{u}}, \tilde{\varrho}) \mapsto \varrho\} : X \to C^{1+\kappa/2,2+\kappa}(Q)$ is continuous. Combining the results obtained we conclude that the mapping

$$\Xi_\tau : X \to C^1(0, T; W^{8,2}(\Omega)) \times C^{1+\kappa/2,2+\kappa}(Q)$$

is continuous. Since the embedding $C^1(0, T; W^{8,2}(\Omega)) \times C^{1+\kappa/2,2+\kappa}(Q) \hookrightarrow X$ is compact, it follows that the mapping $\Xi_\tau : X \to X$ is continuous and compact. The continuity of this mapping with respect to $\tau$ is obvious. □

Next, we have to prove that all fixed points of the operators $\Xi_\tau$ belong to some bounded set in $X$. This means that the set of all solutions to equations (5.2.12) in $X$ (under the assumption that they exist) is uniformly bounded in $X$. In this way we also establish the desired estimates (5.2.2)–(5.2.10). In fact all these estimates can be derived from the first energy estimate which is a consequence of the energy conservation law. We begin by proving (5.2.10) and split the proof into three lemmas.

**Lemma 5.2.3.** *Under the assumptions of Theorem 5.2.1, let $(\mathbf{u}, \varrho) \in X$ be a fixed point of the operator $\Xi_\tau$ with $0 < \tau \le 1$. Then there is a constant $c_e$ as in Theorem 5.2.1 and $\varepsilon_0 > 0$, depending only on $c_e$, such that for all $\varepsilon < \varepsilon_0$,*

$$\begin{aligned}
&\frac{1}{2}\int_\Omega \big(b|\Delta^2(\mathbf{u} - \mathbf{U})|^2 + b|\mathbf{u} - \mathbf{U}|^2 + \varrho|\mathbf{u} - \mathbf{U}|^2\big)(x, t)\, dx + \frac{1}{2}\int_0^t\int_\Omega |\nabla \mathbf{u}|^2\, dxds \\
&\quad \le \int_0^t\int_\Omega p(\varrho) \operatorname{div} \mathbf{u}\, dxds + c_e + \int_0^t\int_\Omega \Big(\frac{1}{2}\varepsilon^{4/3}|\nabla\varrho|^2 + c_e P(\varrho) + c_e\varrho|\mathbf{u}|^2\Big)\, dxds.
\end{aligned} \tag{5.2.20}$$

*Proof.* We have already proved that each fixed point $(\mathbf{u}, \varrho) \in X$ of $\Xi_\tau$ belongs to $C^{1,2}(Q)$ and $\partial_t\mathbf{u} \in C(0, T; W^{8,2}(\Omega))$. Hence the fixed point is a strong solution to the equation

$$\frac{1}{\tau}\Delta^2(b\Delta^2\partial_t(\mathbf{u} - \mathbf{U})) + \frac{b}{\tau}\partial_t(\mathbf{u} - \mathbf{U}) + \varrho\partial_t(\mathbf{u} - \mathbf{U}) = \Psi[\mathbf{u}, \varrho] \quad \text{in } Q.$$

Multiplying this equation by $\mathbf{u} - \mathbf{U}$ and integrating the result over the cylinder

$\Omega \times [0, t]$ we arrive at the identity

$$\int_0^t \int_\Omega \Big( \frac{1}{\tau} \Delta^2 (b \Delta^2 \partial_t (\mathbf{u} - \mathbf{U})) + \frac{b}{\tau} \partial_t (\mathbf{u} - \mathbf{U}) \Big) \cdot (\mathbf{u} - \mathbf{U})\, dxdt$$
$$+ \frac{1}{2} \int_0^t \int_\Omega \{ \varrho \partial_t |\mathbf{u} - \mathbf{U}|^2 + (\varrho \mathbf{u} - \varepsilon \nabla \varrho) \nabla |\mathbf{u} - \mathbf{U}|^2 \}\, dxdt$$
$$+ \int_0^t \int_\Omega (\nabla p - (\Delta \mathbf{u} + \lambda \nabla \operatorname{div} \mathbf{u})) \cdot (\mathbf{u} - \mathbf{U})\, dxdt$$
$$+ \int_0^t \int_\Omega \{ \varrho \partial_t \mathbf{U} + (\varrho \mathbf{u} - \varepsilon \nabla \varrho) \nabla \mathbf{U} \} \cdot (\mathbf{u} - \mathbf{U})\, dxdt = \int_0^t \int_\Omega \varrho \mathbf{f} \cdot (\mathbf{u} - \mathbf{U})\, dxdt. \tag{5.2.21}$$

Since $\partial_n^k (\mathbf{u} - \mathbf{U}) = 0$ on the lateral surface $S_T$ for all integers $0 \le k \le 3$, and $\mathbf{u} - \mathbf{U} = 0$ at the bottom $\Omega \times \{t = 0\}$, and

$$\partial_t \varrho + \operatorname{div}(\varrho \mathbf{u} - \varepsilon \nabla \varrho) = 0,$$

we can integrate by parts in (5.2.21) to obtain

$$\frac{1}{2} \int_\Omega \Big( \frac{b}{\tau} (|\Delta^2 (\mathbf{u} - \mathbf{U})|^2 + |\mathbf{u} - \mathbf{U}|^2) + \varrho |\mathbf{u} - \mathbf{U}|^2 \Big)(x, t)\, dx$$
$$+ \int_0^t \int_\Omega (|\mathbf{u}|^2 + \lambda |\operatorname{div} \mathbf{u}|^2)\, dxdt + I_1 + I_2 + I_3 = \int_0^t \int_\Omega p \operatorname{div} \mathbf{u}\, dxdt, \tag{5.2.22}$$

where

$$I_1 = \int_0^t \int_\Omega (\varrho \partial_t \mathbf{U} \cdot \mathbf{u} + \varrho \mathbf{u} \nabla \mathbf{U} \cdot \mathbf{u} - \varepsilon \nabla \varrho \nabla \mathbf{U} \cdot \mathbf{u} - \varrho \mathbf{f} \cdot \mathbf{u})\, dxdt,$$
$$I_2 = -\int_0^t \int_\Omega (\varrho \partial_t \mathbf{U} \cdot \mathbf{U} + \varrho \mathbf{u} \nabla \mathbf{U} \cdot \mathbf{U} - \varepsilon \nabla \varrho \nabla \mathbf{U} \cdot \mathbf{U} - \varrho \mathbf{f} \cdot \mathbf{U})\, dxdt,$$
$$I_3 = -\int_0^t \int_\Omega (\nabla \mathbf{u} : \nabla \mathbf{U} + \lambda \operatorname{div} \mathbf{u} \operatorname{div} \mathbf{U} - p \operatorname{div} \mathbf{U})\, dxdt.$$

Recalling that the absolute values of $\mathbf{U}$, $f$, and the first derivatives of $\mathbf{U}$ are bounded by the constant $c_e$ we get the estimate

$$|I_1 + I_2 + I_3| \le c_e \int_0^t \int_\Omega (p + \varrho + \varrho |\mathbf{u}| + \varrho |\mathbf{u}|^2 + \varepsilon |\nabla \varrho|\, |\mathbf{u}| + \varepsilon |\nabla \varrho|)\, dxdt$$
$$+ c_e \int_0^t \int_\Omega |\nabla \mathbf{u}|\, dxdt. \tag{5.2.23}$$

The Cauchy inequality implies

$$\varrho |\mathbf{u}| \le \varrho + \varrho |\mathbf{u}|^2, \quad |\nabla \mathbf{u}| \le \delta |\nabla \mathbf{u}|^2 + \delta^{-1},$$
$$c_e \varepsilon |\nabla \varrho|\, |\mathbf{u}| \le \frac{1}{4} \varepsilon^{4/3} |\nabla \varrho|^2 + 4 c_e^2 \varepsilon^{2/3} |\mathbf{u}|^2, \quad \varepsilon c_e |\nabla \varrho| \le \frac{1}{4} \varepsilon^2 |\nabla \varrho|^2 + 4 c_e^2,$$

for every $\delta \in (0,1)$. Substituting these estimates into (5.2.23) we obtain

$$|I_1+I_2+I_3| \le \frac{c_e}{\delta} + c_e \int_0^t \int_\Omega \big(p+\varrho+\varrho|\mathbf{u}|^2 + (\varepsilon^{2/3}+\delta)|\mathbf{u}|^2\big)\,dxdt \\ + \frac{1}{2}\varepsilon^{4/3}\int_0^t\int_\Omega |\nabla\varrho|^2\,dxdt + \delta c_e \int_0^t\int_\Omega |\nabla\mathbf{u}|^2\,dxdt. \tag{5.2.24}$$

Since $\mathbf{u}-\mathbf{U}$ vanishes at $S_T$, the Poincaré inequality yields

$$\int_0^t\int_\Omega |\mathbf{u}|^2\,dxdt \le \int_0^t\int_\Omega |\mathbf{u}-\mathbf{U}|^2\,dxdt + \int_0^t\int_\Omega |\mathbf{U}|^2\,dxdt \\ \le c_e + \int_0^t\int_\Omega |\mathbf{u}-\mathbf{U}|^2\,dxdt \le c_e + c_e\int_0^t\int_\Omega |\nabla\mathbf{u}-\nabla\mathbf{U}|^2\,dxdt \\ \le c_e + c_e\int_0^t\int_\Omega |\nabla\mathbf{u}|^2\,dxdt + c_e\int_0^t\int_\Omega |\nabla\mathbf{U}|^2\,dxdt \le c_e + c_e\int_0^t\int_\Omega |\nabla\mathbf{u}|^2\,dxdt.$$

Inserting this into (5.2.24) we obtain

$$|I_1+I_2+I_3| \\ \le \frac{c_e}{\delta} + \int_0^t\int_\Omega \Big(c_e p + c_e\varrho + c_e\varrho|\mathbf{u}|^2 + \frac{1}{2}\varepsilon^{4/3}|\nabla\varrho|^2 + c_e(2\varepsilon^{2/3}+\delta)|\nabla\mathbf{u}|^2\Big)\,dxdt.$$

In turn, substituting this into (5.2.22) and recalling that $\tau \le 1$ we arrive at

$$\frac{1}{2}\int_\Omega \big(b|\Delta^2(\mathbf{u}-\mathbf{U})|^2 + b|(\mathbf{u}-\mathbf{U})|^2 + \varrho|\mathbf{u}-\mathbf{U}|^2\big)(x,t)\,dx \\ + (1-c_e\delta - c_e\varepsilon^{2/3})\int_0^t\int_\Omega |\nabla\mathbf{u}|^2\,dxdt \\ \le \frac{c_e}{\delta} + \int_0^t\int_\Omega \Big(c_e p + c_e\varrho + c_e\varrho|\mathbf{u}|^2 + \frac{1}{2}\varepsilon^{4/3}|\nabla\varrho|^2\Big)\,dxdt + \int_0^t\int_\Omega p \,\mathrm{div}\,\mathbf{u}\,dxdt. \tag{5.2.25}$$

Next, by (5.1.8) we have

$$P(\varrho) \ge c_p^{-1}p - c_p{}^{-1} \ge c_p^{-2}\varrho^{\upsilon} - c_p^{-2} - c_p^{-1} \ge c_p^{-2}\varrho - c_e.$$

Inserting this into (5.2.25) we obtain

$$\frac{1}{2}\int_\Omega \big(b|\Delta^2(\mathbf{u}-\mathbf{U})|^2 + b|(\mathbf{u}-\mathbf{U})|^2 + \varrho|\mathbf{u}-\mathbf{U}|^2\big)(x,t)\,dx \\ + (1-c_e\delta - c_e\varepsilon^{2/3})\int_0^t\int_\Omega |\nabla\mathbf{u}|^2\,dxdt \\ \le \frac{c_e}{\delta} + \int_0^t\int_\Omega \Big(c_e P(\varrho) + c_e\varrho|\mathbf{u}|^2 + \frac{1}{2}\varepsilon^{4/3}|\nabla\varrho|^2\Big)\,dxdt + \int_0^t\int_\Omega p\,\mathrm{div}\,\mathbf{u}\,dxdt.$$

Finally choosing $\varepsilon_0$ and $\delta$, depending only on $c_e$, such that $c_e\delta + c_e\varepsilon_0^{2/3} \le 1/2$ we arrive at the desired inequality (5.2.20). $\square$

**Lemma 5.2.4.** *There is a constant $c_e$ as in Remark 5.1.2 such that for all $t \in (0,T)$ and $\tau \in (0,1]$, any fixed point $(\mathbf{u}, \varrho)$ of the operator $\Xi_\tau$ satisfies*

$$\frac{1}{2}\int_\Omega P(\varrho)(x,t)\,dx + \varepsilon\int_0^t\int_\Omega\Big(P''(\varrho) - \frac{1}{2}\varepsilon^{1/3}\Big)|\nabla\varrho|^2\,dxdt$$
$$\le c_e + 2\varepsilon^{2/3}\int_0^t\int_\Omega|\nabla P'(\varrho_\infty)|^2\,dxdt$$
$$-\int_0^t\int_\Omega\varrho\big(\partial_t P'(\varrho_\infty) + \mathbf{u}\cdot\nabla P'(\varrho_\infty)\big)\,dxdt - \int_0^t\int_\Omega p\,\mathrm{div}\,\mathbf{u}\,dxdt. \quad (5.2.26)$$

*Proof.* If $(\mathbf{u}, \varrho)$ is a fixed point of $\Xi_\tau$, then $\varrho$ satisfies

$$\begin{aligned} &\partial_t\varrho - \varepsilon\Delta\varrho + \mathbf{u}\cdot\nabla\varrho + \mathrm{div}\,\mathbf{u}\,\varrho = 0 \quad \text{in } Q,\\ &\varrho = \varrho_\infty \quad \text{on } \sqcup_T. \end{aligned} \quad (5.2.27)$$

Since $\varrho_\infty \in C^\infty(Q)$ satisfies the compatibility conditions (5.1.11), the boundary value problem for $\varrho$ meets all requirements of Theorem 1.7.12 and hence has a unique solution $\varrho \in C^{1+\alpha/2,2+\alpha}(Q)$ which is strictly positive in $Q$. Thus we can multiply both sides of the diffusion equation in (5.2.27) by $P'(\varrho) - P'(\varrho_\infty)$ and use the identity $P'(\varrho)\varrho - P(\varrho) = p(\varrho)$ to obtain

$$\begin{aligned} 0 &= (\partial_t\varrho - \varepsilon\Delta\varrho + \mathbf{u}\cdot\nabla\varrho + \mathrm{div}\,\mathbf{u}\,\varrho)(P'(\varrho) - P'(\varrho_\infty))\\ &= \partial_t(P(\varrho) - P'(\varrho_\infty)\varrho) + \mathrm{div}\,\big((P(\varrho) - P'(\varrho_\infty)\varrho)\mathbf{u}\big)\\ &\quad - \varepsilon\,\mathrm{div}\,\big((P'(\varrho) - P'(\varrho_\infty))\nabla\varrho\big)\\ &\quad + \varepsilon P''(\varrho)|\nabla\varrho|^2 - \varepsilon\nabla\varrho\cdot\nabla P'(\varrho_\infty)\\ &\quad + \varrho\big(\partial_t P'(\varrho_\infty) + \nabla P'(\varrho_\infty)\cdot\mathbf{u}\big) + p(\varrho)\,\mathrm{div}\,\mathbf{u}. \end{aligned}$$

Integrating this equality by parts over the cylinder $\Omega\times(0,t)$ and using the identity $P'(\varrho_\infty)\varrho_\infty - P'(\varrho_\infty) = p(\varrho_\infty)$ we arrive at

$$\int_\Omega\big(P(\varrho) - \varrho P'(\varrho_\infty)\big)(x,t)\,dx + \varepsilon\int_0^t\int_\Omega P''(\varrho)|\nabla\varrho|^2\,dxdt$$
$$+\int_0^t\int_\Omega\varrho\big(\partial_t P'(\varrho_\infty) + \mathbf{u}\cdot\nabla P'(\varrho_\infty)\big)\,dxdt - \varepsilon\int_0^t\int_\Omega\nabla\varrho\cdot\nabla P'(\varrho_\infty)\,dxdt$$
$$+\int_\Omega p(\varrho_\infty)(x,0)\,dx + \int_0^t\int_{\partial\Omega}p(\varrho_\infty)\mathbf{U}\cdot\mathbf{n}\,dSdt = -\int_0^t\int_\Omega p\,\mathrm{div}\,\mathbf{u}\,dxdt,$$

which leads to the inequality

$$\int_\Omega (P(\varrho) - c_e\varrho)(x,t)\,dx + \int_0^t\int_\Omega \{\varepsilon P''(\varrho)|\nabla\varrho|^2 + \varrho(\partial_t P'(\varrho_\infty) + \mathbf{u}\cdot\nabla P'(\varrho_\infty))\}\,dxdt - \varepsilon\int_0^t\int_\Omega \nabla\varrho\cdot\nabla P'(\varrho_\infty)\,dxdt \le c_e - \int_0^t\int_\Omega p\operatorname{div}\mathbf{u}\,dxdt.$$

Next notice that

$$\varepsilon|\nabla P'(\varrho_\infty)|\,|\nabla\varrho| \le \frac{1}{2}\varepsilon^{4/3}|\nabla\varrho|^2 + 2\varepsilon^{2/3}|\nabla P'(\varrho_\infty)|^2,$$

which leads to

$$\int_\Omega (P(\varrho) - c_e\varrho)(x,t)\,dx + \varepsilon\int_0^t\int_\Omega \Big(P''(\varrho) - \frac{1}{2}\varepsilon^{1/3}\Big)|\nabla\varrho|^2\,dxdt + \int_0^t\int_\Omega \varrho(\partial_t P'(\varrho_\infty) + \mathbf{u}\cdot\nabla P'(\varrho_\infty))\,dxdt \le c_e + 2\varepsilon^{2/3}\int_0^t\int_\Omega |\nabla P'(\varrho_\infty)|^2\,dxdt - \int_0^t\int_\Omega p\operatorname{div}\mathbf{u}\,dxdt. \quad (5.2.28)$$

Next, condition (5.1.8) and the Young inequality imply

$$\varrho^\upsilon \le c_p P(\varrho) + c_p, \quad c_e\varrho \le \delta\varrho^\upsilon + \delta^{-\upsilon/(\upsilon-1)}c_e^{\upsilon/(\upsilon-1)},$$

where $\delta > 0$ is an arbitrary number. In particular, for sufficiently small $\delta > 0$ and a suitable choice of the constant $c_e$, we have

$$c_e\varrho \le P(\varrho)/2 + c_e. \quad (5.2.29)$$

Inserting (5.2.29) into (5.2.28) we obtain the desired estimate (5.2.26). □

**Lemma 5.2.5.** *There is a constant $c_e$ as in Remark 5.1.2 and $\varepsilon_0 > 0$, depending only on $c_{E,p}$, such that for all $\varepsilon \in (0,\varepsilon_0)$, $t \in (0,T)$, and $\tau \in (0,1]$, any fixed point $(\mathbf{u},\varrho)$ of the operator $\Xi_\tau$ satisfies*

$$\int_\Omega \big(b|\Delta^2\mathbf{u}|^2 + b|\mathbf{u}|^2 + \varrho|\mathbf{u}|^2 + P(\varrho)\big)(x,t)\,dx + \int_0^t\int_\Omega \big(|\nabla\mathbf{u}|^2 + \varepsilon(P''(\varrho) - \varepsilon^{1/3})|\nabla\varrho|^2\big)\,dxdt \le c_e + c_e\int_0^t\int_\Omega (P(\varrho) + \varrho|\mathbf{u}|^2)\,dxdt + M(t), \quad (5.2.30)$$

$$\int_\Omega \big(b|\Delta^2\mathbf{u}|^2 + b|\mathbf{u}|^2 + \varrho|\mathbf{u}|^2 + P(\varrho)\big)(x,t)\,dx$$
$$+\int_0^t\int_\Omega \big(|\nabla\mathbf{u}|^2 + \varepsilon P''(\varrho)|\nabla\varrho|^2\big)\,dxdt$$
$$\le c_{E,p} + c_{E,p}\int_0^t\int_\Omega (P(\varrho) + \varrho|\mathbf{u}|^2)\,dxdt, \quad (5.2.31)$$

*where $M(t)$ is defined by* (5.2.11).

*Proof.* Adding (5.2.20) and (5.2.26) we arrive at

$$\frac{1}{2}\int_\Omega \big(b|\Delta^2(\mathbf{u}-\mathbf{U})|^2 + b|\mathbf{u}-\mathbf{U}|^2 + \varrho|\mathbf{u}-\mathbf{U}|^2 + P(\varrho)\big)(x,t)\,dx$$
$$+\frac{1}{2}\int_0^t\int_\Omega \big(|\nabla\mathbf{u}|^2 + \varepsilon P''(\varrho)|\nabla\varrho|^2 - \varepsilon^{4/3}|\nabla\varrho|^2\big)\,dxdt$$
$$\le c_e + c_e\int_0^t\int_\Omega (P(\varrho) + \varrho|\mathbf{u}|^2)\,dxdt$$
$$+ 2\varepsilon^{2/3}\int_0^t\int_\Omega |\nabla P'(\varrho_\infty)|^2\,dxdt - \int_0^t\int_\Omega \varrho\big(\partial_t P'(\varrho_\infty) + \mathbf{u}\cdot\nabla P'(\varrho_\infty)\big)\,dxdt. \quad (5.2.32)$$

Next, using the inequality $(a-b)^2 \ge 2^{-1}a^2 - b^2$ and noting that $|\mathbf{U}|, b \le c_e$ we obtain

$$\int_\Omega \Big(\frac{b}{2}|\Delta^2(\mathbf{u}-\mathbf{U})|^2 + \frac{b}{2}|\mathbf{u}-\mathbf{U}|^2 + \frac{\varrho}{2}|\mathbf{u}-\mathbf{U}|^2\Big)(x,t)\,dx$$
$$\ge \frac{1}{4}\int_\Omega \big(b|\Delta^2\mathbf{u}|^2 + b|\mathbf{u}|^2 + \varrho|\mathbf{u}|^2\big)(x,t)\,dx - \frac{1}{2}\int_\Omega \big(b|\Delta^2\mathbf{U}|^2 + c_e\varrho + c_e\big)(x,t)\,dx.$$

From this and inequality (5.2.29) we obtain

$$\frac{1}{2}\int_\Omega \big(b|\Delta^2(\mathbf{u}-\mathbf{U})|^2 + b|\mathbf{u}-\mathbf{U}|^2 + \varrho|\mathbf{u}-\mathbf{U}|^2 + P\big)(x,t)\,dx$$
$$\ge \frac{1}{4}\int_\Omega \big(b|\Delta^2\mathbf{u}|^2 + b|\mathbf{u}|^2 + \varrho|\mathbf{u}|^2 + 2P - c_e\varrho\big)(x,t)\,dx$$
$$-\frac{1}{2}\int_\Omega (b|\Delta^2\mathbf{U}|^2)(x,t)\,dx - c_e$$
$$\ge \frac{1}{4}\int_\Omega \big(b|\Delta^2\mathbf{u}|^2 + b|\mathbf{u}|^2 + \varrho|\mathbf{u}|^2 + P\big)(x,t)\,dx$$
$$-\frac{1}{2}\int_\Omega (b|\Delta^2\mathbf{U}|^2)(x,t)\,dx - c_e. \quad (5.2.33)$$

Next, substituting (5.2.33) into (5.2.32) we finally obtain

$$\int_\Omega \big(b|\Delta^2\mathbf{u}|^2 + b|\mathbf{u}|^2 + \varrho|\mathbf{u}|^2 + P(\varrho)\big)(x,t)\,dx$$
$$+ \int_0^t \int_\Omega \big(|\nabla\mathbf{u}|^2 + \varepsilon(P''(\varrho) - \varepsilon^{1/3})|\nabla\varrho|^2\big)\,dxdt$$
$$\leq c_e + c_e\int_0^t \int_\Omega (P(\varrho) + \varrho|\mathbf{u}|^2)\,dxdt + 2\int_\Omega (b|\Delta^2\mathbf{U}|^2)(x,t)\,dx$$
$$+ 4\int_0^t \int_\Omega \big(2\varepsilon^{2/3}|\nabla P'(\varrho_\infty)|^2\,dxdt - \varrho\,(\partial_t P'(\varrho_\infty) + \mathbf{u}\cdot\nabla P'(\varrho_\infty))\big)\,dxdt. \tag{5.2.34}$$

Recalling formula (5.2.11) for $M(t)$ we conclude that (5.2.34) coincides with claim (5.2.30). To prove (5.2.31) notice that in view of (5.1.7) there is $\varepsilon_0$, depending only on $c_E$ and hence only on $c_{E,p}$, such that for all $\varepsilon \in (0, \varepsilon_0)$ we have

$$P''(\varrho) - \varepsilon^{1/3} \geq \frac{1}{2}P''(\varrho). \tag{5.2.35}$$

On the other hand, in view of Remark 5.1.2 we have

$$|M| \leq c_{E,p} + c_{E,p}\int_\Omega (\varrho + \varrho|\mathbf{u}|^2)\,dx \leq c_{E,p} + c_{E,p}\int_\Omega (P(\varrho) + \varrho|\mathbf{u}|^2)\,dx.$$

Inserting this into (5.2.34) leads to estimate (5.2.31). □

**Lemma 5.2.6.** *There is a constant $c_e$ as in Remark 5.1.2 and $\varepsilon_0 > 0$, depending only on $c_{E,p}$, such that for all $\varepsilon \in (0, \varepsilon_0)$, $t \in (0, T)$, and $\tau \in (0, 1]$, any fixed point $(\mathbf{u}, \varrho)$ of the operator $\Xi_\tau$ satisfies inequalities* (5.2.2) *and* (5.2.10).

*Proof.* Choose $\varepsilon_0 > 0$ that meets all requirements of Lemma 5.2.5. We begin by proving (5.2.10). It follows from (5.2.30) that for all $t \in (0, T)$,

$$\int_\Omega (\varrho|\mathbf{u}|^2 + P(\varrho))(x,t)\,dx - c_e\int_0^t \int_\Omega (P(\varrho) + \varrho|\mathbf{u}|^2)\,dxdt \leq c_e + M(t).$$

Multiplying by $\exp(-c_e t)$ we can rewrite this inequality in the equivalent form

$$\frac{d}{dt}\Big\{e^{-c_e t}\int_0^t \int_\Omega (P(\varrho) + \varrho|\mathbf{u}|^2)\,dxdt\Big\} \leq e^{-c_e t}(c_e + M(t)).$$

Integrating over $(0, t)$ and multiplying the result by $\exp(c_e t)$ we arrive at

$$\int_0^t \int_\Omega (P(\varrho) + \varrho|\mathbf{u}|^2)\,dxdt \leq c_e + \int_0^t e^{c_e(t-s)}M(s)ds.$$

Inserting this into the right hand side of (5.2.30) we obtain (5.2.10). It remains to prove (5.2.2). To this end notice that (5.2.31) implies, for all $t \in (0, T)$,

$$\int_\Omega (\varrho|\mathbf{u}|^2 + P(\varrho))(x,t)\,dx - c_e\int_0^t \int_\Omega (P(\varrho) + \varrho|\mathbf{u}|^2)\,dxdt \leq c_{E,p}.$$

Arguing as before we obtain

$$\int_0^t \int_\Omega (P(\varrho) + \varrho|\mathbf{u}|^2)\, dxdt \le \int_0^t e^{c_e(t-s)} c_{E,p} ds \le c_{E,p}.$$

Inserting this into the right hand side of (5.2.31) we obtain (5.2.2). □

**Lemma 5.2.7.** *Let $\varepsilon_0$ be defined by Lemma 5.2.5. Then for any $\varepsilon \in (0, \varepsilon_0)$ and $\tau \in (0,1]$, every fixed point $(\mathbf{u}, \varrho) \in X$ of the operator $\Xi_\tau$ satisfies inequalities (5.2.3)–(5.2.8).*

*Proof.* We begin by proving (5.2.3). The Poincaré inequality yields

$$\begin{aligned}\|\mathbf{u}(t)\|_{L^2(\Omega)} &\le \|\mathbf{u}(t) - \mathbf{U}(t)\|_{L^2(\Omega)} + \|\mathbf{U}(t)\|_{L^2(\Omega)}\\ &\le c_e\|\nabla(\mathbf{u}(t) - \mathbf{U}(t))\|_{L^2(\Omega)} + \|\mathbf{U}(t)\|_{L^2(\Omega)}\\ &\le c_e\|\nabla\mathbf{u}(t)\|_{L^2(\Omega)} + c_e\|\nabla\mathbf{U}(t))\|_{L^2(\Omega)} + \|\mathbf{U}(t)\|_{L^2(\Omega)}\\ &\le c_e\|\nabla\mathbf{u}(t)\|_{L^2(\Omega)} + c_e.\end{aligned}$$

Thus we get

$$\begin{aligned}\|\mathbf{u}\|^2_{L^2(0,T;W^{1,2}(\Omega))} &= \int_0^T (\|\mathbf{u}(t)\|^2_{L^2(\Omega)} + \|\nabla\mathbf{u}(t)\|^2_{L^2(\Omega)})\, dt\\ &\le c_e \int_0^T \|\nabla\mathbf{u}(t)\|^2_{L^2(\Omega)}\, dt + c_e,\end{aligned}$$

which along with (5.2.2) gives the first inequality in (5.2.3); the second is a straightforward consequence of (5.2.2). Notice that by conditions (5.1.5)–(5.1.7) and formula (5.1.3) we have

$$P(\varrho) \ge c_E^{-1}\varrho^\gamma - c_E^{-1}, \quad P''(\varrho) \ge (2c_E)^{-1}(1+\varrho)^{\gamma-2}. \tag{5.2.36}$$

On the other hand, (5.2.2) implies

$$\sup_{t\in(0,T)} \int_\Omega P(x,t)\, dx + \varepsilon \int_Q P''(\varrho)|\nabla\varrho|^2\, dxdt \le c_{E,p}.$$

Combining this with (5.2.36) we arrive at (5.2.4). It remains to prove (5.2.5)–(5.2.8).

It follows from (5.2.3) that the fixed point $(\mathbf{u}, \varrho)$ is a bounded energy function. Moreover, its energy is uniformly bounded by a constant $c_{E,p}$. Therefore, the fixed point $(\mathbf{u}, \varrho)$ satisfies all hypotheses of Proposition 4.2.1 with $E$ replaced by $c_{E,p}$. Hence we can apply Corollaries 4.2.2 and 4.2.3 to obtain estimates (5.2.5) and (5.2.6).

Since $\gamma > 2d$, it follows from estimates (5.2.3)–(5.2.4) that the functions $(\mathbf{u}, \varrho)$ have the dissipation rate

$$\varepsilon \int_Q (1+\varrho)^{\gamma-2}|\nabla\varrho(x,t)|^2\, dxdt \le c_{E,p}.$$

In particular, these functions meet all requirements of Proposition 4.3.1 with $E$ replaced by $c_{E,p}$. It remains to notice that inequalities (5.2.7)–(5.2.8) coincide with (4.3.2)–(4.3.3) in that proposition. $\square$

Lemmas 5.2.6 and 5.2.7 give uniform estimates for the fixed points of the operator $\Xi_\tau$ in the energy space norm. The next lemma shows that the boundedness of the fixed points of $\Xi_\tau$ in the Banach space $X$ is a straightforward consequence of the energy estimates.

**Lemma 5.2.8.** *There is a constant $c$ independent of $\tau \in (0,1]$ such that $\|(\mathbf{u},\varrho)\|_X < c$ for each fixed point $(\mathbf{u},\varrho) \in X$ of the operator $\Xi_\tau$.*

*Proof.* We begin by analysing the nonlinear operator $\Psi$ which is part of the operator $\Xi_\tau$. It follows from the definition (5.2.13) of $\Psi$ that for every $\zeta \in C_0^2(\Omega)$ and $t \in (0,T)$,

$$\int_\Omega \Psi[\mathbf{u},\varrho](x,t)\zeta(x)\,dx = -\int_\Omega \mathbb{S}(\mathbf{u}) : \nabla\zeta\,dx + \int_\Omega p(\varrho)\operatorname{div}\zeta\,dx - \int_\Omega \big((\varrho\,\mathbf{u} - \varepsilon\nabla\varrho)\nabla\mathbf{u} - \varrho\mathbf{f} + \varrho\,\partial_t\mathbf{U}\big)\zeta\,dx.$$

We have

$$|\mathbb{S}(\mathbf{u})| + p + \big|(\varrho\,\mathbf{u} - \varepsilon\nabla\varrho)\nabla\mathbf{u} - \varrho\mathbf{f} + \varrho\,\partial_t\mathbf{U}\big| \le c_{E,p}(|\nabla\mathbf{u}|^2 + |\mathbf{u}|^2 + \varrho^\gamma + |\varrho\mathbf{u}|^2 + \varepsilon^2|\nabla\varrho|^2 + 1),$$

which gives

$$\Big|\int_\Omega \Psi[\mathbf{u},\varrho](x,t)\zeta(x)\,dx\Big| \le c_{E,p}\|\zeta\|_{C^1(\Omega)}\big(\|\mathbf{u}(t)\|^2_{W^{1,2}(\Omega)} + \|\varrho^\gamma(t)\|_{L^1(\Omega)} + \|\varrho(t)\mathbf{u}(t)\|^2_{L^2(\Omega)} + \varepsilon^2\|\nabla\varrho(t)\|^2_{L^2(\Omega)}\big)$$
$$\le c_{E,p}\|\zeta\|_{C^1(\Omega)}\big(1 + \varepsilon^2\|\nabla\varrho(t)\|^2_{L^2(\Omega)} + \|\mathbf{u}(t)\|^2_{W^{1,2}(\Omega)} + \|\varrho(t)\mathbf{u}(t)\|^2_{L^2(\Omega)}\big).$$

Here, we use the estimate $\|\varrho^\gamma(t)\|_{L^1(\Omega)} \le c_{E,p}$ that follows from (5.2.4). Since the embedding $W^{4,2}(\Omega) \hookrightarrow C^1(\Omega)$ is bounded, we conclude that

$$\|\Psi[\mathbf{u},\varrho](\cdot,t)\|_{W^{-4,2}(\Omega)} \le c_{E,p}\big(1 + \varepsilon^2\|\nabla\varrho(t)\|^2_{L^2(\Omega)} + \|\mathbf{u}(t)\|^2_{W^{1,2}(\Omega)} + \|\varrho(t)\mathbf{u}(t)\|^2_{L^2(\Omega)}\big).$$

Applying Lemma 1.7.7 to problem (5.2.14a) we obtain

$$\|\mathbf{v}(t)\|_{W^{4,2}(\Omega)} \le c_{E,p}\big(1 + \varepsilon^2\|\nabla\varrho(t)\|^2_{L^2(\Omega)} + \|\mathbf{u}(t)\|^2_{W^{1,2}(\Omega)} + \|\varrho(t)\mathbf{u}(t)\|^2_{L^2(\Omega)}\big). \tag{5.2.37}$$

Noting that

$$\mathbf{u}(x,t) = \mathbf{U}(x,t) + \int_0^t \mathbf{v}(x,s)\,ds$$

we arrive at

$$\|\mathbf{u}\|_{C(0,T;W^{4,2}(\Omega))} \le c_{E,p}\big(1+\varepsilon^2\|\nabla\varrho\|^2_{L^2(Q)}+\|\mathbf{u}\|^2_{L^2(0,T;W^{1,2}(\Omega))}+\|\varrho\mathbf{u}\|^2_{L^2(0,T;L^2(\Omega))}\big).$$

Now estimates (5.2.3)–(5.2.5) yield $\|\mathbf{u}\|_{C(0,T;W^{4,2}(\Omega))} \le c_{E,p}$. Thus we get

$$\|\mathbf{u}\|_{C(0,T;C^2(\Omega))} \le \|\mathbf{u}\|_{C(0,T;W^{4,2}(\Omega))} \le c, \tag{5.2.38}$$

where $c$ is independent of $\varrho$, $\mathbf{u}$ and $\tau$. From this and (5.2.4) we obtain

$$\|\operatorname{div}(\varrho\mathbf{u})\|_{L^2(Q)} \le c.$$

Next, we apply Theorem 1.7.13 to the mass diffusion equation (5.2.14c) to obtain

$$\|\varrho\|_{C(0,T;W^{1,2}(\Omega))} \le c\|\operatorname{div}(\varrho\mathbf{u})\|_{L^2(Q)} \le c,$$

where $c$ is independent of $\mathbf{u}$, $\varrho$ and $\tau$. Combining this with (5.2.38) we conclude that for a.e. $t \in (0,T)$,

$$\|\nabla\varrho(\cdot,t)\|_{L^2(\Omega)} + \|\mathbf{u}(\cdot,t)\|_{W^{1,2}(\Omega)} \le c,$$

which along with (5.2.37) yields $\|\mathbf{v}(\cdot,t)\|_{W^{4,2}(\Omega)} \le c$. Hence

$$\|\mathbf{u}\|_{C^1(0,T;C^2(\Omega))} \le \|\mathbf{u}\|_{C^1(0,T;W^{4,2}(\Omega))} \le c + \|\mathbf{v}\|_{C(0,T;W^{4,2}(\Omega))} \le c. \tag{5.2.39}$$

In particular, we have

$$\|\mathbf{u}\|_{C^1(Q)} + \|\operatorname{div}\mathbf{u}\|_{C^1(Q)} \le c.$$

Thus we can apply Theorem 1.7.12 to the diffusion equation (5.2.12c) to obtain for all $\alpha \in (0,1)$ the estimate $\|\varrho\|_{C^{1+\alpha/2,2+\alpha}(Q)} \le c$. Combining this with (5.2.39) we conclude that the norms of all fixed points of the operator $\Xi_\tau$ in $X$ are bounded by a constant $c$ independent of the fixed points and of $\tau$. □

**Proof of Theorem 5.2.1.** We are now in a position to complete the proof of Theorem 5.2.1. First of all we notice that by Lemma 5.2.2 the mapping

$$X \times [0,1] \ni (\mathbf{u},\varrho,\tau) \mapsto \Xi_\tau(\mathbf{u},\varrho) \in X$$

is continuous. Moreover, for any $\tau \in [0,1]$ the mapping $\Xi_\tau : X \to X$ is compact. Set $G = \{(\mathbf{u},\varrho) \in X : \|(\mathbf{u},\varrho)\|_X < c\}$, where $c$ is the constant of Lemma 5.2.8. By that lemma, $\Xi_\tau$ for $\tau \in (0,1]$ has no fixed points on $\partial G = \{\|(\mathbf{u},\varrho)\|_X = c\}$. Next, for $\tau = 0$ we have $\Xi_0(\mathbf{u},\varrho) = (\mathbf{U},\varrho_0)$, where $\varrho_0 \in C^{1+\alpha/2,2+\alpha}(Q)$ is a solution to the boundary value problem

$$\partial_t\varrho_0 + \operatorname{div}(\varrho_0\mathbf{U}) = \varepsilon\Delta\varrho_0 \quad \text{in } Q, \qquad \varrho_0 = \varrho_\infty \quad \text{on } \sqcup_T.$$

It is clear that the homeomorphism $\mathbb{I} - \Xi_0 : X \to X$ has the only zero $(\mathbf{U}, \varrho_0)$ and we can assume that $c$ is so large that the ball $G$ includes the point $(\mathbf{U}, \varrho_0)$. It follows that the mappings $\Xi_\tau$ and the set $G$ meet all requirements of the Leray-Schauder theorem 1.1.16, so $\Xi_1$ has a fixed point $(\mathbf{u}, \varrho)$ in the ball $G$. It follows from the definition (5.2.16) of the Banach space $X$ that

$$(\mathbf{u}, \varrho) \in C^\kappa(0, T; C^2(\Omega)) \times C^{1+\kappa/2, 2+\kappa}(Q)$$

for any $\kappa \in (0, 1)$. Moreover, as is proved in Lemma 5.2.2, the mapping

$$X \ni (\mathbf{u}, \varrho) \mapsto \Xi_\tau(\mathbf{u}, \varrho) \in C^1(0, T; W^{4,2}(\Omega)) \times C^{1+\kappa/2, 2+\kappa}(Q)$$

is continuous. Therefore, for any fixed point $(\mathbf{u}, \varrho)$, $\mathbf{u}$ belongs to $C^1(0, T; W^{4,2}(\Omega))$. Since the embedding $W^{4,2}(\Omega) \hookrightarrow C^1(\Omega)$ is continuous, it follows from (5.2.13) that $\Psi[\mathbf{u}, \varrho] \in C^1(0, T; C(\Omega))$. We also have $\mathbf{v} \equiv \partial_t \mathbf{u} \in C(0, T; C^1(\Omega))$ and hence

$$\tau\Psi[\mathbf{u}, \varrho] - b\mathbf{v} - \tau\varrho\mathbf{v} \in C(0, T; L^2(\Omega)). \tag{5.2.40}$$

In view of (5.2.14a) the fixed point $(\mathbf{u}, \varrho)$ satisfies the equation

$$\Delta^2(b\Delta^2\mathbf{v}) = \tau\Psi[\mathbf{u}, \varrho] - b\mathbf{v} - \tau\varrho\mathbf{v}.$$

Applying Theorem 1.7.5 to this equation and recalling (5.2.40) we conclude that $u \in C^1(0, T; W^{8,2}(\Omega))$. It remains to note that the energy estimate (5.2.2) in Theorem 5.2.1 is given by Lemma 5.2.6. Estimates (5.2.3)–(5.2.8) are given by Lemma 5.2.7, and (5.2.10) is established in Lemma 5.2.6. The proof of Theorem 5.2.1 is complete.

## 5.3 Passage to the limit. The first level

In this section we perform the first step on the way from the regularized equations to the compressible Navier-Stokes equations. We investigate the behavior of solutions when the coefficient $b$ tends to zero. At this stage we let the coefficient $b$ tend to zero inside the cylinder $Q$ while the diffusion coefficient $\varepsilon$ remains strictly positive. More precisely, we assume that

$$b(x) = a(x) + \sigma \quad \text{in } \Omega, \tag{5.3.1}$$

where $\sigma$ is a positive parameter and $a \in C^\infty(\Omega)$ is a nonnegative function with

$$a(x) = 1 \quad \text{for } \operatorname{dist}(x, \partial\Omega) \le \delta/2, \quad a(x) = 0 \quad \text{for } \operatorname{dist}(x, \partial\Omega) \ge \delta. \tag{5.3.2}$$

Here, $\delta$ is a fixed small positive number. For given $\delta$ we denote by $\Omega_\delta$ a subdomain in $\Omega$ and by $Q_\delta$ the corresponding subcylinder in $Q$, defined by

$$\Omega_\delta = \{x \in \Omega : \operatorname{dist}(x, \partial\Omega) > \delta\}, \quad Q_\delta = \Omega_\delta \times (0, T). \tag{5.3.3}$$

With this notation the function $a$ takes the value 0 in $Q_\delta$ and 1 in $Q \setminus Q_{\delta/2}$. It is clear, in view of (5.3.2), that for this choice of $b$ equations (5.1.20) meet all requirements of Theorem 5.2.1 and so have a solution satisfying inequalities (5.2.2)–(5.2.10). Fix $a$ and choose an arbitrary sequence $\sigma_n \to 0$ as $n \to \infty$. Denote by $(\mathbf{u}_n, \varrho_n)$ solutions to problem (5.1.20) corresponding to $\sigma_n$. It follows from estimates (5.2.2)–(5.2.8) that there is a subsequence, still denoted by $(\mathbf{u}_n, \varrho_n)$, and functions $(\mathbf{u}, \varrho)$, $\mathbf{X}$, with the following properties:

$$\begin{aligned} \varrho_n &\rightharpoonup \varrho \quad \text{weakly in } L^{3\gamma/2}(Q), \\ \nabla \varrho_n &\rightharpoonup \nabla \varrho \quad \text{weakly in } L^2(Q), \\ \mathbf{u}_n &\rightharpoonup \mathbf{u} \quad \text{weakly in } L^2(0,T; W^{1,2}(\Omega)), \\ \sqrt{a+\sigma_n}\,\Delta^2 \mathbf{u}_n &\rightharpoonup \mathbf{X} \quad \text{weakly}^\star \text{ in } L^\infty(0,T; L^2(\Omega)). \end{aligned} \tag{5.3.4}$$

Our task is to pass to the limit as $\sigma_n \to 0$ in the regularized equations (5.1.20) and to obtain a system of equations and boundary conditions for the limits $\varrho$ and $\mathbf{u}$. The corresponding result is given by the following theorems. The first gives estimates for the couple $(\mathbf{u}, \varrho)$ and shows that $(\mathbf{u}, \varrho)$ is a weak solution to the momentum balance equation.

**Theorem 5.3.1.** *Let the pressure function $p(\varrho)$, domain $\Omega$, and functions $\mathbf{U}$, $\varrho_\infty$ satisfy Condition 5.1.1 with the adiabatic exponent $\gamma > 2d$. Let $a$ be a given function satisfying condition (5.3.2) and $\sigma_n$ a sequence with $\sigma_n \to 0$ as $n \to \infty$. Furthermore assume that $(\mathbf{u}_n, \varrho_n)$ are solutions to problem (5.1.20) with $b = a + \sigma_n$ that meet all requirements of Theorem 5.2.1 and satisfy conditions (5.3.4). Then there exists a subsequence, still denoted by $(\mathbf{u}_n, \varrho_n)$, and a matrix-valued function $\mathbb{K}$ with the following properties:*

*(i) $\varrho_n \to \varrho$ strongly in $L^s(Q)$ for $1 < s < 3\gamma/2$ and*

$$\varrho_n \mathbf{u}_n \otimes \mathbf{u}_n \rightharpoonup \mathbb{K} \quad \text{weakly in } L^2(0,T; L^z(\Omega)) \text{ for } 1 < z < (1 - 2^{-1}\mathfrak{a})^{-1} \tag{5.3.5}$$

*with $\mathfrak{a} = 2d^{-1} - \gamma^{-1}$. Moreover, we can characterize the limits as follows:*

$$\mathbb{K} = \varrho \mathbf{u} \otimes \mathbf{u} \quad \text{in } Q_\delta, \tag{5.3.6}$$

$$\mathbf{X} = \sqrt{a}\,\Delta^2 \mathbf{u} \quad \text{in } Q \setminus Q_\delta. \tag{5.3.7}$$

*(ii) The functions $\varrho$, $\mathbf{u}$ satisfy the energy inequality*

$$\operatorname*{ess\,sup}_{t\in(0,T)} \int_\Omega \{a|\Delta^2 \mathbf{u}|^2 + a|\mathbf{u}|^2 + \varrho|\mathbf{u}|^2 + P(\varrho)\}(x,t)\,dx + \int_0^T \int_\Omega \{|\nabla \mathbf{u}|^2 + \varepsilon P''(\varrho)|\nabla \varrho|^2\}\,dxdt \le c_{E,p}. \tag{5.3.8}$$

*Moreover, for almost every* $t \in (0, T)$,

$$\int_\Omega \{a|\Delta^2\mathbf{u}|^2 + a|\mathbf{u}|^2 + \varrho|\mathbf{u}|^2 + P(\varrho)\}(x,t)\,dx \\ + \int_0^t \int_\Omega \{|\nabla\mathbf{u}|^2 + \varepsilon(P''(\varrho) - \varepsilon^{1/3})|\nabla\varrho|^2\}\,dxdt \\ \leq c_e + M(t) + c_e \int_0^t e^{c_e(t-s)} M(s)\,ds, \quad (5.3.9)$$

*where* $M(t)$ *is given by* (5.2.11) *with* $b$ *replaced by* $a$. *The constants* $c_e$ *and* $c_{E,p}$ *are as in Remark 5.1.2.*

*(iii) The functions* $\varrho$, $\mathbf{u}$ *satisfy inequalities* (5.2.3)–(5.2.8).

*(iv) The integral identity*

$$\int_Q \big(a\Delta^2(\mathbf{u} - \mathbf{U}) \cdot \partial_t \Delta^2 \boldsymbol{\zeta} + a(\mathbf{u} - \mathbf{U}) \cdot \partial_t \boldsymbol{\zeta} + \varrho\mathbf{u} \cdot \partial_t \boldsymbol{\zeta}\big)\,dxdt \\ + \int_Q \big((\varrho\mathbf{u} - \varepsilon\nabla\varrho) \otimes \mathbf{u} + p(\varrho)\,\mathbb{I} - \mathbb{S}(\mathbf{u})\big) : \nabla\boldsymbol{\zeta}\,dxdt + \int_Q \varrho\mathbf{f} \cdot \boldsymbol{\zeta}\,dxdt \\ + \int_\Omega \varrho_\infty(x,0)\mathbf{U}(x,0) \cdot \boldsymbol{\zeta}(x,0)\,dx = 0 \quad (5.3.10)$$

*holds for all vector fields* $\boldsymbol{\zeta} \in C^\infty(Q)$ *satisfying*

$$\boldsymbol{\zeta}(x,T) = 0 \quad \text{in } \Omega, \quad \boldsymbol{\zeta}(x,t) = 0 \quad \text{in } Q \setminus Q_\delta. \quad (5.3.11)$$

The second theorem describes the behavior of $(\mathbf{u}, \varrho)$ near $S_T$ and shows that $\varrho$ is a weak renormalized solution to the mass diffusion equation.

**Theorem 5.3.2.** *Under the assumptions of Theorem 5.3.1, set* $A^\delta := \Omega \setminus \Omega_{\delta/3}$. *Then*

$$\|\mathbf{u}\|_{L^\infty(0,T;C^2(\Omega\setminus\Omega_{\delta/2}))} \leq c(\delta) c_{E,p}. \quad (5.3.12)$$

*There is a constant* $c(\delta, \varepsilon)$, *depending only on* $\delta$ *and* $\varepsilon$, *such that*

$$\|\partial_t\varrho\|_{L^2(0,T;L^2(A^\delta))} + \|\varrho\|_{L^2(0,T;W^{2,2}(A^\delta))} \\ + \|\varrho\|_{C(0,T;W^{1,2}(A^\delta))} \leq c(\delta,\varepsilon)\,c_{E,p}. \quad (5.3.13)$$

*In particular, the normal derivative* $\partial_n\varrho$ *is well defined and square-integrable on the lateral boundary* $S_T$. *The integral identity*

$$\int_Q (\varrho\partial_t\psi - \varepsilon\nabla\varrho \cdot \nabla\psi + \varrho\mathbf{u} \cdot \nabla\psi)\,dxdt \\ + \int_{S_T} \psi(\varepsilon\nabla\varrho - \varrho_\infty\mathbf{U}) \cdot \mathbf{n}\,dSdt + \int_\Omega (\psi\varrho_\infty)(x,0)\,dx = 0 \quad (5.3.14)$$

*holds for all* $\psi \in C^\infty(Q)$ *vanishing at* $t = T$. *Moreover,*

$$\int_Q \Big(\varphi(\varrho)\partial_t\psi + (\varphi(\varrho)\mathbf{u} - \varepsilon\nabla\varphi(\varrho)) \cdot \nabla\psi - \psi(\varphi'(\varrho)\varrho - \varphi(\varrho)) \operatorname{div}\mathbf{u}\Big)\, dxdt$$
$$+ \int_{S_T} \psi(\varepsilon\varphi'(\varrho_\infty)\nabla\varrho - \varphi(\varrho_\infty)\mathbf{U}) \cdot \mathbf{n}\, dSdt + \int_\Omega (\psi\varphi(\varrho_\infty))(x,0)\, dx \geq 0 \quad (5.3.15)$$

*for all smooth nonnegative functions* $\psi$, *vanishing in a neighborhood of* $S_T \setminus \Sigma_{\text{in}}$ *and in a neighborhood of the top* $\Omega \times \{t = T\}$, *and for all convex functions* $\varphi \in C^2[0,\infty)$ *satisfying the growth condition*

$$|\varphi(\varrho)| + |\varphi'(\varrho)\varrho| + |\varphi''(\varrho)\varrho^2| \leq C\varrho^2.$$

The proof of Theorem 5.3.1 naturally falls into four steps given in Sections 5.3.1–5.3.5. The proof of Theorem 5.3.2 appears in Section 5.3.6.

### 5.3.1 Step 1. Convergence of densities and momenta

In this section we show that the sequence $\varrho_n$ converges almost everywhere in the cylinder $Q$. This fact results from the following lemma due to Yu. Dubinski and J.-L. Lions [77].

**Lemma 5.3.3.** *Let* $X \hookrightarrow Y \hookrightarrow Z$ *be continuously embedded Banach spaces such that* $Y$ *is reflexive and the embedding* $X \hookrightarrow Y$ *is compact. Furthermore, assume that* $\varrho_n$ *is a bounded sequence in* $L^r(0,T;X)$ *for some* $1 < r \leq \infty$, *such that the sequence* $\partial_t\varrho_n$ *is bounded in* $L^1(0,T;Z)$. *Then there is a subsequence, still denoted by* $\varrho_n$, *and a function* $\varrho \in L^r(0,T;Y)$ *such that for all* $s \in (1,r)$,

$$\varrho_n \to \varrho \quad \text{in } L^s(0,T;Y) \quad \text{as } n \to \infty.$$

This lemma implies the following proposition.

**Proposition 5.3.4.** *Under the assumptions of Theorem 5.3.1, as* $n \to \infty$,

$$\varrho_n \to \varrho \quad \text{in } L^s(Q) \text{ for all } s \in [1, 3\gamma/2), \quad (5.3.16)$$
$$p(\varrho_n) \to p(\varrho) \quad \text{in } L^s(Q) \text{ for all } s \in [1, 3/2), \quad (5.3.17)$$
$$\varrho_n\mathbf{u}_n \rightharpoonup \varrho\mathbf{u} \quad \text{weakly in } L^2(Q). \quad (5.3.18)$$

*Proof.* Let us prove that $\varrho_n$ converges a.e. in $Q$. Consider the triple of Banach spaces

$$X = W^{1,2}(\Omega) \subset Y = L^2(\Omega) \subset Z = W^{-4,2}(\Omega).$$

Obviously, it meets the requirements of Lemma 5.3.3. The sequence $\varrho_n$ is bounded in $L^2(0,T;X)$ by (5.2.4).To estimate the time derivative of $\varrho_n$ in $Z$, choose $\zeta \in W^{4,2}_0(\Omega) \subset C^2(\Omega)$. Recall that by Theorem 5.2.1 the function $\varrho_n \in C^{1,2}(Q)$ is a solution to the diffusion equation (5.1.20b) with $\mathbf{u}$ replaced by $\mathbf{u}_n$. Multiplying this equation by $\zeta$ and integrating over $\Omega$ we obtain

$$|\langle \partial_t \varrho_n, \zeta\rangle| = \Big|\int_\Omega (\varepsilon \varrho_n \Delta\zeta + \varrho_n \mathbf{u}_n \cdot \nabla\zeta)\, dx\Big|.$$

By the energy estimate (5.2.2) and inequality (5.1.5) we conclude that

$$\begin{aligned}|\langle \partial_t \varrho_n, \zeta\rangle| &\le c\|\zeta\|_{C^2(\Omega)} \int_\Omega (\varrho_n + \varrho_n|\mathbf{u}_n|)\, dx \\ &\le c\|\zeta\|_{W^{2,4}(\Omega)} \int_\Omega (\varrho_n^\gamma + \varrho_n|\mathbf{u}_n|^2 + 1)\, dx \le c,\end{aligned}$$

where the constant $c$ is independent of $n$. Hence the sequence $\partial_t \varrho_n$ is bounded in $L^\infty(0,T;Z)$, so Lemma 5.3.3 yields the compactness of the sequence $\varrho_n$ in any $L^s(Q)$ with $1 \le s < 2$. On the other hand, by (5.3.4) this sequence converges weakly to $\varrho$ in $L^{3\gamma/2}(Q)$. Hence it converges to $\varrho$ strongly in $L^s(Q)$. On the other hand, in view of estimate (5.2.8), the sequence $\varrho_n$ is bounded in $L^{3\gamma/2}(Q)$. Thus Lemma 1.3.2 shows that $\varrho_n$ converges strongly to $\varrho$ in all spaces $L^r(Q)$ with $r < 3\gamma/2$, which yields (5.3.16). Relation (5.3.17) obviously follows from (5.3.16) and the restrictions (5.1.6) imposed on $p(\varrho)$. Next, $\mathbf{u}_n$ converges weakly in $L^2(Q)$ in view of (5.3.4). By the strong convergence of $\varrho_n$ in $L^2(Q)$ we conclude that

$$\lim_{n\to\infty} \int_Q \varrho_n \mathbf{u}_n \cdot \boldsymbol{\xi}\, dxdt = \int_Q \varrho \mathbf{u} \cdot \boldsymbol{\xi}\, dxdt \quad \text{for all } \boldsymbol{\xi} \in C_0^\infty(Q).$$

Since by (5.2.5) the sequence $\varrho_n \mathbf{u}_n$ is bounded in $L^2(Q)$, we see that the momentum $\varrho_n \mathbf{u}_n$ converges to $\varrho\mathbf{u}$ weakly in $L^2(Q)$. □

### 5.3.2 Step 2. Weak convergence of the kinetic energy tensor

In this section the weak convergence of the matrix-valued functions $\varrho_n \mathbf{u}_n \otimes \mathbf{u}_n$ is investigated.

**Proposition 5.3.5.** *Under the assumptions of Theorem 5.3.1,*

$$\varrho_n \mathbf{u}_n \otimes \mathbf{u}_n \rightharpoonup \mathbb{K} \in L^2(0,T; L^z(\Omega)) \quad \textit{weakly in } L^2(0,T;L^z(\Omega)) \textit{as } n \to \infty$$

*for all $z$ satisfying $1 < z < (1 - 2^{-1}\mathfrak{a})^{-1}$, where $\mathfrak{a} = 2d^{-1} - \gamma^{-1}$. Moreover, $\mathbb{K} = \varrho\mathbf{u} \otimes \mathbf{u}$ in $Q_\delta$.*

*Proof.* We begin with the observation that $(\mathbf{u}_n, \varrho_n)$ is a solution to the regularized problem (5.1.20), meets all requirements of Theorem 5.2.1, and admits the energy estimate (5.2.2). Notice that in view of condition (5.1.5) the estimate (5.2.2) coincides with (4.2.1), and hence $(\mathbf{u}_n, \varrho_n)$ are functions of bounded energy in the sense of Section 4.2. Moreover, their energy is bounded by a constant $c_{E,p}$ independent of $n$. Hence the conclusion of Proposition 4.2.1 holds for $(\mathbf{u}_n, \varrho_n)$. In particular, we can apply estimate (4.2.21) in Corollary 4.2.3 to deduce that the sequence $\varrho_n \mathbf{u}_n \otimes \mathbf{u}_n$ is bounded in $L^2(0,T;L^{z}(\Omega))$. Hence, after passing to a subsequence, we may assume that this sequence converges weakly in $L^2(0,T;L^{z}(\Omega))$ to a matrix-valued function $\mathbb{K}$. It remains to prove $\mathbb{K}$ coincides with $\varrho\mathbf{u}\otimes\mathbf{u}$ in the smaller cylinder $Q_\delta$. It suffices to show that

$$\lim_{n\to\infty}\int_{Q_\delta}\varrho_n\mathbf{u}_n\otimes\mathbf{u}_n : \mathbb{H}\,dxdt = \int_{Q_\delta}\varrho\mathbf{u}\otimes\mathbf{u} : \mathbb{H}\,dxdt \tag{5.3.19}$$

for any matrix-valued function $\mathbb{H}\in C_0^\infty(Q_\delta)$. To do this we prove that the momentum $\varrho_n\mathbf{u}_n(x,t)$ converges weakly for almost every $t\in(0,T)$, and next apply Lemma 4.4.5. Let us consider the vector functions $\mathcal{H}_n = \varrho_n\mathbf{u}_n$ and $\mathcal{S}_n = \sigma_n\Delta^4\mathbf{u}_n + \sigma_n\mathbf{u}_n$. It follows from representations (5.3.1)–(5.3.2) with $\sigma=\sigma_n$ that the coefficient $b(x)$ is equal to $\sigma_n$ in the subdomain $\Omega_\delta$. From this and the energy estimate (5.2.2) we conclude that

$$\|\sqrt{\sigma_n}(\Delta^2\mathbf{u}_n+\mathbf{u}_n)\|_{L^\infty(0,T;L^2(\Omega_\delta))}\le c_{E,p},$$

where the constant $c_{E,p}$ is independent of $n$. This leads to

$$\|\mathcal{S}_n\|_{L^\infty(0,T;W^{-4,2}(\Omega))}\le c_{E,p}\sqrt{\sigma_n}. \tag{5.3.20}$$

Since $\sigma_n\to0$ as $n\to\infty$, we can assume, passing to a subsequence if necessary, that

$$\|\mathcal{S}_n(t)\|_{W^{-4,2}(\Omega_\delta)}\to 0 \quad\text{as } n\to\infty \quad \text{for a.e. } t\in(0,T). \tag{5.3.21}$$

On the other hand, inequality (5.2.5) yields the estimate

$$\|\varrho_n\mathbf{u}_n\|_{L^\infty(0,T;L^\beta(\Omega))}\le c_{E,p}\quad \text{for } \beta=2\gamma/(\gamma+1),$$

which gives

$$\|\mathcal{H}_n\|_{L^\infty(0,T;L^\beta(\Omega))}\le c_{E,p}. \tag{5.3.22}$$

As $b=\sigma_n$ in $Q_\delta$, the first equation (5.1.20a) in the system of regularized equations (5.1.20) reads

$$\frac{\partial}{\partial t}(\mathcal{S}_n+\mathcal{H}_n)=\operatorname{div}\mathfrak{Q}_n+\mathfrak{q}_n \quad \text{in } Q_\delta,$$

where

$$\mathfrak{Q}_n=\mathbb{S}(\mathbf{u}_n)-p(\varrho_n)\mathbb{I}-\varrho_n\mathbf{u}_n\otimes\mathbf{u}_n+\varepsilon\nabla\varrho_n\otimes\mathbf{u}_n,\quad \mathfrak{q}_n=\varrho_n\mathbf{f}.$$

Since $(\mathbf{u}_n,\varrho_n)$ are solutions to the regularized equations (5.1.20) and meet all requirements of Theorem 5.2.1, they satisfy estimates (5.2.3)–(5.2.8), which combined with the constraints on the pressure growth rate imposed by (5.1.6) leads to

$$\|\mathfrak{Q}_n\|_{L^1(Q)}+\|\mathfrak{q}_n\|_{L^1(Q)}\le c_{E,p}$$

with a constant $c_{E,p}$ independent of $n$. It now follows from (5.3.20)–(5.3.21) that the sequence $\mathcal{H}_n$ meets all requirements of Lemma 4.4.4 with $r = \beta$, $k = 4$, $h = 2$ BK', and with $\Omega$ replaced by $\Omega_\delta$, so

$$\mathcal{H}_n(t) = \varrho_n \mathbf{u}_n \rightharpoonup \mathcal{H}(t) \quad \text{weakly in } L^\beta(\Omega_\delta) \quad \text{for a.e. } t \in [0,T].$$

In view of (5.3.18) we have $\mathcal{H} = \varrho\mathbf{u}$. Since $\beta > 2d/(d+2)$, it now follows from (5.3.4) that the sequences $\mathcal{H}_n = \varrho_n \mathbf{u}_n$ and $\mathbf{u}_n$ satisfy all hypotheses of Lemma 4.4.5, and hence

$$\varrho_n \mathbf{u}_n \otimes \mathbf{u}_n = \mathcal{H}_n \otimes \mathbf{u}_n \rightharpoonup \mathcal{H} \otimes \mathbf{u} = \varrho\mathbf{u} \otimes \mathbf{u}$$

weakly in $L^2(0,T; L^m(\Omega_\delta))$ for $m^{-1} > \beta^{-1} + (d-2)(2d)^{-1} = 1 - 2^{-1}\mathfrak{a}$, which yields the desired relation (5.3.19). □

### 5.3.3 Step 3. Weak limits of $\varepsilon\nabla\varrho_n \otimes \mathbf{u}_n$

**Lemma 5.3.6.** *Under the assumptions of Theorem 5.3.1, the couple $(\varrho, \mathbf{u})$ satisfies estimate (5.2.7), and the equality*

$$\lim_{n\to\infty} \varepsilon \int_Q (\nabla\varrho_n \otimes \mathbf{u}_n) : \nabla\boldsymbol{\zeta}\, dxdt = \varepsilon \int_Q (\nabla\varrho \otimes \mathbf{u}) : \nabla\boldsymbol{\zeta}\, dxdt \tag{5.3.23}$$

*holds for any vector field $\boldsymbol{\zeta} \in C^\infty(Q)$ vanishing in a neighborhood of $S_T$.*

*Proof.* First observe that by (5.2.4), we have $\varepsilon^{1/2}\|\nabla\varrho_n\|_{L^2(Q)} \le c_{E,p}$. Hence $\nabla\varrho_n \rightharpoonup \nabla\varrho$ weakly in $L^2(Q)$ and so $\varepsilon^{1/2}\|\nabla\varrho\|_{L^2(Q)} \le c_{E,p}$. The Hölder inequality implies

$$\begin{aligned}\||\nabla\varrho|\,|\mathbf{u}|\|_{L^1(0,T;L^{\gamma/(\gamma-1)}(\Omega))} &\le \|\nabla\varrho\|_{L^2(0,T;L^2(\Omega))}\|\mathbf{u}\|_{L^2(0,T;L^r(\Omega))}\\ &\le \varepsilon^{-1/2} c_{E,p}\|\mathbf{u}\|_{L^2(0,T;L^r(\Omega))}\end{aligned}$$

for $r^{-1} = (\gamma-1)\gamma^{-1} - 2^{-1}$ or equivalently $r^{-1} = 2^{-1} - \gamma^{-1}$. Since $\gamma > d$, we have $r < 2d/(d-2)$. Hence the embedding $W^{1,2}(\Omega) \hookrightarrow L^r(\Omega)$ is bounded. Thus we get

$$\||\nabla\varrho|\,|\mathbf{u}|\|_{L^1(0,T;L^{\gamma/(\gamma-1)}(\Omega))} \le \varepsilon^{-1/2} c_{E,p}\|\mathbf{u}\|_{L^2(0,T;W^{1,2}(\Omega))} \le c_{E,p}\varepsilon^{-1/2},$$

which is the desired estimate (5.2.7). It remains to prove (5.3.23). To this end notice that $\mathbf{u}_n$ and $\varrho_n$ are continuously differentiable in $x$. Hence we can integrate by parts to obtain

$$\varepsilon \int_Q \nabla\varrho_n \otimes \mathbf{u}_n : \nabla\boldsymbol{\zeta}\, dxdt = -\varepsilon \int_Q \varrho_n(\nabla\mathbf{u}_n : \nabla\boldsymbol{\zeta} + \mathbf{u}_n \cdot \Delta\boldsymbol{\zeta})\, dxdt. \tag{5.3.24}$$

Since $2 < 3\gamma/2$, it follows from Proposition 5.3.4 that $\varrho_n \to \varrho$ strongly in $L^2(Q)$. On the other hand, (5.3.4) implies $\nabla\mathbf{u}_n \rightharpoonup \nabla\mathbf{u}$ and $\mathbf{u}_n \rightharpoonup \mathbf{u}$ weakly in $L^2(Q)$. Letting $n \to \infty$ in (5.3.24) leads to

$$\lim_{n\to\infty} \varepsilon \int_Q \nabla\varrho_n \otimes \mathbf{u}_n : \nabla\boldsymbol{\zeta}\, dxdt = -\varepsilon \int_Q \varrho(\nabla\mathbf{u} : \nabla\boldsymbol{\zeta} + \mathbf{u} \cdot \Delta\boldsymbol{\zeta})\, dxdt. \tag{5.3.25}$$

Integrating by parts on the right hand side we obtain (5.3.23). □

### 5.3.4 Step 4. Proof of estimates (5.3.8–5.3.9)

Now we are in a position to establish the basic energy estimates (5.3.8)–(5.3.9). Our considerations are based on the energy estimates (5.2.2), (5.2.10) and the following auxiliary lemmas.

**Lemma 5.3.7.** *Let* $\mathcal{X}$ *be defined by the limit relation* (5.3.4). *Then*

$$\mathcal{X} = \sqrt{a}\Delta^2\mathbf{u} \quad \text{in } Q\setminus Q_\delta, \quad \mathcal{X} = 0 \quad \text{in } Q_\delta. \tag{5.3.26}$$

*Proof.* Set $\mathcal{X}_n = (a+\sigma_n)^{1/2}\Delta^2\mathbf{u}_n$. It follows from Theorem 5.2.1 that the sequence $\mathcal{X}_n$ is bounded in $L^\infty(0,T;L^2(\Omega))$. Hence, passing to a subsequence if necessary, we can assume that $\mathcal{X}_n$ converges weakly$^\star$ in $L^\infty(0,T;L^2(\Omega))$ to a function $\mathcal{X} \in L^\infty(0,T;L^2(\Omega))$. Since $\sigma_n \to 0$, we conclude that

$$(a+\sigma_n)\Delta^2\mathbf{u}_n \rightharpoonup a^{1/2}\mathcal{X} \quad \text{weakly}^\star \text{ in } L^\infty(0,T;L^2(\Omega)).$$

Next, notice that for any $\zeta \in C_0^\infty(\mathbb{R}^{d+1})$ compactly supported in $Q\setminus Q_\delta$, we have

$$\int_Q (a+\sigma_n)^{-1/2}\mathcal{X}_n\zeta\, dxdt = \int_Q \mathbf{u}_n\Delta^2\zeta\, dxdt \to \int_Q \mathbf{u}\Delta^2\zeta\, dxdt.$$

On the other hand, since $a$ is strictly positive on the support of $\zeta$,

$$\int_Q (a+\sigma_n)^{-1/2}\mathcal{X}_n\zeta\, dxdt \to \int_Q a^{-1/2}\mathcal{X}\zeta\, dxdt.$$

Thus we obtain the integral identity

$$\int_{Q\setminus Q_\delta} a^{-1/2}\mathcal{X}\zeta\, dxdt = \int_{Q\setminus Q_\delta} \mathbf{u}\Delta^2\zeta\, dxdt \quad \text{for all } \zeta \in C_0^\infty(Q\setminus Q_\delta),$$

which yields $\Delta^2\mathbf{u} = a^{-1/2}\mathcal{X}$ in $Q\setminus Q_\delta$. Next, since $\Delta^2\mathbf{u}_n$ converges to $\Delta^2\mathbf{u}$ in the sense of distributions in $Q$, the sequence $\sigma_n^{1/2}\Delta^2\mathbf{u}_n$ converges to zero in the sense of distributions in the cylinder $Q_\delta$. Hence $\mathcal{X} = 0$ in $Q_\delta$. □

**Lemma 5.3.8.** *Under the assumption of Theorem 5.3.1, for any* $0 \le T' < T'' \le T$,

$$\int_{T'}^{T''}\int_\Omega a|\Delta^2\mathbf{u}|^2\, dxdt \le \liminf_{n\to\infty}\int_{T'}^{T''}\int_\Omega (a+\sigma_n)|\Delta^2\mathbf{u}_n|^2\, dxdt, \tag{5.3.27a}$$

$$\int_{T'}^{T''}\int_\Omega P(\varrho)\, dxdt \le \liminf_{n\to\infty}\int_{T'}^{T''}\int_\Omega P(\varrho_n)\, dxdt, \tag{5.3.27b}$$

$$\int_{T'}^{T''}\int_\Omega \varrho|\mathbf{u}|^2\, dxdt \le \liminf_{n\to\infty}\int_{T'}^{T''}\int_\Omega \varrho_n|\mathbf{u}_n|^2\, dxdt, \tag{5.3.27c}$$

$$\int_{T'}^{T''}\int_\Omega |\nabla\mathbf{u}|^2\, dxdt \le \liminf_{n\to\infty}\int_{T'}^{T''}\int_\Omega |\nabla\mathbf{u}_n|^2\, dxdt, \tag{5.3.27d}$$

$$\int_{T'}^{T''}\int_\Omega P''(\varrho)|\nabla\varrho|^2\, dxdt \le \liminf_{n\to\infty}\int_{T'}^{T''}\int_\Omega P''(\varrho_n)|\nabla\varrho_n|^2\, dxdt. \tag{5.3.27e}$$

*Proof.* Since the sequence $\mathcal{X}_n = \sqrt{a+\sigma_n}\Delta^2\mathbf{u}_n$ converges to $\mathcal{X}$ in $L^2(Q)$, we have

$$\int_{T'}^{T''}\int_\Omega |\mathcal{X}|^2\,dxdt \le \liminf_{n\to\infty}\int_{T'}^{T''}\int_\Omega (a+\sigma_n)|\Delta^2\mathbf{u}_n|^2\,dxdt,$$

which along with Lemma 5.3.7 yields (5.3.27a). Next notice that the function $P$ is nonnegative and continuous. Moreover, in view of (5.3.16), $\varrho_n$ converges to $\varrho$ in $L^1(Q)$. Passing to a subsequence we can assume that $\varrho_n$ converges to $\varrho$ a.e. in $Q$. Hence the sequence of nonnegative functions $P(\varrho_n)$ converges a.e. to $P(\varrho)$, and the Levi Theorem yields (5.3.27b).

Let us prove (5.3.27c). Since $(\mathbf{u}_n, \varrho_n)$ satisfy the energy inequality (5.2.2), the sequence $\sqrt{\varrho_n}\mathbf{u}_n$ is bounded in $L^2(Q)$. On the other hand, $\mathbf{u}_n \rightharpoonup \mathbf{u}$ weakly in $L^2(Q)$, $\sqrt{\varrho_n}\to\sqrt{\varrho}$ strongly in $L^{2\gamma}(Q)$, and $2\gamma > 2$. Thus we get

$$\int_Q \sqrt{\varrho_n}\mathbf{u}_n\xi\,dxdt \to \int_Q \sqrt{\varrho}\mathbf{u}\xi\,dxdt \quad \text{as } n\to\infty \quad \text{for all } \xi\in C(Q).$$

Hence $\sqrt{\varrho_n}\mathbf{u}_n \rightharpoonup \sqrt{\varrho}\mathbf{u}$ weakly in $L^2(Q)$. Recalling that the norm in a Banach space is lower semicontinuous in the weak convergence, we arrive at (5.3.27c). Inequality (5.3.27d) obviously follows from the weak convergence of $\nabla\mathbf{u}_n$ to $\nabla\mathbf{u}$ in $L^2(Q)$. It remains to prove (5.3.27e). To this end introduce the function

$$\Phi(s) = \int_0^s \sqrt{P''(\tau)}\,d\tau.$$

It follows from the constitutive relations (5.1.3) and (5.1.6) that $\Phi$ is continuously differentiable and $\Phi(\varrho)\le c_E(1+\varrho)^{\gamma/2}$. By (5.2.8), the sequence $\varrho_n$ is bounded in $L^{3\gamma/2}(Q)$ and hence $\Phi(\varrho_n)$ is bounded in $L^{3\gamma/(\gamma+1)}(Q)$. Noting that $\varrho_n\to\varrho$ a.e. in $Q$ and $3\gamma/(\gamma+1)>2$, we conclude that $\Phi(\varrho_n)\to\Phi(\varrho)$ in $L^2(Q)$. On the other hand, inequality (5.2.2) yields the boundedness of the sequence

$$\sqrt{P''(\varrho_n)}\nabla\varrho_n = \nabla\Phi(\varrho_n)$$

in $L^2(Q)$. Moreover, for any vector field $\boldsymbol{\xi}\in C_0^\infty(Q)$, we have

$$\begin{aligned}\int_Q \sqrt{P''(\varrho_n)}\nabla\varrho_n\cdot\boldsymbol{\xi}\,dxdt &= \int_Q \nabla\Phi(\varrho_n)\cdot\boldsymbol{\xi}\,dxdt = -\int_Q \Phi(\varrho_n)\operatorname{div}\boldsymbol{\xi}\,dxdt\\ \to -\int_Q \Phi(\varrho)\operatorname{div}\boldsymbol{\xi}\,dxdt &= \int_Q \nabla\Phi(\varrho)\cdot\boldsymbol{\xi}\,dxdt = \int_Q \sqrt{P''(\varrho)}\nabla\varrho\cdot\boldsymbol{\xi}\,dxdt.\end{aligned}$$

It follows that $\sqrt{P''(\varrho_n)}\nabla\varrho_n \rightharpoonup \sqrt{P''(\varrho)}\nabla\varrho$ weakly in $L^2(Q)$, which leads to (5.3.27e). □

The next lemma ensures the validity of inequalities (5.3.8)–(5.3.9).

**Lemma 5.3.9.** *Under the assumptions of Theorem 5.3.1 the couple* $(\mathbf{u},\varrho)$ *satisfies estimates* (5.3.8)–(5.3.9).

*Proof.* We begin by proving (5.3.8). Choose $t_0 \in (0,T)$, and $h > 0$ so small that $(t_0 - h, t_0 + h) \subset (0,T)$. It follows from (5.2.2) with $(\mathbf{u}, \varrho)$ replaced by $(\mathbf{u}_n, \varrho_n)$ that

$$\frac{1}{2h}\int_{t_0-h}^{t_0+h}\int_\Omega (a|\Delta^2\mathbf{u}_n|^2 + a|\mathbf{u}|^2 + \varrho|\mathbf{u}_n|^2 + P(\varrho_n))(x,t)\,dxdt \le c_{E,p}.$$

Letting $n \to \infty$ and applying (5.3.27a)–(5.3.27c) with $T' = t_0 - h$ and $T'' = t_0 + h$ leads to

$$\frac{1}{2h}\int_{t_0-h}^{t_0+h}\int_\Omega (a|\Delta^2\mathbf{u}|^2 + a|\mathbf{u}|^2 + \varrho|\mathbf{u}|^2 + P(\varrho))(x,t)\,dxdt \le c_{E,p}.$$

Letting $h \to 0$ gives

$$\int_\Omega (a|\Delta^2\mathbf{u}|^2 + a|\mathbf{u}|^2 + \varrho|\mathbf{u}|^2 + P(\varrho))(x,t_0)\,dx \le c_{E,p} \quad \text{for a.e. } t_0 \in (0,T). \tag{5.3.28}$$

Next, (5.2.2) implies

$$\int_Q (|\nabla \mathbf{u}_n|^2 + \varepsilon P''(\varrho)|\nabla\varrho|^2)\,dxdt \le c_{E,p}.$$

Letting $n \to \infty$ and applying (5.3.27a)–(5.3.27c) with $T' = 0$ and $T'' = T$ gives

$$\int_Q (|\nabla \mathbf{u}|^2 + \varepsilon P''(\varrho)|\nabla\varrho|^2)\,dxdt \le c_{E,p}. \tag{5.3.29}$$

Combining (5.3.28) and (5.3.29) leads to the desired estimate (5.3.8).

To prove (5.3.9) integrate both sides of (5.2.10), with $(\mathbf{u}, \varrho)$ replaced by $(\mathbf{u}_n, \varrho_n)$ and $b$ replaced by $a + \sigma_n$, to obtain

$$\begin{aligned}
&\int_{t_0-h}^{t_0+h}\int_\Omega \big((a+\sigma_n)|\Delta^2\mathbf{u}_n|^2 + (a+\sigma_n)|\mathbf{u}_n|^2 + \varrho_n|\mathbf{u}_n|^2 + P(\varrho_n)\big)(x,t)\,dxdt \\
&\quad + \int_{t_0-h}^{t_0+h}\Big\{\int_0^t\int_\Omega (|\nabla\mathbf{u}_n|^2 + \varepsilon(P''(\varrho_n) - \varepsilon^{1/3})|\nabla\varrho_n|^2)(x,s)\,dxds\Big\}\,dt \\
&\quad \le \int_{t_0-h}^{t_0+h}\Big\{c_e + M_n(t) + c_e\int_0^t e^{c_e(t-s)}M_n(s)\,ds\Big\}\,dt,
\end{aligned} \tag{5.3.30}$$

where

$$\begin{aligned}
M_n(t) = {}& 2\int_\Omega (a+\sigma_n)|\Delta^2\mathbf{U}|^2(x,t)\,dx + 8\varepsilon^{2/3}\int_0^t\int_\Omega |\nabla P(\varrho_\infty)|^2\,dxdt \\
& - 4\int_0^t\int_\Omega \varrho_n\big(\partial_t P'(\varrho_\infty) + \nabla P'(\varrho_\infty)\cdot\mathbf{u}_n\big)\,dxdt.
\end{aligned} \tag{5.3.31}$$

Letting $n \to \infty$ in (5.3.30) and applying Lemma 5.3.8 with $T' = t_0 - h$ and $T'' = t_0 + h$ we arrive at

$$\int_{t_0-h}^{t_0+h} \int_\Omega (a|\Delta^2 \mathbf{u}|^2 + a|\mathbf{u}|^2 + \varrho|\mathbf{u}|^2 + P(\varrho))(x,t)\,dxdt$$
$$+ \int_{t_0-h}^{t_0+h} \Big\{ \int_0^t \int_\Omega (|\nabla \mathbf{u}|^2 + \varepsilon(P''(\varrho) - \varepsilon^{1/2})|\nabla\varrho|^2)(x,s)\,dxds \Big\}\,dt$$
$$\le \lim_{n\to\infty} \int_{t_0-h}^{t_0+h} \Big\{ c_e + M_n(t) + c_e \int_0^t e^{c_e(t-s)} M_n(s)\,ds \Big\}\,dt. \quad (5.3.32)$$

It follows from estimates (5.2.3)–(5.2.4) that the sequences $\varrho_n$ and $\varrho_n \mathbf{u}_n$ are bounded in $L^\infty(0,T;L^1(\Omega))$. Hence $M_n(t)$ is uniformly bounded on $(0,T)$. In fact the functions $M_n$ have uniformly bounded derivatives on $(0,T)$. On the other hand, Proposition 5.3.4 implies that $\varrho_n \to \varrho$ in $L^2(Q)$ and $\varrho_n \mathbf{u}_n \rightharpoonup \varrho\mathbf{u}$ weakly in $L^2(Q)$, which yields for all $t \in (0,T)$,

$$\int_0^t \int_\Omega \varrho_n\big(\partial_t P'(\varrho_\infty) + \nabla P'(\varrho_\infty)\cdot \mathbf{u}_n\big)\,dxdt \to \int_0^t \int_\Omega \varrho\big(\partial_t P'(\varrho_\infty) + \nabla P'(\varrho_\infty)\cdot \mathbf{u}\big)\,dxdt.$$

It now follows from (5.3.31) that

$$M_n(t) \to 2\int_\Omega \left(a|\Delta^2 \mathbf{U}|^2\right)(x,t)\,dx + 8\varepsilon^{2/3} \int_0^t \int_\Omega |\nabla P(\varrho_\infty)|^2\,dxdt$$
$$- 4\int_0^t \int_\Omega \varrho\big(\partial_t P'(\varrho_\infty) + \nabla P'(\varrho_\infty)\cdot \mathbf{u}\big)\,dxdt =: M(t)$$

for all $t \in (0,T)$. Recall that the functions $M_n(t)$ are uniformly bounded on $(0,T)$. Applying the Lebesgue dominated convergence theorem we arrive at

$$\lim_{n\to\infty} \int_{t_0-h}^{t_0+h} \Big\{ M_n(t) + c_e \int_0^t e^{c_e(t-s)} M_n(s)\,ds \Big\}\,dt$$
$$= \int_{t_0-h}^{t_0+h} \Big\{ M(t) + c_e \int_0^t e^{c_e(t-s)} M(s)\,ds \Big\}\,dt.$$

Substituting this into (5.3.32) we finally obtain

$$\frac{1}{2h}\int_{t_0-h}^{t_0+h} \int_\Omega (a|\Delta^2 \mathbf{u}|^2 + a|\mathbf{u}|^2 + \varrho|\mathbf{u}|^2 + P(\varrho))\,dxdt$$
$$+ \frac{1}{2h}\int_{t_0-h}^{t_0+h} \Big\{ \int_0^t \int_\Omega (|\nabla \mathbf{u}|^2 + \varepsilon(P''(\varrho) - \varepsilon^{1/3})|\nabla\varrho|^2)(x,s)\,dxds \Big\}\,dt$$
$$\le \frac{1}{2h}\int_{t_0-h}^{t_0+h} \Big\{ c_e + M(t) + c_e \int_0^t e^{c_e(t-s)} M(s)\,ds \Big\}\,dt. \quad (5.3.33)$$

Letting $h \to 0$ we arrive at the desired inequality (5.3.9). □

### 5.3.5 Step 5. Proof of Theorem 5.3.1

We are now in a position to prove Theorem 5.3.1.

**Proof of (i)–(ii).** First note that the strong convergence of $\varrho_n$ is exactly the statement of Proposition 5.3.4. Relation (5.3.5) is proved in Proposition 5.3.5. Next, representation (5.3.6) for the matrix-valued function $\mathbb{K}$ is a direct consequence of Proposition 5.3.5, and Lemma 5.3.7 gives (5.3.7). This proves (i). To prove (ii), notice that the energy estimates (5.3.8) and (5.3.9) are established in Lemma 5.3.9.

**Proof of (iii).** The energy estimate (5.3.8) gives

$$\operatorname*{ess\,sup}_{t\in(0,T)} \int_\Omega (\varrho|\mathbf{u}|^2 + P(\varrho))(x,t)\,dx + \int_0^T \int_\Omega \{|\nabla \mathbf{u}|^2 + \varepsilon P''(\varrho)|\nabla\varrho|^2\}\,dxdt \le c_{E,p}.$$

On the other hand, the constitutive relations (5.1.3)–(5.1.6) imply

$$P(\varrho) \ge c_{E,p}(\varrho^\gamma - 1), \quad P''(\varrho) \ge c_{E,p}(1+\varrho^{\gamma-2}).$$

It follows that

$$\|\mathbf{u}\|_{L^2(0,T;W^{1,2}(\Omega))} \le c_{E,p}, \quad \|\varrho|\mathbf{u}|^2\|_{L^\infty(0,T;L^1(\Omega))} \le c_{E,p}, \tag{5.3.34}$$

$$\|\varrho^\gamma\|_{L^\infty(0,T;L^1(\Omega))} + \varepsilon\|(1+\varrho)^{\gamma-2}|\nabla\varrho|^2\|_{L^1(Q)} \le c_{E,p}, \tag{5.3.35}$$

which yields (5.2.3)–(5.2.4). Moreover, by (5.3.34)–(5.3.35), $(\mathbf{u},\varrho)$ are bounded energy functions and meet all requirements of Proposition 4.2.1 with $E$ replaced by $c_{E,p}$. Hence, Corollary 4.2.2 implies $\|\varrho\mathbf{u}\|_{L^\infty(0,T;L^{2\gamma/(\gamma+1)}(\Omega))} \le c_{E,p}$, which immediately yields (5.2.5). In its turn, applying Corollary 4.2.3 we get the estimate $\|\varrho|\mathbf{u}|^2\|_{L^2(0,T;L^z(\Omega_\delta))} \le c_{E,p}$, which gives (5.2.6). Since $\gamma > 2d$, it follows from (5.3.34) and (5.3.35) that $(\mathbf{u},\varrho)$ are bounded dissipation rate functions and satisfy all hypotheses of Proposition 4.3.1. Estimate (4.3.3) in that proposition yields $\|\varrho^{3\gamma/2}\|_{L^1(Q)} \le c_{E,p}\,\varepsilon^{-1}$, which is exactly the required estimate (5.2.8). It remains to note that estimate (5.2.7) is given by Lemma 5.3.6.

**Proof of (iv).** Now choose $\boldsymbol{\zeta} \in C^\infty(Q)$ which vanishes at the top of $Q$ and outside of $Q_\delta$. Multiplying by $\boldsymbol{\zeta}$ the regularized momentum equation (5.1.20a) with $b$, $\mathbf{u}$, $\varrho$ replaced by $a+\sigma_n$, $\mathbf{u}_n$, and $\varrho_n$, respectively, and integrating the result by parts we arrive at

$$\int_Q (\sigma_n\Delta^2(\mathbf{u}_n - \mathbf{U})\cdot\partial_t\Delta^2\boldsymbol{\zeta} + \sigma_n(\mathbf{u}_n - \mathbf{U})\cdot\partial_t\boldsymbol{\zeta} + \varrho\mathbf{u}\cdot\partial_t\boldsymbol{\zeta})\,dxdt$$

$$+ \int_Q (\varrho_n\mathbf{u}_n\otimes\mathbf{u}_n - \varepsilon\nabla\varrho_n\otimes\mathbf{u}_n + p(\varrho_n)\mathbb{I} - \mathbb{S}(\mathbf{u}_n)) : \nabla\boldsymbol{\zeta}\,dxdt$$

$$+ \int_Q \varrho_n\mathbf{f}\cdot\boldsymbol{\zeta}\,dxdt + \int_\Omega (\varrho_\infty\mathbf{U})(x,0)\,dx = 0.$$

Letting $n \to \infty$ and using (5.3.4) and Propositions 5.3.4, 5.3.5 and Lemma 5.3.6 we obtain the integral identity (5.3.10). The proof of Theorem 5.3.1 is complete.

### 5.3.6 Proof of Theorem 5.3.2

Since the coefficient $a(x)$ equals 1 in the annulus $\Omega \setminus \Omega_{\delta/2}$, estimate (5.3.8) implies

$$\operatorname*{ess\,sup}_{t \in (0,T)} \int_{\Omega \setminus \Omega_{\delta/2}} \left(|\Delta^2 \mathbf{u}(x,t)|^2 + |\mathbf{u}(x,t)|^2\right) dx \leq c_{E,p}.$$

Since the boundary of $\Omega \setminus \Omega_{\delta/2}$ is a surface of class $C^3$ depending on $\delta$ only, we conclude that

$$\|\mathbf{u}\|_{L^\infty(0,T,W^{4,2}(\Omega \setminus \Omega_{\delta/2}))} \leq c(\delta) c_{E,p}.$$

Next, since the embedding $W^{4,2}(\Omega \setminus \Omega_{\delta/2}) \hookrightarrow C^{2+\alpha}(\Omega \setminus \Omega_{\delta/2})$ is bounded for $0 < \alpha < 1/2$ with a bound depending on $\delta$, it follows that

$$\|\mathbf{u}\|_{L^\infty(0,T;C^{2+\alpha}(\Omega \setminus \Omega_{\delta/2}))} \leq c(\delta) c_{E,p} \quad \text{for all } \alpha \in (0, 1/2), \tag{5.3.36}$$

which leads to (5.3.12). Let us derive an estimate for $\varrho$. Choose $\eta \in C^\infty(\Omega)$ with

$$\eta = 1 \quad \text{in } \Omega \setminus \Omega_{\delta/3}, \quad \eta = 0 \quad \text{outside of } \Omega \setminus \Omega_{\delta/2}.$$

Next, set $\rho_n = \eta \varrho_n$. Recall that $\varrho_n$ satisfies the diffusion equation (5.2.12c) with $(\mathbf{u}, \varrho)$ replaced by $(\mathbf{u}_n, \varrho_n)$. It follows that

$$\partial_t \rho_n - \varepsilon \Delta \rho_n = f_n \quad \text{in } Q, \quad \rho_n = \eta \varrho_\infty \quad \text{on } \sqcup_T, \tag{5.3.37}$$

where

$$f_n = -\eta \mathbf{u}_n \cdot \nabla \varrho_n - \eta \operatorname{div} \mathbf{u}_n \varrho_n - \varepsilon \Delta \eta \varrho_n - 2\varepsilon \nabla \eta \nabla \varrho_n.$$

Notice that by (5.3.36) we have

$$|f_n| \leq c(\eta, \delta) c_{e,p}(|\varrho_n| + |\nabla \varrho_n| + 1).$$

Since $\eta$ depends only on $\delta$, it now follows from the energy estimate (5.2.4) that $\|f_n\|_{L^2(Q)} \leq c(\varepsilon, \delta) c_{E,p}$. Thus we can apply Theorem 1.7.13 to the initial-value problem (5.3.37) to obtain the estimate

$$\|\partial_t \rho_n\|_{L^2(0,T;L^2(\Omega))} + \|\rho_n\|_{L^2(0,T;W^{2,2}(\Omega))} + \|\rho_n\|_{C(0,T;W^{1,2}(\Omega))} \leq c(\varepsilon, \delta) c_{E,p}.$$

Recall the notation $A^\delta$ for the annulus $\Omega \setminus \Omega_{\delta/3}$. Since $\varrho_n = \rho_n$ in $A^\delta \times (0,T)$, we obtain

$$\|\partial_t \varrho_n\|_{L^2(0,T;L^2(A^\delta))} + \|\varrho_n\|_{L^2(0,T;W^{2,2}(A^\delta))} + \|\varrho_n\|_{C(0,T;W^{1,2}(A^\delta))} \leq c(\varepsilon, \delta) c_{E,p}. \tag{5.3.38}$$

As $\varrho_n \rightharpoonup \varrho$ weakly in $L^2(0,T;L^2(A^\delta))$ we can assume by passing to a subsequence that $\varrho_n \rightharpoonup \varrho$ weakly in $L^2(0,T;W^{2,2}(A^\delta))$ and weakly$^\star$ in $L^\infty(0,T;W^{1,2}(A^\delta))$.

Moreover we can assume that $\partial_t \varrho_n \rightharpoonup \partial_t \varrho$ weakly in $L^2(0,T;L^2(A^\delta))$. Since the norm is lower semicontinuous in the weak convergence, the limit $\varrho$ also satisfies estimate (5.3.38), which leads to the desired estimate (5.3.13).

As the embedding $W^{1,2}(\Omega \setminus \Omega_{\delta/3}) \hookrightarrow L^2(\partial\Omega)$ is continuous (see [2, Thm. 5.22]) we have $\|\nabla\varrho\|_{L^2(\partial\Omega)} \le c(\delta)\|\varrho\|_{W^{2,2}(A^\delta)}$. From this and estimates (5.3.38), (5.3.13) we obtain

$$\|\nabla\varrho_n\|_{L^2(S_T)} \le c(\delta)\|\varrho_n\|_{L^2(0,T;W^{2,2}(A^\delta))} \le c(\varepsilon,\delta)c_{E,p}, \tag{5.3.39}$$

$$\|\nabla\varrho\|_{L^2(S_T)} \le c(\delta,\varepsilon)\|\varrho\|_{L^2(0,T;W^{2,2}(A^\delta))} \le c(\varepsilon,\delta)c_{E,p}. \tag{5.3.40}$$

Moreover, the weak convergence $\varrho_n \rightharpoonup \varrho$ in $L^2(0,T;W^{2,2}(A^\delta))$ yields

$$\nabla\varrho_n \rightharpoonup \nabla\varrho \quad \text{weakly in } L^2(S_T). \tag{5.3.41}$$

Now, fix $\psi \in C^\infty(Q)$ vanishing in a neighborhood of $\Sigma_{\text{in}}$ and in a neighborhood of $\Omega \times \{t = T\}$. Multiplying (5.1.20b) by $\psi$ and integrating the result by parts in $Q$ we arrive at the integral identity

$$\int_Q (\varrho_n\partial_t\psi - \varepsilon\nabla\varrho_n \cdot \nabla\psi + \varrho_n\mathbf{u}_n \cdot \nabla\psi)\, dxdt$$
$$+ \int_{S_T} \psi(\varepsilon\nabla\varrho_n - \varrho_\infty\mathbf{U}) \cdot \mathbf{n}\, dSdt + \int_{\partial\Omega} (\psi\varrho_\infty)(x,0)\, dx = 0. \tag{5.3.42}$$

Recall that by Theorem 5.3.1, $\varrho_n \to \varrho$ strongly in $L^2(Q)$, $\nabla\varrho_n \to \nabla\varrho$ weakly in $L^2(Q)$, and $\mathbf{u}_n \rightharpoonup \mathbf{u}$ weakly in $L^2(0,T;W^{1,2}(\Omega))$. It follows that

$$\int_Q (\varrho_n\partial_t\psi - \varepsilon\nabla\varrho_n \cdot \nabla\psi + \varrho_n\mathbf{u}_n \cdot \nabla\psi)\, dxdt$$
$$\to \int_Q (\varrho\partial_t\psi - \varepsilon\nabla\varrho \cdot \nabla\psi + \varrho\mathbf{u} \cdot \nabla\psi)\, dxdt.$$

On the other hand, relation (5.3.41) yields

$$\int_{S_T} \psi\nabla\varrho_n \cdot \mathbf{n}\, dSdt \to \int_{S_T} \psi\nabla\varrho \cdot \mathbf{n}\, dSdt. \tag{5.3.43}$$

Letting $n \to \infty$ in (5.3.42) we obtain the desired integral identity (5.3.14). It remains to prove (5.3.15). To this end fix a convex function $\varphi$ satisfying

$$\varphi \in C^2[0,\infty), \quad |\varphi(\varrho)| + |\varphi'(\varrho)\varrho| + |\varphi''(\varrho)\varrho^2| \le C\varrho^2, \quad \varphi''(\varrho) \ge 0. \tag{5.3.44}$$

Next, choose a smooth nonnegative function $\psi$ vanishing in a neighborhood of $\Sigma_{\text{in}}$ and in a neighborhood of $\Omega \times \{t = T\}$. Multiplying (5.1.20b) by $\varphi'(\varrho_n)\psi$ and integrating the result by parts over $Q$ we obtain

$$\int_Q \varphi(\varrho_n)(\partial_t\psi + \nabla\psi\cdot\mathbf{u}_n)\,dxdt - \varepsilon\int_Q \varphi'(\varrho_n)\nabla\varrho_n\nabla\psi\,dxdt$$
$$-\int_Q \big(\varphi'(\varrho_n)\varrho_n - \varphi(\varrho_n)\big)\operatorname{div}\mathbf{u}_n\psi\,dxdt + \int_\Omega (\varphi(\varrho_\infty)\psi)(x,0)\,dx$$
$$-\int_{S_T}\varphi(\varrho_\infty)\mathbf{U}\cdot\mathbf{n}\,dSdt + \varepsilon\int_{S_T}\varphi'(\varrho_\infty)\nabla\varrho_n\cdot\mathbf{n}\,\psi\,dSdt = \varepsilon\int_Q \psi\varphi''(\varrho_n)|\nabla\varrho_n|^2\,dxdt \geq 0. \tag{5.3.45}$$

It follows from the growth conditions (5.3.44) and estimate (5.2.8) that the sequences $\varphi(\varrho_n)$ and $\varphi'(\varrho_n)\varrho_n$ are bounded in $L^{3\gamma/4}(Q)$. Moreover, we have already shown that $\varrho_n \to \varrho$ a.e. in $Q$. Since $3\gamma/4 > 2$ these sequences converge strongly in $L^2(Q)$ to $\varphi(\varrho)$ and $\varphi'(\varrho)\varrho$ respectively. On the other hand, $\operatorname{div}\mathbf{u}_n$ converges weakly in $L^2(Q)$ to $\operatorname{div}\mathbf{u}$. Hence we can pass to the limit to obtain

$$\int_Q \varphi(\varrho_n)(\partial_t\psi + \nabla\psi\cdot\mathbf{u}_n)\,dxdt - \varepsilon\int_Q \varphi'(\varrho_n)\nabla\varrho_n\nabla\psi\,dxdt$$
$$-\int_Q \big(\varphi'(\varrho_n)\varrho_n - \varphi(\varrho_n)\big)\operatorname{div}\mathbf{u}_n\psi\,dxdt$$
$$\to \int_Q \varphi(\varrho)(\partial_t\psi + \nabla\psi\cdot\mathbf{u})\,dxdt - \varepsilon\int_Q \varphi'(\varrho)\nabla\varrho\nabla\psi\,dxdt$$
$$-\int_Q \big(\varphi'(\varrho_n)\varrho - \varphi(\varrho)\big)\operatorname{div}\mathbf{u}\psi\,dxdt. \tag{5.3.46}$$

Letting $n \to \infty$ in (5.3.45) and applying (5.3.43), (5.3.46) we obtain inequality (5.3.15). The proof of Theorem 5.3.2 is complete.

### 5.3.7 Local pressure estimate

Theorem 5.3.1 guarantees the existence of the limits $(\mathbf{u}, \varrho)$ of a sequence of solutions $(\mathbf{u}_n\varrho_n)$ to the regularized equations (5.1.20) with $b = a + \sigma_n$, where $\sigma_n \to 0$ and $a$ is concentrated near the lateral boundary $S_T$ of the cylinder $Q$. Recall that Theorem 5.3.1 provides the estimate $\|p(\varrho)\|_{L^\infty(0,T;L^1(\Omega))} \leq c_{E,p}$, where a constant $c_{E,p}$ as in Remark 5.1.2 depends only on the diameter of the flow domain $\Omega$, length $T$ of the time interval, and the given data and elementary properties of the constitutive law (5.1.4), (5.1.10). Estimate (5.2.8) for $\varrho$ established in Theorem 5.3.1 gives an estimate for the pressure function $p(\varrho)$ in $L^{3/2}(Q)$, but this estimate depends on the small parameter $\varepsilon$ and it is useless. Here, we show that the pressure function satisfies $L^{(1+\gamma)/\gamma}$ estimates on compact subsets of the cylinder $Q$. The corresponding result is given in the following proposition.

**Proposition 5.3.10.** *Under the assumptions of Theorem 5.3.1 there exists $\delta_0 > 0$ independent of $\varepsilon$ such that for for all $\delta \in (0, \delta_0]$ and all compact subsets $Q'$ of the cylinder $Q_\delta$,*

$$\int_{Q'} \varrho^{\gamma+1}\, dxdt + \int_{Q'} p(\varrho)^{(\gamma+1)/\gamma}\, dxdt \le c_{E,p}(Q'), \tag{5.3.47}$$

*where the constant $c_{E,p}(Q')$ depends only on the constant $c_{E,p}$ and $Q'$, $\delta_0$. In particular, $c_{E,p}(Q')$ is independent of $\delta \le \delta_0$.*

*Proof.* The proof is based on the basic integral identity (4.5.10) of Theorem 4.5.2. First, by the integral identities (5.3.10) and (5.3.14), the couple $(\mathbf{u}, \varrho)$ satisfies the mass and momentum balance equations

$$\begin{aligned} &\partial_t \varrho + \operatorname{div}(\varrho\mathbf{u} - \mathbf{g}) = 0 \quad \text{in } Q_\delta, \\ &\partial_t(\varrho\mathbf{u}) + \operatorname{div}(\varrho\mathbf{u}\otimes\mathbf{u}) = \operatorname{div}(\mathbb{T} + \mathbb{G}) + \rho\mathbf{f} \quad \text{in } Q_\delta, \end{aligned}$$

which are understood in the sense of distributions. Here, the matrix-valued functions $\mathbb{T}$, $\mathbb{G}$ and the vector-valued function $\mathbf{g}$ are defined by

$$\begin{aligned} &\mathbb{G} = \varepsilon\nabla\varrho\otimes\mathbf{u}, \quad \mathbb{T} = \mathbb{S}(\mathbf{u}) - p(\varrho)\,\mathbb{I}, \quad \mathbf{g} = \varepsilon\nabla\varrho, \\ &\mathbb{S}(\mathbf{u}) = \nabla\mathbf{u} + \nabla\mathbf{u}^\top + (\lambda - 1)\operatorname{div}\mathbf{u}\,\mathbb{I}. \end{aligned} \tag{5.3.48}$$

Next, set

$$r = 3, \quad s = \gamma, \quad q = s/(s-1), \quad \varphi = \varrho, \quad \mathbf{g}^\varphi = \mathbf{g} = \varepsilon\nabla\varrho, \quad \varpi = 0. \tag{5.3.49}$$

Let us show that the functions $\varrho$, $\varphi$, the vector field $\mathbf{u}$, and the matrix-valued functions $\mathbb{T}, \mathbb{G}$ meet all requirements of Theorem 4.5.2 in the cylinder $Q_\delta \subset Q$. To this end we have to check that $\varrho$, $\varphi$, $\mathbf{u}$, and $\mathbb{T}$ satisfy conditions (4.5.4)–(4.5.7), and then to prove that $\mathbf{g}$ and $\mathbb{G}$ belong to $L^2(Q)$ and $L^1(0,T;L^q(\Omega_T))$, respectively.

By Theorem 5.3.1, the functions $\varrho$, $\mathbf{u}$, and $\varphi = \varrho$ satisfy estimates (5.2.3) and (5.2.4). It is easily seen that these imply (4.5.4)–(4.5.6) with $E$ replaced by $c_{E,p}$. It follows from (5.2.3) and (5.2.4) that

$$\|\mathbb{T}\|_{L^1(Q)} + \|\mathbf{g}\|_{L^2(Q)} \le c_{E,p}. \tag{5.3.50}$$

It also follows from Theorem 5.3.1 that $\varrho$ and $\mathbf{u}$ satisfy inequalities (5.2.7)–(5.2.8), which leads to the estimate

$$\varepsilon\|\varrho^{3\gamma/2}\|_{L^1(Q)} + \varepsilon^{1/2}\||\nabla\varrho||\mathbf{u}|\|_{L^1(0,T;L^{d/(d-1)}(\Omega))} \le c_{E,p}.$$

Since $q < d/(d-1)$ we deduce that $\mathbb{G} \in L^1(0,T;L^q(\Omega))$. As $p(\varrho) \le c_{E,p}(\varrho^\gamma + 1)$, we find from (5.2.3) and (5.2.8) that

$$\|\mathbb{T}\|_{L^{3/2}(Q)} \le \|\nabla\mathbf{u}\|_{L^{3/2}(Q)} + \|p\|_{L^{3/2}(Q)} \le c_{E,p}/\varepsilon.$$

Since $r' = 3/2$, it follows that the matrix-valued function $\mathbb{T} \in L^r(Q)$ satisfies (4.5.7). Hence the functions $\varrho, \varphi, \varpi$, the vector fields $\mathbf{u}, \mathbf{g}, \mathbf{g}^\varphi$, and the tensors $\mathbb{T}$, $\mathbb{G}$ satisfy Condition 4.5.1 of Theorem 4.5.2. Finally, the exponents $\gamma$ and $s$ obviously satisfy condition (4.5.8) of that theorem. Thus

$$\int_{Q_\delta} \xi\varrho\mathbf{R} : [\zeta(\mathbb{T}+\mathbb{G})]\, dxdt = \int_{Q_\delta} \big(\mathbf{H}\cdot\mathbf{u}+\zeta\varrho(\mathbf{N}+\mathbf{P}+\mathbf{Q})\cdot\mathbf{u}+S\big)\, dxdt \tag{5.3.51}$$

for all $\xi \in C_0^\infty(Q_\delta)$ and $\zeta \in C_0^\infty(\Omega_\delta)$. Here,

$$\begin{aligned}
&\mathbf{H} = \mathbf{R}[\xi\varrho]\,(\zeta\varrho\mathbf{u}) - (\xi\varrho)\,\mathbf{R}[\zeta\varrho\mathbf{u}],\\
&S = (\nabla\zeta\otimes\mathbf{A}[\xi\varrho]) : (\varrho\mathbf{u}\otimes\mathbf{u} - \mathbb{T} - \mathbb{G}) + \zeta\mathbf{A}[\xi\varrho]\cdot(\varrho\mathbf{f}),\\
&\mathbf{N} = \mathbf{R}[\xi\,\mathbf{g}], \quad \mathbf{P} = \mathbf{A}[\varrho\partial_t\xi + \varrho\mathbf{u}\cdot\nabla\xi], \quad \mathbf{Q} = -\mathbf{A}[\mathbf{g}\cdot\nabla\xi].
\end{aligned}$$

Using the relation $\mathbf{R} : [\zeta p\mathbb{I}] = \zeta p$, we can rewrite equality (5.3.51) in the form

$$\begin{aligned}
\int_{Q_\delta} \xi\zeta p(\varrho)\varrho\, dxdt = &\int_{Q_\delta} \xi\varrho\mathbf{R} : [\zeta(\mathbb{S}(\mathbf{u}) + \mathbb{G})]\, dxdt\\
&- \int_{Q_\delta} \big(\mathbf{H}\cdot\mathbf{u} + \zeta\varrho(\mathbf{N}+\mathbf{P}+\mathbf{Q})\cdot\mathbf{u} + S\big)\, dxdt.
\end{aligned} \tag{5.3.52}$$

Now our task is to estimate all terms on the right hand side of (5.3.52). Let us estimate $\mathbf{N}$, $\mathbf{P}$, and $\mathbf{Q}$. By (5.2.3) and (5.2.4), $(\mathbf{u}, \varrho)$ are finite energy functions, and the energy is bounded by a constant $c_{E,p}$. Hence, $(\mathbf{u}, \varrho)$ satisfies all hypotheses of Proposition 4.2.1 with $E$ replaced by $c_{E,p}$. We apply Corollary 4.2.2 of this proposition to obtain

$$\|\varrho\mathbf{u}\|_{L^2(0,T;L^\kappa(\Omega))} \le c_{E,p} \quad \text{for all } \kappa^{-1} > 2^{-1} + \gamma^{-1} - d^{-1}. \tag{5.3.53}$$

Since $\gamma > 2d$, we can take $\kappa = 2$ to obtain

$$\|\varrho\mathbf{u}\|_{L^2(Q)} \le c_{E,p}. \tag{5.3.54}$$

Notice that $\xi, \zeta \in C_0^\infty(\mathbb{R}^d)$ are compactly supported in $\Omega_\delta$. It follows that the vector field $\zeta\mathbf{A}[\xi\varrho\mathbf{u}]$ is compactly supported in $Q_\delta$ and satisfies

$$\|\zeta\mathbf{A}[\xi\varrho\mathbf{u}]\|_{L^2(0,T;W^{1,2}(\Omega))} \le c(\xi)c_{E,p}. \tag{5.3.55}$$

Next, since $\varrho$ is in $L^\infty(0,T;L^\gamma(\Omega))$ with $\gamma > 2d > 2$, we have

$$\|\mathbf{A}[\partial_t\xi\varrho]\|_{L^\infty(0,T;W^{1,2}(\Omega))} \le c(\xi)c_{E,p}. \tag{5.3.56}$$

Combining (5.3.55) and (5.3.56) we obtain

$$\|\mathbf{P}\|_{L^2(0,T;W^{1,2}(\Omega))} \le c(\xi)c_{E,p}. \tag{5.3.57}$$

Since the vector field $\mathbf{g} = \varepsilon\nabla\varrho \in L^2(0,T;L^2(\Omega))$ satisfies inequality (5.2.4), the same arguments as before lead to

$$\|\mathbf{A}[\xi\mathbf{g}]\|_{L^2(0,T;W^{1,2}(\Omega))} + \|\mathbf{R}[\xi\mathbf{g}]\|_{L^2(0,T;L^2(\Omega))} \le c(\xi)c_{E,p}. \tag{5.3.58}$$

This gives

$$\|\mathbf{Q}\|_{L^2(0,T;W^{1,2}(\Omega))} \le c(\xi)c_{E,p}, \quad \|\mathbf{N}\|_{L^2(0,T;L^2(\Omega))} \le c(\xi)c_{E,p}. \tag{5.3.59}$$

Combining estimates (5.3.57)–(5.3.59) we get

$$\|\mathbf{N}+\mathbf{P}+\mathbf{Q}\|_{L^2(Q)} \le c(\xi,\zeta)c_{E,p},$$

which along with (5.3.54) leads to

$$\|\zeta\varrho(\mathbf{N}+\mathbf{P}+\mathbf{Q})\cdot\mathbf{u}_\varepsilon\|_{L^1(Q)} \le c(\xi,\zeta)c_{E,p}. \tag{5.3.60}$$

Recall that for $\gamma > d$ the embedding $W^{1,\gamma}(\Omega) \hookrightarrow L^\infty(\Omega)$ is compact. Thus we get

$$\begin{aligned}\|\mathbf{A}[\xi\varrho]\|_{L^\infty(Q)} &\le c\|\mathbf{A}[\xi\varrho]\|_{L^\infty(0,T;W^{1,\gamma}(\Omega))} \le c\|\xi\varrho\|_{L^\infty(0,T;L^\gamma(\Omega))}\\ &\le c(\xi,\zeta)c_{E,p}.\end{aligned}$$

Moreover, by (5.2.7) we have

$$\|\mathbb{G}\|_{L^1(Q)} \le \|\mathbb{G}\|_{L^1(0,T;L^{\gamma/(\gamma-1)}(\Omega))} \le \varepsilon c_{E,p}. \tag{5.3.61}$$

This along with (5.3.50) leads to

$$\begin{aligned}\|S\|_{L^1(Q)} &\le c(\xi,\zeta)c_{E,p}\big(\|\varrho|\mathbf{u}|^2\|_{L^1(Q)} + \|\mathbb{T}+\mathbb{G}\|_{L^1(Q)} + \|\varrho\mathbf{f}\|_{L^1(Q)}\big)\\ &\le c(\xi,\zeta)c_{E,p}.\end{aligned} \tag{5.3.62}$$

Let us estimate the vector field $\mathbf{H}$. Notice that the operator $\mathbf{R}$ is bounded in $L^\kappa(\mathbb{R}^d)$ for all $\kappa \in (1,\infty)$. From this and the Hölder inequality we obtain

$$\begin{aligned}\|\mathbf{H}\|_{L^2(Q)} &\le \|\mathbf{R}[\xi\varrho]\|_{L^\infty(0,T;L^\gamma(\Omega))}\|\zeta\varrho\mathbf{u}\|_{L^2(0,T;L^\sigma(\Omega))}\\ &\quad + \|\xi\varrho\|_{L^\infty(0,T;L^\gamma(\Omega))}\|\mathbf{R}[\zeta\varrho\mathbf{u}]\,\|_{L^2(0,T;L^\sigma(\Omega))}\\ &\le c\|\xi\varrho\|_{L^\infty(0,T;L^\gamma(\Omega))}\|\zeta\varrho\mathbf{u}\|_{L^2(0,T;L^\sigma(\Omega))}\\ &\le c_{E,p}\|\zeta\varrho\mathbf{u}\|_{L^2(0,T;L^\sigma(\Omega))}, \quad \text{where} \quad \sigma^{-1} = 2^{-1} - \gamma^{-1}.\end{aligned}$$

It is easy to check that with $\gamma > 2d$, we have

$$\sigma^{-1} = 2^{-1} - \gamma^{-1} > 2^{-1} + \gamma^{-1} - d^{-1} = \kappa,$$

which along with (5.3.53) leads to

$$\|\zeta\varrho\mathbf{u}\|_{L^2(0,T;L^\sigma(\Omega))} \le c_{E,p}c(\zeta).$$

We thus get $\|\mathbf{H}\|_{L^2(Q)} \le c(\xi,\zeta)c_{E,p}$. This inequality along with the estimate $\|\mathbf{u}\|_{L^2(0,T;W^{1,2}(\Omega))} \le c_{E,p}$ implies

$$\|\mathbf{H}\cdot\mathbf{u}\|_{L^1(Q)} \le c(\xi,\zeta)c_{E,p}. \tag{5.3.63}$$

Next, the boundedness of $\mathbf{R}$ in Lebesgue spaces yields

$$\begin{aligned}
&\|\varrho\mathbf{R} : [\zeta(\mathbb{S}(\mathbf{u}) + \mathbb{G})]\|_{L^1(Q)} \\
&\quad \le c(\zeta)\|\varrho\|_{L^\infty(0,T;L^\gamma(\Omega))}\big(\|\mathbf{R}[\zeta\mathbb{S}]\|_{L^1(0,T;L^{\gamma/(\gamma-1)}(\Omega))} + \|\mathbf{R}[\zeta\mathbb{G}]\|_{L^1(0,T;L^{\gamma/(\gamma-1)}(\Omega))}\big) \\
&\qquad \le c(\zeta)c_{E,p}\big(\|\mathbb{S}\|_{L^1(0,T;L^{\gamma/(\gamma-1)}(\Omega))} + \|\mathbb{G}\|_{L^1(0,T;L^{\gamma/(\gamma-1)}(\Omega))}\big).
\end{aligned}$$

Since $\gamma/(\gamma-1) < 2$ we have

$$\|\mathbb{S}(\mathbf{u})\|_{L^1(0,T;L^{\gamma/(\gamma-1)}(\Omega))} \le c\|\mathbf{u}\|_{L^2(0,T;W^{1,2}(\Omega))} \le c_{E,p}.$$

Combining this result with (5.3.61) we get

$$\|\varrho\mathbf{R} : [\zeta(\mathbb{S}(\mathbf{u}) + \mathbb{G})]\|_{L^1(Q)} \le c(\zeta)c_{E,p}. \tag{5.3.64}$$

Finally, combining estimates (5.3.60), (5.3.62), (5.3.63), and (5.3.64) with identity (5.3.52) we arrive at

$$\int_{Q_\delta} \xi\zeta p(\varrho)\varrho\, dxdt \le c(\xi,\zeta)c_{E,p}.$$

It remains to note that for any compact $Q' \Subset Q_\delta$ there are nonnegative functions $\zeta \in C_0^\infty(\Omega)$ and $\xi \in C_0^\infty(Q)$ such that $\zeta\xi = 1$ on $Q'$. □

### 5.3.8 Normal derivative of the density

In this section we derive an estimate for the normal derivative of the density on the boundary. The result is given by the following proposition.

**Proposition 5.3.11.** *Under the assumptions of Theorem 5.3.1, for any compact set $\Sigma' \Subset \Sigma_{\text{in}} \subset S_T$, there is a constant $C$ independent of $\varepsilon$ such that with $\varkappa = 1/2 - d/(2\gamma)$,*

$$\varepsilon\left|\frac{\partial\varrho}{\partial n}\right| \le C\varepsilon^{\varkappa}(\|\varrho\|_{L^\gamma(Q)} + \|\varrho_\infty\|_{C^2(Q)}) \quad \text{on } \Sigma'. \tag{5.3.65}$$

Before the proof of Proposition 5.3.11, we provide a general result on the normal derivative of the solution to the diffusion equation with vanishing viscosity. This result is of independent interest, and it is necessary to justify our regularization procedure.

**Normal derivative of a singularly perturbed transport equation.** Let $(z, t^*) \in \Sigma_{\rm in}$ and let $c_0$, $\iota$, $l$ be positive constants. We introduce the following condition on the velocity field $\mathbf{u}$:

**Condition 5.3.12.** • Denote by $\mathcal{Q}_\iota$ the cylinder

$$\mathcal{Q}_\iota := B(z, \iota) \times (t^* - \iota, t^* + \iota), \tag{5.3.66}$$

where $B(z, \iota) \subset \mathbb{R}^d$ is the ball of radius $\iota$ centered at $z$. The velocity field $\mathbf{u}$ admits the representation

$$\begin{aligned} \mathbf{u} = \nabla H + \mathbf{v} \quad &\text{in } Q \cap \mathcal{Q}_\iota, \\ H = 0, \quad \mathbf{v} \cdot \mathbf{n} = 0 \quad &\text{on } S_T \cap \mathcal{Q}_\iota, \end{aligned} \tag{5.3.67}$$

where the function $H$ and the vector field $\mathbf{v}$ satisfy

$$\begin{aligned} &\|H\|_{C^2(Q \cap \mathcal{Q}_\iota)} \leq c_0, \\ &\operatorname*{ess\,sup}_{t \in (t^* - \iota, t^* + \iota)} \|\mathbf{v}(\cdot, t)\|_{C^2(B(z,\iota) \cap \Omega)} \leq c_0. \end{aligned} \tag{5.3.68}$$

• The "potential" $H$ satisfies

$$\begin{aligned} \frac{1}{2}\partial_t H(x,t) + \frac{1}{4}|\nabla H(x,t)|^2 > l > 0 \quad &\text{in } \mathcal{Q}_\iota \cap Q, \\ -\nabla H(x,t) \cdot \mathbf{n}(x) > l > 0 \quad &\text{on } \mathcal{Q}_\iota \cap S_T. \end{aligned} \tag{5.3.69}$$

Consider the following boundary value problem in the cylinder $\mathcal{Q}_\iota \cap Q$:

$$\partial_t \varrho + \operatorname{div}(\varrho \mathbf{u}) - \varepsilon \Delta \varrho = 0 \quad \text{in } \mathcal{Q}_\iota \cap Q, \tag{5.3.70a}$$

$$\varrho = \varrho_\infty \quad \text{on } S_T \cap \mathcal{Q}_\iota. \tag{5.3.70b}$$

The following theorem whose proof is given in Section 13.4 plays a key role in our further considerations.

**Theorem 5.3.13.** *Let $\Omega \subset \mathbb{R}^d$ be a bounded domain with $C^3$ boundary and suppose there are $(z, t^*) \in S_T$ and $\iota > 0$ such that $\mathbf{u}$ satisfies Condition 5.3.12 in a cylinder $Q_\iota$. Let $\varrho \in L^\gamma(\mathcal{Q}_\iota \cap Q)$, $\gamma > d + 2$, be a strong solution to problem (5.3.70) with $\partial_t \varrho, \partial_x^2 \varrho \in L^2(\mathcal{Q}_\iota)$. Then there are constants $C$ and $a_0 < \iota$, depending only on $d$, $\gamma$, $\Omega$, $\iota$, and on the constants $c_0$, $l$ in Condition 5.3.12, such that*

$$\varepsilon \left| \frac{\partial \varrho}{\partial n} \right| \leq C \varepsilon^{\varkappa} (\|\varrho\|_{L^\gamma(\mathcal{Q}_\iota \cap Q)} + \|\varrho_\infty\|_{C^2(\mathcal{Q}_\iota \cap Q)}) \quad \text{on } S_T \cap \mathcal{Q}_{a_0},$$

*where $\varkappa = 2^{-1} - d(2\gamma)^{-1}$, $\mathcal{Q}_{a_0}$ is defined by (5.3.66) for $\iota := a_0$.*

**Proof of Proposition 5.3.11.** In order to apply Theorem 5.3.13 we must check that $\varrho, \mathbf{u}$ satisfy its hypotheses. The proof of this fact is based on the following lemma which gives the construction of $H$.

**Lemma 5.3.14.** *Let $\Sigma' \Subset \Sigma_{\text{in}}$ be a compact subset of $\Sigma_{\text{in}}$ in the topology of $S_T = \partial\Omega \times (0,T)$ and $(z,t^*) \in \Sigma'$. Furthermore, assume that $\varrho$ and $\mathbf{u}$ satisfy all conditions of Theorem 5.3.1. Then there are positive constants $\iota$, $l$, $c_0$, depending only on $\Sigma'$, $\Omega$, $\delta$ and $c_{E,p}$, such that Condition 5.3.12 is fulfilled for all $(z,t^*) \in \Sigma'$. We stress that the constants $\iota$, $l$, and $c_0$ do not depend on the choice of $(z,t^*) \in \Sigma'$.*

*Proof.* Choose $\iota < \delta/3$ so small that the closure of the $\iota$-neighborhood of $\Sigma'$ in the topology of $S_T$ is contained in $\Sigma_{\text{in}}$. This means that there is a compact set $\Sigma'' \subset \Sigma_{\text{in}}$ such that $\mathcal{Q}_\iota \cap S_T \Subset \Sigma''$ for all $(z,t^*) \in \Sigma'$. For such a choice there is $l \in (0,1)$ depending only on $\iota$ and $\partial\Omega$, $\mathbf{U}$ such that

$$-\mathbf{U}\cdot\mathbf{n} > 4l^{1/2} \quad \text{on } \mathcal{Q}_\iota \subset S_T \quad \text{for all } (z,t^*) \in \Sigma'.$$

More precisely, the constant $l > 0$ depends on the second order derivatives of the normal coordinates, i.e., on the curvatures of $\partial\Omega$ and their derivatives.

Next choose $\iota > 0$ depending on $\partial\Omega$ such that for any $z \in \partial\Omega$ the intersection $B(z,\iota) \cap \Omega$ is a normal neighborhood of $z$ (see Section 13.2 for definitions and details), i.e. every point $x \in B(z,\iota)$ has a representation

$$x = \omega(x) + \tau(x)\mathbf{n}(\omega(x)), \quad \omega, \tau \in C^2(B(z,\iota)), \quad \omega(x) \in \partial\Omega.$$

Notice that $\tau(x) = \operatorname{dist}(x,\partial\Omega)$ and $\nabla\tau = \mathbf{n}$ on $\partial\Omega$. It follows from Theorem 5.3.2 that $\operatorname{ess\,sup}_{(0,T)} \|\mathbf{u}\|_{C^{2+\alpha}(\Omega\setminus\Omega_{\delta/3})} \le c(\delta)c_{E,p}$. By the choice of $\iota$ we have $\mathcal{Q}_\iota \cap Q \subset (\Omega \setminus \Omega_{\delta/3}) \times (0,T)$, which leads to

$$\operatorname*{ess\,sup}_{(t^*-\iota,t^*+\iota)} \|\mathbf{u}\|_{C^{2+\alpha}(B(z,\iota)\cap\Omega)} \le c(\varepsilon,\delta)c_{E,p}. \tag{5.3.71}$$

Set $H(x,t) = -\mathbf{U}(x,t)\cdot\mathbf{n}(\omega(x))\tau(x)$. Since $\tau\mathbf{n}(\omega) \in C^2\big(B(z,\iota)\big)$ for all $z \in \partial\Omega$, the function $H$ belongs to $C^2(B(z,\iota)\times(0,T))$. Hence $H \in C^2(\mathcal{Q}_\iota)$ for all cylinders $\mathcal{Q}_\iota$ centered at points $(z,t^*) \in \Sigma'$. Moreover, the norm $\|H\|_{C^2(\mathcal{Q}_\iota)}$ is bounded by a constant depending only on $\iota$, $\mathbf{U}$ and the curvature of $\partial\Omega$. It is clear that

$$|\partial_t H(x,t)| \le c_{E,p}\operatorname{dist}(x,\partial\Omega) \le c_{E,p}\iota \quad \text{in } \mathcal{Q}_\iota \quad \text{for all } (z,t^*) \in \Sigma',$$

and

$$-\nabla H\cdot\mathbf{n} = -\mathbf{U}\cdot\mathbf{n} > 4l^{1/2} > l > 0 \quad \text{on } S_T \cap \mathcal{Q}_\iota \quad \text{for all } (z,t^*) \in \Sigma'.$$

Choosing $\iota$, depending on $\mathbf{U}$, $\partial\Omega$, and $l$, sufficiently small we deduce that

$$|\nabla H| > 3l^{1/2} > 0 \quad \text{on } Q \cap \mathcal{Q}_\iota \quad \text{for all } (z,t^*) \in \Sigma'.$$

In particular, we have

$$\frac{1}{2}\,\partial_t H(x,t) + \frac{1}{4}|\nabla H(x,t)|^2 \ge 2l - c_{E,p}\iota.$$

Choosing $\iota < c_{E,p}^{-1} l$ we see that $H$ satisfies Condition 5.3.12 at each $(z,t^*) \in \Sigma'$. □

*Proof of Proposition 5.3.11.* In view of Lemma 5.3.14 there are positive constants $\iota$, $c_0$, $l$ and a function $H$, depending only on $\Sigma'$, $\Omega$, $\delta$, such that $\mathbf{u}$ and $H$ satisfy Condition 5.3.12 for all $(z,t^*) \in \Sigma'$. Then $\mathcal{Q}_\iota \cap Q \subset (\Omega \setminus \Omega_{\delta/3}) \times (0,T)$. By Theorem 5.3.2, $\varrho$ is a strong solution to problem (5.3.70) in the cylindrical annulus $(\Omega \setminus \Omega_{\delta/3}) \times (0,T)$ and hence in $\mathcal{Q}_\iota \cap Q$. Hence $\varrho$ and $\mathbf{u}$ satisfy Condition 5.3.12 with constants $\iota, l, c_0$ depending only on $\delta$, $c_{E,p}$, $\Omega$ and $\Sigma'$. Thus the couple $(\mathbf{u}, \varrho)$ meets all requirements of Theorem 5.3.13, which completes the proof. □

## 5.4 Passage to the limit. The second level

Theorem 5.3.1 guarantees the existence of the limits $(\mathbf{u}\varrho,)$ of a sequence of solutions $(\mathbf{u}_n\varrho_n,)$ to the regularized equations (5.1.20) with $b = a + \sigma_n$, where $\sigma_n \to 0$ and $a$ is concentrated in the $\delta$-neighborhood of the lateral boundary $S_T = \partial\Omega \times (0,T)$. These limits depend of course on a small parameter $\varepsilon$. It is convenient to manifest this dependence explicitly and give the following definition:

**Definition 5.4.1.** If $(\varrho, \mathbf{u})$ are weak limits of solutions to the regularized equations defined by Theorem 5.3.1, $\mathbb{K}$ is a weak limit of the kinetic energy tensors $\varrho_n \mathbf{u}_n \otimes \mathbf{u}_n$, and $M(t)$ is the function in the energy estimate (5.3.9), then

$$\varrho_\varepsilon := \varrho, \quad \mathbf{u}_\varepsilon := \mathbf{u}, \quad \mathbb{K}_\varepsilon := \mathbb{K}, \quad M_\varepsilon := M.$$

Now for fixed $a$ and $\delta$ our task is to pass to the limit $\varepsilon \to 0$ in the governing equations and boundary conditions. We begin with the observation that by Theorem 5.3.1, the functions $\varrho_\varepsilon$ and $\mathbf{u}_\varepsilon$ satisfy the estimates, independent of $\varepsilon$,

$$\|\mathbf{u}_\varepsilon\|_{L^2(0,T;W^{1,2}(\Omega))} \le c_{E,p}, \quad \|\varrho_\varepsilon|\mathbf{u}_\varepsilon|^2\|_{L^\infty(0,T;L^1(\Omega))} \le c_{E,p}, \tag{5.4.1}$$

$$\|\varrho_\varepsilon^\gamma\|_{L^\infty(0,T;L^1(\Omega))} + \varepsilon\|(1+\varrho_\varepsilon)^{\gamma-2}|\nabla\varrho_\varepsilon|^2\|_{L^1(Q)} \le c_{E,p}, \tag{5.4.2}$$

$$\|\varrho_\varepsilon\mathbf{u}_\varepsilon\|_{L^\infty(0,T;L^\beta(\Omega))} \le c_{E,p}, \tag{5.4.3}$$

$$\|\varrho_\varepsilon|\mathbf{u}_\varepsilon|^2\|_{L^2(0,T;L^z(\Omega_\delta))} \le c_{E,p}, \tag{5.4.4}$$

$$\||\nabla\varrho_\varepsilon|\mathbf{u}_\varepsilon\|_{L^1(0,T;L^s(\Omega))} \le \varepsilon^{-1/2}c_{E,p}, \tag{5.4.5}$$

$$\varepsilon\|\varrho_\varepsilon^{3\gamma/2}\|_{L^1(Q)} \le c_{E,p}, \tag{5.4.6}$$

where $\beta = 2\gamma/(\gamma+1)$, $s = d/(d-1)$, and $z \in (1, (1-2^{-1}\mathfrak{a})^{-1})$, $\mathfrak{a} = 2d^{-1} - \gamma^{-1}$. Recall that the constant $c_{E,p}$ is independent of $\delta$, $a$ and $\varepsilon$. It also follows from Proposition 5.3.5 that

$$\|\mathbb{K}_\varepsilon\|_{L^2(0,T;L^z(\Omega))} + \|\mathbb{K}_\varepsilon\|_{L^\infty(0,T;L^1(\Omega))} \le c_{E,p}. \tag{5.4.7}$$

The representation $\mathbb{K}_\varepsilon = \varrho_\varepsilon\mathbf{u}_\varepsilon \otimes \mathbf{u}_\varepsilon$ holds in the subcylinder $Q_\delta$. Next, by Proposition 5.3.10, for all $Q' \Subset Q_\delta$ and $\delta \in (0,\delta_0)$, with $\delta_0$ independent of $\varepsilon$ and $a$,

$$\|\varrho_\varepsilon\|_{L^{\gamma+1}(Q')} \le c(Q')_{E,p} \tag{5.4.8}$$

where the constant $c(Q')_{E,p}$ is independent of $\varepsilon$ and $\delta \in (0, \delta_0)$. Moreover, for almost every $t \in (0,T)$,

$$\begin{aligned}\int_\Omega \{a|\Delta^2\mathbf{u}_\varepsilon|^2 + a|\mathbf{u}_\varepsilon|^2 + \varrho_\varepsilon|\mathbf{u}_\varepsilon|^2 + P(\varrho_\varepsilon)\}(x,t)\,dx \\ + \int_0^t\int_\Omega \{|\nabla\mathbf{u}_\varepsilon|^2 + \varepsilon(P''(\varrho_\varepsilon) - \varepsilon^{1/3})|\nabla\varrho_\varepsilon|^2\}\,dxdt \\ \le c_e + M_\varepsilon(t) + c_e\int_0^t e^{c_e(t-s)}M_\varepsilon(s)\,ds, \quad (5.4.9)\end{aligned}$$

where

$$\begin{aligned}M_\varepsilon(t) = 2\int_\Omega (a|\Delta^2\mathbf{U}|^2)\,dxdt + 8\varepsilon^{2/3}\int_0^t\int_\Omega |\nabla P(\varrho_\infty)|^2\,dxdt \\ - 4\int_0^t\int_\Omega \varrho_\varepsilon\big(\partial_t P'(\varrho_\infty) + \nabla P'(\varrho_\infty)\cdot\mathbf{u}_\varepsilon\big)\,dxdt.\end{aligned} \quad (5.4.10)$$

Recall that the constants $c_e$ and $c_{E,p}$ are as in Remark 5.1.2. Hence, passing to a subsequence if necessary, we can assume that the sequences $\varrho_\varepsilon, \mathbf{u}_\varepsilon$ tend to $\varrho_\delta$, $\mathbf{u}_\delta$ as $\varepsilon \to 0$, and

$$\begin{aligned}\varrho_\varepsilon &\rightharpoonup \varrho_\delta \quad \text{weakly}^\star \text{ in } L^\infty(0,T;L^\gamma(\Omega)),\\ \mathbf{u}_\varepsilon &\rightharpoonup \mathbf{u}_\delta \quad \text{weakly in } L^2(0,T;W^{1,2}(\Omega)),\\ \varrho_\varepsilon &\rightharpoonup \varrho_\delta \quad \text{weakly in } L^{\gamma+1}(Q'), \quad \text{for every compact set } Q' \Subset Q_\delta.\end{aligned} \quad (5.4.11)$$

The natural question arises on convergence of the sequences of kinetic energy tensors, momenta, and pressures as $\varepsilon \to 0$. The weak convergence of the momenta and the kinetic energy tensors easily follows from Theorem 4.4.2:

**Lemma 5.4.2.** *Let $\varrho_\delta$ and $\mathbf{u}_\delta$ be defined by* (5.4.11)*. Then there exist subsequences, still denoted by $\varrho_\varepsilon$, $\mathbf{u}_\varepsilon$, such that for $\varepsilon \to 0$,*

$$\begin{aligned}\varrho_\varepsilon\mathbf{u}_\varepsilon &\rightharpoonup \varrho_\delta\mathbf{u}_\delta \quad \text{weakly}^\star \text{ in } L^\infty(0,T;L^{2\gamma/(\gamma+1)}(\Omega)),\\ \varrho_\varepsilon\mathbf{u}_\varepsilon\otimes\mathbf{u}_\varepsilon &\rightharpoonup \varrho_\delta\mathbf{u}_\delta\otimes\mathbf{u}_\delta \quad \text{weakly in } L^2(0,T;L^z(\Omega)),\end{aligned} \quad (5.4.12)$$

*for every $z \in (1, (1-2^{-1}\mathfrak{a})^{-1})$, where $\mathfrak{a}$ is given by* (4.2.2).

*Proof.* By Theorem 5.3.1 the functions $\varrho_\varepsilon$ and $\mathbf{u}_\varepsilon$ satisfy the mass and momentum balance equations

$$\begin{aligned}\partial_t\varrho_\varepsilon &= \operatorname{div}\mathfrak{u}_\varepsilon \quad \text{in } Q,\\ \partial_t(\varrho_\varepsilon\mathbf{u}_\varepsilon) &= \operatorname{div}\mathfrak{V}_\varepsilon + \mathfrak{w}_\varepsilon \quad \text{in } Q_\delta,\end{aligned}$$

where

$$\begin{aligned}\mathfrak{u}_\varepsilon &= -\varrho_\varepsilon\mathbf{u}_\varepsilon + \varepsilon\nabla\varrho_\varepsilon, \quad \mathfrak{w}_\varepsilon = \varrho_\varepsilon\mathbf{f},\\ \mathfrak{V}_\varepsilon &= \mathbb{S}(\mathbf{u}_\varepsilon) - p(\varrho_\varepsilon)\,\mathbb{I} - \varrho_\varepsilon\mathbf{u}_\varepsilon\otimes\mathbf{u}_\varepsilon + \varepsilon\nabla\varrho_\varepsilon\otimes\mathbf{u}_\varepsilon.\end{aligned}$$

It follows from estimates (5.4.1) and (5.4.2) that the sequences $\mathbf{u}_\varepsilon$ and $\varrho_\varepsilon$ are bounded in $L^2(0,T;W^{1,2}(\Omega))$ and $L^\infty(0,T;L^\gamma(\Omega))$ respectively. On the other hand, in view of estimates (5.4.2) and (5.4.3), the sequence $\mathfrak{u}_\varepsilon$ is bounded in $L^1(Q)$. Hence the functions $\varrho_\varepsilon$ and the vector fields $\mathfrak{u}_\varepsilon$ satisfy Condition 4.4.1 of Theorem 4.4.2 with $n$ replaced by $\varepsilon$. Therefore

$$\varrho_\varepsilon \mathbf{u}_\varepsilon \rightharpoonup \varrho_\delta \mathbf{u}_\delta \quad \text{weakly in } L^2(0,T;L^m(\Omega)) \quad \text{for } m^{-1} > 2^{-1} + \gamma^{-1} - d^{-1}.$$

Since the sequence $\varrho_\varepsilon \mathbf{u}_\varepsilon$ is bounded in $L^\infty(0,T;L^{2\gamma/(\gamma+1)}(\Omega))$, it also converges to $\varrho_\delta \mathbf{u}_\delta$ weakly$^\star$ in $L^\infty(0,T;L^{2\gamma/(\gamma+1)}(\Omega))$. This leads to the first relation in (5.4.12). To prove the second, notice that by (5.4.1), (5.4.2), and (5.4.5) the sequence $\mathfrak{V}_\varepsilon$ is bounded in $L^1(Q)$. On the other hand, since $\mathbf{f}$ is bounded, it follows from (5.4.2) that the sequence $\mathfrak{w}_\varepsilon$ is bounded in $L^1(Q_\delta)$. Therefore the sequences $\varrho_\varepsilon$, $\mathbf{u}_\varepsilon$, $\mathfrak{V}_\varepsilon$, and $\mathfrak{w}_\varepsilon$ satisfy all hypotheses of Theorem 4.4.2 with $n$ replaced by $\varepsilon$ and $Q$ replaced by $Q_\delta$. This yields the second relation in (5.4.12). □

However, in contrast to the first level limit, we are unable to prove the strong convergence of $\varrho_\varepsilon$ and $p_\varepsilon$ at this stage. Instead we show the weak convergence of the pressure and obtain the representation of the weak limits of $p(\varrho_\varepsilon)$, and, more generally, composite functions $\varphi(\varrho_\varepsilon)$, in terms of the Young measure associated with the sequence $\varrho_\varepsilon$ (see Section 1.4 for definitions and basic statements of Young measure theory). In addition, we prove that the limit function $\varrho_\delta$ serves as a renormalized solution to the mass transport equation.

To obtain the representation for the weak limits of $\varrho_\varepsilon$ we recall that, in view of (5.4.2) and (5.4.8), the sequence $\varrho_\varepsilon$ is bounded in $L^\gamma(Q)$ and in $L^{\gamma+1}(Q')$ for every subcylinder $Q' \subset Q$. Hence we can apply the fundamental theorem on Young measures (Theorem 1.4.5) to obtain

**Theorem 5.4.3.** *There is a subsequence, still denoted by $\varrho_\varepsilon$, and a Young measure $\mu^\delta \in L^\infty_w(Q;\mathcal{M}(\mathbb{R}))$ such that for any $\varphi \in C_0(\mathbb{R})$:*

- *As $\varepsilon \to 0$,*

$$\varphi(\varrho_\varepsilon) \rightharpoonup \overline{\varphi} \quad \textit{weakly}^\star \textit{ in } L^\infty(Q), \quad \textit{where} \quad \overline{\varphi}(x) = \langle \mu^\delta_{xt}, \varphi \rangle.$$

- *For a.e. $(x,t) \in Q$ the measures $\mu^\delta_{xt}$ are supported on $[0,\infty)$.*
- *For any Carathéodory function $f : Q \times \mathbb{R} \to \mathbb{R}$ bounded from below,*

$$\liminf_{\varepsilon \to 0} \int_Q f(x,t,\varrho_\varepsilon(x,t))\,dx \geq \int_Q \overline{f}(x,t)\,dx, \quad \textit{where } \overline{f}(x,t) = \langle \mu_{xt}, f(x,t,\cdot)\rangle.$$

*Moreover, if in addition*

$$|f(x,t,\lambda)|\lambda^{-\gamma} \to 0 \quad \textit{as } \lambda \to \infty, \quad \textit{uniformly in } (x,t) \in Q,$$

*then*

$$\lim_{\varepsilon \to 0} \int_Q f(x,t,\varrho_\varepsilon(x,t))\,dx = \int_Q \overline{f}(x,t)\,dx.$$

*If $Q' \Subset Q$ and*

$$|f(x,t,\lambda)|\lambda^{-\gamma-1} \to 0 \quad \text{as } \lambda \to \infty, \quad \text{uniformly in } (x,t) \in Q',$$

*then*

$$\lim_{\varepsilon \to 0} \int_{Q'} f(x,t,\varrho_\varepsilon(x,t))\,dx = \int_{Q'} \overline{f}(x,t)\,dx.$$

Recall that a function $f : Q \times \mathbb{R} \to \mathbb{R}$ is said to be a Carathéodory function if for a.e. $(x,t) \in Q$ the function $f(x,t,\cdot)$ is continuous and for every $y \in \mathbb{R}$ the function $f(\cdot,\cdot,y)$ is measurable. It is worth noting that the weak limits $\varrho_\delta$ of the sequence $\varrho_\varepsilon$ have the representation

$$\varrho_\delta(x,t) = \langle \mu^\delta_{xt}, \lambda \rangle := \int_{[0,\infty)} \lambda\, d\mu^\delta_{xt}(\lambda). \tag{5.4.13}$$

Since $p(\lambda)$, $P(\lambda)$ are continuous functions with asymptotic behavior $p(\lambda), P(\lambda) \sim \lambda^\gamma$ at infinity, the application of Theorem 5.4.3 to the pressure function $p(\varrho_\varepsilon)$ and the internal energy function $P(\varrho_\varepsilon)$ leads to the following result:

**Lemma 5.4.4.** *The functions $\overline{p}_\delta, \overline{P}_\delta : Q \to \mathbb{R}^+$ defined by*

$$\overline{p}_\delta(x,t) = \langle \mu^\delta_{xt}, p(\lambda) \rangle, \quad \overline{P}_\delta(x,t) = \langle \mu^\delta_{xt}, P(\lambda) \rangle$$

*belong to $L^\infty(0,T;L^1(\Omega))$. For any measurable function $\psi : Q \to \mathbb{R}^+$, we have*

$$\begin{aligned} \int_Q \psi \overline{p}_\delta(x,t)\,dxdt &\le \liminf_{\epsilon \to 0} \int_Q \psi p(\varrho_\epsilon)\,dxdt, \\ \int_Q \psi \overline{P}_\delta(x,t)\,dxdt &\le \liminf_{\epsilon \to 0} \int_Q \psi P(\varrho_\epsilon)\,dxdt. \end{aligned} \tag{5.4.14}$$

*For every $Q' \Subset Q$, the function $\overline{p\varrho}_\delta : Q \to \mathbb{R}^+$ given by*

$$\overline{p\varrho}_\delta(x,t) = \langle \mu^\delta_{xt}, p(\lambda)\lambda \rangle$$

*belongs to $L^1(Q')$ and for any continuous function $\psi : Q \to \mathbb{R}^+$,*

$$\int_{Q'} \psi \overline{p\varrho}_\delta(x,t)\,dxdt \le \liminf_{\epsilon \to 0} \int_{Q'} \psi p(\varrho_\epsilon)\varrho_\epsilon\,dxdt. \tag{5.4.15}$$

*For all $\psi \in L^\infty(Q')$,*

$$\begin{aligned} \int_{Q'} \psi \overline{p}_\delta(x,t)\,dxdt &= \lim_{\epsilon \to 0} \int_{Q'} \psi p(\varrho_\epsilon)\,dxdt, \\ \int_{Q'} \psi \overline{P}_\delta(x,t)\,dxdt &= \lim_{\epsilon \to 0} \int_{Q'} \psi P(\varrho_\epsilon)\,dxdt. \end{aligned} \tag{5.4.16}$$

*Moreover,*

$$\overline{p\varrho}_\delta(x,t) - \overline{p}_\delta(x,t)\varrho_\delta(x,t) \ge 0 \quad \text{a.e. in } Q. \tag{5.4.17}$$

*Equality holds if and only if $\mu^\delta_{xt}$ is the Dirac measure concentrated at $\varrho_\delta(x,t)$, i.e., $\langle \mu^\delta_{xt}, \varphi \rangle = \varphi(\varrho_\delta(x,t))$ for every $\varphi \in C_0(\mathbb{R})$.*

*Proof.* The lemma is a straightforward consequence of Theorem 5.4.3, applied to the Carathéodory functions $\psi(x,t)p(\lambda)$ and $\psi(x,t)P(\lambda)$. To prove (5.4.17) notice that

$$\begin{aligned}
&\langle \mu_{xt}^{\delta}, (p(\lambda) - p(\varrho_\delta(x,t)))(\lambda - \varrho_\delta(x,t))\rangle \\
&\quad = \langle \mu_{xt}^{\delta}, p(\lambda)\lambda\rangle - \langle \mu_{xt}, p(\lambda)\rangle \varrho_\delta(x,t) \\
&\qquad - p(\varrho_\delta(x,t))\langle \mu_{xt}^{\delta}, \lambda\rangle + p(\varrho_\delta(x,t))\varrho_\delta(x,t)\langle \mu_{xt}^{\delta}, 1\rangle \\
&\quad = \langle \mu_{xt}^{\delta}, p(\lambda)\lambda\rangle - \langle \mu_{xt}^{\delta}, p(\lambda)\rangle \varrho_\delta(x,t) - p(\varrho_\delta(x,t))\varrho_\delta(x,t) + p(\varrho_\delta(x,t))\varrho_\delta(x,t) \\
&\quad = \langle \mu_{xt}^{\delta}, p(\lambda)\lambda\rangle - \langle \mu_{xt}^{\delta}, p(\lambda)\rangle \varrho_\delta(x,t) = \overline{p\varrho}_\delta(x,t) - \overline{p}_\delta(x,t)\varrho_\delta(x,t).
\end{aligned}$$

Since $p$ is strictly monotone, the left hand side of this identity is positive and vanishes if and only if $\mu_{xt}^{\delta}$ is the Dirac measure concentrated at $\varrho_\delta(x,t)$. □

The following Theorem 5.4.5 is the main result of this section. It gives estimates, independent of $\delta$, for the weak limits $(\varrho_\delta, \mathbf{u}_\delta)$. It also ensures that $\varrho_\delta$ and $\mathbf{u}_\delta$ serve as a renormalized solution to the mass balance equation. Next, the theorem shows that in the subcylinder $Q_{2\delta}$ the functions $(\varrho_\delta, \mathbf{u}_\delta)$ satisfy the modified moment balance equation with the pressure function $p(\varrho)$ replaced with the function $\overline{p}_\delta$ defined in Lemma 5.4.4. We do not claim the equality $p(\varrho_\delta) = \overline{p}_\delta$ at this stage. However, we make a step forward towards proving this equality by showing that the difference of $\overline{p\varrho}_\delta$ and $\overline{p}_\delta\varrho$ vanishes as $\delta \to 0$.

**Theorem 5.4.5.** *Under the above assumptions, the couple $(\varrho_\delta, \mathbf{u}_\delta)$ and the Young measure $\mu^\delta$ have the following properties:*

*(i) The functions $(\varrho_\delta, \mathbf{u}_\delta)$ satisfy the estimates*

$$\|\mathbf{u}_\delta\|_{L^2(0,T;W^{1,2}(\Omega))} \le c_{E,p}, \quad \|\varrho_\delta|\mathbf{u}_\delta|^2\|_{L^\infty(0,T;L^1(\Omega_\delta))} \le c_{E,p}, \tag{5.4.18}$$

$$\|\varrho_\delta^\gamma\|_{L^\infty(0,T;L^1(\Omega))} + \|\overline{p}_\delta\|_{L^\infty(0,T;L^1(\Omega))} \le c_{E,p}, \tag{5.4.19}$$

$$\|\varrho_\delta \mathbf{u}_\delta\|_{L^\infty(0,T;L^\beta(\Omega))} \le c_{E,p}, \tag{5.4.20}$$

$$\|\varrho_\delta|\mathbf{u}_\delta|^2\|_{L^2(0,T;L^z(\Omega_\delta))} \le c_{E,p}, \tag{5.4.21}$$

*where $\beta = 2\gamma/(\gamma+1)$, $s = d/(d-1)$, and $z$ is an arbitrary number in $(1, (1-2^{-1}\mathfrak{a})^{-1})$. Recall that the constant $c_{E,p}$ is as in Remark 5.1.2 and it is independent of $\delta$.*

*For any compact set $Q' \Subset Q_{\delta_0}$ and for all $\delta \in (0, \delta_0]$, where $\delta_0$ is defined in Proposition 5.1.18, there is a constant $c(Q')$ independent of $\delta$ such that*

$$\|\varrho_\delta\|_{L^{\gamma+1}(Q')} + \|\overline{p\varrho}_\delta\|_{L^1(Q')} \le c(Q'). \tag{5.4.22}$$

*(ii) The integral identity*

$$\int_Q \varrho_\delta \mathbf{u}_\delta \cdot \partial_t \boldsymbol{\zeta}\, dxdt + \int_Q \big(\varrho_\delta \mathbf{u}_\delta \otimes \mathbf{u}_\delta + \overline{p}_\delta\, \mathbb{I} - \mathbb{S}(\mathbf{u}_\delta)\big) : \nabla \boldsymbol{\zeta}\, dxdt$$
$$+ \int_Q \varrho_\delta \mathbf{f} \cdot \boldsymbol{\zeta}\, dxdt + \int_\Omega (\varrho_\infty \mathbf{U} \cdot \boldsymbol{\zeta})(x,0)\, dx = 0 \quad (5.4.23)$$

*holds for any vector field* $\boldsymbol{\zeta} \in C^\infty(Q)$ *such that*

$$\boldsymbol{\zeta}(x,t) = 0 \quad \textit{if}\ x \in \Omega \setminus \Omega_{2\delta}\ \textit{or}\ t = T.$$

*(iii) The integral identity*

$$\int_Q \big(\varphi(\varrho_\delta)\partial_t \psi + \varphi(\varrho_\delta)\mathbf{u}_\delta \cdot \nabla \psi - (\varphi'(\varrho_\delta)\varrho_\delta - \varphi(\varrho_\delta))\,\mathrm{div}\,\mathbf{u}_\delta\big)\, dxdt$$
$$- \int_{\Sigma_{\mathrm{in}}} \psi \varphi(\varrho_\infty)\mathbf{U} \cdot \mathbf{n}\, d\Sigma + \int_\Omega (\psi\varphi(\varrho_\infty))(x,0)\, dx = 0 \quad (5.4.24)$$

*holds for all smooth functions* $\psi$ *vanishing in a neighborhood of* $\Omega \times \{t = T\}$ *and in a neighborhood of* $S_T \setminus \Sigma_{\mathrm{in}}$*, and for all* $\varphi \in C^2[0,\infty)$ *satisfying the growth condition*

$$|\varphi(\varrho)| + |\varphi'(\varrho)\varrho| + |\varphi''(\varrho)\varrho^2| \le C\varrho^2. \quad (5.4.25)$$

*(iv) For any compact set* $Q' \Subset Q_\delta$*,*

$$\int_{Q'} (\overline{p\varrho}_\delta - \overline{p}_\delta \varrho_\delta)\, dxdt \le c_{E,p}\, \mathrm{meas}(Q \setminus Q')^{(\gamma-2)/(2\gamma)}. \quad (5.4.26)$$

*(v) For almost every* $t \in (0,T)$*,*

$$\int_\Omega \{\varrho_\delta |\mathbf{u}_\delta|^2 + \overline{P}_\delta\}(x,t)\, dx + \int_0^t \int_\Omega |\nabla \mathbf{u}_\delta|^2\, dxdt$$
$$\le c_e + M_\delta(t) + c_e \int_0^t e^{c_e(t-s)} M_\delta(s)\, ds, \quad (5.4.27)$$

*where*

$$M_\delta(t) = 2\int_\Omega (a|\Delta^2 \mathbf{U}|^2)(x,t)\, dx. \quad (5.4.28)$$

The remainder of this section is devoted to the proof of Theorem 5.4.5. By abuse of notation the subscript $\delta$ is omitted, and we write simply $\varrho$, $\mathbf{u}$ and $\mu$ instead of $\varrho_\delta$, $\mathbf{u}_\delta$ and $\mu^\delta$. We split the proof of each claim (i)–(v) into a sequence of lemmas.

**Proof of claim (i).** In view of (5.4.11) the functions $\varrho_\varepsilon$ converge to $\varrho_\delta$ weakly$^\star$ in $L^\infty(0,T;L^\gamma(\Omega))$ and weakly in $L^{\gamma+1}(Q')$ for every $Q' \Subset Q$. Next, the functions $\mathbf{u}_\varepsilon$ converge weakly in $L^2(0,T;W^{1,2}(\Omega))$. Finally, by Lemma 5.4.2 the momenta $\varrho_\varepsilon \mathbf{u}_\varepsilon$ converge weakly$^\star$ in $L^\infty(0,T;L^\beta(\Omega))$ and the kinetic energy tensors $\varrho_\varepsilon \mathbf{u}_\varepsilon \otimes \mathbf{u}_\varepsilon$ converge to $\varrho_\delta \mathbf{u}_\delta \otimes \mathbf{u}_\delta$ weakly in $L^2(0,T;L^z(\Omega_\delta))$. Therefore, (5.4.18)–(5.4.22) follow from (5.4.1)–(5.4.8) and from the lower semicontinuity of the norm in the weak topology. Next, (5.4.22) is a straightforward consequence of (5.4.8) and (5.4.15). It remains to prove (5.4.19) for $\overline{p} := \overline{p}_\delta$. To this end choose $t_0 \in (0,T)$ and $h > 0$. Consider the nonnegative Carathéodory function

$$f(x,t,\lambda) = p(\lambda) \quad \text{for } |t - t_0| < h, \quad \text{and} \quad f(x,t) = 0 \quad \text{otherwise.}$$

From Lemma 5.4.4, (5.4.14), (5.4.2) and the inequality $p(\varrho) \le c_e(1+\varrho^\gamma)$ we obtain

$$\frac{1}{2h}\int_{t_0-h}^{t_0+h}\int_\Omega \overline{p}_\delta \, dxdt \le \lim_{\varepsilon\to 0} \frac{1}{2h}\int_{t_0-h}^{t_0+h}\int_\Omega p(\varrho_\varepsilon)\, dxdt \le c_{E,p} \quad \text{for all } h > 0,$$

which obviously implies (5.4.19).

**Proof of claim (ii).** Our next task is to prove (5.4.23). Choose a vector field $\boldsymbol{\zeta}$ that vanishes in $Q \setminus Q_{2\delta}$ and in a neighborhood of the top $\Omega \times \{t = T\}$. Since $a(x) = 0$ in $Q_\delta$, the integral identity (5.3.10) with $\varrho$ and $\mathbf{u}$ replaced by $\varrho_\varepsilon$ and $\mathbf{u}_\varepsilon$ implies

$$\int_Q \Big( \varrho_\varepsilon \mathbf{u}_\varepsilon \cdot \partial_t \boldsymbol{\zeta} + ((\varrho_\varepsilon \mathbf{u}_\varepsilon - \varepsilon \nabla \varrho_\varepsilon) \otimes \mathbf{u}_\varepsilon + p(\varrho_\varepsilon)\mathbb{I} - \mathbb{S}(\mathbf{u}_\varepsilon)) : \nabla \boldsymbol{\zeta} \Big)\, dxdt$$
$$+ \int_Q \varrho_\varepsilon \mathbf{f} \cdot \boldsymbol{\zeta}\, dxdt + \int_\Omega (\varrho_\infty \mathbf{U} \cdot \boldsymbol{\zeta})(x,0)\, dx = 0. \quad (5.4.29)$$

It follows from (5.4.12) that

$$\lim_{\varepsilon\to 0} \int_Q \{\varrho_\varepsilon \mathbf{u}_\varepsilon \cdot \partial_t \boldsymbol{\zeta} + (\varrho_\varepsilon \mathbf{u}_\varepsilon \otimes \mathbf{u}_\varepsilon - \mathbb{S}(\mathbf{u}_\varepsilon)) : \nabla \boldsymbol{\zeta} + \varrho_\varepsilon \mathbf{f} \cdot \boldsymbol{\zeta}\}\, dxdt$$
$$= \int_Q \{\varrho_\delta \mathbf{u}_\delta \cdot \partial_t \boldsymbol{\zeta} + (\varrho_\delta \mathbf{u}_\delta \otimes \mathbf{u}_\delta - \mathbb{S}(\mathbf{u}_\delta)) : \nabla \boldsymbol{\zeta} + \varrho_\delta \mathbf{f} \cdot \boldsymbol{\zeta}\}\, dxdt. \quad (5.4.30)$$

Next, estimate (5.4.5) yields

$$\lim_{\varepsilon\to 0} \int_Q \varepsilon \nabla \varrho_\varepsilon \otimes \mathbf{u}_\varepsilon : \nabla \boldsymbol{\zeta}\, dxdt = 0. \quad (5.4.31)$$

Now choose $h > 0$ and notice that $\Omega_{2\delta} \times (h, T-h) \Subset Q$. Then (5.4.16) yields

$$\lim_{\varepsilon\to 0} \int_h^{T-h} \int_{\Omega_{2\delta}} (p(\varrho_\varepsilon) - \overline{p}_\delta)\, \mathrm{div}\, \boldsymbol{\zeta}\, dxdt = 0.$$

From this, (5.4.1), the inequality $c(1+\varrho_\varepsilon^\gamma) \geq p(\varrho_\varepsilon)$, and estimate (5.4.19) we obtain

$$\begin{aligned}\limsup_{\varepsilon\to 0}\Big|\int_{Q_{2\delta}} (p(\varrho_\varepsilon)-\overline{p}_\delta)\operatorname{div}\boldsymbol{\zeta}\,dxdt\Big| &\leq \limsup_{\varepsilon\to 0}\Big|\int_h^{T-h}\int_{\Omega_{2\delta}} (p(\varrho_\varepsilon)-\overline{p}_\delta)\operatorname{div}\boldsymbol{\zeta}\,dxdt\Big|\\ &+ c(\zeta)\limsup_{\varepsilon\to 0}\Big\{\int_0^h + \int_{T-h}^T\Big\}\int_{\Omega_{2\delta}} (p(\varrho_\varepsilon)+\overline{p}_\delta))\,dxdt\\ &\leq c(\zeta)c_{E,p}\Big\{\int_0^h+\int_{T-h}^T\Big\}\,dt = 2c(\zeta)c_{E,p}h \to 0 \quad \text{as } h\to 0. \quad (5.4.32)\end{aligned}$$

Letting $\varepsilon \to 0$ in (5.4.29) and using (5.4.30)–(5.4.32) we obtain (5.4.23).

**Proof of claim (iii).** Now our task is to prove that the limits $\varrho := \varrho_\delta$ and $\mathbf{u} := \mathbf{u}_\delta$ determine a renormalized solution to the limit transport equation and satisfy the integral identity (5.4.24). The following lemma shows that $\varrho$ is a weak solution to the limit transport equation.

**Lemma 5.4.6.** *Under the assumptions of Theorem 5.4.5, the integral identity*

$$\int_Q \varrho(\partial_t\psi + \nabla\psi\cdot\mathbf{u})\,dxdt + \int_\Omega (\varrho_\infty\psi)(x,0)\,dx = \int_{S_T}\psi\varrho_\infty\mathbf{U}\cdot\mathbf{n}\,dSdt \quad (5.4.33)$$

*holds for any $\psi \in C^1(Q)$ which vanishes in the vicinity of $S_T\setminus\Sigma_{\text{in}}$ and of the top $\Omega\times\{T\}$.*

*Proof.* Choose an arbitrary smooth function $\psi$ which is zero in the vicinity of $S_T\setminus\Sigma_{\text{in}}$ and of $\Omega\times\{t=T\}$. This means that $\operatorname{supp}\psi\cap S_T \Subset \Sigma_{\text{in}}$. It follows from the integral identity (5.3.14) with $\varrho$ and $\mathbf{u}$ replaced by $\varrho_\varepsilon$ and $\mathbf{u}_\varepsilon$ that

$$\begin{aligned}&\int_Q (\varrho_\varepsilon\partial_t\psi - \varepsilon\nabla\varrho_\varepsilon\cdot\nabla\psi + \varrho_\varepsilon\mathbf{u}_\varepsilon\cdot\nabla\psi)\,dxdt\\ &\qquad + \int_{S_T}\psi(\varepsilon\nabla\varrho_\varepsilon - \varrho_\infty\mathbf{U})\cdot\mathbf{n}\,dSdt + \int_\Omega(\psi\varrho_\infty)(x,0)\,dx = 0. \quad (5.4.34)\end{aligned}$$

Proposition 5.3.11 implies that $\varepsilon|\partial_n\varrho_\varepsilon| \leq c\varepsilon^\varkappa$ at all $(z,t)\in\operatorname{supp}\psi\cap S_T\Subset\Sigma_{\text{in}}$, where $\varkappa>0$ and $c$ are independent of $(z,t)$ and $\varepsilon$. Hence

$$\Big|\varepsilon\int_{S_T}\psi\nabla\varrho_\varepsilon\cdot\mathbf{n}\,dSdt\Big| \leq C\varepsilon^\varkappa \to 0 \quad \text{as } \varepsilon\to 0. \quad (5.4.35)$$

On the other hand, estimate (5.4.2) implies

$$\Big|\varepsilon\int_Q \nabla_\varepsilon\varrho\cdot\nabla\psi\,dxdt\Big| \leq c(\psi)c_{E,p}\,\varepsilon^{1/2} \to 0 \quad \text{as } \varepsilon\searrow 0. \quad (5.4.36)$$

Letting $\varepsilon$ tend to 0 in the integral identity (5.4.34), and using (5.4.12), (5.4.35) and (5.4.36) we obtain (5.4.33). □

It follows from Lemma 5.4.6 that $\varrho := \varrho_\delta$ is a weak solution to the boundary value problem

$$\partial_t \varrho + \operatorname{div}(\varrho \mathbf{u}) = 0 \quad \text{in } Q, \quad \varrho = \varrho_\infty \quad \text{on } \Sigma_{\text{in}}, \quad \varrho = \varrho_\infty \quad \text{on } \Omega \times \{0\}.$$

The next lemma justifies the renormalization procedure (see Section 1.6) and shows that $\varrho$ and $\mathbf{u}$ determine a renormalized solution to the mass balance equation.

**Lemma 5.4.7.** *Under the assumptions of Theorem 5.4.5, the integral identity*

$$\int_Q \varphi(\varrho)(\partial_t \psi + \nabla \psi \cdot \mathbf{u})\, dxdt - \int_Q (\varphi'(\varrho)\varrho - \varphi(\varrho)) \operatorname{div} \mathbf{u}\, dxdt \\ + \int_\Omega \varphi(\varrho_\infty(\cdot, 0))\psi(\cdot, 0)\, dx = \int_{S_T} \psi \varrho_\infty \mathbf{U} \cdot \mathbf{n}\, dSdt \quad (5.4.37)$$

*holds for all functions $\psi \in C^1(Q)$ which vanish in the vicinity of $S_T \setminus \Sigma_{\text{in}}$ and of the top $\Omega \times \{T\}$, and for any $C^2$ function $\varphi : \mathbb{R} \to \mathbb{R}$ such that*

$$|\varphi(s)| + |(1 + |s|)\varphi'(s)| + |(1 + s^2)\varphi''(s)| \le c(1 + |s|)^2, \quad s \in \mathbb{R}. \quad (5.4.38)$$

*Proof.* First we assume that $\varphi$ satisfies the stronger condition

$$|\varphi(s)| + |(1 + |s|)\varphi'(s)| + |(1 + s^2)\varphi''(s)| \le c, \quad s \in \mathbb{R}. \quad (5.4.39)$$

Choose a test function $\psi$ as in the lemma. Without loss of generality assume that $(\psi, \varrho_\infty)$ and $\mathbf{U}$ are extended to $C^1(\mathbb{R}^d \times \mathbb{R}^+)$ and $C^\infty(\mathbb{R}^d \times \mathbb{R}^+)$ functions respectively. Moreover, we can assume that the extensions vanish for large $|x|$ and $t$. Next, fix an open set $D \subset \mathbb{R}^d \times \mathbb{R}^+$ such that $\operatorname{cl} D \cap (\operatorname{cl} S_T \setminus \Sigma_{\text{in}}) = \emptyset$. Notice that $S_T \cap \operatorname{cl} D$ is a compact subset of $\Sigma_{\text{in}}$. Finally, choose cut-off functions $\eta, \zeta \in C^\infty(\mathbb{R}^d \times \mathbb{R}^+)$ such that

$$0 \le \eta, \zeta \le 1, \quad \zeta = \eta = 0 \quad \text{in the vicinity of } S_T \setminus \Sigma_{\text{in}}, \\ \eta = 1 \quad \text{in } D, \quad \zeta = 1 \quad \text{on } \operatorname{supp} \eta.$$

Now set

$$\begin{aligned} \mathbf{u}^* &= \zeta \mathbf{u} \quad \text{in } Q \quad \text{and} \quad \mathbf{u}^* = \zeta \mathbf{U} \quad \text{in } \mathbb{R}^d \times (0, T) \setminus Q, \\ \omega_\infty &= \eta \varrho_\infty \quad \text{in } \mathbb{R}^d \times (0, T), \\ \omega &= \eta \varrho \quad \text{in } Q, \quad \omega = \omega_\infty \quad \text{in } \mathbb{R}^d \times (0, T) \setminus Q, \\ F^+ &= (\zeta \varrho)(\partial_t \eta + \nabla \eta \cdot \mathbf{u}) \quad \text{in } Q, \\ F^- &= \partial_t \omega_\infty + \operatorname{div}(\omega_\infty \mathbf{u}^*) \quad \text{in } \mathbb{R}^d \times (0, T) \setminus Q, \end{aligned}$$

and finally

$$F = F^+ \quad \text{in } Q, \quad F = F^- \quad \text{in } \mathbb{R}^d \times (0, T) \setminus Q.$$

With this notation and $\psi$ replaced by $\eta\psi$, the integral identity (5.4.33) reads

$$\int_Q \big(\omega(\partial_t\psi + \nabla\psi\cdot\mathbf{u}^*) + \psi F^+\big)\, dxdt + \int_\Omega (\omega_\infty\psi)(x,0)\, dx = \int_{S_T} \omega_\infty \mathbf{u}^*\cdot\mathbf{n}\psi\, dSdt. \tag{5.4.40}$$

Notice that this holds for all $\psi \in C^1(\mathbb{R}^d\times(0,T))$ vanishing for $t=T$, since $\omega$ vanishes in a neighborhood of $S_T\setminus\Sigma_{\text{in}}$. Direct calculations show that for every $\psi\in C^1(\mathbb{R}^d\times(0,T))$ vanishing at $\{t=T\}$,

$$\int_{(\mathbb{R}^d\times(0,T))\setminus Q} \big(\omega_\infty(\partial_t\psi + \nabla\psi\cdot\mathbf{u}^*) + \psi F^-\big)\, dxdt + \int_{\mathbb{R}^d\setminus\Omega} (\omega_\infty\psi)(x,0)\, dx$$
$$= -\int_{S_T} \omega_\infty \mathbf{u}^*\cdot\mathbf{n}\psi\, dSdt. \tag{5.4.41}$$

It is important to note that the sign of the right hand side is opposite to the sign of the right hand side in (5.4.40), since the direction of the outward normal to $\mathbb{R}^d\setminus\Omega$ is opposite to the direction of $\mathbf{n}$. Combining (5.4.40) and (5.4.41) yields

$$\int_{\mathbb{R}^d\times(0,T)} \big(\omega(\partial_t\psi + \nabla\psi\cdot\mathbf{u}^*) + \psi F\big)\, dxdt + \int_{\mathbb{R}^d} (\omega_\infty\psi)(x,0)\, dx = 0$$

for any $\psi\in C^1(\mathbb{R}^d\times(0,T))$ vanishing for $t=T$. It follows that $\omega$ is a weak solution to the following Cauchy problem in the strip $\mathbb{R}^d\times(0,T)$:

$$\partial_t\omega + \operatorname{div}(\omega\mathbf{u}^*) = F \quad \text{in } \mathbb{R}^d\times(0,T), \qquad \omega=\omega_\infty \quad \text{for } t=0.$$

By (5.4.18)–(5.4.20) with $\varrho_\delta := \varrho$ and $\mathbf{u}_\delta := \mathbf{u}$ we have

$$\omega\in L^\infty(0,T;L^\gamma(\mathbb{R}^d)),\quad \mathbf{u}^*\in L^2(0,T;W^{1,2}(\mathbb{R}^d)),\quad F\in L^\infty(0,T;L^{2\gamma/(\gamma+2)}(\mathbb{R}^d)).$$

Moreover, these functions vanish for large $|x|$. Since $\gamma>2d$, we can apply the renormalization Lemma 1.6.2 to conclude that the composite function $\varphi(\omega)$ is a weak solution to the Cauchy problem in the strip $\mathbb{R}^d\times(0,T)$,

$$\begin{aligned}&\partial_t\varphi(\omega) + \operatorname{div}(\varphi(\omega)\mathbf{u}^*) + (\varphi'(\omega)\omega - \varphi(\omega))\operatorname{div}\mathbf{u}^* = \varphi'(\omega)F,\\ &\varphi(\omega) = \varphi(\eta\varrho_\infty)\equiv\varphi(\omega_\infty) \quad \text{for } t=0.\end{aligned}$$

This means that for any $C^1$ function $\psi$ with $\psi(\cdot,T)=0$,

$$\int_{\mathbb{R}^d\times(0,T)} \varphi(\omega)(\partial_t\psi + \nabla\psi\cdot\mathbf{u}^*)\, dxdt - \int_{\mathbb{R}^d\times(0,T)} \psi(\varphi'(\omega)\omega - \varphi(\omega))\operatorname{div}\mathbf{u}^*\, dxdt$$
$$+ \int_{\mathbb{R}^d} (\varphi(\eta\varrho_\infty)\psi)(x,0)\, dx + \int_{\mathbb{R}^d\times(0,T)} \psi\varphi'(\omega)F dxdt = 0.$$

On the other hand, straightforward calculations lead to the identity

$$\begin{aligned}\int_{\mathbb{R}^d\times(0,T)\setminus Q}&\varphi(\omega)(\partial_t\psi+\nabla\psi\cdot\mathbf{u}^*)\,dxdt\\&-\int_{\mathbb{R}^d\times(0,T)\setminus Q}(\varphi'(\omega)\omega-\varphi(\omega))\operatorname{div}\mathbf{u}^*\,dxdt\\&+\int_{\mathbb{R}^d\setminus\Omega}(\varphi(\eta\varrho_\infty)\psi)(x,0)\,dx+\int_{\mathbb{R}^d\times(0,T)\setminus Q}\psi\varphi'(\omega)F^-\,dxdt\\&\qquad\qquad=-\int_{S_T}\psi\varphi(\omega_\infty)\mathbf{u}^*\cdot\mathbf{n}\,dSdt,\end{aligned}$$

which gives

$$\begin{aligned}\int_Q\varphi(\omega)(\partial_t\psi+\nabla\psi\cdot\mathbf{u}^*)\,dxdt&-\int_Q\psi(\varphi'(\omega)\omega-\varphi(\omega))\operatorname{div}\mathbf{u}^*\,dxdt\\+\int_\Omega(\varphi(\eta\varrho_\infty)\psi)(x,0)\,dx&+\int_Q\psi\varphi'(\omega)F^+dxdt=\int_{S_T}\psi\varphi(\omega_\infty)\mathbf{u}^*\cdot\mathbf{n}\,dSdt,\end{aligned}\tag{5.4.42}$$

for every $C^1$ function $\psi$ in $Q$ vanishing for $t=T$. Now, choose $\psi\in C^1(\mathbb{R}^d\times(0,T))$ equal to 0 in a neighborhood of $S_T\setminus\Sigma_{\text{in}}$. The intersection $\operatorname{supp}\psi\cap S_T$ is a compact subset of $\Sigma_{\text{in}}$. Hence there is an open set $D\subset\mathbb{R}^d\times(0,T)$ with

$$\operatorname{supp}\psi\Subset D,\quad D\cap(\operatorname{cl}S_T\setminus\Sigma_{\text{in}})=\emptyset.$$

For such a $D$ and the corresponding function $\eta$, we have

$$\omega=\varrho,\quad \omega_\infty:=\eta\varrho_\infty=\varrho_\infty,\quad \mathbf{u}^*=\mathbf{u}\quad\text{on }\operatorname{supp}\psi\cap\operatorname{cl}Q.$$

Inserting these into (5.4.42) we obtain (5.4.37).

To complete the proof we approximate an arbitrary function $\varphi$ satisfying (5.4.38) by the functions

$$\varphi_n(s)=\begin{cases}\varphi(s) & \text{if }|\varphi(s)|\le n,\\ n+1 & \text{if }\varphi(s)\ge n+1,\\ -n-1 & \text{if }\varphi(s)\le -n-1,\end{cases}$$

which are $C^2$ on $\mathbb{R}$ and satisfy (5.4.39). It is clear that

$$\varphi_n(\varrho)\to\varphi(\varrho),\quad \varphi'(\varrho)\to\varphi'(\varrho)$$

everywhere in $Q$. Moreover, it follows from (5.4.38) and $\varrho^2\in L^{\gamma/2}(Q)\subset L^2(Q)$ that

$$\|\varphi_n(\varrho)\|_{L^{\gamma/2}(Q)}+\|\varrho\varphi_n'(\varrho)\|_{L^{\gamma/2}(Q)}\le c.$$

Since $\gamma > 4$, it now follows from Lemma 1.3.2 that

$$\varphi_n(\varrho) \to \varphi(\varrho), \quad \varrho\varphi_n'(\varrho) \to \varrho\varphi'(\varrho) \quad \text{in } L^2(Q) \quad \text{as } n \to \infty. \tag{5.4.43}$$

Substituting $\varphi_n(\varrho)$ for $\varphi(\varrho)$ into (5.4.33), letting $n \to \infty$ and recalling that $\mathbf{u}, \operatorname{div}\mathbf{u} \in L^2(Q))$ we conclude that the integral identity (5.4.37) holds for all $\varphi$ satisfying (5.4.38). □

It remains to note that claim (iii) is a straightforward consequence of Lemmas 5.4.6 and 5.4.7.

**Proof of claim (iv).** We obtain the desired inequality (5.4.26) in a few steps.

**Lemma 5.4.8.** *Let the assumptions of Theorem 5.4.5 be satisfied and $\varphi$ be an arbitrary $C^2$ function on $\mathbb{R}$ satisfying condition* (5.4.38). *Then with the notation $\varrho := \varrho_\delta$ and $\mathbf{u} := \mathbf{u}_\delta$ for the limits, we obtain*

$$\varphi(\varrho_\varepsilon(t)) \rightharpoonup \overline{\varphi}(t) \quad \textit{weakly in } L^{\gamma/2}(\Omega), \tag{5.4.44}$$

$$\varphi(\varrho_\varepsilon)\,\mathbf{u}_\varepsilon \rightharpoonup \overline{\varphi}\,\mathbf{u} \quad \textit{weakly in } L^m(Q) \quad \textit{with} \quad m = 2\gamma/(\gamma+4), \tag{5.4.45}$$

$$(\varphi'(\varrho_\varepsilon)\varrho_\varepsilon - \varphi(\varrho_\varepsilon))\operatorname{div}\mathbf{u}_\varepsilon \rightharpoonup \overline{(\varphi'\varrho - \varphi)\operatorname{div}\mathbf{u}} \quad \textit{weakly in } L^m(Q). \tag{5.4.46}$$

*Here $\overline{(\varphi'\varrho - \varphi)\operatorname{div}\mathbf{u}}$ is just a notation for a weak limit, and $\overline{\varphi}$ is defined by the Young measure $\mu^\delta_{xt}$,*

$$\overline{\varphi}(x,t) = \langle \mu^\delta_{xt}, \varphi(\lambda)\rangle. \tag{5.4.47}$$

*Proof.* First, by (5.4.38) and (5.4.2) we have

$$\|\varphi'(\varrho_\varepsilon)\varrho_\varepsilon - \varphi(\varrho_\varepsilon)\|_{L^\infty(0,T;L^{\gamma/2}(\Omega))} + \|\varphi(\varrho_\varepsilon)\|_{L^\infty(0,T;L^{\gamma/2}(\Omega))} \le c, \tag{5.4.48}$$

where $c$ is independent of $\varepsilon$. Hence, passing to a subsequence, we can assume that $\varphi(\varrho_\varepsilon)$ converges to some $\overline{\varphi}$ weakly$^\star$ in $L^\infty(0,T;L^{\gamma/2}(\Omega))$. Moreover, since $\varphi(\varrho)\varrho^{-\gamma} \to 0$ as $\varrho \to 0$, we can apply Theorem 5.4.3 to obtain (5.4.47). Estimate (5.4.48) combined with the Hölder inequality and (5.4.1) gives

$$\|(\varphi'(\varrho_\varepsilon)\varrho_\varepsilon - \varphi(\varrho_\varepsilon))\operatorname{div}\mathbf{u}_\varepsilon\|_{L^m(Q)} \le \|\varphi'(\varrho_\varepsilon)\varrho_\varepsilon - \varphi(\varrho_\varepsilon)\|_{L^{\gamma/2}(Q)}\|\operatorname{div}\mathbf{u}_\varepsilon\|_{L^2(Q)} \le c. \tag{5.4.49}$$

Here $m = 2\gamma/(4+\gamma) > 1$ satisfies $m^{-1} = 2^{-1} + 2\gamma^{-1}$. Hence, passing to a subsequence, we can assume that $(\varphi'(\varrho_\varepsilon)\varrho_\varepsilon - \varphi(\varrho_\varepsilon))\operatorname{div}\mathbf{u}_\varepsilon$ converges weakly in $L^m(Q)$ as $\varepsilon \to 0$, which leads to (5.4.46).

Next, notice that $\varphi(\varrho_\varepsilon)$ is a solution to the equation

$$\partial_t\varphi(\varrho_\varepsilon) = \operatorname{div}\mathfrak{v}_\varepsilon + \mathfrak{h}_\varepsilon,$$

where

$$\mathfrak{v}_\varepsilon = \varepsilon\varphi'(\varrho_\varepsilon)\nabla\varrho_\varepsilon - \varphi(\varrho_\varepsilon)\mathbf{u}_\varepsilon,$$
$$\mathfrak{h}_\varepsilon = -(\varphi'(\varrho_\varepsilon)\varrho_\varepsilon - \varphi(\varrho_\varepsilon))\operatorname{div}\mathbf{u}_\varepsilon - \varepsilon\varphi''(\varrho_\varepsilon)|\nabla\varrho_\varepsilon|^2 = 0.$$

By (5.4.48), $\varphi(\varrho_\varepsilon)$ and $\varphi'(\varrho_\varepsilon)\varrho_\varepsilon$ are bounded in $L^2(Q)$. Moreover, in view of (5.4.1), $\mathbf{u}_\varepsilon$ and $\operatorname{div}\mathbf{u}_\varepsilon$ are bounded in $L^2(Q)$. Therefore, $(\varphi'(\varrho_\varepsilon)\varrho_\varepsilon - \varphi(\varrho_\varepsilon))\operatorname{div}\mathbf{u}_\varepsilon$ and $\varphi(\varrho_\varepsilon)\mathbf{u}_\varepsilon$ are bounded in $L^1(Q)$.

It also follows from (5.4.2) that $\sqrt{\varepsilon}\nabla\varrho_\varepsilon$ is bounded in $L^2(Q)$. Since $\varphi''$ is bounded it follows that $\varepsilon\varphi'(\varrho_\varepsilon)\nabla\varrho_\varepsilon$ and $\varepsilon\varphi''|\nabla\varrho_\varepsilon|^2$ are bounded in $L^1(Q)$.

Therefore the sequences $\mathfrak{v}_\varepsilon$ and $\mathfrak{g}_\varepsilon$ are bounded in $L^1(Q)$. Since $\varphi(\varrho_\varepsilon)$ converges to $\overline{\varphi}$ weakly$^\star$ in $L^\infty(0,T;L^{\gamma/2})$, we can apply Theorem 4.4.2 with $n$ replaced by $\varepsilon$, $\varphi_n$ replaced by $\varphi(\varrho_\varepsilon)$, and $s$ replaced by $\gamma/2$ to obtain (5.4.44) and (5.4.45). Here we use the inequality $m^{-1} > 2/\gamma + 2^{-1} - d^{-1}$. □

**Lemma 5.4.9.** *Let*

$$\mathfrak{D}_\varphi = (\varphi'(\varrho)\varrho - \varphi(\varrho))\operatorname{div}\mathbf{u} - \overline{(\varphi'\varrho - \varphi)\operatorname{div}\mathbf{u}}, \quad \Phi = \overline{\varphi} - \varphi(\varrho).$$

*Furthermore, assume that $\varphi$ is a $C^2$ convex function satisfying the growth condition (5.4.38), and $\psi \in C^1(Q)$ is a nonnegative function vanishing in a neighborhood of $S_T \setminus \Sigma_{\mathrm{in}}$ and in a neighborhood of $\Omega \times \{t = T\}$. Then*

$$\int_Q \Phi(\partial_t\psi + \nabla\psi\cdot\mathbf{u})\,dxdt + \int_Q \mathcal{D}_\varphi\psi\,dxdt \geq 0. \tag{5.4.50}$$

Notice that $\Phi \geq 0$ due the convexity of $\varphi$.

*Proof.* Recall that by Theorem 5.3.2 the functions $\varrho_\varepsilon$ and $\mathbf{u}_\varepsilon$ satisfy inequality (5.3.15), which reads

$$\begin{aligned}&\int_Q \big(\varphi(\varrho_\varepsilon)\partial_t\psi + (\varphi(\varrho_\varepsilon)\mathbf{u}_\varepsilon - \varepsilon\varphi'(\varrho_\varepsilon)\nabla\varrho_\varepsilon)\cdot\nabla\psi - \psi(\varphi'(\varrho_\varepsilon)\varrho_\varepsilon - \varphi(\varrho_\varepsilon))\operatorname{div}\mathbf{u}_\varepsilon\big)\,dxdt\\ &+ \int_{S_T}\psi\big(\varepsilon\varphi'(\varrho_\infty)\nabla\varrho - \varphi(\varrho_\infty)\mathbf{U}\big)\cdot\mathbf{n}\,dSdt + \int_\Omega(\psi\varphi(\varrho_\infty))(x,0)\,dx \geq 0.\end{aligned} \tag{5.4.51}$$

By Proposition 5.3.11, $\varepsilon|\partial_n\varrho_\varepsilon| \leq c\varepsilon^{\varkappa}$ for all $(z,t) \in \operatorname{supp}\psi \cap S_T \Subset \Sigma_{\mathrm{in}}$, where $\varkappa > 0$ and $c$ are independent of $(z,t)$ and $\varepsilon$. Hence

$$\Big|\varepsilon\int_{S_T}\psi\varphi'(\varrho_\infty)\nabla\varrho_\varepsilon\cdot\mathbf{n}\,dSdt\Big| \leq c\varepsilon^{\varkappa} \to 0 \quad \text{as } \varepsilon \to 0. \tag{5.4.52}$$

On the other hand, we have $(\gamma - 2)/2 > 1$, which along with the growth condition (5.4.25) and estimate (5.4.2) implies

$$\begin{aligned}\varepsilon\int_Q|\varphi'(\varrho_\varepsilon)\nabla\varrho_\varepsilon|\,dxdt &\leq \varepsilon c\int_Q|(1+\varrho_\varepsilon)\nabla\varrho_\varepsilon|\,dxdt\\ &\leq \varepsilon c\int_Q|(1+\varrho_\varepsilon)^{(\gamma-2)/2}\nabla\varrho_\varepsilon|\,dxdt \leq c\varepsilon^{1/2} \to 0 \quad \text{as } \varepsilon \to 0.\end{aligned} \tag{5.4.53}$$

Letting $\varepsilon \to 0$ in (5.4.51) and using (5.4.45), (5.4.46), (5.4.52), and (5.4.53) we conclude that

$$\int_Q \left(\overline{\varphi}\partial_t\psi + \overline{\varphi}\mathbf{u}\cdot\nabla\psi - \overline{(\varphi'\varrho-\varphi)\operatorname{div}\mathbf{u}}\right) dxdt$$
$$- \int_{\Sigma_{\rm in}} \psi\varphi(\varrho_\infty)\mathbf{U}\cdot\mathbf{n}\, d\Sigma + \int_\Omega (\psi\varphi(\varrho_\infty))(x,0)\, dx \geq 0 \quad (5.4.54)$$

for all smooth nonnegative $\psi$, vanishing in a neighborhood of $\Omega\times\{t=T\}$ and of $S_T\setminus\Sigma_{\rm in}$. Next, we subtract (5.4.37) from (5.4.54) to obtain (5.4.50). □

Our further considerations are based on the following geometric lemma:

**Lemma 5.4.10.** *Let $\partial\Omega$ be a $C^{1+\alpha}$ surface, $0<\alpha<1$, $\mathbf{U}\in C^1(\mathbb{R}^d\times(0,T))$, and $\mathbf{u}-\mathbf{U}\in L^2(0,T;W^{1,2}_0(\Omega))$. Furthermore, assume that a compact "interface" $\Gamma=(S_T\setminus\Sigma_{\rm in})\cap \operatorname{cl}\Sigma_{\rm in}$ satisfies condition* (5.1.9)*:*

$$\lim_{\sigma\to 0}\sigma^{-d}\operatorname{meas}\mathcal{O}_\sigma<\infty, \quad (5.4.55)$$

*where $\mathcal{O}_\sigma$ is a tubular neighborhood of $\Gamma$,*

$$\mathcal{O}_\sigma=\{(x,t)\in\mathbb{R}^{d+1}:\operatorname{dist}((x,t),\Gamma)\leq\sigma\}.$$

*Then there is a sequence of Lipschitz functions $\psi_n: Q\to[0,1]$, $n\geq 1$, with the following properties: $\psi_n$ vanishes in some neighborhood of $S_T\setminus\Sigma_{\rm in}$ and of $\{t=T\}$, $\psi_n\nearrow 1$ everywhere in $Q$, and*

$$\limsup_{n\to\infty}\int_Q \Phi(\partial_t\psi_n+\nabla\psi_n\cdot\mathbf{u})\, dxdt\leq 0 \quad \textit{for all nonnegative } \Phi\in L^q(Q),\ q>2.$$

*Proof.* This is a particular case of Theorem 13.3.3. □

**Lemma 5.4.11.** *The functions $\varrho:=\varrho_\delta$ and $\mathbf{u}:=\mathbf{u}_\delta$ satisfy the inequality*

$$\int_Q(\overline{\varrho\operatorname{div}\mathbf{u}}-\varrho\operatorname{div}\mathbf{u})\, dxdt\leq 0. \quad (5.4.56)$$

*Proof.* Set

$$\varphi_k=-\left(\frac{3}{2}+\log k\right)\varrho+2k\varrho^2-\frac{1}{2}k^2\varrho^3 \quad \text{for } \varrho\leq k^{-1}, \quad \varphi_k=\varrho\log\varrho \quad \text{for } \varrho>k^{-1}.$$

Obviously $\varphi_k\in C^2[0,\infty)$ satisfies the growth condition (5.4.25). Substituting $\varphi_k$ and the functions $\psi_n$ from Lemma 5.4.10 into (5.4.50), letting $n\to\infty$ and using Lemma 5.4.10 we obtain

$$\int_Q \mathfrak{D}_{\varphi_k}\, dxdt\geq 0. \quad (5.4.57)$$

Next note that

$$\varrho\varphi_k'(\varrho)-\varphi_k(\varrho)-\varrho = 2k\varrho^2-k^2\varrho^3 \quad \text{for } \varrho < k^{-1}, \quad \varrho\varphi_k'(\varrho)-\varphi_k(\varrho) = \varrho \quad \text{for } \varrho \geq k^{-1}.$$

Hence the sequence $\varphi_k'\varrho - \varphi_k$ converges to $\varrho$ uniformly on $\mathbb{R}$. Letting $k \to \infty$ in (5.4.57) we obtain (5.4.56). □

**Lemma 5.4.12.** *For any subcylinder $Q' \Subset Q_\delta$ and $\psi \in C_0(Q')$,*

$$(1+\lambda)\int_{Q'} \psi\big(\overline{\varrho \operatorname{div} \mathbf{u}} - \varrho \operatorname{div} \mathbf{u}\big)\, dxdt = \lim_{\varepsilon\to 0}\int_{Q'} \psi(p(\varrho_\varepsilon)\varrho_\varepsilon - \overline{p}\varrho)\, dxdt.$$

*Proof.* Notice that $\varrho_\varepsilon$ and $\mathbf{u}_\varepsilon$ satisfy the differential equations

$$\begin{aligned} &\partial_t\varrho_\varepsilon + \operatorname{div}(\varrho_\varepsilon\mathbf{u}_\varepsilon - \mathbf{g}_\varepsilon) = 0 \quad \text{in } Q_\delta,\\ &\partial_t(\varrho_\varepsilon\mathbf{u}_\varepsilon) + \operatorname{div}(\varrho_\varepsilon\mathbf{u}_\varepsilon \otimes \mathbf{u}_\varepsilon) = \operatorname{div}(\mathbb{T}_\varepsilon + \mathbb{G}_\varepsilon) + \rho_\varepsilon\mathbf{f} \quad \text{in } Q_\delta, \end{aligned}$$

where

$$\mathbb{T}_\varepsilon = \mathbb{S}(\mathbf{u}_\varepsilon) - p(\varrho_\varepsilon)\,\mathbb{I}, \quad \mathbf{g}_\varepsilon = \varepsilon\nabla\varrho_\varepsilon, \quad G_\varepsilon = \varepsilon\nabla\varrho_\varepsilon \otimes \mathbf{u}_\varepsilon.$$

Set

$$r = \gamma+1, \quad s = \gamma, \quad q = s/(s-1), \quad r' = (\gamma+1)/\gamma,$$
$$\varphi_\varepsilon = \varrho_\varepsilon, \quad \mathbf{g}_\varepsilon^\varphi = \mathbf{g}_\varepsilon, \quad \varpi_\varepsilon = 0,$$

Let us prove that $\varrho_\varepsilon$, $\mathbf{u}_\varepsilon$, $\varphi_\varepsilon$, and $\mathbb{T}_\varepsilon$, $\mathbb{G}_\varepsilon$, $\mathbf{g}_\varepsilon$ meet all requirements of Theorem 4.7.1 with $n$ replaced by $\varepsilon$ and $Q$ replaced by $Q_\delta$. It follows from estimates (5.4.1)–(5.4.2) that the functions $(\varrho_\varepsilon, \mathbf{u}_\varepsilon, \varphi_\varepsilon)$ satisfy inequalities (4.6.4)–(4.6.6). Estimate (5.4.8) implies that $\varphi_\varepsilon = \varrho_\varepsilon$ satisfies (4.6.8). Next, estimates (5.4.1)–(5.4.2) yield the $L^1$-estimate (4.6.7) for $\mathbb{T}_n$. On the other hand, the inequalities $p \leq c(\varrho^\gamma + 1)$ and $(\gamma+1)/\gamma \leq 2$ along with (5.4.2) and (5.4.8) give, for every $Q' \Subset Q_\delta$,

$$\|\mathbb{T}_\varepsilon\|_{L^{(\gamma+1)/\gamma}(Q')} \leq \|\mathcal{S}(\mathbf{u}_\varepsilon)\|_{L^{(\gamma+1)/\gamma}(Q')} + \|p(\varrho_\varepsilon)\|_{L^{(\gamma+1)/\gamma}(Q')} \leq c(Q').$$

It follows that $\mathbb{T}_\varepsilon$ satisfies (4.6.8). Next (5.4.2) and (5.4.5) imply

$$\|\mathbf{g}_\varepsilon\|_{L^2(Q)} = \varepsilon\|\nabla\varrho_\varepsilon\|_{L^2(Q)} \leq \varepsilon^{1/2}c \to 0 \quad \text{as } \varepsilon \to 0,$$
$$\|\mathbb{G}_\varepsilon\|_{L^1(0,T;L^q(\Omega))} \leq \varepsilon\|\nabla\varrho_\varepsilon|\mathbf{u}_\varepsilon|\|_{L^1(0,T;L^{d/(d-1)}(\Omega))} \leq \varepsilon^{1/2}c \to 0 \quad \text{as } \varepsilon \to 0,$$

since $q < d/(d-1)$. Hence $\mathbf{g}_\varepsilon$ and $\mathbb{G}_\varepsilon$ satisfy (4.6.9)–(4.6.10) with $n \to \infty$ replaced by $\varepsilon \to 0$. Therefore, $\varrho_\varepsilon$, $\mathbf{u}_\varepsilon$, $\varphi_\varepsilon$, and $\mathbb{T}_\varepsilon$, $\mathbb{G}_\varepsilon$, $\mathbf{g}_\varepsilon$ satisfy Condition 4.6.1. Applying Theorem 4.7.1 and recalling that $\varphi_\varepsilon = \varrho_\varepsilon$ we obtain

$$\lim_{\varepsilon\to 0}\int_{Q_\delta} \psi(p(\varrho_\varepsilon) - (\lambda+1)\operatorname{div}\mathbf{u}_\varepsilon)\varrho_\varepsilon\, dxdt = \int_{Q_\delta} \psi(\overline{p} - (\lambda+1)\operatorname{div}\mathbf{u})\varrho\, dxdt.$$

This completes the proof. □

**Lemma 5.4.13.** *Under the assumptions of Theorem 5.4.5 inequality* (5.4.26) *holds for any subcylinder* $Q' \Subset Q_\delta$.

*Proof.* Choose $Q' \Subset Q_\delta$ and a nonnegative function $\psi \in C_0^\infty(Q_\delta)$ such that $\psi = 1$ on $Q'$. It follows from Lemmas 5.4.4 with and 5.4.12 and (5.4.56) that with

$$\overline{p} := \overline{p}_\delta, \quad \overline{\varrho} =:= \overline{\varrho}_\delta, \quad ,\overline{p\varrho} := \overline{p\varrho}_\delta, \quad \mathbf{u} := \mathbf{u}_\delta,$$

we have

$$\begin{aligned}
\int_{Q'} (\overline{p\varrho} - \overline{p}\varrho)\, dxdt &\le \int_{Q_\delta} \psi(\overline{p\varrho} - \overline{p}\varrho)\, dxdt \le \lim_{\varepsilon\to 0} \int_{Q'} \psi(p(\varrho_\varepsilon)\varrho_\varepsilon - \overline{p}\varrho)\, dxdt \\
&= (1+\lambda)\int_{Q_\delta} \psi(\overline{\varrho \operatorname{div} \mathbf{u}} - \varrho \operatorname{div} \mathbf{u})\, dxdt = (1+\lambda)\int_{Q} (\overline{\varrho \operatorname{div} \mathbf{u}} - \varrho \operatorname{div} \mathbf{u})\, dxdt \\
&\qquad + (1+\lambda)\int_{Q} (\psi - 1)(\overline{\varrho \operatorname{div} \mathbf{u}} - \varrho \operatorname{div} \mathbf{u})\, dxdt \\
&\qquad \le (1+\lambda)\int_{Q} (\psi - 1)(\overline{\varrho \operatorname{div} \mathbf{u}} - \varrho \operatorname{div} \mathbf{u})\, dxdt \\
&\qquad \le (1+\lambda)\int_{Q\setminus Q'} |\overline{\varrho \operatorname{div} \mathbf{u}} - \varrho \operatorname{div} \mathbf{u}|\, dxdt.
\end{aligned} \tag{5.4.58}$$

Next, estimates (5.4.1)–(5.4.2) and the Hölder inequality imply that for $m = 2\gamma/(\gamma+2)$,

$$\begin{aligned}
\|\overline{\varrho \operatorname{div} \mathbf{u}}\|_{L^m(Q)} &\le \liminf_{\varepsilon\to 0} \|\varrho_\varepsilon \operatorname{div} \mathbf{u}_\varepsilon\|_{L^m(Q)} \\
&\le \liminf_{\varepsilon\to 0} \|\varrho_\varepsilon\|_{L^\gamma(Q)} \|\operatorname{div} \mathbf{u}_\varepsilon\|_{L^2(Q)} \le c_{E,p}.
\end{aligned}$$

The same arguments combined with (5.4.18)–(5.4.19) give $\|\varrho \operatorname{div} \mathbf{u}\|_{L^m(Q)} \le c_{E,p}$, hence

$$\begin{aligned}
&\int_{Q\setminus Q'} |\overline{\varrho \operatorname{div} \mathbf{u}} - \varrho \operatorname{div} \mathbf{u}|\, dxdt \\
&\qquad \le \operatorname{meas}(Q\setminus Q')^{1/m'} \|\overline{\varrho \operatorname{div} \mathbf{u}} - \varrho \operatorname{div} \mathbf{u}\|_{L^m(Q)} \le c_{E,p} \operatorname{meas}(Q\setminus Q')^{1/m'},
\end{aligned}$$

where $1/m' = (\gamma-2)/(2\gamma)$. Inserting this into (5.4.58) we obtain (5.4.26). □

It remains to note that the statement of Lemma 5.4.13 is exactly claim (iv) of Theorem 5.4.5.

**Proof of claim (v).** The proof is based on the following lemma analogous to Lemma 5.3.8. This lemma is valid for any weak limit, however it is formulated for the weak limit under considerations.

**Lemma 5.4.14.** *For any* $0 \le T' < T'' \le T$, *the weak limit* $(\mathbf{u}, \varrho)$ *of the sequence* $(\mathbf{u}_\varepsilon, \varrho_\varepsilon)$ *satisfies*

$$\int_{T'}^{T''}\int_\Omega \overline{P}\, dxdt \le \liminf_{\varepsilon\to 0}\int_{T'}^{T''}\int_\Omega P(\varrho_\varepsilon)\, dxdt, \tag{5.4.59a}$$

$$\int_{T'}^{T''}\int_\Omega \varrho|\mathbf{u}|^2\, dxdt = \liminf_{\varepsilon\to 0}\int_{T'}^{T''}\int_\Omega \varrho_\varepsilon|\mathbf{u}_\varepsilon|^2\, dxdt, \tag{5.4.59b}$$

$$\int_{T'}^{T''}\int_\Omega |\nabla\mathbf{u}|^2\, dxdt \le \liminf_{\varepsilon\to 0}\int_{T'}^{T''}\int_\Omega |\nabla\mathbf{u}_\varepsilon|^2\, dxdt. \tag{5.4.59c}$$

*Proof.* First observe that if $\psi(x,t)$ is the characteristic function of $\Omega\times[T',T'']$, then (5.4.59a) turns into (5.4.14). Let us prove (5.4.59b). By Lemma 5.4.2 the sequence $\varrho_\varepsilon\mathbf{u}_\varepsilon\otimes\mathbf{u}_\varepsilon$ converges weakly in $L^2(0,T;L^z(\Omega))$. Hence it also converges weakly in $L^z(Q)$, whence in $L^z(\Omega\times(T',T''))$. Recalling that $\varrho|\mathbf{u}|^2 = \operatorname{tr}\varrho\mathbf{u}\otimes\mathbf{u}$ we obtain (5.4.59b). Inequality (5.4.59c) obviously follows from the weak convergence of $\nabla\mathbf{u}_\varepsilon$ to $\nabla\mathbf{u}$ in $L^2(Q)$. □

**Lemma 5.4.15.** *The functions* $M_\varepsilon : [0,T]\to\mathbb{R}$ *defined by* (5.4.10) *are uniformly bounded and*

$$\lim_{\varepsilon\to 0} M_\varepsilon(t) \le M_\delta(t) + c_e = 2\int_\Omega (a|\Delta^2\mathbf{U}|^2)(x,t)\, dx + c_e \tag{5.4.60}$$

*where* $c_e$ *is a constant as in Remark 5.1.2.*

*Proof.* By (5.4.2) and (5.4.3) the sequences $\varrho_\varepsilon$ and $\varrho_\varepsilon\mathbf{u}_\varepsilon$ are uniformly bounded in $L^\infty(0,T;L^1(\Omega))$. Hence the boundedness of $M_\varepsilon(t)$ is a straightforward consequence of (5.4.10):

$$M_\varepsilon(t) = M_\delta(t) + 4\int_0^t\int_\Omega \Big(\varepsilon^{2/3}|\nabla P(\varrho_\infty)|^2 - \varrho_\varepsilon\big(\partial_t P'(\varrho_\infty) + \nabla P'(\varrho_\infty)\cdot\mathbf{u}_\varepsilon\big)\Big)\, dxds.$$

Thus we get $\lim_{\varepsilon\to 0} M_\varepsilon = M_\delta + 4\lim_{\varepsilon\to 0} I_\varepsilon$, where

$$I_\varepsilon(t) = -\int_0^t\int_\Omega \varrho_\varepsilon\big(\partial_t P'(\varrho_\infty) + \nabla P'(\varrho_\infty)\mathbf{u}_\varepsilon\big)\, dxds.$$

Next, by relation (5.4.11) and Lemma 5.4.2 we have $\varrho_\varepsilon \rightharpoonup \varrho_\delta =: \varrho$ weakly$^\star$ in $L^\infty(0,T;L^\gamma(\Omega))$ and $\varrho_\varepsilon\mathbf{u}_\varepsilon \rightharpoonup \varrho_\delta\mathbf{u}_\delta =: \varrho\mathbf{u}$ weakly$^\star$ in $L^\infty(0,T;L^{2\gamma/(\gamma+1)}(\Omega))$. Hence letting $\varepsilon\to 0$ we obtain

$$I := \lim_{\varepsilon\to 0} I_\varepsilon(t) = -\int_0^t\int_\Omega \varrho(\partial_t P'(\varrho_\infty) + \nabla P'(\varrho_\infty)\mathbf{u})\, dxds.$$

It follows from Theorem 13.3.3 (see Appendix) that there is a sequence of Lipschitz functions $\psi_n : \Omega\times(0,t)\to\mathbb{R}$ such that

- $0 \le \psi_n \le 1$, $\psi_n \nearrow 1$ in $\Omega \times (0,t)$ as $n \to \infty$;
- $\psi_n$ vanishes in a neighborhood of $\partial\Omega \times (0,t) \setminus \Sigma_{\text{in}}$ and of $\Omega \times \{t\}$;
- for any nonnegative function $\Phi \in L^2(\Omega \times (0,t))$,

$$\liminf_{n\to\infty} \int_{\Omega\times(0,t)} \Phi(\partial_t\psi_n + \nabla\psi_n \cdot \mathbf{u})\, dxds \le 0. \tag{5.4.61}$$

We have $I = \lim_{n\to\infty} I_n$, where

$$\begin{aligned} I_n &= -\int_0^t \int_\Omega \psi_n \varrho(\partial_t P'(\varrho_\infty) + \nabla P'(\varrho_\infty)\mathbf{u})\, dxds \\ &= \int_0^t \int_\Omega \varrho P'(\varrho_\infty)(\partial_t\psi_n + \nabla\psi_n \cdot \mathbf{u})\, dxds \\ &\quad - \int_0^t \int_\Omega \varrho\big(\partial_t(P'(\varrho_\infty)\psi_n) + \nabla(P'(\varrho_\infty)\psi_n)\cdot\mathbf{u}\big)\, dxds. \end{aligned}$$

Notice that $\psi_n P'(\varrho_\infty)$ is a Lipschitz function vanishing in a neighborhood of $\Omega\times\{t\}$ and in a neighborhood of $\partial\Omega \times (0,t) \setminus \Sigma_{\text{in}}$. Extended by zero over $Q$, $\psi_n P'(\varrho_\infty)$ becomes a Lipschitz function vanishing in a neighborhood of $S_T \setminus \Sigma_{\text{in}}$ and in a neighborhood of $\Omega \times \{t = T\}$. Hence $\psi_n P'(\varrho_\infty)$ can be regarded as a test function in the mass balance equation. In particular, the integral identity (5.4.33) with $\psi$ replaced by $\psi_n P'(\varrho_\infty)$ implies

$$\begin{aligned} &\Big|\int_0^t \int_\Omega \varrho\big(\partial_t(P'(\varrho_\infty)\psi_n) + \nabla(P'(\varrho_\infty)\psi_n)\cdot\mathbf{u}\big)\, dxds\Big| \\ &\quad = \Big|\int_0^t \int_{\partial\Omega} \psi_n\varrho_\infty P'(\varrho_\infty)\mathbf{U}\cdot\mathbf{n}\, dS(x)ds - \int_\Omega (\psi_n\varrho_\infty P'(\varrho_\infty))(x,0)\, dx\Big| \le c_e. \end{aligned}$$

Thus we get

$$I_n(t) \le \int_0^t \int_\Omega \varrho_\delta P'(\varrho_\infty)(\partial_t\psi_n + \nabla\psi_n \cdot \mathbf{u})\, dxds + c_e.$$

Since $\Phi := \varrho P'(\varrho_\infty)$ is nonnegative and belongs to $L^2(Q)$ we can apply (5.4.61) to obtain

$$\lim_{n\to\infty} \int_0^t \int_\Omega \varrho_\delta P'(\varrho_\infty)(\partial_t\psi_n + \nabla\psi_n \cdot \mathbf{u})\, dxds \le 0,$$

which leads to $I = \lim_{n\to\infty} I_n \le c_e$. Combining this result with the identity $\lim_{\varepsilon\to 0} M_\varepsilon = M_\delta + 4I$ we obtain (5.4.60). □

**Lemma 5.4.16.** *Let* $\mathbf{u} = \mathbf{u}_\delta$, $\varrho = \varrho_\delta$ *and* $\overline{P} = \overline{P}_\delta$ *be defined by* (5.4.11) *and Lemma 5.4.4, respectively. Then for almost every* $t \in (0,T)$ *the functions* $(\mathbf{u}, \varrho)$ *and* $\overline{P}$ *satisfy inequality* (5.4.27).

*Proof.* Choose $t_0 \in (0,T)$ and $h > 0$ such that $(t_0 - h, t_0 + h) \subset (0,T)$. Recall that $P'' \geq 0$ in view of (5.1.7). Integrate both sides of (5.4.9) over $(t_0 - h, t_0 + h)$ to obtain

$$\int_{t_0-h}^{t_0+h} \int_\Omega (\varrho_\varepsilon |\mathbf{u}_\varepsilon|^2 + P(\varrho_\varepsilon))(x,t)\,dxdt + \int_{t_0-h}^{t_0+h} \Big\{ \int_0^t \int_\Omega (|\nabla \mathbf{u}_\varepsilon|^2)(x,s)\,dxds \Big\}\,dt$$
$$\leq \int_{t_0-h}^{t_0+h} \Big\{ \int_0^t \int_\Omega \varepsilon^{4/3} |\nabla \varrho_\varepsilon|^2\,dxds \Big\}\,dt$$
$$+ \int_{t_0-h}^{t_0+h} \Big\{ c_e + M_\varepsilon(t) + c_e \int_0^t e^{c_e(t-s)} M_\varepsilon(s)\,ds \Big\}\,dt.$$

It follows from (5.2.4) that $\|\varepsilon^{4/3} |\nabla \varrho_\varepsilon|^2\|_{L^1(Q)} \leq c_{E,p} \varepsilon^{1/3} \to 0$ as $\varepsilon \to 0$. Letting $\varepsilon \to 0$ and applying Lemma 5.4.14 with $T' = t_0 - h$ and $T'' = t_0 + h$ and Lemma 5.4.15 we obtain

$$\frac{1}{2h} \int_{t_0-h}^{t_0+h} \int_\Omega (\varrho |\mathbf{u}|^2 + \overline{P})(x,t)\,dxdt + \frac{1}{2h} \int_{t_0-h}^{t_0+h} \Big\{ \int_0^t \int_\Omega (|\nabla \mathbf{u}|^2)(x,s)\,dxds \Big\}\,dt$$
$$\leq \frac{1}{2h} \int_{t_0-h}^{t_0+h} \Big\{ c_e + M_\delta(t) + c_e \int_0^t e^{c_e(t-s)} M_\delta(s)\,ds \Big\}\,dt.$$

Letting $h \to 0$ yields (5.4.27). □

It remains to notice that the last claim (v) coincides with the statement of Lemma 5.4.16. This completes the proof of Theorem 5.4.5.

## 5.5 Passage to the limit. The third level

In this section we pass to the limit as $\delta \to 0$ and thus we complete the proof of the main Theorem 5.1.6. We begin with the observation that the functions $\varrho_\delta$ and $\mathbf{u}_\delta$ satisfy estimates (5.4.18)–(5.4.22) of Theorem 5.4.5. Hence, passing to a subsequence, we can assume that there exist $\varrho \in L^\infty(0,T; L^\gamma(\Omega)) \cap L^{\gamma+1}_{\text{loc}}(Q)$ and $\mathbf{u} \in L^\infty(0,T; W^{1,2}(\Omega))$ with the properties

$$\begin{aligned} &\varrho_\delta \rightharpoonup \varrho \quad \text{weakly}^\star \text{ in } L^\infty(0,T; L^\gamma(\Omega)), \\ &\mathbf{u}_\delta \rightharpoonup \mathbf{u} \quad \text{weakly in } L^2(0,T; W^{1,2}(\Omega)), \\ &\varrho_\delta \rightharpoonup \varrho \quad \text{weakly in } L^{\gamma+1}(Q') \quad \text{for any compact } Q' \Subset Q. \end{aligned} \tag{5.5.1}$$

Finally arguing as in the proof of Lemma 5.4.2 we can assume that

$$\begin{aligned} &\varrho_\delta \mathbf{u}_\delta \rightharpoonup \varrho \mathbf{u} \quad \text{weakly}^\star \text{ in } L^\infty(0,T; L^{2\gamma/(\gamma+1)}(\Omega)), \\ &\varrho_\delta \mathbf{u}_\delta \otimes \mathbf{u}_\delta \rightharpoonup \varrho \mathbf{u} \otimes \mathbf{u} \quad \text{weakly in } L^2(0,T; L^z(\Omega)). \end{aligned} \tag{5.5.2}$$

In the previous section we have proved the existence of a Young measure $\mu^\delta$ associated with the sequence $\varrho_\varepsilon$. Recall that $\mu^\delta$ is an element of the Banach space $L^\infty_w(Q; \mathcal{M}(\mathbb{R}))$, where $\mathcal{M}(\mathbb{R})$ is the space of Borel measures on $\mathbb{R}$. By Theorem 1.2.24, the latter space is the dual to the Banach space $C_0(\mathbb{R})$ of all continuous functions on the real axis vanishing at infinity. Since $C_0(\mathbb{R})$ is separable, we can apply the generalized Riesz Theorem 1.3.24 to deduce that $L^\infty_w(Q; \mathcal{M}(\mathbb{R}))$ is the dual of $L^1(Q; C_0(\mathbb{R}))$, and for every functional $V \in L^1(Q; C_0(\mathbb{R}))'$, there exists a unique parameterized measure $\nu \in L^\infty(Q; \mathcal{M}(\mathbb{R}))$ such that

$$\langle V, v \rangle = \int_Q \langle \nu_{xt}, v(x,t,\cdot) \rangle \, dxdt \quad \text{for all } v \in L^1(Q; C_0(\mathbb{R})).$$

The converse is true in the sense that every $\nu \in L^\infty(Q; \mathcal{M}(\mathbb{R}))$ determines an element of the dual space $L^1(Q; C_0(\mathbb{R}))'$. Since $L^1(Q; C_0(\mathbb{R}))$ is separable, it follows from the Alaoglu theorem that every bounded sequence in $L^1(Q; C_0(\mathbb{R}))'$ contains a weakly$^\star$ convergent subsequence. By definition, the Young measures $\mu^\delta$, $\delta > 0$, belong to the space $L^\infty_w(Q; \mathcal{M}(\mathbb{R}))$ and the norm of $\mu^\delta$ in this space is equal to 1. Hence, passing to a subsequence, we can assume that $\mu^\delta$ converges weakly$^\star$ in $L^\infty_w(Q; \mathcal{M}(\mathbb{R}))$ to some $\mu \in L^\infty_w(Q; \mathcal{M}(\mathbb{R}))$. This means that for all $v \in L^1(Q; C_0(\mathbb{R}))$,

$$\int_Q \langle \mu^\delta_{xt}, v(x,t,\cdot) \rangle \, dxdt \to \int_Q \langle \mu_{xt}, v(x,t,\cdot) \rangle \, dxdt \quad \text{as } \delta \to 0. \tag{5.5.3}$$

It is worth noting that this convergence alone does not imply that $\mu$ is a Young measure.

Now our task is to show that the limit functions $(\mathbf{u}, \varrho)$ determine a renormalized solution to problem (5.1.12) in the sense of Definition 5.1.4 and meet all requirements of Theorem 5.1.6. We split the proof into several lemmas. The first shows that the parameterized measure $\mu$ defined by (5.5.3) is a Young measure.

**Lemma 5.5.1.** *(i) The parameterized measure $\mu$ defined by* (5.5.3) *is a Young measure. The limit function $\varrho$ defined by* (5.5.1) *has the representation $\varrho(x,t) = \langle \mu_{xt}, \lambda \rangle$.*

*(ii) Let $p(\varrho)$ be a pressure function and define the internal energy*

$$P(\varrho) = \varrho \int_0^\varrho s^{-2} p(s) \, ds.$$

*Then the functions*

$$\overline{p}(x,t) := \langle \mu_{xt}, p(\lambda) \rangle, \quad \overline{P}(x,t) := \langle \mu_{xt}, P(\lambda) \rangle$$

*belong to $L^\infty(0,T; L^1(\Omega))$, and $\overline{p\varrho}(x,t) = \langle \mu_{xt}, p(\lambda)\lambda \rangle$ belongs to $L^1(Q')$ for any compact set $Q' \Subset Q$.*

*(iii) For any nonnegative $\psi \in L^\infty(Q)$ and compact $Q' \Subset Q$,*

$$\int_Q \psi\overline{p}\,dxdt \le \liminf_{\delta\to 0}\int_Q \psi\overline{p}_\delta\,dxdt,$$
$$\int_Q \psi\overline{P}\,dxdt \le \liminf_{\delta\to 0}\int_Q \psi\overline{P}_\delta\,dxdt, \tag{5.5.4}$$

$$\int_{Q'} \psi\overline{p\varrho}\,dxdt \le \liminf_{\delta\to 0}\int_{Q'} \psi\overline{p\varrho}_\delta\,dxdt, \quad \int_{Q'} \psi\overline{p}\,dxdt = \lim_{\delta\to 0}\int_{Q'} \psi\overline{p}_\delta\,dxdt, \tag{5.5.5}$$

*where $\overline{p}_\delta$ and $\overline{p\varrho}_\delta$ are defined in Lemma 5.4.4.*

*Proof.* To avoid repetitions we give the proof only for the pressure function $p$. The same proof works for $P$.

*Proof of claim (i).* First observe that by Lemma 5.4.3 and (5.4.19),

$$\int_Q \langle\mu^\delta, p(\lambda)\rangle\,dxdt = \int_Q \overline{p}_\delta\,dxdt \le c_{E,p},$$

and for any fixed compact set $Q' \Subset Q$,

$$\int_{Q'} \langle\mu^\delta, p(\lambda)\lambda\rangle\,dxdt = \int_{Q'} \overline{p\varrho}_\delta\,dxdt \le c_{E,p}(Q').$$

It now follows from the inequality $\varrho^\gamma \le c_e(p(\varrho)+1)$ that

$$\int_Q\int_N^\infty \lambda\,d\mu^\delta_{xt}(\lambda)\,dxdt \le c_{E,p}N^{-\gamma+1}, \quad \int_{Q'}\int_N^\infty p(\lambda)\,d\mu^\delta_{xt}(\lambda)\,dxdt \le c_{E,p}N^{-1}. \tag{5.5.6}$$

Now fix $\eta \in C_0^\infty(\mathbb{R})$ with the properties

$$0 \le \eta \le 1, \quad \eta(\varrho) = 1 \quad \text{if } |\varrho| \le 1, \quad \eta(\varrho) = 0 \quad \text{if } |\varrho| \ge 2.$$

The limit element $\mu$ defines a weakly$^\star$ measurable mapping $Q \ni (x,t) \mapsto \mu_{xt} \in \mathcal{M}(\mathbb{R})$. By the definition of weak$^\star$ limit, the Radon measure $\mu_{xt}$ is nonnegative. Since all $\mu^\delta_{xt}$ are supported on the convex set $[0,\infty)$, so is $\mu_{xt}$. Next, for any $\psi \in L^\infty(Q)$ we have

$$\int_Q \psi\langle\mu_{xt}, \eta(\lambda/N)\lambda\rangle\,dxdt = \lim_{\delta\to 0}\int_Q \psi\langle\mu^\delta_{xt}, \eta(\lambda/N)\lambda\rangle\,dxdt. \tag{5.5.7}$$

Assume now that $\psi \in L^\infty(Q)$ is nonnegative. We have

$$\int_Q \psi\langle\mu^\delta_{xt}, \eta(\lambda/N)\lambda\rangle\,dxdt = \int_Q \psi\varrho_\delta(x,t)\,dxdt + \mathcal{O}_N,$$

where the remainder is of the form

$$\mathcal{O}_N = \int_Q \psi \langle \mu^\delta_{xt}, (\eta(\lambda/N) - 1)\lambda \rangle \, dxdt.$$

By the first inequality in (5.5.6),

$$|\mathcal{O}_N| \leq \|\psi\|_{L^\infty(Q)} \int_Q \Big( \int_N^\infty \lambda \, d\mu^\delta_{xt} \Big) \, dxdt \leq c_{E,p} \|\psi\|_{L^\infty(Q)} N^{-\gamma+1}.$$

Thus we get

$$\limsup_{\delta \to 0} \Big| \int_Q \psi \langle \mu^\delta_{xt}, \eta(\lambda/N)\lambda \rangle \, dxdt - \int_Q \psi \varrho_\delta \, dxdt \Big| \leq c_{E,p} \|\psi\|_{L^\infty(Q)} N^{-\gamma+1}.$$

Since $\varrho_\delta$ converges to $\varrho$ weakly in $L^\gamma(Q)$, we now deduce from (5.5.7) that

$$\Big| \int_Q \psi \langle \mu_{xt}, \eta(\lambda/N)\lambda \rangle \, dxdt - \int_Q \psi \varrho \, dxdt \Big| \leq c_{E,p} \|\psi\|_{L^\infty(Q)} N^{-\gamma+1}. \tag{5.5.8}$$

Notice that $\eta(\lambda/N)\lambda \nearrow \lambda$ as $N \to \infty$ on $[0, \infty)$. Hence by the Fatou Theorem 1.2.16,

$$\langle \mu_{xt}, \eta(\lambda/N)\lambda \rangle \nearrow \langle \mu_{xt}, \lambda \rangle \quad \text{as } N \to \infty.$$

Letting $N \to \infty$ in (5.5.8) we obtain

$$\int_Q \psi(\langle \mu_{xt}, \lambda \rangle - \varrho) \, dxdt = 0 \quad \text{for all nonnegative } \psi \in L^\infty(Q). \tag{5.5.9}$$

Therefore $\langle \mu_{xt}, \lambda \rangle = \varrho(x,t)$ for a.e. $(x,t) \in Q$. Now replace $\langle \mu_{xt}, \eta(\lambda/N)\lambda \rangle$ by $\langle \mu_{xt}, \eta(\lambda/N) \rangle$ in (5.5.7) and repeat the previous arguments. Thus we arrive at $\langle \mu_{xt}, 1 \rangle = 1$ for a.e. $(x,t) \in Q$. Hence $\mu$ is a Young measure. This proves claim (i).

*Proof of claim (ii).* Relation (5.5.3) with $v = \eta(\lambda/N)p(\lambda)$ implies

$$\begin{aligned}\int_Q \langle \mu_{xt}, \eta(\lambda/N)p(\lambda) \rangle \, dxdt &= \lim_{\delta \to 0} \int_Q \langle \mu^\delta_{xt}, \eta(\lambda/N)p(\lambda) \rangle \, dxdt \\ &\leq \liminf_{\delta \to 0} \int_Q \langle \mu^\delta_{xt}, p(\lambda) \rangle \, dxdt = \liminf_{\delta \to 0} \int_Q \overline{p}_\delta \, dxdt.\end{aligned}$$

It now follows from (5.4.19) that

$$\int_Q \langle \mu_{xt}, \eta(\lambda/N)p(\lambda) \rangle \, dxdt \leq c_{E,p}. \tag{5.5.10}$$

Since $\eta(\lambda/N)p(\lambda) \nearrow p(\lambda)$ as $N \to \infty$, by the Fatou Theorem 1.2.16 we have

$$\langle \mu_{xt}, \eta(\lambda/N)p(\lambda) \rangle \nearrow \langle \mu_{xt}, p(\lambda) \rangle = \overline{p}(x,t) \quad \text{as } N \to \infty.$$

Applying once more Theorem 1.2.16 we arrive at

$$c_{E,p} \geq \int_Q \langle \mu_{xt}, \eta(\lambda/N)p(\lambda)\rangle \, dxdt \nearrow \int_Q \langle \mu_{xt}, p(\lambda)\rangle \, dxdt \quad \text{as } N \to \infty.$$

Hence $\overline{p} \in L^1(Q)$. Next choose $Q' \Subset Q$ and define $v(x,t,\lambda) = p(\lambda)\lambda\eta(\lambda/N)$ if $(x,t) \in Q'$ and $v(x,t,\lambda) = 0$ otherwise. By (5.5.3) we have

$$\begin{aligned}\int_{Q'} \langle \mu_{xt}, \eta(\lambda/N)p(\lambda)\lambda\rangle \, dxdt &= \lim_{\delta\to 0} \int_{Q'} \langle \mu^{\delta}_{xt}, \eta(\lambda/N)p(\lambda)\lambda\rangle \, dxdt \\ &\leq \liminf_{\delta\to 0} \int_{Q'} \langle \mu^{\delta}_{xt}, p(\lambda)\lambda\rangle \, dxdt = \liminf_{\delta\to 0} \int_{Q'} \overline{p\varrho}_{\delta} \, dxdt.\end{aligned}$$

From this and (5.4.22) we get

$$\int_{Q'} \langle \mu_{xt}, \eta(\lambda/N)p(\lambda)\lambda\rangle \, dxdt \leq c_{E,p}(Q').$$

Since $\eta(\lambda/N)p(\lambda)\lambda \nearrow p(\lambda)\lambda$ as $N \to \infty$, we apply the Fatou Theorem 1.2.16 twice to obtain

$$\langle \mu_{xt}, \eta(\lambda/N)p(\lambda)\lambda\rangle \nearrow \langle \mu_{xt}, p(\lambda)\lambda\rangle = \overline{p\varrho}(x,t) \quad \text{as } N \to \infty$$

and

$$c_{E,p} \geq \int_{Q'} \langle \mu_{xt}, \eta(\lambda/N)p(\lambda)\lambda\rangle dxdt \nearrow \int_{Q'} \langle \mu_{xt}, p(\lambda)\lambda\rangle \, dxdt \quad \text{as } N \to \infty.$$

Hence $\overline{p\varrho}$ belongs to $L^1(Q)$. This completes the proof of claim (ii).

*Proof of claim (iii).* Let $\psi \in L^{\infty}(Q)$ be nonnegative. By the convergence (5.5.3) with $v = \psi(x,t)\eta(\lambda/N)p(\lambda)$, we have

$$\begin{aligned}\int_Q \psi\langle \mu_{xt}, \eta(\lambda/N)p(\lambda)\rangle \, dxdt &= \lim_{\delta\to 0} \int_Q \psi\langle \mu^{\delta}_{xt}, \eta(\lambda/N)p(\lambda)\rangle \, dxdt \\ &\leq \liminf_{\delta\to 0} \int_Q \psi\langle \mu^{\delta}_{xt}, p(\lambda)\rangle \, dxdt = \liminf_{\delta\to 0} \int_Q \psi\overline{p}_{\delta} \, dxdt. \quad (5.5.11)\end{aligned}$$

Since $\eta(\lambda/N)p(\lambda) \nearrow p(\lambda)$ as $N \to \infty$, the Fatou Theorem 1.2.16 yields

$$\langle \mu_{xt}, \eta(\lambda/N)p(\lambda)\rangle \nearrow \langle \mu_{xt}, p(\lambda)\rangle \quad \text{as } N \to \infty.$$

Applying once more Theorem 1.2.16 we arrive at

$$\int_Q \psi\langle \mu_{xt}, \eta(\lambda/N)p(\lambda)\rangle \, dxdt \nearrow \int_Q \psi\langle \mu_{xt}, p(\lambda)\rangle \, dxdt \quad \text{as } N \to \infty.$$

Therefore, we can let $N \to \infty$ in (5.5.11) to obtain (5.5.4). Now we repeat these arguments to derive (5.5.5). Fix $Q' \Subset Q$. Notice that $Q' \Subset Q_\delta$ for all sufficiently small $\delta$. In view of (5.5.11),

$$\int_{Q'} \psi \langle \mu_{xt}, \eta(\lambda/N) p(\lambda)\rangle \, dxdt = \lim_{\delta \to 0} \int_{Q'} \psi \langle \mu_{xt}^\delta, \eta(\lambda/N) p(\lambda)\rangle \, dxdt \qquad (5.5.12)$$

for any nonnegative $\psi \in L^\infty(Q)$. On the other hand,

$$\int_{Q'} \psi \langle \mu_{xt}^\delta, \eta(\lambda/N) p(\lambda)\rangle \, dxdt = \int_{Q'} \psi \overline{p}_\delta(x,t) + \mathcal{O}'_N,$$

where

$$\mathcal{O}'_N = \int_{Q'} \psi \langle \mu_{xt}^\delta, (\eta(\lambda/N) - 1) p(\lambda)\rangle \, dxdt.$$

By the second inequality in (5.5.6),

$$|\mathcal{O}'_N| \le \|\psi\|_{L^\infty(Q')} \int_{Q'} \Big( \int_N^\infty p(\lambda)\, d\mu_{xt}^\delta \Big) \, dxdt \le c(Q') \|\psi\|_{L^\infty(Q')} N^{-1}.$$

Thus we get

$$\limsup_{\delta \to 0} \Big| \int_{Q'} \psi \langle \mu_{xt}^\delta, \eta(\lambda/N) p(\lambda)\rangle \, dxdt - \int_{Q'} \psi \overline{p}_\delta \, dxdt \Big| \le c(Q') \|\psi\|_{L^\infty(Q')} N^{-1}.$$

We now find from (5.5.12) that

$$\begin{aligned} - cN^{-1} + \int_{Q'} \psi \langle \mu_{xt}, \eta(\lambda/N) p(\lambda)\rangle \, dxdt &\le \liminf_{\delta \to 0} \int_{Q'} \psi \overline{p}_\delta \, dxdt \\ \le \limsup_{\delta \to 0} \int_{Q'} \psi \overline{p}_\delta \, dxdt &\le cN^{-1} + \int_{Q'} \psi \langle \mu_{xt}, \eta(\lambda/N) p(\lambda)\rangle \, dxdt. \qquad (5.5.13) \end{aligned}$$

On the other hand, applying the Fatou theorem twice we obtain

$$\lim_{N \to \infty} \int_{Q'} \psi \langle \mu_{xt}, \eta(\lambda/N) p(\lambda)\rangle \, dxdt = \int_{Q'} \psi \langle \mu_{xt}, p(\lambda)\rangle \, dxdt = \int_Q \psi \overline{p} \, dxdt.$$

Letting $N \to \infty$ in (5.5.13) we obtain

$$\int_{Q'} \psi \langle \mu_{xt}, p(\lambda)\rangle = \lim_{\delta \to 0} \int_{Q'} \psi \overline{p}_\delta \, dxdt,$$

which yields the equality in (5.5.5). It remains to prove the inequality in (5.5.5). For any nonnegative $\psi \in L^\infty(Q)$ we have

$$\begin{aligned} \int_{Q'} \psi \langle \mu_{xt}, \eta(\lambda/N) p(\lambda) \lambda\rangle \, dxdt &= \lim_{\delta \to 0} \int_{Q'} \psi \langle \mu_{xt}^\delta, \eta(\lambda/N) p(\lambda)\lambda\rangle \, dxdt \\ &\le \liminf_{\delta \to 0} \int_{Q'} \psi \langle \mu_{xt}^\delta, p(\lambda)\lambda\rangle \, dxdt = \liminf_{\delta \to 0} \int_{Q'} \psi \overline{p\varrho}_\delta \, dxdt. \qquad (5.5.14) \end{aligned}$$

Since $\eta(\lambda/N)p(\lambda)\lambda \nearrow p(\lambda)\lambda$ as $N \to \infty$, arguing as before, we apply the Fatou theorem and let $N \to \infty$ to obtain the inequality in (5.5.5). □

Lemma 5.5.1 implies the strong convergence of the sequence $\varrho_\delta$.

**Lemma 5.5.2.** *The limits $\varrho$, $\overline{p}$, and $\overline{P}$ satisfy $\overline{p} = p(\varrho)$ and $\overline{P} = P(\varrho)$. Moreover, $\varrho_\delta \to \varrho$ a.e. in $Q$.*

*Proof.* First we show that for a.e. $(x,t) \in Q$, the Young measure $\mu_{xt}$ is the Dirac measure concentrated at $\varrho(x,t)$. We begin with the observation that the conclusion of Lemma 5.4.4 obviously holds for $\mu_{xt}$ in place of $\mu^\delta_{xt}$. By that lemma it suffices to prove that $\overline{p\varrho} - \overline{p}\varrho = 0$ a.e. in $Q$. Notice that $\overline{p\varrho} - \overline{p}\varrho \ge 0$ by (5.4.17).

Next, fix a compact subset $Q' \Subset Q$ and let $\psi$ be the characteristic function of $Q'$. Then (5.5.5) implies

$$\int_{Q'} \overline{p\varrho}\, dxdt \le \liminf_{\delta\to 0} \int_{Q'} \overline{p\varrho_\delta}\, dxdt \quad \text{and} \quad \int_{Q'} \overline{p}\, dxdt = \lim_{\delta\to 0} \int_{Q'} \overline{p}_\delta\, dxdt,$$

which leads to

$$\int_{Q'} (\overline{p\varrho} - \overline{p}\varrho)\, dxdt \le \liminf_{\delta\to 0} \int_{Q'} (\overline{p\varrho_\delta} - \overline{p}_\delta\varrho)\, dxdt.$$

Combining this with (5.4.26) we obtain

$$\int_{Q'} (\overline{p\varrho} - \overline{p}\varrho)\, dxdt \le c_{E,p}\, \mathrm{meas}(Q \setminus Q')^{(\gamma-2)/(2\gamma)}.$$

Choosing an increasing sequence of sets $Q'$ whose union covers $Q$ and recalling that $\overline{p\varrho} - \overline{p}\varrho \ge 0$ we arrive at

$$\int_{Q} (\overline{p\varrho} - \overline{p}\varrho)\, dxdt = 0, \quad \text{and so} \quad \overline{p\varrho} - \overline{p}\varrho = 0 \quad \text{a.e. in } Q.$$

Applying Lemma 5.4.4 we conclude that $\mu_{xt}$ is the Dirac measure concentrated at $\varrho(x,t)$ for a.e. $(x,t) \in Q$. In particular,

$$\overline{p}(x,t) = \langle \mu_{xt}, p(\lambda)\rangle = p(\varrho(x,t)), \quad \overline{P}(x,t) = \langle \mu_{xt}, P(\lambda)\rangle = P(\varrho(x,t)).$$

This leads to the desired equalities in Lemma 5.5.2. Next notice that the sequence $\varrho_\delta^2$ is bounded in $L^{\gamma/2}(Q)$ and converges weakly in this space to $\overline{\varrho^2}$. We have

$$\overline{\varrho^2}(x,t) = \langle \mu_{xt}, \lambda^2\rangle = \varrho(x,t)^2,$$

since $\mu_{xt}$ is the Dirac measure concentrated at $\varrho(x,t)$. This yields

$$\lim_{\delta\to 0} \int_Q \varrho_\delta(x,t)^2\, dxdt = \int_Q \varrho(x,t)^2\, dxdt.$$

Therefore, $\varrho_\delta$ converges to $\varrho$ weakly in $L^2(Q)$ and $\|\varrho_\delta\|_{L^2(Q)} \to \|\varrho\|_{L^2(Q)}$. This is possible if and only if $\varrho_\delta \to \varrho$ strongly in $L^2(Q)$. Passing to a subsequence we can assume that $\varrho_\delta$ converges to $\varrho$ a.e. in $Q$. □

Our next task is to derive an energy estimate for $(\mathbf{u}, \varrho)$.

**Lemma 5.5.3.** *Let $(\mathbf{u}, \varrho)$ be defined by (5.5.1). Then*

$$\int_\Omega (P(\varrho) + \varrho|\mathbf{u}|^2)(x,t)\,dx + \int_0^t \int_\Omega |\nabla \mathbf{u}(x,s)|^2\,dxds \le c_e \tag{5.5.15}$$

*for a.e. $t \in (0,T)$, where the constant $c_e$ is as in Remark 5.1.2.*

*Proof.* Choose $t_0 \in (0,T)$ and $h > 0$ such that $(t_0 - h, t_0 + h) \subset (0,T)$. Let $\psi(x,t)$ be the characteristic function of $\Omega \times (t_0 - h, t_0 + h)$. By the choice of $\psi$ and (5.5.4),

$$\int_{t_0-h}^{t_0+h} \int_\Omega \overline{P}(x,t)\,dxdt \le \liminf_{\delta\to 0} \int_{t_0-h}^{t_0+h} \int_\Omega \overline{P}_\delta(x,t)\,dxdt.$$

By (5.5.2), the sequence $\varrho_\delta \mathbf{u}_\delta \otimes \mathbf{u}_\delta$ converges to $\varrho\mathbf{u} \otimes \mathbf{u}$ weakly in $L^2(0,T; L^z(\Omega))$. Hence it also converges weakly in $L^1(Q)$, whence in $L^1(\Omega \times (t_0 - h, t_0 + h))$. Recalling that $\varrho|\mathbf{u}|^2 = \operatorname{tr} \varrho\mathbf{u} \otimes \mathbf{u}$ we obtain

$$\int_{t_0-h}^{t_0+h} \int_\Omega \varrho|\mathbf{u}|^2\,dxdt = \liminf_{\delta\to 0} \int_{t_0-h}^{t_0+h} \int_\Omega \varrho_\delta|\mathbf{u}_\delta|^2\,dxdt.$$

Since $\nabla \mathbf{u}_\delta$ converges to $\nabla \mathbf{u}$ weakly in $L^2(Q)$, we also have

$$\int_0^t \int_\Omega |\nabla \mathbf{u}|^2\,dxds \le \liminf_{\delta\to 0} \int_0^t \int_\Omega |\nabla \mathbf{u}_\delta|^2\,dxds.$$

It follows that

$$\begin{aligned}\frac{1}{2h}\int_{t_0-h}^{t_0+h} &\Big\{ \int_\Omega (\overline{P} + \varrho|\mathbf{u}|^2)(x,t)\,dx + \int_0^t \int_\Omega |\nabla \mathbf{u}|^2\,dxds \Big\}\,dt \\ &\le \liminf_{\delta\to 0} \frac{1}{2h}\int_{t_0-h}^{t_0+h} \Big\{ \int_\Omega (\overline{P}_\delta + \varrho_\delta|\mathbf{u}_\delta|^2)(x,t)\,dx + \int_0^t \int_\Omega |\nabla \mathbf{u}_\delta|^2\,dxds \Big\}\,dt.\end{aligned} \tag{5.5.16}$$

Inequality (5.4.27) implies

$$\begin{aligned}\frac{1}{2h}\int_{t_0-h}^{t_0+h} &\Big\{ \int_\Omega (\overline{P}_\delta + \varrho_\delta|\mathbf{u}_\delta|^2)(x,t)\,dx + \int_0^t \int_\Omega |\nabla \mathbf{u}_\delta|^2\,dxds \Big\}\,dt \\ &\le \frac{1}{2h}\int_{t_0-h}^{t_0+h} \Big\{ c_e + M_\delta(t) + c_e \int_0^t e^{c_e(t-s)} M_\delta(s)\,ds \Big\}\,dt \le c_e + c_e \sup_{t\in[0,T]} M_\delta(t),\end{aligned} \tag{5.5.17}$$

where $M_\delta(t) = 2\|\sqrt{a}\Delta^2 \mathbf{U}(t)\|^2_{L^2(\Omega)}$. Notice that $\mathbf{U} \in C^\infty(Q)$. The nonnegative function $a$ does not exceed 1 and is supported in the $\delta$-neighborhood of $\partial\Omega$. This leads to the estimate $M_\delta \le c_{E,p}\delta$. Inserting this in (5.5.17), and applying (5.5.16), we obtain

$$\frac{1}{2h}\int_{t_0-h}^{t_0+h}\Big\{\int_\Omega (\overline{P}+\varrho|\mathbf{u}|^2)(x,t)\,dx + \int_0^t\int_\Omega |\nabla\mathbf{u}|^2\,dxdt\Big\}\,dt \le c_e.$$

Letting $h\to 0$ and noting that $\overline{P}=P(\varrho)$ we arrive at (5.5.15). □

The next lemma shows that the limit functions $(\mathbf{u},\varrho)$ satisfy the mass balance equation.

**Lemma 5.5.4.** *Let $(\mathbf{u},\varrho)$ be defined by (5.5.1). Then the integral identity*

$$\int_Q (\varrho\mathbf{u}\cdot\partial_t\boldsymbol{\xi} + \varrho\mathbf{u}\otimes\mathbf{u}:\nabla\boldsymbol{\xi} + p(\varrho)\operatorname{div}\boldsymbol{\xi} - \mathbb{S}(\mathbf{u}):\nabla\boldsymbol{\xi})\,dxdt$$
$$+\int_Q \varrho\mathbf{f}\cdot\boldsymbol{\xi}\,dxdt + \int_\Omega (\varrho_\infty\mathbf{U}\cdot\boldsymbol{\xi})(x,0)\,dx = 0 \quad (5.5.18)$$

*holds for all vector fields $\boldsymbol{\xi}\in C^\infty(Q)$ equal to 0 in a neighborhood of the lateral side $S_T=\partial\Omega\times(0,T)$ and of the top $\Omega\times\{T\}$.*

*Proof.* Choose a vector field $\boldsymbol{\zeta}$ which vanishes in a neighborhood of $S_T$ and of $\Omega\times\{T\}$. There is a $\delta_0>0$ such that $\operatorname{supp}\boldsymbol{\zeta}\subset Q_{2\delta_0}$. It follows from (5.5.1)–(5.5.2) that

$$\lim_{\delta\to 0}\int_Q \{\varrho_\delta\mathbf{u}_\delta\cdot\partial_t\boldsymbol{\zeta} + (\varrho_\delta\mathbf{u}_\delta\otimes\mathbf{u}_\delta - \mathbb{S}(\mathbf{u}_\delta)):\nabla\boldsymbol{\zeta} + \varrho_\delta\mathbf{f}\cdot\boldsymbol{\zeta}\}\,dxdt$$
$$=\int_Q \{\varrho\mathbf{u}\cdot\partial_t\boldsymbol{\zeta} + (\varrho\mathbf{u}\otimes\mathbf{u} - \mathbb{S}(\mathbf{u})):\nabla\boldsymbol{\zeta} + \varrho\mathbf{f}\cdot\boldsymbol{\zeta}\}\,dxdt. \quad (5.5.19)$$

Since for any $h>0$ the set $Q'=\Omega_{2\delta_0}\times[h,T-h]$ is a compact subset of $Q_{\delta_0}$, the equality in (5.5.5) with $\psi=\operatorname{div}\zeta$ implies

$$\limsup_{\delta\to 0}\Big|\int_{Q_{2\delta_0}}\operatorname{div}\boldsymbol{\zeta}\,(\overline{p}-\overline{p}_\delta)\,dxdt\Big| \le \limsup_{\delta\to 0}\Big|\int_h^{T-h}\int_{\Omega_{2\delta_0}}\operatorname{div}\boldsymbol{\zeta}(\overline{p}-\overline{p}_\delta)\,dxdt\Big|$$
$$+c\limsup_{\delta\to 0}\Big\{\int_0^h+\int_{T-h}^T\Big\}\int_{\Omega_{2\delta}}(\overline{p}+\overline{p}_\delta)\,dxdt \le c\Big\{\int_0^h+\int_{T-h}^T\Big\}\,dt = 2ch\to 0$$
$$(5.5.20)$$

as $h\to 0$. Letting $\delta\to 0$ in (5.4.23), using (5.5.19)–(5.5.20) and recalling the equality $\overline{p}=p(\varrho)$ in Lemma 5.5.2 we obtain the integral identity (5.5.18). Hence the limit functions $\varrho$ and $\mathbf{u}$ satisfy the momentum balance equation. □

Finally we show that $\varrho$ is a renormalized solution to the mass balance equation.

**Lemma 5.5.5.** *Let* $(\mathbf{u}, \varrho)$ *be defined by* (5.5.1). *Then the integral identity*

$$\int_Q \Big(\varphi(\varrho)\partial_t\psi + \varphi(\varrho)\mathbf{u}\cdot\nabla\psi + \psi\big(\varphi(\varrho) - \varphi'(\varrho)\varrho\big)\operatorname{div}\mathbf{u}\Big)\,dxdt$$
$$= \int_{\Sigma_{\text{in}}} \psi\varphi(\varrho_\infty)\mathbf{U}\cdot\mathbf{n}\,d\Sigma - \int_\Omega (\varphi(\varrho_\infty)\psi)(x,0)\,dx \quad (5.5.21)$$

*holds for all* $\psi \in C^\infty(Q)$ *vanishing in a neighborhood of the surface* $S_T \setminus \Sigma_{\text{in}}$ *and of the top* $\Omega \times \{T\}$, *and for all smooth functions* $\varphi : [0,\infty) \to \mathbb{R}$ *such that*

$$\limsup_{\varrho\to\infty}(\varrho^{-1}|\varphi(\varrho)| + |\varphi'(\varrho)|) < \infty. \quad (5.5.22)$$

*Proof.* Fix $\psi$ and $\varphi$ satisfying the conditions of this lemma. Recall that by the integral identity (5.4.24) we have

$$\int_Q \Big(\varphi(\varrho_\delta)\partial_t\psi + \varphi(\varrho_\delta)\mathbf{u}_\delta\cdot\nabla\psi - \big(\varphi'(\varrho_\delta)\varrho_\delta - \varphi(\varrho_\delta)\big)\operatorname{div}\mathbf{u}_\delta\Big)\,dxdt$$
$$- \int_{\Sigma_{\text{in}}} \psi\varphi(\varrho_\infty)\mathbf{U}\cdot\mathbf{n}\,d\Sigma + \int_\Omega (\psi\varphi(\varrho_\infty))(x,0)\,dx = 0. \quad (5.5.23)$$

In view of Lemma 5.5.2, the sequence $\varrho_\delta$ converges to $\varrho$ a.e. in $Q$. Since $\varphi$ and $\varphi'$ are continuous, the sequences $\varphi(\varrho_\delta)$ and $\varphi'(\varrho_\delta)\varrho_\delta$ converge to $\varphi(\varrho)$ and $\varphi'(\varrho)\varrho$, respectively, a.e. in $Q$. On the other hand, it follows from (5.4.19) and the growth condition (5.5.22) imposed on $\varphi$ that the sequences $\varphi(\varrho_\delta)$ and $\varphi'(\varrho_\delta)\varrho_\delta$ are bounded in $L^\gamma(Q)$. By the Lebesgue dominated convergence theorem, these sequences converge to $\varphi(\varrho)$ and $\varphi'(\varrho)\varrho$, respectively, in $L^1(Q)$. Letting $\delta \to 0$ in (5.5.23) we obtain the integral identity (5.5.21). □

We are now in a position to complete the proof of the main Theorem 5.1.6. It follows from Lemmas 5.5.4 and 5.5.5 that the functions $(\mathbf{u}, \varrho)$ defined by (5.5.1) satisfy the integral identities (5.1.13), (5.1.14) and meet all requirements of Definition 5.1.4 of a renormalized solution to problem (5.1.12). On the other hand, in view of Lemma 5.5.3, $(\mathbf{u}, \varrho)$ satisfy the energy estimate (5.1.17). This completes the proof of Theorem 5.1.6.

# Chapter 6

# Pressure estimate

In Chapter 5 we have proved the existence of renormalized solutions of problem (5.1.12) under the assumptions that the flow domain, the given data $\mathbf{U}$, $\varrho_\infty$, and the constitutive relation $p = p(\varrho)$ satisfy Condition 5.1.1. Moreover, it was also assumed that the adiabatic exponent $\gamma$ in Condition 5.1.1 satisfies $\gamma > 2d$. In this case Theorem 5.1.6 gives the existence of a renormalized solution $(\mathbf{u}, \varrho)$ satisfying the energy estimate (5.1.17). It follows from that estimate that the energy of $(\mathbf{u}, \varrho)$ is bounded by a constant $c_e$ as in Remark 5.1.2, that is, depending only on $\operatorname{diam}\Omega$, $\|\mathbf{U}\|_{C^1(Q)}$, $\|\varrho_\infty\|_{L^\infty(\sqcup_T)}$, and the constant $c_p$ which depends only on $p(\varrho)$. In addition, Proposition 5.3.10 ensures that the pressure function $p(\varrho)$ is integrable with exponent $(\gamma+1)/\gamma > 1$ on every compact subset of the cylinder $Q$. This local estimate plays a crucial role in the proof of the main Theorem 5.1.6. Unfortunately, this estimate involves the constant $c_{E,p}$, which depends on higher order derivatives of the boundary data.

In this chapter we show that for a suitable $l = l(\gamma) > 1$ and $\gamma > d/2$ the $L^l$-norm of the pressure function $p(\varrho)$ is bounded on every compact subset $Q' \Subset Q$ by a constant depending only on $c_e$ and $Q'$. In Chapter 7 this result is applied in order to establish existence theory for the adiabatic exponent $\gamma > d/2$ and for fast oscillating boundary data.

**Problem 6.0.1.** *For given* $\mathbf{f} \in C(Q)$, *find functions* $(\mathbf{u}, \varrho)$ *satisfying the system of equations*

$$\partial_t(\varrho \mathbf{u}) + \operatorname{div}(\varrho\, \mathbf{u} \otimes \mathbf{u}) + \nabla p(\rho) = \operatorname{div} \mathbb{S}(\mathbf{u}) + \varrho \mathbf{f} \quad \textit{in } Q, \tag{6.0.1a}$$

$$\partial_t \varrho + \operatorname{div}(\varrho \mathbf{u}) = 0 \quad \textit{in } \Omega. \tag{6.0.1b}$$

We introduce the following condition on the pressure function $p(\varrho)$.

**Condition 6.0.2.** For the general constitutive law $p = p(\varrho)$ and the internal energy density

$$P(\varrho) = \varrho \int_0^\varrho s^{-2} p(s)\, ds,$$

the functions $p, P : [0,\infty) \to \mathbb{R}^+$ are continuous and there are an exponent $\gamma > 1$ and a constant $c_p > 0$ such that

$$\varrho^\gamma \le c_p(P(\varrho)+1), \quad \varrho^\gamma \le c_p(p(\varrho)+1), \quad p(\varrho) \le c_p(P(\varrho)+1). \tag{6.0.2}$$

**Proposition 6.0.3.** *Let $\Omega$ be a bounded domain with boundary of class $C^\infty$ in $\mathbb{R}^d$, $d = 2,3$, and suppose the pressure function $p$ satisfies Condition 6.0.2 with $\gamma \in (d/2, \infty)$. Let $\varrho \in L^\infty(0,T; L^\gamma(\Omega))$, $\mathbf{u} \in L^2(0,T; W^{1,2}(\Omega))$ be a renormalized solution to equations (6.0.1) such that $p \in L^\infty(0,T; L^1(\Omega)) \cap L^{r'}_{\mathrm{loc}}(Q)$, $1 < r' < 2$, and*

$$\|\mathbf{u}\|_{L^2(0,T;W^{1,2}(\Omega))} + \|\varrho|\mathbf{u}|^2\|_{L^\infty(0,T;L^1(\Omega))} + \|P(\varrho)\|_{L^\infty(0,T;L^1(\Omega))} \le c_e. \tag{6.0.3}$$

*Then for any exponent $\theta$ from the interval*

$$0 \le \theta < \min\{2\gamma/d - 1, \, \gamma/2\}$$

*and for any cylinder $Q' \Subset Q$, there is a constant $c$, depending only on $c_p$, $c_e$, $Q'$, $\gamma$ and $\theta$, such that*

$$\int_{Q'} p(\varrho)\varrho^\theta \, dxdt \le c. \tag{6.0.4}$$

*Proof.* First, fix $\theta \in (0, \min\{2\gamma/d - 1, \gamma/2\})$ and $K > 1$. Set

$$\varphi(\varrho) = \begin{cases} \varrho^\theta & \text{for } 0 \le \varrho < K, \\ K^\theta - \theta K^{\theta-1}\{\varrho^2/2 - (1+K)\varrho + K + K^2/2\} & \text{for } K \le \varrho \le K+1, \\ \varphi(K+1) & \text{for } \varrho > K+1. \end{cases}$$

The function $\varphi$ is continuously differentiable, with the second order derivative piecewise continuous on $(0,\infty)$. It is easily seen that $\varphi(\varrho) \le \varrho^\theta$ on this interval. Set $\varphi := \varphi(\varrho)$. Since $K > 1$ is arbitrary, it suffices to prove that

$$\int_{Q'} p(\varrho)\varphi(\varrho) \, dxdt \le c(c_e, Q'), \tag{6.0.5}$$

where $c(c_e, Q')$ is independent of $K$. It follows from Definition 5.1.4 of renormalized solutions that the functions $(\mathbf{u}, \varrho, \varphi)$ satisfy the integral identities

$$\int_Q (\varrho\mathbf{u} \cdot \partial_t \boldsymbol{\xi} + \varrho\mathbf{u} \otimes \mathbf{u} : \nabla\boldsymbol{\xi} + p \operatorname{div}\boldsymbol{\xi} - \mathbb{S}(\mathbf{u}) : \nabla\boldsymbol{\xi} + \varrho\mathbf{f} \cdot \boldsymbol{\xi}) \, dxdt = 0,$$

$$\int_Q (\varrho\partial_t\psi + \varrho\mathbf{u} \cdot \nabla\psi) \, dxdt = 0,$$

$$\int_Q \Big(\varphi\partial_t\psi + \varphi\mathbf{u} \cdot \nabla\psi + \psi\big(\varphi(\varrho) - \varphi'(\varrho)\varrho\big) \operatorname{div}\mathbf{u}\Big) \, dxdt = 0,$$

which hold for all vector fields $\boldsymbol{\xi} \in C_0^\infty(Q)$ and functions $\psi \in C_0^\infty(Q)$. This means that $(\mathbf{u}, \varrho, \varphi)$ satisfy in the cylinder $Q$ the equations

$$\begin{aligned} &\partial_t(\varrho\mathbf{u}) + \operatorname{div}(\varrho\mathbf{u}\otimes\mathbf{u}) = \operatorname{div}\mathbb{T} + \varrho\mathbf{f},\\ &\partial_t\varrho + \operatorname{div}(\varrho\mathbf{u}) = 0,\\ &\partial_t\varphi + \operatorname{div}(\varphi\mathbf{u}) + \varpi = 0, \end{aligned} \tag{6.0.6}$$

where

$$\mathbb{T} = \mathbb{S}(\mathbf{u}) - p(\varrho)\,\mathbb{I}, \quad \varpi = \big(\varphi(\varrho) - \varphi'(\varrho)\varrho\big)\operatorname{div}\mathbf{u}. \tag{6.0.7}$$

Next, it follows from (6.0.3) and (6.0.2) that

$$\|\mathbb{T}\|_{L^1(Q)} \le c_e, \quad \mathbb{T} \in L^{r'}(Q'). \tag{6.0.8}$$

Moreover, the boundedness of $\varphi$ and estimate (6.0.3) imply

$$\varphi \in L^\infty(Q), \quad \varpi \in L^2(Q).$$

It follows that the functions $(\mathbf{u}, \varrho, \varphi)$, $\varpi$ and tensors $\mathbb{T}$ satisfy Condition 4.5.1 with $\mathbf{g} = \mathbf{g}^\varphi = 0$, $\mathbb{G} = 0$. Hence they meet all requirements of Theorem 4.5.2 with $s = \infty$, $\gamma$. Now fix a cylinder $Q' \Subset Q$ and two nonnegative functions $\xi \in C_0^\infty(Q)$ and $\zeta \in C_0^\infty(\Omega)$ such that

$$\xi = \zeta = 1 \quad \text{on } Q'.$$

Now, it is convenient to consider $\zeta \in C_0^\infty(\Omega)$ as a function in $C^\infty(\Omega\times\mathbb{R})$ which is constant with respect to the time variable. By Theorem 4.5.2 the functions $(\mathbf{u}, \varrho, \varphi)$ satisfy the integral identity (4.5.10), which in our case reads

$$\int_Q \xi\varphi\mathbf{R} : [\zeta\mathbb{T}]\,dxdt = \int_Q \big(\mathbf{H}\cdot\mathbf{u} + \zeta\varrho(\mathbf{P}+\mathbf{Q})\cdot\mathbf{u} + S\big)\,dxdt. \tag{6.0.9}$$

Here, we denote

$$\mathbf{H} = \mathbf{R}[\xi\varphi](\zeta\varrho\mathbf{u}) - (\xi\varphi)\,\mathbf{R}[\zeta\varrho\mathbf{u}], \tag{6.0.10}$$

$$S = (\nabla\zeta\otimes\mathbf{A}[\xi\varphi]) : (\varrho\mathbf{u}\otimes\mathbf{u} - \mathbb{T}) + \zeta\mathbf{A}[\xi\varphi]\cdot(\varrho\mathbf{f}), \tag{6.0.11}$$

$$\mathbf{P} = \mathbf{A}[\varphi\partial_t\xi + \varphi\mathbf{u}\cdot\nabla\xi], \quad \mathbf{Q} = -\mathbf{A}[\varpi\xi]. \tag{6.0.12}$$

Using the identity $\mathbf{R} : [\zeta p(\varrho)\,\mathbb{I}] = \zeta p(\varrho)\mathbb{I}$ we obtain from (6.0.9)

$$\begin{aligned} \int_Q \xi\zeta\varphi p(\varrho)\,dxdt &= \int_Q \xi\varphi\mathbf{R} : [\zeta\mathbb{S}(\mathbf{u})]\,dxdt\\ &\quad - \int_Q \big(\mathbf{H}\cdot\mathbf{u} + \zeta\varrho(\mathbf{P}+\mathbf{Q})\cdot\mathbf{u} + S\big)\,dxdt. \end{aligned}$$

Since $\zeta = \xi = 1$ on $Q'$, we have

$$\int_{Q'} \varphi p(\varrho)\,dxdt \le \int_Q |\xi\varphi\mathbf{R} : [\zeta\mathbb{S}(\mathbf{u})]|\,dxdt$$
$$+ \int_Q \big(|\mathbf{H}\cdot\mathbf{u}| + |\zeta\varrho|(|\mathbf{P}| + |\mathbf{Q}|)|\mathbf{u}| + |S|\big)\,dxdt. \qquad (6.0.13)$$

Let us estimate all terms on the right hand side. We have

$$\|\mathbf{R} : [\zeta\mathbb{S}(\mathbf{u})]\|_{L^2(0,T;L^2(\mathbb{R}^d))} \le \|\zeta\mathbb{S}(\mathbf{u})\|_{L^2(0,T;L^2(\mathbb{R}^d))} \le c\|\mathbb{S}(\mathbf{u})\|_{L^2(0,T;L^2(\Omega))}$$
$$\le c\|\nabla\mathbf{u}\|_{L^2(0,T;L^2(\Omega))} \le c(c_e,\zeta).$$

On the other hand, the inequality $\varphi \le \varrho^\theta$ along with (6.0.2) and (6.0.3) implies

$$\|\xi\varphi\|_{L^\infty(0,T;L^{\gamma/\theta}(\mathbb{R}^d))} \le \|\varrho^\theta\|_{L^\infty(0,T;L^{\gamma/\theta}(\Omega))} \le \|\varrho\|^\theta_{L^\infty(0,T;L^\gamma(\Omega))} \le c(c_e,\xi).$$

Recalling that $1 \ge 1/2 + \theta/\gamma$ and applying the Hölder inequality we arrive at

$$\|\xi\varphi\mathbf{R} : [\zeta\mathbb{S}(\mathbf{u})]\|_{L^1(Q)}$$
$$\le c\|\mathbf{R} : [\zeta\mathbb{S}(\mathbf{u})]\|_{L^2(0,T;L^2(\mathbb{R}^d))}\|\xi\varphi\|_{L^\infty(0,T;L^{\gamma/\theta}(\mathbb{R}^d))} \le c(c_e,\xi,\zeta). \qquad (6.0.14)$$

Next, it follows from the Hölder inequality that

$$\|\mathbf{u}\cdot\mathbf{H}\|_{L^1(Q)} \le c\|\mathbf{u}\|_{L^2(0,T;L^m(\Omega))}\,\|\zeta\varrho\mathbf{u}\|_{L^2(0,T;L^n(\Omega))}\,\|\mathbf{R}[\xi\varphi]\|_{L^\infty(0,T;L^\kappa(\Omega))}$$
$$+ c\|\mathbf{u}\|_{L^2(0,T;L^m(\Omega))}\,\|\mathbf{R}[\zeta\varrho\mathbf{u}]\|_{L^2(0,T;L^n(\Omega))}\,\|\xi\varphi\|_{L^\infty(0,T;L^\kappa(\Omega))}$$

for all exponents satisfying

$$n, m, \kappa \in (1,\infty), \quad n^{-1} + m^{-1} + \kappa^{-1} \le 1.$$

By the boundedness of the operator $\mathbf{R}$ in Lebesgue spaces on $\mathbb{R}^d$, we have

$$\|\mathbf{R}[\xi\varphi]\|_{L^\infty(0,T;L^\kappa(\mathbb{R}^d))} \le c\|\xi\varphi\|_{L^\infty(0,T;L^\kappa(\mathbb{R}^d))} \le c\|\varphi\|_{L^\infty(0,T;L^\kappa(\Omega))}$$

and

$$\|\mathbf{R}[\zeta\varrho\mathbf{u}]\|_{L^2(0,T;L^n(\mathbb{R}^d))} \le c\|\zeta\varrho\mathbf{u}\|_{L^2(0,T;L^n(\mathbb{R}^d))} \le c\|\varrho\mathbf{u}\|_{L^2(0,T;L^n(\Omega))}.$$

Thus we get

$$\|\mathbf{u}\cdot\mathbf{H}\|_{L^1(Q)} \le c\|\mathbf{u}\|_{L^2(0,T;L^m(\Omega))}\|\varrho\mathbf{u}\|_{L^2(0,T;L^n(\Omega))}\|\varphi\|_{L^\infty(0,T;L^\kappa(\Omega))}. \qquad (6.0.15)$$

Now set $\kappa = \gamma/\theta$, which gives $\varphi^\kappa \le \varrho^\gamma$, hence (by (6.0.2) and (6.0.3))

$$\|\varphi\|_{L^\infty(0,T;L^\kappa(\Omega))} \le c_e.$$

It follows that

$$\|\mathbf{u}\cdot\mathbf{H}\|_{L^1(Q)} \le c\|\mathbf{u}\|_{L^2(0,T;L^m(\Omega))}\|\varrho\mathbf{u}\|_{L^2(0,T;L^n(\Omega))} \tag{6.0.16}$$

for

$$n,m\in(1,\infty),\quad n^{-1}+m^{-1}\le 1-\theta/\gamma. \tag{6.0.17}$$

By the energy estimate (6.0.3) the functions $(\mathbf{u},\varrho)$ meet all requirements of Proposition 4.2.1. Applying Corollary 4.2.2 of that proposition we conclude that

$$\|\varrho\mathbf{u}\|_{L^2(0,T;L^m(\Omega))}\le c_e \quad \text{if } m^{-1}>2^{-1}+\gamma^{-1}-d^{-1}. \tag{6.0.18}$$

On the other hand, the embedding theorem implies

$$\|\mathbf{u}\|_{L^2(0,T;L^n(\Omega))}\le c\|\mathbf{u}\|_{L^2(0,T;W^{1,2}(\Omega))}\le c_e \quad \text{if } n^{-1}>2^{-1}-d^{-1}. \tag{6.0.19}$$

Since $\theta/\gamma<2d^{-1}-\gamma^{-1}$, we have

$$(2^{-1}+\gamma^{-1}-d^{-1})+(2^{-1}-d^{-1})=1+\gamma^{-1}-2d^{-1}<1-\theta/\gamma.$$

Hence there exist $m,n\in(1,\infty)$ satisfying

$$m^{-1}>2^{-1}+\gamma^{-1}-d^{-1},\quad n^{-1}>2^{-1}-d^{-1},\quad \text{and}\quad n^{-1}+m^{-1}\le 1-\theta/\gamma.$$

For such $m$ and $n$ substituting (6.0.18)–(6.0.19) into (6.0.16) leads to

$$\|\mathbf{u}\cdot\mathbf{H}\|_{L^1(Q)}\le c(c_e,\zeta,\xi). \tag{6.0.20}$$

Let us estimate the vector field $\mathbf{P}$. For any $\sigma\in[1,\infty)$ we have

$$\|\varphi\partial_t\xi+\varphi\mathbf{u}\cdot\nabla\xi\|_{L^2(0,T;L^\sigma(\Omega))}\le c(\xi)\|\varphi\|_{L^\infty(0,T;L^\sigma(\Omega))}+c(\xi)\|\varphi\mathbf{u}\|_{L^2(0,T;L^\sigma(\Omega))}.$$

Moreover, by the Hölder inequality we have

$$\|\varphi\mathbf{u}\|_{L^2(0,T;L^\sigma(\Omega))}\le c(\Omega)\|\varphi\|_{L^\infty(0,T;L^n(\Omega))}\|\mathbf{u}\|_{L^2(0,T;L^m(\Omega))} \tag{6.0.21}$$

for all exponents $m$ and $n$ satisfying

$$n^{-1}+m^{-1}\le\sigma^{-1},\quad m,n\in[1,\infty].$$

Set $n=\gamma/\theta$ and note that for such $n$ we have $\varphi^n\le\varrho^\gamma$. Thus we get

$$\|\varphi\|_{L^\infty(0,T;L^n(\Omega))}\le\|\varrho^\gamma\|^\theta_{L^\infty(0,T;L^1(\Omega))}\le c_e,$$

hence

$$\|\varphi\mathbf{u}\|_{L^2(0,T;L^\sigma(\Omega))}\le c_e\|\mathbf{u}\|_{L^2(0,T;L^m(\Omega))} \tag{6.0.22}$$

for any exponent $m\in(1,\infty)$ satisfying $m^{-1}\le\sigma^{-1}-\theta\gamma^{-1}$. On the other hand, for $m^{-1}>2^{-1}-d^{-1}$ the embedding theorem implies

$$\|\mathbf{u}\|_{L^2(0,T;L^m(\Omega))}\le c\|\mathbf{u}\|_{L^2(0,T;W^{1,2}(\Omega))}\le c_e. \tag{6.0.23}$$

Combining (6.0.21)–(6.0.23) yields

$$\|\varphi \mathbf{u}\|_{L^2(0,T;L^\sigma(\Omega))} \le c_e \tag{6.0.24}$$

for all $\sigma$ such that there exists $m$ with

$$2^{-1} - d^{-1} < m^{-1} < \sigma^{-1} - \theta\gamma^{-1}.$$

In other words, inequality (6.0.24) holds for all $\sigma$ such that

$$\sigma \in (1,\infty), \quad \sigma^{-1} > \theta\gamma^{-1} + 2^{-1} - d^{-1}. \tag{6.0.25}$$

Notice that $\sigma < \gamma\theta^{-1}$, which yields

$$\|\varphi\|_{L^\infty(0,T;L^\sigma(\Omega))} \le c\|\varphi\|_{L^\infty(0,T;L^{\gamma/\theta}(\Omega))} \le c_e\|\varrho^\gamma\|^{\theta/\gamma}_{L^\infty(0,T;L^1(\Omega))} \le c_e. \tag{6.0.26}$$

In view of (6.0.24) and (6.0.26) we get

$$\|\varphi\partial_t\xi + \varphi\mathbf{u}_\epsilon\cdot\nabla\xi\|_{L^2(0,T;L^\sigma(\Omega))} \le c(\xi)c_e \quad \text{when } \sigma^{-1} > \theta\gamma^{-1} + 2^{-1} - d^{-1}. \tag{6.0.27}$$

Since $\xi$ is compactly supported in $Q$, we have

$$\|\mathbf{A}[\varphi\partial_t\xi + \varphi\mathbf{u}\cdot\nabla\xi]\|_{L^2(0,T;W^{1,\sigma}(\Omega))} \le c(\xi)c_e.$$

Recalling that the embedding $W^{1,\sigma}(\Omega) \hookrightarrow L^\iota(\Omega)$ is bounded if $\iota^{-1} > \sigma^{-1} - d^{-1}$, we deduce that

$$\|\mathbf{P}\|_{L^2(0,T;L^\iota(\Omega))} = \|\mathbf{A}[\varphi\partial_t\xi + \varphi\mathbf{u}\cdot\nabla\xi]\|_{L^2(0,T;L^\iota(\Omega))} \le c(\xi)c_e \tag{6.0.28}$$

for all $\iota$ with $\sigma^{-1} - d^{-1} < \iota^{-1} \le 1$, where $\sigma$ is an arbitrary number satisfying (6.0.25). From this and (6.0.25) we conclude that (6.0.28) holds for all exponents $\iota$ satisfying

$$2^{-1} + \theta\gamma^{-1} - 2d^{-1} < \iota^{-1} \le 1. \tag{6.0.29}$$

Next, the inequality $\theta\gamma^{-1} < 2d^{-1} - \gamma^{-1}$ yields

$$2^{-1} + \gamma^{-1} - d^{-1} < 1 - (2^{-1} + \theta\gamma^{-1} - 2d^{-1}). \tag{6.0.30}$$

It follows that there are $m, \iota \in (1,\infty)$ such that

$$2^{-1} + \gamma^{-1} - d^{-1} < m^{-1} < 1 - \iota^{-1} \quad \text{and} \quad \iota > 2^{-1} + \theta\gamma^{-1} - 2d^{-1}.$$

For such $m$ and $\iota$, the Hölder inequality along with (6.0.18) and (6.0.28) implies the desired estimate for $\mathbf{P}$:

$$\begin{aligned}\|\zeta\varrho\mathbf{u}\cdot\mathbf{P}\|_{L^1(Q)} &\le c(\xi,\zeta)\|\varrho\mathbf{u}\|_{L^2(0,T;L^m(\Omega))}\|\mathbf{P}\|_{L^2(0,T;L^\iota(\Omega))}\\ &\le c(\xi,\zeta)c_e\|\varrho\mathbf{u}\|_{L^2(0,T;L^m(\Omega))} \le c(\xi,\zeta)c_e.\end{aligned} \tag{6.0.31}$$

We use similar arguments to estimate $\mathbf{Q}$. Notice that $|\varpi| \le c\varrho^\theta |\operatorname{div}\mathbf{u}|$. By the Hölder inequality, we have

$$\begin{aligned}\|\xi\varpi\|_{L^2(0,T;L^\tau(\Omega))} &\le c(\xi)\|\nabla\mathbf{u}\|_{L^2(0,T;L^2(\Omega))}\|\varrho^\theta\|_{L^\infty(0,T;L^{\gamma/\theta}(\Omega))}\\ &= c(\xi)\|\nabla\mathbf{u}\|_{L^2(0,T;L^2(\Omega))}\|\varrho\|^\theta_{L^\infty(0,T;L^\gamma(\Omega))} \le c(\xi)c_e\end{aligned}$$

for $\tau^{-1} = \theta\gamma^{-1} + 2^{-1}$. Notice that $\tau \in (1,\infty)$ since $\theta\gamma^{-1} < 2^{-1}$. Since $\xi\varpi$ is compactly supported in $Q$, we get

$$\|\mathbf{Q}\|_{L^2(0,T;W^{1,\tau}(\Omega))} = \|\mathbf{A}[\xi\varpi]\|_{L^2(0,T;W^{1,\tau}(\Omega))} \le c(\xi)c_e.$$

Recall that the embedding $W^{1,\tau}(\Omega) \hookrightarrow L^\omega(\Omega)$ is bounded if $\omega^{-1} > \tau^{-1} - d^{-1}$. Thus we get

$$\|\mathbf{Q}\|_{L^2(0,T;L^\omega(\Omega))} \le c(\xi)c_e \quad \text{if } \omega^{-1} > \theta\gamma^{-1} + 2^{-1} - d^{-1},\ \omega \ge 1. \tag{6.0.32}$$

Next, it follows from (6.0.30) that

$$2^{-1} + \gamma^{-1} - d^{-1} < 1 - (\theta\gamma^{-1} + 2^{-1} - d^{-1}).$$

Hence there are $m, \omega \in (1,\infty)$ with

$$2^{-1} + \gamma^{-1} - d^{-1} < m^{-1} = 1 - \omega^{-1}, \quad \omega^{-1} > \theta\gamma^{-1} + 2^{-1} - d^{-1}.$$

For such $m$ and $\omega$, the Hölder inequality along with (6.0.18) and (6.0.32) implies the estimate for $\mathbf{Q}$:

$$\begin{aligned}\|\zeta\varrho\mathbf{u}\cdot\mathbf{Q}\|_{L^1(Q)} &\le c(\Omega,\xi,\zeta)\|\varrho\mathbf{u}\|_{L^2(0,T;L^m(\Omega))}\|\mathbf{Q}\|_{L^2(0,T;L^\omega(\Omega))}\\ &\le c(\xi,\zeta)c_e\|\varrho\mathbf{u}\|_{L^2(0,T;L^m(\Omega))} \le c(\xi,\zeta)c_e.\end{aligned} \tag{6.0.33}$$

It remains to estimate $S$. First, we observe that $\xi\varphi$ is compactly supported in $Q$ and satisfies

$$\|\xi\varphi\|_{L^\infty(0,T;L^{\gamma/\theta}(\Omega))} \le c\|\varrho^\gamma\|^{\theta/\gamma}_{L^\infty(0,T;L^1(\Omega))} \le c_e,$$

hence

$$\|\mathbf{A}[\xi\varphi]\|_{L^\infty(0,T;W^{1,\gamma/\theta}(\Omega))} \le c(\xi)c_e.$$

Since $\gamma/\theta > d$, the embedding $W^{1,\gamma/\theta}(\Omega) \hookrightarrow C(\Omega)$ is bounded, whence

$$\|\mathbf{A}[\xi\varphi]\|_{L^\infty(Q)} \le c(\xi)c_e.$$

It follows that

$$\|S\|_{L^1(Q)} \le c(\xi,\zeta)c_e(\|\varrho|\mathbf{u}|^2\|_{L^1(Q)} + \|\mathbb{T}\|_{L^1(Q)} + \|\varrho\|_{L^1(Q)} + 1) \le c(\xi,\zeta)c_e.$$

Inserting this estimate along with (6.0.20), (6.0.31), and (6.0.33) into (6.0.13) and noting that $\zeta$, $\xi$ depend only on $Q'$ we obtain (6.0.5), and the proposition follows. □

# Chapter 7

# Kinetic theory. Fast density oscillations

## 7.1 Problem formulation. Main results

In Chapter 5 the existence of renormalized solutions to problem (5.1.12) is proved under restrictive regularity assumptions on the flow domain, the given data $\mathbf{U}$, $\varrho_\infty$, and the constitutive relation $p = p(\varrho)$. In particular, the adiabatic exponent $\gamma$ satisfies $\gamma > 2d$. The restrictions are used mainly in the proof of the existence of strong solutions to the regularized equations and to justify the limit passages when the regularization parameters go to zero. As the existence of a renormalized solution is proved, the restrictions on the data and the constitutive relations should be weakened. It is most important to relax the restriction $\gamma > 2d$ and extend the existence theory to the range $\gamma > d/2$, common for homogeneous boundary value problems. Another problem is to extend the solvability of the nonhomogeneous problem to nonsmooth densities $\varrho_\infty$.

These problems are closely related to the question of compactness of the set of solutions to compressible Navier-Stokes equations corresponding to various pressure functions and boundary data. Indeed, we can approximate the pressure function $p(\varrho)$ with a fast growing artificial pressure $p_\varepsilon$ and next approximate the bounded function $\varrho_\infty$ with a sequence of smooth functions $\varrho_\infty^\epsilon$, $\epsilon \in (0, 1]$. Then we can apply Theorem 5.1.6 to obtain a sequence of solutions $(\mathbf{u}_\epsilon, \varrho_\epsilon)$ to problem (5.1.12) with $p$ replaced by $p_\epsilon$ and $\varrho_\infty$ replaced by $\varrho_\infty^\epsilon$. By estimate (5.1.17) in Theorem 5.1.6 the solutions $(\mathbf{u}_\epsilon, \varrho_\epsilon)$ have uniformly bounded energies. Hence, passing to a subsequence if necessary, we can assume that $(\mathbf{u}_\epsilon, \varrho_\epsilon)$ converge weakly in the energy space to some limit functions $(\mathbf{u}, \varrho)$.

Our goal now is to derive equations for the limit $(\mathbf{u}, \varrho)$. We can expect that the equations depend on the approximation methods. Strongly convergent approximation should lead to Navier-Stokes equations, but weakly convergent approximation may lead to some other equations. In the case of weakly convergent approximation we encounter the problem of *fast oscillating data* arising in nonlinear acoustics. In the most general case such data appear as weak limits of sequences of smooth data. To give a rigorous meaning to the above discussion we formulate explicit assumptions on the artificial pressure function $p(\varrho)$, its approximations $p_\epsilon$, and the boundary data: the densities $\varrho_\infty^\epsilon$ and the velocity $\mathbf{U}$.

**Condition 7.1.1.** The function $p : [0,\infty) \to \mathbb{R}^+$ has continuous first derivative and

$$p'(\varrho) \geq 0, \quad p(0) = p'(0) = 0. \tag{7.1.1}$$

There are an exponent $\gamma > d/2$ and a constant $c_p > 0$ such that for $\varrho \in [0,\infty)$,

$$\varrho^\gamma \leq c_p(P(\varrho)+1), \quad \varrho^\gamma \leq c_p(p(\varrho)+1), \quad p(\varrho) \leq c_p(P(\varrho)+1), \tag{7.1.2}$$

and for $\varrho \in [1,\infty)$,

$$c_p^{-1}\varrho^\gamma \leq p(\varrho) \leq c_p\varrho^\gamma, \quad c_p^{-1}\varrho^{\gamma-1} \leq p'(\varrho) \leq c_p\varrho^{\gamma-1}, \quad p(1) \leq c_p. \tag{7.1.3}$$

In particular,

$$p(\varrho) \leq c_p(\varrho^\gamma + 1). \tag{7.1.4}$$

Therefore, the energy density $P = \varrho \int_0^\varrho s^{-1}p(s)\,ds$ is twice continuously differentiable on $(0,\infty)$, but $P''(\varrho)$ can develop a singularity at $\varrho = 0$.

**Condition 7.1.2.**

- We take the approximation of the pressure function in the form

$$p_\epsilon(\varrho) = \chi\Big(\frac{\varrho}{\epsilon}\Big)p(\varrho) + \epsilon(\varrho^2 + \varrho^n), \quad \epsilon \in (0,1], \tag{7.1.5}$$

where $n > \max\{\gamma, 4d\}$ and $\chi \in C^\infty(\mathbb{R})$ is a nondecreasing function such that

$$\chi(s) = 0 \quad \text{for } s \leq 1/2, \quad \chi(s) = 1 \quad \text{for } s \geq 1. \tag{7.1.6}$$

- The function $\mathbf{U} \in C^\infty(\mathbb{R}^d \times (0,T))$ and the $C^\infty$ boundary $\partial\Omega$ satisfy the geometric condition (5.1.9) in Condition 5.1.1.
- The nonnegative function $\varrho_\epsilon^\infty \in C^\infty(Q)$, $\epsilon \in (0,1]$, vanishes in a neighborhood of the edge $\partial\Omega \times \{0\}$.
- The functions $\varrho_\infty^\epsilon$ are uniformly bounded on the parabolic boundary $\sqcup_T = S_T \cup (\mathrm{cl}\,\Omega \times \{t=0\})$ of the cylinder $Q$ and

$$\varrho_\infty^\epsilon \rightharpoonup \varrho_\infty \in L^\infty(\sqcup_T) \quad \text{weakly}^\star \text{ in } L^\infty(\sqcup_T) \quad \text{as } \epsilon \to 0. \tag{7.1.7}$$

The weak$^\star$ convergence in (7.1.7) implies *fast oscillations of solutions* in the model.

We denote by $c_e$, as in Remark 5.1.2, generic constants depending only on

$$\lambda, \quad \gamma, \quad \mathrm{diam}\,\Omega, \quad T, \quad \|\varrho_\infty\|_{L^\infty(\sqcup_T)}, \quad \|\mathbf{U}\|_{C^1(Q)}, \quad \|\mathbf{f}\|_{C(Q)}, \quad c_p.$$

**Regularized problem.** Let us consider the regularized boundary value problem

$$\partial_t(\varrho_\epsilon \mathbf{u}_\epsilon) + \operatorname{div}(\varrho_\epsilon \mathbf{u}_\epsilon \otimes \mathbf{u}_\epsilon) + \nabla p_\epsilon(\varrho_\epsilon) = \operatorname{div} \mathbb{S}(\mathbf{u}_\epsilon) + \varrho_\epsilon \mathbf{f} \quad \text{in } Q, \tag{7.1.8a}$$

$$\partial_t \varrho_\epsilon + \operatorname{div}(\varrho_\epsilon \mathbf{u}_\epsilon) = 0 \quad \text{in } Q, \tag{7.1.8b}$$

$$\begin{aligned} &\mathbf{u}_\epsilon = \mathbf{U} \quad \text{on } S_T, \quad \varrho_\epsilon = \varrho_\infty^\epsilon \quad \text{on } \Sigma_{\text{in}}, \\ &\mathbf{u}_\epsilon|_{t=0} = \mathbf{U}, \quad \varrho_\epsilon|_{t=0} = \varrho_\infty^\epsilon \quad \text{in } \Omega. \end{aligned} \tag{7.1.8c}$$

The following result is a direct consequence of Theorem 5.1.6:

**Theorem 7.1.3.** *Let Condition 7.1.2 be satisfied. Then problem* (7.1.8) *has a renormalized solution which satisfies the estimate*

$$\|\mathbf{u}_\epsilon\|_{L^2(0,T;W^{1,2}(\Omega))} + \|\varrho_\epsilon|\mathbf{u}_\epsilon|^2\|_{L^\infty(0,T;L^1(\Omega))} + \|P_\epsilon(\varrho_\epsilon)\|_{L^\infty(0,T;L^1(\Omega))} \le c_e, \tag{7.1.9}$$

*where $c_e$ is independent of $\epsilon$. For any $\theta$, $0 < \theta < \min\{2\gamma d^{-1} - 1, 2^{-1}\gamma\}$ and a compact set $Q' \Subset Q$, there is a constant $c_e(Q')$ independent of $\epsilon$ such that*

$$\int_{Q'} p_\epsilon(\varrho_\epsilon)\varrho_\epsilon^\theta \, dxdt \le c_e(Q'), \quad \int_{Q'} (\varrho_\epsilon^{\gamma+\theta} + \varepsilon \varrho_\epsilon^{n+\theta}) \, dxdt \le c_e(Q'). \tag{7.1.10}$$

*Proof.* It suffices to check that $\Omega$, $p_\epsilon$, $\mathbf{U}$ and $\varrho_\infty^\epsilon$ satisfy Condition 5.1.1 with $(\upsilon, \gamma)$ replaced by $(\gamma, n)$. It follows from Condition 7.1.2 that $\Omega$, $\mathbf{U}$, and $\varrho_\infty^\epsilon$ obviously satisfy Condition 5.1.1. It remains to check that the pressure function $p(\varrho_\epsilon)$ and the internal energy density

$$P_\epsilon := \varrho \int_0^\varrho s^{-1} p(s) \, ds = \varrho \int_0^\varrho s^{-1} \chi\Big(\frac{s}{\epsilon}\Big) p(s) \, ds + \epsilon\Big(\frac{1}{2}\varrho^2 + \frac{1}{n}\varrho^n\Big). \tag{7.1.11}$$

satisfy (5.1.4)–(5.1.8). First we show that $p_\epsilon$ and $P_\epsilon$ satisfy (5.1.4)–(5.1.7). Since $p$ is continuously differentiable and $\chi(s/\epsilon)$ vanishes on $[0, \epsilon/2]$, the function $P_\epsilon$ is twice continuously differentiable on $[0, \infty)$. Since all functions in representations (7.1.5) and (7.1.11) are nonnegative and monotone, we have

$$\varrho^n \le \epsilon^{-1} p_\epsilon(\varrho), \quad \varrho^n \le n\epsilon^{-1} P_\epsilon(\varrho). \tag{7.1.12}$$

Notice that $p_\epsilon$ are uniformly bounded on $[0, 1]$. Next, (7.1.3) yields, for $\varrho \ge 1$,

$$p_\epsilon(\varrho) \le c_p \varrho^\gamma + \epsilon(\varrho^2 + \varrho^n) \le c(\epsilon)\varrho^n, \quad p_\epsilon'(\varrho) \le c_p \varrho^{\gamma-1} + \epsilon(2\varrho + n\varrho^{n-1}) \le c(\epsilon)\varrho^{n-1}. \tag{7.1.13}$$

This leads to the inequality $p_\epsilon(\varrho) \le c(\epsilon)(\varrho^n + 1)$, which along with (7.1.12) implies

$$p_\epsilon(\varrho) \le c(\epsilon)(P_\epsilon(\varrho) + 1). \tag{7.1.14}$$

Finally note that $P_\epsilon''(\varrho) = \varrho^{-1} p_\epsilon' \ge 2\epsilon$. It follows from (7.1.12)–(7.1.14) that $p_\varepsilon$ and $P_\epsilon$ satisfy the desired relations (5.1.5) with $\gamma$ replaced by $n$ and $C_E$ replaced by a constant $c(\epsilon)$, depending on $c_p$ and $\epsilon$.

It remains to show that $p_\epsilon$ and $P_\epsilon$ satisfy (5.1.8). We first observe that (7.1.11) can be rewritten in the form

$$P_\epsilon(\varrho) = P(\varrho) - P^*_\epsilon(\varrho) + \epsilon\Big(\frac{1}{2}\varrho^2 + \frac{1}{n}\varrho^n\Big), \quad P^*_\epsilon(\varrho) = \varrho\int_0^\varrho s^{-1}\Big(1-\chi\Big(\frac{s}{\epsilon}\Big)\Big)p(s)\,ds. \tag{7.1.15}$$

It is easy to check that

$$0 \le P^*_\epsilon(\varrho) \le \varrho\int_0^\epsilon s^{-1}p(s)\,ds = \eta(\epsilon)\varrho, \quad \text{where} \quad \eta(\epsilon)\to 0 \quad \text{as } \epsilon\to 0. \tag{7.1.16}$$

Now (7.1.2) yields

$$P_\epsilon(\varrho) \ge P(\varrho) - P^*_\epsilon(\varrho) \ge c_p^{-1}\varrho^\gamma - c_p^{-1} - \eta(1)\varrho \ge 2^{-1}c_p^{-1}\varrho^\gamma - C, \tag{7.1.17}$$

where $C$ depends only on $c_p$ and $\gamma$. Hence $\varrho^\gamma \le c_e(P_\epsilon+1)$ for some constant $c_e$ depending only on $c_p$ and $\gamma$. Next, in view of (7.1.4) we have

$$p_\epsilon(\varrho) \le p(\varrho) + \epsilon(\varrho^2+\varrho^n) \le p(\varrho) + nP_\epsilon(\varrho) \le c_p\varrho^\gamma + c_p + nP_\varepsilon(\varrho).$$

Combining this with (7.1.17) we arrive at

$$p_\epsilon(\varrho) \le (2c_p^2+n)P_\epsilon(\varrho) + C,$$

where $C$ depends only on $c_p$ and $\gamma$. Hence

$$p_\epsilon \le c_e(P_\epsilon+1) \tag{7.1.18}$$

for some constant $c_e$. Recall that $\chi(\varrho/\epsilon)-1$ vanishes for $\varrho\ge 1$, which yields $(1-\chi(\varrho/\epsilon))p(\varrho)\le c_p$. Applying (7.1.2) we obtain

$$p_\epsilon(\varrho) \ge p(\varrho) - (1-\chi(\varrho/\epsilon))p(\varrho) \ge c_p^{-1}\varrho^\gamma - 1 - c_p. \tag{7.1.19}$$

Hence $\varrho^\gamma \le c_e(p_\epsilon+1)$ for a suitable constant $c_e$. Now we assemble the results obtained and arrive at

$$\varrho^\gamma \le c_e(P_\epsilon+1), \quad \varrho^\gamma\le c_e(p_\epsilon+1), \quad p_\epsilon\le c_e(P_\epsilon+1) \quad \text{for } \varrho\ge 0. \tag{7.1.20}$$

Therefore, $p_\epsilon$ and $P_\epsilon$ satisfy (5.1.8) with $\upsilon$ and $c_p$ replaced by $\gamma$ and $c_e$. It follows that $\Omega$, $\mathbf{U}$, $\varrho^\epsilon_\infty$, and $p_\epsilon$ satisfy Condition 5.1.1 with exponents $(\upsilon,\gamma)$ with $\upsilon\in(1,\gamma)$ replaced by $(\gamma,n)$ with $\gamma\in(1,n)$. Thus we can apply Theorem 5.1.6 to conclude that problem (7.1.8) has a renormalized solution $(\mathbf{u}_\epsilon,\varrho_\epsilon)$ which satisfies estimate (7.1.9) with a constant $c_e$ depending only on $\operatorname{diam}\Omega$, $\|\mathbf{U}\|_{C^1(Q)}$, the constant $c_p$ in Condition 7.1.1, and $\|\varrho^\epsilon_\infty\|_{L^\infty(\sqcup_T)}$.

Next, notice that in view of (7.1.20), the function $p_\epsilon$ satisfies Condition 6.0.2. It then follows from estimate (7.1.9) that the solution $(\mathbf{u}_\epsilon,\varrho_\epsilon)$ meets all requirements of Proposition 6.0.3 with $p$ replaced by $p_\epsilon$ and $c_p$ replaced by $c_e$. Hence

$$\int_{Q'} p_\epsilon(\varrho_\epsilon)\varrho_\epsilon^\theta\,dxdt \le c_e(Q'), \tag{7.1.21}$$

which is exactly the first inequality in (7.1.10). In order to prove the second, notice that (7.1.19) and the inequality $p_\epsilon(\varrho) \geq \epsilon\varrho^n$ imply

$$c_p\varrho_\epsilon^\gamma + \epsilon\varrho_\epsilon^n \leq 2p_\epsilon(\varrho_\epsilon) + c_p.$$

Since $\theta \leq \gamma$, estimate (7.1.9) implies

$$\int_{Q'} \varrho_\epsilon^\theta \, dxdt \leq c_e(Q') \int_{Q'} (\varrho_\epsilon^\gamma + 1) \, dxdt \leq c_e(Q') \int_{Q'} (P_\epsilon + 1) \, dxdt \leq c_e(Q'),$$

which gives

$$\int_{Q'} (c_p\varrho_\epsilon^{\gamma+\theta} + \epsilon\varrho_\epsilon^{n+\theta}) \, dxdt \leq 2 \int_{Q'} p_\epsilon \, \varrho_\epsilon^\theta dxdt + c_p \int_{Q'} \varrho_\epsilon^\theta \, dxdt \leq c_e(Q').$$

Together with (7.1.21), this leads to the second estimate in (7.1.10). □

Let us consider the sequence of solutions to problem (7.1.8) defined by Theorem 7.1.3. Passing to a subsequence we can assume that there exist $\mathbf{u} \in L^2(0,T;W^{1,2}(\Omega))$ and $\overline{\varrho} \in L^\infty(0,T;L^\gamma(\Omega))$ such that

$$\begin{aligned} \mathbf{u}_\epsilon &\rightharpoonup \mathbf{u} \quad \text{weakly in } L^2(0,T;W^{1,2}(\Omega)), \\ \varrho_\epsilon &\rightharpoonup \overline{\varrho} \quad \text{weakly}^\star \text{ in } L^\infty(0,T;L^\gamma(\Omega)), \\ \varrho_\epsilon &\rightharpoonup \overline{\varrho} \quad \text{weakly in } L^{\gamma+\theta}(Q') \quad \text{for all } Q' \Subset Q. \end{aligned} \tag{7.1.22}$$

Notice that in Condition 7.1.2 it is assumed that the sequence of boundary and initial data $\varrho_\infty^\varepsilon$ is only weakly convergent, and it may be rapidly oscillating as $\epsilon \to 0$. The well known example of a rapidly oscillating sequence is

$$\varrho_\infty^\epsilon = R\Big(x, t, \frac{x}{\epsilon}, \frac{t}{\epsilon}\Big), \tag{7.1.23}$$

where $R(x,t,y,\tau)$ is a smooth function periodic in $y$ and $\tau$. The theory of weak convergence for such sequences is called *two-scale convergence* theory. In view of the mass transport equation (7.1.8c), oscillations in boundary and initial data are transferred inside the flow domain along fluid particle trajectories. Hence we can expect that these oscillations induce rapid oscillations of the density in the flow domain, and the propagation of such oscillations will now be under discussion. It is worth noting that density oscillations can be regarded as sound waves, studied in acoustics. Rapid oscillations appear if the wavelength of the sound is small compared to the diameter of the flow domain. In that case, the sound waves can be considered in the same way as light rays and wave fronts in optics. For example, in classical linear acoustics, sound propagation is described by the wave equation

$$\partial_t^2\varrho - c^2\Delta\varrho = 0. \tag{7.1.24}$$

The passage from the wave equation to geometric acoustics can be performed if approximate solutions of the wave equation (7.1.24) in the form

$$\varrho = A(x,t)e^{\frac{i}{\epsilon}\mathfrak{s}(t,x,\epsilon)} \tag{7.1.25}$$

are considered. Substitution of this formula into (7.1.24) and further formal asymptotic analysis leads to the Hamilton-Jacobi equation for the phase function $\mathfrak{s}$ which describes the position of the wave front (see e.g. [20] for details). This discussion shows that rapid oscillations appear in a natural way in modeling a viscous gas motion in an acoustic field. However, the method of asymptotic expansions in the linear theory of partial differential equations is difficult to apply to nonlinear problems of viscous gas dynamics. The difficulty is the lack of information on smoothness properties of solutions to compressible Navier-Stokes equations.

It seems that Bakhvalov and Eglit have been the first to consider the problem of averaged equations of the one-dimensional motion of a viscous compressible medium with rapidly oscillating boundary and initial data. In [8] new equations of motion are formally derived. It is discovered that the averaging description for this problem cannot be given in terms of the averaging characteristics and the resulting new equations have a substantially different form compared to the original equations. These results have been rigorously justified and generalized by Amosov and Zlotnik [5], [6]. Notice that all those results only apply to the one-dimensional case and with data functions of the form (7.1.23).

In this chapter another approach based on Young measure theory is developed. It follows from relations (7.1.22) and the fundamental theorem on Young measures (Theorem 1.4.5) that there exist subsequences, still denoted by $\varrho_\varepsilon$, $\varrho^\epsilon_\infty$, and Young measures $\mu \in L^\infty_w(Q; \mathcal{M}(\mathbb{R}))$, $\mu^\infty \in L^\infty_w(\sqcup_T; \mathcal{M}(\mathbb{R}))$, with the following properties:

For any continuous function $\varphi \in C_0(\mathbb{R})$ we have

$$\varphi(\varrho_\epsilon) \rightharpoonup \overline{\varphi} \quad \text{weakly}^\star \text{ in } L^\infty(Q), \tag{7.1.26a}$$

$$\varphi(\varrho^\epsilon_\infty) \rightharpoonup \overline{\varphi}_\infty \quad \text{weakly}^\star \text{ in } L^\infty(\sqcup_T), \tag{7.1.26b}$$

where

$$\overline{\varphi}(x,t) = \langle \mu_{xt}, \varphi \rangle, \quad \overline{\varphi}_\infty(x,t) = \langle \mu^\infty_{xt}, \varphi \rangle. \tag{7.1.26c}$$

In this framework rapidly oscillating sequences are associated with some Young measures. At this point it is worth noting that a Young measure determines a random function (depending on the spatial variables and the time variable), according to the following definition ([40]):

**Definition 7.1.4.** Let $(\mathcal{E}, \mathcal{A}, \pi)$ be a probability space, i.e., $\mathcal{A}$ is a $\sigma$-algebra on the set $\mathcal{E}$ and $\pi : \mathcal{A} \to \mathbb{R}$ is a probability measure, $\pi(\mathcal{E}) = 1$. A *random variable* is a Borel map $\varsigma : \mathcal{E} \to \mathbb{R}$. Recall that $\varsigma$ is *Borel* if $\varsigma^{-1}(B) \in \mathcal{A}$ for any Borel set $B \subset \mathbb{R}$. The *probability distribution* of the random variable $\varsigma$ is the probability measure $\mu$ on the real line defined by $\mu(B) = \pi(\varsigma^{-1}(B))$ for all Borel sets $B \subset \mathbb{R}$.

The *distribution function* (cumulative distribution function) of $\varsigma$ is defined by $f(\lambda) = \mu(-\infty, \lambda]$, $\lambda \in \mathbb{R}$. A family of random variables $\varsigma_{xt}$ labeled by points $(x,t) \in Q$ is called a *random function* (random field) on $Q$.

Therefore, we can consider rapidly oscillating sequences $\varrho_\epsilon$ and $\varrho^\epsilon_\infty$ as random functions on the cylinder $Q$ and on the surface $\sqcup_T$ with the associated probability distributions $\mu_{xt}$ and $\mu^\infty_{xt}$, respectively. Introduce the corresponding cumulative distribution functions

$$f(x,t,s) = \mu_{xt}(-\infty, s], \quad f_\infty(x,t,s) = \mu^\infty_{xt}(-\infty, s]. \tag{7.1.27}$$

For a.e. $(x,t) \in Q$ (resp. $(x,t) \in \sqcup_T$), the functions $f(x,t,s)$ and $f_\infty(x,t,s)$ are monotone and right continuous in $s$, tend to 1 as $s \to \infty$ and vanish for $s < 0$. In accordance with Example 1.3.6, for any $\varphi \in C_0(\mathbb{R})$ integrals with respect to Young measures can be written in the form of Stieltjes integrals (see Section 1.3.1 for definition):

$$\langle \mu_{xt}, \varphi \rangle = \int_{\mathbb{R}} \varphi(s)\, d_s f(x,t,s), \quad \langle \mu^\infty_{xt}, \varphi \rangle = \int_{\mathbb{R}} \varphi(s)\, d_s f_\infty(x,t,s). \tag{7.1.28}$$

In particular, the weak limits $\overline{\varrho}$ and $\overline{p}$ of the densities $\varrho_\epsilon$ and pressure functions $p(\varrho_\epsilon)$ are determined as the expectation values

$$\overline{\varrho} = \int_{\mathbb{R}} s\, d_s f(x,t,s), \quad \overline{p} = \int_{\mathbb{R}} p(s)\, d_s f(x,t,s). \tag{7.1.29}$$

In other words the density and the pressure can be considered as random functions on $Q$ with the common probability measure $\mu_{xt}$.

Now, the main problem is to derive the so-called *kinetic equation* for the cumulative probability. In this way the system of governing equations along with boundary conditions for the limit velocity field $\mathbf{u}$ and the distribution function $f$ can be obtained. To make the presentation clear, first, the kinetic equation is postulated, and then rigorously derived.

**Problem 7.1.5** (Kinetic problem). *For given $\mathbf{U} \in C^\infty(Q)$, $\mathbf{f} \in C(Q)$ and a distribution function $f_\infty : \sqcup_T \times \mathbb{R} \to [0,1]$, find a velocity field $\mathbf{u} \in L^2(0,T; W^{1,2}(\Omega))$ and a distribution function $f : Q \times \mathbb{R} \to [0,1]$ such that:*

*(i) $f$ is measurable and satisfies*

$$f(x,t,s) = \lim_{h \searrow 0} f(x,t,s+h), \quad f(x,t,s') \le f(x,t,s'') \quad \textit{for } s' \le s'', \tag{7.1.30}$$

$$f(x,t,s) = 0 \quad \textit{for } s < 0, \quad \lim_{s\to\infty} f(x,t,s) = 1, \tag{7.1.31}$$

$$\int_Q \Big\{ \int_{\mathbb{R}} |s|^\gamma\, d_s f(x,t,s) \Big\}\, dxdt < \infty. \tag{7.1.32}$$

*(ii)* $\mathbf{u}$ *and* $f$ *satisfy the following equations and boundary conditions:*

$$\partial_t(\overline{\varrho}\mathbf{u}) + \operatorname{div}(\overline{\varrho}\mathbf{u} \otimes \mathbf{u}) + \nabla \overline{p} = \operatorname{div} \mathbb{S}(\mathbf{u}) + \overline{\varrho}\,\mathbf{f} \quad \text{in } Q, \tag{7.1.33a}$$

$$\partial_t f + \operatorname{div}(f\mathbf{u}) - \partial_s(sf \operatorname{div} \mathbf{u} + s\,\mathfrak{C}[f]) = 0 \quad \text{in } Q \times \mathbb{R}, \tag{7.1.33b}$$

$$\mathbf{u} = \mathbf{U} \quad \text{on } \sqcup_T, \tag{7.1.33c}$$

$$f = f_\infty \quad \text{on } \Sigma_{\text{in}} \times \mathbb{R}, \quad f(x,0,s) = f_\infty(x,0,s) \quad \text{on } \Omega \times \mathbb{R}. \tag{7.1.33d}$$

*Here the nonlinear operator* $\mathfrak{C}[f]$ *is defined by*

$$\mathfrak{C}[f] = \frac{1}{\lambda+1} \int_{(-\infty,s]} (p(\tau) - \overline{p})\, d_\tau f(x,t,\tau), \tag{7.1.33e}$$

*and the functions* $\overline{\varrho}$ *and* $\overline{p}$ *are given by* (7.1.29). *Since* $0 \le p \le c(1+\varrho^\gamma)$, *it follows from* (7.1.32) *that* $\overline{p} \in L^1(Q)$.

**Remark 7.1.6.** The equality $f(x,t,s) = 0$ on $(-\infty, 0)$ simply means that the corresponding probability measure vanishes on this interval, i.e. $\mu_{xt}(-\infty,0) = 0$. Therefore if a function $\varphi$ is integrable with respect to the measure $\mu_{xt} = d_s f(x,t,s)$, then

$$\int_{\mathbb{R}} \varphi\, d_s f(x,t,s) = \int_{[0,\infty)} \varphi\, d_s f(x,t,s). \tag{7.1.34}$$

In particular,

$$\begin{aligned} &\overline{\varrho} = \int_{[0,\infty)} s\, d_s f(x,t,s), \quad \overline{p} = \int_{[0,\infty)} p(s)\, d_s f(x,t,s), \\ &\mathfrak{C}[f](x,t,s) = \frac{1}{\lambda+1} \int_{[0,s]} (p(\tau) - \overline{p}(x,t))\, d_\tau f(x,t,\tau) \quad \text{for } s \ge 0, \\ &\mathfrak{C}[f](x,t,s) = 0 \quad \text{for } s < 0. \end{aligned} \tag{7.1.35}$$

**Remark 7.1.7.** Since $0 \le \varrho_\infty^\epsilon \le c_e$, it follows from Theorem 1.4.5 that $\operatorname{supp} \mu_{xt}^\infty \subset [0, c_e]$. Therefore, the distribution function $f_\infty$ has the properties

$$f_\infty(x,t,s) = 0 \quad \text{for } s < 0, \quad f_\infty(x,t,s) = 1 \quad \text{for } s > c_e, \tag{7.1.36}$$

$$f_\infty(x,t,s) \text{ is monotone and right continuous in } s. \tag{7.1.37}$$

Further we will consider solutions to Problem 7.1.5 satisfying additional regularity requirements concerning the behavior of the distribution function at infinity. To this end, the notion of regular solution to the kinetic problem is introduced.

**Definition 7.1.8.** A solution $(\mathbf{u}, f)$ to Problem 7.1.5 is said to be *regular* if

$$\|\mathfrak{H}\|_{L^{1+\gamma}(Q)} + \sup_{v \in \mathbb{R}^+} \|\mathfrak{V}_v\|_{L^1(Q)} < \infty \tag{7.1.38}$$

where

$$\begin{aligned} \mathfrak{V}_v(x,t) &= \int_{[0,\infty)} \min\{s,v\}(p(s) - \overline{p})\, d_s f(x,t,s), \\ \mathfrak{H}(x,t) &= \int_{[0,\infty)} f(x,t,s)(1 - f(x,t,s))\, ds. \end{aligned} \tag{7.1.39}$$

**Results: Convergence of fast oscillating flow to a solution of the kinetic problem.** The following theorem which is the first main result of this section shows that the vector field $\mathbf{u}$ and the distribution function $f$ defined by (7.1.26) and (7.1.27) are a regular weak solution to Problem 7.1.5.

**Theorem 7.1.9.** *Assume that the pressure function $p$, the functions $\mathbf{U}$, $\varrho_\infty^\epsilon$, and the domain $\Omega$ satisfy Conditions 7.1.1 and 7.1.2. Furthermore assume that $(\mathbf{u}_\epsilon, \varrho_\epsilon)$ are solutions to problem (7.1.8) defined by Theorem 7.1.3 and $\mathbf{u}$, $f$, $f_\infty$ are defined by (7.1.22), (7.1.26), and (7.1.27). Then:*

*(i) The distribution function $f$ satisfies conditions (7.1.30)–(7.1.31). There is a constant $c_e$, depending only on $\operatorname{diam}\Omega$, $\|\mathbf{U}\|_{C^1(Q)}$, the function $p(\varrho)$, the time period $T$, and $\|\varphi_\infty^\epsilon\|_{L^\infty(\sqcup_T)}$, such that*

$$\operatorname*{ess\,sup}_{t\in(0,T)} \int_\Omega \int_{[0,\infty)} s^\gamma \, d_s f(x,t,s)\, dx \le c_e . \tag{7.1.40}$$

*Moreover, for any $Q' \Subset Q$ and $0 < \theta < \min\{2\gamma d^{-1} - 1, \gamma/2\}$, there is a constant $c$ depending only on $c_e$, $Q'$ and $\theta$ such that*

$$\int_{Q'} \int_{[0,\infty)} s^{\gamma+\theta}\, d_s f(x,t,s)\, dx \le c. \tag{7.1.41}$$

*(ii) The integral identity for momentum balance*

$$\int_Q \big(\overline{\varrho}\mathbf{u}\cdot\partial_t\boldsymbol{\xi} + \overline{\varrho}\mathbf{u}\otimes\mathbf{u} : \nabla\boldsymbol{\xi} + \overline{p}\operatorname{div}\boldsymbol{\xi} - \mathbb{S}(\mathbf{u}) : \nabla\boldsymbol{\xi}\big)\, dxdt$$
$$+ \int_Q \overline{\varrho}\mathbf{f}\cdot\boldsymbol{\xi}\, dxdt + \int_\Omega (\varrho_\infty \mathbf{U}\cdot\boldsymbol{\xi})(x,0)\, dx = 0 \tag{7.1.42}$$

*holds for all vector fields $\boldsymbol{\xi} \in C^\infty(Q)$ equal to 0 in a neighborhood of the lateral boundary $S_T$ and of the top $\Omega\times\{t=T\}$. Here the functions $\overline{\varrho}$ and $\overline{p}$ are defined by (7.1.29).*

*(iii) The integral identity for the distribution function*

$$\int_Q\int_{\mathbb{R}} \big(f\partial_t\psi + f\nabla_x\psi\cdot\mathbf{u} - fs\partial_s\psi\operatorname{div}\mathbf{u} - s\partial_s\psi\,\mathfrak{C}[f]\big)\, dsdxdt$$
$$+ \int_\Omega\int_{\mathbb{R}} f_\infty(x,0,s)\psi(x,0,s)\, dsdx - \int_{\Sigma_{\text{in}}}\int_{\mathbb{R}} f_\infty(x,t,s)\psi(x,t,s)\mathbf{U}\cdot\mathbf{n}\, dsd\Sigma = 0 \tag{7.1.43}$$

*holds for all $\psi \in C^\infty(Q\times\mathbb{R})$ which satisfy the conditions*

$$\begin{aligned} &\psi = 0 \quad \text{in a neighborhood of } (S_T\setminus\Sigma_{\text{in}})\times\mathbb{R} \text{ and of } \Omega\times\{t=T\}\times\mathbb{R},\\ &\psi(x,t,s) = 0 \quad \text{for all large } |s| \text{ uniformly in } (x,t). \end{aligned} \tag{7.1.44}$$

*The functions $\mathfrak{C}[f]$ and $f_\infty$ are defined by (7.1.33e), (7.1.27), respectively.*

*(iv) Assume, in addition, that the pressure function admits the representation $p(\varrho) = p_c(\varrho) + p_b(\varrho)$ where $p_c : [0, \infty) \to \mathbb{R}$ is convex and $p_b$ is uniformly bounded. Then the function $f$ satisfies condition* (7.1.38), *i.e., $f$ is a regular solution to the kinetic equation.*

For the proof, see Section 7.2.

**Results: Main theorem on the kinetic equation. Deterministic data.** The theory of the kinetic equation (7.1.33b) is of independent interest aside from the theory of Navier-Stokes equations itself. The following theorem on kinetic equations with deterministic data, which is the second main result of this chapter, makes it possible, among other things, to prove compactness properties of solutions to compressible Navier-Stokes equations and to investigate the domain dependence of solutions to these equations. Before formulating the result we give the following

**Definition 7.1.10.** The random function $\mu_{xt}$ is *deterministic* if for a.e. $(x, t)$ the probability measure $\mu_{xt}$ is concentrated at a single point. This means that there is a measurable function $\varrho : Q \to \mathbb{R}$ such that the distribution function $f(x, t, s) = 0$ for $s < \varrho(x, t)$ and $f(x, t, s) = 1$ for $s \geq \varrho(x, t)$.

Let us consider the boundary value problem for the kinetic equation

$$\partial_t f + \operatorname{div}(f\mathbf{u}) - \partial_s(sf \operatorname{div} \mathbf{u} + s\,\mathfrak{C}[f]) = 0 \quad \text{in } Q \times \mathbb{R}, \tag{7.1.45a}$$

$$f = f_\infty \quad \text{on } \Sigma_{\text{in}} \times \mathbb{R}, \quad f(x, 0, s) = f_\infty(x, 0, s) \quad \text{on } \Omega \times \mathbb{R}. \tag{7.1.45b}$$

Here the nonlinear operator $\mathfrak{C}$ is defined by

$$\mathfrak{C}[f] = \frac{1}{\lambda + 1} \int_{(-\infty, s]} (p(\tau) - \overline{p})\, d_\tau f(x, t, \tau), \quad \overline{p} = \int_{\mathbb{R}} p(s)\, d_s f(x, t, s). \tag{7.1.45c}$$

We emphasize that here $\mathbf{u}$ is a given vector field that has nothing to do with Navier-Stokes equations. The next question concerns the deterministic case. If the boundary data $f_\infty$ is deterministic, is a solution to Problem 7.1.5 deterministic? This question is important because if $f$ is deterministic, then obviously $\overline{p} = p(\overline{\varrho})$ and a solution of the kinetic equation becomes a weak renormalized solution of the mass balance equation. The following theorem which is the main result of this chapter gives a positive answer to this question.

Our goal is to derive conditions under which any solution to problem (7.1.45) with deterministic boundary and initial distribution function $f_\infty$ is deterministic. Throughout this chapter we assume that the domain $\Omega$, the given vector field $\mathbf{u}$, and $f$, $f_\infty$ satisfy the following conditions:

**Condition 7.1.11.** • $\Omega \subset \mathbb{R}^d$ is a bounded domain with $C^\infty$ boundary. The vector field $\mathbf{u} \in L^2(0, T; W^{1,2}(\Omega))$ satisfies the following boundary condition on the parabolic boundary $\sqcup_T = S_T \cup (\operatorname{cl} \Omega \times \{t = 0\})$:

$$\mathbf{u} = \mathbf{U} \quad \text{on } \sqcup_T. \tag{7.1.46}$$

- The lateral surface $S_T = \partial\Omega \times (0,T)$ and the function $\mathbf{U} \in C^\infty(\mathbb{R}^d \times (0,T))$ satisfy the geometric condition (5.1.9).
- The distribution functions $f \in L^\infty(Q \times \mathbb{R})$ and $f_\infty \in L^\infty(\sqcup_T \times \mathbb{R})$ are monotone and right continuous in $s$ for a.e. $(x,t)$ in $Q$ and $\sqcup_T$, respectively. This means that

$$f(x,t,s') \le f(x,t,s'') \quad \text{for } s' \le s'', \quad \lim_{h\searrow 0} f(x,t,s+h) = f(x,t,s),$$

and the same condition holds for $f_\infty$. Moreover,

$$\begin{aligned} &f(x,t,s) \to 1, \quad f_\infty(x,t,s) \to 1 \quad \text{as } s \to \infty, \\ &f(x,t,s) = 0, \quad f_\infty(x,t,s) = 0 \quad \text{for } s < 0. \end{aligned}$$

- $p : \mathbb{R}^+ \to \mathbb{R}^+$ satisfies Condition 7.1.1 with $\gamma > d/2$, and there is $c > 0$ such that

$$\operatorname*{ess\,sup}_{t\in(0,T)} \int_\Omega \Big\{ \int_{[0,\infty)} s^\gamma \, d_s f(x,t,s) \Big\} \, dx < c < \infty, \tag{7.1.47}$$

- There is $\theta > 0$ such that for every $Q' \Subset Q$,

$$\int_{Q'} \Big\{ \int_{[0,\infty)} s^{\gamma+\theta} \, d_s f(x,t,s) \Big\} \, dx < c(Q') < \infty.$$

- There is a positive $\varrho_\infty \in C^1(\mathbb{R}^d \times (0,T))$ such that for a.e. $(x,t) \in \sqcup_T$,

$$f_\infty(x,t,s) = 0 \text{ for } s < \varrho_\infty(x,t), \; f_\infty(x,t,s) = 1 \text{ for } s \ge \varrho_\infty(x,t). \tag{7.1.48}$$

**Theorem 7.1.12.** *Let Condition 7.1.11 be satisfied. Furthermore assume that $f$ satisfies equations (7.1.45) and inequality (7.1.38), i.e., $f$ is a regular solution to problem (7.1.45). Then there is $\varrho \in L^\infty(0,T; L^\gamma(\Omega))$ such that for a.e. $(x,t) \in Q$,*

$$f(x,t,s) = 0 \quad \text{for } s < \varrho(x,t), \quad f(x,t,s) = 1 \quad \text{for } s \ge \varrho(x,t). \tag{7.1.49}$$

## 7.2 Proof of Theorem 7.1.9

**Proof of claim (i).** We split the proof into a sequence of lemmas.

**Lemma 7.2.1.** *Under the assumptions of Theorem 7.1.9,*

$$\int_Q \psi(x,t) \Big\{ \int_{[0,\infty)} s^\gamma \, d_s f(x,t,s) \Big\} \, dxdt \le \limsup_{\epsilon\to 0} \int_Q \psi(x,t) (\varrho_\epsilon)^\gamma \, dxdt, \tag{7.2.1}$$

$$\int_{Q'} \psi(x,t) \Big\{ \int_{[0,\infty)} s^{\theta+\gamma} \, d_s f(x,t,s) \Big\} \, dxdt \le \limsup_{\epsilon\to 0} \int_{Q'} \psi(x,t) (\varrho_\epsilon)^{\theta+\gamma} \, dxdt, \tag{7.2.2}$$

*for all nonnegative $\psi \in L^\infty(Q)$ and all $Q' \Subset Q$. Moreover for all $\psi \in L^\infty(Q')$,*

$$\int_{Q'} \overline{p}\psi \,dxdt = \int_{Q'} \psi(x,t)\Big\{ \int_{[0,\infty)} p(s)\,d_s f(x,t,s)\Big\}\,dxdt$$
$$= \lim_{\epsilon\to 0}\int_{Q'} \psi(x,t)p(\varrho_\epsilon)\,dxdt. \tag{7.2.3}$$

*Proof.* Consider the nonnegative Carathéodory function $F(x,t,s) = \psi(x,t)|s|^\gamma$. It follows from estimate (7.1.9) and the inequality $s^\gamma \le c_p(1+P(s))$ that

$$\int_Q F(x,t,\varrho_\epsilon)\,dxdt \le c_e\|\psi\|_{L^\infty(Q)} \int_Q (1+P(\varrho_\epsilon))\,dxdt \le c_e\|\psi\|_{L^\infty(Q)}.$$

Hence the integrals of $F(x,t,\varrho_\epsilon)$ are uniformly bounded. Recall that $f$ is the distribution function of the Young measure $\mu_{xt}$ associated with the sequence $\varrho_\epsilon$, i.e., $d\mu_{xt}(s) = d_s f(x,t,s)$. Hence we can apply Theorem 1.4.5 with $\Omega$ replaced by $Q$ and $v_n$ replaced by $\varrho_\epsilon$ to obtain (7.2.1). Next consider the integrand $F(x,t,s) = \psi(x,t)|s|^{\theta+\gamma}$ on a compact set $Q' \Subset Q$. It follows from estimate (7.1.10) that

$$\int_{Q'} F(x,t,\varrho_\epsilon)\,dxdt \le c_e(Q')\|\psi\|_{L^\infty(Q)} \int_{Q'} \varrho_\epsilon^{\theta+\gamma}\,dxdt \le c_e(Q')\|\psi\|_{L^\infty(Q)}.$$

Applying now Theorem 1.4.5 with $\Omega$ replaced by $Q'$ and $v_n$ replaced by $\varrho_\epsilon$ we obtain (7.2.2). Finally choose an arbitrary $Q' \Subset Q$ and $\psi \in L^\infty(Q')$. Consider the integrand $F(x,t,s) = \psi(x,t)p(s)$. Since $p(s) \le c(1+s^\gamma)$ for $s \ge 0$, we have

$$|F(x,t,\varrho_\epsilon)|^{1+\theta/\gamma} \le \|\psi\|^{1+\theta/\gamma}_{L^\infty(Q')}(1+\varrho_\epsilon^{\gamma+\theta}).$$

It now follows from estimate (7.1.10) that

$$\int_{Q'} |F(x,t,\varrho_\epsilon)|^{1+\theta/\gamma}\,dxdt \le c_e\|\psi\|^{1+\theta/\gamma}_{L^\infty(Q')}\Big(1+\int_{Q'}\varrho_\epsilon^{\gamma+\theta}\,dxdt\Big) \le c,$$

where $c$ is independent of $\epsilon$. Since $\theta/\gamma > 0$, it follows that the sequence $F(x,t,\varrho_\epsilon)$ is equi-integrable in $Q'$. Applying Theorem 1.4.5 we conclude that

$$\int_{Q'} \langle \mu_{xt}, F(x,t,\cdot)\rangle\,dxdt = \lim_{\epsilon\to 0}\int_{Q'} F(x,t,\varrho_\epsilon)\,dxdt,$$

which along with the identity

$$\langle \mu_{xt}, F(x,t,\cdot)\rangle = \psi(x,t)\int_{[0,\infty)} p(s)\,d_s f(x,t,s)$$

yields (7.2.3). □

**Lemma 7.2.2.** *Under the assumptions of Theorem 7.1.9, the distribution function $f$ satisfies inequalities* (7.1.40) *and* (7.1.41).

*Proof.* Choose $t_0 \in (0,T)$ and $h > 0$ such that $(t_0 - h, t_0 + h) \subset (0,T)$. Next set $\psi(x,t) = 1$ if $t \in (t_0 - h, t_0 + h)$ and $\psi(x,t) = 0$ otherwise. It follows from (7.2.1), (7.1.9), and the inequality $s^\gamma \le c_e(1 + P(s))$ that

$$\frac{1}{2h}\int_{t_0-h}^{t_0+h}\int_\Omega\Big\{\int_{[0,\infty)} s^\gamma\, d_s f(x,t,s)\Big\}\,dxdt \le \limsup_{\epsilon\to 0}\frac{1}{2h}\int_{t_0-h}^{t_0+h}\int_\Omega \varrho_\epsilon^\gamma\,dxdt$$
$$\le c_e \limsup_{\epsilon\to 0}\frac{1}{2h}\int_{t_0-h}^{t_0+h}\int_\Omega (1 + P(\varrho_\epsilon))\,dxdt \le c_e.$$

Letting $h \to 0$ we obtain (7.1.40). Next choose $Q' \Subset Q$ and set $\psi(x,t) = 1$ in $Q'$ and $\psi = 0$ elsewhere. Combining (7.1.10) and (7.2.2) we obtain (7.1.41). □

**Proof of claim (ii)**

**Lemma 7.2.3.** *There exist subsequences, still denoted by $\varrho_\epsilon$, $\mathbf{u}_\epsilon$, such that as $\epsilon \to 0$,*

$$\begin{aligned} &\varrho_\epsilon\mathbf{u}_\epsilon \rightharpoonup \overline{\varrho}\mathbf{u} \quad \textit{weakly}^\star \textit{ in } L^\infty(0,T;L^{2\gamma/(\gamma+1)}(\Omega)),\\ &\varrho_\epsilon\mathbf{u}_\epsilon\otimes\mathbf{u}_\epsilon \rightharpoonup \overline{\varrho}\mathbf{u}\otimes\mathbf{u} \quad \textit{weakly in } L^2(0,T;L^{z}(\Omega)),\end{aligned} \tag{7.2.4}$$

*where $z$ is an arbitrary number from the interval* $(1,(1-2^{-1}\mathfrak{a})^{-1})$, $\mathfrak{a} = 2d^{-1} - \gamma^{-1}$.

*Proof.* By (7.1.8) the functions $\varrho_\epsilon$ and $\mathbf{u}_\epsilon$ satisfy the mass and momentum balance equations

$$\begin{aligned} \partial_t\varrho_\epsilon &= \operatorname{div}\mathfrak{u}_\epsilon \quad \text{in } Q,\\ \partial_t(\varrho_\epsilon\mathbf{u}_\epsilon) &= \operatorname{div}\mathfrak{V}_\epsilon + \mathfrak{w}_\epsilon \quad \text{in } Q,\end{aligned}$$

which are understood in the sense of distributions. Here,

$$\mathfrak{u}_\epsilon = -\varrho_\epsilon\mathbf{u}_\epsilon, \quad \mathfrak{w}_\epsilon = \varrho_\epsilon\mathbf{f}, \quad \mathfrak{V}_\epsilon = \mathbb{S}(\mathbf{u}_\epsilon) - p_\epsilon(\varrho_\epsilon)\,\mathbb{I} - \varrho_\epsilon\mathbf{u}_\epsilon\otimes\mathbf{u}_\epsilon.$$

It follows from estimate (7.1.9) that the sequences $\mathbf{u}_\epsilon$ and $\varrho_\epsilon$ are bounded in $L^2(0,T;W^{1,2}(\Omega))$ and $L^\infty(0,T;L^\gamma(\Omega))$, respectively. Moreover, that estimate implies that $(\mathbf{u}_\epsilon,\varrho_\epsilon)$ are bounded energy functions in the sense of Section 4.2, and their energies are bounded by the constant $c_e$ in (7.1.9) independent of $\epsilon$. Hence we can apply Corollary 4.2.2 to conclude that

$$\|\varrho_\epsilon\mathbf{u}_\epsilon\|_{L^\infty(0,T;L^{2\gamma/(\gamma+1)}(\Omega))} \le c.$$

Hence the sequence $\mathfrak{u}_\epsilon$ is bounded in $L^1(Q)$ and the functions $\varrho_\epsilon$, $\mathfrak{u}_\epsilon$ satisfy Condition 4.4.1 of Theorem 4.4.2 with $n$ replaced by $\epsilon$. Therefore

$$\varrho_\epsilon\mathbf{u}_\epsilon \rightharpoonup \overline{\varrho}\mathbf{u} \quad \text{weakly in } L^2(0,T;L^m(\Omega)) \text{ for } m^{-1} > 2^{-1} + \gamma^{-1} - d^{-1}.$$

Since the sequence $\varrho_\epsilon \mathbf{u}_\epsilon$ is bounded in $L^\infty(0,T;L^{2\gamma/(\gamma+1)}(\Omega))$, it also converges weakly* in that space. This leads to the first relation in (7.2.4). To prove the second, notice that by (7.1.9) and the inequalities $0 \le p \le c(1+P)$, the sequence $\mathfrak{V}_\epsilon$ is bounded in $L^1(Q)$. On the other hand, since $\mathbf{f}$ is bounded, it follows from (7.1.9) that $\mathfrak{w}_\epsilon$ is bounded in $L^1(Q)$. Therefore the sequences $\varrho_\epsilon$, $\mathbf{u}_\epsilon$, $\mathfrak{V}_\epsilon$, and $\mathfrak{w}_\epsilon$ satisfy all hypotheses of Theorem 4.4.2 with $n$ replaced by $\epsilon$. This yields the second relation in (7.2.4). □

**Lemma 7.2.4.** *Under the assumptions of Theorem 7.1.3 for any $Q' \Subset Q$ there is a constant $c(Q')$ independent of $\epsilon$ such that*

$$\|p_\epsilon(\varrho_\epsilon)\|_{L^r(Q')} \le c(r,Q') \quad \text{for } 1 \le r < 1 + \theta(n\gamma)^{-1} \tag{7.2.5}$$

*and*

$$\lim_{\epsilon\to 0} \int_{Q'} |p(\varrho_\epsilon) - p_\epsilon(\varrho_\epsilon)|\, dxdt = 0. \tag{7.2.6}$$

*Proof.* Since $p(\varrho) \le c(1+\varrho^\gamma)$ and $\epsilon\varrho^2 \le 1 + \epsilon\varrho^n$, we have

$$\begin{aligned}\int_{Q'} p_\epsilon(\varrho_\epsilon)^r\, dxdt &\le c\int_{Q'} (1+\varrho_\epsilon^\gamma + \epsilon\varrho_\epsilon^n)^r\, dxdt \\ &\le c + c\int_{Q'} (\varrho_\epsilon^{r\gamma} + \epsilon\varrho_\epsilon^{nr})\, dxdt \le c + c\int_{Q'} (\varrho_\epsilon^{\theta+\gamma} + \epsilon\varrho_\epsilon^{n+\theta})\, dxdt,\end{aligned}$$

which along with (7.1.10) yields (7.2.5). Next, we have

$$p_\epsilon(\varrho) = p(\varrho) + p_\epsilon^*(\varrho) + \epsilon(\varrho^2 + \varrho^n), \quad \text{where} \quad p_\epsilon^*(\varrho) = (1-\chi(\varrho/\epsilon))p(\varrho).$$

Since $1-\chi(\varrho/\epsilon)$ vanishes for $\varrho > \epsilon$, we have

$$0 \le p_\epsilon^*(\varrho) \le \max_{\varrho\in[0,\epsilon]} p(\varrho) = \eta(\epsilon) \to 0 \quad \text{as } \epsilon \to 0.$$

From this and the inequality $\varrho^2 \le 1 + \varrho^n$ we obtain

$$|p_\epsilon(\varrho) - p(\varrho)| \le \eta(\epsilon) + \epsilon + 2\epsilon\varrho^n. \tag{7.2.7}$$

Next, the Young inequality implies

$$\epsilon^{n/(n+\theta)}\varrho_\epsilon^n \le \epsilon\varrho_\epsilon^{n+\theta} + 1.$$

Thus from (7.1.10) we get

$$\begin{aligned}\epsilon\int_{Q'} \varrho_\epsilon^n\, dxdt &= \epsilon^{\theta/(n+\theta)}\int_{Q'} \epsilon^{n/(n+\theta)}\varrho_\epsilon^n\, dxdt \le \epsilon^{\theta/(n+\theta)}\int_{Q'} (\epsilon\varrho_\epsilon^{n+\theta}+1)\, dxdt \\ &\le c_e(Q')\epsilon^{\theta/(n+\theta)},\end{aligned}$$

Combining this with (7.2.7) we obtain

$$\int_{Q'} |p(\varrho_\epsilon) - p_\epsilon(\varrho_\epsilon)|\, dxdt \le c_e\epsilon + c_e\eta(\epsilon) + c_e(Q')\epsilon^{\theta/(n+\theta)} \to 0 \quad \text{as } \epsilon \to 0. \qquad \square$$

**Lemma 7.2.5.** *Under the assumptions of Theorem 7.1.9 the identity (7.1.42) holds for all vector fields $\boldsymbol{\xi} \in C^\infty(Q)$ equal to 0 in a neighborhood of $S_T = \partial\Omega \times (0,T)$ and of the top $\Omega \times \{t = T\}$.*

*Proof.* Choose $\boldsymbol{\xi} \in C^\infty(Q)$ as in the statement. Since $(\varrho_\epsilon, \mathbf{u}_\epsilon)$ is a weak solution to the momentum balance equation (7.1.8a), we have

$$\int_Q \varrho_\epsilon \mathbf{u}_\epsilon \cdot \partial_t \boldsymbol{\zeta}\, dxdt + \int_Q \big(\varrho_\epsilon \mathbf{u}_\epsilon \otimes \mathbf{u}_\epsilon + p_\epsilon(\varrho_\epsilon)\mathbb{I} - \mathbb{S}(\mathbf{u}_\epsilon)\big) : \nabla \boldsymbol{\zeta}\, dxdt \\ + \int_Q \varrho_\epsilon \mathbf{f} \cdot \boldsymbol{\zeta}\, dxdt + \int_\Omega \varrho_\infty^\epsilon(\cdot,0)\mathbf{U}(\cdot,0)\cdot\boldsymbol{\zeta}(\cdot,0)\, dx = 0. \quad (7.2.8)$$

It follows from Lemma 7.2.3 that

$$\lim_{\epsilon\to 0} \int_Q \{\varrho_\epsilon \mathbf{u}_\epsilon \cdot \partial_t \boldsymbol{\zeta} + \big(\varrho_\epsilon \mathbf{u}_\epsilon \otimes \mathbf{u}_\epsilon - \mathbb{S}(\mathbf{u}_\epsilon)\big) : \nabla\boldsymbol{\zeta} + \varrho_\epsilon \mathbf{f}\cdot\boldsymbol{\zeta}\} dxdt \\ = \int_Q \{\overline{\varrho}\mathbf{u}\cdot\partial_t\boldsymbol{\zeta} + \big(\overline{\varrho}\mathbf{u}\otimes\mathbf{u} - \mathbb{S}(\mathbf{u})\big) : \nabla\boldsymbol{\zeta} + \overline{\varrho}\mathbf{f}\cdot\boldsymbol{\zeta}\}\, dxdt. \quad (7.2.9)$$

Next notice that $\boldsymbol{\zeta}$ vanishes in a neighborhood of $S_T = \partial\Omega \times (0,T)$. Hence there is a cylinder

$$Q_\delta = \Omega_\delta \times (0,T), \quad \text{where} \quad \Omega_\delta = \{x \in \Omega : \operatorname{dist}(x, \partial\Omega) > \delta\},$$

such that $\operatorname{supp}\boldsymbol{\zeta} \subset Q_\delta$. Now choose $h > 0$ and notice that $\Omega_\delta \times (h, T-h) \Subset Q$. It follows from (7.2.3) with $\psi = \operatorname{div}\boldsymbol{\zeta}$ that

$$\lim_{\epsilon\to 0} \int_h^{T-h} \int_{\Omega_\delta} p(\varrho_\epsilon) \operatorname{div}\boldsymbol{\zeta}\, dxdt = \int_h^{T-h} \int_{\Omega_\delta} \overline{p} \operatorname{div}\boldsymbol{\zeta}\, dxdt. \quad (7.2.10)$$

On the other hand, (7.2.6) with $Q' = \Omega_\delta \times (h, T-h)$ implies

$$\lim_{\epsilon\to 0} \int_h^{T-h} \int_{\Omega_\delta} |p(\varrho_\epsilon) - p_\epsilon(\varrho_\epsilon)|\, dxdt = 0.$$

Combining this with (7.2.10) we obtain

$$\lim_{\epsilon\to 0} \int_h^{T-h} \int_{\Omega_\delta} p_\epsilon(\varrho_\epsilon) \operatorname{div}\boldsymbol{\zeta}\, dxdt = \int_h^{T-h} \int_{\Omega_\delta} \overline{p} \operatorname{div}\boldsymbol{\zeta}\, dxdt. \quad (7.2.11)$$

Next, in view of (7.1.18) we have $0 \le p_\epsilon \le c(1 + P_\epsilon)$. Along with (7.1.9) this implies

$$\Big|\Big\{\int_0^h + \int_{T-h}^T\Big\} \int_{\Omega_\delta} p(\varrho_\epsilon) \operatorname{div}\boldsymbol{\zeta}\, dxdt\Big| \le c\Big\{\int_0^h + \int_{T-h}^T\Big\} \int_\Omega (1 + P_\epsilon)\, dxdt \le ch. \quad (7.2.12)$$

Notice that $p(s) \le c(1+s^\gamma)$ for $s \ge 0$ in view of (7.1.4). Thus

$$\overline{p}(x,t) := \int_{[0,\infty)} p(s)\, d_s f(x,t,s) \le c\Big(1 + \int_{[0,\infty)} s^\gamma\, d_s f(x,t,s)\Big).$$

From (7.1.40) we conclude that

$$\operatorname*{ess\,sup}_{t\in(0,T)} \int_\Omega \overline{p}(x,t)\, dx \le c,$$

which leads to

$$\Big|\Big\{\int_0^h + \int_{T-h}^T\Big\}\int_{\Omega_\delta} \overline{p}\operatorname{div}\boldsymbol{\zeta}\, dxdt\Big| \le ch. \tag{7.2.13}$$

Combining (7.2.11)–(7.2.13), and noting that $\boldsymbol{\zeta}$ vanishes outside $\Omega_\delta \times (0,T)$, we arrive at

$$\limsup_{\epsilon\to 0}\Big|\int_Q p_\epsilon(\varrho_\epsilon)\operatorname{div}\boldsymbol{\zeta}\, dxdt - \int_Q \overline{p}\operatorname{div}\boldsymbol{\zeta}\, dxdt\Big| \le ch.$$

Finally, letting $h \to 0$ we obtain

$$\lim_{\epsilon\to 0}\int_Q p_\epsilon(\varrho_\epsilon)\operatorname{div}\boldsymbol{\zeta}\, dxdt = \int_Q \overline{p}\operatorname{div}\boldsymbol{\zeta}\, dxdt. \tag{7.2.14}$$

Letting $\epsilon \to 0$ in (7.2.8) and recalling (7.2.9), (7.2.14) we obtain the integral identity (7.1.42). □

**Proof of claim (iii)**

**Lemma 7.2.6.** *Let the hypotheses of Theorem 7.1.9 be satisfied and* $\Phi \in C(\mathbb{R}^+)$ *vanish near* $+\infty$. *Then passing to a subsequence we can assume that*

$$\Phi_\epsilon \operatorname{div}\mathbf{u}_\epsilon \rightharpoonup \overline{\Phi\operatorname{div}\mathbf{u}} \quad \text{weakly in } L^2(Q), \tag{7.2.15}$$

$$V_\epsilon\Phi_\epsilon \rightharpoonup -(\lambda+1)\overline{\Phi\operatorname{div}\mathbf{u}} + \overline{\Phi p} \quad \text{weakly in } L^2(Q), \tag{7.2.16}$$

$$\overline{\Phi\operatorname{div}\mathbf{u}} = \overline{\Phi}\operatorname{div}\mathbf{u} + \frac{1}{\lambda+1}\big(\overline{\Phi p} - \overline{\Phi}\overline{p}\big), \tag{7.2.17}$$

*where* $\Phi_\epsilon = \Phi(\varrho_\epsilon)$, $V_\epsilon = -(1+\lambda)\operatorname{div}\mathbf{u}_\epsilon + p_\epsilon(\varrho_\epsilon)$. *Notice that* $\overline{\Phi\operatorname{div}\mathbf{u}}$ *is just a notation for the weak limit, whereas* $\overline{p}$, $\overline{\Phi}$, *and* $\overline{\Phi p}$ *are represented in terms of the Young measure* $\mu_{xt}$.

*Proof.* Assume that $\Phi(s)$ is extended to a function in $C_c(\mathbb{R})$. Moreover, it suffices to prove the lemma for $\Phi \in C_0^\infty(\mathbb{R})$ since $C_0^\infty(\mathbb{R})$ is dense in $C_0(\mathbb{R})$. Observe that the governing equations (7.1.8) can be rewritten in the form

$$\partial_t(\varrho_\epsilon\mathbf{u}_\epsilon) + \operatorname{div}(\varrho_\epsilon\mathbf{u}_\epsilon\otimes\mathbf{u}_\epsilon) = \operatorname{div}\mathbb{T}_\epsilon + \varrho_\epsilon\mathbf{f} \quad \text{in } Q, \tag{7.2.18a}$$

$$\partial_t\varrho_\epsilon + \operatorname{div}(\varrho_\epsilon\mathbf{u}_\epsilon) = 0 \quad \text{in } Q. \tag{7.2.18b}$$

Here

$$\mathbb{T}_\epsilon = \mathbb{S}(u_\epsilon) - p_\epsilon(\varrho_\epsilon)\,\mathbb{I}. \tag{7.2.18c}$$

Moreover, since $(\mathbf{u}_\epsilon, \varrho_\epsilon)$ is a renormalized solution to problem (7.1.8) we have the following equation for $\Phi_\epsilon = \Phi(\varrho_\epsilon)$:

$$\partial\Phi_\epsilon + \operatorname{div}(\Phi_\epsilon \mathbf{u}_\epsilon) + \varpi_\epsilon = 0. \tag{7.2.18d}$$

Here

$$\varpi_\epsilon = (\Phi'(\varrho_\epsilon)\varrho_\epsilon - \Phi(\varrho_\epsilon))\mathbf{u}_\epsilon. \tag{7.2.19}$$

Now choose the exponents

$$1 < r' \le 1 + \frac{\theta}{n\gamma}, \quad r = r'/(r'-1), \quad 1 < q \le 2, \quad s = \infty, \tag{7.2.20}$$

and set

$$\mathbf{g}_\epsilon = \mathbf{g}_\epsilon^\varphi = 0, \quad \mathbb{G}_\epsilon = 0. \tag{7.2.21}$$

Let us check that the functions $\varrho_\epsilon$, $\varphi_\epsilon := \Phi_\epsilon$, the vector fields $\mathbf{u}_\epsilon$, $\mathbf{g}_\epsilon$, $\mathbf{g}_\epsilon^\varphi$, the matrix-valued functions $\mathbb{T}_\epsilon$, $\mathbb{G}_\epsilon$, and the exponents $\gamma$, $s$, $r$, $q$ meet all requirements of Theorem 4.7.1 with $n \to \infty$ replaced by $\epsilon \to 0$. Since the stress tensor $\mathbb{T}_\epsilon$ admits representation (7.2.18c) it suffices to prove that all these quantities satisfy Condition 4.6.1 and (4.6.16).

First, by (7.1.9) and the inequality $\varrho^\gamma \le c(1+P)$ in Condition 7.1.1, the functions $\varrho_\epsilon$, $\varphi_\epsilon = \Phi_\epsilon$, and the vector fields $\mathbf{u}_\epsilon$ satisfy (4.6.4)–(4.6.6) with $n$ and $E$ replaced by $\epsilon$ and a constant $c$ independent of $\epsilon$, respectively. Next, since $\Phi'\varrho - \Phi$ is bounded, it follows from (7.1.9) and (7.2.5) that for any $Q' \Subset Q$,

$$\|\mathbb{T}_\epsilon\|_{L^1(Q)} \le c, \quad \|\varpi_\epsilon\|_{L^2(Q)} \le c, \quad \|\mathbb{T}_\epsilon\|_{L^{r'}(Q')} \le c(Q'), \tag{7.2.22}$$

where $c$ is independent of $\epsilon$. Hence $\mathbb{T}_\epsilon$ and $\Phi_\epsilon$ satisfy (4.6.7)–(4.6.8) with $n$, $E$, and $\varphi_\epsilon$ replaced by $\epsilon$, $c$, and $\Phi_\epsilon$, respectively. Next, (7.2.21) and (7.2.22) obviously imply that $\varpi_\epsilon$, $\mathbf{g}_\epsilon$, $\mathbf{g}_\epsilon^\varphi$, and $\mathbb{G}_\epsilon$ satisfy (4.6.9)–(4.6.11). On the other hand, by (7.1.22) and (7.1.26a),

$$\begin{aligned}
\mathbf{u}_\epsilon &\rightharpoonup \mathbf{u} \quad \text{weakly in } L^2(0,T;W^{1,2}(\Omega)),\\
\varrho_\epsilon &\rightharpoonup \overline{\varrho} \quad \text{weakly}^\star \text{ in } L^\infty(0,T;L^\gamma(\Omega)),\\
\Phi_\epsilon &\rightharpoonup \overline{\Phi} \quad \text{weakly}^\star \text{ in } L^\infty(Q).
\end{aligned}$$

Hence $\varrho_\epsilon$, $\mathbf{u}_\epsilon$, and $\Phi_\epsilon$ satisfy (4.6.12) with $n \to \infty$ replaced by $\epsilon \to 0$.

Next, by (7.2.5) the functions $p_\epsilon(\varrho_\epsilon)$ are bounded in $L^{r'}(Q')$ for every $Q' \Subset Q$. Hence passing to a subsequence we can assume that $p_\epsilon(\varrho_\epsilon)$ converges weakly in $L^{r'}(Q')$ to some $p^* \in L^1(Q) \cap L^{r'}_{\mathrm{loc}}(Q)$. Since $r' \le 2$, we have

$$\begin{aligned} \mathbb{T}_\epsilon &= \mathbb{S}(\mathbf{u}_\epsilon) + p_\epsilon(\varrho_\epsilon) \rightharpoonup \mathbb{S}(\mathbf{u}) + p^* = \overline{\mathbb{T}}, \\ V_\epsilon &= -(\lambda+1)\operatorname{div}\mathbf{u}_\epsilon + p_\epsilon(\varrho_\epsilon) \rightharpoonup -(\lambda+1)\operatorname{div}\mathbf{u} + p^* = \overline{V} \end{aligned} \tag{7.2.23}$$

weakly in $L^{r'}(Q')$. Hence the sequence $\mathbb{T}_\epsilon$ also satisfies (4.6.12). Therefore, the functions $\varrho_\epsilon$, $\varphi_\epsilon = \Phi_\epsilon$, the vector fields $\mathbf{u}_\epsilon$, $\mathbf{g}_\epsilon$, $\mathbf{g}_\epsilon^\varphi$, and the matrix-valued functions $\mathbb{T}_\epsilon$, $\mathbb{G}_\epsilon$ satisfy Condition 4.6.1 of Theorem 4.6.3 with exponents $\gamma$, $s$, $r$, $q$ given by (7.2.20). Obviously the exponents and the functions $\mathbb{G}_\epsilon$, $\mathbf{g}_\epsilon^\varphi$ satisfy (4.6.16). In summary, all these quantities meet all requirements of Theorems 4.6.3 and 4.7.1. Applying Theorem 4.7.1 we obtain

$$\lim_{\epsilon\to 0}\int_Q V_\epsilon \Phi_\epsilon \eta\, dxdt = \int_Q \overline{V}\,\overline{\Phi}\,\eta\, dxdt \quad \text{for all } \eta \in C_0^\infty(Q), \tag{7.2.24}$$

where $V_\epsilon$ and $\overline{V}$ are defined by (7.2.23). Now we specify the integrands in (7.2.24). By (7.2.3) and (7.2.6) we have

$$\lim_{\epsilon\to 0}\int_{Q'} \psi(x,t) p_\epsilon(\varrho_\epsilon)\, dxdt = \int_{Q'} \psi(x,t)\overline{p}\, dxdt \quad \text{for all } \psi \in L^\infty(Q').$$

It follows that $p^* = \overline{p}$ and hence

$$\overline{V} = -(\lambda+1)\operatorname{div}\mathbf{u} + \overline{p}. \tag{7.2.25}$$

Since $\Phi \in C_0^\infty(\mathbb{R})$, the sequence $\Phi_\epsilon \operatorname{div}\mathbf{u}_\epsilon$ is bounded in $L^2(Q)$. Passing to a subsequence we can assume that there is a function $\overline{\Phi\operatorname{div}\mathbf{u}} \in L^2(Q)$ such that

$$\Phi_\epsilon \operatorname{div}\mathbf{u}_\epsilon \rightharpoonup \overline{\Phi\operatorname{div}\mathbf{u}} \quad \text{weakly in } L^2(Q). \tag{7.2.26}$$

This is (7.2.15). Notice that $\overline{\Phi\operatorname{div}\mathbf{u}}$ is just a notation for the weak limit. Next, since $\Phi \in C_0^\infty(\mathbb{R})$, the sequences $\Phi(\varrho_\epsilon)p_\epsilon(\varrho_\epsilon)$ and $\Phi(\varrho_\epsilon)p(\varrho_\epsilon)$ are bounded in $L^\infty(Q)$. Hence there are functions $(\Phi p)^*$ and $\overline{\Phi p}$ such that for any $\psi \in L^1(Q)$,

$$\lim_{\epsilon\to 0}\int_Q \psi p_\epsilon(\varrho_\epsilon)\Phi(\varrho_\epsilon)\, dxdt = \int_Q \psi(x,t)(\Phi p)^*\, dxdt, \tag{7.2.27}$$

$$\lim_{\epsilon\to 0}\int_Q \psi p(\varrho_\epsilon)\Phi(\varrho_\epsilon)\, dxdt = \int_Q \psi(x,t)\overline{\Phi p}\, dxdt. \tag{7.2.28}$$

Moreover, since $\Phi$ is compactly supported, from (7.1.5) we have

$$(p_\epsilon(\varrho) - p(\varrho))\Phi(\varrho) \to 0 \quad \text{as } \epsilon \to 0 \quad \text{uniformly in } \varrho.$$

Hence $(\Phi p)^* = \overline{\Phi p}$. From (7.2.24), (7.2.27), and (7.2.26) we now obtain

$$\lim_{\epsilon\to 0}\int_Q V_\epsilon \Phi_\epsilon \eta\, dxdt = \int_Q \left(-(\lambda+1)\overline{\Phi\operatorname{div}\mathbf{u}} + \overline{\Phi p}\right)\eta\, dxdt, \tag{7.2.29}$$

which yields (7.2.16). Finally combining (7.2.24), (7.2.29), and (7.2.25) we get

$$-(\lambda+1)\overline{\Phi\operatorname{div}\mathbf{u}}+\overline{\Phi p}=-(\lambda+1)\overline{\Phi}\operatorname{div}\mathbf{u}+\overline{\Phi}\overline{p},$$

which yields (7.2.17). □

**Lemma 7.2.7.** *Let the hypotheses of Theorem 7.1.9 be satisfied. Furthermore assume that $\varphi \in C_0^\infty(\mathbb{R})$ and $\Phi(s)=\varphi'(s)s-\varphi(s)$. Then*

$$\int_Q \overline{\varphi}(\partial_t\psi+\nabla\psi\cdot\mathbf{u})\,dxdt-\frac{1}{\lambda+1}\int_Q \psi(\overline{\Phi p}-\overline{\Phi}\overline{p})\,dxdt \\ -\int_Q \psi\overline{\Phi}\operatorname{div}\mathbf{u}\,dxdt+\int_\Omega(\psi\overline{\varphi}_\infty)(x,0)\,dx-\int_{S_T}\psi\overline{\varphi}_\infty\mathbf{U}\cdot\mathbf{n}\,dSdt=0 \quad (7.2.30)$$

*for all $\psi\in C^\infty(Q)$ vanishing in a neighborhood of $S_T\setminus\Sigma_{\text{in}}$ and of $\Omega\times\{t=T\}$. Here*

$$\overline{\varphi}=\int_{\mathbb{R}}\varphi(s)\,d_sf(x,t,s),\quad \overline{p}=\int_{\mathbb{R}}p(s)\,d_sf(x,t,s),\quad \overline{\varphi}_\infty=\int_{\mathbb{R}}\varphi(s)\,d_sf_\infty(x,t,s),$$
$$\overline{\Phi p}=\int_{\mathbb{R}}\Phi(s)p(s)\,d_sf(x,t,s),\quad \overline{\Phi}=\int_{\mathbb{R}}\Phi(s)\,d_sf(x,t,s).$$

*Proof.* Choose $\varphi$ and $\psi$ as in the statement. Set $\varphi_\epsilon=\varphi(\varrho_\epsilon)$, $\varphi_\infty^\epsilon=\varphi(\varrho_\infty^\epsilon)$, and $\Phi_\epsilon=\Phi(\varrho_\epsilon)$. Since $\varrho_\epsilon$ is a renormalized solution to the boundary value problem

$$\partial_t\varrho_\epsilon+\operatorname{div}(\varrho_\epsilon\mathbf{u}_\epsilon)=0\quad\text{in } Q,$$
$$\varrho_\epsilon=\varrho_\infty^\epsilon\quad\text{on }\Sigma_{\text{in}},\quad \varrho_\epsilon|_{t=0}=\varrho_\infty^\epsilon\quad\text{in }\Omega,$$

we have

$$\int_Q\varphi_\epsilon(\partial_t\psi+\nabla\psi\cdot\mathbf{u}_\epsilon)\,dxdt-\int_Q\psi\Phi_\epsilon\operatorname{div}\mathbf{u}_\epsilon\,dxdt \\ +\int_\Omega(\psi\varphi_\infty^\epsilon)(x,0)\,dx-\int_{S_T}\psi\varphi_\infty^\epsilon\mathbf{U}\cdot\mathbf{n}\,dSdt=0.$$

Letting $\epsilon\to 0$ and using (7.1.26a)–(7.1.26b) and (7.2.26) we obtain

$$\int_Q\overline{\varphi}(\partial_t\psi+\nabla\psi\cdot\mathbf{u})\,dxdt-\int_Q\psi\overline{\Phi\operatorname{div}\mathbf{u}}\,dxdt \\ +\int_\Omega(\psi\overline{\varphi}_\infty)(x,0)\,dx-\int_{S_T}\psi\overline{\varphi}_\infty\mathbf{U}\cdot\mathbf{n}\,dSdt=0.$$

Inserting (7.2.17) we obtain (7.2.30). □

**Lemma 7.2.8.** *Under the assumptions of Theorem 7.1.9 the distribution function $f$ defined by (7.1.26) and (7.1.27) satisfies the integral identity (7.1.43).*

*Proof.* Choose $\eta \in C_0^\infty(\mathbb{R})$ and set

$$\varphi(s) = \int_s^\infty \eta(\tau)\, d\tau, \quad \Phi(s) = s\varphi'(s) - \varphi(s) = -s\eta(s) - \int_s^\infty \eta(\tau)\, d\tau. \quad (7.2.31)$$

First, we derive formulae for $\overline{\varphi}$, $\overline{\Phi}$, and $\overline{\Phi p}$. We have

$$\overline{\varphi}(x,t) = \int_{\mathbb{R}} \varphi(s)\, d_s f(x,t,s) = \int_{\mathbb{R}} \Big( \int_s^\infty \eta(\tau)\, d\tau \Big)\, d_s f(x,t,s).$$

Since the distribution function $f$ vanishes for $s < 0$ and the smooth function $\varphi$ vanishes for all large $s$ and $d\varphi = -\eta\, ds$, we can use Lemma 1.3.7 and integrate by parts in the Stieltjes integral to obtain

$$\overline{\varphi}(x,t) = \int_{\mathbb{R}} \eta(s) f(x,t,s)\, ds. \quad (7.2.32)$$

The same arguments give

$$\overline{\varphi}_\infty(x,t) = \int_{\mathbb{R}} \eta(s) f_\infty(x,t,s)\, ds. \quad (7.2.33)$$

As the smooth function $s\varphi'(s)$ is compactly supported and $d(s\varphi'(s)) = -(s\eta)'\, ds$, integration by parts gives

$$\overline{\varrho\varphi'} = \int_{\mathbb{R}} s\varphi'(s)\, d_s f(x,t,s) = \int_{\mathbb{R}} (s\eta)' f(x,t,s)\, ds.$$

Combining this with (7.2.32) we obtain

$$\overline{\Phi}(x,t) = \overline{\varrho\varphi'}(x,t) - \overline{\varphi}(x,t) = \int_{\mathbb{R}} s\eta'(s) f(x,t,s)\, ds. \quad (7.2.34)$$

Next, for a.e. $(x,t) \in Q$ denote by $F$ the distribution function of the Borel measure $d\nu_{xt}(s) = p(s) d\mu_{xt}(s)$ on $\mathbb{R}$,

$$F(x,t,s) = \int_{(-\infty, s]} p(\tau)\, d_\tau f(x,t,\tau). \quad (7.2.35)$$

Since $p(s)$ is integrable with respect to $\mu_{xt}$ for a.e. $(x,t) \in Q$, the function $F(x,t,s)$ is monotone, right continuous and

$$F(x,t,s) = 0 \quad \text{for } s < 0, \quad F(x,t,s) \to \overline{p}(x,t) \quad \text{as } s \to \infty.$$

Repeating the previous arguments with $f$ replaced by $F$ we obtain the following analogues of formulae (7.2.32)–(7.2.34):

$$\int_{\mathbb{R}} \varphi(s) p(s)\, d_s f(x,t,s) = \int_{\mathbb{R}} \varphi(s)\, d_s F(x,t,s) = \int_{\mathbb{R}} \eta(s) F(x,t,s)\, ds,$$

$$\int_{\mathbb{R}} \varphi'(s) s p(s)\, d_s f(x,t,s) = \int_{\mathbb{R}} \varphi'(s) s\, d_s F(x,t,s) = \int_{\mathbb{R}} (s\eta)' F(x,t,s)\, ds,$$

which gives

$$\overline{\Phi p} = \int_{\mathbb{R}} (\varphi'(s)s - \varphi(s))p(s)\, d_s f(x,t,s) = \int_{\mathbb{R}} s\eta' F(x,t,s)\, ds$$
$$= \int_{\mathbb{R}} \eta'(s)s\Big\{ \int_{(-\infty,s]} p(\tau)\, d_\tau f(x,t,\tau)\Big\}\, ds.$$

Combining this with (7.2.34) and recalling formula (7.1.33e) for $\mathfrak{C}[f]$ we obtain, for a.e. $(x,t) \in Q$,

$$\overline{\Phi p} - \overline{\Phi}\overline{p} = \int_{\mathbb{R}} \eta'(s)s\Big\{ \int_{(-\infty,s]} (p(\tau) - \overline{p})\, d_\tau f(x,t,\tau)\Big\}\, ds$$
$$= (\lambda+1)\int_{\mathbb{R}} s\eta'(s)\mathfrak{C}[f])\, ds. \tag{7.2.36}$$

Let $\varsigma \in C^\infty(Q)$ vanish in a neighborhood of $S_T \setminus \Sigma_{\text{in}}$ and of $\Omega \times \{t = T\}$. Then $\varsigma$ and $\varphi$ meet all requirements of Lemma 7.2.7 and hence satisfy the integral identity

$$\int_Q \overline{\varphi}(\partial_t\varsigma + \nabla\varsigma\cdot\mathbf{u})\, dxdt - \frac{1}{\lambda+1}\int_Q \varsigma\big(\overline{\Phi p} - \overline{\Phi}\overline{p}\big)\, dxdt$$
$$- \int_Q \varsigma\overline{\Phi}\,\operatorname{div}\mathbf{u}\, dxdt + \int_\Omega (\varsigma\overline{\varphi}_\infty)(x,0)\, dx - \int_{S_T} \varsigma\overline{\varphi}_\infty \mathbf{U}\cdot\mathbf{n}\, dSdt = 0. \tag{7.2.37}$$

It now follows from representations (7.2.32) and (7.2.33) that

$$\int_Q \overline{\varphi}(\partial_t\varsigma + \nabla\varsigma\cdot\mathbf{u})\, dxdt = \int_Q\int_{\mathbb{R}} \eta(s)(\partial_t\varsigma + \nabla\varsigma\cdot\mathbf{u})f(x,t,s)\, ds\, dxdt \tag{7.2.38}$$

and

$$\int_\Omega (\varsigma\overline{\varphi}_\infty)(x,0)\, dx - \int_{S_T} \varsigma\overline{\varphi}_\infty \mathbf{U}\cdot\mathbf{n}\, dSdt$$
$$= \int_\Omega\int_{\mathbb{R}} \varsigma(x,0)\eta(s)f_\infty(x,0,s)\, dsdx - \int_{S_T}\int_{\mathbb{R}} \varsigma\eta(s)\mathbf{U}\cdot\mathbf{n}\, f_\infty(x,t,s)\, ds\, dSdt.$$

Next, (7.2.34) and (7.2.36) imply

$$\begin{aligned} \frac{1}{\lambda+1}\int_Q \varsigma\big(\overline{\Phi p} - \overline{\Phi}\overline{p}\big)\, dxdt &= \int_Q\int_{\mathbb{R}} \varsigma s\eta'(s)\mathfrak{C}[f](x,t,s)\, ds\, dxdt,\\ \int_Q \varsigma\overline{\Phi}\,\operatorname{div}\mathbf{u}\, dxdt &= \int_Q\int_{\mathbb{R}} \varsigma s\eta'(s)f(x,t,s)\,\operatorname{div}\mathbf{u}\, dsdxdt. \end{aligned} \tag{7.2.39}$$

Inserting (7.2.38)–(7.2.39) into (7.2.37) we finally obtain

$$\int_{Q\times\mathbb{R}} \Big(f\big(\partial_t(\varsigma\eta) + \nabla_x(\varsigma\eta)\cdot\mathbf{u}\big) - s\partial_s(\varsigma\eta)\big(f\,\operatorname{div}\mathbf{u} + \mathfrak{C}[f]\big)\Big)\, dxdtds$$
$$+ \int_{\Omega\times\mathbb{R}} (\varsigma\eta)(x,0,s)f_\infty(x,0,s)\, dxds - \int_{S_T\times\mathbb{R}} (\varsigma\eta)f_\infty\mathbf{U}\cdot\mathbf{n}\, dSdt\, ds = 0. \tag{7.2.40}$$

Thus we have proved the desired identity (7.1.43) for $\psi(x,t,s) = \varsigma(x,t)\eta(s)$. It remains to extend this result to all smooth functions $\psi(x,t,s)$ satisfying (7.1.44). Fix such a $\psi \in C^\infty(Q \times \mathbb{R})$. Denote by $\hat{\psi}$ the Fourier transform of $\psi$ with respect to $s$,

$$\hat{\psi}(x,t,\xi) = \frac{1}{\sqrt{2\pi}} \int_{\mathbb{R}} \psi(x,t,s) e^{-i\xi s}\, ds.$$

Next choose $\zeta \in C_0^\infty(\mathbb{R})$ nonnegative with $\zeta(0) = 1$ and for $\xi \in \mathbb{R}$ set

$$\varsigma(x,t,\xi) = \hat{\psi}(x,t,\xi), \quad \eta(s,\xi) = \zeta(s/k)e^{i\xi s}.$$

Since $\psi$ is $C^\infty$ with respect to all arguments and compactly supported in $s$, it follows that $\hat{\psi}$ is smooth in $x,t,\xi$ and

$$|\hat{\psi}(x,t,\xi)| + |\nabla_x \hat{\psi}(x,t,\xi)| + |\partial_t \hat{\psi}(x,t,\xi)| \le c(m)(1+|\xi|)^{-m} \tag{7.2.41}$$

for any $m \ge 0$. Inserting the definitions of $\varsigma$ and $\eta$ in (7.2.40) we arrive at

$$\begin{aligned}\int_{Q\times\mathbb{R}} e^{i\xi s}\Big((\partial_t\hat{\psi} + \nabla_x\hat{\psi}\cdot\mathbf{u})\zeta(s/k)f \\ - \hat{\psi}s\big(\zeta(s/k)' + i\xi\zeta(s/k)\big)(f \operatorname{div}\mathbf{u}(x,t) + \mathfrak{C}[f])\Big)\, dxdtds \\ + \int_{\Omega\times\mathbb{R}} e^{i\xi s}\hat{\psi}(x,0,\xi)\zeta(s/k)f_\infty(x,0,s)\, dxds \\ = \int_{S_T\times\mathbb{R}} e^{i\xi s}\hat{\psi}(x,t,\xi)\zeta(s/k)f_\infty\mathbf{U}\cdot\mathbf{n}\, dSdtds.\end{aligned} \tag{7.2.42}$$

Here $\hat{\psi} = \hat{\psi}(x,t,\xi)$ and the integrals depend on $\xi$. For every positive $N$ define

$$\psi_N(x,t,s) = \frac{1}{\sqrt{2\pi}} \int_{-N}^{N} e^{i\xi s}\hat{\psi}(x,t,\xi)\, d\xi.$$

Integrating (7.2.42) with respect to $\xi$ over $[-N,N]$ we obtain

$$\begin{aligned}\int_{Q\times\mathbb{R}} \Big((\partial_t\psi_N + \nabla_x\psi_N\cdot\mathbf{u})\zeta(s/k)f \\ - s\big(\psi_N\zeta(s/k)' + \partial_s\psi_N\zeta(s/k)\big)(f \operatorname{div}\mathbf{u}(x,t) + \mathfrak{C}[f])\Big)\, dxdtds \\ + \int_{\Omega\times\mathbb{R}} \psi_N(x,0,s)\zeta(s/k)f_\infty(x,0,s)\, dxds \\ = \int_{S_T\times\mathbb{R}} \psi_N(x,t,s)\zeta(s/k)f_\infty\mathbf{U}\cdot\mathbf{n}\, dSdtds.\end{aligned} \tag{7.2.43}$$

Notice that the nonnegative functions $f, f_\infty$ are bounded by 1. Obviously we have $|\mathfrak{C}[f]| \le 2\overline{p}$, a majorant independent of $s$ and integrable on $Q$. Next, $\operatorname{div}\mathbf{u} \in L^2(Q)$.

Since $\zeta$ is compactly supported, it follows that

$$\begin{aligned} &s(f\operatorname{div}\mathbf{u}+\mathfrak{C}[f])\zeta(s/k),\ s(f\operatorname{div}\mathbf{u}+\mathfrak{C}[f])(\zeta(s/k))'\in L^1(Q\times\mathbb{R}),\\ &\zeta(s/k)f\in L^1(Q\times\mathbb{R}),\quad \zeta(s/k)f_\infty\in L^1(\sqcup_T\times\mathbb{R}). \end{aligned}\tag{7.2.44}$$

The formula for the inverse Fourier transform and estimates (7.2.41) imply that $\psi_N(x,t,s)\to\psi(x,t,s)$ uniformly in $x,t,s$ as $N\to\infty$. The same holds for $\partial_t\psi_N$, $\nabla_x\psi_N$ and $\partial_s\psi_N$. Letting $N\to\infty$ in (7.2.43), recalling (7.2.44), and applying the Lebesgue dominated convergence theorem we obtain, for every $k>0$,

$$\begin{aligned} \int_{Q\times\mathbb{R}}\Big((\partial_t\psi+\nabla_x\psi\cdot\mathbf{u})\zeta(s/k)f\\ -s\big(\zeta(s/k)'\psi+\zeta(s/k)\partial_s\psi\big)(f\operatorname{div}\mathbf{u}(x,t)+\mathfrak{C}[f])\Big)\,dxdtds\\ +\int_{\Omega\times\mathbb{R}}\psi(x,0,s)\zeta(s/k)f_\infty(x,0,s)\,dxds\\ =\int_{S_T\times\mathbb{R}}\psi(x,t,s)\zeta(s/k)f_\infty\mathbf{U}\cdot\mathbf{n}\,dSdtds. \end{aligned}\tag{7.2.45}$$

Finally notice that the smooth function $\psi(x,t,s)$ is compactly supported in $s$ and $\zeta(s/k)\to1$, $\zeta(s/k)'=k^{-1}\zeta'(s/k)\to0$ as $k\to\infty$. Letting $k\to\infty$ in (7.2.45) we obtain (7.1.43). □

**Proof of claim (iv).** We have already proved claims (i)–(iii) of Theorem 7.1.9 which ensure that $(\mathbf{u},f)$ is a solution to Problem 7.1.5. It remains to prove that this solution is regular in the sense of Definition 7.1.8, i.e. satisfies estimate (7.1.38). We split the proof into a sequence of lemmas. For any $\vartheta:Q\to\mathbb{R}$ and for any $\varrho\in\mathbb{R}$ set

$$\phi_\vartheta(x,t,\varrho)=\min\{\varrho,\ \vartheta(x,t)\}.\tag{7.2.46}$$

It follows from the fundamental theorem on Young measures (Theorem 1.4.5) and the convergences (7.1.26) that for any $\vartheta\in C(Q)$,

$$\begin{aligned} &\phi_\vartheta(\cdot,\varrho_\epsilon)\rightharpoonup\overline{\phi}_\vartheta\quad\text{weakly}^\star\text{ in }L^\infty(Q),\quad\text{where}\\ &\overline{\phi}_\vartheta(x,t)=\int_{[0,\infty)}\min\{s,\vartheta(x,t)\}\,d_sf(x,t,s). \end{aligned}\tag{7.2.47}$$

**Lemma 7.2.9.** *Let $\vartheta\in C(Q)$. Then for any $h\in C_c(Q)$,*

$$\begin{aligned} \lim_{\epsilon\to0}\int_Q h\big(\phi_\vartheta(\varrho_\epsilon)\operatorname{div}\mathbf{u}_\epsilon-\overline{\phi}_\vartheta\operatorname{div}\mathbf{u}\big)\,dxdt\\ =\frac{1}{\lambda+1}\lim_{\epsilon\to0}\int_Q h\big(\phi_\vartheta(\varrho_\epsilon)p(\varrho_\epsilon)-\overline{\phi}_\vartheta\overline{p}\big)\,dxdt. \end{aligned}\tag{7.2.48}$$

*Proof.* Choose $N > \sup_Q \vartheta(x,t)$ and $\delta > 0$. Since $\phi_\vartheta(x,t,\varrho)$ is uniformly continuous in $Q \times [0,N]$, there are smooth functions $\varphi_k : [0,N] \to \mathbb{R}$ and $\vartheta_k : Q \to \mathbb{R}$, $1 \le k \le n$, such that the function

$$\phi_\delta(x,t,\varrho) = \sum_{k=1}^{n} \vartheta_k(x,t)\varphi_k(\varrho) \tag{7.2.49}$$

approximates $\phi_\vartheta$ with accuracy $\delta$, i.e.,

$$\sup_{(x,t,\varrho)\in Q\times[0,N]} |\phi_\vartheta(x,t,\varrho) - \phi_\delta(x,t,\varrho)| \le \delta.$$

Extend $\varphi_k$ to $[0,\infty)$ by setting $\varphi_k(s) = \varphi_k(N)$ for $s > N$. This yields an extension of $\phi_\delta$ to $Q \times [0,\infty)$. We keep the same notation for the extended functions. It follows from (7.2.46) that

$$\sup_{(x,t,\varrho)\in Q\times[0,\infty)} |\phi_\vartheta(x,t,\varrho) - \phi_\delta(x,t,\varrho)| \le \delta. \tag{7.2.50}$$

Let $h \in C_c(Q)$. Fix a compact set $Q' \Subset Q$ such that $\operatorname{supp} h \subset Q'$. Let us prove that the desired equality (7.2.48) holds for $\phi_\delta$ in place of $\phi_\vartheta$. By linearity it suffices to prove that for every $k$,

$$\lim_{\epsilon\to 0}\int_Q h\vartheta_k\big(\varphi_k(\varrho_\epsilon)\operatorname{div}\mathbf{u}_\epsilon - \overline{\varphi}_k \operatorname{div}\mathbf{u}\big)\,dxdt$$
$$= \frac{1}{\lambda+1}\lim_{\epsilon\to 0}\int_Q h\big(\varphi_k(\varrho_\epsilon)p(\varrho_\epsilon) - \overline{\varphi_k p}\big)\,dxdt. \tag{7.2.51}$$

To this end notice that the Lipschitz function $\varphi_k(\varrho) - \varphi_k(N)$ vanishes for $\varrho \ge N$ and hence, by (7.2.17),

$$\lim_{\epsilon\to 0}\int_Q h\vartheta_k\big((\varphi_k(\varrho_\epsilon) - \varphi_k(N))\operatorname{div}\mathbf{u}_\epsilon - \overline{\varphi_k - \varphi_k(N)}\operatorname{div}\mathbf{u}\big)\,dxdt$$
$$= \frac{1}{\lambda+1}\lim_{\epsilon\to 0}\int_Q h\vartheta_k\big((\varphi_k(\varrho_\epsilon) - \varphi_k(N))p(\varrho_\epsilon) - \overline{\varphi_k - \varphi_k(N)}\,\overline{p}\big)\,dxdt. \tag{7.2.52}$$

Recalling that $\operatorname{div}\mathbf{u}_\epsilon$ converges weakly to $\operatorname{div}\mathbf{u}$ in $L^2(Q)$ and using (7.2.3) we obtain

$$0 = \lim_{\epsilon\to 0}\int_Q h\vartheta_k\big(\varphi_k(N)\operatorname{div}\mathbf{u}_\epsilon - \varphi_k(N)\operatorname{div}\mathbf{u}\big)\,dxdt,$$

$$0 = \lim_{\epsilon\to 0}\int_Q h\vartheta_k\big(\varphi_k(N)p(\varrho_\epsilon) - \varphi_k(N)\overline{p}\big)\,dxdt.$$

Combining this with (7.2.52) we arrive at (7.2.51). Summing (7.2.51) with respect to $k$ we obtain

$$\lim_{\epsilon\to 0}\int_Q h\big(\phi_\delta(\varrho_\epsilon)\operatorname{div}\mathbf{u}_\epsilon - \overline{\phi}_\delta\operatorname{div}\mathbf{u}\big)\,dxdt$$
$$= \frac{1}{\lambda+1}\lim_{\epsilon\to 0}\int_Q h\big(\phi_\delta(\varrho_\epsilon)p(\varrho_\epsilon) - \overline{\phi}_\delta\overline{p}\big)\,dxdt. \quad (7.2.53)$$

Next, by (7.1.9) we have

$$\|\operatorname{div}\mathbf{u}_\epsilon\|_{L^2(Q)} + \|p(\varrho_\epsilon)\|_{L^1(Q)} \le c,$$

which gives

$$\Big|\int_Q h\big(\phi_\vartheta(\varrho_\epsilon) - \phi_\delta(\varrho_\epsilon)\big)\operatorname{div}\mathbf{u}_\epsilon\,dxdt\Big| + \Big|\int_Q h\big(\phi_\vartheta(\varrho_\epsilon) - \phi_\delta(\varrho_\epsilon)\big)p(\varrho_\epsilon)\,dxdt\Big| \le c\delta. \quad (7.2.54)$$

It follows from the obvious inequalities $|\overline{\phi}_\vartheta - \overline{\phi}_\delta| = |\overline{\phi_\vartheta - \phi_\delta}| \le \delta$ and the inclusion $\overline{p}\in L^1(Q)$ that

$$\Big|\int_Q h\big(\overline{\phi}_\vartheta - \overline{\phi}_\delta\big)\operatorname{div}\mathbf{u}\,dxdt\Big| + \Big|\int_Q h\big(\overline{\phi}_\vartheta - \overline{\phi}_\delta\big)\overline{p}\,dxdt\Big| \le c\delta. \quad (7.2.55)$$

Combining (7.2.53)–(7.2.55) we obtain

$$-c\delta \le \liminf_{\epsilon\to 0}\int_Q \Big\{\phi_\vartheta(\varrho_\epsilon)\operatorname{div}\mathbf{u}_\epsilon - \overline{\phi}_\vartheta\operatorname{div}\mathbf{u} - \frac{1}{\lambda+1}\big(\phi_\vartheta(\varrho_\epsilon)p(\varrho_\epsilon) - \overline{\phi}_\vartheta\overline{p}\big)\Big\}h\,dxdt$$
$$\le \limsup_{\epsilon\to 0}\int_Q \Big\{\phi_\vartheta(\varrho_\epsilon)\operatorname{div}\mathbf{u}_\epsilon - \overline{\phi}_\vartheta\operatorname{div}\mathbf{u} - \frac{1}{\lambda+1}\big(\phi_\vartheta(\varrho_\epsilon)p(\varrho_\epsilon) - \overline{\phi}_\vartheta\overline{p}\big)\Big\}h\,dxdt \le c\delta.$$

Letting $\delta\to 0$ we obtain equality (7.2.48). □

Now define the function

$$\mathcal{T}_\vartheta(x,t) = \phi_\vartheta(x,t,\overline{\varrho}) - \overline{\phi}_\vartheta(x,t). \quad (7.2.56)$$

Recall

$$\begin{aligned}\phi_\vartheta(x,t,\overline{\varrho}) &= \min\Big\{\int_{[0,\infty)} s\,d_s f(x,t,s),\ \vartheta(x,t)\Big\},\\ \overline{\phi}_\vartheta(x,t) &= \int_{[0,\infty)} \min\{s,\vartheta(x,t)\}\,d_s f(x,t,s).\end{aligned} \quad (7.2.57)$$

**Lemma 7.2.10.** *Under the assumptions of Theorem 7.1.9, for every $\vartheta\in C(Q)$ and every compact set $Q'\Subset Q$ there is a constant $c$ independent of $\vartheta$ and $Q'$ such that*

$$\|\mathcal{T}_\vartheta\|^{\gamma+1}_{L^{\gamma+1}(Q')} \le \limsup_{\epsilon\to 0}\int_{Q'} |\min\{\varrho_\epsilon,\vartheta\} - \min\{\overline{\varrho},\vartheta\}|^{\gamma+1}\,dxdt \le c. \quad (7.2.58)$$

*Proof.* As the function $s \mapsto \phi_\vartheta(x,t,s)$ defined by (7.2.46) is monotone and its derivative does not exceed 1, it follows that for any $s'' \le s'$,

$$|\phi_\vartheta(x,t,s') - \phi_\vartheta(x,t,s'')|^{1+\gamma} \le (\phi_\vartheta(x,t,s') - \phi_\vartheta(x,t,s''))(s'^\gamma - s''^\gamma). \quad (7.2.59)$$

Next, (7.1.3) implies

$$s'^\gamma - s''^\gamma \le c(p(s') - p(s'')) \quad \text{for } 1 \le s'' \le s'. \quad (7.2.60)$$

By (5.1.4), the pressure function $p$ satisfies $s'^\gamma \le c(p(s')+1)$ and is bounded on $[0,1]$, which leads to

$$s'^\gamma - s''^\gamma \le c(p(s') - p(s'')) + c \quad \text{for } s'' \le 1 \le s'. \quad (7.2.61)$$

Combining (7.2.60) and (7.2.61) yields

$$s'^\gamma - s''^\gamma \le c(p(s') - p(s'')) + c \quad \text{for } 0 \le s'' \le s'. \quad (7.2.62)$$

From this and (7.2.59) we finally obtain

$$\begin{aligned} &|\phi_\vartheta(x,t,s') - \phi_\vartheta(x,t,s'')|^{1+\gamma} \\ &\qquad \le c_p(\phi_\vartheta(x,t,s') - \phi_\vartheta(x,t,s''))(p(s') - p(s'') + c). \end{aligned} \quad (7.2.63)$$

Now choose $Q' \Subset Q$ and $h \in C_c(Q)$ such that $0 \le h \le 1$, and $h(x,t) = 1$ in $Q'$. Notice that $\mathcal{T}_\vartheta$ is the weak limit of the sequence $\phi_\vartheta(x,t,\varrho_\epsilon) - \phi_\vartheta(x,t,\overline{\varrho})$. It follows that

$$\begin{aligned} \int_{Q'} |\mathcal{T}_\vartheta|^{\gamma+1}\,dxdt &\le \int_Q h|\mathcal{T}_\vartheta|^{\gamma+1}\,dxdt \\ &\le \lim_{\epsilon\to 0}\int_Q h|\phi_\vartheta(x,t,\varrho_\epsilon) - \phi_\vartheta(x,t,\overline{\varrho})|^{1+\gamma}\,dxdt. \end{aligned} \quad (7.2.64)$$

Now our task is to estimate the right hand side of this inequality. First we use (7.2.63) to obtain

$$\begin{aligned} &\lim_{\epsilon\to 0}\int_Q h|\phi_\vartheta(x,t,\varrho_\epsilon) - \phi_\vartheta(x,t,\overline{\varrho})|^{1+\gamma}\,dxdt \\ &\qquad \le c_p \lim_{\epsilon\to 0}\int_Q h\big(\phi_\vartheta(x,t,\varrho_\epsilon) - \phi_\vartheta(x,t,\overline{\varrho})\big)\big(p(\varrho_\epsilon) - p(\overline{\varrho})\big)\,dxdt \\ &\qquad\quad + c\lim_{\epsilon\to 0}\int_Q h\big(\phi_\vartheta(x,t,\varrho_\epsilon) - \phi_\vartheta(x,t,\overline{\varrho})\big)\,dxdt \\ &\qquad = c_p\int_Q h\,I(x,t)\,dxdt + c\int_Q h(\overline{\phi}_\vartheta - \phi_\vartheta(x,t,\overline{\varrho}))\,dxdt, \end{aligned} \quad (7.2.65)$$

where

$$I = \overline{\phi_\vartheta p} - \overline{\phi}_\vartheta\, p(\overline{\varrho}) - \phi_\vartheta(\cdot,\overline{\varrho})\,\overline{p} + \phi_\vartheta(\cdot,\overline{\varrho})\,p(\overline{\varrho}).$$

Now rewrite the expression for $I$ in the form

$$I = \overline{\phi_\vartheta p} - \overline{\phi}_\vartheta\, \overline{p} + \big(\overline{\phi}_\vartheta - \phi_\vartheta(\cdot, \overline{\varrho})\big)\big(\overline{p} - p(\overline{\varrho})\big). \tag{7.2.66}$$

Recall that, by the assumptions of Theorem 7.1.9 (claim (iv)), the pressure function has a representation $p(\varrho) = p_c(\varrho) + p_b(\varrho)$ where $p_c$ is convex and $p_b$ is bounded, i.e. there exists $c$ such $|p_b| \le c$. Since $p_c$ is convex, we have $\overline{p_c} \ge p_c(\overline{\varrho})$. It follows that

$$\overline{p} \ge \overline{p_c} - c \ge p_c(\overline{\varrho}) - c \ge p(\overline{\varrho}) - 2c, \quad \text{so} \quad \overline{p} - p(\overline{\varrho}) \ge -2c.$$

On the other hand, since $\phi_\vartheta(x, t, \cdot)$ is concave, we have $\overline{\phi}_\vartheta \le \phi_\vartheta(\overline{\varrho})$. Thus

$$\big(\overline{\phi}_\vartheta - \phi_\vartheta(\overline{\varrho})\big)\big(\overline{p} - p(\overline{\varrho})\big) \le 2c\big(\phi_\vartheta(\overline{\varrho}) - \overline{\phi}_\vartheta\big),$$

which along with (7.2.66) leads to

$$I \le \overline{\phi_\vartheta p} - \overline{\phi}_\vartheta\, \overline{p} + c\big(\phi_\vartheta(\overline{\varrho}) - \overline{\phi}_\vartheta\big).$$

Combining this with (7.2.65) we arrive at

$$\lim_{\epsilon \to 0} \int_Q h|\phi_\vartheta(x, t, \varrho_\epsilon) - \phi_\vartheta(x, t, \overline{\varrho})|^{1+\gamma}\, dxdt$$
$$\le c \int_Q h\big(\overline{\phi_\vartheta p} - \overline{\phi}_\vartheta\, \overline{p}\big)\, dxdt + c \int_Q h\big(\phi_\vartheta(x, t, \overline{\varrho}) - \overline{\phi}_\vartheta\big)\, dxdt. \tag{7.2.67}$$

Next, the obvious inequality $0 \le \phi_\vartheta(x, t, \varrho) \le \varrho$ implies $0 \le \overline{\phi}_\vartheta \le \overline{\varrho}$ and $0 \le \phi_\vartheta(x, t, \overline{\varrho}) \le \overline{\varrho}$. From this, (7.1.29) and (7.1.40), we obtain

$$\int_Q h\big(\phi_\vartheta(x, t, \overline{\varrho}) - \overline{\phi}_\vartheta\big)\, dxdt \le 2 \int_Q h\overline{\varrho}\, dxdt \le c.$$

Combining this with (7.2.67) we arrive at

$$\lim_{\epsilon \to 0} \int_Q h|\phi_\vartheta(x, t, \varrho_\epsilon) - \phi_\vartheta(x, t, \overline{\varrho})|^{1+\gamma}\, dxdt \le c \int_Q h\big(\overline{\phi_\vartheta p} - \overline{\phi}_\vartheta\, \overline{p}\big)\, dxdt + c.$$

From this and (7.2.48) we obtain

$$\lim_{\epsilon \to 0} \int_Q h|\phi_\vartheta(x, t, \varrho_\epsilon) - \phi_\vartheta(x, t, \overline{\varrho})|^{1+\gamma}\, dxdt$$
$$\le c(\lambda + 1) \int_Q h\big(\overline{\phi_\vartheta \operatorname{div} \mathbf{u}} - \overline{\phi}_\vartheta \operatorname{div} \mathbf{u}\big)\, dxdt + c. \tag{7.2.68}$$

By the definition of weak convergence, we have

$$\int_Q h\big(\overline{\phi_\vartheta \operatorname{div} \mathbf{u}} - \overline{\phi}_\vartheta \operatorname{div} \mathbf{u}\big)\, dxdt = \lim_{\epsilon \to 0} \int_Q h\phi_\vartheta(x, t, \varrho_\epsilon)(\operatorname{div} \mathbf{u}_\epsilon - \operatorname{div} \mathbf{u})\, dxdt.$$

Since $\phi(x,t,\overline{\varrho})$ is bounded and $\nabla\mathbf{u}_\epsilon \rightharpoonup \nabla\mathbf{u}$ weakly in $L^2(Q)$, it follows that

$$\lim_{\epsilon\to 0}\int_Q h\phi_\vartheta(x,t,\overline{\varrho})(\operatorname{div}\mathbf{u}_\epsilon - \operatorname{div}\mathbf{u})\,dxdt = 0.$$

Combining the above results we arrive at

$$\int_Q h\big(\overline{\phi_\vartheta \operatorname{div}\mathbf{u}} - \overline{\phi}_\vartheta \operatorname{div}\mathbf{u}\big)\,dxdt$$
$$= \lim_{\epsilon\to 0}\int_Q h\big(\phi_\vartheta(x,t,\varrho_\epsilon) - \phi_\vartheta(x,t,\overline{\varrho})\big)\operatorname{div}(\mathbf{u}_\epsilon - \mathbf{u})\,dxdt. \quad (7.2.69)$$

Next, the Young inequality implies

$$c(\lambda+1)h\big(\phi_\vartheta(x,t,\varrho_\epsilon) - \phi_\vartheta(x,t,\overline{\varrho})\big)\operatorname{div}(\mathbf{u}_\epsilon - \mathbf{u})$$
$$\le 2^{-1}h|\phi_\vartheta(x,t,\varrho_\epsilon) - \phi_\vartheta(x,t,\overline{\varrho})|^{\gamma+1} + 2^{\frac{1}{\gamma}}h\{c(\lambda+1)|\operatorname{div}(\mathbf{u}_\epsilon - \mathbf{u})|\}^{\frac{\gamma+1}{\gamma}}.$$

Since $\mathbf{u}_\epsilon$ satisfies (7.1.9), (7.1.22), the sequence $\operatorname{div}(\mathbf{u}_\epsilon - \mathbf{u})$ is bounded in $L^2(Q)$. From this and the inequality $(\gamma+1)/\gamma \le 2$ we obtain

$$\limsup_{\epsilon\to 0}\int_Q |\operatorname{div}(\mathbf{u}_\epsilon - \mathbf{u})|^{\frac{\gamma+1}{\gamma}}\,dxdt \le c,$$

which gives

$$c(\lambda+1)\lim_{\epsilon\to 0}\int_Q h\big(\phi_\vartheta(x,t,\varrho_\epsilon) - \phi_\vartheta(x,t,\overline{\varrho})\big)\operatorname{div}(\mathbf{u}_\epsilon - \mathbf{u})\,dxdt$$
$$\le 2^{-1}\lim_{\epsilon\to 0}\int_Q h|\phi_\vartheta(x,t,\varrho_\epsilon) - \phi_\vartheta(x,t,\overline{\varrho})|^{\gamma+1}\,dxdt + c.$$

Recalling identity (7.2.69) we arrive at

$$c(\lambda+1)\int_Q h\big(\overline{\phi_\vartheta \operatorname{div}\mathbf{u}} - \overline{\phi_\vartheta}\operatorname{div}\mathbf{u}\big)\,dxdt$$
$$\le 2^{-1}\lim_{\epsilon\to 0}\int_Q h|\phi_\vartheta(x,t,\varrho_\epsilon) - \phi_\vartheta(x,t,\overline{\varrho})|^{\gamma+1}\,dxdt + c.$$

Inserting this into (7.2.68) we get

$$2^{-1}\lim_{\epsilon\to 0}\int_Q h|\phi_\vartheta(x,t,\varrho_\epsilon) - \phi_\vartheta(x,t,\overline{\varrho})|^{1+\gamma}\,dxdt \le c,$$

which along with (7.2.64) leads to (7.2.58). □

**Lemma 7.2.11.** *Under the assumptions of Theorem 7.1.9, there is a constant c such that*

$$\|\mathfrak{H}\|_{L^{1+\gamma}(Q)} + \sup_{\upsilon\in\mathbb{R}^+} \|\mathfrak{V}_\upsilon\|_{L^1(Q)} < c, \tag{7.2.70}$$

*where*

$$\begin{aligned} \mathfrak{V}_\upsilon(x,t) &= \int_{[0,\infty)} \min\{s,\upsilon\}(p(s)-\overline{p})\, d_s f(x,t,s), \\ \mathfrak{H}(x,t) &= \int_{[0,\infty)} f(x,t,s)(1-f(x,t,s))\, ds. \end{aligned} \tag{7.2.71}$$

*Proof.* Pick $\vartheta \in C(Q)$ nonnegative. Recall the definition of $\mathcal{T}_\vartheta$:

$$\mathcal{T}_\vartheta(x,t) = \min\left\{\overline{\varrho}(x,t),\, \vartheta(x,t)\right\} - \int_{[0,\infty)} \min\{s,\vartheta(x,t)\}\, d_s f(x,t,s).$$

By integration by parts (Lemma 1.3.7) we obtain

$$\begin{aligned} &\int_{[0,\infty)} \min\{s,\vartheta(x,t)\}\, d_s f(x,t,s) \\ &\qquad = \int_{[0,\vartheta(x,t)]} s\, d_s f(x,t,s) + \vartheta(x,t) \int_{(\vartheta(x,t),\infty)} d_s f(x,t,s) \\ &\qquad = \vartheta - \int_0^\vartheta f(x,t,s)\, ds = \int_0^\vartheta (1-f(x,t,s))\, ds. \end{aligned}$$

Next, for a.e. $(x,t) \in Q$ and every $N > 0$ we have

$$N^\gamma(1-f(x,t,N)) = N^\gamma \int_{[N,\infty)} d_s f(x,t) \le \int_{[0,\infty)} s^\gamma\, d_s f(x,t) < \infty,$$

which leads to $\lim_{N\to\infty} N(1-f(x,t,N)) = 0$. Integrating by parts we obtain

$$\begin{aligned} \int_{[0,N]} s d_s f(x,t,s)) &= -\int_{[0,N]} s d_s(1-f(x,t,s)) \\ &= \int_{[0,N]} (1-f(x,t,s))\, ds - N(1-f(x,t,N)). \end{aligned}$$

Letting $N \to \infty$ we arrive at

$$\overline{\varrho}(x,t) = \int_{[0,\infty)} s\, d_s f(x,t,s) = \int_0^\infty (1-f(x,t,s))\, ds \tag{7.2.72}$$

for a.e. $(x,t) \in Q$. It follows that

$$\begin{aligned} \mathcal{T}_\vartheta(x,t) &= \int_0^{\vartheta(x,t)} f(x,t,s)\, ds \qquad \text{for } \overline{\varrho}(x,t) \ge \vartheta(x,t), \\ \mathcal{T}_\vartheta(x,t) &= \int_{\vartheta(x,t)}^\infty (1-f(x,t,s))\, ds \quad \text{for } \overline{\varrho}(x,t) \le \vartheta(x,t). \end{aligned} \tag{7.2.73}$$

Next choose a sequence of nonnegative continuous functions $\vartheta_n$ such that $\vartheta_n \to \overline{\varrho}$ a.e. in $Q$. Substituting $\vartheta_n$ into (7.2.73) and letting $n \to \infty$ we conclude that for a.e. $(x,t) \in Q$,

$$\mathcal{T}_{\vartheta_n}(x,t) \to \mathcal{T}_{\varrho} = \int_0^{\overline{\varrho}(x,t)} f(x,t,s)\,ds = \int_{\overline{\varrho}(x,t)}^{\infty} (1 - f(x,t,s))\,ds. \qquad (7.2.74)$$

It follows from (7.2.58) that there is a constant $c$ independent of $Q'$ and $n$ such that $\|\mathcal{T}_{\vartheta_n}\|_{L^{\gamma+1}(Q')} \le c$ and hence $\|\mathcal{T}_{\vartheta_n}\|_{L^{\gamma+1}(Q)} \le c$. Thus $\|\mathcal{T}_{\varrho}\|_{L^{\gamma+1}(Q)} \le c$. Noting that $\mathfrak{H} \le \mathcal{T}_{\varrho}$ we arrive at the desired estimate for $\mathfrak{H}$ in (7.2.70). To estimate $\mathfrak{V}_v$ notice that by (7.2.48), for any $Q' \Subset Q$,

$$\frac{1}{\lambda+1}\mathfrak{V}_v = \operatorname*{w\text{-}lim}_{\epsilon\to 0}(\phi_v(\varrho_\epsilon)\operatorname{div}\mathbf{u}_\epsilon) - \overline{\phi}_v \operatorname{div}\mathbf{u},$$

where $w$-lim denotes the weak limit in $L^2(Q')$. Thus we get

$$\frac{1}{\lambda+1}\mathfrak{V}_v = \operatorname*{w\text{-}lim}_{\epsilon\to 0}(\phi_v(\varrho_\epsilon) - \phi_v(\overline{\varrho}))\operatorname{div}\mathbf{u}_\epsilon - \Big(\operatorname*{w\text{-}lim}_{\epsilon\to 0}\phi_v(\varrho_\epsilon) - \phi_v(\overline{\varrho})\Big)\operatorname{div}\mathbf{u}.$$

From this and (7.2.58) we obtain, for every $Q' \Subset Q$,

$$\begin{aligned}\|\mathfrak{V}_v\|_{L^1(Q')} &\le c \limsup_{\epsilon\to 0}\big\{(\|\operatorname{div}\mathbf{u}_\epsilon\|_{L^2(Q')} + \|\operatorname{div}\mathbf{u}\|_{L^2(Q')})\|\phi_v(\varrho_\epsilon) - \phi_v(\overline{\varrho})\|_{L^2(Q')}\big\}\\ &\le c \limsup_{\epsilon\to 0}\|\phi_v(\varrho_\epsilon) - \phi_v(\overline{\varrho})\|_{L^2(Q')} \le c,\end{aligned}$$

where $c$ is independent of $Q'$. Thus we arrive at the estimate for $\mathfrak{V}_v$ in (7.2.70). □

It remains to note that claim (iv) is a direct consequence of Lemma 7.2.11. This completes the proof of Theorem 7.1.9.

## 7.3 Proof of Theorem 7.1.12

We split the proof into a sequence of lemmas. The first lemma determines the extension of a solution to problem (7.1.8) outside of the domain $\Omega$.

**Lemma 7.3.1.** *Let $f_\infty$ be defined by (7.1.48). Then*

$$\begin{aligned}&\int_{(\mathbb{R}^d\setminus\Omega)\times(0,T)}\int_{\mathbb{R}} f_\infty\big(\partial_t\psi + \nabla_x\psi\cdot\mathbf{U} - s\partial_s\psi\operatorname{div}\mathbf{U} + \partial_s\psi F_\infty\big)\,dsdxdt\\ &+ \int_{\mathbb{R}^d\setminus\Omega}\int_{\mathbb{R}} f_\infty(x,0,s)\psi(x,0,s)\,dsdx + \int_{S_T}\int_{\mathbb{R}} f_\infty(x,t,s)\psi(x,t,s)\mathbf{U}\cdot\mathbf{n}\,ds\,dSdt = 0\end{aligned} \qquad (7.3.1)$$

*for all $\psi \in C^\infty(\mathbb{R}^d \times (0,T) \times \mathbb{R})$ vanishing for all sufficiently large $|x|$, $s$, and for $t = T$. Here*

$$F_\infty = \partial_t\varrho_\infty + \operatorname{div}(\varrho_\infty\mathbf{U}), \qquad (7.3.2)$$

*and* $\mathbf{n}$ *is the outward normal vector to* $\partial\Omega$*. Notice that* (7.3.1) *is simply an identity which holds for any deterministic distribution* $f_\infty$*.*

*Proof.* As this is a simpler version of Lemma 7.2.8, we only give some comments. Choose $\eta \in C_0^\infty(\mathbb{R})$ and set

$$\varphi(s) = \int_s^\infty \eta(\tau)\,d\tau, \quad \Phi(s) = s\varphi'(s) - \varphi(s) = -s\eta(s) - \int_s^\infty \eta(\tau)\,d\tau.$$

It easily follows from (7.1.48) that

$$\varphi(\varrho_\infty)(x,t) = \int_{\mathbb{R}} \eta(s) f_\infty(x,t,s)\,ds \tag{7.3.3}$$

and

$$\Phi(\varrho_\infty)(x,t) = \int_{\mathbb{R}} s\eta'(s) f_\infty(x,t,s)\,ds, \quad \varphi'(\varrho_\infty)(x,t) = \int_{\mathbb{R}} \eta'(s) f_\infty(x,t,s)\,ds. \tag{7.3.4}$$

Multiplying the identity $\partial_t \varrho_\infty + \operatorname{div}(\varrho_\infty \mathbf{U}) = F_\infty$ by $\varphi'(\varrho_\infty)$ we obtain

$$\partial_t \varphi(\varrho_\infty) + \operatorname{div}(\varphi(\varrho_\infty)\mathbf{U}) + \Phi(\varrho_\infty)\operatorname{div}\mathbf{U} = \varphi'(\varrho_\infty) F_\infty.$$

Pick $\varsigma \in C^\infty(\mathbb{R}^d \times (0,T) \times \mathbb{R}))$ vanishing for large $|x|+|s|$ and for $t = T$. Multiplying the last identity by $\varsigma$ and integrating by parts we arrive at

$$\begin{aligned}\int_{(\mathbb{R}^d\setminus\Omega)\times(0,T)} &\{\varphi(\varrho_\infty)(\partial_t\varsigma + \nabla\varsigma\cdot\mathbf{U}) - \varsigma\Phi(\varrho_\infty)\operatorname{div}\mathbf{U}\}\,dxdt \\ &+ \int_{(\mathbb{R}^d\setminus\Omega)\times(0,T)} \varphi'(\varrho_\infty)F_\infty\,dxdt + \int_{\mathbb{R}^d\setminus\Omega} (\varsigma\varphi(\varrho_\infty))(x,0)\,dx \\ &\qquad + \int_{S_T} \varsigma\varphi(\varrho_\infty)\mathbf{U}\cdot\mathbf{n}\,dSdt = 0. \end{aligned} \tag{7.3.5}$$

Substituting (7.3.3) and (7.3.4) into (7.3.5) yields (7.3.1) with $\psi = \varsigma(x,t)\eta(s)$. Arguing as in the proof of Lemma 7.2.8 we conclude that (7.3.1) holds for all test functions as in the hypotheses of Lemma 7.3.1. □

We now conclude from the integral identity (7.3.1) that $f_\infty$ is a weak solution to the equation

$$\partial_t f_\infty + \operatorname{div}(f_\infty \mathbf{U}) - \partial_s(s f_\infty \operatorname{div}\mathbf{U}) + \partial_s(F_\infty f_\infty) = 0 \tag{7.3.6}$$

in $\big((\mathbb{R}^d \times (0,T)) \setminus Q\big) \times \mathbb{R}$. Now our task is to combine this equation with (7.1.33b) in $Q \times \mathbb{R}$ and next using cut-off functions to reduce the boundary value problem (7.1.45) to the Cauchy problem in the strip $\mathbb{R}^d \times (0,T) \times \mathbb{R}$. To this end introduce a cut-off function $\eta \in C^\infty(\mathbb{R}^d \times (0,T))$ with $0 \le \eta \le 1$ and vanishing in some neighborhood of $S_T \setminus \Sigma_{\text{in}}$ and for all sufficiently large $|x|$. Next we set

$$\begin{aligned} &\mathbf{v} = \mathbf{u} \quad \text{in } Q, \quad \mathbf{v} = \mathbf{U} \quad \text{in } (\mathbb{R}^d \times (0,T)) \setminus Q, \\ &f^*(x,t,s) = \eta(x,t) f(x,t,s) \quad \text{in } Q \times \mathbb{R}, \\ &f^*(x,t,s) = \eta(x,t) f_\infty(x,t,s) \quad \text{in } \big((\mathbb{R}^d \times (0,T)) \setminus Q\big) \times \mathbb{R}, \\ &H = -s f^* \operatorname{div} \mathbf{u}(x,t) - s\eta(x,t)\mathfrak{C}[f] \quad \text{in } Q \times \mathbb{R}, \\ &H = -s f^* \operatorname{div} \mathbf{U}(x,t) + f^* F_\infty(x,t) \quad \text{in } \big((\mathbb{R}^d \times (0,T)) \setminus Q\big) \times \mathbb{R}, \end{aligned} \tag{7.3.7}$$

**Lemma 7.3.2.** *The integral identity*

$$\int_{\mathbb{R}^d \times (0,T) \times \mathbb{R}} \{ f^*(\partial_t \psi + \mathbf{v}\nabla\psi) + H\partial_s\psi + P\psi \} \, dxdtds + \int_{\mathbb{R}^d \times \mathbb{R}} \eta f_\infty(x,0,s)\psi(x,0,s)\, dxds = 0 \tag{7.3.8}$$

*holds for all $\psi \in C^\infty(\mathbb{R}^d \times (0,T) \times \mathbb{R})$ vanishing for $t = T$ and for sufficiently large $|x|$, $s$. Here*

$$P = (\partial_t\eta + \nabla\eta\cdot\mathbf{u}) f \text{ in } Q\times\mathbb{R}, \quad P = (\partial_t\eta + \nabla\eta\cdot\mathbf{U}) f_\infty \text{ in } \big((\mathbb{R}^d \times (0,T)) \setminus Q\big) \times \mathbb{R}. \tag{7.3.9}$$

*Proof.* The integral identity (7.3.1) with $\psi$ replaced by $\psi\eta$ implies

$$\begin{aligned} \int_{((\mathbb{R}^d \times (0,T))\setminus Q)\times\mathbb{R}} &\{ (\eta f_\infty)\big(\partial_t\psi + \nabla_x\psi\cdot\mathbf{U} - s\partial_s\psi \operatorname{div}\mathbf{U} + \partial_s\psi F_\infty\big) + P\psi \}\, dxdtds \\ &+ \int_{\mathbb{R}^d\setminus\Omega}\int_{\mathbb{R}} \eta f_\infty(x,0,s)\psi(x,0,s)\, dsdx \\ &+ \int_{S_T}\int_{\mathbb{R}} \eta f_\infty(x,t,s)\psi(x,t,s)\mathbf{U}\cdot\mathbf{n}\, ds\, dSdt = 0. \end{aligned} \tag{7.3.10}$$

On the other hand substituting $\psi\eta$ in (7.1.43) we obtain

$$\begin{aligned} \int_{Q\times\mathbb{R}} &\{ (f\eta)\big(\partial_t\psi + \mathbf{u}\nabla\psi - \partial_s\psi s \operatorname{div}\mathbf{u}\big) + P\psi \}\, dxdtds \\ &- \int_{Q\times\mathbb{R}} \partial_s\psi s\eta\, \mathfrak{C}[f]\, dxdtds - \int_{S_T}\int_{\mathbb{R}} \eta f_\infty(x,t,s)\psi(x,t,s)\mathbf{U}\cdot\mathbf{n}\, ds\, dSdt \\ &+ \int_{\Omega\times\mathbb{R}} \eta f_\infty(x,0,s)\psi(x,0,s)\, dxds = 0. \end{aligned} \tag{7.3.11}$$

Summing (7.3.10) and (7.3.11) gives (7.3.8). □

The integral identity (7.3.8) means that $f^*$ is a weak solution to the Cauchy problem

$$\begin{aligned} &\partial_t f^* + \operatorname{div}(f^*\mathbf{v}) + \partial_s H = P \quad \text{in } \mathbb{R}^d \times (0,T) \times \mathbb{R}, \\ &f^*(x,0,s) = \eta(x,0,s) f_\infty(x,0,s) \quad \text{in } \mathbb{R}^d \times \mathbb{R}. \end{aligned} \tag{7.3.12}$$

Recall that $f_\infty$ is given by Condition 7.1.11. The next step is the renormalization of this problem. In other words we intend to derive an equation for a composite function $\Psi(f^*)$. We choose $\Psi$ in such a way that it is concave and $\Psi(f)$ vanishes for every deterministic distribution function $f$, i.e., for $f$ which takes only two values 0 and 1. The simplest choice is

$$\Psi(f) = f(1-f), \quad \Psi'(f) = 1 - 2f. \tag{7.3.13}$$

The justification of the renormalization procedure is based on the following lemma on the properties of the mollifying operator. Recall the definition (1.6.1)–(1.6.3) of the mollifiers in Section 1.6:

$$[f]_{,k}(x,t,\tau) := k \int_{\mathbb{R}} \theta(k(s-\tau)) f(x,t,s)\, ds, \tag{7.3.14}$$

$$[f]_{m,}(y,t,s) := m^d \int_{\mathbb{R}^d} \Theta(m(x-y)) f(x,t,s)\, dx, \quad \Theta(x) = \prod_{i=1}^d \theta(x_i), \tag{7.3.15}$$

and $[f]_{m,k} = [[f]_{,k}]_{m,}$. Here the mollifying kernel $\theta$ has the properties

$$\theta \in C_0^\infty(\mathbb{R}), \quad \int_{\mathbb{R}} \theta(s)\, ds = 1, \quad \operatorname{supp}\theta \subset (-1,1), \quad \theta(-s) = \theta(s). \tag{7.3.16}$$

**Lemma 7.3.3.** *Suppose $g : \mathbb{R}^d \times (0,T) \times \mathbb{R} \to \mathbb{R}$ admits a representation*

$$g(x,t,s) = s g_1(x,t,s) + g_0(x,t,s), \quad \textit{where} \quad \|g_i(\cdot,t,s)\|_{L^1(\mathbb{R}^d)} \le M_i(t). \tag{7.3.17}$$

*Furthermore assume that $\chi(s)$ is continuous, satisfies $0 \le \chi \le 1$ and vanishes for $|s| \ge N$. Then for any $m > 0$ and $k \ge 1$, and for all $s \in \mathbb{R}$,*

$$\begin{aligned} &\|[\chi g]_{,k}(\cdot,t,s)\|_{L^1(\mathbb{R}^d)} \le N\, M_1(t) + M_0(t), \\ &\|[\chi g]_{m,k}(\cdot,t,s)\|_{L^1(\mathbb{R}^d)} \le N\, M_1(t) + cM_0(t). \end{aligned} \tag{7.3.18}$$

*Moreover, for every $m > 0$, $N > 0$ and $k \ge 1$, there is a constant $c(m,k,N)$ such that for all $x \in \mathbb{R}^d$, $t \in [0,T]$, and $s \in [-N,N]$,*

$$|\partial_x [g]_{m,}(x,t,s)| + |\partial_s [g]_{,k}(x,t,s)| \le c(N+2) M_1(t) + cM_0(t). \tag{7.3.19}$$

*Proof.* The inequalities $k \geq 1$ and $\chi(\tau)|\tau| \leq N$ imply

$$\|[\chi g]_{,k}(\cdot,t,s)\|_{L^1(\mathbb{R}^d)} = \int_{\mathbb{R}^d}\Big|\int_{\mathbb{R}} k\theta(k(\tau-s))\chi(\tau)g(x,t,\tau)\,d\tau\Big|\,dx$$
$$\leq \int_{\mathbb{R}^{d+1}} k\theta(k(\tau-s))\chi(\tau)|g|(x,t,\tau)\,dxd\tau = \int_{\mathbb{R}} k\theta(k(\tau-s))\chi(\tau)\|g(\cdot,t,\tau)\|_{L^1(\mathbb{R}^d)}\,d\tau$$
$$\leq M_1(t)\int_{\mathbb{R}} k\theta(k(\tau-s))\chi(\tau)|\tau|\,d\tau + M_0(t)\int_{\mathbb{R}} k\theta(k(\tau-s))\,d\tau \leq NM_1(t)+M_0(t),$$

which yields the first estimate in (7.3.18). To prove the second, we note that in view of the properties of mollifiers we have

$$\|[\chi g]_{m,k}(\cdot,t,s)\|_{L^1(\mathbb{R}^d)} = \|[[\chi g]_{,k}]_{m,}(\cdot,t,s)\|_{L^1(\mathbb{R}^d)} \leq \|[\chi g]_{,k}(\cdot,t,s)\|_{L^1(\mathbb{R}^d)}.$$

Hence the second inequality in (7.3.18) follows from the first. It remains to prove (7.3.19). To this end choose a continuous function $\chi(s)$ such that $\chi(s) = 1$ for $|s| \leq N+1$ and $\chi(s) = 0$ for $|s| \geq N+2$. Obviously $[g]_{m,k} = [\chi g]_{m,k}$ for $|s| \leq N$. Hence it suffices to prove (7.3.19) for $\chi g$. It follows from (7.3.17) that

$$\|\chi g(\cdot,t,\cdot)\|_{L^1(\mathbb{R}^{d+1})} \leq (N+2)(M_1(t)+M_0(t)).$$

By the general properties of mollifiers,

$$\|[\chi g(\cdot,t,\cdot)]_{m,k}\|_{C^1(\mathbb{R}^{d+1})} \leq c\|\chi g(\cdot,t,\cdot)\|_{L^1(\mathbb{R}^{d+1})} \leq c(N+2)(M_1(t)+M_0(t)),$$

which gives (7.3.19). □

The next proposition justifies the renormalization procedure of the Cauchy problem (7.3.12).

**Proposition 7.3.4.** *Let $\Psi$ be given by (7.3.13) and suppose the function $f^*$ defined by (7.3.7) satisfies (7.3.8). Then for any $\psi \in C^\infty(\mathbb{R}^d \times (0,T) \times \mathbb{R})$ vanishing for $t = T$ and for sufficiently large $|x| + |s|$, we have*

$$\int_{Q\times\mathbb{R}} \{\Psi(f^*)(\partial_t\psi + \nabla\psi\cdot\mathbf{u} - s\partial_s\psi\,\operatorname{div}\mathbf{u}) - s\eta\partial_s\psi\,\Psi'(f^*)\,\mathfrak{C}[f]\}\,dxdtds$$
$$+2\int_Q \eta\Big\{\int_{\mathbb{R}} s\psi\mathfrak{M}\,d_sf^*\Big\}\,dxdt + \mathfrak{N} = 0. \quad (7.3.20)$$

*Here*

$$\mathfrak{M}(x,t,s) = \frac{1}{2}\lim_{h\searrow 0}\mathfrak{C}[f](x,t,s-h) + \frac{1}{2}\lim_{h\searrow 0}\mathfrak{C}[f](x,t,s+h), \quad (7.3.21)$$

$$\mathfrak{N} = \int_{((\mathbb{R}^d\times(0,T))\setminus Q)\times\mathbb{R}} \{\Psi(\eta f_\infty)\big(\partial_t\psi + \nabla\psi\cdot\mathbf{U} - \partial_s\psi\,(s\operatorname{div}\mathbf{U} - F_\infty)\big)\}\,dxdtds$$
$$+\int_{\mathbb{R}^d\times(0,T)\times\mathbb{R}} \psi\Psi'(f^*)P\,dxdtds + \int_{\mathbb{R}^d\times\mathbb{R}} \Psi(\eta f_\infty(x,0,s))\psi(x,0,s)\,dxds.$$
$$(7.3.22)$$

*Proof.* The proof naturally falls into five steps.

**Step 1. Integral identity for mollifiers.** Set

$$K(y,\tau;x,s) = km^d\theta(k(s-\tau))\Theta(m(x-y))$$

for integers $m, k \geq 1$. We have

$$[f^*]_{m,k}(y,t,\tau) \equiv \int_{\mathbb{R}^{d+1}} K(y,\tau;x,s) f^*(x,t,s)\,dxds. \tag{7.3.23}$$

For a fixed $(y,\tau)$ the function $K(y,\tau;\cdot,\cdot)$ belongs to the class $C_0^\infty(\mathbb{R}^{d+1})$. Now choose $h \in C^\infty[0,T]$ such that $h(T) = 0$. Fix $(y,\tau) \in \mathbb{R}^{d+1}$ and set $\psi(x,t,s) = h(t)K(y,\tau;x,s)$. Substituting $\psi$ in the integral identity (7.3.8) gives

$$\begin{aligned}
&\int_0^T h'(t)\Big\{\int_{\mathbb{R}^{d+1}} K(y,\tau;x,s) f^*(x,t,s)\,dxds\Big\}\,dt \\
&+\int_0^T h(t)\Big\{\int_{\mathbb{R}^{d+1}} \big(\nabla_x K(y,\tau;x,s)\cdot\mathbf{v}(x,t,s) f^*(x,t,s) + \partial_s K(y,\tau;x,s) H(x,t,s) \\
&\qquad\qquad + K(y,\tau;x,s) P(x,t,s)\big)\,dxds\Big\}\,dt \\
&\qquad\qquad + \int_{\mathbb{R}^{d+1}} h(0)\eta K(y,\tau;x,s) f_\infty(x,0,s)\eta(x,0,s)\,dxds = 0.
\end{aligned}$$

From this, the identities $\nabla_x K(y,\tau;x,s) = -\nabla_y K(y,\tau;x,s)$, $\partial_s K(y,\tau;x,s) = -\partial_\tau K(y,\tau;x,s)$, and (7.3.23) we obtain

$$\begin{aligned}
&\int_0^T h'(t)[f^*]_{m,k}(y,t,\tau)\,dt \\
&\quad = \int_0^T h(t)\big(\operatorname{div}_y[\mathbf{v}f^*]_{m,k}(y,t,\tau) + \partial_\tau[H]_{m,k}(y,t,\tau) - [P]_{m,k}(y,t,\tau)\big)\,dt \\
&\qquad\qquad - h(0)[\eta f_\infty]_{m,k}(y,0,\tau).
\end{aligned} \tag{7.3.24}$$

Let us return to notation $(x,s)$ instead of $(y,\tau)$ in this integral identity. It follows from (7.3.24) that the function $[f^*]_{m,k}(x,t,s)$ satisfies the differential equation

$$\partial_t[f^*]_{m,k} = \operatorname{div}[\mathbf{g}^{(0)}]_{m,k} + \partial_s[g^{(1)}]_{m,k} + [g^{(2)}]_{m,k} \tag{7.3.25}$$

in the sense of distributions. Here

$$\mathbf{g}^{(0)} = -f^*\mathbf{v}, \quad g^{(1)} = -H, \quad g^{(2)} = P.$$

Recall that $\mathbf{v} = \mathbf{u}$ in the cylinder $Q$ and $\mathbf{v} = \mathbf{U}$ outside of $Q$. The function $\mathbf{U} \in C^1(\mathbb{R}^d \times [0,T])$ vanishes for all large $|x|$ and $\mathbf{u} \in L^2(0,T;W^{1,2}(\Omega))$, which gives $\|\mathbf{v}\|_{L^2(0,T;W^{1,2}(\mathbb{R}^d)} \leq c$. Since $\mathbf{v}$ vanishes for all large $|x|$, we have

$$\|\mathbf{v}(\cdot,t)\|_{L^1(\mathbb{R}^d)} + \|\nabla\mathbf{v}\|_{L^1(\mathbb{R}^2)} \leq c + cM(t), \tag{7.3.26}$$

where

$$M(t) = \|\mathbf{u}(\cdot,t)\|_{W^{1,2}(\Omega)}, \quad \|M(t)\|_{L^2(0,T)} \le c. \tag{7.3.27}$$

Let us prove that the functions $\mathbf{g}^{(0)}$ and $g^{(2)}$ satisfy the inequalities

$$\|\mathbf{g}^{(0)}(\cdot,t,s)\|_{L^1(\mathbb{R}^d)} + \|g^{(2)}(\cdot,t,s)\|_{L^1(\mathbb{R}^d)} \le cM(t) + c \quad \text{for a.e. } (t,s) \in [0,T] \times \mathbb{R}. \tag{7.3.28}$$

Since $f^*$ does not exceed 1, estimate (7.3.28) for $\mathbf{g}^{(0)}$ obviously follows from (7.3.26). Next, since the smooth function $\eta$ vanishes for all sufficiently large $|x|$, estimate (7.3.28) for $g^{(2)}$ follows from formula (7.3.9) for $P$ and estimate (7.3.26) for $\mathbf{v}$. It remains to estimate $g^{(1)}$. The expression (7.3.7) for $H$ implies

$$\begin{aligned} |g^{(1)}| &= |H| \le |s|\,|\mathfrak{C}[f]| + |s|\,|\operatorname{div}\mathbf{v}| \quad \text{in } Q \times \mathbb{R}, \\ |g^{(1)}| &= |H| \le |s|\,|\operatorname{div}\mathbf{v}| + |F_\infty| \quad \text{in } \big((\mathbb{R}^d \times (0,T)) \setminus Q\big) \times \mathbb{R}. \end{aligned} \tag{7.3.29}$$

Since $\varrho_\infty \in C^1(\mathbb{R}^d \times (0,T))$ vanishes for large $|x|$, the expression (7.3.2) for $F_\infty$ yields the estimate $\|F_\infty(\cdot,t,s)\|_{L^1(\mathbb{R}^d\setminus\Omega)} \le c$ for a.e. $(t,s) \in [0,T] \times \mathbb{R}$. From (7.3.26) we obtain

$$\begin{aligned} \|H(\cdot,t,s)\|_{L^1(\Omega)} &\le |s|\,\|\mathfrak{C}[f](\cdot,t,s)\,\|_{L^1(\Omega)} + |s|M(t), \\ \|H(\cdot,t,s)\|_{L^1(\mathbb{R}^d\setminus\Omega)} &\le |s|M(t) + c, \end{aligned} \tag{7.3.30}$$

for a.e. $(t,s) \in [0,T] \times \mathbb{R}$. Next, the function $\mathfrak{C}[f](x,t,s)$, defined by (7.1.33e), vanishes for $s < 0$ and satisfies

$$\begin{aligned} |\mathfrak{C}[f](x,t,s)| &\le \int_{\mathbb{R}} p(s)\, d_s f(x,t,s) + \overline{p}(x,t) \int_{\mathbb{R}} d_s f(x,t,s) \\ &= 2\int_{\mathbb{R}} p(s)\, d_s f(x,t,s) \equiv 2\overline{p}(x,t). \end{aligned} \tag{7.3.31}$$

Inequality (7.1.3) in Condition 7.1.1 and inequality (7.1.47) in Condition 7.1.11 imply

$$\begin{aligned} \|\overline{p}\|_{L^\infty(0,T;L^1(\Omega))} &\equiv \operatorname*{ess\,sup}_{t\in(0,T)} \int_\Omega \Big\{ \int_{[0,\infty)} p(s)\, d_s f(x,t,s) \Big\}\, dx \\ &\le c \operatorname*{ess\,sup}_{t\in(0,T)} \int_\Omega \Big\{ \int_{[0,\infty)} (1+s^\gamma)\, d_s f(x,t,s) \Big\}\, dx \le c. \end{aligned} \tag{7.3.32}$$

Combining this with (7.3.30) and (7.3.31) gives the estimate

$$\|H(\cdot,t,s)\|_{L^1(\mathbb{R}^d)} \equiv \|g^{(1)}(\cdot,t,s)\|_{L^1(\mathbb{R}^d)} \le c|s|M(t) + c|s| + c \tag{7.3.33}$$

for a.e. $(t,s) \in [0,T] \times \mathbb{R}$. In view of (7.3.28) and (7.3.33) the functions $\mathbf{g}^{(0)}$, $g^{(i)}$ in (7.3.25) meet all requirements of Lemma 7.3.3. Applying inequality (7.3.19) of that lemma with $g$ replaced by $\mathbf{g}^{(0)}$ and $g^{(i)}$ we conclude that the right hand side of (7.3.25) admits the estimate

$$\begin{aligned}\big|\operatorname{div}[\mathbf{g}^{(0)}]_{m,k} + \partial_s[g^{(1)}]_{m,k} + [g^{(2)}]_{m,k}\big| \\ \le c(m,k)(N+2)(|s|+1)(M(t)+1) \qquad (7.3.34)\end{aligned}$$

for all $x \in \mathbb{R}^d$, $s \in [-N, N]$, and all $N \ge 1$. It now follows from (7.3.27) and (7.3.25) that $\|\partial_t[f^*]_{m,k}(x,\cdot,s)\|_{L^2(0,T)}$ is uniformly bounded on every compact subset of $\mathbb{R}^{d+1}$. Since the embedding $W^{1,2}(0,T) \hookrightarrow C(0,T)$ is compact, the functions $[f^*]_{m,k}(\cdot,t,\cdot)$ converge to some limit as $t \searrow 0$ uniformly on every compact subset of $\mathbb{R}^{d+1}$. By (7.3.24) this limit coincides with $[\eta f_\infty]_{m,k}(x,0,s)$. Thus we get

$$[f^*]_{m,k}(x,t,s) \to [\eta f_\infty]_{m,k}(x,0,s) \quad \text{as } t \searrow 0 \qquad (7.3.35)$$

uniformly on every compact set. Now rewrite equation (7.3.25) in the form

$$\partial_t[f^*]_{m,k} + \operatorname{div}([f^*]_{m,k}\mathbf{v}) + \partial_s[H]_{m,k} = [P]_{m,k} + \operatorname{div} I_{m,k}, \qquad (7.3.36)$$

where

$$I_{m,k} = \operatorname{div}([f^*]_{m,k}\mathbf{v} - [f^*\mathbf{v}]_{m,k}). \qquad (7.3.37)$$

Recall that the function $[f^*]_{m,k}$ is infinitely differentiable with respect to $x$, $s$ and has the time derivative locally square integrable. Hence it is a strong solution to (7.3.36). Multiplying this equation by the smooth bounded function $\Psi'([f^*]_{k,m})$ and noting that $\Psi'(s)s - \Psi(s) = -s^2$ we arrive at

$$\begin{aligned}\partial_t\Psi([f^*]_{k,m}) + \operatorname{div}(\Psi([f^*]_{k,m})\mathbf{v}) - [f^*]_{k,m}^2 \operatorname{div}\mathbf{v} + \Psi'([f^*]_{k,m})\partial_s[H]_{k,m} \\ = \Psi'([f^*]_{k,m})[P]_{k,m} + \Psi'([f^*]_{k,m})I_{m,k}. \qquad (7.3.38)\end{aligned}$$

Choose $\psi \in C^\infty(\mathbb{R}^d \times (0,T) \times \mathbb{R})$ such that $\psi$ vanishes for $t = T$ and for large $|x|$ and $|s|$, say

$$\psi(x,t,s) = 0 \quad \text{for } |x| \ge N-1, \quad \psi(x,t,s) = 0 \quad \text{for } |s| \ge N-1. \qquad (7.3.39)$$

Next choose $\chi \in C^\infty(\mathbb{R})$ such that

$$\chi(s) = 1 \quad \text{for } |s| \le N, \quad \chi(s) = 0 \quad \text{for } |s| \ge N+1. \qquad (7.3.40)$$

It is clear that for $k, m \ge 1$,

$$\chi H = H, \quad [\chi H]_{,k} = [H]_{,k}, \quad [\chi H]_{m,k} = [H]_{m,k} \quad \text{on the support of } \psi. \qquad (7.3.41)$$

The same conclusion can be drawn for $P$. Multiplying (7.3.38) by $\psi$, integrating the result over $\mathbb{R}^d \times (0,T) \times \mathbb{R}$, and using (7.3.35) and (7.3.41) we arrive at

$$\begin{aligned}
&\int_{\mathbb{R}^d\times(0,T)\times\mathbb{R}} \big\{\Psi([f^*]_{m,k})(\partial_t\psi + \nabla\psi\cdot\mathbf{v}) + \psi[f^*]_{m,k}^2 \operatorname{div}\mathbf{v}\big\}\,dxdtds \\
&\quad + \int_{\mathbb{R}^d\times(0,T)\times\mathbb{R}} \partial_s\psi\,\Psi'([f^*]_{k,m})\,[\chi H]_{k,m} dxdtds \\
&\quad + \int_{\mathbb{R}^d\times(0,T)\times\mathbb{R}} \psi\Psi'([f^*]_{k,m})([\chi P]_{m,k} + I_{m,k})\,dxdtds \\
&\quad + \int_{\mathbb{R}^d\times\mathbb{R}} \psi(x,0,s)\Psi\big([\eta f_\infty]_{m,k}(x,0,s)\big)\,dxds - 2J_{m,k} = 0, \qquad (7.3.42)
\end{aligned}$$

where

$$J_{m,k} = \int_{\mathbb{R}^d\times(0,T)\times\mathbb{R}} \psi[\chi H]_{m,k}\partial_s[f^*]_{m,k}\,dxdtds.$$

Here we use the equality $\Psi'' = -2$.

**Step 2. First level limit.** It follows from (7.3.33), (7.3.27) and (7.3.40) that

$$\begin{aligned}
\|\chi H\|_{L^1(\mathbb{R}^d\times(0,T)\times\mathbb{R})} &\le c\int_{(0,T)\times\mathbb{R}} \chi(s)(|s|+1)\,(M(t)+1)\,dtds \\
&\le c(N+1)^2\int_{(0,T)} (M(t)+1)\,dt \\
&\le c.
\end{aligned}$$

Next estimate (7.3.28) for $g^{(2)} = P$ gives

$$\begin{aligned}
\|\chi P\|_{L^1(\mathbb{R}^d\times(0,T)\times\mathbb{R})} &\le c\int_{(0,T)\times\mathbb{R}} \chi(s)(M(t)+1)\,dtds \\
&\le c(N+1)\int_{(0,T)} (M(t)+1)\,dt \\
&\le c.
\end{aligned}$$

By the general properties of mollifiers, it follows that

$$[\chi H]_{m,k} \to \chi H, \quad [\chi P]_{m,k} \to \chi P \quad \text{in } L^1(\mathbb{R}^d\times(0,T)\times\mathbb{R}) \quad \text{as } (m,k)\to\infty.$$

On the other hand, the functions $[f^*]_{m,k}$ are uniformly bounded and converge to $f^*$ a.e. in $\mathbb{R}^d \times (0,T) \times \mathbb{R}$ as $(m,k) \to \infty$. This leads to

$$\begin{aligned}
&\int_{\mathbb{R}^d \times (0,T)\times\mathbb{R}} \{\Psi([f^*]_{m,k})(\partial_t\psi + \nabla\psi\cdot\mathbf{v}) + \psi[f^*]^2_{m,k}\,\mathrm{div}\,\mathbf{v}\}\,dxdtds \\
&\quad + \int_{\mathbb{R}^d \times (0,T)\times\mathbb{R}} \{\partial_s\psi\,\Psi'([f^*]_{k,m})\,[\chi H]_{k,m} + \psi\Psi'([f^*]_{k,m})[\chi P]_{m,k}\}\,dxdtds \\
&\qquad \to \int_{\mathbb{R}^d \times (0,T)\times\mathbb{R}} \{\Psi(f^*)(\partial_t\psi + \nabla\psi\cdot\mathbf{v}) + \psi(f^*)^2\,\mathrm{div}\,\mathbf{v}\}\,dxdtds \\
&\qquad\qquad + \int_{\mathbb{R}^d \times (0,T)\times\mathbb{R}} \{\partial_s\psi\,\Psi'(f^*)H + \psi\Psi'(f^*)P\}\,dxdtds \qquad (7.3.43)
\end{aligned}$$

as $(m,k) \to \infty$. Here we use the equalities $\psi\chi = \psi$ and $\partial_s\psi\chi = \partial_s\psi$, which are direct consequences of (7.3.39) and (7.3.40). Repeating this argument we obtain

$$\int_{\mathbb{R}^d\times\mathbb{R}} \psi(x,0,s)\Psi\big([\eta f_\infty]_{m,k}(x,0,s)\big)\,dxds \to \int_{\mathbb{R}^d\times\mathbb{R}} \psi(x,0,s)\Psi\big(\eta f_\infty(x,0,s)\big)\,dxds \tag{7.3.44}$$

as $(m,k) \to \infty$. Next, applying the Lions-Di Perna commutator Lemma 1.6.1 and noting that $f^*$ is bounded and $\psi$ is bounded and compactly supported we obtain

$$\lim_{k\to\infty}\lim_{m\to\infty} \int_{\mathbb{R}^d \times (0,T)\times\mathbb{R}} \psi\Psi'([f^*]_{k,m})I_{m,k}\,dxdtds = 0. \tag{7.3.45}$$

**Step 3. Second level limit.** Now our task is to pass to the limit in the integral

$$J_{m,k} = \int_{\mathbb{R}^d \times (0,T)\times\mathbb{R}} \psi[\chi H]_{m,k}\partial_s[f^*]_{m,k}\,dxdtds$$

By abuse of notation we will write simply $H_k$ instead of $[\chi H]_{,k}$ and $f_k$ instead of $[f^*]_{,k}$. Notice that the functions $H_k$ and their derivatives with respect to $s$ of any order belong to $L^1(\mathbb{R}^d \times (0,T)\times\mathbb{R})$. In turn, the functions $f_k$ and their derivatives with respect to $s$ are bounded in $\mathbb{R}^d \times (0,T)\times\mathbb{R}$. It now follows from the symmetry of the mollifying operator that

$$J_{m,k} \equiv \int_{\mathbb{R}^d \times (0,T)\times\mathbb{R}} \psi[H_k]_{m,}\partial_s[f_k]_{m,}\,dxdtds = \int_{\mathbb{R}^d \times (0,T)\times\mathbb{R}} [\psi[H_k]_{m,}]_{m,}\partial_s f_k\,dxdtds.$$

First we pass to the limit in the latter integral as $m \to \infty$. To this end consider the function

$$\Phi_{m,k}(t,s) = \|[H_k]_{m,}(\cdot,t,s) - H_k(\cdot,t,s)\|_{L^1(\mathbb{R}^d)}.$$

In view of (7.3.40), $\Phi_{m,k}(t,s) = 0$ for $|s| \ge N+2$. Since $H$ satisfies (7.3.33), it meets all requirements of Lemma 7.3.3 with $g$ replaced by $H$, and $M_1$, $M_2$ replaced by $M$, 1. It follows from estimate (7.3.18) in that lemma that

$$\begin{aligned}\Phi_{m,k}(t,s) &\leq \|[H_k]_{m,}(\cdot,t,s)\|_{L^1(\mathbb{R}^d)} + \|H_k(\cdot,t,s)\|_{L^1(\mathbb{R}^d)}\\ &= \|[\chi H]_{m,k}(\cdot,t,s)\|_{L^1(\mathbb{R}^d)} + \|[\chi H]_{,k}(\cdot,t,s)\|_{L^1(\mathbb{R}^d)}\\ &\leq c(N+2)(M(t)+1) \end{aligned} \tag{7.3.46}$$

for a.e. $(t,s) \in [0,T] \times \mathbb{R}$. By the general properties of the mollifying operators, $\Phi_{m,k}(t,s) \to 0$ as $m \to \infty$ for a.e. $(t,s) \in [0,T] \times \mathbb{R}$. On the other hand, in view of (7.3.27), the function $M(t)+1$ is integrable on $[0,T]$. Hence $\Phi_{m,k}(t,s)$ has an integrable majorant on the rectangle $[0,T] \times [-N-2, N+2]$. Applying the Lebesgue dominated convergence theorem we obtain

$$\int_0^T \int_{-N-2}^{N+2} \Phi_{m,k}\, dsdt0 \quad \text{as } m \to \infty.$$

Since $\Phi_{m,k}$ vanishes for $|s| \geq N+2$, it follows that $\Phi_{m,k} \to 0$ in $L^1((0,T)\times\mathbb{R})$, which gives

$$[H_k]_{m,} - H_k \to 0 \quad \text{in } L^1(\mathbb{R}^d \times (0,T) \times \mathbb{R}) \quad \text{as } m \to \infty. \tag{7.3.47}$$

It follows that

$$\left\| \left[[H_k]_{m,} - H_k\right]_{m,} \right\|_{L^1(\mathbb{R}^d\times(0,T)\times\mathbb{R})} \leq \|[H_k]_{m,} - H_k\|_{L^1(\mathbb{R}^d\times(0,T)\times\mathbb{R})} \to 0 \tag{7.3.48}$$

as $m \to \infty$. Combining (7.3.47)–(7.3.48) with the identity

$$\left[[H_k]_{m,}\right]_{m,} - H_k = \left[[H_k]_{m,} - H_k\right]_{m,} + [H_k]_{m,} - H_k$$

implies

$$\left[[H_k]_{m,}\right]_{m,} - H_k \to 0 \quad \text{in } L^1(\mathbb{R}^d \times (0,T)\times\mathbb{R}) \quad \text{as } m \to \infty.$$

Letting $m \to \infty$ in the expression for $J_{m,k}$ with fixed $k \geq 1$ and noting that the functions $\partial_s f_k$, $\psi$ are bounded we arrive at

$$J_{m,k} = \int_{\mathbb{R}^d\times(0,T)\times\mathbb{R}} \left[\psi[H_k]_{m,}\right]_{m,} \partial_s f_k \, dxdtds \to \int_{\mathbb{R}^d\times(0,T)\times\mathbb{R}} \psi H_k \partial_s f_k\, dxdtds$$

as $m \to \infty$. Recalling $f_k = [f^*]_{,k}$ and $H_k = [\chi H]_{,k}$ we finally obtain

$$\lim_{m\to\infty} J_{m,k} = \int_{\mathbb{R}^d\times(0,T)\times\mathbb{R}} \psi[\chi H]_{,k}\partial_s[f^*]_{,k}\, dxdtds. \tag{7.3.49}$$

**Step 4. Third level limit.** Now our task is to pass to the limit in (7.3.49) as $k \to \infty$. Let us consider the sequence of functions

$$\Phi_k(x,t) = \int_{\mathbb{R}} \psi[\mathsf{X}H]_{,k}(x,t,s)\partial_s[f^*]_{,k}(x,t,s)\,ds, \quad (x,t) \in \mathbb{R}^d \times (0,T).$$

Let us show that they have an integrable majorant independent of $k$. Inequalities (7.3.29) and (7.3.31) imply

$$|\mathsf{X}H| \leq 2\mathsf{X}(s)|s|\,\overline{p}(x,t) + \mathsf{X}(s)|s|\,|\operatorname{div}\mathbf{v}| \quad \text{in } Q \times \mathbb{R},$$
$$|\mathsf{X}H| \leq \mathsf{X}(s)|s|\,|\operatorname{div}\mathbf{v}| + \mathsf{X}(s)|F_\infty| \quad \text{in } \big((\mathbb{R}^d \times (0,T)) \setminus Q\big) \times \mathbb{R}$$

Now set

$$p^*(x,t) = \overline{p}(x,t) \quad \text{in } Q, \quad p^*(x,t) = 0 \quad \text{elsewhere.}$$

Since the given vector field $\mathbf{U}$ belongs to $C^1(\mathbb{R}^d \times (0,T))$, formula (7.3.2) for $F_\infty$ yields

$$|F_\infty| \leq c(|\partial_t \varrho_\infty| + |\nabla \varrho_\infty| + \varrho_\infty). \tag{7.3.50}$$

It follows that

$$|\mathsf{X}H| \leq c(N+1)\big(p^* + |\operatorname{div}\mathbf{v}| + |\partial_t \varrho_\infty| + |\nabla \varrho_\infty| + \varrho_\infty\big) =: \mathcal{M}(x,t). \tag{7.3.51}$$

Since the right hand side is independent of $s$, the same holds for $[\mathsf{X}H(x,t,s)]_{,k}$. Thus we get

$$\begin{aligned}|\Phi_k(x,t)| &\leq \int_{\mathbb{R}} |\psi[\mathsf{X}H]_{,k}|(x,t,s)\partial_s[f^*]_{,k}(x,t,s)\,ds \leq \mathcal{M}(x,t) \int_{\mathbb{R}} \partial_s[f^*]_{,k}(x,t,s)\,ds \\ &= \mathcal{M}(x,t) \lim_{s\to\infty} [f^*]_{,k}(x,t,s) \leq \mathcal{M}(x,t).\end{aligned} \tag{7.3.52}$$

Here we use the relation $[f^*]_{,k}(x,t,s) \nearrow \eta(x,t) \leq 1$ as $s \to \infty$, which follows from the similar relation for $f^*$. To estimate $\mathcal{M}$, notice that $\varrho_\infty \in C^1(\mathbb{R}^d \times (0,T))$ vanishes for all large $|x|$. Inequalities (7.3.32) imply

$$\|p^*\|_{L^\infty(0,T;L^1(\Omega))} \leq c.$$

From this and estimates (7.3.26)–(7.3.27) for $\mathbf{v}$ we obtain

$$\|\mathcal{M}\|_{L^1(\mathbb{R}^d \times (0,T))} \leq c. \tag{7.3.53}$$

Our next task is to calculate $\lim_{k\to\infty} \Phi_k(x,t)$. To this end notice that $f^*$ is monotone in $s$, and $\mathfrak{C}[f]$ is a difference of two functions monotone in $s$. Hence $\mathsf{X}(s)H(x,t,s)$ as a function of $s$ has the right and left limits at each $s \in \mathbb{R}$. Therefore, $\mathsf{X}H$ and $f^*$, as functions of $s$, meet all requirements of Lemma 1.6.3. Applying relation (1.6.11) in that lemma and noting that $\mathsf{X}$, $\psi$ are continuous we obtain

$$\lim_{k\to\infty} \Phi_k(x,t) = \int_{\mathbb{R}} \psi(x,t,s)\mathsf{X}(s)\tilde{H}(x,t,s)\,d_s f^*(x,t,s), \tag{7.3.54}$$

where $\tilde{H}(x,t,s) = 2^{-1}\lim_{h\searrow 0} H(x,t,s-h) + 2^{-1}\lim_{h\searrow 0} H(x,t,s+h)$. The expression in (7.3.7) for $H$ implies

$$\tilde{H}(x,t,s) = -s\eta\mathfrak{M}(x,t,s) - s\operatorname{div}\mathbf{u}(x,t)\tilde{f}^*(x,t,s) \quad \text{in } Q\times\mathbb{R},$$

where $\mathfrak{M}$ is given by (7.3.21), and

$$\tilde{H}(x,t,s) = \big(F_\infty(x,t) - s\operatorname{div}\mathbf{U}(x,t)\big)\tilde{f}^*(x,t,s) \quad \text{in } \big((\mathbb{R}^d\times(0,T))\setminus Q\big)\times\mathbb{R}.$$

Next, identity (1.6.12) yields

$$\int_{\mathbb{R}} \psi s\tilde{f}^*\, d_s f^* = \frac{1}{2}\int_{\mathbb{R}} \psi s\, d_s(f^*)^2 = -\frac{1}{2}\int_{\mathbb{R}} \partial_s(s\psi)(f^*)^2\, ds,$$
$$\int_{\mathbb{R}} \psi\tilde{f}^*\, d_s f^* = \frac{1}{2}\int_{\mathbb{R}} \psi\, d_s(f^*)^2 = -\frac{1}{2}\int_{\mathbb{R}} \partial_s\psi(f^*)^2\, ds.$$

From this and the expression for $\tilde{H}$ we obtain the existence of the pointwise limit

$$\lim_{k\to\infty} \Phi_k(x,t) = \Phi(x,t) \tag{7.3.55}$$

given by

$$\Phi = \frac{1}{2}\operatorname{div}\mathbf{u}(x,t)\int_{\mathbb{R}} \partial_s(\psi s)(f^*)^2\, ds - \int_{\mathbb{R}} \psi s\eta\,\mathfrak{M}\, d_s f^* \quad \text{in } Q,$$
$$\Phi = \frac{1}{2}\operatorname{div}\mathbf{U}\int_{\mathbb{R}} \partial_s(\psi s)(f^*)^2\, ds - \frac{1}{2}F_\infty\int_{\mathbb{R}} \partial_s\psi\,(f^*)^2\, ds \quad \text{in } (\mathbb{R}^d\times(0,T))\setminus Q. \tag{7.3.56}$$

Here we use the fact that $\chi = 1$ on the support of $\psi$. Since by (7.3.52) and (7.3.53) the sequence $\Phi_k$ has an integrable majorant, Lebesgue's dominated convergence theorem yields

$$\lim_{k\to\infty}\lim_{m\to\infty} J_{m,k} = \int_{\mathbb{R}^d\times(0,T)} \Phi(x,t)\, dxdt. \tag{7.3.57}$$

**Step 5. Renormalization.** Letting first $m\to\infty$ and then $k\to\infty$ in the integral identity (7.3.42) and using (7.3.43)–(7.3.45) and (7.3.57) we arrive at

$$\begin{aligned}
&\int_{\mathbb{R}^d\times(0,T)\times\mathbb{R}} \big\{\Psi(f^*)(\partial_t\psi + \nabla\psi\cdot\mathbf{v}) + \psi(f^*)^2\operatorname{div}\mathbf{v}\big\}\, dxdtds \\
&\qquad + \int_{\mathbb{R}^d\times(0,T)\times\mathbb{R}} \partial_s\psi\,\Psi'(f^*)H\, dxdtds - 2\int_{\mathbb{R}^d\times(0,T)} \Phi\, dxdt \\
&\quad + \int_{\mathbb{R}^d\times(0,T)\times\mathbb{R}} \psi\Psi'(f^*)P\, dxdtds + \int_{\mathbb{R}^d\times\mathbb{R}} \psi(x,0,s)\Psi(\eta f_\infty(x,0,s))\, dxds = 0.
\end{aligned} \tag{7.3.58}$$

It follows from formulae (7.3.56) for $\Phi$ and (7.3.7) for $H$ that

$$\int_{Q\times\mathbb{R}} \partial_s\psi\,\Psi'(f^*)H\,dxdtds - 2\int_Q \Phi\,dxdt = 2\int_Q \eta\Big\{\int_{\mathbb{R}} s\psi\,\mathfrak{M}\,d_sf^*\Big\}\,dxdt$$
$$-\int_{Q\times\mathbb{R}} \Big\{\Big(s\partial_s\psi\big(\Psi'(f^*)f^*+(f^*)^2\big)+(f^*)^2\psi\Big)\operatorname{div}\mathbf{u}+s\eta\partial_s\psi\,\Psi'(f^*)\,\mathfrak{C}[f]\Big\}\,dxdtds \tag{7.3.59}$$

and

$$\int_{((\mathbb{R}^d\times(0,T))\setminus Q)\times\mathbb{R}} \partial_s\psi\,\Psi'(f^*)\,H\,dxdtds - 2\int_{(\mathbb{R}^d\times(0,T))\setminus Q}\Phi\,dxdt =$$
$$-\int_{((\mathbb{R}^d\times(0,T))\setminus Q)\times\mathbb{R}} \Big\{\partial_s\psi\big(\Psi'(f^*)f^*+(f^*)^2\big)\big(s\operatorname{div}\mathbf{U}-F_\infty\big)+(f^*)^2\psi\operatorname{div}\mathbf{U}\Big\}\,dxdtds. \tag{7.3.60}$$

Next notice that $\Psi'(f^*)f^*+(f^*)^2\equiv\Psi(f^*)$ and

$$\mathbf{v}=\mathbf{u}\quad\text{in } Q,\qquad \mathbf{v}=\mathbf{U}\text{ and } f^*=\eta f_\infty\quad\text{in } ((\mathbb{R}^d\times(0,T))\setminus Q.$$

Inserting (7.3.59)–(7.3.60) into (7.3.58) we obtain the integral identity (7.3.20) which is our claim. □

The next step is to eliminate the dependence of $\psi$ on $x$ and to replace the integral identity (7.3.20) by an integral *inequality* containing an arbitrary nonnegative test function depending only on $t$ and $s$.

**Lemma 7.3.5.** *The inequality*

$$\int_{Q\times\mathbb{R}} \{\Psi(f)(\partial_t\Phi - s\partial_s\Phi\operatorname{div}\mathbf{u}) - s\partial_s\Phi\,\Psi'(f)\,\mathfrak{C}[f]\}\,dxdtds$$
$$+2\int_Q\Big\{\int_{\mathbb{R}} s\Phi\mathfrak{M}\,d_sf\Big\}\,dxdt \ge 0 \tag{7.3.61}$$

*holds for any nonnegative function $\Phi\in C^\infty((0,T)\times\mathbb{R})$ vanishing for $t=T$ and for all sufficiently large $|s|$.*

*Proof.* First observe that for every Lipschitz function $\psi:\mathbb{R}^d\times(0,T)\times\mathbb{R}\to\mathbb{R}$ which vanishes in a neighborhood of the plane $\{t=T\}$ and for all sufficiently large $|x|$, $|s|$, there is a sequence of functions $\psi_n$ such that

$$\psi_n\in C^\infty(\mathbb{R}^d\times(0,T)\times\mathbb{R}),\quad \psi_n=0\quad\text{for } t=T\text{ and large } |x|,|s|,$$

$\psi_n\to\psi$ uniformly in $\mathbb{R}^d\times(0,T)\times\mathbb{R}$, and the first derivatives of $\psi_n$ are uniformly bounded and converge to the derivatives of $\psi$ a.e. in $\mathbb{R}^d\times(0,T)\times\mathbb{R}$. Such a sequence can be obtained by mollification of $\psi$. Substituting $\psi_n$ into (7.3.20) and letting $n \to \infty$ we conclude that the integral identity (7.3.20) holds for all Lipschitz functions $\psi$ defined in $\mathbb{R}^d \times (0, T) \times \mathbb{R}$ such that $\psi$ vanishes in a neighborhood of the plane $\{t = T\}$ and for all sufficiently large $|x|$, $|s|$.

Let us consider the following construction. Introduce the set $K = (S_T \setminus \Sigma_{\text{in}}) \cup \{t = T\} \subset \mathbb{R}^{d+1}$. For every $h > 0$ denote by $O(h)$ the $h$-neighborhood of $K$,

$$O(h) = \{(x, t) : \operatorname{dist}((x, t), K) < h\}$$

Next, choose a Lipschitz function $\phi : Q \to \mathbb{R}$ vanishing in a neighborhood of $K$. By the sharp form of Whitney's extension theorem [31], the function $\phi$ has a Lipschitz extension over $\mathbb{R}^{d+1}$, still denoted by $\phi$. In particular, $\phi$ vanishes in $O(h_1)$ for some $h_1 > 0$. We can choose $0 < h_3 < h_2 < h_1$ so small that $B_i = \{(x, t) : |x| + |t| < h_i^{-1}\}$ contain the cylinder $Q$. Thus we get

$$B_1 \setminus O(h_1) \Subset B_2 \setminus O(h_2) \Subset B_3 \setminus O(h_3).$$

Now choose $\xi, \eta \in C_0^\infty(\mathbb{R}^{d+1})$ with the properties

$$\begin{aligned} \xi &= 1 \quad \text{in } B_1 \setminus O(h_1), \quad & \xi &= 0 \quad \text{outside of } B_2 \setminus O(h_2), \\ \eta &= 1 \quad \text{in } B_2 \setminus O(h_2), \quad & \eta &= 0 \quad \text{outside of } B_3 \setminus O(h_3). \end{aligned}$$

Finally, choose $\Phi \in C^\infty((0, T) \times \mathbb{R})$ satisfying the hypotheses of the lemma and set $\psi(x, t, s) = \Phi(t, s)\phi(x, t)\xi(x, t)$. Then $\psi$ is Lipschitz in $\mathbb{R}^d \times (0, T) \times \mathbb{R}^d$ and vanishes for $t = T$ and for sufficiently large $|x|$, $|s|$. Moreover, by the choice of $\xi$, $\psi = \Phi\phi$ in $Q \times \mathbb{R}$. On the other hand, $\eta = 1$ and $f^* = f$ on the support of $\psi$. Substituting $\psi$ into the integral identity (7.3.20) we obtain

$$\begin{aligned} &\int_{Q \times \mathbb{R}} \phi \{\Psi(f)(\partial_t \Phi - s \partial_s \Phi \operatorname{div} \mathbf{u}) - s\, \partial_s \Phi\, \Psi'(f)\, \mathfrak{C}[f]\}\, dx dt ds \\ &\quad + \int_{Q \times \mathbb{R}} \Phi \Psi(f)(\partial_t \phi + \nabla \phi \cdot \mathbf{u})\, dx dt ds + 2 \int_Q \phi \Big\{ \int_{\mathbb{R}} s \Phi \mathfrak{M}\, d_s f \Big\}\, dx dt + \mathfrak{N} = 0, \end{aligned}$$

where $\mathfrak{N}$ is given by (7.3.22). Since $\eta = 1$ on the support of $\psi$, it follows from the identity $f_\infty^2 = f_\infty$ and formula (7.3.9) for $P$ that

$$\Psi(\eta f_\infty) = \Psi(f_\infty) = 0, \quad P = 0 \quad \text{on the support of } \psi.$$

Hence $\mathfrak{N} = 0$ and

$$\begin{aligned} &\int_{Q \times \mathbb{R}} \phi \{\Psi(f)(\partial_t \Phi - s \partial_s \Phi \operatorname{div} \mathbf{u}) - s \partial_s \Phi\, \Psi'(f)\, \mathfrak{C}[f]\}\, dx dt ds \\ &\quad + \int_{Q \times \mathbb{R}} \Phi \Psi(f)(\partial_t \phi + \nabla \phi \cdot \mathbf{u})\, dx dt ds + 2 \int_Q \phi \Big\{ \int_{\mathbb{R}} s \Phi \mathfrak{M}\, d_s f \Big\}\, dx dt = 0. \end{aligned} \tag{7.3.62}$$

Observe that by Condition 7.1.11, the vector field $\mathbf{U}$ and the surface $S_T = \partial\Omega \times (0, T)$ meet the hypotheses of the geometric Theorem 13.3.3. Hence there exists a sequence of Lipschitz functions $\phi_n : Q \to [0,1]$ such that each $\phi_n$ vanishes in some neighborhood of $S_T \setminus \Sigma_{\text{in}}$ and of $\{t = T\}$, $\phi_n \to 1$ everywhere in $Q$ and

$$\int_Q G(\partial_t \phi_n + \nabla \phi_n \cdot \mathbf{u})\, dxdt \le 0$$

for all bounded nonnegative functions $G$. Substituting $\phi = \phi_n$ in (7.3.62) and letting $n \to \infty$ we arrive at (7.3.61). □

Our next task is to obtain an analog of inequality (7.3.61) with the test function $\Phi$ independent of $s$. The derivation of such an inequality is based on the following lemma concerning the properties of $\mathfrak{C}$ and $\mathfrak{M}$:

**Lemma 7.3.6.** *For a.e. $(x,t) \in Q$, the function $\mathfrak{C}[f](x,t,\cdot)$ is nonpositive. Moreover, for all $g \in C_0^\infty(0,\infty)$,*

$$(\lambda+1) \int_{\mathbb{R}} g(s) \mathfrak{C}[f](x,t,s)\, ds = \int_{[0,\infty)} g'(\tau) \mathfrak{V}_\tau(x,t)\, d\tau \tag{7.3.63}$$

*where $\mathfrak{V}_\tau$ is defined by* (7.1.39).

*Proof.* Since $p(s)$ is monotone, for almost every $(x,t) \in Q$ there exists $\overline{s} \in [0,\infty)$, depending on $(x,t)$, such that $p(s) \le \overline{p}(x,t)$ for $s \le \overline{s}$ and $p(s) \ge \overline{p}(x,t)$ otherwise. Since $f(x,t,s) = 0$ for $s < 0$ we have for any $s' \le s'' \le \overline{p}(x,t)$,

$$\mathfrak{C}[f](x,t,s'') - \mathfrak{C}[f](x,t,s') = \frac{1}{\lambda+1} \int_{[s',s'')} (p(s) - \overline{p})\, d_s f(x,t,s) \le 0.$$

The same arguments give $\mathfrak{C}[f](x,t,s'') \ge \mathfrak{C}[f](x,t,s')$ for $\overline{s} \le s' \le s''$. Hence $\mathfrak{C}[f](x,t,\cdot)$ does not increase on $(-\infty,\overline{s})$ and does not decrease on $(\overline{s},\infty)$. Since $\mathfrak{C}[f](x,t,\cdot)$ vanishes at $\pm\infty$, it is nonpositive. The same conclusion can be drawn for $\mathfrak{M}$. Next, it follows from the identity

$$\int_{[0,s)} (p(z) - \overline{p}(x,t))\, d_z f(x,t,z) = -\int_{[s,\infty)} (p(z) - \overline{p}(x,t))\, d_z f(x,t,z)$$

that

$$\begin{aligned}
(\lambda+1) &\int_{\mathbb{R}} g(s)\, \mathfrak{C}[f](x,t,s)\, ds \\
&= \int_{[0,\infty)} \Big( \int_{[s,\infty)} g'(\tau)\, d\tau \Big) \Big( \int_{[s,\infty)} (p(z) - \overline{p})\, d_z f(x,t,z) \Big)\, ds \\
&= \int_{[0,\infty)} g'(\tau) \Big( \int_{[0,\tau)} ds \int_{[s,\infty)} (p(z) - \overline{p})\, d_z f(x,t,z) \Big)\, d\tau \\
&= \int_{[0,\infty)} g'(\tau) \Big( \int_{[0,\infty)} F(s)\, dG(s) \Big)\, d\tau
\end{aligned}$$

where

$$G(s) = \min\{s, \tau\}, \quad F(s) = \int_{[s,\infty)} (p(z) - \overline{p})\, d_z f(x, t, z).$$

It remains to note that in view of Lemma 1.3.7,

$$\int_{[0,\infty)} F(s)\, dG(s) = -\lim_{s\to -0} F(s)G(s) - \int_{[0,\infty)} G(s)\, dF(s)$$
$$= \int_{[0,\infty)} G(s)(p(s) - \overline{p})\, d_s f = \int_{[0,\infty)} \min\{s,\tau\}(p(s) - \overline{p})\, d_s f \equiv \mathfrak{V}_\tau(x, t). \qquad \square$$

Now we are in a position to eliminate the variable $s$ from inequality (7.3.61).

**Lemma 7.3.7.** *Under the assumption of Theorem 7.1.12,*

$$\int_{Q\times\mathbb{R}} \Psi(f)\, dxdtds \le 0. \tag{7.3.64}$$

*Proof.* Introduce a function $\omega \in C_0^\infty(\mathbb{R})$ with the properties

$$\omega \ge 0, \quad \operatorname{supp}\omega \subset (0,1), \quad \int_{\mathbb{R}} \omega(s)\, ds = 1.$$

Fix $\xi \ge 1$ and set

$$\upsilon(s) = \int_s^\infty \omega(\tau - \xi)\, d\tau, \quad h(t) = T - t.$$

Substituting $\Phi(t,s) = h(t)\upsilon(s)$ into the integral inequality (7.3.61) we obtain

$$-\int_{Q\times\mathbb{R}} \Psi(f)\upsilon(s)\, dxdtds + 2\int_Q \Big\{ \int_{\mathbb{R}} h(t)\upsilon(s)(s\mathfrak{M})\, d_s f(x,t,s) \Big\}\, dxdt$$
$$\ge \int_{Q\times\mathbb{R}} s\upsilon'(s)h(t)\Psi(f)\operatorname{div}\mathbf{u}\, dxdtds + \int_{Q\times\mathbb{R}} s\upsilon'(s)h(t)\Psi'(f)\,\mathfrak{C}[f]\, dxdtds. \tag{7.3.65}$$

Recall that $f$, $\Psi(f)$ and $\mathfrak{C}[f]$, $\mathfrak{M}$ vanish for $s < 0$. In view of Lemma 7.3.6, $\mathfrak{M}$ is nonpositive, which leads to

$$\int_Q \Big\{ \int_{\mathbb{R}} h(t)\upsilon(s)(s\mathfrak{M})\, d_s f(x,t,s) \Big\}\, dxdt \le 0. \tag{7.3.66}$$

Since $\upsilon' \le 0$, we have

$$\int_{Q\times\mathbb{R}} s\upsilon'(s)\Psi(f)h(t)\operatorname{div}\mathbf{u}\, dxdtds \ge T\int_{Q\times\mathbb{R}} s\upsilon'(s)\Psi(f)|\operatorname{div}\mathbf{u}|\, dxdtds$$
$$\ge (\xi+1)T\int_{Q\times\mathbb{R}} \upsilon'(s)\Psi(f)|\operatorname{div}\mathbf{u}|\, dxdtds. \tag{7.3.67}$$

Let us introduce the function

$$\wp_1(s) = T\int_Q |\operatorname{div}\mathbf{u}|\Big\{\int_{[0,s]} \Psi(f)(x,t,\tau)\,d\tau\Big\}\,dxdt \tag{7.3.68}$$

and note that

$$T\int_{Q\times\mathbb{R}} \upsilon'(s)\Psi(f)|\operatorname{div}\mathbf{u}|\,dxdtds = -\int_{[0,\infty)} \upsilon''(s)\wp_1(s)\,ds.$$

Then we can write inequality (7.3.67) in the form

$$\int_{Q\times\mathbb{R}} \Psi(f)h(t)\upsilon'(s)\operatorname{div}\mathbf{u}\,dxdtds \geq -(\xi+1)\int_{[0,\infty)} \upsilon''(s)\wp_1(s)\,ds. \tag{7.3.69}$$

Next, by Lemma 7.3.6 we have $\upsilon'(s)\,\mathfrak{C}[f] \geq 0$ and $\upsilon'(s)\,\mathfrak{C}[f] = 00$ for $s \geq \xi+1$, which along with the inequality $\Psi'(f) = 1-2f \geq -1$ gives

$$\int_{Q\times\mathbb{R}} s\upsilon'(s)h(t)\Psi'(f)\,\mathfrak{C}[f]\,dxdtds \geq -(\xi+1)T\int_{Q\times\mathbb{R}} \upsilon'(s)\,\mathfrak{C}[f]\,dxdtds.$$

On the other hand, in view of (7.3.63), we have

$$\int_{Q\times\mathbb{R}} \upsilon'(s)\,\mathfrak{C}[f]\,dxdtds = \frac{1}{\lambda+1}\int_{Q\times\mathbb{R}} \upsilon''(s)\,\mathfrak{V}_s(x,t)\,dxdtds.$$

Thus we get

$$\int_{Q\times\mathbb{R}} s\upsilon'(s)h(t)\Psi'(f)\,\mathfrak{C}[f]\,dxdtds \geq -c(\xi+1)\int_{Q\times\mathbb{R}} \upsilon''(s)\,\mathfrak{V}_s(x,t)\,dxdtds, \tag{7.3.70}$$

where $c = T/(\lambda+1)$. Let us introduce the function

$$\wp_2(s) = c\int_Q \mathfrak{V}_s\,dxdt \tag{7.3.71}$$

and note that

$$c\int_{Q\times\mathbb{R}} \upsilon''(s)\,\mathfrak{V}_s(x,t)\,dxdtds = \int_{[0,\infty)} \upsilon''(s)\wp_2(s)\,ds;$$

hence we can rewrite inequality (7.3.70) in the form

$$\int_{Q\times\mathbb{R}} s\upsilon'(s)h(t)\Psi'(f)\,\mathfrak{C}[f]\,dxdtds \geq -(\xi+1)\int_{[0,\infty)} \upsilon''(s)\wp_2(s)\,ds. \tag{7.3.72}$$

Inserting inequalities (7.3.66), (7.3.69) and (7.3.72) into (7.3.65) we obtain

$$-\int_{Q\times\mathbb{R}} \Psi(f)\upsilon(s)\,dxdtds \geq -(\xi+1)\int_{[0,\infty)} \upsilon''(s)(\wp_1(s)+\wp_2(s))\,ds. \tag{7.3.73}$$

Next recalling the definition of $v$, we get

$$\int_{[0,\infty)} v''(s)\wp_i(s)\,ds = -\int_{[0,\infty)} \omega'(s-\xi)\wp_i(s)\,ds = \frac{d}{d\xi}\int_{[0,\infty)} \omega(s-\xi)\wp_i(s)\,ds,$$

which along with (7.3.73) yields the inequality

$$\int_{Q\times\mathbb{R}} \Psi(f)v(s)\,dxdtds \le (1+\xi)\frac{d}{d\xi}\int_{[0,\infty)} \omega(s-\xi)(\wp_1(s)+\wp_2(s))\,ds.$$

Since $v(s)=1$ for $s\le\xi$ and $\Psi\ge 0$ we have

$$\int_{Q\times[0,\xi]} \Psi(f)\,dxdtds \le (1+\xi)\frac{d}{d\xi}\int_{[0,\infty)} \omega(s-\xi)(\wp_1(s)+\wp_2(s))\,ds. \tag{7.3.74}$$

Let us prove that the functions $\wp_i$, $i=1,2$, are bounded on the positive semi-axis. At this point we use Condition 7.1.11 which says that $f$ satisfies inequality (7.1.38), i.e. $f$ is a regular solution to the kinetic equation (7.1.45). Notice that the obvious inequality

$$\int_0^s \Psi(f(x,t,s))\,ds \le \int_0^\infty \Psi(f(x,t,s))\,ds \equiv \mathfrak{H}(x,t)$$

along with (7.1.38) implies

$$|\wp_1(s)| \le c\|\operatorname{div}\mathbf{u}\|_{L^2(Q)}\,\|\mathfrak{H}\|_{L^2(Q)} \le c < \infty \quad \text{for all} \quad s\ge 0, \tag{7.3.75}$$

which yields the boundedness of $\wp_1$. Using estimate (7.1.38) for $\mathfrak{V}_s$ we conclude that $|\wp_2|\le c$ with a constant $c$ independent of $\xi$. Therefore, the smooth function

$$\xi \mapsto \int_{[0,\infty)} \omega(s-\xi)(\wp_1(s)+\wp_2(s))\,ds$$

is uniformly bounded on the positive semi-axis. Hence there exists a sequence $\xi_k\to\infty$ such that

$$(1+\xi_k)\frac{d}{d\xi}\int_{[0,\infty)} [\omega(s-\xi)(\wp_1(s)+\wp_2(s))]\Big|_{\xi=\xi_k} ds \to 0 \quad \text{as } k\to\infty.$$

Substituting $\xi=\xi_k$ into (7.3.74) and letting $k\to\infty$, we arrive at (7.3.64). □

It follows from Lemma 7.3.7 and $\Psi(f)=f(1-f)\ge 0$ that $f(1-f)=0$ a.e. in $Q\times\mathbb{R}$. Hence the distribution function $f$ is deterministic in the sense of Definition 7.1.10. This completes the proof of Theorem 7.1.12.

## 7.4 Compactness and existence of solutions for adiabatic exponent $\gamma > 3/2$

Now we are in a position to answer the questions formulated at the beginning of this chapter. In this section the restriction $\gamma > 2d$ on the adiabatic exponent $\gamma$ and the compatibility conditions imposed in Conditions 5.1.1 on the data and the constitutive relations, obeying in Theorem 5.1.6, are relaxed. This is important since the range $\gamma > 2d$, assumed in Theorem 5.1.6, in applications corresponds only to liquids and metals, while $\gamma > d/2$ covers all gases in the case of space dimension $d = 2$ and mono-atomic gases in the case $d = 3$. The compatibility conditions also cannot be accepted from the physical standpoint, since they require the existence of a vacuum zone near $\partial\Omega \cap \{t = 0\}$.

The main result of this section is the following theorem:

**Theorem 7.4.1.** *Suppose the pressure function $p(\varrho)$ satisfies Condition 7.1.1 and has the representation $p = p_c(\varrho) + p_b(\varrho)$, where $p_c$ is convex and $p_b$ is bounded. Furthermore, assume that the given functions*

$$\varrho_\infty \in C^1(\mathbb{R}^d \times [0,T]), \quad \mathbf{U} \in C^\infty(\mathbb{R}^d \times [0,T]), \tag{7.4.1}$$

*and the bounded domain $\Omega \subset \mathbb{R}^d$ with $C^\infty$ boundary satisfy the geometric Condition 7.1.2. Then problem* (5.1.12) *has a weak renormalized solution* $(\mathbf{u}, \varrho)$ *which meets all requirements of Definition 5.1.4 and satisfies the estimate*

$$\|\mathbf{u}\|_{L^2(0,T;W^{1,2}(\Omega))} + \|\varrho|\mathbf{u}|^2\|_{L^\infty(0,T;L^1(\Omega))} + \|\varrho_\epsilon\|_{L^\infty(0,T;L^\gamma(\Omega))} \le c_e, \tag{7.4.2}$$

*where the constant $c_e$ depends only on $\gamma > d/2$, $\|\mathbf{U}\|_{C^1(\mathbb{R}^d\times[0,T])}$, $\|\varrho_\infty\|_{L^\infty(\mathbb{R}^d\times[0,T])}$,* $d$, $\operatorname{diam}\Omega$, $T$, *and the constant $c_p$ in Condition 7.1.1. Moreover, for any cylinder $Q' \Subset Q$, and any $\theta$ with $0 < \theta < \min\{2\gamma d^{-1} - 1, 2^{-1}\gamma\}$, there is a constant $c$, depending only on $\theta$, $c_e$, and $Q'$, such that*

$$\int_{Q'} \varrho^{\gamma+\theta}\,dxdt \le c. \tag{7.4.3}$$

Theorem 7.4.1 is a starting point for further investigations of properties of solutions to in/out flow problems for compressible Navier-Stokes equations, important for applications. In particular, as a result of such investigations, the existence of solutions in nonsmooth domains can be shown, and the continuous domain dependence of solutions can be established.

*Proof.* Denote by $\Upsilon$ the bottom edge of the cylinder $Q$,

$$\Upsilon = \{(x,0) : x \in \partial\Omega\}.$$

Introduce $\eta^\epsilon \in C^\infty(\mathbb{R}^d \times [0,T])$ such that

$$0 \le \eta^\epsilon \le 1, \quad \eta^\epsilon(x,t) = 1 \quad \text{for } \operatorname{dist}\big((x,t),\Upsilon\big) > \epsilon, \tag{7.4.4}$$

$$\eta^\epsilon = 0 \quad \text{in a neighborhood of } \Upsilon. \tag{7.4.5}$$

Next choose a sequence of functions $\rho_\infty^\epsilon \in C^\infty(\mathbb{R}^d \times [0,T])$ such that $\rho_\infty^\epsilon \to \varrho_\infty$ in $C(Q)$ as $\epsilon \to 0$. Set $\varrho_\infty^\epsilon = \eta^\epsilon \rho_\infty^\epsilon$. The functions $\varrho_\infty^\epsilon$ belong to $C^\infty(\mathbb{R}^d \times (0,T))$, and vanish in a neighborhood of the edge $\Upsilon$. Therefore, the sequence $\varrho_\infty^\epsilon$ satisfies Condition 7.1.2 and hence meets the requirements of Theorem 7.1.9. Notice that under the hypotheses of Theorem 7.4.1 the artificial pressure function $p_\epsilon$ given by (7.1.5) and the vector field $\mathbf{U}$ also satisfy Condition 7.1.2. Applying Theorem 7.1.3 we conclude that the modified problem (7.1.8) admits a family of renormalized solutions $(\mathbf{u}_\epsilon, \varrho_\epsilon)$, $\epsilon > 0$, which meets all requirements of Theorem 7.1.3 and, in particular, satisfies estimates (7.1.9), (7.1.10). After passing to a subsequence we can assume that there exist $\mathbf{u} \in L^2(0,T;W^{1,2}(\Omega))$ and $\overline{\varrho} \in L^\infty(0,T;L^\gamma(\Omega))$ such that

$$\begin{aligned} \mathbf{u}_\epsilon &\rightharpoonup \mathbf{u} \quad \text{weakly in } L^2(0,T;W^{1,2}(\Omega)), \\ \varrho_\epsilon &\rightharpoonup \overline{\varrho} \quad \text{weakly}^\star \text{ in } L^\infty(0,T;L^\gamma(\Omega)), \\ \varrho_\epsilon &\rightharpoonup \overline{\varrho} \quad \text{weakly in } L^{\gamma+\theta}(Q') \text{ for all } Q' \Subset Q. \end{aligned} \tag{7.4.6}$$

From this and the fundamental Young measure theorem 1.4.5 we conclude that there exist subsequences, still denoted by $\varrho_\epsilon$ and $\varrho_\infty^\epsilon$, and Young measures $\mu \in L_w^\infty(Q;\mathcal{M}(\mathbb{R}))$ and $\mu^\infty \in L_w^\infty(\sqcup_T;\mathcal{M}(\mathbb{R}))$, with the following properties: For any $\varphi \in C_0(\mathbb{R})$ we have

$$\varphi(\varrho_\epsilon) \rightharpoonup \overline{\varphi} \quad \text{weakly}^\star \text{ in } L^\infty(Q), \tag{7.4.7}$$

$$\varphi(\varrho_\infty^\epsilon) \rightharpoonup \overline{\varphi}_\infty \quad \text{weakly}^\star \text{ in } L^\infty(\sqcup_T), \tag{7.4.8}$$

where

$$\overline{\varphi}(x,t) = \langle \mu_{xt}, \varphi \rangle, \quad \overline{\varphi}_\infty(x,t) = \langle \mu_{xt}^\infty, \varphi \rangle. \tag{7.4.9}$$

Moreover, since $\varphi(\varrho_\infty)$ converges to $\varrho_\infty$ a.e. in $\sqcup_T$, we have $\mu_{xt}^\infty = \delta(\cdot - \varrho_\infty(x,t))$ and the corresponding distribution function $f_\infty$ is deterministic in the sense of Definition 7.1.10.

Applying Theorem 7.1.9 we conclude that the limit vector field $\mathbf{u}$ and the distribution function $f$ of the measure $\mu_{xt}$ serve as a solution to the kinetic Problem 7.1.5 and this solution is regular in the sense of Definition 7.1.8. Since the boundary distribution function $f_\infty$ is deterministic, we can apply Theorem 7.1.12 to deduce that so is $f$, and the Young measure $\mu_{xt}$, associated with the sequence $\varrho_\epsilon$, is the Dirac measure concentrated at some point $\varrho(x,t)$. This means that

$$\overline{\varrho} = \varrho, \quad \overline{p} = p(\varrho), \quad \overline{\varphi} = \varphi(\varrho).$$

The integral identity (7.1.42) along with the equality $\overline{p} = p(\varrho)$ implies that the couple $(\mathbf{u}, \varrho)$ satisfies the momentum balance equation. Since in the deterministic case $\overline{p\varphi} = p(\varrho)\varphi(\varrho) = \overline{p}\,\overline{\varphi}$, it follows from the integral identity (7.2.30) and from

$$\overline{\varphi} = \varphi(\varrho), \quad \overline{\Phi} = \Phi(\varrho) \equiv \varphi'(\varrho)\varrho - \varphi(\varrho)$$

that $\varrho$ is a renormalized solution to the mass balance equations. It remains to note that estimates (7.4.2)–(7.4.3) are consequences of (7.1.9)–(7.1.10). □

# Chapter 8

# Domain convergence

The question of domain stability of solutions to compressible Navier-Stokes equations is important for many applications. It can be formulated as follows. Assume that domains $\Omega_n \subset \mathbb{R}^d$ converge in some sense to a domain $\Omega$. The question is whether a sequence of solutions to a boundary value problem for compressible Navier-Stokes equation in $\Omega_n$ contains a subsequence which converges weakly to a solution of the same boundary value problem for compressible Navier-Stokes equations in the limiting domain $\Omega$. In this auxiliary chapter we investigate in detail the question of convergence of subsets of Euclidean space $\mathbb{R}^d$. We discuss two types of convergence, the Hausdorff convergence of compact sets in $\mathbb{R}^d$ and the Kuratowski-Mosco convergence of subspaces of Banach spaces. For the purposes of shape optimization we introduce the $\mathcal{S}$-convergence of compact sets, which combines the Hausdorff convergence with the Kuratowski-Mosco convergence. We formulate sufficient geometrical conditions which provide the compactness of classes of compact sets in the $\mathcal{S}$ topology.

## 8.1 Hausdorff and Kuratowski-Mosco convergences

**Hausdorff convergence.** If $K$ and $S$ are compact subsets of $\mathbb{R}^d$, then the Hausdorff distance $d_H$ between them is defined as follows:

$$d_H(K, S) = \max\Big\{ \sup_{x \in S} \operatorname{dist}(x, K), \sup_{y \in K} \operatorname{dist}(y, S)\Big\}. \tag{8.1.1}$$

Equipped with this metric, the totality of compact subsets of $\mathbb{R}^d$ becomes a complete metric space. A remarkable property of this space is that it is locally compact. In particular, we have the following lemma:

**Lemma 8.1.1.** *Let $B$ be a bounded domain in $\mathbb{R}^d$. Then any sequence of compact sets $S_n \subset B$ contains a subsequence which converges to some compact set $S \subset \overline{B}$ in the Hausdorff metric.*

The notion of the Hausdorff metric is not as useful as it seems at first glance, since the topological and metric properties of the Hausdorff limit are not related to the properties of the members of the convergent sequence.

**Kuratowski-Mosco convergence.** Kuratowski-Mosco convergence is widely applied in the analysis of domain dependence of solutions to elliptic equations. It deals with the properties of sequences of functions $u_n$ from Sobolev spaces $W^{1,p}(\Omega_n)$ on a family of domains $\Omega_n$. However, this convergence is related to sequences of Banach spaces rather than of domains. For simplicity of presentation we restrict our considerations to Kuratowski-Mosco convergence of subspaces of a "hold-all" Banach space.

**Definition 8.1.2.** Let $Y$ be a Banach space and let $X_n$, $n \geq 1$, $X$, be closed subspaces of $Y$. The sequence $X_n$ *converges to* $X$ *in the sense of Kuratowski-Mosco* if

- for any sequence $u_n \rightharpoonup u$ weakly convergent in $Y$ with $u_n \in X_n$, the limit element $u$ belongs to $X$;
- whenever $u \in X$, there is a sequence $u_n \in X_n$ with $u_n \to u$ strongly in $Y$.

With applications to nonstationary problems in mind, we investigate the properties of Kuratowski-Mosco convergence for spaces of functions depending on the time variable. The following proposition is used in the proof of the domain stability of solutions to initial-boundary value problems for compressible Navier-Stokes equations.

**Proposition 8.1.3.** *Let $Y$ be a separable reflexive Banach space, and suppose that a sequence of closed subspaces $X_n \subset Y$ converges to a closed subspace $X \subset Y$ in the sense of Kuratowski-Mosco. Furthermore, assume that a sequence $u_n \in L^r(0,T;X_n)$, $n \geq 1$, $1 < r < \infty$, converges weakly in $L^r(0,T;Y)$ to some $u \in L^r(0,T;Y)$. Then $u \in L^r(0,T;X)$.*

*Proof.* First we prove the result under the assumption that the mappings $t \mapsto u_n(t)$ are uniformly bounded, i.e.,

$$\|u_n(t)\|_Y \leq M < \infty \quad \text{for all } t \in (0,T),\ n \geq 1. \tag{8.1.2}$$

For any interval $I \subset (0,T)$, denote by $\mathcal{L}_I$ the set of all weak limit points in $Y$ of sequences $u_n(t)$, $t \in I$. A point $u \in Y$ belongs to $\mathcal{L}_I$ if and only if there exists $t \in I$ and a subsequence $u_{n_m}(t)$ such that $u_{n_m}(t) \rightharpoonup y$ weakly in $Y$ as $m \to \infty$.

Denote by $\mathcal{C}_I$ the closure in $Y$ of the convex hull of $\mathcal{L}_I$. Since $u_n(t) \in X_n$, Definition 8.1.2 implies that $\mathcal{L}_I \subset X$. Hence $\mathcal{C}_I$ is a closed convex subset of $X$.

Let us show that there exists a set $\mathcal{E}$ of full measure in $(0,T)$ such that for every $t_0 \in \mathcal{E}$,

$$\lim_{n\to\infty} n \int_{I_n} \|u(t) - u(t_0)\|_Y \, dt = 0, \quad \text{where} \quad I_n = (t_0 - 1/2n, t_0 + 1/2n). \tag{8.1.3}$$

By the Lusin Theorem 1.3.13, for any $\epsilon > 0$, there is a closed set $\mathcal{T}_\epsilon \subset (0,T)$ such that the mapping $u : \mathcal{T}_\epsilon \to Y$ is continuous and $\operatorname{meas}((0,T) \setminus \mathcal{T}_\epsilon) < \epsilon$. Denote by $\mathcal{E}_\epsilon$ the set of all Lebesgue points of $\mathcal{T}_\epsilon$. Recall that $\mathcal{E}_\epsilon$ is of full measure in $\mathcal{T}_\epsilon$. By the definition of the Lebesgue point, for any $t_0 \in \mathcal{E}_\epsilon$,

$$\lim_{n\to\infty} n \operatorname{meas}(I_n \setminus \mathcal{E}_\epsilon) = 0.$$

Notice that by (8.1.2) the mapping $u : t \mapsto u(t)$ is bounded. Thus we get

$$\begin{aligned} n \int_{I_n} \|u(t) - u(t_0)\|_Y \, dt \\ &\leq n \int_{I_n \cap \mathcal{E}_\epsilon} \|u(t) - u(t_0)\|_Y \, dt + n \int_{I_n \setminus \mathcal{E}_\epsilon} \|u(t) - u(t_0)\|_Y \, dt \\ &\leq \sup_{I_n \cap \mathcal{E}_\epsilon} \|u(t) - u(t_0)\|_Y + 2Mn \operatorname{meas}(I_n \setminus \mathcal{E}_\epsilon) \to 0 \quad \text{as } n \to \infty. \end{aligned}$$

Therefore, (8.1.3) holds for all $t_0 \in \mathcal{E}_\epsilon$ and hence for all $t_0 \in \mathcal{E} = \bigcup_\epsilon \mathcal{E}_\epsilon$. Since $\operatorname{meas}((0,T) \setminus \mathcal{E}_\epsilon) < \epsilon$, the set $\mathcal{E}$ is of full measure in $(0,T)$. This proves (8.1.3). Next we show that for all $t_0 \in \mathcal{E}$,

$$u(t_0) \in \bigcap_n \mathcal{C}_{I_n}, \quad \text{where} \quad I_n = (t_0 - 1/2n, t_0 + 1/2n). \tag{8.1.4}$$

Assume that, contrary to our claim, $u(t_0) \notin \bigcap_n \mathcal{C}_{I_n}$. Since $\mathcal{C}_{I_n}$ decreases, there is an integer $N$ such that $u(t_0) \notin \mathcal{C}_{I_n}$ for all $n \geq N$. Since $\mathcal{C}_{I_n}$ is a closed convex subset of the Banach space $Y$, it follows from the Hahn-Banach theorem that there exist a functional $f \in Y'$ and a constant $e$ such that

$$f(u(t_0)) > e > \sup_{u \in \mathcal{C}_{I_N}} f(u). \tag{8.1.5}$$

In order to obtain a contradiction we will prove that

$$\limsup_{k\to\infty} f(u_k(t)) \leq e \quad \text{for any } t \in I_N. \tag{8.1.6}$$

To this end fix $t \in I_N$ and a subsequence $u_{k_i}(t)$ with

$$\lim_{i\to\infty} f(u_{k_i}(t)) = \limsup_{k\to\infty} f(u_k(t)). \tag{8.1.7}$$

By (8.1.2), the sequence $u_{k_i}(t)$ is bounded in $Y$. Passing to a subsequence we may assume that $u_{k_i}(t)$ converges weakly to some $v \in Y$. Thus $v \in \mathcal{L}_{I_N} \subset \mathcal{C}_{I_N}$. It now follows from (8.1.5) that

$$\lim_{i\to\infty} f(u_{k_i}(t)) = f(v) \leq e,$$

which proves (8.1.6). Next, since $u_k(t)$ converges to $u(t)$ weakly in $L^r(0,T;Y)$, it follows from (8.1.6) that for any $n \geq N$,

$$n\int_{I_n} f(u(t))\,dt = \lim_{k\to\infty} n\int_{I_n} f(u_k(t))\,dt \leq n\int_{I_n} \limsup_{k\to\infty} f(u_k(t))\,dt \leq e. \tag{8.1.8}$$

On the other hand, (8.1.3) and (8.1.5) imply

$$\begin{aligned}\liminf_{n\to\infty} n\int_{I_n} f(u(t))\,dt &= \lim_{n\to\infty} n\int_{I_n} f(u(t_0))\,dt + \liminf_{n\to\infty} n\int_{I_n} f(u(t)-u(t_0))\,dt\\ &= f(u(t_0)) + \liminf_{n\to\infty} n\int_{I_n} f(u(t)-u(t_0))\,dt\\ &\geq f(u(t_0)) - \|f\|_{Y'}\lim_{n\to\infty}\int_{I_n}\|u(t)-u(t_0)\|_Y\,dt\\ &= f(x(t_0)) > e,\end{aligned} \tag{8.1.9}$$

which contradicts (8.1.8). This proves the inclusion (8.1.4). Since $\mathcal{E}$ is a set of full measure in $(0,T)$, it follows from (8.1.4) that

$$u(t_0) \in \bigcap_n \mathcal{C}_{I_n} \subset X \quad \text{for a.e. } t_0 \in (0,T), \tag{8.1.10}$$

which implies the inclusion $u \in L^r(0,T;X)$.

Hence the assertion of Proposition 8.1.3 holds for all bounded sequences $u_n(t)$ satisfying (8.1.2). Let us consider the general case. Choose an arbitrary sequence $u_n(t)$ with the properties

$$u_n \rightharpoonup u \quad \text{weakly in } L^r(0,T;Y), \quad u_n(t) \in X_n \quad \text{for a.e. } t \in (0,T)$$

and set

$$u_n^M(t) = u_n(t) \quad \text{if } \|u_n(t)\|_Y \leq M \quad \text{and} \quad u_n^M(t) = 0 \quad \text{otherwise,}$$

where $M$ is an arbitrary fixed positive number. It is clear that the elements $u_n^M$ satisfy (8.1.2). By Corollary 1.3.25 the space $L^r(0,T;Y)$ is reflexive. Hence passing to a subsequence we can assume that $u_n^M$, $n \geq 1$, converges weakly in $L^r(0,T;Y)$ to some $u^M \in L^r(0,T;Y)$. Moreover, as proved above, $u^M \in L^r(0,T;X)$. Since $r > 1$, the sequences $u_n$ and $u_n^M$ converge weakly in $L^1(0,T;Y)$ to $u$ and $u^M$, respectively. Thus we get

$$\begin{aligned}\int_0^T \|u^M(t)-u(t)\|_Y\,dt &\leq \liminf_{n\to\infty}\int_0^T \|u_n^M(t)-u_n(t)\|_Y\,dt\\ &= \liminf_{n\to\infty}\int_{\{t:\|u_n(t)\|_Y\geq M\}} \|u_n(t)\|_Y\,dt\\ &\leq \liminf_{n\to\infty}\Big(\int_{\{t:\|u_n(t)\|_Y\geq M\}} dt\Big)^{1/r'}\Big(\int_{\{t:\|u_n(t)\|_Y\geq M\}} \|u_n(t)\|_Y^r\,dt\Big)^{1/r}\\ &\leq c\liminf_{n\to\infty}\big(\operatorname{meas}\{t : \|u_n(t)\|_Y \geq M\}\big)^{1/r'}.\end{aligned}$$

On the other hand, we have

$$\begin{aligned}\operatorname{meas}\{t : \|u_n(t)\|_Y \geq M\} &\leq \frac{1}{M}\int_{\{t:\|u_n(t)\|_Y\geq M\}} \|u_n(t)\|_Y\,dt \\ &\leq \frac{1}{M}\int_0^T \|u_n\|_Y\,dt \leq \frac{c}{M}.\end{aligned}$$

Combining the results obtained we arrive at

$$\int_0^T \|u^M(t) - u(t)\|_Y\,dt \leq cM^{-1/r'}.$$

Therefore, $u^M$ converges to $u$ in $L^1(0,T;Y)$ as $M \to \infty$. Passing to a subsequence we conclude that $\|u^M(t) - u(t)\|_Y \to 0$ as $M \to \infty$ for a.e. $t$. Since $u^M(t) \in X$, the values of the limit function $u(t)$ also belong to $X$ for a.e. $t \in (0,T)$, which proves Proposition 8.1.3. □

## 8.2 Capacity, quasicontinuity and fine topology

In this section we collect basic facts on fine properties of sets and Sobolev functions. The main relevant sources are [1], [28], [76]. Our presentation is mainly based on the monograph [1].

**Capacity.** Sobolev functions when compared with Lebesgue integrable functions enjoy many additional useful properties. In particular, Sobolev functions are well defined on some sets of zero Lebesgue measure, for instance, on smooth manifolds of codimension one. The properties that characterize pointwise behavior of Sobolev functions are called fine properties. Capacity is a very useful technical tool in the fine theory of Sobolev functions. It can be regarded as a nonnegative function defined on subsets of Euclidean space $\mathbb{R}^d$, and hence capacity theory has many common features with measure theory. But there is one important difference: capacity is not an additive function of sets but only subadditive. The explicit definition of capacity (Bessel capacity) is the following (cf. Definitions 2.2.1, 2.2.2, 2.2.4 in [1]):

**Definition 8.2.1.** Let $K \subset \mathbb{R}^d$ be compact. Then

$$\operatorname{Cap}(K) = \inf\{\|\varphi\|^2_{W^{1,2}(\mathbb{R}^d)} : \varphi \in C_0^\infty(\mathbb{R}^d),\ \varphi \geq 1 \text{ on } K\}.$$

**Definition 8.2.2.** Let $G \subset \mathbb{R}^d$ be open. Then

$$\operatorname{Cap}(G) = \sup\{\operatorname{Cap}(K) : K \subset G,\ K \text{ compact}\}.$$

**Definition 8.2.3.** Let $A \subset \mathbb{R}^d$ be arbitrary. Then

$$\operatorname{Cap}(A) = \inf\{\operatorname{Cap}(G) : G \supset A,\ G \text{ open}\}.$$

An equivalent definition can be given in terms of the Bessel kernel $G_{\mathcal{B}}$ which is the inverse Fourier transform of the function $\widehat{G_{\mathcal{B}}}(\xi) = (1 + |\xi|^2)^{-1/2}$, $\xi \in \mathbb{R}^d$. It is well known (see [133]) that a function $\varphi$ is in $W^{1,2}(\mathbb{R}^d)$ if and only if it is a convolution $\varphi = G_{\mathcal{B}} * g$, $g \in L^2(\mathbb{R}^d)$, and

$$c^{-1}\|g\|_{L^2(\mathbb{R}^d)} \le \|\varphi\|_{W^{1,2}(\mathbb{R}^d)} \le c\|g\|_{L^2(\mathbb{R}^d)},$$

where $c$ depends only on $d$. This leads to

$$\mathrm{Cap}(A) = \inf\{\|g\|_{L^2(\mathbb{R}^d)} : g \ge 0,\, G_{\mathcal{B}} * g \ge 1 \text{ on } A\}.$$

Capacity enjoys many properties similar to properties of measures. It is nonnegative, monotone, and subadditive, i.e.

$$A \subset B \Rightarrow 0 \le \mathrm{Cap}(A) \le \mathrm{Cap}(B),$$
$$\mathrm{Cap}(A \cup B) \le \mathrm{Cap}(A) + \mathrm{Cap}(B),$$

For any compact set $K \subset \mathbb{R}^d$,

$$\mathrm{Cap}(K) = \inf\{\mathrm{Cap}(G) : G \supset K,\, G \text{ is open}\}.$$

Capacity also enjoys nice metric properties. In particular, it is invariant with respect to translations and rotations and more generally, with respect to all affine isometries. The following lemma given in [1] shows that capacity has some invariance properties with respect to bi-Lipschitz transformations of the ambient Euclidean space.

**Lemma 8.2.4.** *Let $\chi : \mathbb{R}^d \to \mathbb{R}^d$ be a bijection that is bi-Lipschitz:*

$$k^{-1}|x' - x''| \le |\chi(x') - \chi(x'')| \le k|x' - x''| \quad \text{for all } x', x'' \in \mathbb{R}^d. \tag{8.2.1}$$

*Then there is a constant $c > 0$, depending only on $k$ and $d$, such that for any compact set $K \subset \mathbb{R}^d$,*

$$c^{-1}\,\mathrm{Cap}(K) \le \mathrm{Cap}(\chi(K)) \le c\,\mathrm{Cap}(K).$$

In many important cases, capacity admits robust estimates. In particular, in the three dimensional case we have

$$c^{-1}a \le \mathrm{Cap}(B_a) \le ca \tag{8.2.2a}$$

for the ball $B_a = \{|x| \le a\} \subset \mathbb{R}^3$, and

$$c^{-1}a \le \mathrm{Cap}(D_a) \le ca \tag{8.2.2b}$$

for the disk $D_a = \{x = (x_1, x_2, 0) : |x| \le a\} \subset \mathbb{R}^3$, where $c$ is an absolute constant.

In the two-dimensional case, calculating the capacity is more complicated because of the peculiarities of the logarithmic potential. For $a \in (0,1)$, we have (see [76])

$$c^{-1}/\log(2/a)^{-1} \le \mathrm{Cap}(B_a) \le c/\log(2/a)^{-1} \tag{8.2.2c}$$

for $B_a = \{|x| \le a\} \subset \mathbb{R}^2$, and

$$c^{-1}/\log(2/a)^{-1} \le \mathrm{Cap}(I_a) \le c/\log(2/a)^{-1} \tag{8.2.2d}$$

for $I_a = \{x = (x_1, 0) : |x_1| \le a\} \subset \mathbb{R}^2$. Here, the constant $c$ is independent of $a$.

If $K$ is a locally Lipschitz manifold in $\mathbb{R}^d$ then

$$\begin{aligned} \mathrm{Cap}(K) &= 0 \quad \text{if } \dim K \le d-2, \\ \mathrm{Cap}(K) &> 0 \quad \text{if } \dim K \ge d-1. \end{aligned}$$

**Thickness.** The notion of thickness of a set at some point arises in potential theory in connection with the continuity properties of harmonic functions. Recall the classical result on solvability of the Dirichlet problem for the Laplace equation (see [76]).

**Theorem 8.2.5.** *Let $\Omega \subset \mathbb{R}^d$ be a bounded domain and $f : \mathbb{R}^d \to \mathbb{R}$ be a continuous function. Assume that for every $x \in A := \mathbb{R}^d \setminus \Omega$,*

$$\sum_{n=0}^{\infty} 2^{n(d-2)}\,\mathrm{Cap}(B_{2^{-n}}(x) \cap A) = \infty, \tag{8.2.3}$$

*where $B_a(x)$ is the ball of radius $a$ centered at $x$. Then the Dirichlet problem*

$$\Delta v = 0 \quad \textit{in } \Omega, \qquad v = f \quad \textit{on } \partial\Omega, \tag{8.2.4}$$

*has a unique solution $v \in C(\overline{\Omega})$ such that*

$$v(y) \to f(x) \quad \textit{as } \Omega \ni y \to x \quad \textit{for all } x \in \partial\Omega. \tag{8.2.5}$$

Condition (8.2.3) is known as the Wiener regularity condition and it ensures that the closed set $\mathbb{R}^d \setminus \Omega$ is in some sense "thick" at the point $x$. Thus we come to the following definition of thickness ([1, Def. 6.3.7]):

**Definition 8.2.6.** We say that a set $E \subset \mathbb{R}^d$ is *thin at a point* $x$ if

$$\int_0^1 \mathrm{Cap}(B_s(x) \cap E)\frac{ds}{s^{d-1}} < \infty, \tag{8.2.6}$$

or equivalently,

$$\sum_{n=0}^{\infty} 2^{n(d-2)}\mathrm{Cap}(B_{2^{-n}}(x) \cap E) < \infty, \tag{8.2.7}$$

where $B_s(x) = \{y \in \mathbb{R}^d : |y - x| \le s\}$. If a set is not thin at $x$ it is *thick at* $x$. A set $E$ is *thick* if it is thick at every point $x \in E$.

Recall some simple properties of thick and thin sets.

**Lemma 8.2.7.** *If $F \subset \mathbb{R}^d$ is thick at $x$ and $E \subset \mathbb{R}^d$ is thin at $x$ then $F \setminus E$ is thick at $x$. Moreover, $x$ is a limit point of $F$.*

*Proof.* As capacity is subadditive, we have $\operatorname{Cap}(B_s(x) \cap (F \setminus E)) \geq \operatorname{Cap}(B_s(x) \cap F) - \operatorname{Cap}(B_s(x) \cap E)$ and

$$\int_0^1 \operatorname{Cap}(B_s(x) \cap (F \setminus E)) \frac{ds}{s^{d-1}} \geq \lim_{\epsilon \to 0} \int_\epsilon^1 \operatorname{Cap}(B_s(x) \cap F) \frac{ds}{s^{d-1}} - \lim_{\epsilon \to 0} \int_\epsilon^1 \operatorname{Cap}(B_s(x) \cap E) \frac{ds}{s^{d-1}}.$$

It remains to note that the first limit on the right hand side is infinite while the second is finite. To prove that $x$ is a limit point of $F$ it suffices to note that in the opposite case, $\operatorname{Cap}(B_s(x) \cap F) = 0$ for all sufficiently small $s$, and the integral

$$\int_0^1 \operatorname{Cap}(B_s(x) \cap F) \frac{ds}{s^{d-1}}$$

converges at 0. This contradicts the thickness of $F$ at $x$. □

**Lemma 8.2.8.** *Suppose a bi-Lipschitz bijection $\mathcal{X} : \mathbb{R}^d \to \mathbb{R}^d$ satisfies condition (8.2.1) and a compact set $\mathcal{T}$ is thick at 0. Then for any $x \in \mathbb{R}^d$, the set $x + \mathcal{X}(\mathcal{T})$ is thick at $x$.*

*Proof.* If $y \in B(0, s/k)$ then, by (8.2.1), $\mathcal{X}(y) \in B(0, s)$. This leads to the inclusion $B(0, s) \cap \mathcal{X}(\mathcal{T}) \supset \mathcal{X}(B(0, s/k) \cap \mathcal{T})$. Since capacity is invariant with respect to translations of $\mathbb{R}^d$, we obtain

$$\operatorname{Cap}\big(B(x, s) \cap (x + \mathcal{X}(\mathcal{T}))\big) = \operatorname{Cap}\big(B(0, s) \cap \mathcal{X}(\mathcal{T})\big) \geq \operatorname{Cap}\big(\mathcal{X}(B(0, s/k) \cap \mathcal{T})\big).$$

On the other hand, Lemma 8.2.4 with $K$ replaced by $\mathcal{X}(B(0, s/k) \cap \mathcal{T})$ implies $\operatorname{Cap}\big(\mathcal{X}(B(0, s/k) \cap \mathcal{T})\big) \geq c(k) \operatorname{Cap}(B(0, s/k) \cap \mathcal{T})$. Thus we get

$$\operatorname{Cap}\big(B(x, s) \cap (x + \mathcal{X}(\mathcal{T}))\big) \geq c(k) \operatorname{Cap}(B(0, s/k) \cap \mathcal{T}).$$

From this we conclude that

$$\int_0^1 \operatorname{Cap}\big(B(x, s) \cap (x + \mathcal{X}(\mathcal{T}))\big) \frac{ds}{s^{d-1}} \geq c(k) \int_0^1 \operatorname{Cap}(B(0, s/k) \cap \mathcal{T}) \frac{ds}{s^{d-1}}$$
$$= c(k) \int_0^{1/k} \operatorname{Cap}(B(0, s) \cap \mathcal{T}) \frac{ds}{s^{d-1}} = \infty. \quad □$$

The following examples illustrate the properties of thick sets; the examples are also used in the next sections.

**Example 8.2.9** (Flat spearhead). For $\omega \in (0, \pi/2]$ consider the plane sector

$$\triangle(\omega) = \{x \in \mathbb{R}^3 : x = (r\cos\alpha, r\sin\alpha, 0),\ |\alpha| \le \omega,\ r \in (0,1)\} \subset \mathbb{R}^3. \qquad (8.2.8)$$

**Lemma 8.2.10.** *The set* $\triangle(\omega)$ *is thick at* 0.

*Proof.* Choose an integer $N \ge [\pi/\omega]+1$. Note that the disk $D_s = \{x \in \mathbb{R}^3 : x_3 = 0, x_1^2 + x_2^2 \le s^2\}$ can be covered by the $N$ sectors obtained from $B(0,s) \cap \triangle(\omega)$ by rotation through $i\pi/\omega$, $1 \le i \le N$, about the $x_3$ axis. Since capacity is invariant under rotations, it follows that $N\,\mathrm{Cap}(B(0,s) \cap \triangle(\omega)) \ge \mathrm{Cap}\, D_s$. Combining this with (8.2.2b) we arrive at $\mathrm{Cap}(B(0,s) \cap \triangle(\omega)) \ge cs/N$, which yields

$$\int_0^1 \mathrm{Cap}(B(x,s) \cap \triangle(\omega))\, \frac{ds}{s^2} \ge \frac{c}{N} \int_0^1 \frac{ds}{s} = \infty. \qquad \square$$

**Example 8.2.11** (Interval). Let us consider the one-dimensional interval

$$I = \{x \in \mathbb{R}^2 : x = (r,0),\ r \in [0,1]\} \subset \mathbb{R}^2. \qquad (8.2.9)$$

**Lemma 8.2.12.** *The set* $I$ *is thick at* 0.

*Proof.* Notice that $B(0,s) \cap I = [0,s]$ and by (8.2.2d) we have $\mathrm{Cap}(B(0,s) \cap I) \ge c|\log s|^{-1}$. Hence

$$\int_0^1 \mathrm{Cap}(B(x,s) \cap I)\, \frac{ds}{s^1} \ge c \int_0^1 \frac{ds}{s \log|s|} = \infty. \qquad \square$$

**Fine topology, fine continuity.** For given $x \in \mathbb{R}^d$, we can regard the sets which are thin at $x$ as "small". This leads to the concepts of fine topology and fine continuity, based on the following definition (see [1, Defs. 6.4.1–6.4.2]):

**Definition 8.2.13.** Let $x \in \mathbb{R}^d$. Then a set $U$ is called a *fine neighborhood* of $x$ if $x \in U$ and $\mathbb{R}^d \setminus U$ is thin at $x$. A set $G \subset \mathbb{R}^d$ is *finely open* if it is a fine neighborhood of each of its points. A set $F \subset \mathbb{R}^d$ is *finely closed* if $\mathbb{R}^d \setminus F$ is finely open.

Finely open sets define the fine topology in $\mathbb{R}^d$, which leads to the notion of a finely continuous function.

**Definition 8.2.14.** A real-valued function $f$ defined on a set $F \subset \mathbb{R}^d$ is *finely continuous* at $x \in F$ if the set $\{y \in F : |f(y) - f(x)| \ge \epsilon\}$ is thin at $x$ for all $\epsilon > 0$.

As a consequence, we obtain the following proposition (see [1, Prop. 6.4.3]):

**Proposition 8.2.15.** *Let $f$ be a real-valued function defined on a set $F \subset \mathbb{R}^d$ and finely continuous at a point $x \in F$ such that $F$ is thick at $x$. Then there is a set $E \subset F$ such that $E$ is thin at $x$ and*

$$\lim_{F \setminus E \ni y \to x} f(y) = f(x).$$

**Quasicontinuity.** One of the most important fine properties is quasicontinuity, which is defined as follows ([1, Def. 6.1.1]):

**Definition 8.2.16.** Let $B \subset \mathbb{R}^d$ be a bounded set. A function $f : B \to \mathbb{R}$ is said to be *quasicontinuous* on $B$ if for any $\epsilon > 0$ there exists an open set $G$ with $\mathrm{Cap}(G) < \epsilon$ such that $f$ is continuous on $B \setminus G$. A function $f : \mathbb{R}^d \to \mathbb{R}$ is quasicontinuous if it is quasicontinuous on every bounded subset of $\mathbb{R}^d$.

**Remark 8.2.17.** Further if a statement is true for all $x$ except for a set $E$ with $\mathrm{Cap}(E) = 0$, then we say that it is true *quasieverywhere*, written q.e.

The following result (see [1, Thm. 6.4.5, Thm. 6.4.6]) establishes the relation between quasicontinuity and fine continuity.

**Theorem 8.2.18.** *A function defined on $\mathbb{R}^d$ is quasicontinuous if and only if it is finely continuous q.e. in $\mathbb{R}^d$.*

**Quasicontinuity of elements of Sobolev spaces.** Every function $f \in W^{1,2}(\mathbb{R}^d)$ is, in fact, an equivalence class of measurable functions which coincide off a set of zero measure. Each element of such a class is called a *representative* of $f$. The following statement ([1, Prop. 6.1.2]) states that every Sobolev function becomes quasicontinuous after possible changes of its values on a set of zero measure.

**Proposition 8.2.19.** *Every $f \in W^{1,2}(\mathbb{R}^d)$ has a quasicontinuous representative.*

It follows from Theorem 8.2.18 that the quasicontinuous representative is finely continuous q.e. in $\mathbb{R}^d$. The next statement ([1, Thm. 6.1.4]) ensures that the quasicontinuous representative is uniquely defined.

**Proposition 8.2.20.** *Let $f_1$ and $f_2$ be quasicontinuous functions, and suppose that $f_1 = f_2$ almost everywhere. Then $f_1 = f_2$ quasieverywhere.*

This leads to a new notion of the trace of a Sobolev function on an arbitrary subset of $\mathbb{R}^d$ (see [1, Def. 6.1.6]).

**Definition 8.2.21.** Let $f \in W^{1,2}(\mathbb{R}^d)$ and let $E \subset \mathbb{R}^d$ be an arbitrary set. Then the *trace* of $f$ on $E$, denoted $f|_E$, is the restriction to $E$ of any quasicontinuous representative of $f$. Thus the trace is uniquely defined q.e. on $E$.

The following approximation theorem due to Hedberg (see [1, Thm. 9.1.3], [53]) plays an important role in our theory.

**Theorem 8.2.22.** *Let $f \in W^{1,2}(\mathbb{R}^d)$. Let $K \subset \mathbb{R}^d$ be a closed set. Then the following statements are equivalent:*

*(a)* $f|_K = 0$;

*(b)* $f \in W_0^{1,2}(\mathbb{R}^d \setminus K)$;

*(c) for any $\epsilon > 0$ and any compact $F \subset \mathbb{R}^d \setminus K$, there is an $\eta \in C_0^\infty(\mathbb{R}^d \setminus K)$ such that $\eta = 1$ on $F$, $0 \le \eta \le 1$ and $\|f - \eta f\|_{W^{1,2}(\mathbb{R}^d)} < \epsilon$.*

## 8.3 Applications to flow around an obstacle

In this section we apply the general results to convergence of condenser type domains. This question is important for our purposes since such a geometry is typical for flow modeling around an obstacle. Let $B \subset \mathbb{R}^d$ be a bounded hold-all domain with $C^3$ boundary. Assume that the flow occupies a condenser type domain $\Omega = B \setminus S$, with a compact obstacle $S \Subset B$ inside of $\Omega$.

**Definition 8.3.1.** Let $S$ be a compact subset of $B$. Denote by $C^\infty_S(B)$ the linear space of all $C^\infty(B)$ functions $u$ that vanish in some neighborhood of $S$ (depending on $u$). Denote by $W^{1,r}_S(B)$ the completion of $C^\infty_S(B)$ in the norm of $W^{1,r}(B)$. In particular, $W^{1,r}_S(B)$ is a closed subspace of $W^{1,r}(B)$.

The Banach spaces $W^{1,2}_S(B)$, $S \Subset B$, form a collection of closed subspaces of the Banach space $W^{1,2}(B)$. Thus applying Definition 8.1.2 of Kuratowski-Mosco convergence to this particular case we arrive at the following definition:

**Definition 8.3.2.** A sequence of compact sets $S_n \Subset B$, $n \geq 1$, *converges in the sense of Kuratowski-Mosco* to a compact set $S \Subset B$ if $W^{1,2}_{S_n}(B)$ converges in the sense of Kuratowski-Mosco to $W^{1,2}_S(B)$, i.e.,

- for every sequence $u_n \rightharpoonup u$ weakly convergent in $W^{1,2}(B)$ with $u_n \in W^{1,2}_{S_n}(B)$, the limit function $u$ is in $W^{1,2}_S(B)$;
- conversely, whenever $u \in W^{1,2}_S(B)$, there exists a sequence $u_n \in W^{1,2}_{S_n}(B)$ such that $u_n \to u$ strongly in $W^{1,2}(B)$.

Now Proposition 8.1.3 implies the following corollary:

**Corollary 8.3.3.** *Let compact sets $S_n$, $n \geq 1$, converge to a compact set $S$ in the sense of Kuratowski-Mosco and let $u_n \in L^2(0,T; W^{1,2}_{S_n}(B))$ converge to $u$ weakly in $L^2(0,T; W^{1,2}(B))$. Then $u \in L^2(0,T; W^{1,2}_S(B))$.*

Theorem 8.2.22 implies the following property of elements of $W^{1,2}_S(B)$:

**Corollary 8.3.4.** *Let $f \in W^{1,2}(B)$. Let $S \Subset B$ be a compact set. Then the following statements are equivalent:*

*(a) the quasicontinuous representative of $f$ is equal to $0$ q.e. on $S$;*

*(b) $f \in W^{1,2}_S(B)$.*

*Proof.* It suffices to note that $f$ has an extension $\overline{F} \in W^{1,2}(\mathbb{R}^d)$ and apply Theorem 8.2.22. □

**Combined $\mathcal{S}$-convergence.** In order to use the advantages of both Hausdorff and Kuratowski-Mosco convergences we introduce the notion of combined convergence of compact subsets of the hold-all domain $B$.

**Definition 8.3.5.** We say that compact sets $S_n \Subset B$, $n \geq 1$, *$\mathcal{S}$-converge* to a compact set $S \Subset B$ if

- there is a compact set $B' \Subset B$ such that $S_n, S \Subset B'$;
- the sets $S_n$ converge to $S$ in the Hausdorff metric;
- the spaces $W^{1,2}_{S_n}(B)$ converge to $W^{1,2}_S(B)$ in the sense of Kuratowski-Mosco.

In this case we write $S_n \xrightarrow{\mathcal{S}} S$ and we say that the compact set $S$ is the *$\mathcal{S}$-limit* of the sequence $S_n$.

The following proposition shows that every compact obstacle $S$ is the $\mathcal{S}$-limit of a sequence of smooth obstacles $S_n \supset S$. In Chapter 9 this result is applied to existence theory for compressible Navier-Stokes equations in nonsmooth domains.

**Proposition 8.3.6.** *Let $B \subset \mathbb{R}^d$ be a bounded domain with $C^1$ boundary and $S \Subset B$ be a compact set. Then there is a sequence of compact sets $S_n \supset S$ with $C^\infty$ boundaries such that $S_n \xrightarrow{\mathcal{S}} S$.*

The proof of Proposition 8.3.6 is based on two simple but useful lemmas.

**Lemma 8.3.7.** *Let $B \subset \mathbb{R}^d$ be a bounded domain with $C^1$ boundary and $S \Subset B$ be a compact set. Let $u \in W^{1,2}(B)$ have the property that there is a sequence $u_n \in W^{1,2}(B)$ such that $u_n \to u$ strongly in $W^{1,2}(B)$ and each $u_n$ vanishes in some neighborhood of $S$. Then there exists a sequence $\varphi_n \in C^\infty(B)$ such that each $\varphi_n$ vanishes in a neighborhood of $S$ and $\varphi_n \to u$ strongly in $W^{1,2}(B)$, therefore, $u \in W^{1,2}_S(B)$.*

*Proof.* Since $B$ is bounded there exists a ball $B_N = \{|x| \leq N\}$ such that $B \Subset B_N$. Since $\partial B$ is a $C^1$ closed surface, there is an extension $\overline{u}_n \in W^{1,2}(\mathbb{R}^d)$ of $u_n$ to $\mathbb{R}^d$ such that

$$\|\overline{u}_n\|_{W^{1,2}(\mathbb{R}^d)} \leq c\|u_n\|_{W^{1,2}(B)} \quad \text{and} \quad \overline{u}_n = 0 \quad \text{in } \mathbb{R}^d \setminus B_N.$$

Here the constant $c$ depends only on $B$. Next choose a sequence of integers $m_n$ with the properties

$$1/m_n < 2^{-d-1}\operatorname{dist}(S, \operatorname{supp} u_n) \quad \text{and} \quad \|[\overline{u}_n]_{m_n}, - \overline{u}_n\|_{W^{1,2}(\mathbb{R}^d)} \leq 1/n.$$

Here $[\cdot]_{m_n}$, is a mollifier defined by (1.6.3),

$$[\overline{u}_n]_{m_n},(x) = m_n^d \int_{\mathbb{R}^d} \Theta(m_n(x-y))\overline{u}_n(y)\, dy.$$

Since $\Theta(z)$ is supported in the cube $[-1,1]^d$, it follows from the choice of $m_n$ that the functions $\varphi_n := [\overline{u}_n]_{m_n}, \in C_0^\infty(\mathbb{R}^d)$ vanish in a neighborhood of $S$ and

$$\begin{aligned}\|\varphi_n - u\|_{W^{1,2}(B)} &\leq c\|u_n - u\|_{W^{1,2}(B)} + \|\varphi_n - \overline{u}_n\|_{W^{1,2}(\mathbb{R}^d)} \\ &\leq c\|u_n - u\|_{W^{1,2}(B)} + 1/n \to 0 \quad \text{as } n \to \infty.\end{aligned}$$

□

**Lemma 8.3.8.** *Let $S \subset \mathbb{R}^d$ be an arbitrary compact set. Then there is a decreasing sequence of compact sets $S_n \subset \mathbb{R}^d$ with $C^\infty$ boundaries such that*

$$S_n \supset S, \quad d_H(S, S_n) \searrow 0 \quad \text{as } n \to \infty,$$

*where the Hausdorff distance $d_H$ is defined by* (8.1.1).

*Proof.* Choose $\eta \in C^\infty(\mathbb{R})$ satisfying

$$\eta(s) = 0 \quad \text{for } s \le 1, \quad \eta(s) = 1 \quad \text{for } s \ge 2, \quad \eta'(s) \ge 0,$$

and set

$$\varphi_n(x) = \eta\big(2^{3n} \operatorname{dist}(x, S)\big), \quad n \ge 1.$$

It is clear that $0 \le \varphi_n \le 1$ and

$$\varphi_n(x) = 0 \quad \text{for } \operatorname{dist}(x, S) \le 2^{-3n}, \quad \varphi_n(x) = 1 \quad \text{for } \operatorname{dist}(x, S) \ge 2^{-3n+1}.$$

Finally set

$$f_n(x) = 2^{3n+1} \int_{\mathbb{R}^d} \Theta\big(2^{3n+1}(x-y)\big)\varphi_n(y)\, dy$$

with a nonnegative mollifying kernel $\Theta$ such that

$$\Theta \in C_0^\infty(\mathbb{R}^d), \quad \Theta(x) = 0 \quad \text{for } |x| \ge 1, \quad \int_{\mathbb{R}^d} \Theta(x)\, dx = 1.$$

The function $f_n$ belongs to $C^\infty(\mathbb{R}^d)$ and satisfies

$$\begin{aligned} f_n(x) = 0 \quad &\text{for } \operatorname{dist}(x, S) \le 2^{-3n} - 2^{-3n-1} = 2^{-3n-1}, \\ f_n(x) = 1 \quad &\text{for } \operatorname{dist}(x, S) \ge 2^{-3n+1} + 2^{-3n-1} = 5 \cdot 2^{-3n-1}. \end{aligned}$$

By the Sard Theorem [118], the level set $\{f_n = c\}$ does not contain critical points of $f_n$ for almost every $c \in [0,1]$. Choose $c_n \in (0,1)$ with this property and set

$$S_n := \{x : f_n(x) \le c_n\}.$$

Then $S_n$ is contained in the $5 \cdot 2^{-3n-1}$-neighborhood of $S$, and it contains the $2^{-3n-1}$-neighborhood of $S$. It follows that $d_H(S, S_n) \le 5 \cdot 2^{-3n-1} \to 0$ as $n \to \infty$. Since $5 \cdot 2^{-3(n+1)-1} < \cdot 2^{-3n-1}$, we also have $S_{n+1} \subset S_n$. By the choice of $c_n$, $\nabla f_n \ne 0$ on $\partial S_n = \{f_n = c_n\}$, so $\partial S_n$ is a $C^\infty$ surface. □

*Proof of Proposition 8.3.6.* It suffices to show that the sequence of compact sets $S_n$ defined in Lemma 8.3.8 $\mathcal{S}$-converges to $S$. By that lemma, $S_n \to S$ in the Hausdorff metric. Hence it remains to show that $W^{1,2}_{S_n}(B)$ converges to $W^{1,2}_S(B)$ in the sense of Kuratowski-Mosco. To this end we have to prove that

(a) if $W^{1,2}_{S_n}(B) \ni u_n \rightharpoonup u$ weakly in $W^{1,2}(B)$, then $u \in W^{1,2}_S(B)$;

(b) for every $u \in W^{1,2}_S(B)$ there exists a sequence $u_n \in W^{1,2}_{S_n}(B)$ such that $u_n \to u$ in $W^{1,2}(B)$.

To prove (a) note that by the Mazur theorem, the weak convergence implies that there is a sequence of finite convex combinations

$$w_N = \sum_{k=1}^{m(N)} \alpha_k^N u_{n_k}, \quad \alpha_k^N \in [0,1], \quad \sum_{k=1}^{m(N)} \alpha_k^N = 1,$$

such that $w_N \to u$ converges strongly in $W^{1,2}(B)$ as $N \to \infty$. It follows from Lemma 8.3.7 that for every $k$, there exists $\varphi_{n_k} \in C^\infty(B)$ with

$$\|\varphi_{n_k} - u_{n_k}\|_{W^{1,2}(B)} \le 1/N, \quad \varphi_{n_k} = 0 \text{ in a neighborhood of } S_{n_k}.$$

Since $S \subset S_{n_k}$, each $\varphi_{n_k}$ vanishes in a neighborhood of $S$. Now set

$$\phi_N = \sum_{k=1}^{m(N)} \alpha_k^N \varphi_{n_k}.$$

Obviously $\phi_N \in C^\infty(B)$ and vanishes in a neighborhood of $S$. Next,

$$\begin{aligned}\|\phi_N - u\|_{W^{1,2}(B)} &\le \|w_N - u\|_{W^{1,2}(B)} + \sum_{k=1}^{m(N)} \alpha_k^N \|\varphi_{n_k} - u_{n_k}\|_{W^{1,2}(B)}\\ &\le \|w_N - u\|_{W^{1,2}(B)} + 1/N \to 0 \quad \text{as } N \to \infty.\end{aligned}$$

Hence $u \in W^{1,2}_S(B)$. This proves claim (a).

To prove (b), fix $u \in W^{1,2}_S(B)$. Then there exists a sequence $\varphi_n \in C^\infty(B)$ such that

$$\|\varphi_n - u\|_{W^{1,2}(B)} \le 1/n, \quad \varphi_n = 0 \quad \text{in a neighborhood of } S.$$

Since the distance between $S_m$ and $S$ tends to 0 as $m \to \infty$, for every $n$ there is $m_n$ such that $\varphi_n = 0$ in a neighborhood of $S_{m_n}$. Without loss of generality we can assume that $m_n \to \infty$ as $n \to \infty$. It is clear that $\varphi_n \in W^{1,2}_{S_{m_n}}(B)$. Moreover, since the sequence $S_n$ decreases, $\varphi_n \in W^{1,2}_{S_j}(B)$ for al $j \ge m_n$. Now set $u_j = \varphi_n$ for $m_n \le j \le m_{n+1}$. We have $u_j \in W^{1,2}_{S_j}(B)$ and

$$\|u_j - u\|_{W^{1,2}(B)} \le 1/n \to 0 \quad \text{as } j \to \infty,$$

which proves claim (b) and Proposition 8.3.6. □

The last proposition in this section is an extension of the approximation Theorem 8.2.22 to functions depending on time. This result plays a crucial role in the proof of the domain stability of solutions to the initial-boundary value problem for compressible Navier-Stokes equations.

**Proposition 8.3.9.** *Let $B \subset \mathbb{R}^d$ be a bounded domain with $C^1$ boundary and $S \Subset B$ be a compact set. Furthermore assume that $\mathbf{u} \in L^2(0,T; W^{1,2}_S(B))$. Then there is a sequence of functions $\psi_n \in C^\infty(B)$ with the following properties:*

*Each $\psi_n$ vanishes in a neighborhood of $S$, satisfies $0 \le \psi_n \le 1$, and*

$$\int_0^T \int_{B\setminus S} |\mathbf{u}(x,t)|\, |\nabla \psi_n(x)|\, dx dt \to 0 \quad \text{as } n \to \infty, \tag{8.3.1}$$

$$\psi_n(x) \to 1 \quad \text{as } n \to \infty \quad \text{for all } x \in B \setminus S. \tag{8.3.2}$$

We split the proof into a sequence of lemmas.

**Lemma 8.3.10.** *Let $\mathbf{u}_n \to \mathbf{u}$ strongly in $W^{1,2}(B)$. Then*

$$\big\| |\mathbf{u}| - |\mathbf{u}_n| \big\|^2_{W^{1,2}(B)} \le \|\mathbf{u} - \mathbf{u}_n\|^2_{W^{1,2}(B)} + \int_B F_n^2(x)\, dx \tag{8.3.3}$$

*for some $F_n \in L^2(B)$, where $F_n \to 0$ a.e. in $B$ and $|F_n| \le 2|\nabla \mathbf{u}|$, so in particular $F_n \to 0$ in $L^2(B)$.*

*Proof.* Since the mapping $z \mapsto |z|$ is Lipschitz, $|\mathbf{u}|$ belongs to $W^{1,2}(B)$ and

$$\nabla |\mathbf{u}| = |\mathbf{u}|^{-1} \mathbf{u} \nabla \mathbf{u} \quad \text{for } \mathbf{u} \ne 0, \quad \nabla |\mathbf{u}| = 0 \quad \text{for } \mathbf{u} = 0$$

(see [133, Thm. 2.1.11]). It is clear that

$$\big\| |\mathbf{u}| \big\|_{W^{1,2}(B)} \le \|\mathbf{u}\|_{W^{1,2}(B)}, \quad \big\| |\mathbf{u}_n| \big\|_{W^{1,2}(B)} \le \|\mathbf{u}_n\|_{W^{1,2}(B)}.$$

Next we have

$$\nabla|\mathbf{u}| - \nabla|\mathbf{u}_n| = \begin{cases} |\mathbf{u}_n|^{-1}\mathbf{u}_n(\nabla\mathbf{u} - \nabla\mathbf{u}_n) + (|\mathbf{u}|^{-1}\mathbf{u} - |\mathbf{u}_n|^{-1}\mathbf{u}_n)\nabla\mathbf{u} & \\ \qquad\qquad \text{for } \mathbf{u} \ne 0,\ \mathbf{u}_n \ne 0, \\ |\mathbf{u}|^{-1}\mathbf{u}\nabla\mathbf{u} & \text{for } \mathbf{u} \ne 0,\ \mathbf{u}_n = 0, \\ -|\mathbf{u}_n|^{-1}\mathbf{u}_n\nabla\mathbf{u}_n & \text{for } \mathbf{u} = 0,\ \mathbf{u}_n \ne 0. \end{cases}$$

It follows that

$$\big|\nabla|\mathbf{u}| - \nabla|\mathbf{u}_n|\big| \le |\nabla\mathbf{u} - \nabla\mathbf{u}_n| + F_n, \tag{8.3.4}$$

where

$$F_n = \big|\, |\mathbf{u}|^{-1}\mathbf{u} - |\mathbf{u}_n|^{-1}\mathbf{u}_n \big|\, |\nabla\mathbf{u}| \quad \text{for } \mathbf{u} \ne 0 \text{ and } \mathbf{u}_n \ne 0,$$

and $F_n = 0$ otherwise. Since $\mathbf{u}_n$ converges to $\mathbf{u}$ a.e. in $B$ we have

$$\big|\, |\mathbf{u}|^{-1}\mathbf{u} - |\mathbf{u}_n|^{-1}\mathbf{u}_n \big| \to 0 \quad \text{a.e. on } \{\mathbf{u} \ne 0\}.$$

Since $F_n$ vanishes on $\{\mathbf{u} = 0\}$, we conclude that $F_n \to 0$ a.e. in $B$. On the other hand obviously $F_n \le 2|\nabla \mathbf{u}|$. Since $F_n^2$ has the integrable majorant $4|\nabla\mathbf{u}|^2$ and converges to 0 as $n \to \infty$, the Lebesgue dominated convergence theorem implies the strong convergence of $F_n$ to 0 in $L^2(B)$. It follows from (8.3.4) and the inequality $\big|\,|\mathbf{u}| - |\mathbf{u}_n|\,\big| \le |\mathbf{u} - \mathbf{u}_n|$ that

$$\big\|\,|\mathbf{u}| - |\mathbf{u}_n|\,\big\|^2_{W^{1,2}(B)} = \int_B \Big(\big|\nabla|\mathbf{u}| - \nabla|\mathbf{u}_n|\,\big|^2 + \big|\,|\mathbf{u}| - |\mathbf{u}_n|\,\big|^2\Big)\,dx$$
$$\le \|\mathbf{u} - \mathbf{u}_n\|^2_{W^{1,2}(B)} + \int_B F_n^2\,dx.$$

Hence $F_n$ meets all requirements of the lemma. □

**Lemma 8.3.11.** *If* $\mathbf{v} \in W^{1,2}_S(B)$ *then* $|\mathbf{v}| \in W^{1,2}_S(B)$.

*Proof.* There is a sequence of vector functions $\boldsymbol{\varphi}_n \in C^\infty(B)$ such that

$$\boldsymbol{\varphi}_n \text{ vanishes in a neighborhood of } S \text{ and } \boldsymbol{\varphi}_n \to \mathbf{v} \text{ strongly in } W^{1,2}(B). \tag{8.3.5}$$

By Lemma 8.3.10 we have

$$\big\|\,|\mathbf{v}| - |\boldsymbol{\varphi}_n|\,\big\|^2_{W^{1,2}(B)} \le \|\mathbf{v} - \boldsymbol{\varphi}_n\|^2_{W^{1,2}(B)} + \int_B F_n^2(x)\,dx, \tag{8.3.6}$$

where $F_n \to 0$ in $L^2(B)$. Letting $n \to \infty$ shows that $|\boldsymbol{\varphi}_n|$ converges to $|\mathbf{v}|$ in $W^{1,2}(B)$. Since each $|\boldsymbol{\varphi}_n|$ vanishes in a neighborhood of $S$, we have $|\mathbf{v}| \in W^{1,2}_S(B)$. □

**Lemma 8.3.12.** *Let* $\mathbf{u} \in L^2(0,T; W^{1,2}_S(B))$. *Then for any* $\epsilon > 0$ *there exists a finite collection of disjoint measurable sets* $E_i \subset [0,T]$, $1 \le i \le N$, *with* $\bigcup_i E_i = [0,T]$, *and functions* $\varphi_i \in C^\infty(B)$ *vanishing in a neighborhood of* $S$ *and such that*

$$\int_0^T \big\|\varphi - |\mathbf{u}|\,\big\|^2_{W^{1,2}(B)}\,dt \le \epsilon^2/T, \tag{8.3.7}$$

*where* $\varphi \in L^2(0,T; W^{1,2}(B))$ *is the simple function defined by*

$$\varphi(x,t) = \varphi_i(x) \quad \text{for } t \in E_i, \quad i = 1,\dots,N. \tag{8.3.8}$$

*Proof.* By Theorem 1.3.20 there is a sequence $\mathbf{s}_n \in L^2(0,T; W^{1,2}_S(B))$ of simple functions (see Definition 1.3.8) such that

$$\int_0^T \|\mathbf{s}_n - \mathbf{u}\|^2_{W^{1,2}_S(B)}\,dt \to 0 \quad \text{as } n \to \infty. \tag{8.3.9}$$

Passing to a subsequence we can assume that

$$\|\mathbf{s}_n(t) - \mathbf{u}(t)\|_{W^{1,2}_S(B)} \to 0 \quad \text{as } n \to \infty \quad \text{for a.e. } t \in [0,T].$$

Applying Lemma 8.3.10 we obtain, for a.e. $t \in [0,T]$,

$$\big\|\,|\mathbf{s}_n(t)| - |\mathbf{u}(t)|\,\big\|^2_{W^{1,2}(B)} \le \|\mathbf{s}_n(t) - \mathbf{u}(t)\|^2_{W^{1,2}(B)} + \|F_n(t)\|^2_{L^2(B)}, \tag{8.3.10}$$

where $\|F_n(t)\|_{L^2(B)} \to 0$ as $n \to \infty$ and $\|F_n(t)\|_{L^2(B)} \le 2\|\mathbf{u}(t)\|_{W^{1,2}(B)}$, so by the Lebesgue dominated convergence theorem,

$$\int_0^T \|F_n(t)\|^2_{L^2(B)}\, dt \to 0 \quad \text{as } n \to \infty. \tag{8.3.11}$$

Next (8.3.10) yields

$$\int_0^T \|\, |\mathbf{s}_n(t)| - |\mathbf{u}(t)|\, \|^2_{W^{1,2}(B)}$$
$$\le \int_0^T \|\mathbf{s}_n(t) - \mathbf{u}(t)\|^2_{W^{1,2}(B)}\, dt + \int_0^T \|F_n(t)\|^2_{L^2(B)}\, dt.$$

Letting $n \to \infty$ and using (8.3.9), (8.3.11) we conclude that the left hand side of the last inequality tends to 0 as $n \to \infty$. Hence for given $\epsilon > 0$, there is $n > 1$ such that

$$\int_0^T \|\, |\mathbf{s}_n| - |\mathbf{u}|\, \|^2_{W^{1,2}(B)}\, dt \le \epsilon^2/(4T). \tag{8.3.12}$$

By Definition 1.3.8, the simple function $\mathbf{s}_n : [0,T] \to W^{1,2}_S(B)$ has a representation

$$\mathbf{s}_n = \sum_{i=1}^N \mathbf{c}_i \chi_{E_i},$$

where $\chi_{E_i}(t)$ are the characteristic functions of the pairwise disjoint measurable sets $E_i \subset [0,T]$, and $\mathbf{c}_i \in W^{1,2}_S(B)$. Thus we get

$$|\mathbf{s}_n| = \sum_{i=1}^N |\mathbf{c}_i| \chi_{E_i}.$$

By Lemma 8.3.11 we have $|\mathbf{c}_i| \in W^{1,2}_S(B)$. Hence for every $i$ there is a function $\varphi_i \in C^\infty_S(B)$ such that

$$\|\varphi_i - |\mathbf{c}_i|\, \|_{W^{1,2}(B)} \le \epsilon/(2T). \tag{8.3.13}$$

Now set

$$\varphi(x,t) = \sum_i \varphi_i(x) \chi_{E_i}(t).$$

It follows from (8.3.12), (8.3.13) and the inequality

$$\|\varphi - |\mathbf{u}|\, \|^2_{W^{1,2}(B)} \le 2\|\, |\mathbf{s}_n| - |\mathbf{u}|\, \|^2_{W^{1,2}(B)} + 2\|\varphi - |\mathbf{s}_n|\, \|^2_{W^{1,2}(B)}$$

that

$$\begin{aligned}\int_0^T &\|\varphi - |\mathbf{u}|\,\|^2_{W^{1,2}(B)}\,dt\\ &\le 2\int_0^T \|\,|\mathbf{s}_n| - |\mathbf{u}|\,\|^2_{W^{1,2}(B)}\,dt + 2\int_0^T \|\,|\mathbf{s}_n| - \varphi\|^2_{W^{1,2}(B)}\,dt\\ &= 2\int_0^T \|\,|\mathbf{s}_n| - |\mathbf{u}|\,\|^2_{W^{1,2}(B)}\,dt + 2\sum_i \|\varphi_i - |\mathbf{c}_i|\,\|^2_{W^{1,2}(B)}\,\mathrm{meas}\,E_i\\ &\le \frac{\epsilon^2}{2T} + \frac{\epsilon^2}{2T^2}\sum_i \mathrm{meas}\,E_i \le \epsilon^2/T. \qquad \square\end{aligned}$$

**Lemma 8.3.13.** *If $\mathbf{u} \in L^2(0,T;W^{1,2}_S(B))$, then the function*

$$W(x) = \int_0^T |\mathbf{u}(x,t)|\,dt \tag{8.3.14}$$

*belongs to $W^{1,2}_S(B)$.*

*Proof.* It suffices to prove that for any $\epsilon > 0$ there is $\Phi \in C^\infty(B)$ that vanishes in a neighborhood of $S$ and $\|\Phi - W\|_{W^{1,2}(B)} \le \epsilon$. Let $\varphi$ be given by Lemma 8.3.12. Define

$$\Phi = \int_0^T \varphi\,dt = \sum_{i=1}^N \varphi_i\,\mathrm{meas}\,E_i.$$

Since each $\varphi_i \in C^\infty(B)$ vanishes in a neighborhood of $S$, $\Phi$ belongs to $C^\infty(B)$ and vanishes in a neighborhood of $S$. Next, by Theorem 1.3.15,

$$\|\Phi - W\|_{W^{1,2}(B)} \le \int_0^T \|\varphi - |\mathbf{u}|\,\|_{W^{1,2}(B)}\,dt \le \sqrt{T}\Big(\int_0^T \|\varphi - |\mathbf{u}|\,\|^2_{W^{1,2}(B)}\,dt\Big)^{1/2},$$

and (8.3.7) yields $\|\Phi - W\|_{W^{1,2}(B)} \le \epsilon$. $\square$

**Lemma 8.3.14.** *Let $W$ be defined in Lemma 8.3.13. Then there exists a sequence $\psi_n \in C^\infty(B)$, $n \ge 1$, such that each $\psi_n$ vanishes in a neighborhood of $S$ and satisfies $0 \le \psi_n \le 1$ and*

$$\int_{B\setminus S} W(x)|\nabla\psi_n(x)|\,dx \to 0 \quad \text{as } n\to\infty, \tag{8.3.15}$$

$$\psi_n(x) \to 1 \quad \text{as } n\to\infty \quad \text{for all } x \in B\setminus S. \tag{8.3.16}$$

*Proof.* Consider the sequence of open sets

$$V_n = \{x : \mathrm{dist}(x,S) < 1/n\}.$$

We have $V_n \setminus S \subset V_{n-1} \setminus S$ and $\bigcap_n (V_n \setminus S) = \emptyset$. Hence

$$\operatorname{meas}(V_n \setminus S) \to 0 \quad \text{as } n \to \infty.$$

Notice that $W \in W^{1,2}_S(B)$ can be extended to a function $\overline{W} \in W^{1,2}(\mathbb{R}^d)$. Since $W$ vanishes q.e. on $S$, i.e. $W|_S = 0$, it follows from Theorem 8.2.22 that $\overline{W} \in W^{1,2}_0(\mathbb{R}^d \setminus S)$. Denote by $F$ the compact set $\operatorname{cl} B \setminus V_n$. Applying to $\overline{W}$ statement (c) of Theorem 8.2.22 with $K$ replaced by $S$ and $\epsilon = 1/n$ we conclude that there is $\psi_n \in C^\infty(\mathbb{R}^d)$ such that

- $0 \le \psi_n \le 1$ in $\mathbb{R}^d$,
- $\psi_n = 1$ on $F$, $\psi_n = 0$ in a neighborhood of $S$,
- $\|\overline{W}(1-\psi_n)\|_{W^{1,2}(\mathbb{R}^d)} \le 1/n$.

It immediately follows that

$$\psi_n = 1 \quad \text{in } B \setminus V_n, \quad \|W(1-\psi_n)\|_{W^{1,2}(B)} \le 1/n. \tag{8.3.17}$$

Next we have

$$W|\nabla\psi_n| = W|\nabla(1-\psi_n)| \le |\nabla W|(1-\psi_n) + |\nabla(W(1-\psi_n))|. \tag{8.3.18}$$

Since $1 - \psi_n = 0$ on $B \setminus V_n$ and $\nabla W = 0$ a.e. on $S$ we have

$$\int_B |\nabla W|(1-\psi_n)\,dx \le \int_{V_n\setminus S} |\nabla W| dx \le \Big(\int_B |\nabla W|^2\,dx\Big)^{1/2} \big(\operatorname{meas}(V_n \setminus S)\big)^{1/2}.$$

On the other hand, (8.3.17) implies

$$\begin{aligned}\int_B |\nabla(W(1-\psi_n))|\,dx &\le \Big(\int |\nabla(W(1-\psi_n))|^2\,dx\Big)^{1/2} (\operatorname{meas} B)^{1,2}\\ &\le c\|W(1-\psi_n)\|_{W^{1,2}(B)} \le c/\sqrt{n}.\end{aligned}$$

Combining the results obtained with (8.3.18) we finally obtain

$$\int_B W|\nabla\psi_n|\,dx \le \|W\|_{W^{1,2}(B)} \big(\operatorname{meas}(V_n \setminus S)\big)^{1/2} + cn^{-1/2} \to 0$$

as $n \to \infty$. □

**Proof of Proposition 8.3.9.** It suffices to note that

$$\int_0^T \int_{B\setminus S} |\mathbf{u}(x,t)|\,|\nabla\psi_n(x)|\,dxdt = \int_{B\setminus S} W(x)|\nabla\psi_n(x)|\,dx,$$

and apply Lemma 8.3.14.

## 8.4 $\mathcal{S}$-compact classes of admissible obstacles

In this section we present an example of an $\mathcal{S}$-compact family of admissible obstacles. Such a class must satisfy two conditions. First, it should be sufficiently large in order to meet practical requirements arising in real shape optimization problems. In particular, it should include the $d$-dimensional obstacles satisfying the double cone condition and the $(d-1)$-dimensional Lipschitz manifolds. On the other hand a family of admissible obstacles should be defined by clear geometrical constrains. The main difficulty is that there is no geometrical criterion for Kuratowski-Mosco convergence. Intuitively, it is clear that a compact set in the Kuratowski-Mosco topology should be in some sense "equi-thick", but specifying such a requirement is a delicate task.

A possible way to construct a family of admissible obstacles is the following. Let us assume that for every point $x$ of an admissible obstacle $S$, there is a test set which is contained in $S$ and thick at $x$. If we choose the test set as a Lipschitz image of a fixed compact $\mathcal{T} \subset \mathbb{R}^d$ which is thick at 0, then we arrive at the following definition.

**Definition 8.4.1.** Let $\mathcal{T} \subset \mathbb{R}^d$ be a compact set thick at 0, and $B' \subset \mathbb{R}^d$ an arbitrary compact set. The family $\mathcal{O}(k, \mathcal{T}, B')$ consists of all compact sets $S \subset B'$ with the following property: for every $x_0 \in S$, there exists a bi-Lipschitz bijection $\chi : \mathbb{R}^d \to \mathbb{R}^d$ such that

$$\chi(0) = 0, \quad k^{-1}|y' - y''| \le |\chi(y') - \chi(y'')| \le k|y' - y''| \tag{8.4.1}$$

for all $y', y'' \in \mathbb{R}^d$, and $x_0 + \chi(\mathcal{T}) \subset S$.

Notice that in this definition the bijection $\chi$ depends on $x_0$, but the constant $k$ does not. By Lemma 8.2.8 every obstacle in $\mathcal{O}(k, \mathcal{T}, B')$ is thick at each of its points. In accordance with Lemmas 8.2.10 and 8.2.12 we can choose $\mathcal{T} = \triangle(\omega)$ in dimension three and $\mathcal{T} = I$ in dimension two. For this choice the class $\mathcal{T}$ includes all $(d-1)$-dimensional compact manifolds with Lipschitz boundaries and all compact sets $S$ which are starlike with respect to three noncollinear points. The following theorem shows that $\mathcal{O}(k, \mathcal{T}, B')$ is compact with respect to $\mathcal{S}$-convergence.

**Theorem 8.4.2.** *Every sequence $S_n \in \mathcal{O}(k, \mathcal{T}, B')$ contains a subsequence which $\mathcal{S}$-converges to a set $S \in \mathcal{O}(k, \mathcal{T}, B')$.*

We split the proof into a sequence of lemmas.

**Lemma 8.4.3.** *If $S_n \in \mathcal{O}(k, \mathcal{T}, B')$ converges in the Hausdorff metric to a compact set $S$, then $S \in \mathcal{O}(k, \mathcal{T}, B')$.*

*Proof.* Obviously $S \subset B'$. Next choose $x_0 \in S$. By the definition of the Hausdorff convergence, there is a sequence $x_n \in S_n$ such that $x_n \to x_0$. Since $S_n \in \mathcal{O}(k, \mathcal{T}, B')$, there is a bi-Lipschitz bijection $\chi_n : \mathbb{R}^d \to \mathbb{R}^d$ satisfying (8.4.1) such that $x_n + \chi_n(\mathcal{T}) \subset S_n$. Passing to a subsequence we can assume that $\chi_n$ converges uniformly on every compact set in $\mathbb{R}^d$ to a mapping $\mathcal{X} : \mathbb{R}^d \to \mathbb{R}^d$. Obviously $\mathcal{X}$ satisfies (8.4.1). Choose $y \in \mathcal{T}$. We have $S_n \ni x_n + \mathcal{X}_n(y) \to x_0 + \mathcal{X}(y)$. Since $S_n \to S$ in the Hausdorff metric, we conclude that $x_0 + \mathcal{X}(y) \in S$ and hence $x_0 + \mathcal{X}(\mathcal{T}) \subset S$. Therefore, $S$ belongs to $\mathcal{O}(k, \mathcal{T}, B')$. □

**Lemma 8.4.4.** *Let $Z^n : \mathbb{R}^d \to \mathbb{R}^d$ be bi-Lipschitz bijections satisfying*

$$k^{-1}|x' - x''| \le |Z^n(x') - Z^n(x'')| \le k|x' - x''| \quad \textit{for all } x', x'' \in \mathbb{R}^d. \tag{8.4.2}$$

*Assume that $Z : \mathbb{R}^d \to \mathbb{R}^d$ is a map such that $Z^n \to Z$ uniformly on every bounded set in $\mathbb{R}^d$. Then $Z$ is a bijection satisfying (8.4.2) and the Jacobians $J_n = \det(D_x Z^n)$ converge weakly in $L^r(D)$ to $J = \det(D_x Z)$ for all $r \in [1, \infty)$ and all bounded measurable sets $D \subset \mathbb{R}^d$.*

*Proof.* Letting $n \to \infty$ in (8.4.2) we find that $Z$ satisfies (8.4.2) and hence it is a bi-Lipschitz bijection. In particular, the Jacobian matrices $D_x Z^n$ and $D_x Z$ are uniformly bounded in $L^\infty(\mathbb{R}^d)$. Hence it suffices to prove that

$$\lim_{n\to\infty} \int_{\mathbb{R}^d} J_n(x)\varphi(x)\,dx = \int_{\mathbb{R}^d} J(x)\varphi(x)\,dx \quad \text{for all } \varphi \in C_0^\infty(\mathbb{R}^d). \tag{8.4.3}$$

Notice that the mappings $Z^n$ and $Z$ take homeomorphically $\mathbb{R}^d$ onto itself and have the bi-Lipschitz inverses $(Z^n)^{-1}$ and $Z^{-1}$. Obviously $(Z^n)^{-1} \to Z^{-1}$ uniformly on every bounded set. Thus we get

$$\int_{\mathbb{R}^d} J_n(x)\varphi(x)\,dx = \int_{\mathbb{R}^d} \varphi \circ (Z^n)^{-1}(z)\,dz$$
$$\to \int_{\mathbb{R}^d} \varphi \circ Z^{-1}(z)\,dz = \int_{\mathbb{R}^d} J(x)\varphi(x)\,dx.$$ □

**Lemma 8.4.5.** *Let $\overline{u}_n \rightharpoonup \overline{u}$ weakly in $W^{1,2}(\mathbb{R}^d)$. Let $\mathcal{X}_n$ be a sequence of bi-Lipschitz bijections satisfying (8.4.1) and converging uniformly on every compact set to a bijection $\mathcal{X}$. Let $\mathbb{R}^d \ni x_n \to x_0$. Then the sequence $v_n(y) = \overline{u}_n(x_n + \mathcal{X}_n(y))$ converges weakly in $W^{1,2}(\mathbb{R}^d)$ to $v(y) = \overline{u}(x_0 + \mathcal{X}(y))$.*

*Proof.* Since the sequence $v_n$ is obviously bounded in $W^{1,2}(\mathbb{R}^d)$, it contains a subsequence which converges weakly in $W^{1,2}(\mathbb{R}^d)$ to an element $v^* \in W^{1,2}(\mathbb{R}^d)$. Of course, this subsequence converges to $v^*$ weakly in $L^2(\mathbb{R}^d)$. Notice that $v_n \rightharpoonup v$ weakly in $W^{1,2}(\mathbb{R}^3)$ if and only if all such $v^*$, i.e. all limits of subsequences, coincide and are equal to $v$. Hence it suffices to show that $v_n \rightharpoonup v$ weakly in $L^2(\mathbb{R}^d)$. Since $v_n$ is uniformly bounded in $L^2(\mathbb{R}^d)$, it suffices to prove that

$$\lim_{n\to\infty} \int_{\mathbb{R}^d} v_n \cdot \varphi\,dy = \int_{\mathbb{R}^d} v \cdot \varphi\,dy \quad \text{for all } \varphi \in C_0^\infty(\mathbb{R}^d). \tag{8.4.4}$$

Denote by $y = Z^n(x)$ and $y = Z(x)$ the inverse mappings of $x = x_n + \mathcal{X}_n(y)$ and $x = x_0 + \mathcal{X}(y)$, respectively. Since $\mathcal{X}_n$ and $\mathcal{X}$ satisfy (8.4.1) and $\mathcal{X}_n \to \mathcal{X}$ uniformly on every compact set, the mappings $Z^n$ and $Z$ meet all requirements of Lemma 8.4.4. On the other hand, we have $\overline{u}_n(x) = v_n(Z^n(x))$ and $\overline{u}(x) = v(Z(x))$, which leads to

$$\begin{aligned} \int_{\mathbb{R}^d} v_n \cdot \varphi \, dy &= \int_{\mathbb{R}^d} \overline{u}_n(x) \cdot \varphi(Z^n(x)) J_n(x) \, dy, \\ \int_{\mathbb{R}^d} v \cdot \varphi \, dy &= \int_{\mathbb{R}^d} \overline{u}(x) \cdot \varphi(Z(x)) J(x) \, dy, \end{aligned} \tag{8.4.5}$$

where $J_n$ and $J$ are the Jacobians of $Z^n$ and $Z$. Since

$$|Z^n(x)| \geq k^{-1}|x| - c, \quad |Z(x)| \geq k^{-1}|x| - c, \quad c = \sup |x_n|,$$

and $\varphi(y)$ is compactly supported in $\mathbb{R}^d$, there is a bounded domain $D \subset \mathbb{R}^d$ such that the vector fields $\varphi \circ Z^n$ and $\varphi \circ Z$ vanish outside of $D$. Moreover, $\varphi \circ Z^n \to \varphi \circ Z$ uniformly on $D$. Recall that the embedding $W^{1,2}(\mathbb{R}^d) \hookrightarrow L^2(D)$ is compact. Hence $\overline{u}_n \to \overline{u}$ strongly in $L^2(D)$. Obviously $\overline{u}_n \cdot \varphi \circ Z^n \to \overline{u} \cdot \varphi \circ Z$ strongly in $L^2(D)$. On the other hand, by Lemma 8.4.4, $J_n \rightharpoonup J$ weakly in $L^2(D)$. Thus

$$\lim_{n \to \infty} \int_{\mathbb{R}^d} \overline{u}_n(x) \cdot \varphi(Z^n(x)) J_n(x) \, dy = \int_{\mathbb{R}^d} \overline{u}(x) \cdot \varphi(Z(x)) J(x) \, dy.$$

Now (8.4.5) yields the desired equality (8.4.4). □

**Lemma 8.4.6.** *Let $S_n \in \mathcal{O}(k, \mathcal{T}, B')$ and let $\overline{u}_n \in W^{1,2}(\mathbb{R}^d)$ satisfy*

$$\overline{u}_n \in W_0^{1,2}(\mathbb{R}^d \setminus S_n), \quad \overline{u}_n \rightharpoonup \overline{u} \quad \textit{weakly in } W^{1,2}(\mathbb{R}^d).$$

*Let $\mathcal{X}_n : \mathbb{R}^d \to \mathbb{R}^d$ be a sequence of bi-Lipschitz bijections satisfying* (8.4.1) *and converging to a bi-Lipschitz bijection $\mathcal{X}$ uniformly on every bounded subset of $\mathbb{R}^d$. Furthermore assume that there is a sequence $\mathbb{R}^d \ni x_n \to x_0$ with $x_n + \mathcal{X}_n(\mathcal{T}) \subset S_n$. Then the quasicontinuous representative of $\overline{u}$ vanishes q.e. on $x_0 + \mathcal{X}(\mathcal{T})$, i.e., in accordance with Definition 8.2.21,*

$$\overline{u}|_{x_0 + \mathcal{X}(\mathcal{T})} = 0.$$

*Proof.* First we show that $v_n(y) = \overline{u}_n(x_n + \mathcal{X}_n(y))$ belongs to $W_)^{1,2}(\mathbb{R}^d \setminus \mathcal{T})$. Observe that since $\overline{u}_n \in W_0^{1,2}(\mathbb{R}^d \setminus S_n)$, there is a sequence $\varphi_k \in C_0^\infty(\mathbb{R}^d)$ with the following properties. Every $\varphi_k$ vanishes in a neighborhood $U_k$ of $S_n$ and $\varphi_k \to \overline{u}_n$ in $W^{1,2}(\mathbb{R}^d)$ as $k \to \infty$. Notice that $U_k \ni x_n + \mathcal{X}_n(\mathcal{T})$. The inverse bi-Lipschitz homeomorphism $Z^n = (x_n + \mathcal{X}_n)^{-1}$ takes $U_k$ onto the open set $Z^n(U_k)$ and $x_n + \mathcal{X}_n(\mathcal{T})$ onto $\mathcal{T}$. Obviously $Z^n(U_k)$ is a neighborhood of $\mathcal{T}$. Hence the Lipschitz function $\varphi_k(x_n + \mathcal{X}_n(y))$ vanishes in a neighborhood of $\mathcal{T}$. On the other hand,

$$\varphi_k(x_n + \mathcal{X}_n) \to \overline{u}_n(x_n + \mathcal{X}_n) = v_n \text{ as } k \to \infty \text{ in } W^{1,2}(\mathbb{R}^d)$$

since $x_n + \mathcal{X}_n$ is a bi-Lipschitz homeomorphism. It follows that $v_n \in W_0^{1,2}(\mathbb{R}^d \setminus \mathcal{T})$.

By Lemma 8.4.5, the sequence $v_n$ converges weakly in $W^{1,2}(\mathbb{R}^d)$ to $v = \overline{u}\circ(x_0+\mathcal{X})$. Every closed subspace of the Hilbert space $W^{1,2}(\mathbb{R}^d)$ is weakly closed, since all Hilbert spaces are reflexive. In particular, $W_0^{1,2}(\mathbb{R}^d \setminus \mathcal{T})$ is weakly closed. Therefore, $v \in W_0^{1,2}(\mathbb{R}^d \setminus \mathcal{T})$. From Theorem 8.2.22 we now conclude that the quasicontinuous representative of $v$ is equal to zero q.e. on $\mathcal{T}$.

Next, we show that if $v$ is quasicontinuous, then so is $\overline{u} = v \circ Z$, where $Z = (x_0 + \mathcal{X})^{-1}$. First, as $\mathcal{X}$ satisfies (8.4.1), Lemma 8.2.4 with $\mathcal{X}$ replaced by $x_0 + \mathcal{X}$ shows that

$$\operatorname{Cap}(x_0 + \mathcal{X}(G)) \le c(k) \operatorname{Cap} G \quad \text{for all open } G \subset \mathbb{R}^d.$$

Next, choose an open set $D \subset \mathbb{R}^d$ and $\epsilon > 0$. Since $v$ is quasicontinuous, there is an open set $G$ with $\operatorname{Cap} G < \epsilon/c(k)$ such that $v$ is continuous on $Z(D) \setminus G$. It follows that $\overline{u}$ is continuous on $D \setminus G^*$, where $G^* = x_0 + \mathcal{X}(G)$. We also have $\operatorname{Cap} G^* < \epsilon$. Hence from Definition 8.2.16 we deduce that $\overline{u} = v \circ Z$ is quasicontinuous.

Finally notice that by Lemma 8.2.4, the bi-Lipschitz homeomorphism $Z$ takes zero capacity sets onto zero capacity sets. Since $v = 0$ q.e. on $\mathcal{T}$ and $Z$ takes $\mathcal{T}$ onto $x_0 + \mathcal{X}(\mathcal{T})$, it follows that $\overline{u} = v \circ Z$ is zero q.e. on $x_0 + \mathcal{X}(\mathcal{T})$. □

**Lemma 8.4.7.** *Let $B' \Subset B$ and suppose $S_n \in \mathcal{O}(k, \mathcal{T}, B')$ converges in the Hausdorff metric to a compact set $S$. Furthermore assume that $u_n \in W_{S_n}^{1,2}(B)$ converges weakly in $W^{1,2}(B)$ to some $u \in W^{1,2}(B)$. Then $u \in W_S^{1,2}(B)$.*

*Proof.* By Corollary 8.3.4 it suffices to prove that the quasicontinuous representative of $u$ is equal to zero q.e. on $S$.

Let us consider the following construction. Since $B \subset \mathbb{R}^d$ is a bounded domain with $C^1$ boundary, there exists a bounded linear extension operator $\Xi : W^{1,2}(B) \to W^{1,2}(\mathbb{R}^d)$ which assigns to every $u \in W^{1,2}(B)$ its extension $\overline{u} \in W^{1,2}(\mathbb{R}^d)$, such that

$$\|\overline{u}\|_{W^{1,2}(\mathbb{R}^d)} \le c\|u\|_{W^{1,2}(B)}, \quad \overline{u} = u \quad \text{on } B.$$

If $u \in W_S^{1,2}(B)$ then $\overline{u} \in W_0^{1,2}(\mathbb{R}^d \setminus S)$, so we can set $\overline{u}_n = \Xi u_n \in W_0^{1,2}(\mathbb{R}^d \setminus S_n)$. Since the operator $\Xi : W^{1,2}(B) \to W^{1,2}(\mathbb{R}^d)$ is bounded, $\overline{u}_n \rightharpoonup \overline{u} = \Xi u$ weakly in $W^{1,2}(\mathbb{R}^d)$. The lemma will be proved if we show that the quasicontinuous representative of $\overline{u}$ is zero q.e. on $S$. To this end notice that in view of Theorem 8.2.18, the quasicontinuous representative of $\overline{u}$ is finely continuous q.e. in $\mathbb{R}^d$. Hence it suffices to show that $\overline{u}(x_0) = 0$ at every point $x_0 \in S$ such that the quasicontinuous representative of $\overline{u}$ is finely continuous at $x_0$.

Fix such an $x_0 \in S$. Since $S_n \to S$ in the Hausdorff metric, there is a sequence $x_n \in S_n$ which converges to $x_0$. Recall that $S_n \in \mathcal{O}(k, \mathcal{T}, B')$, so by Definition 8.4.1 there exists a bi-Lipschitz homeomorphism $\mathcal{X}_n : \mathbb{R}^d \to \mathbb{R}^d$ satisfying (8.4.1) such that $x_n + \mathcal{X}_n(\mathcal{T}) \subset S_n$. Passing to a subsequence we can assume that $\mathcal{X}_n$ converges to a bi-Lipschitz bijection $\mathcal{X}$ uniformly on every bounded subset of $\mathbb{R}^d$. Hence the sequences $\overline{u}_n$, $\overline{u}$, and $\mathcal{X}_n$ meet all requirements of Lemma 8.4.6. Thus, the quasicontinuous representative of $\overline{u}$ is equal to zero q.e. on $x_0 + \mathcal{X}(\mathcal{T})$.

Since $\overline{u}$ is finely continuous at $x_0$, by Proposition 8.2.15 there exists a set $E$ such that

$$\lim_{\mathbb{R}^d \setminus E \ni x \to x_0} \overline{u}(x) = \overline{u}(x_0) \quad \text{and} \quad E \text{ is thin at } x_0. \tag{8.4.6}$$

Next, by Lemma 8.2.8, the set $x_0 + X(\mathcal{T})$ is thick at $x_0$. Lemma 8.2.7 shows that the set $F = (x_0 + X(\mathcal{T})) \setminus E$ is thick at $x_0$. Since $\overline{u} = 0$ q.e. on $x_0 + X(\mathcal{T})$, there is a set $X$ of zero capacity such that $\overline{u} = 0$ on $F \setminus X$. Since $F$ is thick at $x_0$, Lemma 8.2.7 implies that $F \setminus X$ is thick at $x_0$, and $x_0$ is a limit point of $F \setminus X$. On the other hand, $(F \setminus X) \cap E = \emptyset$, so (8.4.6) yields the desired equality $\overline{u}(x_0) = \lim_{F \setminus X \ni x \to x_0} \overline{u}(x) = 0$. □

**Lemma 8.4.8.** *Let $S \in \mathcal{O}(k, \mathcal{T}, B')$ and suppose a sequence of compact sets $S_n \subset B'$ converges to $S$ in the Hausdorff metric. Let $u \in W^{1,2}_S(B)$. Then there is a sequence $u_n \in W^{1,2}_{S_n}(B)$ which converges to $u$ in $W^{1,2}(B)$.*

*Proof.* Since $u \in W^{1,2}_S(B)$, there is a sequence $\varphi_k \in C^\infty(B)$ such that $\varphi_k \to u$ in $W^{1,2}(B)$ and $\varphi_k$ vanishes in a neighborhood $G_k$ of $S$. Without loss of generality we can assume that $G_{k+1} \subset G_k$ and $\bigcap_k G_k = S$. By the definition (8.1.1) of the Hausdorff metric, for every $k$ there is $N_k$ such that $S_n \Subset G_k$ for all $n \geq N_k$. Hence we can take $u_n = \varphi_k$ for all $n \in [N_k, N_{k+1})$. □

**Proof of Theorem 8.4.2.** Take any sequence $S_n \in \mathcal{O}(k, \mathcal{T}, B')$. Since the sets $S_n$ are compact and contained in the compact set $B'$, Lemma 8.1.1 yields the existence of a subsequence, still denoted by $S_n$, and a compact set $S \subset B'$ such that $S_n \to S$ in the Hausdorff metric. By Lemma 8.4.7 for any sequence $W^{1,2}_{S_n}(B) \ni u_n \rightharpoonup u$ weakly in $W^{1,2}(B)$, the limit element $u$ belongs to $W^{1,2}_S(B)$. Conversely, Lemma 8.4.8 implies that for any $u \in W^{1,2}_S(B)$ there exists a sequence $u_n \in W^{1,2}_{S_n}(B)$ which converges to $u$ strongly in $W^{1,2}(B)$. From Definition 8.3.2 we thus conclude that $S_n \to S$ in the Kuratowski-Mosco sense. The Kuratowski-Mosco convergence combined with the Hausdorff convergence yields the $\mathcal{S}$-convergence of $S_n$. It remains to note that in view of Lemma 8.4.3, the limit set $S$ belongs to $\mathcal{O}(k, \mathcal{T}, B')$.

# Chapter 9

# Flow around an obstacle. Domain dependence

## 9.1 Preliminaries

There are two applications of domain stability results for the governing equations of the in/out flow problem:

- The first is to the existence of weak renormalized solutions with reasonable regularity assumptions on the data from the point of view of applications. This means that at the first stage the existence is obtained under restrictive assumptions, in particular on the domain. Next, the assumptions are relaxed by applying the stability result. In this important direction we have a nice result: a closed obstacle is perfectly admissible for the existence of weak renormalized solutions for compressible Navier-Stokes equations. In other words, there is no need for artificial domain constraints—solutions for the governing equations exist even in the case of a closed obstacle.
- The second is to shape optimization problems, in order to establish the existence of an optimal shape within a given family of admissible domains or obstacles, for a specific cost functional, with the state equation in the form of compressible Navier-Stokes equations. In this case the expected result would be a broad family of admissible obstacles such that the associated shape optimization problems enjoy solutions. But even for the simplest case of the Laplacian there is no satisfactory answer to the question, therefore we cannot expect to find a better solution. We establish the existence of an optimal domain in the same class as in the case of the Laplacian. The result seems to be optimal, and unexpected.

In this chapter it is shown that weak renormalized solutions to compressible Navier-Stokes equations are stable with respect to domain perturbations. In this way the existence results are extended to more realistic data. The general framework is used for the analysis of shape optimization problems.

## 9.2 Existence theory

In this chapter we consider in more detail the problem of a flow around an obstacle, which is a particular case of problem (5.1.12) and is formulated as follows:

**Condition 9.2.1.** Let $B$ be an open, bounded hold-all domain in $\mathbb{R}^d$ with $C^\infty$ boundary $\partial B$ and

$$\Sigma = \partial B \times (0, T).$$

Furthermore a compact "obstacle" $S$ is placed inside $B$, i.e., $S$ is a compact subset of $B$. A given *exterior vector field* $\mathbf{U} \in C^\infty(\mathbb{R}^d \times [0, T])$ splits the surface $\Sigma$ into three disjoint parts (subsets) $\Sigma_{\text{in}}$, $\Sigma_{\text{out}}$ and $\Gamma$ defined by the sign of the normal component of $\mathbf{U}$,

$$\begin{aligned} \Sigma_{\text{in}} &= \{(x, t) \in \Sigma : \mathbf{n}(x) \cdot \mathbf{U}(x, t) < 0\}, \\ \Gamma &= \{(x, t) \in \Sigma : \mathbf{n}(x) \cdot \mathbf{U}(x, t) = 0\}, \\ \Sigma_{\text{out}} &= \{(x, t) \in \Sigma : \mathbf{n}(x) \cdot \mathbf{U}(x, t) > 0\}, \end{aligned} \tag{9.2.1}$$

where $\mathbf{n}$ is the outward normal to $\partial B$. Hence in the topology of $\Sigma$, the compact set $\Gamma$ separates the open sets $\Sigma_{\text{in}}$ and $\Sigma_{\text{out}}$. With this notation we have

$$\Omega = B \setminus S, \quad Q = \Omega \times (0, T), \quad S_T = \Sigma \cup (\partial S \times (0, T)). \tag{9.2.2}$$

We also assume that the given vector field $\mathbf{U}$ vanishes near the obstacle,

$$\mathbf{U}(x, t) = 0 \quad \text{in a neighborhood of } S \times (0, T). \tag{9.2.3}$$

Under these assumptions the problem of the flow around an obstacle can be formulated in the same way as problem (5.1.12).

**Problem 9.2.2.** *Find a velocity field* $\mathbf{u}$ *and a density function* $\varrho$ *satisfying*

$$\partial_t(\varrho \mathbf{u}) + \operatorname{div}(\varrho\, \mathbf{u} \otimes \mathbf{u}) + \nabla p(\rho) = \operatorname{div} \mathbb{S}(\mathbf{u}) + \varrho \mathbf{f} \quad \textit{in } Q, \tag{9.2.4a}$$

$$\partial_t \varrho + \operatorname{div}(\varrho \mathbf{u}) = 0 \quad \textit{in } Q, \tag{9.2.4b}$$

$$\begin{aligned} &\mathbf{u} = \mathbf{U} \quad \textit{on } S_T, \\ &\varrho = \varrho_\infty \quad \textit{on } \Sigma_{\text{in}} \times (0, T), \\ &\mathbf{u}|_{t=0} = \mathbf{U}, \quad \varrho|_{t=0} = \varrho_\infty \quad \textit{in } \Omega. \end{aligned} \tag{9.2.4c}$$

*Notice that* $\mathbf{u} = 0$ *on* $\partial S \times (0, T)$.

In Chapter 5 we studied weak renormalized solutions to this problem, but in this particular case we use a definition of renormalized solution which is stronger compared to Definition 5.1.4. In order to treat the case of nonsmooth obstacles we recall Definition 8.3.1 of the Sobolev space $W^{1,2}_S(B)$ of functions vanishing around an obstacle.

**Definition 9.2.3.** A couple

$$\mathbf{u} \in L^2(0,T;W^{1,2}_S(B)), \quad \varrho \in L^\infty(0,T;L^\gamma(B\setminus S))$$

is said to be a *weak renormalized solution* to problem (9.2.4) if it satisfies:

- The kinetic energy is bounded, i.e., $\varrho|\mathbf{u}|^2 \in L^\infty(0,T;L^1(B\setminus S))$.
- The velocity $\mathbf{u}$ satisfies the Dirichlet boundary condition on the lateral side of the cylinder,
$$\mathbf{u} = \mathbf{U} \quad \text{on } \Sigma.$$
This condition makes sense since $\mathbf{u} \in L^2(0,T;W^{1,2}(\Omega))$ has a trace on the smooth surface $\Sigma$.
- The extended velocity satisfies $\mathbf{u} \in L^2(0,T;W^{1,2}_S(B))$.
- The integral identity
$$\int_{(B\setminus S)\times(0,T)} \big(\varrho\mathbf{u}\cdot\partial_t\boldsymbol{\xi} + \varrho\mathbf{u}\otimes\mathbf{u} : \nabla\boldsymbol{\xi} + p\,\mathrm{div}\,\boldsymbol{\xi} - \mathbb{S}(\mathbf{u}) : \nabla\boldsymbol{\xi}\big)\,dxdt \\ + \int_{(B\setminus S)\times(0,T)} \varrho\mathbf{f}\cdot\boldsymbol{\xi}\,dxdt + \int_{(B\setminus S)\times\{t=0\}} \mathbf{U}\cdot\boldsymbol{\xi}\,dx = 0 \quad (9.2.5)$$
holds for all vector fields $\boldsymbol{\xi} \in C^\infty((B\setminus S)\times(0,T))$ equal to 0 in a neighborhood of the lateral side $(\partial B\cup\partial S)\times(0,T)$ and of the top $(B\setminus S)\times\{t=T\}$.
- The integral identity
$$\int_{(B\setminus S)\times(0,T)} \big(\varphi(\varrho)\partial_t\psi + \varphi(\varrho)\mathbf{u}\cdot\nabla\psi + \psi(\varphi(\varrho) - \varphi'(\varrho)\varrho)\,\mathrm{div}\,\mathbf{u}\big)\,dxdt \\ = \int_{\Sigma_{\mathrm{in}}} \psi\varphi(\varrho_\infty)\mathbf{U}\cdot\mathbf{n}\,d\Sigma - \int_{(B\setminus S)\times\{t=0\}} \varphi(\varrho_\infty)\psi\,dx \quad (9.2.6)$$
holds for all $\psi \in C^\infty((B\setminus S)\times(0,T))$ vanishing in a neighborhood of the surface $\Sigma\setminus\Sigma_{\mathrm{in}}$ and in a neighborhood of the top $(B\setminus S)\times\{t=T\}$, and for all $C^1$ functions $\varphi : [0,\infty) \to \mathbb{R}$ such that
$$\limsup_{\varrho\to\infty}(|\varphi(\varrho)| + |\varrho\varphi'(\varrho)|) < \infty. \quad (9.2.7)$$
Condition (9.2.7) means that $\varphi$ has minimal admissible smoothness and it is bounded at infinity.

The only difference between Definitions 5.1.4 and 9.2.3 is that in the latter we require the integral identity (9.2.6) to hold for all test functions $\psi$ that vanish in a neighborhood of $\Sigma \setminus \Sigma_{\text{in}}$, while in Definition 5.1.4, only for those that vanish in a neighborhood of $\Sigma \setminus \Sigma_{\text{in}}$ and of $\partial S \times (0,T)$. Hence Definition 9.2.3 is stronger.

Renormalized solutions in the sense of Definition 9.2.3 have the additional property that a solution $\varrho$ to the mass balance equation (9.2.4b) can be extended to the obstacle $S$ and the extended function also satisfies the transport equation. This property is used throughout this chapter, and it is specified by the following lemma.

**Lemma 9.2.4.** *Let* $\mathbf{u}$ *and* $\varrho$ *meet all requirements of Definition 9.2.3, and define* $\mathbf{u}^* : B \times (0,T) \to \mathbb{R}^d$ *and* $\varrho^* : B \times (0,T) \to \mathbb{R}^+$ *by*

$$\mathbf{u}^* = \mathbf{u} \quad \textit{in } (B \setminus S) \times (0,T), \qquad \mathbf{u}^* = 0 \quad \textit{in } S \times (0,T),$$

$$\varrho^* = \varrho \quad \textit{in } (B \setminus S) \times (0,T), \qquad \varrho^*(x,t) = \varrho_\infty(x,0) \quad \textit{in } S \times (0,T).$$

*Then the integral identity*

$$\int_{B\times(0,T)} \big(\varphi(\varrho^*)\partial_t\psi + \varphi(\varrho^*)\mathbf{u}^* \cdot \nabla\psi + \psi(\varphi(\varrho^*) - \varphi'(\varrho^*)\varrho^*)\operatorname{div}\mathbf{u}^*\big)\,dxdt$$
$$= \int_{\Sigma_{\text{in}}} \psi\varphi(\varrho_\infty)\mathbf{U}\cdot\mathbf{n}\,d\Sigma - \int_B (\varphi(\varrho_\infty)\psi)(x,0)\,dx \tag{9.2.8}$$

*holds for all* $\psi \in C^\infty(Q)$ *vanishing in a neighborhood of the surface* $\Sigma \setminus \Sigma_{\text{in}}$ *and in a neighborhood of the top* $\Omega \times \{T\}$*, and for all smooth functions* $\varphi : [0,\infty) \to \mathbb{R}$ *such that*

$$\limsup_{\varrho\to\infty}(|\varphi(\varrho)| + |\varrho\varphi'(\varrho)|) < \infty. \tag{9.2.9}$$

*Proof.* It is easy to see that

$$\int_{S\times(0,T)} \varphi(\varrho^*)\partial_t\psi\,dxdt = -\int_{S\times\{t=0\}} \varphi(\varrho_\infty)\psi\,dx. \tag{9.2.10}$$

Notice that $\mathbf{u}^* \in L^2(0,T; W^{1,2}_S(B))$, in fact $\mathbf{u}^*$ is simply $\mathbf{u}$ if $\mathbf{u}$ is considered as an element of $W^{1,2}_S(B)$. In particular, $\mathbf{u}^* = 0$ and $\nabla\mathbf{u}^* = 0$ on $S \times (0,T)$. Now from (9.2.6), (9.2.10), and the definition of $\varrho^*$ we obtain (9.2.8). □

We are now in a position to prove the main result of this section, on existence of a weak renormalized solution to the boundary value problem (9.2.4). We assume that the flow domain, the given data and the constitutive function $p(\varrho)$ satisfy the following conditions:

**Condition 9.2.5.** • The function $p : [0,\infty) \to \mathbb{R}^+$ has continuous first derivative and

$$p'(\varrho) \ge 0, \quad p(0) = p'(0) = 0. \tag{9.2.11}$$

There are an exponent $\gamma > d/2$ and a constant $c_p > 0$ such that for $\varrho \in [0,\infty)$,

$$\varrho^\gamma \le c_p(P(\varrho)+1), \quad \varrho^\gamma \le c_p(p(\varrho)+1), \quad p(\varrho) \le c_p(P(\varrho)+1), \tag{9.2.12}$$

and for $\varrho \in [1,\infty)$,

$$c_p^{-1}\varrho^\gamma \le p(\varrho) \le c_p\varrho^\gamma, \quad c_p^{-1}\varrho^{\gamma-1} \le p'(\varrho) \le c_p\varrho^{\gamma-1}. \tag{9.2.13}$$

In particular,

$$p(\varrho) \le c_p(\varrho^\gamma + 1). \tag{9.2.14}$$

Hence the energy density $P(\varrho) = \varrho \int_0^\varrho s^{-1}p(s)\,ds$ is twice continuously differentiable on $(0,\infty)$.

- The pressure function has the representation $p(\varrho) = p_c(\varrho) + p_b(\varrho)$, where $p_c$ is convex and $p_b$ is bounded.
- $B \subset \mathbb{R}^d$ is a bounded domain with $C^\infty$ boundary $\partial B$. The characteristic set $\Gamma$ defined in (9.2.1) and the given vector field $\mathbf{U}$ satisfy the geometric condition

$$\limsup_{\sigma\to 0} \sigma^{-d}\,\mathrm{meas}\,\mathcal{O}_\sigma < \infty, \quad \mathcal{O}_\sigma = \{(x,t) : \mathrm{dist}((x,t),\Gamma) \le \sigma\}. \tag{9.2.15}$$

- The initial and boundary data satisfy the conditions

$$\begin{aligned} &\mathbf{U} \in C^\infty(\mathbb{R}^d \times (0,T)), \quad \varrho_\infty \in C^1(\mathbb{R}^d \times (0,T)), \\ &\mathbf{f} \in C(B \times (0,T)), \quad \mathbf{U} = 0 \quad \text{in a neighborhood of } S \times (0,T). \end{aligned} \tag{9.2.16}$$

**Remark 9.2.6.** Throughout, we denote by $c_e$ a generic constant depending only on

$$\gamma, \quad \mathrm{diam}\,\Omega, \quad T, \quad \|\varrho_\infty\|_{L^\infty(S_T\cup\Omega\times\{0\})}, \quad \|\mathbf{U}\|_{C^1(Q)}, \quad \|\mathbf{f}\|_{C(Q)}, \quad c_p.$$

The following theorem, which is an analog of Theorem 7.4.1, gives the existence of renormalized solutions to problem (9.2.4).

**Theorem 9.2.7.** *Let $\gamma > d/2$ be given and suppose Condition 9.2.5 is satisfied. Furthermore, assume that $\partial S$ is a $C^\infty$ surface. Then problem (9.2.4) has a weak renormalized solution $(\mathbf{u},\varrho)$ which meets all requirements of Definition 9.2.3 and satisfies the estimate*

$$\|\mathbf{u}\|_{L^2(0,T;W^{1,2}(\Omega))} + \|\varrho|\mathbf{u}|^2\|_{L^\infty(0,T;L^1(\Omega))} + \|\varrho\|_{L^\infty(0,T;L^\gamma(\Omega))} \le c_e. \tag{9.2.17}$$

*Moreover, for any cylinder $Q' \Subset Q$, and for $\vartheta \in (0, \min\{2\gamma d^{-1} - 1, 2^{-1}\gamma\})$, there is a constant $c$, depending only on $\vartheta$, $c_e$, and $Q'$, such that*

$$\int_{Q'} \varrho^{\gamma+\vartheta}\,dxdt \le c. \tag{9.2.18}$$

*Proof.* First observe that under the present assumptions, the flow domain $\Omega$ and the given data $\mathbf{U}$, $\varrho_\infty$ satisfy all hypotheses of Theorem 7.4.1 and hence problem (9.2.4) has a weak solution which meets all requirements of Definition 5.1.4 and satisfies estimates (9.2.17), (9.2.18). Hence it remains to prove that $\varrho$ satisfies the integral identity (9.2.6) for any $\psi \in C^\infty(B \times (0,T))$ vanishing in a neighborhood of $\Sigma \setminus \Sigma_{\text{in}}$ and of $\Omega \times \{T\}$. Fix such a $\psi$, and choose $\eta \in C^\infty(\mathbb{R})$ such that $\eta(t) = 1$ for $t > 2$ and $\eta(t) = 0$ for $t < 1$. Set

$$\eta_n(x) = \eta(n \operatorname{dist}(x,S)) \quad \text{for } x \in \Omega = B \setminus S, \quad \eta_n(x) = 0 \quad \text{for } x \in S.$$

Since $\partial S$ is $C^3$, for sufficiently large $n$ the function $\eta_n$ belongs to $C^3(B)$, vanishes in the $(1/n)$-neighborhood of $S$ and is 1 outside the $(2/n)$-neighborhood of $S$. In particular $\psi_n = \psi\eta_n$ vanishes in a neighborhood of $\partial\Omega \setminus \Sigma_{\text{in}}$ and hence, by Definition 5.1.4, the integral identity

$$\int_Q \big(\varphi(\varrho)\partial_t\psi_n + \varphi(\varrho)\mathbf{u}\cdot\nabla\psi_n + \psi_n(\varphi(\varrho) - \varphi'(\varrho)\varrho)\operatorname{div}\mathbf{u}\big)\,dxdt$$
$$= \int_{\Sigma_{\text{in}}} \psi_n\varphi(\varrho_\infty)\mathbf{U}\cdot\mathbf{n}\,d\Sigma - \int_\Omega (\varphi(\varrho_\infty)\psi_n)(x,0)\,dx \quad (9.2.19)$$

holds for all sufficiently large $n$ and for all smooth $\varphi : [0,\infty) \to \mathbb{R}$ such that

$$\limsup_{\varrho\to\infty}(|\varphi(\varrho)| + |\varrho\varphi'(\varrho)|) < \infty.$$

Since $\eta_n$ is independent of $t$, the functions $\psi_n$ and $\partial_t\psi_n$ are uniformly bounded and converge to $\psi$ and $\partial_t\psi$ a.e. in $Q = \Omega \times (0,T)$. From the Lebesgue dominated convergence theorem we conclude that

$$\lim_{n\to\infty}\int_Q \big(\varphi(\varrho)\partial_t\psi_n + \psi_n(\varphi(\varrho) - \varphi'(\varrho)\varrho)\operatorname{div}\mathbf{u}\big)\,dxdt$$
$$= \int_Q \big(\varphi(\varrho)\partial_t\psi + \psi(\varphi(\varrho) - \varphi'(\varrho)\varrho)\operatorname{div}\mathbf{u}\big)\,dxdt, \quad (9.2.20)$$

and

$$\lim_{n\to\infty}\int_{\Sigma_{\text{in}}} \psi_n\varphi(\varrho_\infty)\mathbf{U}\cdot\mathbf{n}\,d\Sigma - \lim_{n\to\infty}\int_\Omega (\varphi(\varrho_\infty)\psi_n)(x,0)\,dx$$
$$= \int_{\Sigma_{\text{in}}} \psi\varphi(\varrho_\infty)\mathbf{U}\cdot\mathbf{n}\,d\Sigma - \int_\Omega (\varphi(\varrho_\infty)\psi)(x,0)\,dx. \quad (9.2.21)$$

Next we have

$$\int_Q \varphi(\varrho)\mathbf{u}\cdot\nabla\psi_n\,dxdt = \int_Q \varphi(\varrho)\eta_n\mathbf{u}\cdot\nabla\psi\,dxdt + \int_Q \varphi(\varrho)\psi\mathbf{u}\cdot\nabla\eta_n\,dxdt. \quad (9.2.22)$$

The Lebesgue dominated convergence theorem implies

$$\lim_{n\to\infty}\int_Q \varphi(\varrho)\eta_n\mathbf{u}\cdot\nabla\psi\,dxdt=\int_Q\varphi(\varrho)\mathbf{u}\cdot\nabla\psi\,dxdt. \tag{9.2.23}$$

On the other hand, since $\mathbf{u}$ vanishes at $\partial S$ we have

$$\int_Q \operatorname{dist}(x,S)^{-1}|\mathbf{u}|^2\,dxdt\le c(\Omega)\|\mathbf{u}\|^2_{L^2(0,T;W^{1,2}(\Omega))},$$

hence

$$\lim_{n\to\infty}\int_{\operatorname{dist}(x,\partial S)\le 2/n}\operatorname{dist}(x,S)^{-1}|\mathbf{u}|^2\,dxdt=0.$$

This leads to

$$\begin{aligned}\int_Q|\varphi(\varrho)\psi\mathbf{u}\cdot\nabla\eta_n|\,dxdt&\le c(\varphi,\psi)\int_Q|\mathbf{u}|\,|\nabla\eta_n|\,dxdt\le cn\int_{\operatorname{dist}(x,\partial S)\le 2/n}|\mathbf{u}|\,dxdt\\&\le c\Big(n\int_{\operatorname{dist}(x,\partial S)\le 2/n}dxdt\Big)^{1/2}\Big(\int_{\operatorname{dist}(x,\partial S)\le 2/n}\operatorname{dist}(x,S)^{-1}|\mathbf{u}|^2\,dxdt\Big)^{1/2}\\&\le c\Big(\int_{\operatorname{dist}(x,\partial S)\le 2/n}\operatorname{dist}(x,S)^{-1}|\mathbf{u}|^2\,dxdt\Big)^{1/2}\to 0\quad\text{as } n\to\infty,\end{aligned}$$

which along with (9.2.23) and (9.2.22) implies

$$\lim_{n\to\infty}\int_Q\varphi(\varrho)\mathbf{u}\cdot\nabla\psi_n\,dxdt=\int_Q\varphi(\varrho)\mathbf{u}\cdot\nabla\psi\,dxdt. \tag{9.2.24}$$

Letting $n\to\infty$ in identity (9.2.19) and invoking (9.2.20), (9.2.21) and (9.2.24) yields (9.2.6). □

## 9.3 Main stability theorem

We are now in a position to prove the main result of this chapter on stability of solutions to the in/out flow problem for compressible Navier-Stokes equations.

**Theorem 9.3.1.** *Let $B\subset\mathbb{R}^d$ be a bounded domain with $C^\infty$ boundary, $S\Subset B$ a compact set, and $S_n$ a sequence of compact sets with $S_n\to S$ in the sense of $\mathcal{S}$-convergence defined in Definition 8.3.5. Suppose the pressure function $p(\varrho)$ and the boundary and initial data $(\mathbf{U},\varrho_\infty)$ satisfy Condition 9.2.5. Let $\mathbf{u}_n: B\times(0,T)\to\mathbb{R}^d$ be vector fields and $\varrho_n: B\times(0,T)\to\mathbb{R}$, $n\ge 1$, be nonnegative functions satisfying*

$$\mathbf{u}_n(x,t)=0,\quad \varrho_n(x,t)=\varrho_\infty(x,0)\quad\text{in } S_n\times(0,T), \tag{9.3.1}$$
$$\mathbf{u}_n\in L^2(0,T;W^{1,2}_{S_n}(B)),\quad \varrho_n\in L^\infty(0,T;L^\gamma(B)). \tag{9.3.2}$$

*Assume that the couples* $(\mathbf{u}_n, \varrho_n)$, $n \geq 1$, *are renormalized solutions to problem* (9.2.4) (*with fixed initial and boundary data* $(\mathbf{U}, \varrho_\infty)$) *in* $(B \setminus S_n) \times (0,T)$, *and meet all requirements of Definition 9.2.3. Furthermore, assume that*

$$\|\mathbf{u}_n\|_{L^2(0,T;W^{1,2}_{S_n}(B))} + \|\varrho_n|\mathbf{u}_n|^2\|_{L^\infty(0,T;L^1(B))} + \|\varrho_n\|_{L^\infty(0,T;L^\gamma(B))} \leq c_e, \tag{9.3.3}$$

*for all sufficiently large* $n$, *and for any compact* $Q' \Subset (B \setminus S) \times (0,T)$ *and* $\theta \in (0, \min\{2\gamma d^{-1} - 1, 2^{-1}\gamma\})$, *there is a constant* $c$, *depending only on* $\theta$, $c_e$, *and* $Q'$, *such that*

$$\int_{Q'} \varrho_n^{\gamma+\theta}\, dxdt \leq c. \tag{9.3.4}$$

*Then there are* $\mathbf{u} \in L^2(0,T;W^{1,2}_S(B))$ *and* $\varrho \in L^\infty(0,T;L^\gamma(\Omega))$ *and subsequences, still denoted by* $\mathbf{u}_n$ *and* $\varrho_n$, *with the properties*

$$\begin{aligned} &\mathbf{u}_n \rightharpoonup \mathbf{u} \quad \textit{weakly in } L^2(0,T;W^{1,2}(B)),\\ &\varrho_n \rightharpoonup \varrho \quad \textit{weakly}^\star \textit{ in } L^\infty(0,T;L^\gamma(B)),\\ &\varrho_n \to \varrho \quad \textit{strongly in } L^r((B\setminus S)\times(0,T)) \quad \textit{for all } 1 \leq r < \gamma. \end{aligned} \tag{9.3.5}$$

*Moreover, there is* $b > 1$ *such that for all open* $S' \Supset S$,

$$\begin{aligned} &\varrho_n \mathbf{u}_n \otimes \mathbf{u}_n \rightharpoonup \varrho\mathbf{u}\otimes\mathbf{u} \quad \textit{weakly in } L^2(0,T;L^b(B\setminus S')),\\ &\varrho_n\mathbf{u}_n \rightharpoonup \varrho\mathbf{u} \quad \textit{weakly}^\star \textit{ in } L^\infty(0,T;L^{2\gamma/(\gamma+1)}(B)). \end{aligned} \tag{9.3.6}$$

*For any compact set* $\Omega' \Subset B \setminus S$,

$$p(\varrho_n) \to p(\varrho) \quad \textit{in } L^1(\Omega' \times (0,T)). \tag{9.3.7}$$

*The couple* $(\mathbf{u}, \varrho)$ *is a renormalized solution to problem* (9.2.4), *meets all requirements of Definition 9.2.3, and satisfies estimates* (9.3.3)–(9.3.4).

The proof is based on the kinetic equation method. Our strategy is the following. First we prove that the distribution function $f(x,t,s)$ associated with the sequence $\varrho_n$ satisfies the kinetic equations in the cylinder $(B \setminus S) \times (0,T)$. This part of the proof imitates the proof of Theorem 7.4.1. Next we employ Proposition 8.3.9 to prove that the kinetic equations can be extended to the larger cylinder $B \times (0,T)$. Then we apply Theorem 7.1.12 to ensure that $f(x,t,\cdot)$ is a Heaviside step function, which yields the strong convergence of the densities $\varrho_n$. Once the convergence of the densities is established, it is a routine task to prove that the limit $(\mathbf{u}, \varrho)$ is a renormalized solution to compressible Navier-Stokes equations.

**Part 1. Kinetic equation.** First of all we adopt the convention that the functions $\varrho_n$ are extended to $B \times (0,T)$ by

$$\varrho_n(x,t) = \varrho_\infty(x,0) \quad \text{in } S_n \times (0,T). \tag{9.3.8}$$

**Lemma 9.3.2.** *There are subsequences, still denoted by $(\mathbf{u}_n, \varrho_n)$, a vector field $\mathbf{u} \in L^2(0,T; W^{1,2}_S(B))$, a function $\overline{\varrho} \in L^\infty(0,T; L^\gamma(B))$, and a Young measure $\mu \in L^\infty_w(Q; \mathcal{M}(\mathbb{R}))$ such that for any function $\varphi \in C_0(\mathbb{R})$,*

$$\begin{aligned} \mathbf{u}_n &\rightharpoonup \mathbf{u} \quad \text{weakly in } L^2(0,T; W^{1,2}(B)), \\ \varrho_n &\rightharpoonup \varrho \quad \text{weakly}^\star \text{ in } L^\infty(0,T; L^\gamma(B)), \\ \varphi(\varrho_n) &\rightharpoonup \overline{\varphi} \quad \text{weakly}^\star \text{ in } L^\infty(B \times (0,T)), \end{aligned} \tag{9.3.9}$$

*where $\overline{\varphi}(x,t) = \langle \mu_{xt}, \varphi \rangle$.*

*Proof.* Since $\mathbf{u}_n$ is bounded in $L^2(0,T; W^{1,2}(\Omega))$, passing to a subsequence we can assume that $\mathbf{u}_n \in L^2(0,T; W^{1,2}_{S_n}(\Omega))$ converges weakly in $L^2(0,T; W^{1,2}(B))$ to some vector field $\mathbf{u}$. It remains to note that by Proposition 8.1.3 the limit $\mathbf{u}$ belongs to $L^2(0,T; W^{1,2}_S(\Omega))$. The existence of $\varrho$ follows from the boundedness of $\varrho_n$ in the space $L^\infty(0,T; L^\gamma(B))$ and weak$^\star$ compactness of bounded subsets of this space. The existence of the Young measure $\mu$ is a straightforward consequence of the fundamental Young measure theorem (Theorem 1.4.5). □

Denote as before by $f(x,t,s)$ the distribution function of the Young measure $\mu_{xt}$ defined by Lemma 9.3.2,

$$f(x,t,s) = \mu_{xt}(-\infty, s].$$

Since the $\varrho_n$ are nonnegative, we have

$$f(x,t,s) = 0 \quad \text{for } s < 0 \quad \text{and} \quad f(x,t,s) = \mu_{xt}[0,s] \quad \text{for } s \geq 0. \tag{9.3.10}$$

The next lemma establishes general properties of the distribution function as well as the weak limit of the pressure function. Denote by $\overline{p} : B \times (0,T) \to [0,\infty]$ the function defined by

$$\overline{p}(x,t) = \int_{[0,\infty)} p(s)\, d_s f(x,t,s), \quad (x,t) \in B \times (0,T).$$

**Lemma 9.3.3.** *Under the assumptions of Theorem 9.3.1, we have*

$$\int_{B\times(0,T)} \psi(x,t) \Big\{ \int_{[0,\infty)} s^\gamma\, d_s f(x,t,s) \Big\}\, dxdt \leq \lim_{n\to\infty} \int_{B\times(0,T)} \psi \varrho_n^\gamma\, dxdt, \tag{9.3.11}$$

$$\int_{Q'} \psi(x,t) \Big\{ \int_{[0,\infty)} s^{\theta+\gamma}\, d_s f(x,t,s) \Big\}\, dxdt \leq \lim_{n\to\infty} \int_{Q'} \psi \varrho_n^{\gamma+\theta}\, dxdt, \tag{9.3.12}$$

*for all nonnegative $\psi \in L^\infty(B \times (0,T))$ and all $Q' \Subset Q$, $Q = (B \setminus S) \times (0,T)$. Moreover, for all $\psi \in L^\infty(Q')$,*

$$\int_{Q'} \psi \overline{p}\, dxdt = \lim_{n\to\infty} \int_{Q'} \psi p(\varrho_n)\, dxdt. \tag{9.3.13}$$

*Proof.* Consider the nonnegative Carathéodory function $F(x,t,s) = \psi(x,t)|s|^\gamma$. It follows from estimate (9.3.3) that

$$\int_{B\times(0,T)} F(x,t,\varrho_n)\,dxdt \le c_e\|\psi\|_{L^\infty(B\times(0,T))}.$$

Hence the integrals of $F(x,t,\varrho_n)$ over $B\times(0,T)$ are uniformly bounded for $n = 1,2,\dots$. Recall that $f$ is the distribution function of the Young measure $\mu_{xt}$ associated with the sequence $\varrho_n$, i.e. $d\mu_{xt}(s) = d_s f(x,t,s)$. Hence we can apply the fundamental theorem on Young measures (Theorem 1.4.5) with $\Omega$ replaced by $B\times(0,T)$ and $v_n$ replaced by $\varrho_n$ to obtain (9.3.11). Next consider the integrand $F(x,t,s) = \psi(x,t)|s|^{\theta+\gamma}$ on a compact set $Q' \Subset Q$. It follows from estimate (9.3.4) that

$$\int_{Q'} F(x,t,\varrho_n(x,t))\,dxdt \le c\|\psi\|_{L^\infty(Q)}\int_{Q'} \varrho_n^{\theta+\gamma}\,dxdt \le c_e\|\psi\|_{L^\infty(Q)}.$$

Applying Theorem 1.4.5 with $\Omega$ replaced by $Q'$ and $v_n$ replaced by $\varrho_n$ we obtain (9.3.12). Finally choose $Q' \Subset Q$ and $\psi \in L^\infty(Q')$. Consider the integrand $F(x,t,s) = \psi(x,t)p(s)$. Since $p \le c(1+s^\gamma)$ for $s \ge 0$, we have

$$|F(x,t,\varrho_n)|^{1+\theta/\gamma} \le c\|\psi\|^{1+\theta/\gamma}_{L^\infty(Q')}(1+\varrho_n^{\gamma+\theta}).$$

Now estimate (9.3.4) yields

$$\int_{Q'} |F(x,t,\varrho_n)|^{1+\theta/\gamma}\,dxdt \le c_e\|\psi\|^{1+\vartheta/\gamma}_{L^\infty(Q')}\Big(1+\int_{Q'}\varrho_n^{\gamma+\vartheta}\,dxdt\Big) \le c,$$

where $c$ is independent of $n$. Since $\theta/\gamma > 0$, the sequence $F(x,t,\varrho_n)$ is equi-integrable in $Q'$. Applying Theorem 1.4.5 we conclude that

$$\int_{Q'} \langle \mu_{xt}, F(x,t,\cdot)\rangle\,dxdt = \lim_{n\to\infty}\int_{Q'} F(x,t,\varrho_n(x,t))\,dxdt,$$

which along with the identity

$$\langle \mu_{xt}, F(x,t,\cdot)\rangle = \psi(x,t)\int_{[0,\infty)} p(s)\,d_s f(x,t,s)$$

yields (9.3.13). □

In particular, the weak limits $\varrho$ and $\overline{p}$ of the densities $\varrho_n$ and pressure functions $p(\varrho_n)$ are determined as the expectation values

$$\varrho(x,t) = \int_{\mathbb{R}} s\,d_s f(x,t,s), \quad \overline{p}(x,t) = \int_{\mathbb{R}} p(s)\,d_s f(x,t,s). \tag{9.3.14}$$

**Lemma 9.3.4.** *Let $\varphi \in C_0^1(\mathbb{R})$, let $S_\delta$ be the $\delta$-neighborhood of $S$ and let $Q_\delta = (B \setminus S_\delta) \times (0,T) \subset Q$. Then there exist subsequences, still denoted by $\varrho_n$, $\mathbf{u}_n$, such that as $n \to \infty$,*

$$\begin{aligned} \varrho_n \mathbf{u}_n &\rightharpoonup \varrho \mathbf{u} \quad \textit{weakly}^\star \textit{ in } L^\infty(0,T; L^{2\gamma/(\gamma+1)}(B)), \\ \varphi(\varrho_n)\mathbf{u}_n &\rightharpoonup \overline{\varphi}\mathbf{u} \quad \textit{weakly in } L^2(0,T; L^2(B)), \\ \varrho_n \mathbf{u}_n \otimes \mathbf{u}_n &\rightharpoonup \varrho \mathbf{u} \otimes \mathbf{u} \quad \textit{weakly in } L^2(0,T; L^b(B \setminus S_\delta)). \end{aligned} \tag{9.3.15}$$

*Here $b$ is an arbitrary number from the interval $(1, (1-2^{-1}\mathfrak{a})^{-1})$, $\mathfrak{a} = 2d^{-1} - \gamma^{-1}$.*

*Proof.* By the extension lemma (Lemma 9.2.4), the functions $\varrho_n$ and $\mathbf{u}_n$ satisfy the mass and momentum balance equations

$$\begin{aligned} &\partial_t \varrho_n = \operatorname{div} \mathfrak{u}_n \quad \text{in } B \times (0,T), \quad \partial_t \varphi_n = \operatorname{div} \mathfrak{f}_n + \mathfrak{h}_n \quad \text{in } B \times (0,T), \\ &\partial_t(\varrho_n \mathbf{u}_n) = \operatorname{div} \mathfrak{V}_n + \mathfrak{w}_n \quad \text{in } (B \setminus S_n) \times (0,T), \end{aligned}$$

understood in the sense of distributions. Recall that $\mathbf{u}_n \in L^2(0,T; W^{1,2}_{S_n}(B))$ vanishes on $S_n \times (0,T)$ and $S_n \subset S_\delta$ for all sufficiently large $n$. Here,

$$\begin{aligned} &\mathfrak{u}_n = -\varrho_n \mathbf{u}_n, \quad \mathfrak{f}_n = -\varphi(\varrho_n)\, \mathbf{u}_n, \quad \mathfrak{w}_n = \varrho_n \mathbf{f}, \\ &\mathfrak{h}_n = (\varphi(\varrho_n) - \varrho_n \varphi'(\varrho_n)) \operatorname{div} \mathbf{u}_n, \quad \mathfrak{V}_n = \mathbb{S}(\mathbf{u}_n) - p(\varrho_n)\, \mathbb{I} - \varrho_n \mathbf{u}_n \otimes \mathbf{u}_n. \end{aligned}$$

It follows from estimates (9.3.3) that the sequences $\mathbf{u}_n$ and $\varrho_n$ are bounded in $L^2(0,T; W^{1,2}(B))$ and $L^\infty(0,T; L^\gamma(B))$, respectively. On the other hand, (9.3.3) implies that $(\mathbf{u}_n, \varrho_n)$ are bounded energy functions in the sense of Section 4.2. Moreover their energies are bounded by the constant $c_e$ of (9.3.3) independent of $n$. Hence we can apply Corollary 4.2.2 to conclude that

$$\|\varrho_n \mathbf{u}_n\|_{L^\infty(0,T; L^{2\gamma/(\gamma+1)}(B))} \le c_e.$$

Consequently, the sequence $\mathfrak{u}_n$ is bounded in $L^1(B \times (0,T))$. Thus the functions $\varrho_n$ and the vector fields $\mathfrak{u}_n$ satisfy Condition 4.4.1 of Theorem 4.4.2, and so

$$\varrho_n \mathbf{u}_n \rightharpoonup \varrho \mathbf{u} \quad \text{weakly in } L^2(0,T; L^m(\Omega)) \quad \text{for } m^{-1} > 2^{-1} + \gamma^{-1} - d^{-1}.$$

Since the sequence $\varrho_n \mathbf{u}_n$ is bounded in $L^\infty(0,T; L^{2\gamma/(\gamma+1)}(\Omega))$, it also converges weakly$^\star$ in $L^\infty(0,T; L^{2\gamma/(\gamma+1)}(\Omega))$. This leads to the first relation in (9.3.15). Repeating these arguments we obtain the second relation. To prove the third relation, notice that $B \setminus S_\delta \subset B \setminus S_n$ for all sufficiently large $n$. Next, by estimates (9.3.3) and the inequalities $0 \le p \le c(1 + \varrho^\gamma)$, the sequence $\mathfrak{V}_n$ is bounded in $L^1(Q_\delta)$. On the other hand, since $\mathbf{f}$ is bounded, it follows from (9.3.3) that $\mathfrak{w}_n$ is bounded in $L^1(Q_\delta)$. Therefore, the sequences $\varrho_n$, $\mathbf{u}_n$, $\mathfrak{V}_n$, and $\mathfrak{w}_n$ satisfy all hypotheses of Theorem 4.4.2, which yields the last relation of (9.3.15). □

Now our task is to derive the kinetic equation for the distribution function $f(x,t,s)$ in the cylinder $(B \setminus S) \times (0,T)$. This is a consequence of two lemmas which can be considered as modifications of Lemmas 7.2.6–7.2.7.

**Lemma 9.3.5.** *Let the hypotheses of Theorem 9.3.1 be satisfied and suppose* $\Phi \in C(\mathbb{R}^+)$ *vanishes for all sufficiently large* $s$. *Then, possibly passing to a subsequence, we can assume that*

$$\Phi_n \operatorname{div} \mathbf{u}_n \rightharpoonup \overline{\Phi \operatorname{div} \mathbf{u}} \quad \textit{weakly in } L^2(Q), \tag{9.3.16}$$

$$V_n \Phi_n \rightharpoonup -(\lambda+1)\overline{\Phi \operatorname{div} \mathbf{u}} + \overline{\Phi p} \quad \textit{weakly in } L^2(Q), \tag{9.3.17}$$

$$\overline{\Phi \operatorname{div} \mathbf{u}} = \overline{\Phi} \operatorname{div} \mathbf{u} + \frac{1}{\lambda+1}\big(\overline{\Phi p} - \overline{\Phi}\overline{p}\big). \tag{9.3.18}$$

*Here*

$$\Phi_n = \Phi(\varrho_n), \quad V_n = -(1+\lambda) \operatorname{div} \mathbf{u}_n + p(\varrho_n).$$

*Proof.* Assume that $\Phi$ is extended to the semiaxis $(-\infty, 0)$, and the extended function belongs to $C_c(\mathbb{R})$. It suffices to prove the lemma for $\Phi \in C_0^\infty(\mathbb{R})$ since $C_0^\infty(\mathbb{R})$ is dense in $C_0(\mathbb{R})$. First observe that the governing equations (9.2.4) for $(\mathbf{u}_n, \varrho_n)$ can be rewritten in the form

$$\partial_t(\varrho_n \mathbf{u}_n) + \operatorname{div}(\varrho_n \mathbf{u}_n \otimes \mathbf{u}_n) = \operatorname{div} \mathbb{T}_n + \varrho_n \mathbf{f} \quad \text{in } (B \setminus S_n) \times (0,T), \tag{9.3.19a}$$

$$\partial_t \varrho_n + \operatorname{div}(\varrho_n \mathbf{u}_n) = 0 \quad \text{in } (B \setminus S_n) \times (0,T), \tag{9.3.19b}$$

where

$$\mathbb{T}_n = \mathbb{S}(\mathbf{u}_n) - p(\varrho_n)\,\mathbb{I},. \tag{9.3.19c}$$

Moreover, since the couple $(\mathbf{u}_n, \varrho_n)$ is a renormalized solution to problem (9.2.4) we have the following equation for $\Phi_n = \Phi(\varrho_n)$:

$$\partial \Phi_n + \operatorname{div}(\Phi_n \mathbf{u}_n) + \varpi_n = 0 \quad \text{in } (B \setminus S_n) \times (0,T). \tag{9.3.19d}$$

Here

$$\varpi_n = (\Phi'(\varrho_n)\varrho_n - \Phi(\varrho_n)) \operatorname{div} \mathbf{u}_n. \tag{9.3.19e}$$

Now choose $r'$ satisfying $1 < r' \le 1 + \theta\gamma^{-1}$, $q \in (1,2)$ and take $s = \infty$. Next set $\mathbb{G}_n = 0$, $\mathbf{g}_n = \mathbf{g}_n^\varphi = 0$. It follows from estimates (9.3.3) and (9.3.4) and the inequality $p \le c(1 + \varrho^\gamma)$ that for any $Q' \Subset (B \setminus S_n) \times (0,T)$,

$$\|\mathbb{T}_n\|_{L^1(B\times(0,T))} \le c, \quad \|\mathbb{T}_n\|_{L^{r'}(Q')} \le c(Q'), \quad \|\varpi_n\|_{L^2(B\times(0,T))} \le c_e. \tag{9.3.20}$$

The latter inequality is a consequence of estimate (9.3.3) for $\mathbf{u}_n$ and the boundedness of $\Phi'\varrho - \Phi$. Now choose a sufficiently small $\delta > 0$ and denote by $S_\delta$ the $\delta$-neighborhood of $S$. Since $S_n \to S$ in the Hausdorff metric, we have

$$(B \setminus S_\delta) \times (0,T) \subset (B \setminus S_n) \times (0,T)$$

for all sufficiently large $n$. It follows that the triplet $(\varrho_n, \mathbf{u}_n, \Phi_n)$ and the exponents $s, r', q$ satisfy Condition 4.6.1 of Theorem 4.6.3 and, in view of representations

(9.3.19c) and (9.3.19e), they also meet all requirements of Theorem 4.7.1 in the cylinder $(B \setminus S_\delta) \times (0,T)$. Next, by (9.3.4) the functions $p(\varrho_n)$ are bounded in $L^{r'}(Q')$ for any compact subset $Q' \Subset (B \setminus S_\delta) \times (0,T)$. On the other hand, in view of (9.3.13),

$$\lim_{n\to\infty} \int_{Q'} \psi(x,t)p(\varrho_n)\,dxdt = \int_{Q'} \psi(x,t)\overline{p}\,dxdt \quad \text{for all } \psi \in L^\infty(Q').$$

It follows that

$$p(\varrho_n) \rightharpoonup \overline{p} \quad \text{weakly in } L^{r'}(Q'). \tag{9.3.21}$$

Notice that $\overline{p} : B \times (0,T) \to \mathbb{R}$ is already defined and is independent of $Q'$. Thus

$$V_n = -(1+\lambda)\operatorname{div}\mathbf{u}_n + p(\varrho_n) \rightharpoonup \overline{V} = -(1+\lambda)\operatorname{div}\mathbf{u} + \overline{p} \tag{9.3.22}$$

weakly in $L^{r'}(Q')$ for every compact $Q' \Subset (B \setminus S_\delta) \times (0,T)$. Next, since $\Phi \in C_0^\infty(\mathbb{R})$, the convergences (9.3.9) imply

$$\Phi_n \rightharpoonup \overline{\Phi} \quad \text{weakly}^\star \text{ in } L^\infty(B \times (0,T)). \tag{9.3.23}$$

Applying Theorem 4.7.1 we find that

$$\lim_{n\to\infty} \int_{(B\setminus S_\delta)\times(0,T)} V_n \Phi_n \eta\,dxdt = \int_{(B\setminus S_\delta)\times(0,T)} \overline{V}\,\overline{\Phi}\eta\,dxdt \tag{9.3.24}$$

for all $\eta \in C_0^\infty((B \setminus S_\delta) \times (0,T))$. Notice that the integrand on the right hand side of (9.3.24) is defined in $B \times (0,T)$ and is independent of $\delta$.

Recall that $p(s)\Phi(s)$ belongs to $C_c(\mathbb{R})$ and $\Phi_n p(\varrho_n)$ are bounded uniformly in $n$, which along with (9.3.9) implies

$$\Phi_n p(\varrho_n) \rightharpoonup \overline{\Phi p} \quad \text{weakly}^\star \text{ in } L^\infty(B \times (0,T)). \tag{9.3.25}$$

Together with (9.3.24), this implies that for any $\eta \in C_0^\infty((B \setminus S_\delta) \times (0,T))$, the limit

$$\lim_{n\to\infty} \int_Q \eta\Phi_n \operatorname{div}\mathbf{u}_n\,dxdt$$

exists and is independent of $\eta$ and $\delta$. Since the sequence $\Phi_n \operatorname{div}\mathbf{u}_n$ is bounded in $L^2(B \times (0,T))$, this implies the existence of some $\overline{\Phi \operatorname{div}\mathbf{u}} \in L^2((B \setminus S_\delta) \times (0,T))$ such that

$$\Phi_n \operatorname{div}\mathbf{u}_n \rightharpoonup \overline{\Phi \operatorname{div}\mathbf{u}} \quad \text{weakly in } L^2((B \setminus S_\delta) \times (0,T)). \tag{9.3.26}$$

As this limit is independent of $\delta$, this implies (9.3.16). Combining (9.3.25) with (9.3.26) we obtain

$$-(1+\lambda)\Phi(\varrho_n)\operatorname{div}\mathbf{u}_n + \Phi(\varrho_n)p(\varrho_n) \rightharpoonup -(\lambda+1)\overline{\Phi \operatorname{div}\mathbf{u}} + \overline{\Phi p} \quad \text{weakly in } L^2(Q),$$

which yields (9.3.17). It follows from (9.3.24) that the weak limit of $V_n \Phi_n$ exists and is independent of $\delta$. Thus we get

$$\overline{\Phi \operatorname{div} \mathbf{u}} = \overline{\Phi} \operatorname{div} \mathbf{u} + \frac{1}{\lambda + 1} (\overline{\Phi p} - \overline{\Phi} \overline{p}),$$

which is (9.3.18). □

**Lemma 9.3.6.** *Let the hypotheses of Theorem 9.3.1 be satisfied. Furthermore assume that $\varphi \in C_0^\infty(\mathbb{R})$ and $\Phi(s) = \varphi'(s)s - \varphi(s)$. Then the integral identity*

$$\int_Q \overline{\varphi}(\partial_t \psi + \nabla \psi \cdot \mathbf{u})\, dxdt - \frac{1}{\lambda+1} \int_Q \psi (\overline{\Phi p} - \overline{\Phi}\overline{p})\, dxdt$$
$$- \int_Q \psi \overline{\Phi} \operatorname{div} \mathbf{u}\, dxdt + \int_\Omega (\psi \varphi(\varrho_\infty))(x, 0)\, dx - \int_{S_T} \psi \varphi(\varrho_\infty) \mathbf{U} \cdot \mathbf{n}\, dSdt = 0 \tag{9.3.27}$$

*holds for all $\psi \in C^\infty(Q)$ vanishing in a neighborhood of $\Sigma \setminus \Sigma_{\text{in}}$, of $S \times (0, T)$, and of $\Omega \times \{t = T\}$. Here*

$$\overline{\varphi} = \int_{[0,\infty)} \varphi(s)\, d_s f(x, t, s), \quad \overline{p} = \int_{[0,\infty)} p(s)\, d_s f(x, t, s),$$
$$\overline{\Phi p} = \int_{[0,\infty)} \Phi(s) p(s)\, d_s f(x, t, s), \quad \overline{\Phi} = \int_{[0,\infty)} \Phi(s)\, d_s f(x, t, s). \tag{9.3.28}$$

*Proof.* Choose $\varphi \in C_0^\infty(\mathbb{R})$ and a test function $\psi$ as in the statement. Since $\varrho_n$ is a renormalized solution to the boundary value problem (9.2.4) in the cylinder $(B \setminus S_n) \times (0, T)$ in the sense of Definition 9.2.3, we have

$$\int_{(B \setminus S_n) \times (0,T)} \Big( \varphi_n (\partial_t \psi + \nabla \psi \cdot \mathbf{u}_n) - \psi \Phi_n \operatorname{div} \mathbf{u}_n \Big)\, dxdt$$
$$+ \int_{B \setminus S_n} (\psi \varphi(\varrho_\infty))(x, 0)\, dx - \int_{S_T} \psi \varphi(\varrho_\infty) \mathbf{U} \cdot \mathbf{n}\, dSdt = 0,$$

where $\varphi_n = \varphi(\varrho_n)$ and $\Phi_n = \varphi'(\varrho_n)\varrho_n - \varphi(\varrho_n)$. It is important to notice that $\psi$ vanishes in a neighborhood of $\Sigma \setminus \Sigma_{\text{in}}$, of $S_n \times (0, T)$ and of $(B \setminus S_n) \times \{t = T\}$ for all sufficiently large $n$ since $S_n \to S$ in the Hausdorff metric. Letting $n \to \infty$ and using (9.3.15) and (9.3.16) we obtain

$$\int_Q \overline{\varphi}(\partial_t \psi + \nabla \psi \cdot \mathbf{u})\, dxdt - \int_Q \psi \overline{\Phi \operatorname{div} \mathbf{u}}\, dxdt$$
$$+ \int_\Omega (\psi \varphi(\varrho_\infty))(x, 0)\, dx - \int_{S_T} \psi \varphi(\varrho_\infty) \mathbf{U} \cdot \mathbf{n}\, dSdt = 0.$$

Substituting (9.3.18) we obtain (9.3.27). □

**Lemma 9.3.7.** *Under the assumptions of Theorem 9.3.1, let* $\psi \in C^\infty(B\times(0,T)\times\mathbb{R})$ *satisfy*

$$\begin{aligned} &\psi = 0 \quad \text{in a neighborhood of} \\ &\qquad ((\Sigma\setminus\Sigma_{\text{in}})\times\mathbb{R})\cup(B\times\{t=T\}\times\mathbb{R})\cup(S\times(0,T)\times\mathbb{R}), \\ &\psi(x,t,s)=0 \quad \text{for all large } s \text{ uniformly in } (x,t). \end{aligned} \tag{9.3.29}$$

*Then the distribution function* $f$ *defined by* (9.3.10) *satisfies*

$$\begin{aligned} &\int_Q\int_{\mathbb{R}} \big(f\partial_t\psi + f\nabla_x\psi\cdot\mathbf{u} - f\partial_s\psi\,\mathrm{div}\,\mathbf{u} - \partial_s\psi\,\mathfrak{C}[f]\big)\,dsdxdt \\ &\quad + \int_\Omega\int_{\mathbb{R}} f_\infty(x,0,s)\psi(x,0,s)\,dsdx - \int_{\Sigma_{\text{in}}}\int_{\mathbb{R}} f_\infty(x,t,s)\psi(x,t,s)\mathbf{U}\cdot\mathbf{n}\,dsd\Sigma = 0. \end{aligned} \tag{9.3.30}$$

*The nonlinear operator* $\mathfrak{C}[f]$ *is defined by*

$$\mathfrak{C}[f] = \frac{1}{\lambda+1}\int_{(-\infty,s]}(p(\tau)-\overline{p})\,d_\tau f(x,t,\tau), \tag{9.3.31}$$

$$\overline{p} = \int_{\mathbb{R}} p(s)\,d_s f(x,t,s), \tag{9.3.32}$$

*and* $f_\infty$ *is given by*

$$f_\infty(x,t,s)=1 \quad \text{for } s\geq \varrho_\infty(x,t) \quad \text{and} \quad f_\infty(x,t,s)=0 \quad \text{otherwise.} \tag{9.3.33}$$

*Proof.* As the proof uses the same arguments as in the proof of Lemma 7.2.8, we just give an outline. Choose $\eta\in C_0^\infty(\mathbb{R})$ and set

$$\varphi(s)=\int_s^\infty \eta(\tau)\,d\tau, \quad \Phi(s) = s\varphi'(s)-\varphi(s) = -s\eta(s)-\int_s^\infty\eta(\tau)\,d\tau. \tag{9.3.34}$$

First we derive a formula for $\overline{\varphi}$. Repeating the arguments in the proof of Lemma 7.2.8 we obtain

$$\overline{\varphi}(x,t) = \int_{\mathbb{R}} \eta(s)f(x,t,s)\,ds, \quad \overline{\varphi}_\infty(x,t) = \int_{\mathbb{R}}\eta(s)f_\infty(x,t,s)\,ds, \tag{9.3.35}$$

and also

$$\overline{\varrho\varphi'} = \int_{\mathbb{R}}(s\eta)'f(x,t,s)\,ds,\ \overline{\Phi}(x,t) = \overline{\varrho\varphi'}-\overline{\varphi}(x,t) = \int_{\mathbb{R}} s\eta'(s)f(x,t,s)\,ds. \tag{9.3.36}$$

Next for a.e. $(x,t)\in Q$ denote by $F$ the distribution function of the Borel measure $d\nu_{xt}(s)=p(s)\mu_{xt}(s)$ on $\mathbb{R}$,

$$F(x,t,s) = \int_{(-\infty,s]} p(\tau)\,d_\tau f(x,t,\tau).$$

Since $p(s)$ is integrable with respect to the measure $\mu_{xt}$ for a.e. $(x,t) \in Q$, the function $F(x,t,s)$ is monotone, right continuous in $s$ and

$$F(x,t,s) = 0 \quad \text{for } s < 0, \quad F(x,t,s) \to \overline{p}(x,t) \quad \text{as } s \to \infty.$$

We have the following analogues of formulae (9.3.35)–(9.3.36):

$$\int_{\mathbb{R}} \varphi(s)p(s)\, d_s f(x,t,s) = \int_{\mathbb{R}} \varphi(s)\, d_s F(x,t,s) = \int_{\mathbb{R}} \eta(s)F(x,t,s)\, ds,$$
$$\int_{\mathbb{R}} \varphi'(s)sp(s)\, d_s f(x,t,s) = \int_{\mathbb{R}} \varphi'(s)s\, d_s F(x,t,s) = \int_{\mathbb{R}} (s\eta(s))'F(x,t,s)\, ds,$$

which gives

$$\begin{aligned}\overline{\Phi p} &= \int_{\mathbb{R}} (\varphi'(s)s - \varphi(s))p(s)\, d_s f(x,t,s) = \int_{\mathbb{R}} s\eta'(s)F(x,t,s)\, ds \\ &= \int_{\mathbb{R}} \eta'(s)s\Big\{ \int_{(-\infty,s]} p(s)\, d_\tau f(x,t,\tau)\Big\}\, ds.\end{aligned}$$

Combining this formula with (9.3.36) we obtain

$$\begin{aligned}\overline{\Phi p} - \overline{\Phi}\overline{p} &= \int_{\mathbb{R}} \eta'(s)s\Big\{ \int_{(-\infty,s]} (p(s) - \overline{p})\, d_\tau f(x,t,\tau)\Big\}\, ds \\ &= (\lambda+1)\int_{\mathbb{R}} s\eta'(s)\mathfrak{C}[f])\, ds. \end{aligned} \tag{9.3.37}$$

Let $\varsigma \in C^\infty(Q)$ vanish in a neighborhood of the set

$$(\Sigma \setminus \Sigma_{\text{in}}) \cup (S \times (0,T)) \cup (\Omega \times \{t = T\}).$$

Then $\varsigma$ and $\varphi$ meet all requirements of Lemma 9.3.6 and hence

$$\begin{aligned}&\int_Q \overline{\varphi}(\partial_t\varsigma + \nabla\varsigma\cdot\mathbf{u})\, dxdt - \frac{1}{\lambda+1}\int_Q \varsigma(\overline{\Phi p} - \overline{\Phi}\overline{p})\, dxdt \\ &- \int_Q \varsigma\overline{\Phi}\operatorname{div}\mathbf{u}\, dxdt + \int_\Omega (\varsigma\varphi(\varrho_\infty))(x,0)\, dx - \int_{S_T} \varsigma\varphi(\varrho_\infty)\mathbf{U}\cdot\mathbf{n}\, dSdt = 0. \end{aligned} \tag{9.3.38}$$

It follows from (9.3.35) that

$$\int_Q \overline{\varphi}(\partial_t\varsigma + \nabla\varsigma\cdot\mathbf{u})\, dxdt = \int_Q\int_{[0,\infty)} \eta(s)(\partial_t\varsigma + \nabla\varsigma\cdot\mathbf{u})f(x,t,s)\, dsdxdt,$$

$$\begin{aligned}&\int_\Omega (\varsigma\overline{\varphi}_\infty)(x,0)\, dx - \int_{S_T} \varsigma\overline{\varphi}_\infty\mathbf{U}\cdot\mathbf{n}\, dSdt \\ &= \int_\Omega\int_{[0,\infty)} \varsigma(x,0)\eta(s)f_\infty(x,0,s)\, dsdx - \int_{S_T}\int_{[0,\infty)} \varsigma\eta(s)\mathbf{U}\cdot\mathbf{n}\, f_\infty(x,t,s)\, ds\, dSdt.\end{aligned} \tag{9.3.39}$$

Next, formulae (9.3.36) and (9.3.37) imply

$$\begin{aligned}\frac{1}{\lambda+1}\int_Q \varsigma\big(\overline{\Phi p}-\overline{\Phi}\overline{p}\big)\,dxdt &= \int_Q\int_{[0,\infty)} \varsigma s\eta'(s)\mathfrak{C}[f](x,t,s)\,dsdxdt,\\ \int_Q \varsigma\overline{\Phi}\operatorname{div}\mathbf{u}\,dxdt &= \int_Q\int_{[0,\infty)} \varsigma s\eta'(s) f(x,t,s)\operatorname{div}\mathbf{u}\,dsdxdt.\end{aligned} \tag{9.3.40}$$

Substituting (9.3.39)–(9.3.40) into (9.3.38) we finally obtain

$$\begin{aligned}\int_{Q\times[0,\infty)} f\big(\partial_t(\varsigma\eta)+\nabla_x(\varsigma\eta)\cdot\mathbf{u}\big)\,dxdtds - \int_{Q\times[0,\infty)} s\partial_s(\varsigma\eta)\mathfrak{C}[f]\,dxdtds\\ -\int_{Q\times[0,\infty)} s\partial_s(\varsigma\eta) f\operatorname{div}\mathbf{u}\,dxdtds + \int_{\Omega\times[0,\infty)} (\varsigma\eta)(x,0,s)f_\infty(x,0,s)\,dxds\\ -\int_{S_T\times[0,\infty)} (\varsigma\eta)f_\infty\mathbf{U}\cdot\mathbf{n}\,dSdtds = 0.\end{aligned} \tag{9.3.41}$$

Thus we have proved the desired identity (9.3.30) for all test functions $\psi$ having the representation $\psi(x,t,s)=\varsigma(x,t)\eta(s)$.

It remains to extend this result to all $\psi(x,t,s)$ satisfying (9.3.29). To this end, apply the Fourier transform to $\psi(x,t,s)$ with respect to variable $s$ and argue as in the proof of Lemma 7.2.8 to obtain the desired identity (9.3.30). □

It follows from Lemma 9.3.7 that the distribution function $f$ of the Young measure $\mu$ associated with the sequence $\varrho_n$ is a weak solution to the kinetic equation

$$\partial_t f + \operatorname{div}(f\mathbf{u}) - \partial_s(sf\operatorname{div}\mathbf{u}) - \partial_s\mathfrak{C}(f) = 0 \quad \text{in } Q\times\mathbb{R}. \tag{9.3.42}$$

The solution also satisfies in a weak sense the boundary and initial conditions

$$f = f_\infty \quad \text{at } \Sigma_{\text{in}}\times\mathbb{R}, \quad f(x,0,s)=f_\infty(x,0,s) \quad \text{on } (B\setminus S)\times\mathbb{R}. \tag{9.3.43}$$

Our next task is to show that the solution obtained for the kinetic equation (9.3.42) is regular in the sense of Definition 7.1.8, i.e. it has the additional integrability properties pointed out in this definition. This results from the following three lemmas which are modifications of Lemmas 7.2.9–7.2.11. First we prove an identity which is a particular case of (9.3.18). For any continuous bounded function $\vartheta: Q\to\mathbb{R}$ set

$$\phi_\vartheta(\varrho,x,t) = \min\{\varrho,\,\vartheta(x,t)\}. \tag{9.3.44}$$

It follows from the definition of the distribution function $f$ and from the fundamental theorem on Young measures (Theorem 1.4.5) that

$$\begin{aligned}&\phi_\vartheta(\cdot,\varrho_n)\rightharpoonup\overline{\phi}_\vartheta \quad \text{weakly}^\star \text{ in } L^\infty(B\times(0,T)), \quad \text{where}\\ &\overline{\phi}_\vartheta(x,t) = \int_{[0,\infty)} \min\{s,\vartheta(x,t)\}\,d_s f(x,t,s).\end{aligned} \tag{9.3.45}$$

**Lemma 9.3.8.** *Let $\vartheta \in C(Q)$, $Q = (B \setminus S) \times (0,T)$. Then for any $h \in C_c(Q)$,*

$$\lim_{n\to\infty} \int_Q h\big(\phi_\vartheta(\varrho_n) \operatorname{div} \mathbf{u}_n - \overline{\phi}_\vartheta \operatorname{div} \mathbf{u}\big)\, dxdt$$
$$= \frac{1}{\lambda+1} \lim_{n\to\infty} \int_Q h\big(\phi_\vartheta(\varrho_n)p(\varrho_n) - \overline{\phi}_\vartheta \overline{p}\big)\, dxdt. \quad (9.3.46)$$

*Proof.* Let $Q' \Subset Q$ be a compact set such that $\operatorname{supp} h \subset Q'$. Notice that $Q' \Subset (B \setminus S_n) \times (0,T)$ for all sufficiently large $n$. Choose $N > \sup_Q \vartheta(x,t)$ and $\delta > 0$. Since $\phi_\vartheta$ is uniformly continuous in the cylinder $Q \times [0,N]$, there are smooth functions $\varphi_k : [0,N] \to \mathbb{R}$ and $\vartheta_k : Q \to \mathbb{R}$, $1 \le k \le n$, such that the function

$$\phi_\delta(x,t,\varrho) = \sum_{k=1}^{n} \vartheta_k(x,t)\varphi_k(\varrho) \quad (9.3.47)$$

approximates $\phi_\vartheta$ with error bounded by $\delta$,

$$\sup_{Q\times[0,N]} |\phi_\vartheta(x,t,\varrho) - \phi_\delta(x,t,\varrho)| \le \delta.$$

Extend $\varphi_k$ to $[0,\infty)$ by setting $\varphi_k(s) = \varphi_k(N)$ for $s > N$. This gives an extension of $\phi_\delta$ to the cylinder $Q \times [0,\infty)$ We keep the same notation for the extended functions. It follows from (9.3.44) that

$$\sup_{Q\times[0,\infty)} |\phi_\vartheta(x,t,\varrho) - \phi_\delta(x,t,\varrho)| \le \delta. \quad (9.3.48)$$

Let us prove that the desired equality (9.3.46) holds for the approximation $\phi_\delta$. Since the limits on both sides of (9.3.46) are linear in $\phi_\vartheta$, it suffices to prove that for every $k$,

$$\lim_{n\to\infty} \int_Q h\vartheta_k\big(\varphi_k(\varrho_n) \operatorname{div} \mathbf{u}_n - \overline{\varphi}_k \operatorname{div} \mathbf{u}\big)\, dxdt$$
$$= \frac{1}{\lambda+1} \lim_{n\to\infty} \int_Q h\big(\varphi_k(\varrho_n)p(\varrho_n) - \overline{\varphi}_k \overline{p}\big)\, dxdt. \quad (9.3.49)$$

To this end notice that the Lipschitz function $\varphi_k(\varrho) - \varphi_k(N)$ belongs to $C_0(\mathbb{R}^+)$ and hence, by the convergences (9.3.9), we have

$$\lim_{n\to\infty} \int_Q h\vartheta_k\Big(\big(\varphi_k(\varrho_n) - \varphi_k(N)\big) \operatorname{div} \mathbf{u}_n - \big(\overline{\varphi_k - \varphi_k(N)}\big) \operatorname{div} \mathbf{u}\Big)\, dxdt$$
$$= \frac{1}{\lambda+1} \lim_{n\to\infty} \int_Q h\vartheta_k\Big(\big(\varphi_k(\varrho_n) - \varphi_k(N)\big)p(\varrho_n) - \big(\overline{\varphi_k - \varphi_k(N)}\big)\overline{p}\Big)\, dxdt. \quad (9.3.50)$$

Recalling that $\operatorname{div}\mathbf{u}_n$ converges to $\operatorname{div}\mathbf{u}$ weakly in $L^2(B\times(0,T))$ and using (9.3.18) we obtain

$$\lim_{n\to\infty}\int_Q h\vartheta_k\big(\varphi_k(N)\operatorname{div}\mathbf{u}_n-\varphi_k(N)\operatorname{div}\mathbf{u}\big)\,dxdt$$
$$=\frac{1}{\lambda+1}\lim_{n\to\infty}\int_Q h\vartheta_k\big(\varphi_k(N)p(\varrho_n)-\varphi_k(N)\overline{p}\big)\,dxdt.$$

Combining this with (9.3.50) we arrive at (9.3.49). This completes the proof of the desired equality

$$\lim_{n\to\infty}\int_Q h\big(\phi_\delta(\varrho_n)\operatorname{div}\mathbf{u}_n-\overline{\phi}_\delta\operatorname{div}\mathbf{u}\big)\,dxdt$$
$$=\frac{1}{\lambda+1}\lim_{n\to\infty}\int_Q h\big(\phi_\delta(\varrho_n)p(\varrho_n)-\overline{\phi_\delta p}\big)\,dxdt \qquad (9.3.51)$$

for the approximation $\phi_\delta$.

Next, by (9.3.3) we have

$$\|\operatorname{div}\mathbf{u}_n\|_{L^2(Q')}+\|p(\varrho_n)\|_{L^1(Q')}\le c,$$

which gives

$$\Big|\int_Q h(\phi_\vartheta(\varrho_n)-\phi_\delta(\varrho_n))\operatorname{div}\mathbf{u}_n\,dxdt\Big|+\Big|\int_Q h(\phi_\vartheta(\varrho_n)-\phi_\delta(\varrho_n))p(\varrho_n)\,dxdt\Big|\le c\delta. \qquad (9.3.52)$$

Next since $\overline{p}\in L^1(Q)$ and $|\overline{\phi}_\vartheta-\overline{\phi}_\delta|=|\overline{\phi_\vartheta-\phi_\delta}|\le\delta$, we have

$$\Big|\int_Q h(\overline{\phi}_\vartheta-\overline{\phi}_\delta)\operatorname{div}\mathbf{u}\,dxdt\Big|+\Big|\int_Q h(\overline{\phi}_\vartheta-\overline{\phi_\delta})\overline{p}\,dxdt\Big|\le c\delta. \qquad (9.3.53)$$

Writing the desired conclusion (9.3.46) in the form

$$\lim_{n\to\infty}\int_Q H(x,t)\,dxdt=0$$

with appropriate $H$, we observe that combining (9.3.51)–(9.3.53) gives

$$-c\delta\le\liminf_{n\to\infty}\int_Q H(x,t)\,dxdt\le\limsup_{n\to\infty}\int_Q H(x,t)\,dxdt\le c\delta.$$

Letting $\delta\to0$ we obtain (9.3.46). □

Now choose $\vartheta\in C(Q)$ and define

$$\mathcal{T}_\vartheta(x,t)=\phi_\vartheta(x,t,\varrho)-\overline{\phi}_\vartheta,\quad\text{where}\quad \varrho(x,t)=\int_{[0,\infty)}s\,d_sf(x,t,s) \qquad (9.3.54)$$

Recall

$$\overline{\phi}_\vartheta(x,t) = \int_{[0,\infty)} \min\{s, \vartheta(x,t)\}\, d_s f(x,t,s), \tag{9.3.55}$$

$$\phi_\vartheta(x,t,\varrho) = \min\Big\{ \int_{[0,\infty)} s\, d_s f(x,t,s),\, \vartheta(x,t) \Big\}. \tag{9.3.56}$$

**Lemma 9.3.9.** *Under the assumptions of Theorem 9.3.1 for every compact set $Q' \Subset Q = (B \setminus S) \times (0,T)$, there is a constant $c$ independent of $\vartheta$ and $Q'$ such that*

$$\|\mathcal{T}_\vartheta\|^{\gamma+1}_{L^{\gamma+1}(Q')} \leq \limsup_{n\to\infty} \int_{Q'} |\min\{\varrho_n, \vartheta\} - \min\{\varrho, \vartheta\}|^{\gamma+1}\, dxdt \leq c. \tag{9.3.57}$$

*Proof.* Notice that the function $s \mapsto \phi_\vartheta(x,t,s)$ defined by (9.3.54) is monotone and its derivative does not exceed 1. It follows that for any $s'' \leq s'$,

$$|\phi_\vartheta(x,t,s') - \phi_\vartheta(x,t,s'')|^{1+\gamma} \leq (\phi_\vartheta(x,t,s') - \phi_\vartheta(x,t,s''))(s'^\gamma - s''^\gamma). \tag{9.3.58}$$

Next, (9.2.13) implies

$$s'^\gamma - s''^\gamma \leq c(p(s') - p(s'')) \quad \text{for } 1 \leq s'' \leq s'. \tag{9.3.59}$$

By (9.2.12), the pressure function $p$ satisfies the estimate $s'^\gamma \leq c(p(s') + 1)$ and is bounded on $[0,1]$, which leads to

$$s'^\gamma - s''^\gamma \leq c(p(s') - p(s'')) + c \quad \text{for } 0 \leq s'' \leq 1 \leq s'. \tag{9.3.60}$$

Combining (9.3.59) and (9.3.60) we conclude that

$$s'^\gamma - s''^\gamma \leq c(p(s') - p(s'')) + c \quad \text{for } 0 \leq s'' \leq s'. \tag{9.3.61}$$

From (9.3.58) we now obtain

$$\begin{aligned} &|\phi_\vartheta(x,t,s') - \phi_\vartheta(x,t,s'')|^{1+\gamma} \\ &\qquad \leq c_p(\phi_\vartheta(x,t,s') - \phi_\vartheta(x,t,s''))(p(s') - p(s'') + c). \end{aligned} \tag{9.3.62}$$

Now choose $Q' \Subset Q$ and $h \in C_0(Q)$ such that $0 \leq h \leq 1$ and $h = 1$ in $Q'$. It follows from the definition of $\mathcal{T}_\vartheta$ that

$$\begin{aligned} \int_{Q'} |\mathcal{T}_\vartheta|^{\gamma+1}\, dxdt &\leq \int_Q h|\mathcal{T}_\vartheta|^{\gamma+1}\, dxdt \\ &= \lim_{n\to\infty} \int_Q h|\phi_\vartheta(x,t,\varrho_n) - \phi_\vartheta(x,t,\varrho)|^{1+\gamma}\, dxdt. \end{aligned} \tag{9.3.63}$$

Now our task is to estimate the right hand side of this inequality. First we use (9.3.62) to obtain

$$\begin{aligned}
\lim_{n\to\infty}\int_Q h|\phi_\vartheta(x,t,\varrho_n)-\phi_\vartheta(x,t,\varrho)|^{1+\gamma}\,dxdt \\
\le c_p\lim_{n\to\infty}\int_Q h\big(\phi_\vartheta(x,t,\varrho_n)-\phi_\vartheta(x,t,\varrho)\big)\big(p(\varrho_n)-p(\varrho)\big)\,dxdt \\
+c\lim_{n\to\infty}\int_Q h\big(\phi_\vartheta(x,t,\varrho_n)-\phi_\vartheta(x,t,\varrho)\big)\,dxdt \\
=c_p\int_Q h\,I(x,t)\,dxdt+c\int_Q h(\overline{\phi}_\vartheta-\phi_\vartheta(x,t,\varrho))\,dxdt,
\end{aligned}\tag{9.3.64}$$

where

$$I=\overline{\phi_\vartheta p}-\overline{\phi}_\vartheta\,p(\varrho)-\phi_\vartheta(\cdot,\varrho)\,\overline{p}+\phi_\vartheta(\cdot,\varrho)\,p(\varrho).$$

Now rewrite the expression for $I$ in the form

$$I=\overline{\phi_\vartheta p}-\overline{\phi}_\vartheta\,\overline{p}+\big(\overline{\phi}_\vartheta-\phi_\vartheta(\cdot,\varrho)\big)\big(\overline{p}-p(\varrho)\big).\tag{9.3.65}$$

Recall that, by Condition 9.2.5, the pressure function has a representation $p(\varrho)=p_c(\varrho)+p_b(\varrho)$ where $p_c$ is convex and $p_b$ is continuous and bounded, so there exists $c$ such that $|p_b|\le c$. Since $p_c$ is convex, we have $\overline{p}_c\ge p_c(\varrho)$. It follows that

$$\overline{p}\ge\overline{p}_c-c\ge p_c(\varrho)-c\ge p(\varrho)-2c,\quad\text{so}\quad \overline{p}-p(\varrho)\ge-2c.$$

On the other hand, since $\phi_\vartheta(x,t,\cdot)$ is concave, we have $\overline{\phi}_\vartheta\le\phi_\vartheta(\varrho)$. Thus we get

$$\big(\overline{\phi}_\vartheta-\phi_\vartheta(\varrho)\big)\big(\overline{p}-p(\overline{\varrho})\big)\le 2c\big(\phi_\vartheta(\varrho)-\overline{\phi}_\vartheta\big),$$

which along with (9.3.65) leads to

$$I\le\overline{\phi_\vartheta p}-\overline{\phi}_\vartheta\,\overline{p}+c\big(\phi_\vartheta(\varrho)-\overline{\phi}_\vartheta\big).$$

Combining this with (9.3.64) we arrive at

$$\begin{aligned}
\lim_{n\to\infty}\int_Q h|\phi_\vartheta(x,t,\varrho_n)-\phi_\vartheta(x,t,\varrho)|^{1+\gamma}\,dxdt \\
\le c\int_Q h\big(\overline{\phi_\vartheta p}-\overline{\phi}_\vartheta\,\overline{p}\big)\,dxdt+c\int_Q h\big(\phi_\vartheta(\varrho)-\overline{\phi}_\vartheta\big)\,dxdt,
\end{aligned}\tag{9.3.66}$$

Next, the obvious inequality $0\le\phi_\vartheta(x,t,s)\le s$ implies $0\le\overline{\phi}_\vartheta\le\varrho$ and $0\le\phi_\vartheta(x,t,\varrho)\le\varrho$. From this, (9.3.14) and Lemma 9.3.3, we obtain

$$\int_Q h(\phi_\vartheta(x,t,\varrho)-\overline{\phi}_\vartheta)\,dxdt\le 2\int_Q h\varrho\,dxdt\le c.$$

Combining this with (9.3.66) we arrive at

$$\lim_{\epsilon\to 0}\int_Q h|\phi_\vartheta(x,t,\varrho_n)-\phi_\vartheta(x,t,\varrho)|^{1+\gamma}\,dxdt \le c\int_Q h\big(\overline{\phi_\vartheta p}-\overline{\phi}_\vartheta\,\overline{p}\big)\,dxdt + c.$$

From this and Lemma 9.3.8 we obtain

$$\lim_{n\to\infty}\int_Q h|\phi_\vartheta(x,t,\varrho_n)-\phi_\vartheta(x,t,\varrho)|^{1+\gamma}\,dxdt$$
$$\le c(\lambda+1)\int_Q h\big(\overline{\phi_\vartheta \operatorname{div}\mathbf{u}}-\overline{\phi}_\vartheta \operatorname{div}\mathbf{u}\big)\,dxdt + c, \quad (9.3.67)$$

where

$$\int_Q h\big(\overline{\phi_\vartheta \operatorname{div}\mathbf{u}}-\overline{\phi}_\vartheta \operatorname{div}\mathbf{u}\big)\,dxdt = \lim_{n\to\infty}\int_Q h\phi_\vartheta(x,t,\varrho_n)(\operatorname{div}\mathbf{u}_n-\operatorname{div}\mathbf{u})\,dxdt.$$

Since $\phi_\vartheta(x,t,\varrho)$ is bounded and $\nabla\mathbf{u}_n \rightharpoonup \nabla\mathbf{u}$ weakly in $L^2(Q)$, it follows that

$$\lim_{n\to\infty}\int_Q h\phi_\vartheta(x,t,\varrho)(\operatorname{div}\mathbf{u}_n-\operatorname{div}\mathbf{u})\,dxdt = 0.$$

Combining the results obtained we arrive at

$$\int_Q h\big(\overline{\phi_\vartheta \operatorname{div}\mathbf{u}}-\overline{\phi}_\vartheta \operatorname{div}\mathbf{u}\big)\,dxdt$$
$$= \lim_{n\to\infty}\int_Q h\big(\phi_\vartheta(x,t,\varrho_n)-\phi_\vartheta(x,t,\varrho)\big)\operatorname{div}(\mathbf{u}_n-\mathbf{u})\,dxdt. \quad (9.3.68)$$

Next, the Young inequality implies

$$c(\lambda+1)h\big(\phi_\vartheta(x,t,\varrho_n)-\phi_\vartheta(x,t,\varrho)\big)\operatorname{div}(\mathbf{u}_n-\mathbf{u})$$
$$\le 2^{-1}h|\phi_\vartheta(x,t,\varrho_n)-\phi_\vartheta(x,t,\varrho)|^{\gamma+1} + 2^{1/\gamma}h\{c(\lambda+1)|\operatorname{div}(\mathbf{u}_n-\mathbf{u})|\}^{(\gamma+1)/\gamma}.$$

Recall that the sequence $\operatorname{div}(\mathbf{u}_n-\mathbf{u})$ is bounded in $L^2(Q)$. From this and the inequality $(\gamma+1)/\gamma\le 2$ we obtain

$$\limsup_{n\to\infty}\int_Q |\operatorname{div}(\mathbf{u}_n-\mathbf{u})|^{(\gamma+1)/\gamma}\,dxdt \le c,$$

which gives

$$c(\lambda+1)\lim_{n\to\infty}\int_Q h\big(\phi_\vartheta(x,t,\varrho_n)-\phi_\vartheta(x,t,\varrho)\big)\operatorname{div}(\mathbf{u}_n-\mathbf{u})\,dxdt$$
$$\le 2^{-1}\lim_{n\to\infty}\int_Q h|\phi_\vartheta(x,t,\varrho_n)-\phi_\vartheta(x,t,\varrho)|^{\gamma+1}\,dxdt + c.$$

Recalling identity (9.3.68) we arrive at

$$c(\lambda+1)\int_Q h\big(\overline{\phi_\vartheta \operatorname{div}\mathbf{u}} - \overline{\phi}_\vartheta \operatorname{div}\mathbf{u}\big)\,dxdt \le 2^{-1}\lim_{n\to\infty}\int_Q h|\phi_\vartheta(x,t,\varrho_n)-\phi_\vartheta(x,t,\varrho)|^{\gamma+1}\,dxdt + c.$$

Inserting this into (9.3.67) we get

$$2^{-1}\lim_{n\to\infty}\int_Q h|\phi_\vartheta(x,t,\varrho_n)-\phi_\vartheta(x,t,\varrho)|^{1+\gamma}\,dxdt \le c,$$

which along with (9.3.63) leads to (9.3.57). □

**Lemma 9.3.10.** *Under the assumptions of Theorem 9.3.1, there is a constant $c$ such that*

$$\|\mathfrak{H}\|_{L^{1+\gamma}(Q)} + \sup_{\upsilon\in\mathbb{R}^+}\|\mathfrak{V}_\upsilon\|_{L^1(Q)} < c, \tag{9.3.69}$$

*where*

$$\begin{aligned} \mathfrak{V}_\upsilon(x,t) &= \int_{[0,\infty)} \min\{s,\upsilon\}(p(s)-\overline{p})\,d_s f(x,t,s),\\ \mathfrak{H}(x,t) &= \int_{[0,\infty)} f(x,t,s)(1-f(x,t,s))\,ds. \end{aligned} \tag{9.3.70}$$

*Proof.* It suffices to prove that for any compact subset $Q'\Subset Q$,

$$\|\mathfrak{H}\|_{L^{1+\gamma}(Q')} + \sup_{\upsilon\in\mathbb{R}^+}\|\mathfrak{V}_\upsilon\|_{L^1(Q')} < c, \tag{9.3.71}$$

with $c$ independent of $Q'$. Now recall the definitions of the functions $\mathcal{T}_\vartheta$ and $\varrho$:

$$\mathcal{T}_\vartheta(x,t) = \min\Big\{\int_{[0,\infty)} s\,d_s f(x,t,s),\,\vartheta(x,t)\Big\} - \int_{[0,\infty)} \min\{s,\vartheta(x,t)\}\,d_s f(x,t,s),$$

$$\varrho(x,t) = \int_{[0,\infty)} s\,d_s f(x,t,s).$$

Integration by parts in the Stieltjes integral yields

$$\begin{aligned} \int_{[0,\infty)} \min\{s,\vartheta(x,t)\}d_s f(x,t,s) &= \lim_{s\to\infty}\min\{s,\vartheta(x,t)\}f(x,t,s)\\ - \lim_{s\to -0}\min\{s,\vartheta(x,t)\}f(x,t,s) - \int_0^\vartheta f(x,t,s)\,ds &= \vartheta - \int_0^\vartheta f(x,t,s)\,ds \end{aligned} \tag{9.3.72}$$

and

$$\varrho(x,t) = \int_{[0,\infty)} s\,d_s f(x,t,s) = \int_0^\infty (1-f(x,t,s))\,ds. \tag{9.3.73}$$

It follows that

$$\begin{aligned} \mathcal{T}_\vartheta(x,t) &= \int_0^{\vartheta(x,t)} f(x,t,s)\,ds \qquad \text{for } \varrho(x,t) \geq \vartheta(x,t), \\ \mathcal{T}_\vartheta(x,t) &= \int_{\vartheta(x,t)}^{\infty} (1 - f(x,t,s))\,ds \quad \text{for } \varrho(x,t) \leq \vartheta(x,t). \end{aligned} \tag{9.3.74}$$

Next choose a sequence of nonnegative continuous functions $\vartheta_m$ such that $\vartheta_m \to \varrho$ a.e. in $Q$. Substituting $\vartheta_m$ into (9.3.74) and letting $m \to \infty$ we conclude that for a.e. $(x,t) \in Q$,

$$\mathcal{T}_{\vartheta_m}(x,t) \to \mathcal{T}_\varrho = \int_0^{\varrho(x,t)} f(x,t,s)\,ds = \int_{\varrho(x,t)}^{\infty} (1 - f(x,t,s))\,ds \tag{9.3.75}$$

as $m \to \infty$. It now follows from Lemma 9.3.9 that there is a constant $c$ such that

$$\|\mathcal{T}_\varrho\|_{L^{\gamma+1}(Q')} \leq c.$$

Noting that $\mathfrak{H} \leq \mathcal{T}_\varrho$ we arrive at the desired estimate for $\mathfrak{H}$ in (9.3.69). To estimate $\mathfrak{V}_v$, notice that by Lemma 9.3.8,

$$\frac{1}{\lambda+1}\mathfrak{V}_v = w\text{-}\lim_{n\to\infty} \phi_v(\varrho_n)\,\mathrm{div}\,\mathbf{u}_n - \overline{\phi}_v\,\mathrm{div}\,\mathbf{u},$$

where $w$-lim denotes the weak limit in $L^2(Q')$. Thus we get

$$\frac{1}{\lambda+1}\mathfrak{V}_v = w\text{-}\lim_{n\to\infty} (\phi_v(\varrho_n) - \phi_v(\varrho))\,\mathrm{div}\,\mathbf{u}_n - \Big(w\text{-}\lim_{n\to\infty} \phi_v(\varrho_n) - \phi_v(\varrho)\Big)\,\mathrm{div}\,\mathbf{u}.$$

Now Lemma 9.3.9 yields

$$\begin{aligned} \|\mathfrak{V}_v\|_{L^1(Q')} &\leq c \limsup_{n\to\infty} \big\{(\|\mathrm{div}\,\mathbf{u}_n\|_{L^2(Q')} + \|\mathrm{div}\,\mathbf{u}\|_{L^2(Q')})\|\phi_v(\varrho_n) - \phi_v(\varrho)\|_{L^2(Q')}\big\} \\ &\leq c \limsup_{n\to\infty} \|\phi_v(\varrho_n) - \phi_v(\varrho)\|_{L^2(Q')} \leq c. \end{aligned}$$

Thus we have estimated $\mathfrak{V}_v$ as desired in (9.3.69). □

By proving Lemma 9.3.10 we finish the first part of the proof of the main stability theorem (Theorem 9.3.1) and arrive at the following:

**Conclusion.** In view of Lemma 9.3.7 the distribution function $f(x,t,s)$ defined by Lemma 9.3.2 is a solution to the initial boundary value problem for the kinetic equation

$$\partial_t f + \mathrm{div}(f\mathbf{u}) - \partial_s(sf\,\mathrm{div}\,\mathbf{u}) - \partial_s\mathfrak{C}[f] = 0 \quad \text{in } (B \setminus S) \times (0,T) \times \mathbb{R}, \tag{9.3.76}$$

$$f(x,t,s) = f_\infty \quad \text{on } \Sigma_{\text{in}} \times \mathbb{R} \text{ and } (B \setminus S) \times \{t = 0\} \times \mathbb{R}. \tag{9.3.77}$$

Moreover, by Lemma 9.3.10 this solution is regular in the sense of Definition 7.1.8.

**Part 2. Strong convergence of $\varrho_n$.** Now our task is to prove that under the assumptions of Theorem 9.3.1, the sequence of functions $\varrho_n$ converges a.e. in $(B \setminus S) \times (0,T)$. We split the proof into three lemmas. The first lemma shows that the integral identity (9.3.30) in Lemma 9.3.7, which defines a weak solution to the kinetic equation (9.3.76), is satisfied for a wider class of test functions than stated there.

**Lemma 9.3.11.** *Let $\psi \in C^\infty(B \times (0,T) \times \mathbb{R})$ satisfy the conditions*

$$\begin{aligned} &\psi = 0 \quad \text{in a neighborhood of } ((\Sigma \setminus \Sigma_{\text{in}}) \times \mathbb{R}) \cup (B \times \{t = T\} \times \mathbb{R}), \\ &\psi(x,t,s) = 0 \quad \text{for all large } s \text{ uniformly in } (x,t) \in B \times (0,T). \end{aligned} \tag{9.3.78}$$

*Then the distribution function $f$ defined by (9.3.10) satisfies*

$$\begin{aligned} \int_{(B\setminus S)\times(0,T)} \int_{\mathbb{R}} \big(f\partial_t\psi + f\nabla_x\psi \cdot \mathbf{u} - f\partial_s\psi \operatorname{div}\mathbf{u} - \partial_s\psi\, \mathfrak{C}[f]\big)\, dsdxdt& \\ + \int_{B\setminus S}\int_{\mathbb{R}} f_\infty(x,0,s)\psi(x,0,s)\, dsdx& \\ - \int_{\Sigma_{\text{in}}}\int_{\mathbb{R}} f_\infty(x,t,s)\psi(x,t,s)\mathbf{U}\cdot\mathbf{n}\, dsd\Sigma = 0.& \end{aligned} \tag{9.3.79}$$

*Here the operator $f \mapsto \mathfrak{C}[f]$ and the function $f_\infty$ are defined by* (9.3.31)–(9.3.33).

Notice that the only difference between Lemmas 9.3.7 and 9.3.11 is that in Lemma 9.3.11 the test function $\psi$ need not vanish in a neighborhood of $S \times (0,T)$, but this difference is essential. Indeed if the test function is nonnull in a neighborhood of $S\times(0,T)$, then a solution to the kinetic equation in $B\setminus S\times(0,T)$ can be extended to $S\times(0,T)$ in such a way that the extension satisfies the integral identity (9.3.79).

*Proof of Lemma 9.3.11.* By Lemma 9.3.7, (9.3.79) holds for all $\psi$ which satisfy (9.3.78) and vanish in a neighborhood of $S\times(0,T)\times\mathbb{R}$. As $\mathbf{u} \in L^2(0,T,W^{1,2}_S(B))$, Proposition 8.3.9 yields a sequence of functions $\psi_n \in C^\infty(B)$ that vanish in a neighborhood of $S$, satisfy $0 \le \psi_n \le 1$, converge to 1 in $B \setminus S$ and

$$\lim_{n\to\infty} \int_0^T \int_{B\setminus S} |\mathbf{u}|\, |\nabla\psi_n|\, dxdt = 0. \tag{9.3.80}$$

Let now $\psi$ be an arbitrary function satisfying (9.3.78). It is clear that the functions $\psi\psi_n$, $n \ge 1$, also satisfy (9.3.78) and vanish in a neighborhood of $S \times (0,T)$. By Lemma 9.3.7 the identity (9.3.79) holds with $\psi$ replaced by $\psi\psi_n$ for all $n \ge 1$.

Thus, with $\Omega = B \setminus S$ and $Q = \Omega \times (0,T)$, we get

$$\int_Q \int_{\mathbb{R}} (f\partial_t\psi + f\nabla_x\psi \cdot \mathbf{u} - f\partial_s\psi \operatorname{div} \mathbf{u} - \partial_s\psi\, \mathfrak{C}[f])\psi_n \, dsdxdt +$$
$$\int_\Omega \int_{\mathbb{R}} f_\infty(x,0,s)\psi(x,0,s)\psi_n(x)\, dsdx - \int_{\Sigma_{\text{in}}} \int_{\mathbb{R}} \psi_n f_\infty(x,t,s)\psi(x,t,s)\mathbf{U}\cdot\mathbf{n}\, dsd\Sigma$$
$$= - \int_Q \int_{\mathbb{R}} f\psi\nabla_x\psi_n \cdot \mathbf{u}\, dsdxdt. \quad (9.3.81)$$

By (9.3.78), there is $N < \infty$ depending on $\psi$ such that $\psi(x,t,s) = 0$ for $|s| \geq N$. Thus we get

$$\big|f\partial_t\psi + f\nabla_x\psi\cdot\mathbf{u} - f\partial_s\psi \operatorname{div}\mathbf{u}\big| \leq c(1 + |\mathbf{u}| + |\nabla\mathbf{u}|)\chi_N(s), \quad (9.3.82)$$

where $\chi_N$ is the characteristic function of the interval $[-N,N]$, and the constant $c$ depends on $\psi$. It follows from (9.3.31) that

$$|\partial_s\psi\, \mathfrak{C}[f]| \leq c(1+\overline{p})\chi_N(s). \quad (9.3.83)$$

From the boundedness of $\mathbf{U}$ we obtain

$$|f_\infty(x,0,s)\psi(x,0,s)| + |f_\infty(x,t,s)\psi(x,t,s)\mathbf{U}| \leq c\chi_N(s).$$

Since $|\nabla u|$ and $\overline{p}$ are integrable in $B\times(0,T)$, it now follows from (9.3.82)–(9.3.83) that the integrands on the left hand side of (9.3.81) have an integrable majorant independent of $n$. The Lebesgue dominated convergence theorem yields

$$\int_Q \int_{\mathbb{R}} (f\partial_t\psi + f\nabla_x\psi \cdot \mathbf{u} - f\partial_s\psi \operatorname{div} \mathbf{u} - \partial_s\psi\, \mathfrak{C}[f])\psi_n \, dsdxdt$$
$$+ \int_\Omega \int_{\mathbb{R}} f_\infty(x,0,s)\psi(x,0,s)\psi_n(x)\, dsdx - \int_{\Sigma_{\text{in}}} \int_{\mathbb{R}} \psi_n f_\infty(x,t,s)\psi(x,t,s)\mathbf{U}\cdot\mathbf{n}\, dsd\Sigma$$
$$\to \int_Q \int_{\mathbb{R}} (f\partial_t\psi + f\nabla_x\psi \cdot \mathbf{u} - f\partial_s\psi \operatorname{div} \mathbf{u} - \partial_s\psi\, \mathfrak{C}[f]) \, dsdxdt$$
$$+ \int_\Omega \int_{\mathbb{R}} f_\infty(x,0,s)\psi(x,0,s)\, dsdx - \int_{\Sigma_{\text{in}}} \int_{\mathbb{R}} f_\infty(x,t,s)\psi(x,t,s)\mathbf{U}\cdot\mathbf{n}\, dsd\Sigma \quad (9.3.84)$$

as $n\to\infty$. On the other hand, (9.3.80) implies

$$\Big|\int_Q \int_{\mathbb{R}} f\psi\nabla_x\psi_n \cdot \mathbf{u}\, dsdxdt\Big| \leq \int_{\Omega\times(0,T)} \Big\{\int_{\mathbb{R}} |\psi|\, ds\Big\}|\mathbf{u}|\,|\nabla\psi_n(x)|\, dxdt$$
$$\leq c\int_{\Omega\times(0,T)} |\mathbf{u}|\,|\nabla\psi_n(x)|\, dxdt \to 0 \quad \text{as } n\to\infty. \quad (9.3.85)$$

Letting $n\to\infty$ in (9.3.81) and using (9.3.84)–(9.3.85) we obtain (9.3.79). □

Now introduce the distribution function $f^* : B \times (0,T) \times \mathbb{R}$ defined by

$$f^*(x,t,s) = \begin{cases} f(x,t,s) & \text{in } (B \setminus S) \times (0,T) \times \mathbb{R}, \\ f_\infty(x,0,s) & \text{in } S \times (0,T) \times \mathbb{R}. \end{cases} \tag{9.3.86}$$

Recall that in view of (9.3.33),

$$f_\infty(x,0,s) = 0 \quad \text{for } s < \varrho_\infty(x,0), \quad f_\infty(x,0,s) = 1 \quad \text{for } s \geq \varrho_\infty(x,0). \tag{9.3.87}$$

**Lemma 9.3.12.** *Let $\psi \in C^\infty(B \times (0,T) \times \mathbb{R})$ satisfy condition* (9.3.78). *Then*

$$\int_{B\times(0,T)} \int_{\mathbb{R}} \big(f^* \partial_t \psi + f^* \nabla_x \psi \cdot \mathbf{u} - f^* \partial_s \psi \operatorname{div} \mathbf{u} - \partial_s \psi\, \mathfrak{C}[f^*]\big)\, dsdxdt$$
$$+ \int_B \int_{\mathbb{R}} f_\infty(x,0,s)\psi(x,0,s)\, dsdx - \int_{\Sigma_{\text{in}}} \int_{\mathbb{R}} f_\infty(x,t,s)\psi(x,t,s)\mathbf{U} \cdot \mathbf{n}\, dsd\Sigma = 0. \tag{9.3.88}$$

*Proof.* By (9.3.31) and (9.3.32), we have

$$\mathfrak{C}[f^*] = \frac{1}{\lambda+1} \int_{(-\infty,s]} p(\tau)\, d_\tau f^*(x,t,\tau) - \int_{\mathbb{R}} p(\tau)\, d_\tau f^*(x,t,\tau) \int_{(-\infty,s]} d_\tau f^*(x,t,\tau).$$

Notice that for $x \in S$ we have $f^* = f_\infty$, where $f_\infty$ is defined by (9.3.87). It follows that for $x \in S$ and $s < \varrho_\infty(x,0)$,

$$\int_{(-\infty,s]} p(\tau)\, d_\tau f^*(x,t,\tau) = \int_{(-\infty,s]} d_\tau f^*(x,t,\tau) = 0,$$

while for $x \in S$ and $s \geq \varrho_\infty(x,0)$,

$$\int_{(-\infty,s]} p(\tau)\, d_\tau f^*(x,t,\tau) = p(\varrho_\infty(x,0)), \quad \int_{(-\infty,s]} d_\tau f^*(x,t,\tau) = 1.$$

Moreover, for $x \in S$ we have

$$\overline{p} = \int_{\mathbb{R}} p(\tau)\, d_\tau f^*(x,t,\tau) = p(\varrho_\infty(x,0)).$$

Combining the above we find that $\mathfrak{C}[f^*] = 0$ in $S \times (0,T) \times \mathbb{R}$. Noting that $\mathbf{u} = 0$ a.e. in $S \times (0,T)$ and $f^* = f_\infty$ is independent of $t$ on this set, we arrive at

$$\int_{S\times(0,T)} \int_{\mathbb{R}} \big(f^* \partial_t \psi + f^* \nabla_x \psi \cdot \mathbf{u} - f^* \partial_s \psi \operatorname{div} \mathbf{u} - \partial_s \psi\, \mathfrak{C}[f^*]\big)\, dsdxdt$$
$$= \int_{S\times(0,T)} \int_{\mathbb{R}} f^* \partial_t \psi\, dxdtds = -\int_S \int_{\mathbb{R}} f_\infty(x,s)\psi(x,0,t)\, dsdxdt.$$

Combining this with (9.3.79) and noting that $f = f^*$ in $(B \setminus S) \times (0,T) \times \mathbb{R}$ we obtain the desired identity. $\square$

**Lemma 9.3.13.** *The function $f^*$ defined by (9.3.86) satisfies the condition*

$$\|\mathfrak{H}\|_{L^{1+\gamma}(B\times(0,T))} + \sup_{\upsilon\in\mathbb{R}^+} \|\mathfrak{V}_\upsilon\|_{L^1(B\times(0,T))} < \infty, \tag{9.3.89}$$

*where*

$$\mathfrak{V}_\upsilon(x,t) = \int_{[0,\infty)} \min\{s,\upsilon\}(p(s)-\overline{p})\, d_s f^*(x,t,s),$$

$$\mathfrak{H}(x,t) = \int_{[0,\infty)} f^*(x,t,s)(1-f^*(x,t,s))\, ds, \quad \overline{p} = \int_{\mathbb{R}} p(s)\, d_s f^*(x,t,s).$$

*Proof.* It follows from (9.3.86)–(9.3.87) that $f^*(x,t,\cdot)$ is a step function and takes only values 0 or 1 a.e. in $S\times(0,T)$. It easily follows that $\mathfrak{H} = \mathfrak{V}_\upsilon = 0$ in $S\times(0,T)$. Since $f = f^*$ in $(B\setminus S)\times(0,T)$, inequality (9.3.89) is a straightforward consequence of Lemma 9.3.10. □

**Lemma 9.3.14.** *Let* $\mathbf{u}$ *and* $\varrho$ *be defined by Lemma 9.3.2. Then the sequence $\varrho_n$ converges to $\varrho$ strongly in $L^r((B\setminus S)\times(0,T))$ for any $r<\gamma$. Moreover, for any compact set $Q' \Subset (B\setminus S)\times(0,T)$ and $\theta\in[0,\min\{2\gamma/d-1,\gamma/2\})$, the sequence $\varrho_n$ converges to $\varrho$ in $L^{\gamma+\theta}(Q')$, and $p(\varrho_n)$ converges to $p(\varrho)$ in $L^{1+\theta/\gamma}(Q')$. In particular, $\overline{p} = p(\varrho)$ a.e. in $(B\setminus S)\times(0,T)$.*

*Proof.* By Lemma 9.3.12, the distribution function $f^*$ defined by (9.3.86) is a weak solution to the initial boundary problem (7.1.45) in the cylinder $B\times(0,T)$. Moreover, Lemma 9.3.13 implies that this solution is regular in the sense of Definition 7.1.8. It now follows from the hypotheses of Theorem 9.3.1 that $f^*$, $\mathbf{u}$, the data $\varrho_\infty$, $\mathbf{U}$ and the domain $B$ satisfy Condition 7.1.11 (with $\Omega$ replaced by $B$, and $f$ replaced by $f^*$) of Theorem 7.1.12. Hence there is a measurable function $\varrho^* : B\times(0,T)\to\mathbb{R}$ such that

$$f^*(x,t,s) = 0 \quad \text{for } s<\varrho^*(x,t), \quad f^*(x,t,s) = 1 \quad \text{for } s\geq\varrho^*(x,t).$$

Notice that in view of (9.3.86), $f^*$ is *not* equal to the distribution function $f$ of the Young measure associated with the sequence $\varrho_n$ in the whole domain $B\times(0,T)$, but only in $(B\setminus\mathcal{O}(S))\times(0,T)$, where $\mathcal{O}(S)$ is an arbitrary neighborhood of the obstacle $S$. Thus we get

$$\varrho^*(x,t) = \int_{\mathbb{R}} s\, d_s f^*(x,t,s) = \int_{\mathbb{R}} s\, d_s f(x,t,s) = \varrho(x,t) \text{ a.e. in } (B\setminus\mathcal{O}(S))\times(0,T).$$

In other words, the probability distribution generated by the Young measure associated with $\varrho_n$ is deterministic in $(B\setminus\mathcal{O}(S))\times(0,T)$, and the associated Young measure is the Dirac measure concentrated at $\varrho(x,t)$. It now follows from the general properties of Young measures that $\varrho_n\to\varrho$ a.e. in $(B\setminus\mathcal{O}(S))\times(0,T)$. Since $\mathcal{O}(S)$ is an arbitrary neighborhood of $S$, we obtain $\varrho_n\to\varrho$ a.e. in $(B\setminus S)\times(0,T)$. Since the sequence $\varrho_n$ is bounded in $L^\gamma(B\times(0,T))$, Theorem 1.4.7 implies that $\varrho_n\to\varrho$ in $L^r((B\setminus S)\times(0,T))$ for all $1\leq r<\gamma$. It remains to note that the strong convergence of $\varrho_n$ in $L^{\gamma+\theta}(Q')$ is a consequence of estimate (9.3.4). □

**Strong convergence of $p(\varrho_n)$.** The following lemma establishes the limit relation (9.3.7).

**Lemma 9.3.15.** *For any compact set $\Omega' \Subset B \setminus S$,*

$$p(\varrho_n) \to p(\varrho) \quad \textit{in } L^1(\Omega' \times (0,T)). \tag{9.3.90}$$

*Proof.* By Lemma 9.3.14 and the compactness criterion in $L^p$ (Proposition 1.3.1), it suffices to show that the functions $p(\varrho_n)$ are equi-integrable in $\Omega' \times (0,T)$, i.e., for every $\varepsilon > 0$ there is $\delta > 0$ independent of $n$ such that

$$\int_A p(\varrho_n)\, dxdt \le \varepsilon \quad \text{for all } A \subset \Omega' \times (0,T) \quad \text{with meas } A \le \delta. \tag{9.3.91}$$

Choose $0 < h < T$ and set $Q' = \Omega' \times (h, T-h)$. Then

$$\int_A p(\varrho_n)\, dxdt \le \int_{A\cap Q'} p(\varrho_n)\, dxdt + \Big\{ \int_0^h + \int_{T-h}^T \Big\} \int_{B\setminus S} p(\varrho_n)\, dxdt.$$

Since $p \le c(1+\varrho^\gamma)$ it follows from inequalities (9.3.3)–(9.3.4) that

$$\Big\{ \int_0^h + \int_{T-h}^T \Big\} \int_{B\setminus S} p(\varrho_n)\, dxdt \le ch,$$

and

$$\begin{aligned}\int_{A\cap Q'} p(\varrho_n)\, dxdt &\le (\text{meas } A)^{\frac{\theta}{\gamma+\theta}} \Big( \int_{Q'} p(\varrho_n)^{\frac{\gamma+\theta}{\gamma}}\, dxdt \Big)^{\frac{\gamma}{\gamma+\theta}} \\ &\le (\text{meas } A)^{\frac{\theta}{\gamma+\theta}} \Big( \int_{Q'} (1+\varrho_n)^{\gamma+\theta}\, dxdt \Big)^{\frac{\gamma}{\gamma+\theta}} \le C(\Omega', h)(\text{meas } A)^{\frac{\theta}{\gamma+\theta}}.\end{aligned}$$

Here $C = C(\Omega', h)$ denotes a constant depending on $\Omega'$ and $h$. Combining the above we arrive at

$$\int_A p(\varrho_n)\, dxdt \le ch + C(\Omega', h)(\text{meas } A)^{\frac{\theta}{\gamma+\theta}}.$$

Setting $h = \varepsilon/(2c)$ and $\delta = \{\varepsilon/(2C(\Omega', h))\}^{1+\gamma/\theta}$ we obtain (9.3.91). □

**Completion of the proof of Theorem 9.3.1.** Relations (9.3.5) in Theorem 9.3.1 directly follow from Lemmas 9.3.2 and 9.3.14. Lemmas 9.3.4 and 9.3.15 yield (9.3.6) and (9.3.7). It remains to prove that the limit functions $(\mathbf{u}, \varrho)$ solve problem (9.2.4). First we show that they satisfy the momentum balance equations.

**Lemma 9.3.16.** *Let* $(\mathbf{u}, \varrho)$ *be defined by Lemma 9.3.2. Then the integral identity*

$$\int_{B\times(0,T)} \big(\varrho\mathbf{u}\cdot\partial_t\boldsymbol{\xi} + \varrho\mathbf{u}\otimes\mathbf{u} : \nabla\boldsymbol{\xi} + p(\varrho)\operatorname{div}\boldsymbol{\xi} - \mathbb{S}(\mathbf{u}) : \nabla\boldsymbol{\xi}\big)\,dxdt$$
$$+ \int_{B\times(0,T)} \varrho\mathbf{f}\cdot\boldsymbol{\xi}\,dxdt + \int_B (\varrho_\infty\mathbf{U}\cdot\boldsymbol{\xi})(x,0)\,dx = 0 \quad (9.3.92)$$

*holds for all vector fields* $\boldsymbol{\xi} \in C^\infty(B\times(0,T))$ *equal to* 0 *in a neighborhood of the obstacle* $S\times(0,T)$*, of the lateral surface* $\partial B\times(0,T)$*, and of the top* $\Omega\times\{T\}$*.*

*Proof.* Choose $\boldsymbol{\xi} \in C^\infty(B\times(0,T))$ as in the statement. Notice that $\boldsymbol{\xi}$ vanishes in a neighborhood of $S_n\times(0,T)$ for all sufficiently large $n$ since $d_H(S_n, S) \to 0$ as $n\to\infty$. By hypothesis, $\varrho_n$ and $\mathbf{u}_n$ are weak solutions to the momentum balance equation (9.2.4a). The integral identity (9.2.5) yields

$$\int_{B\times(0,T)} \varrho_n\mathbf{u}_n\cdot\partial_t\boldsymbol{\xi}\,dxdt + \int_{B\times(0,T)} \varrho_n\nabla\boldsymbol{\xi} : (\mathbf{u}_n\otimes\mathbf{u}_n)\,dxdt$$
$$+ \int_{B\times(0,T)} \nabla\boldsymbol{\xi} : \big(p(\varrho_n)\,\mathbb{I} - \mathbb{S}(\mathbf{u}_n)\big)\,dxdt + \int_Q \varrho_n\mathbf{f}\cdot\boldsymbol{\xi}\,dxdt$$
$$+ \int_B \varrho_\infty(\cdot,0)\mathbf{U}(\cdot,0)\cdot\boldsymbol{\xi}(\cdot,0)\,dx = 0. \quad (9.3.93)$$

Since $\boldsymbol{\xi}$ vanishes in a neighborhood of $S\times(0,T)$, the convergences (9.3.6) imply

$$\lim_{n\to\infty}\int_{B\times(0,T)} \big\{\varrho_n\mathbf{u}_n\cdot\partial_t\boldsymbol{\xi} + \big(\varrho_n\mathbf{u}_n\otimes\mathbf{u}_n - \mathbb{S}(\mathbf{u}_n)\big) : \nabla\boldsymbol{\xi} + \varrho_n\mathbf{f}\cdot\boldsymbol{\xi}\big\}\,dxdt$$
$$= \int_{B\times(0,T)} \big\{\varrho\mathbf{u}\cdot\partial_t\boldsymbol{\xi} + \big(\varrho\mathbf{u}\otimes\mathbf{u} - \mathbb{S}(\mathbf{u})\big) : \nabla\boldsymbol{\xi} + \varrho\mathbf{f}\cdot\boldsymbol{\xi}\big\}\,dxdt. \quad (9.3.94)$$

By the choice of the test vector field $\boldsymbol{\xi}$, there is a cylinder $\Omega'\times(0,T)$ with $\Omega' \Subset B\backslash S$ such that $\operatorname{supp}\boldsymbol{\xi} \subset \Omega'\times(0,T)$. Now, letting $n\to 0$ and applying Lemma 9.3.15 we obtain

$$\lim_{n\to\infty}\int_{B\times(0,T)} p(\varrho_n)\operatorname{div}\boldsymbol{\xi}\,dxdt = \int_Q p(\varrho)\operatorname{div}\boldsymbol{\xi}\,dxdt. \quad (9.3.95)$$

Finally, letting $n\to\infty$ in (9.3.93) and invoking (9.3.94), (9.3.95) we obtain the desired integral identity (9.3.92). □

**Lemma 9.3.17.** *The integral identity*

$$\int_{(B\backslash S)\times(0,T)} \Big(\varphi(\varrho)\partial_t\psi + \varphi(\varrho)\mathbf{u}\cdot\nabla\psi + \psi\big(\varphi(\varrho) - \varphi'(\varrho)\varrho\big)\operatorname{div}\mathbf{u}\Big)\,dxdt$$
$$= \int_{\Sigma_{\rm in}} \psi\varphi(\varrho_\infty)\mathbf{U}\cdot\mathbf{n}\,d\Sigma - \int_{(B\backslash S)\times\{t=0\}} \varphi(\varrho_\infty)\psi\,dx \quad (9.3.96)$$

*holds for all test functions $\psi \in C^\infty((B \setminus S) \times (0,T))$ vanishing in a neighborhood of the surface $\Sigma \setminus \Sigma_{\text{in}}$ and of the top $(B \setminus S) \times \{t = T\}$, and for all smooth functions $\varphi : [0,\infty) \to \mathbb{R}$ such that*

$$\limsup_{\varrho \to \infty} (|\varphi(\varrho)| + |\varrho\varphi'(\varrho)|) < \infty. \tag{9.3.97}$$

*Proof.* By the hypotheses of Theorem 9.3.1 the functions $(\mathbf{u}_n, \varrho_n)$ are renormalized solutions to the mass balance equations in $(B \setminus S_n) \times (0,T)$, and hence satisfy the identity (9.3.96) with $S$ replaced by $S_n$. Since $d_H(S_n, S) \to 0$ as $n \to \infty$, we have

$$\int_{B \times (0,T)} \Big(\varphi(\varrho_n)\partial_t\psi + \varphi(\varrho_n)\mathbf{u}_n \cdot \nabla\psi + \psi\big(\varphi(\varrho_n) - \varphi'(\varrho_n)\varrho_n\big) \operatorname{div} \mathbf{u}_n\Big)\, dxdt$$
$$= \int_{\Sigma_{\text{in}}} \psi\varphi(\varrho_\infty)\mathbf{U} \cdot \mathbf{n}\, d\Sigma - \int_{B \times \{t=0\}} \varphi(\varrho_\infty)\psi\, dx \tag{9.3.98}$$

for all $\psi \in C^\infty(B \times (0,T))$ vanishing in a neighborhood of $\Sigma \setminus \Sigma_{\text{in}}$, of $B \times \{t = T\}$, and of $S \times (0,T)$ and all sufficiently large $n$. Fix such a $\psi$. It follows from the convergences (9.3.5) and the boundedness condition imposed on $\varphi$ that $\varphi(\varrho_n)$ and $\varphi'(\varrho_n)\varrho_n$ converge to $\varphi(\varrho)$ and $\varphi'(\varrho)\varrho$, respectively, strongly in $L^r(B \times (0,T))$ for all $r \in [1,\infty)$. Letting $n \to \infty$ in (9.3.98) and recalling that $\mathbf{u}_n$ converges weakly in $L^2(0,T; W^{1,2}(B))$ we arrive at the integral identity (9.3.96). It remains to remove the restriction $\psi = 0$ in a neighborhood of $S \times (0,T)$. For this purpose we use Proposition 8.3.9.

Let $\psi_n$ be a sequence of cut-off functions as in Proposition 8.3.9, and let $\psi$ be an arbitrary test function satisfying the hypotheses of the lemma. Since $\psi_n$ vanishes in a neighborhood of $S \times (0,T)$ we may assume $\psi\psi_n \in C^\infty(B \times (0,T))$. Moreover, each $\psi\psi_n$ vanishes in a neighborhood of $\Sigma \setminus \Sigma_{\text{in}}$, of $B \times \{t = T\}$, and of $S \times (0,T)$. Substituting $\psi\psi_n$ for $\psi$ in (9.3.96) we obtain

$$\int_{(B\setminus S) \times (0,T)} \Big(\varphi(\varrho)\partial_t\psi + \varphi(\varrho)\mathbf{u} \cdot \nabla\psi + \psi\big(\varphi(\varrho) - \varphi'(\varrho)\varrho\big) \operatorname{div} \mathbf{u}\Big)\psi_n\, dxdt$$
$$= \int_{\Sigma_{\text{in}}} \psi_n\psi\varphi(\varrho_\infty)\mathbf{U} \cdot \mathbf{n}\, d\Sigma - \int_{(B\setminus S) \times \{t=0\}} \psi_n\varphi(\varrho_\infty)\psi\, dx$$
$$- \int_{(B\setminus S) \times (0,T)} \psi\varphi(\varrho)\mathbf{u} \cdot \nabla\psi_n\, dxdt. \tag{9.3.99}$$

Since $\mathbf{u} \in L^2(0,T; W^{1,2}_S(B))$, we can apply Proposition 8.3.9 to obtain

$$\Big|\int_{(B\setminus S) \times (0,T)} \psi\varphi(\varrho)\mathbf{u} \cdot \nabla\psi_n\, dxdt\Big| \le c\int_{(B\setminus S) \times (0,T)} |\mathbf{u}|\, |\nabla\psi_n|\, dxdt \to 0 \text{ as } n \to \infty.$$

Letting $n \to \infty$ in (9.3.99) and recalling that the uniformly bounded functions $\psi_n$ tend to 1 in $B \setminus S$, we arrive at the integral identity (9.3.96). $\square$

By Lemmas 9.3.16 and 9.3.17, the couple $(\mathbf{u}, \varrho)$ meets all requirements of Definition 9.2.3 and is a renormalized solution to problem (9.2.4). The proof of Theorem 9.3.1 is complete.

# Chapter 10

# Existence theory in nonsmooth domains. Shape optimization, continuity of the work functional

## 10.1 Existence theory in nonsmooth domains

In this section we employ the main stability theorem (Theorem 9.3.1) and Proposition 8.3.6 to prove that the flow problem around an obstacle has a renormalized solution for every compact obstacle. The corresponding result is given by the following theorem, which is one of the main results of this monograph.

**Theorem 10.1.1.** *Let $\gamma > d/2$, assume that the pressure function $p(\varrho)$, the flow domain and the given data satisfy Condition 9.2.5, and $S$ is an arbitrary compact set. Then problem* (9.2.4) *has a weak renormalized solution* $(\mathbf{u}, \varrho)$ *which meets all requirements of Definition 9.2.3 and satisfies the estimate*

$$\|\mathbf{u}\|_{L^2(0,T;W^{1,2}(\Omega))} + \|\varrho|\mathbf{u}|^2\|_{L^\infty(0,T;L^1(\Omega))} + \|\varrho\|_{L^\infty(0,T;L^\gamma(\Omega))} \leq c_e, \tag{10.1.1}$$

*where the constant $c_e$ depends only on $d$, $\|\mathbf{U}\|_{C^1(\mathbb{R}^d\times[0,T])}$, $\|\varrho_\infty\|_{L^\infty(\mathbb{R}^d\times[0,T])}$, $T$,* $\operatorname{diam}\Omega$, *and the constant $c_p$ in Condition 9.2.5. Moreover, for every cylinder $Q' \Subset Q$, and for $\theta \in [0, 2\gamma/d - 1)$, there is a constant $c$, depending only on $\theta$, $c_e$, and $Q'$, such that*

$$\int_{Q'} \varrho^{\gamma+\theta}\, dxdt \leq c. \tag{10.1.2}$$

*Proof.* By Proposition 8.3.6, there exists a sequence of compact subsets $S_n \Subset B$ with $C^\infty$ boundaries such that $S_n \xrightarrow{S} S$ (see Definition 8.3.5). It is easily seen that $B$, $S_n$, and the initial and boundary data $\mathbf{U}$, $\varrho_\infty$ meet all requirements of Theorem 9.2.7. Hence problem (9.2.4) in the cylinder $(B \setminus S_n) \times (0, T)$ has a renormalized solution $(\mathbf{u}_n, \varrho_n)$ in the sense of Definition 9.2.3. Moreover, by Theorem 9.2.7, this solution admits estimates (9.3.3) and (9.3.4). Hence the sequence $(S_n, \mathbf{u}_n, \varrho_n)$ meets all requirements of the stability theorem (Theorem 9.3.1), whose application completes the proof. $\square$

## 10.2 Continuity of the work functional

In this section we show that the work functional $W$ (see (3.3.1)) is well defined on the set of renormalized solutions to problem (9.2.4) in flow domains $\Omega = B \setminus S$ with compact obstacles $S \Subset B$. Moreover, we show that $W$ is continuous with respect to $\mathcal{S}$-convergence. It follows from (3.3.1) that

$$\begin{aligned} W(S, \mathbf{u}, \varrho) = &-\int_{B\setminus S} \eta\{(\varrho\mathbf{u}\cdot\mathbf{W})(x,T) - (\varrho_\infty\mathbf{U}\cdot\mathbf{W})(x,0)\}\,dx \\ &+ \int_0^T\int_{B\setminus S}\{\varrho\eta\mathbf{u}\cdot\partial_t\mathbf{W} + \big(\varrho(\mathbf{u}\otimes\mathbf{u}) - \mathbb{T}\big) : \nabla(\eta\mathbf{W}) + \eta\varrho\mathbf{f}\cdot\mathbf{W}\}\,dxdt, \end{aligned} \tag{10.2.1}$$

where

$$\mathbb{T} = \nabla\mathbf{u} + (\nabla\mathbf{u})^\top + (\lambda - 1)\,\mathrm{div}\,\mathbf{u}\,\mathbb{I} - p(\varrho)\,\mathbb{I}, \tag{10.2.2}$$

and $\mathbf{W} : \mathbb{R}^d \times (0,T) \to \mathbb{R}^d$ is a smooth vector field which characterizes the motion of the obstacle $S$. Recall that $\eta \in C^\infty(B)$ is an arbitrary function such that $\eta = 1$ in a neighborhood of $S$ and $\eta = 0$ in a neighborhood of $\partial B$. We cannot use (10.2.1) directly for renormalized solutions to problem (9.2.4) since the momentum $\varrho\mathbf{u}$ only belongs to $L^\infty(0,T; L^{2\gamma/(\gamma+1)}(B \setminus S))$ and hence the vector field $\varrho(\cdot,T)\mathbf{u}(\cdot,T)$ is not correctly defined as a product of two functions. The element $(\varrho\mathbf{u})(\cdot,T)$ can be defined by using the fact that for a.e. $t$ the time derivative $(\varrho\mathbf{u})_t$ is a distribution on the space $C_0^\infty(B\setminus S)$. The following lemma allows us to cope with this difficulty.

**Lemma 10.2.1.** *Let $\mathbf{U}$, $\varrho_\infty$, and a compact set $S \Subset B$ satisfy all conditions of Theorem 10.1.1. Let $(\mathbf{u}, \varrho)$ be the renormalized solution to problem* (9.2.4) *given by that theorem. Then the functions*

$$\mathbf{m}_h(x,T) := \frac{1}{h}\int_{T-h}^T \varrho(x,t)\mathbf{u}(x,t)\,dt$$

*converge weakly in $L^{2\gamma/(\gamma+1)}(B \setminus S)$ to a function $\mathbf{m}_0(x,T)$ as $h \to 0$. Set*

$$(\varrho\mathbf{u})(x,T) := \mathbf{m}_0(x,T). \tag{10.2.3}$$

*Then*

$$\|(\varrho\mathbf{u})(\cdot,T)\|_{L^{2\gamma/(\gamma+1)}(B\setminus S)} \le c\|\varrho\mathbf{u}\|_{L^\infty(0,T;L^{2\gamma/(\gamma+1)}(B\setminus S))} \le c. \tag{10.2.4}$$

*Proof.* Choose $\varphi \in C_0^\infty(B \setminus S)$ and set

$$v_h(t) = h^{-1}(T-t) \quad \text{for } T-h < t \le T, \quad v_h(t) = 1 \quad \text{for } t \le T-h. \tag{10.2.5}$$

Since $\boldsymbol{\xi} = v_h(t)\boldsymbol{\varphi}$ vanishes near the lateral surface $\partial(B \setminus S) \times [0,T]$ and on the top $\{t = T\}$, we can insert $\boldsymbol{\xi}$ into the integral identity (9.2.5) to obtain

$$\int_{B\setminus S} \mathbf{m}_h \cdot \boldsymbol{\varphi}\, dx = \int_0^T \int_{B\setminus S} v_h(t)\mathfrak{P}_\varphi\, dx\, dt + \int_{B\setminus S} \varrho_\infty \mathbf{U} \cdot \boldsymbol{\varphi}(x,0)\, dx, \tag{10.2.6}$$

where

$$\mathfrak{P}_\varphi = (\varrho\mathbf{u} \otimes \mathbf{u} - \mathbb{S}(\mathbf{u})) : \nabla\boldsymbol{\varphi} + p(\varrho)\operatorname{div}\boldsymbol{\varphi} + \varrho\mathbf{f} \cdot \boldsymbol{\varphi}. \tag{10.2.7}$$

Since the vector field $\mathbf{f}$ is bounded and $p \le c(1 + \varrho^\gamma)$, we have

$$|\mathfrak{P}_\varphi| \le c\|\boldsymbol{\varphi}\|_{C^1(B\setminus S)}(\varrho|\mathbf{u}|^2 + |\nabla\mathbf{u}| + |\mathbf{u}| + \varrho^\gamma + \varrho + 1).$$

It now follows from (10.1.1) that

$$\int_{B\setminus S} |\varrho_\infty \mathbf{U} \cdot \boldsymbol{\varphi}(x,0)|\, dx + \int_0^T \int_{B\setminus S} |\mathfrak{P}_\varphi|\, dx\, dt \le c\|\boldsymbol{\varphi}\|_{C^1(B\setminus S)}.$$

On the other hand, we have $|v_h\mathfrak{P}_\varphi| \le |\mathfrak{P}_\varphi|$, and $v_h\mathfrak{P}_\varphi \to \mathfrak{P}_\varphi$ a.e. in $(B\setminus S)\times(0,T)$ as $h \to 0$. The Lebesgue dominated convergence theorem yields

$$\lim_{h\to 0} \int_{B\setminus S} \mathbf{m}_h \cdot \boldsymbol{\varphi}\, dx = \int_0^T \int_{B\setminus S} \mathfrak{P}_\varphi\, dx\, dt + + \int_{B\setminus S} \varrho_\infty \mathbf{U} \cdot \boldsymbol{\varphi}(x,0)\, dx. \tag{10.2.8}$$

Next, notice that in view of (10.1.1) the bounded energy functions $\varrho$ and $\mathbf{u}$ satisfy all hypotheses of Corollary 4.2.2. Hence $\varrho\mathbf{u} \in L^\infty(0,T; L^{2\gamma/(\gamma+1)}(B \setminus S))$. As

$$\begin{aligned}\|\mathbf{m}_h\|_{L^\infty(0,T;L^{2\gamma/(\gamma+1)}(B\setminus S))} &\le \frac{1}{h}\int_{T-h}^T \|\varrho(t)\mathbf{u}(t)\|_{L^{2\gamma/(\gamma+1)}(B\setminus S)}\, dt \\ &\le \|\varrho\mathbf{u}\|_{L^\infty(0,T;L^{2\gamma/(\gamma+1)}(B\setminus S))},\end{aligned}$$

the sequence $\mathbf{m}_h$ is bounded in $L^{2\gamma/(\gamma+1)}(B \setminus S)$. Combining this with (10.2.8) we conclude that this sequence converges weakly in $L^{2\gamma/(\gamma+1)}(B \setminus S)$ as $h \to 0$. Estimate (10.2.4) obviously follows from the definition of $\mathbf{m}_h$. □

**Remark 10.2.2.** The assertion of Lemma 10.2.1 stems from the fact that the time derivative $\partial_t(\varrho\mathbf{u})$ exists as a distribution defined for a.e. $t \in (0,T)$ on the space $C_0^\infty(B \setminus S)$. This easily follows from the momentum balance equation. Arguing as in the proof of Lemma 10.2.1 we conclude that

$$\frac{1}{h}\int_{t-h}^t \varrho(x,s)\mathbf{u}(x,s)\, ds \rightharpoonup \mathbf{m}_0(x,t) \quad \text{weakly in } L^{2\gamma/(\gamma+1)}(B \setminus S) \quad \text{as } h \to 0$$

for any $t \in (0,T)$. Thus we can define

$$(\varrho\mathbf{u})(x,t) := \mathbf{m}_0(x,t) \tag{10.2.9}$$

for any $t \in (0,T)$. Hence the impulse moment $\varrho\mathbf{u}$ is uniquely defined at each point of the interval $(0,T]$. Moreover, one can prove that the mapping $t \mapsto (\varrho\mathbf{u})(\cdot,t)$ is continuous in the weak $L^{2\gamma/(\gamma+1)}(B\setminus S)$ topology. We point out that $t \mapsto (\varrho\mathbf{u})(\cdot,t)$ in (10.2.9) can be obtained from the product of the functions $t \mapsto \varrho(\cdot,t)$ and $t \mapsto \mathbf{u}(\cdot,t)$ by a possible change of the value of this product on a set of measure zero with respect to the time variable. Therefore, for the fixed value $t = T$ the vector field $(\varrho\mathbf{u})(\cdot,T)$ defined by (10.2.3) is just a notation and it cannot be recovered from just $\varrho(\cdot,T)$ and $\mathbf{u}(\cdot,T)$.

**Remark 10.2.3.** It follows from (10.2.3) and (10.2.8) that

$$(\varrho\mathbf{u})(\cdot,T) = \int_0^T \int_{B\setminus S} \mathfrak{P}_\varphi \, dx\, dt + \int_{B\setminus S} \varrho_\infty \mathbf{U}\cdot\boldsymbol{\varphi}(x,0)\, dx. \tag{10.2.10}$$

It follows from Lemma 10.2.1 that the work functional $W(S,\mathbf{u},\varrho)$ is well defined for all compact obstacles $S$ and all renormalized solutions $(\mathbf{u},\varrho)$ defined by Theorem 10.1.1. However, notice that the function $(\varrho\mathbf{u})(T)$ of the spatial variable in (10.2.1) is defined by Lemma 10.2.1 as a weak trace of the moment $(\varrho\mathbf{u})(x,t)$ on the top $\{t = T\}$. We are now in a position to prove the main result of this section on the continuity properties of the work functional.

Assume that a sequence of compact obstacles $S_n \Subset B$ and functions $(\mathbf{u}_n,\varrho_n) \in L^2(0,T;W^{1,2}_{S_n}(B)) \times L^\infty(0,T;L^\gamma(B\setminus S_n))$ satisfy the following condition:

**Condition 10.2.4.** (i) There is a compact set $S \Subset B$ such that $S_n \xrightarrow{\mathcal{S}} S$ as $n \to \infty$ (see Definition 8.3.5).

(ii) The couples $(\mathbf{u}_n,\varrho_n)$, defined in the cylinder $(B\setminus S_n)\times(0,T)$, are renormalized solutions to problem (9.2.4) with boundary and initial data $\mathbf{U}$, $\varrho_\infty$ given by Theorems 9.3.1 and 10.1.1.

(iii) There are $(\mathbf{u},\varrho) \in L^2(0,T;W^{1,2}_S(B)) \times L^\infty(0,T;L^\gamma(B))$ such that

$$\mathbf{u}_n \rightharpoonup \mathbf{u} \text{ weakly in } L^2(0,T;W^{1,2}(B)),\ \varrho_n \rightharpoonup \varrho \text{ weakly}^\star \text{ in } L^\infty(0,T;L^\gamma(B)).$$

(iv) We have the convergence of volumes,

$$\lim_{n\to\infty} \operatorname{meas} S_n = \operatorname{meas} S. \tag{10.2.11}$$

**Theorem 10.2.5.** *Let $(\mathbf{u}_n,\varrho_n,S_n)$ and $(\mathbf{u},\varrho,S)$ satisfy Condition 10.2.4. Let $\mathbf{W}$ be a divergence free vector field in $C^1((B\setminus S)\times(0,T))$. Then $(\mathbf{u},\varrho)$ is a renormalized solution to problem (9.2.4), and*

$$W(S_n,\mathbf{u}_n,\varrho_n) \to W(S,\mathbf{u},\varrho) \quad \textit{as } n \to \infty. \tag{10.2.12}$$

As the sequence $(\mathbf{u}_n, \varrho_n, S_n)$ meets all requirements of the stability theorem (Theorem 9.3.1), the limit couple $(\mathbf{u}, \varrho)$ is a renormalized solution to problem (9.2.4). Thus it suffices to prove (10.2.12). We split the proof into a sequence of lemmas.

**Lemma 10.2.6.** *Under the assumptions of Theorem 10.2.5,*

$$\lim_{n\to\infty} \operatorname{meas}\big((S \setminus S_n) \cup (S_n \setminus S)\big) = 0.$$

*Proof.* For $\delta > 0$, denote by $S^\delta$ and $S_n^\delta$ the open $\delta$-neighborhoods of the sets $S$ and $S_n$, respectively. Since $S_n$ converges to $S$ in the Hausdorff metric, the inclusion $S_n \subset S^\delta$ holds for sufficiently large $n$. For every such $n$, we have

$$\operatorname{meas}(S \setminus S_n) \le \operatorname{meas}(S^\delta \setminus S_n) = \operatorname{meas} S^\delta - \operatorname{meas} S_n.$$

Letting $n \to \infty$ and using (10.2.11) we obtain

$$\limsup_{n\to\infty} \operatorname{meas}\,(S \setminus S_n) \le \operatorname{meas} S^\delta - \operatorname{meas} S. \tag{10.2.13}$$

Since $S$ is the intersection of the decreasing sequence $\{S^\delta\}$, we have $\lim_{\delta\to 0} \operatorname{meas} S^\delta = \operatorname{meas} S$. Letting $\delta \to 0$ in (10.2.13) we finally get $\limsup_{n\to\infty} \operatorname{meas}(S \setminus S_n) = 0$. It remains to show that $\lim_{n\to\infty} \operatorname{meas}(S_n \setminus S) = 0$. To this end choose $0 < \delta < \varepsilon$. It follows from the definition (8.1.1) of the Hausdorff metric that for all sufficiently large $n$, we have $S_n \subset S_n^\delta \subset S^\varepsilon$, which gives

$$S_n \setminus S \subset S_n^\delta \setminus S \subset S^\varepsilon \setminus S.$$

Thus $\limsup_{n\to\infty} \operatorname{meas}(S_n \setminus S) \le \operatorname{meas} S^\varepsilon - \operatorname{meas} S$. Letting $\varepsilon \to 0$ we obtain $\lim_{n\to\infty} \operatorname{meas}(S_n \setminus S) = 0$. □

**Lemma 10.2.7.** *Under the assumptions of Theorem 10.2.5,*

$$\int_0^T \int_{B\setminus S_n} \eta \varrho_n \mathbf{f} \cdot \mathbf{W}\, dx dt \to \int_0^T \int_{B\setminus S} \eta \varrho \mathbf{f} \cdot \mathbf{W}\, dx dt. \tag{10.2.14}$$

*Proof.* Choose $\varepsilon > 0$ and denote by $S^\varepsilon$ the $\varepsilon$-neighborhood of $S$. Since $S_n \to S$ in the Hausdorff metric, we have $B \setminus S^\varepsilon \subset B \setminus S_n$ for sufficiently large $n$. It follows from Condition 10.2.4 that $(\mathbf{u}_n, \varrho_n, S_n)$ and $(\mathbf{u}, \varrho, S)$ meet all requirements of the stability theorem (Theorem 9.3.1). Hence the functions $\varrho_n$, extended as $\varrho_\infty(\cdot, 0)$ to $S_n \times (0, T)$, are uniformly bounded in $L^\infty(0, T; L^\gamma(B))$ and converge a.e. in $(B \setminus S) \times (0, T)$ to $\varrho$. In particular, $\varrho_n \to \varrho$ in $L^1((B \setminus S) \times (0, T))$ and so

$$\int_0^T \int_{B\setminus S} \eta(\varrho_n - \varrho)\mathbf{f} \cdot \mathbf{W}\, dx dt \to 0 \quad \text{as } n \to \infty. \tag{10.2.15}$$

Notice that $B \setminus S_n = ((B \setminus S) \cup (S \setminus S_n)) \setminus (S_n \setminus S)$, which leads to

$$\int_0^T \int_{B\setminus S_n} \eta \varrho_n \mathbf{f} \cdot \mathbf{W}\, dxdt - \int_0^T \int_{B\setminus S} \eta \varrho \mathbf{f} \cdot \mathbf{W}\, dxdt$$
$$= \int_0^T \int_{B\setminus S} \eta(\varrho_n - \varrho) \mathbf{f} \cdot \mathbf{W}\, dxdt$$
$$+ \int_0^T \int_{S\setminus S_n} \eta \varrho_n \mathbf{f} \cdot \mathbf{W}\, dxdt - \int_0^T \int_{S_n\setminus S} \eta \varrho_n \mathbf{f} \cdot \mathbf{W}\, dxdt.$$

By (10.2.15), the first integral on the right hand side tends to zero as $n \to \infty$, so

$$\limsup_{n\to\infty} \Big| \int_0^T \int_{B\setminus S_n} \eta \varrho_n \mathbf{f} \cdot \mathbf{W}\, dxdt - \int_0^T \int_{B\setminus S} \eta \varrho \mathbf{f} \cdot \mathbf{W}\, dxdt \Big|$$
$$= \limsup_{n\to\infty} \Big| \int_0^T \int_{S\setminus S_n} \eta \varrho_n \mathbf{f} \cdot \mathbf{W}\, dxdt - \int_0^T \int_{S_n\setminus S} \eta \varrho_n \mathbf{f} \cdot \mathbf{W}\, dxdt \Big|$$
$$\leq c \limsup_{n\to\infty} \int_0^T \int_{S\triangle S_n} \varrho_n\, dxdt, \tag{10.2.16}$$

where $c = \|\eta \mathbf{f} \cdot \mathbf{W}\|_{C(B\times(0,T))} < \infty$ and $\triangle$ denotes the symmetric difference of sets, $S \triangle S_n = (S \setminus S_n) \cup (S_n \setminus S)$. Now the Hölder inequality implies

$$\int_0^T \int_{S\triangle S_n} \varrho_n\, dxdt \leq \|\varrho_n\|_{L^\infty(0,T;L^\gamma(S\triangle S_n))} \|1\|_{L^1(0,T;L^{\gamma/(\gamma-1)}(S\triangle S_n))}$$

Since $\|1\|_{L^1(0,T;L^{\gamma/(\gamma-1)}(S\triangle S_n))} = T\,\mathrm{meas}(S \triangle S_n)^{(\gamma-1)/\gamma}$, we obtain

$$\int_0^T \int_{S\triangle S_n} \varrho_n\, dxdt \leq T \|\varrho_n\|_{L^\infty(0,T;L^\gamma(S\triangle S_n))}\, \mathrm{meas}(S \triangle S_n)^{(\gamma-1)/\gamma}.$$

By Condition 10.2.4(ii), the couple $(\mathbf{u}_n, \varrho_n)$ satisfies all conditions of Theorem 9.3.1 and hence admits the energy estimate (9.3.3). It follows that $\|\varrho_n\|_{L^\infty(0,T;L^\gamma(S\triangle S_n))} \leq c_e < \infty$, where $c_e$ is independent of $n$. Applying Lemma 10.2.6 we finally obtain

$$\limsup_{n\to\infty} \int_0^T \int_{S\triangle S_n} \varrho_n\, dxdt \leq T c_e \limsup_{n\to\infty} \mathrm{meas}(S \triangle S_n)^{(\gamma-1)/\gamma} = 0.$$

Substituting this into (10.2.16) we arrive at (10.2.14). □

**Lemma 10.2.8.** *Under the assumptions of Theorem 10.2.5, for $\boldsymbol{\varphi} \in C^1(B \setminus S)$ compactly supported in $B \setminus S$,*

$$\lim_{n\to\infty} \int_{B\setminus S_n} (\varrho_n \mathbf{u}_n)(x,T) \cdot \boldsymbol{\varphi}\, dx = \int_{B\setminus S} (\varrho \mathbf{u})(x,T) \cdot \boldsymbol{\varphi}\, dx. \tag{10.2.17}$$

*Proof.* First we show that for any sequence $h_n \to 0$,

$$\frac{1}{h_n}\int_{T-h_n}^{T}\int_{B\setminus S_n} \varrho_n \mathbf{u}_n \cdot \boldsymbol{\varphi}\, dxdt \to \int_{B\setminus S} (\varrho\mathbf{u})(x,T)\cdot\boldsymbol{\varphi}\, dx \quad \text{as } n\to\infty. \tag{10.2.18}$$

Here the vector field $(\varrho\mathbf{u})(x,T)$ is defined by Lemma 10.2.1. Arguing as in the proof of Lemma 10.2.1 and noting that $\boldsymbol{\varphi}$ is compactly supported in $B\setminus S_n$ for large $n$, we obtain, for all sufficiently large $n$,

$$\begin{aligned}\frac{1}{h_n}\int_{T-h_n}^{T}\int_{B\setminus S_n} \varrho_n \mathbf{u}_n \cdot \boldsymbol{\varphi}\, dxdt &= \int_0^T\int_{B\setminus S} \upsilon_{h_n}(t)\mathfrak{P}^n_{\varphi}\, dx\, dt \\ &+ \int_{B\setminus S} \varrho_\infty \mathbf{U}\cdot\boldsymbol{\varphi}(x,0)\, dx,\end{aligned} \tag{10.2.19}$$

where $\upsilon_h$ is defined by (10.2.5) and

$$\mathfrak{P}^n_{\varphi} = (\varrho_n\mathbf{u}_n\otimes\mathbf{u}_n - \mathbb{S}(\mathbf{u}_n)) : \nabla\boldsymbol{\varphi} + p(\varrho_n)\operatorname{div}\boldsymbol{\varphi} + \boldsymbol{f}\cdot\boldsymbol{\varphi}. \tag{10.2.20}$$

Notice that the functions $\upsilon_{h_n}$ are uniformly bounded and converge everywhere to 1 as $n\to\infty$. In particular, $\upsilon_{h_n}\to 1$ strongly in any $L^r(0,T)$, $r<\infty$. On the other hand, Condition 10.2.4(iii) implies that

$$-\mathbb{S}(\mathbf{u}_n) : \nabla\boldsymbol{\varphi} + \varrho_n\boldsymbol{f}\cdot\boldsymbol{\varphi} \rightharpoonup -\mathbb{S}(\mathbf{u}) : \nabla\boldsymbol{\varphi} + \varrho\boldsymbol{f}\cdot\boldsymbol{\varphi}$$

weakly in $L^{3/2}((B\setminus S)\times(0,T))$. Thus we get

$$\begin{aligned}&\int_0^T\int_{B\setminus S_n} \upsilon_{h_n}(t)\big(-\mathbb{S}(\mathbf{u}_n) : \nabla\boldsymbol{\varphi} + \varrho_n\boldsymbol{f}\cdot\boldsymbol{\varphi}\big)\, dxdt \\ &\qquad \to \int_0^T\int_{B\setminus S}\big(-\mathbb{S}(\mathbf{u}) : \nabla\boldsymbol{\varphi} + \varrho\boldsymbol{f}\cdot\boldsymbol{\varphi}\big)\, dxdt \quad \text{as } n\to\infty.\end{aligned} \tag{10.2.21}$$

Next, as mentioned above, the sequence $(\mathbf{u}_n,\varrho_n,S_n)$ meets all requirements of Theorem 9.3.1. It follows that $p(\varrho_n)\to p(\varrho)$ in $L^1(\Omega'\times(0,T))$ for any measurable set $\Omega'\Subset B\setminus S$. Moreover, the matrix-valued functions $\varrho_n\mathbf{u}_n\otimes\mathbf{u}_n$ converge weakly to $\varrho\mathbf{u}\otimes\mathbf{u}$ in $L^2(0,T;L^b(\Omega'))$ for some $b>1$. Setting $\Omega' = \operatorname{supp}\boldsymbol{\varphi}$ leads to

$$\begin{aligned}&\int_0^T\int_{B\setminus S_n} \upsilon_{h_n}(t)\big(\varrho_n\mathbf{u}_n\otimes\mathbf{u}_n : \nabla\boldsymbol{\varphi} + p(\varrho_n)\operatorname{div}\boldsymbol{\varphi}\big)\, dxdt \\ &\qquad \to \int_0^T\int_{B\setminus S}\big(\varrho\mathbf{u}\otimes\mathbf{u}\nabla : \boldsymbol{\varphi} + p(\varrho)\operatorname{div}\boldsymbol{\varphi}\big)\, dxdt \quad \text{as } n\to\infty.\end{aligned} \tag{10.2.22}$$

Combining (10.2.21) and (10.2.22) we finally obtain

$$\begin{aligned}&\lim_{n\to\infty}\frac{1}{h_n}\int_{T-h_n}^{T}\int_{B\setminus S_n} \varrho_n\mathbf{u}_n\cdot\boldsymbol{\varphi}\, dxdt = \lim_{n\to\infty}\int_0^T\int_{B\setminus S}\upsilon_{h_n}(t)\mathfrak{P}^n_{\varphi}\, dx\, dt \\ &+ \int_{B\setminus S}\varrho_\infty\mathbf{U}\cdot\boldsymbol{\varphi}(x,0)\, dx = \int_0^T\int_{B\setminus S}\mathfrak{P}_{\varphi}\, dx\, dt + \int_{B\setminus S}\varrho_\infty\mathbf{U}\cdot\boldsymbol{\varphi}(x,0)\, dx,\end{aligned} \tag{10.2.23}$$

where $\mathfrak{P}_\varphi$ is defined by (10.2.7). Combining (10.2.23) and identity (10.2.10) we obtain the desired relation (10.2.18). On the other hand, by Lemma 10.2.4 for every $n$ there is $h_n > 0$ such that $h_n < 1/n$ and

$$\left| \int_{B\setminus S_n} \left\{ \frac{1}{h_n} \int_{T-h_n}^{T} \varrho(x,t)\mathbf{u}(x,t)\,dt \right\} \cdot \boldsymbol{\varphi}(x)\,dx - \int_{B\setminus S_n} (\varrho_n \mathbf{u}_n)(x,T) \cdot \boldsymbol{\varphi}(x)\,dx \right| \leq 1/n.$$

Substituting $h_n$ into (10.2.18) we arrive at (10.2.17). □

**Lemma 10.2.9.** *Under the assumptions of Theorem 10.2.5,*

$$\lim_{n\to\infty} \int_{B\setminus S_n} \eta(\varrho_n \mathbf{u}_n)(x,T)\cdot\mathbf{W}(x,T)\,dx = \int_{B\setminus S} \eta(\varrho\mathbf{u})(x,T)\cdot\mathbf{W}(x,T)\,dx. \tag{10.2.24}$$

*Proof.* Choose $0 < \delta < \varepsilon$ and denote the $\delta$-neighborhood and $\varepsilon$-neighborhood of the limit obstacle $S$ by $S^\delta$ and $S^\varepsilon$, respectively. Pick $\varsigma \in C_0^\infty(B)$ such that

$$\varsigma = 1 \quad \text{in } B \setminus S^\varepsilon, \quad \varsigma = 0 \quad \text{in } S^\delta. \tag{10.2.25}$$

Since $\eta$ vanishes near $\partial B$, the vector field $\boldsymbol{\varphi}(\cdot) = \eta\varsigma\mathbf{W}(\cdot,T) \in C^1(B \setminus S)$ is compactly supported in $B \setminus S$, which along with Lemma 10.2.8 implies that for $n \to \infty$,

$$\int_{B\setminus S_n} \varsigma\eta(\varrho_n\mathbf{u}_n)(x,T) \cdot \mathbf{W}(x,T)\,dx - \int_{B\setminus S} \varsigma\eta(\varrho\mathbf{u})(x,T) \cdot \mathbf{W}(x,T)\,dx \to 0. \tag{10.2.26}$$

Hence for $t = T$,

$$\begin{aligned} \limsup_{n\to\infty} &\left| \int_{B\setminus S_n} \eta(\varrho_n\mathbf{u}_n) \cdot \mathbf{W}\,dx - \int_{B\setminus S} \eta(\varrho\mathbf{u}) \cdot \mathbf{W}\,dx \right|_{t=T} \\ &\leq \limsup_{n\to\infty} \left| \int_{B\setminus S_n} (1-\varsigma)\eta(\varrho_n\mathbf{u}_n) \cdot \mathbf{W}\,dx - \int_{B\setminus S} (1-\varsigma)\eta(\varrho\mathbf{u}) \cdot \mathbf{W}\,dx \right|_{t=T}. \end{aligned} \tag{10.2.27}$$

Next notice that $S_n \Subset S^\varepsilon$ for all sufficiently large $n$ and $\varsigma = 1$ outside of $S^\varepsilon$. For all such $n$ we have

$$\begin{aligned} &\left| \int_{B\setminus S_n} (1-\varsigma)\eta(\varrho_n\mathbf{u}_n) \cdot \mathbf{W}\,dx - \int_{B\setminus S} (1-\varsigma)\eta(\varrho\mathbf{u}) \cdot \mathbf{W}\,dx \right|_{t=T} \\ &\quad = \left| \int_{S^\varepsilon\setminus S_n} (1-\varsigma)\eta(\varrho_n\mathbf{u}_n) \cdot \mathbf{W}\,dx - \int_{S^\varepsilon\setminus S} (1-\varsigma)\eta(\varrho\mathbf{u}) \cdot \mathbf{W}\,dx \right|_{t=T} \\ &\quad \leq c \int_{S^\varepsilon\setminus S_n} |(\varrho_n\mathbf{u}_n)(x,T)|\,dx + c \int_{S^\varepsilon\setminus S} |(\varrho\mathbf{u})(x,T)|\,dx, \end{aligned}$$

where $c = \|(1-\varsigma)\eta\mathbf{W}\|_{C(B\times(0,T))}$. Inserting this in (10.2.27) we obtain

$$\limsup_{n\to\infty}\Big|\int_{B\setminus S_n}\eta(\varrho_n\mathbf{u}_n)\cdot\mathbf{W}\,dx-\int_{B\setminus S}\eta(\varrho\mathbf{u})\cdot\mathbf{W}\,dx\Big|_{t=T}$$
$$\le c\limsup_{n\to\infty}\int_{S^\varepsilon\setminus S_n}|(\varrho_n\mathbf{u}_n)(x,T)|\,dx+c\int_{S^\varepsilon\setminus S}|(\varrho\mathbf{u})(x,T)|\,dx. \qquad (10.2.28)$$

Now our task is to show that the right hand side vanishes for $n\to\infty$ and $\varepsilon\to 0$. From Lemma 10.2.1 we conclude that

$$\|(\varrho_n\mathbf{u}_n)(\cdot,T)\|_{L^{2\gamma/(\gamma+1)}(B\setminus S_n)}+\|(\varrho\mathbf{u})(\cdot,T)\|_{L^{2\gamma/(\gamma+1)}(B\setminus S)}\le c,$$

where $c$ is independent of $n$. Next applying the Hölder inequality we obtain

$$\int_{S^\varepsilon\setminus S}|(\varrho\mathbf{u})(x,T)|\,dx\le c\Big(\int_{S^\varepsilon\setminus S}|(\varrho\mathbf{u})|^{2\gamma/(\gamma+1)}\,dx\Big)^{\frac{\gamma+1}{2\gamma}}\operatorname{meas}(S^\varepsilon\setminus S)^{\frac{\gamma-1}{2\gamma}}$$
$$\le c\operatorname{meas}(S^\varepsilon\setminus S)^{(\gamma-1)/(2\gamma)}.$$

The same estimate holds for $(\varrho_n\mathbf{u}_n)$,

$$\int_{S^\varepsilon\setminus S_n}|(\varrho_n\mathbf{u}_n)(x,T)|\,dx\le c\operatorname{meas}(S^\varepsilon\setminus S_n)^{(\gamma-1)/(2\gamma)}.$$

Inserting these into (10.2.28) we finally obtain

$$\limsup_{n\to\infty}\Big|\int_{B\setminus S_n}\eta(\varrho_n\mathbf{u}_n)(x,T)\cdot\mathbf{W}(x,T)\,dx-$$
$$\int_{B\setminus S}\eta(\varrho\mathbf{u})(x,T)\cdot\mathbf{W}(x,T)\,dx\Big|$$
$$\le c\limsup_{n\to\infty}\operatorname{meas}(S^\varepsilon\setminus S_n)^{(\gamma-1)/(2\gamma)}+c\operatorname{meas}(S^\varepsilon\setminus S)^{(\gamma-1)/(2\gamma)}. \qquad (10.2.29)$$

On the other hand, Lemma 10.2.6 and the inclusion $S^\varepsilon\setminus S_n\subset(S^\varepsilon\setminus S)\cup(S\setminus S_n))$ give

$$\limsup_{n\to\infty}\operatorname{meas}(S^\varepsilon\setminus S_n)^{(\gamma-1)/(2\gamma)}$$
$$\le\limsup_{n\to\infty}\big\{\operatorname{meas}(S^\varepsilon\setminus S)+\operatorname{meas}(S\setminus S_n)\big\}^{(\gamma-1)/(2\gamma)}\le\operatorname{meas}(S^\varepsilon\setminus S)^{(\gamma-1)/(2\gamma)}.$$

Inserting this into (10.2.29) yields

$$\limsup_{n\to\infty}\Big|\int_{B\setminus S_n}\eta(\varrho_n\mathbf{u}_n)(x,T)\cdot\mathbf{W}(x,T)\,dx$$
$$-\int_{B\setminus S}\eta(\varrho\mathbf{u})(x,T)\cdot\mathbf{W}(x,T)\,dx\Big|$$
$$\le 2c\operatorname{meas}(S^\varepsilon\setminus S)^{(\gamma-1)/(2\gamma)}.$$

It remains to note that the right hand side tends to 0 as $\varepsilon\to 0$. □

**Remark 10.2.10.** The results of Lemmas 10.2.6–10.2.9 require Condition 10.2.4. It is worth noting that the first three conditions (i)–(iii) in 10.2.4 simply state that $(S_n, \mathbf{u}_n, \varrho_n)$ and $(S, \mathbf{u}, \varrho)$ meet the conditions of the main stability theorem 9.3.1 and the existence theorem 10.1.1. The last condition (iv) is independent of the previous ones. This extra restriction means that the characteristic functions of the sets $S_n$ converge in the $L^1$-norm, which is not in the spirit of the Hausdorff and Kuratowski-Mosco convergences. It is important to notice that Lemmas 10.2.11–10.2.14 below only require Condition 10.2.4(i)–(iii).

**Lemma 10.2.11.** *Let* $S_n \xrightarrow{\mathcal{S}} S \Subset B$ *and let* $\mathbf{u}_n \in L^2(0,T; W^{1,2}_{S_n}(B))$ *converge weakly in* $L^2(0,T;W^{1,2}(B))$ *to* $\mathbf{u} \in L^2(0,T;W^{1,2}(B))$. *Then for any* $q \in [1,2)$ *and* $r \in [1, 2d/(d-2))$,

$$\|\mathbf{u}_n\|_{L^q(0,T;L^r(S))} \to 0 \quad \text{as } n \to \infty. \tag{10.2.30}$$

*If in addition* $(\mathbf{u}_n, \varrho_n)$ *satisfies Condition 10.2.4(i)–(iii), then*

$$\int_0^T \int_{S\setminus S_n} \varrho_n |\mathbf{u}_n|^2 \, dxdt \to 0 \quad \text{as } n \to \infty. \tag{10.2.31}$$

*Proof.* Set

$$\mathfrak{u}_n(x) = \int_0^T |\mathbf{u}_n(x,t)|\, dt \quad \text{for } x \in B. \tag{10.2.32}$$

We have

$$\|\mathfrak{u}_n\|_{L^2(B)} \le \sqrt{T}\|\mathbf{u}_n\|_{L^2(0,T;L^2(B))} \le \sqrt{T}\|\mathbf{u}_n\|_{L^2(0,T;W^{1,2}(B))} \le c. \tag{10.2.33}$$

Recall that a.e. in $B$,

$$\nabla|\mathbf{u}_n| = |\mathbf{u}_n|^{-1}\mathbf{u}_n \nabla \mathbf{u}_n \quad \text{for } \mathbf{u}_n \neq 0, \quad \nabla|\mathbf{u}_n| = 0 \quad \text{otherwise.}$$

Hence

$$\|\nabla \mathfrak{u}_n\|^2_{L^2(B)} \le \int_B \Big(\int_0^T |\nabla \mathbf{u}_n|\, dt\Big)^2 dx \le T \int_B \int_0^T |\nabla \mathbf{u}_n|^2 \, dt\, dx \le c.$$

Altogether, $\mathfrak{u}_n$ is bounded in $W^{1,2}(B)$. Since the embedding $W^{1,2}(B) \hookrightarrow L^2(B)$ is compact, there is a subsequence $\mathfrak{u}_{n_k}$ which converges weakly in $W^{1,2}(B)$ and strongly in $L^2(B)$ to some $\mathfrak{u} \in W^{1,2}(B)$. Moreover, any subsequence of $\mathfrak{u}_n$ contains such a convergent subsequence (with limit depending on the subsequence, of course).

Since $\mathbf{u}_{n_k} \in L^2(0,T;W^{1,2}_{S_{n_k}}(B))$, Lemma 8.3.13 implies that $\mathfrak{u}_{n_k} \in W^{1,2}_{S_{n_k}}(B)$. Next notice that the obstacles $S_n$ converge to $S$ in the sense of Kuratowski-Mosco. It now follows from the weak convergence of $\mathfrak{u}_{n_k}$ that $\mathfrak{u} \in W^{1,2}_S(B)$ and, in particular, $\mathfrak{u} = 0$ on $S$. Recalling that $\mathfrak{u}_{n_k}$ converges to $\mathfrak{u}$ strongly in $L^2(B)$ and hence in $L^2(S)$ we conclude that

$$\int_0^T \int_S |\mathbf{u}_{n_k}|\, dxdt = \|\mathfrak{u}_{n_k}\|_{L^1(S)} \le c\|\mathfrak{u}_{n_k}\|_{L^2(S)} \to 0 \quad \text{as } k \to \infty.$$

Thus $\mathbf{u}_{n_k} \to 0$ in $L^1(S \times (0,T))$. This sequence is bounded in $L^2(0,T;W^{1,2}(B))$, so it is bounded in $L^q(0,T;L^r(S))$ for all $q \le 2$ and $r < 2d/(d-2)$. Hence $\mathbf{u}_{n_k} \to 0$ in $L^q(0,T;L^r(S))$ for all $q \in [1,2)$ and $r \in [1, 2d/(d-2))$. Thus we have proved that any subsequence of $\mathbf{u}_n$ has a subsequence which tends to 0 in $L^q(0,T;L^r(S))$. Hence the whole sequence tends to 0 in this space, which gives (10.2.30).

It remains to prove (10.2.31). To this end notice that by Condition 10.2.4(ii), the functions $(\mathbf{u}_n, \varrho_n)$ satisfy all assumptions of Theorem 9.3.1 and, in particular, the energy estimate (9.3.3) with a constant $c_e$ independent of $n$. Next choose $M > 0$ and denote by $\mathcal{E}_n$ the set of all $t \in (0,T)$ such that $\|\mathbf{u}_n(\cdot,t)\|_{W^{1,2}(B)} \le M$. By (9.3.3) we have

$$\int_{(0,T)\setminus\mathcal{E}_n} \|\mathbf{u}_n(\cdot,t)\|^2_{W^{1,2}(B)}\,dt \le \int_{(0,T)} \|\mathbf{u}_n(\cdot,t)\|^2_{W^{1,2}(B)}\,dt \le c_e^2,$$

which along with $\|\mathbf{u}_n(\cdot,t)\|_{W^{1,2}(B)} \ge M$ on $(0,T)\setminus\mathcal{E}_n$ leads to

$$\mathrm{meas}((0,T)\setminus\mathcal{E}_n) \le c_e^2/M^2.$$

The energy estimate (9.3.3) yields

$$\|\varrho_n|\mathbf{u}_n|^2\|_{L^\infty(0,T;L^1(B\setminus S_n))} \le c_e.$$

Thus we get

$$\begin{aligned}&\int_{(0,T)\setminus\mathcal{E}_n}\int_{B\setminus S_n} \varrho_n|\mathbf{u}_n|^2\,dxdt \\ &\qquad \le c\|\varrho_n|\mathbf{u}_n|^2\|_{L^\infty(0,T;L^1(B\setminus S_n))}\,\mathrm{meas}((0,T)\setminus\mathcal{E}_n) \le c_e\cdot c_e^2/M^2. \qquad (10.2.34)\end{aligned}$$

Next, the Hölder inequality and the energy estimate (see (9.3.3))

$$\|\varrho_n\|_{L^\infty(0,T;L^\gamma(B\setminus S_n))} \le c_e$$

imply

$$\begin{aligned}&\int_{\mathcal{E}_n}\int_{S\setminus S_n} \varrho_n|\mathbf{u}_n|^2\,dxdt \\ &\qquad \le \int_{\mathcal{E}_n}\Big(\int_{B\setminus S_n}\varrho_n(\cdot,t)^\gamma\,dx\Big)^{1/\gamma}\Big(\int_S |\mathbf{u}_n(\cdot,t)|^{(2\gamma)/(\gamma-1)}\,dx\Big)^{1-1/\gamma}dt \\ &\qquad = \int_{\mathcal{E}_n}\|\varrho_n(\cdot,t)\|_{L^\gamma(B\setminus S_n)}\|\mathbf{u}_n(\cdot,t)\|^2_{L^r(S)}\,dt \le c_e\int_{\mathcal{E}_n}\|\mathbf{u}_n(\cdot,t)\|^2_{L^r(S)}\,dt,\end{aligned}$$

where

$$r = 2\gamma/(\gamma-1) < 2d/(d-2) \quad \text{for } \gamma > d/2.$$

Since the embedding $W^{1,2}(B) \hookrightarrow L^r(B)$ is bounded, we have $\|\mathbf{u}_n(\cdot,t)\|_{L^r(S)} \le cM$ for every $t \in \mathcal{E}_n$. This leads to

$$\int_{\mathcal{E}_n}\int_{S\setminus S_n} \varrho_n|\mathbf{u}_n|^2\,dxdt \le c_e M \int_{\mathcal{E}_n} \|\mathbf{u}_n(\cdot,t)\|_{L^r(S)}\,dt. \tag{10.2.35}$$

Combining (10.2.34) and (10.2.35) we deduce

$$\int_0^T\int_{S\setminus S_n} \varrho_n|\mathbf{u}_n|^2\,dxdt \le c_e M\|\mathbf{u}_n\|_{L^1(0,T;L^r(S))} + c_e^3/M.$$

Letting $n \to \infty$ and recalling (10.2.30) we obtain

$$\limsup_{n\to\infty}\int_0^T\int_{S\setminus S_n} \varrho_n|\mathbf{u}_n|^2\,dxdt \le c_e^3/M.$$

Letting $M \to \infty$ we finally arrive at (10.2.31). □

**Lemma 10.2.12.** *Assume that triplets $(\mathbf{u}_n, \varrho_n, S_n)$ and $(\mathbf{u}, \varrho, S)$ satisfy Condition 10.2.4(i)–(iii). Then*

$$\lim_{n\to\infty}\int_0^T\int_{B\setminus S_n} \eta\varrho_n\mathbf{u}_n\cdot\partial_t\mathbf{W}\,dxdt = \int_0^T\int_{B\setminus S} \eta\varrho\mathbf{u}\cdot\partial_t\mathbf{W}\,dxdt. \tag{10.2.36}$$

*Proof.* Recall that the vector fields $\mathbf{u}_n$ belong to $L^2(0,T;W^{1,2}_{S_n}(B))$ and converge to $\mathbf{u}$ weakly in $L^2(0,T;W^{1,2}(B))$. By Corollary 8.3.3 the limit vector field $\mathbf{u}$ belongs to $L^2(0,T;W^{1,2}_S(B))$. It follows that $\mathbf{u}_n$ and $\mathbf{u}$ are zero a.e. on $S_n\times(0,T)$ and $S\times(0,T)$, respectively. Thus (10.2.36) can be equivalently rewritten as

$$\int_0^T\int_B \eta\varrho_n\mathbf{u}_n\cdot\partial_t\mathbf{W}\,dxdt \to \int_0^T\int_B \eta\varrho\mathbf{u}\cdot\partial_t\mathbf{W}\,dxdt. \tag{10.2.37}$$

By Condition 10.2.4(ii), the triplets $(\mathbf{u}_n, \varrho_n, S_n)$ satisfy all hypotheses of Theorem 9.3.1. It follows from relation (9.3.6) in that theorem that the moments $\varrho_n\mathbf{u}_n$ (extended by 0 onto $S_n$) converge weakly$^\star$ in $L^\infty(0,T;L^{2\gamma/(\gamma+1)}(B))$ to $\varrho\mathbf{u}$. This obviously leads to (10.2.37). □

**Lemma 10.2.13.** *Let $(\mathbf{u}_n, \varrho_n, S_n)$ and $(\mathbf{u}, \varrho, S)$ satisfy Condition 10.2.4(i)–(iii). Then*

$$\int_0^T\int_{B\setminus S_n} \mathbb{T}_n : \nabla(\eta\mathbf{W})\,dxdt \to \int_0^T\int_{B\setminus S} \mathbb{T} : \nabla(\eta\mathbf{W})\,dxdt. \tag{10.2.38}$$

*Proof.* Recall that the stress tensor $\mathbb{T}$ is defined by (10.2.2) and consequently

$$\mathbb{T}_n = \nabla\mathbf{u}_n + (\nabla\mathbf{u}_n)^\top + (\lambda-1)\,\mathrm{div}\,\mathbf{u}_n\,\mathbb{I} - p(\varrho_n)\,\mathbb{I}. \tag{10.2.39}$$

Since $\mathbf{u}_n \in L^2(0,T; W^{1,2}_{S_n}(B))$ and $\mathbf{u} \in L^2(0,T; W^{1,2}_S(B))$, the vector fields $\mathbf{u}_n$ and $\mathbf{u}$ vanish a.e. on $S_n \times (0,T)$ and $S \times (0,T)$, respectively. Hence

$$\nabla \mathbf{u}_n = 0 \quad \text{a.e. on } S_n, \qquad \nabla \mathbf{u} = 0 \quad \text{a.e. on } S.$$

Noting that $\nabla \mathbf{u}_n$ converges to $\nabla \mathbf{u}$ weakly in $L^2(0,T; L^2(B))$ we obtain

$$\int_0^T \int_{B\setminus S_n} (\nabla \mathbf{u}_n + (\nabla \mathbf{u}_n)^\top + (\lambda - 1) \operatorname{div} \mathbf{u}_n \, \mathbb{I}) : \nabla(\eta \mathbf{W})\, dxdt$$
$$\to \int_0^T \int_{B\setminus S} (\nabla \mathbf{u} + (\nabla \mathbf{u})^\top + (\lambda - 1) \operatorname{div} \mathbf{u} \, \mathbb{I}) : \nabla(\eta \mathbf{W})\, dxdt. \tag{10.2.40}$$

Recall that by the assumptions of Theorem 10.2.5, we have $\operatorname{div} \mathbf{W} = 0$, which leads to

$$p\,\mathbb{I} : \nabla(\eta \mathbf{W}) = p \nabla \eta \cdot \mathbf{W}.$$

Since $\nabla \eta$ vanishes near $S$ and $\partial B$, $\nabla \eta \cdot \mathbf{W}$ is compactly supported in some cylinder $\Omega' \times (0,T)$ with $\Omega' \Subset B \setminus S$. Now Theorem 9.3.1 shows that $p(\varrho_n) \to p(\varrho)$ in $L^1(\Omega' \times (0,T))$. Thus we get

$$\int_0^T \int_{B\setminus S_n} p(\varrho_n)\, \mathbb{I} : \nabla(\eta \mathbf{W})\, dxdt = \int_0^T \int_{\Omega'} p(\varrho_n) \nabla \eta \cdot \mathbf{W}\, dxdt$$
$$\to \int_0^T \int_{\Omega'} p(\varrho)\, \mathbb{I} : \nabla(\eta \mathbf{W})\, dxdt = \int_0^T \int_{B\setminus S} p(\varrho) \nabla \eta \cdot \mathbf{W}\, dxdt.$$

Combining this relation with (10.2.40) we obtain (10.2.38). □

**Lemma 10.2.14.** *Let $(\mathbf{u}_n, \varrho_n, S_n)$ and $(\mathbf{u}, \varrho, S)$ satisfy Conditions 10.2.4(i)–(iii). Then*

$$\int_0^T \int_{B\setminus S_n} \varrho_n \mathbf{u}_n \otimes \mathbf{u}_n : \nabla(\eta \mathbf{W})\, dxdt \to \int_0^T \int_{B\setminus S} \varrho \mathbf{u} \otimes \mathbf{u} : \nabla(\eta \mathbf{W})\, dxdt \quad \text{as } n \to \infty. \tag{10.2.41}$$

*Proof.* By Condition 10.2.4, the functions $(\mathbf{u}_n, \varrho_n)$ satisfy all hypotheses of Theorem 9.3.1 and hence admit the energy estimate (9.3.3). It follows that the energy of $(\mathbf{u}_n, \varrho_n)$ is bounded from above by a constant $c_e$ independent of $n$. In particular, $(\mathbf{u}_n, \varrho_n)$ meet all requirements of Proposition 4.2.1. Corollary 4.2.3 of that proposition implies that there is an exponent $b > 1$, depending only on the adiabatic exponent $\gamma$ and the space dimension $d$, such that

$$\| \varrho_n |\mathbf{u}_n|^2 \|_{L^2(0,T; L^b(B\setminus S_n))} \le c, \tag{10.2.42}$$

where $c$ depends only on $c_e$, $b$ and the diameter of $B$. Moreover, it follows from the convergences (9.3.6) in Theorem 9.3.1 that for any $\Omega' \Subset B \setminus S$,

$$\varrho_n \mathbf{u}_n \otimes \mathbf{u}_n \rightharpoonup \varrho \mathbf{u} \otimes \mathbf{u} \quad \text{weakly in } L^2(0,T; L^b(\Omega')). \tag{10.2.43}$$

Notice that $\Omega' \Subset B \setminus S_n$ for all large $n$, since $S_n \to S$ in the Hausdorff metric. Now choose $\varepsilon > 0$ and denote by $S^\varepsilon$ the $\varepsilon$-neighborhood of $S$. Then $B \setminus S^\varepsilon \subset B \setminus S_n$ for all sufficiently large $n$. Hence for all such $n$, the sequence $\varrho_n \mathbf{u}_n \otimes \mathbf{u}_n$ is defined in $(B \setminus S^\varepsilon) \times (0,T)$, bounded in $L^2(0,T; L^b(B \setminus S^\varepsilon))$, and converges weakly in $L^2(0,T; L^b(\Omega'))$ to $\varrho \mathbf{u} \otimes \mathbf{u}$ for any compact $\Omega' \Subset B \setminus S^\varepsilon$. Therefore $\varrho_n \mathbf{u}_n \otimes \mathbf{u}_n$ converges weakly in $L^2(0,T; L^b(B \setminus S^\varepsilon))$ to $\varrho \mathbf{u} \otimes \mathbf{u}$. Thus

$$\int_0^T \int_{B\setminus S^\varepsilon} \varrho_n \mathbf{u}_n \otimes \mathbf{u}_n : \nabla(\eta \mathbf{W})\, dxdt \to \int_0^T \int_{B\setminus S^\varepsilon} \varrho \mathbf{u} \otimes \mathbf{u} : \nabla(\eta \mathbf{W})\, dxdt. \tag{10.2.44}$$

Let us consider the identity

$$\begin{aligned}
&\int_0^T \int_{B\setminus S_n} \varrho_n \mathbf{u}_n \otimes \mathbf{u}_n : \nabla(\eta \mathbf{W})\, dxdt - \int_0^T \int_{B\setminus S} \varrho \mathbf{u} \otimes \mathbf{u} : \nabla(\eta \mathbf{W})\, dxdt \\
&\quad = \int_0^T \int_{(S^\varepsilon\setminus S)\setminus S_n} \varrho_n \mathbf{u}_n \otimes \mathbf{u}_n : \nabla(\eta \mathbf{W})\, dxdt - \int_0^T \int_{S^\varepsilon\setminus S} \varrho \mathbf{u} \otimes \mathbf{u} : \nabla(\eta \mathbf{W})\, dxdt \\
&\quad + \int_0^T \int_{B\setminus S^\varepsilon} \varrho_n \mathbf{u}_n \otimes \mathbf{u}_n : \nabla(\eta \mathbf{W})\, dxdt - \int_0^T \int_{B\setminus S^\varepsilon} \varrho \mathbf{u} \otimes \mathbf{u} : \nabla(\eta \mathbf{W})\, dxdt \\
&\quad + \int_0^T \int_{S\setminus S_n} \varrho_n \mathbf{u}_n \otimes \mathbf{u}_n : \nabla(\eta \mathbf{W})\, dxdt.
\end{aligned} \tag{10.2.45}$$

It follows from Lemma 10.2.11 and boundedness of $\nabla(\eta \mathbf{W})$ that

$$\int_0^T \int_{S\setminus S_n} \varrho_n \mathbf{u}_n \otimes \mathbf{u}_n : \nabla(\eta \mathbf{W})\, dxdt \to 0.$$

Inserting this along with (10.2.44) into (10.2.45) we obtain

$$\begin{aligned}
&\limsup_{n\to\infty} \left| \int_0^T \int_{B\setminus S_n} \varrho_n \mathbf{u}_n \otimes \mathbf{u}_n : \nabla(\eta \mathbf{W})\, dxdt - \int_0^T \int_{B\setminus S} \varrho \mathbf{u} \otimes \mathbf{u} : \nabla(\eta \mathbf{W})\, dxdt \right| \\
&\qquad = \limsup_{n\to\infty} \left| \int_0^T \int_{(S^\varepsilon\setminus S)\setminus S_n} \varrho_n \mathbf{u}_n \otimes \mathbf{u}_n : \nabla(\eta \mathbf{W})\, dxdt \right. \\
&\qquad\qquad \left. - \int_0^T \int_{S^\varepsilon\setminus S} \varrho \mathbf{u} \otimes \mathbf{u} : \nabla(\eta \mathbf{W})\, dxdt \right|,
\end{aligned}$$

which leads to

$$\begin{aligned}
&\limsup_{n\to\infty} \left| \int_0^T \int_{B\setminus S_n} \varrho_n \mathbf{u}_n \otimes \mathbf{u}_n : \nabla(\eta \mathbf{W})\, dxdt - \int_0^T \int_{B\setminus S} \varrho \mathbf{u} \otimes \mathbf{u} : \nabla(\eta \mathbf{W})\, dxdt \right| \\
&\quad \le c \limsup_{n\to\infty} \int_0^T \int_{(S^\varepsilon\setminus S)\setminus S_n} \varrho_n |\mathbf{u}_n|^2\, dxdt + c \int_0^T \int_{S^\varepsilon\setminus S} \varrho |\mathbf{u}|^2\, dxdt,
\end{aligned} \tag{10.2.46}$$

where $c = \|\nabla(\eta\mathbf{W})\|_{C(B\times(0,T))}$. Let us estimate the right hand side of this inequality. The Hölder inequality implies

$$\int_0^T\int_{(S^\varepsilon\setminus S)\setminus S_n}\varrho_n|\mathbf{u}_n|^2\,dxdt \\ \le \|\varrho_n|\mathbf{u}_n|^2\|_{L^2(0,T;L^b(B\setminus S_n))}\|1\|_{L^2(0,T;L^{b/(b-1)}((S^\varepsilon\setminus S)\setminus S_n))}.$$

Next, (10.2.42) and

$$\|1\|_{L^2(0,T;L^{b/(b-1)}((S^\varepsilon\setminus S)\setminus S_n))} = \sqrt{T}\,\mathrm{meas}((S^\varepsilon\setminus S)\setminus S_n))^{(b-1)/b} \\ \le \sqrt{T}\,\mathrm{meas}(S^\varepsilon\setminus S)^{(b-1)/b}$$

yield

$$\int_0^T\int_{(S^\varepsilon\setminus S)\setminus S_n}\varrho_n|\mathbf{u}_n|^2\,dxdt \le c\,\mathrm{meas}(S^\varepsilon\setminus S)^{(b-1)/b}. \tag{10.2.47}$$

Similar arguments give

$$\int_0^T\int_{S^\varepsilon\setminus S}\varrho|\mathbf{u}|^2\,dxdt \le c\,\mathrm{meas}(S^\varepsilon\setminus S)^{(b-1)/b}. \tag{10.2.48}$$

Inserting (10.2.47) and (10.2.48) into (10.2.46) we finally arrive at

$$\limsup_{n\to\infty}\Big|\int_{B\setminus S_n}\varrho_n\mathbf{u}_n\otimes\mathbf{u}_n:\nabla(\eta\mathbf{W})\,dxdt - \int_0^T\int_{B\setminus S}\varrho\mathbf{u}\otimes\mathbf{u}:\nabla(\eta\mathbf{W})\,dxdt\Big| \\ \le c\,\mathrm{meas}(S^\varepsilon\setminus S)^{(b-1)/b}.$$

It remains to note that the right hand side tends to 0 as $\varepsilon\to 0$. □

*Proof of Theorem 10.2.5.* We are now in a position to prove the desired relation 10.2.12, which will complete the proof of Theorem 10.2.5. To this end we have to substitute $S_n$, $\mathbf{u}_n$ and $\varrho_n$ into (10.2.1) and let $n\to\infty$. By (10.2.1) we have

$$W(S_n,\mathbf{u}_n,\varrho_n) = W^{(0)}(S_n,\mathbf{u}_n,\varrho_n) + W^{(1)}(S_n,\mathbf{u}_n,\varrho_n), \tag{10.2.49}$$

where

$$W^{(0)}(S,\mathbf{u},\varrho) = -\int_{B\setminus S}\eta(\varrho\mathbf{u})(x,T)\cdot\mathbf{W}(x,T)\,dx + \int_0^T\int_{B\setminus S}\eta\varrho\mathbf{f}\cdot\mathbf{W}\,dxdt,$$

$$W^{(1)}(S,\mathbf{u},\varrho) = \int_{B\setminus S}\eta\varrho_\infty(x,0)\mathbf{U}(x,0)\cdot\mathbf{W}(x,0)\,dx \\ + \int_0^T\int_{B\setminus S}\{\varrho\eta\mathbf{u}\cdot\partial_t\mathbf{W} + (\varrho(\mathbf{u}\otimes\mathbf{u}) - \mathbb{T}) : \nabla(\eta\mathbf{W})\}\,dxdt.$$

Letting $n\to\infty$ and applying Lemmas 10.2.9 and 10.2.7 we obtain $W^{(0)}(S,\mathbf{u},\varrho) = \lim_{n\to\infty}W^{(0)}(S_n,\mathbf{u}_n,\varrho_n)$. The proof is completed by applying Lemmas 10.2.11–10.2.14 to conclude that $W^{(1)}(S,\mathbf{u},\varrho) = \lim_{n\to\infty}W^{(1)}(S_n,\mathbf{u}_n,\varrho_n)$. □

It is important to note that if $\mathbf{W}(\cdot, T) = \mathbf{f} = 0$ then $W^{(0)}(S, \mathbf{u}, \varrho) = 0$. Hence to prove the continuity of $W$ is suffices to show that $W^{(1)}(S, \varrho, \mathbf{u}) = \lim_{n\to\infty} W^{(1)}(S_n, \mathbf{u}_n, \varrho_n)$. As mentioned above, this follows from Lemmas 10.2.11–10.2.14 only. In these lemmas Condition 10.2.4(iv) is not required. Thus we get a stronger version of Theorem 10.2.5.

**Theorem 10.2.15.** *Let $(\mathbf{u}_n, \varrho_n, S_n)$ and $(\mathbf{u}, \varrho, S)$ satisfy Conditions 10.2.4(i)–(iii). Let $\mathbf{W}$ be a divergence free vector field in $C^1((B \setminus S) \times (0,T))$ with $\mathbf{W}(\cdot, T) = 0$. Assume that $\mathbf{f} = 0$. Then $(\mathbf{u}, \varrho)$ is a renormalized solution to problem (9.2.4) and*

$$W(S_n, \mathbf{u}_n, \varrho_n) \to W(S, \mathbf{u}, \varrho) \quad \text{as } n \to \infty. \tag{10.2.50}$$

## 10.3 Applications of the stability theorem to optimization problems

The main stability theorem 9.3.1 and the results on the continuity of the work functional form the theoretical base for studies of shape optimization in aerodynamics. In this section we present existence results for the problem of minimizing the work functional within a family of obstacles with a given volume. In the general case the problem can be formulated as follows:

Assume that the state of the atmosphere and the plan of flight are known, and defined by given vector fields $\mathbf{U}$, $\mathbf{W}$, and the density distribution $\varrho_\infty$, which are fixed data. We also assume that the volume of admissible obstacles is fixed to be a constant $V > 0$. We want to find an obstacle $S$ from a specific admissible class such that the "shape" of the obstacle $S$ minimizes the work of hydrodynamic forces acting on the moving body during the time interval $(0, T)$. In order to give an explicit formulation of this problem we have to define the set of admissible obstacles, some "technological" constraints, and a shape cost functional. The choice of admissible obstacles is restricted by physical and engineering constraints. Otherwise it should be as broad as possible.

It is clear that a physically acceptable obstacle $S$ should have a finite diameter. To bound it we fix a compact set $B' \subset \mathbb{R}^d$ and assume that all admissible obstacles $S$ are contained in $B'$. Next we fix a hold-all domain $B \Supset B'$. The set of admissible obstacles should be compact with respect to $\mathcal{S}$-convergence in order to provide the continuity properties of flow parameters as functions of the "shape" of $S$. One possible way, sufficiently interesting from the mathematical point of view, is to assume that $S \in \mathcal{O}(k, \mathcal{T}, B')$ (see Definition 8.4.1). Moreover, the set of admissible obstacles should have the property that the mapping $S \mapsto \operatorname{meas} S$ is continuous with respect to $\mathcal{S}$-convergence. This condition is fulfilled if we assume that the totality of characteristic functions $\chi_S$ of admissible obstacles is compact in the $L^1(\mathbb{R}^d)$ topology. This obviously holds if the obstacles have uniformly bounded perimeters, or if their characteristic functions are bounded in some Sobolev space $W^{s,r}(\mathbb{R}^d)$ with $s > 0$.

**Definition 10.3.1.** Let $B' \Subset B$ and $V \in (0, \operatorname{meas} B']$ be given. We say that a set $\mathcal{A} \subset 2^{B'}$ is *admissible* if

(i) Every element of $\mathcal{A}$ is a compact subset of $B'$ with $\operatorname{meas} S = V$.

(ii) $\mathcal{A}$ is a closed subset of $\mathcal{O}(k, \mathcal{T}, B')$ with respect to $\mathcal{S}$-convergence.

(iii) The family of characteristic functions $\chi_S$ of $S \in \mathcal{A}$ is compact in $L^1(\mathbb{R}^d)$.

The list of given data for the in/out flow problem in the domain $(B \setminus S) \times (0, T)$ includes a vector field $\mathbf{U} : B \times (0, T) \to \mathbb{R}^d$, a nonnegative function $\varrho_\infty : B \times (0, T) \to \mathbb{R}^+$ and a vector field $\mathbf{W} : \mathbb{R}^d \times (0, T) \to \mathbb{R}^d$. For the flow around a moving obstacle we have (see Section 2.5)

$$\mathbf{W} = \mathbb{U}^\top(t)\big(\dot{\mathbb{U}}(t)x + \dot{\mathbf{a}}(t)\big) \tag{10.3.1}$$

where the mapping $x \mapsto \mathbb{U}(t)x + \mathbf{a}(t)$ defines the evolution of a rigid body. In this case we also have

$$\mathbf{U} = -\mathbf{W} \text{ on } \partial B \times (0, T), \quad \mathbf{U} = 0 \text{ in a neighborhood of } B' \times (0, T). \tag{10.3.2}$$

We can choose an arbitrary field $\mathbf{U}$ satisfying conditions (10.3.2). Furthermore we assume that

$$\mathbf{U}, \mathbf{W} \in C^\infty(\Omega \times [0, T]), \quad \varrho_\infty \in C^1(B \times [0, T]). \tag{10.3.3}$$

By Theorem 10.1.1 for every $S \in \mathcal{A}$ and $\mathbf{U}$, $\mathbf{W}$, $\varrho_\infty$ satisfying conditions (10.3.1)–(10.3.3), the in/out flow problem (9.2.4) has a renormalized solution which meets all requirements of Definition 9.2.3 and satisfies estimates (10.1.1)–(10.1.2). Denote the totality of such solutions by $\mathcal{R}(S)$.

The optimization problem can be formulated as follows:

**Problem 10.3.2.** *For fixed* $\mathbf{U}$, $\mathbf{W}$ *and* $\varrho_\infty$ *satisfying conditions* (10.3.1)–(10.3.3), *find an obstacle* $S \in \mathcal{A}$ *and* $(\mathbf{u}, \varrho) \in \mathcal{R}(S)$ *such that*

$$W(S, \mathbf{u}, \varrho) = \inf_{S' \in \mathcal{A}} \Big\{ \inf_{(\mathbf{u}', \varrho') \in \mathcal{R}(S')} W(S', \mathbf{u}', \varrho') \Big\}. \tag{10.3.4}$$

The following theorem gives an existence result for Problem 10.3.2:

**Theorem 10.3.3.** *Let* $B' \Subset B$ *and* $\mathcal{A} \subset 2^{B'}$ *meet all requirements of Definition 10.3.1. Then Problem 10.3.2 has a solution* $S \in \mathcal{A}$.

*Proof.* Since $\mathbf{U}$, $\mathbf{W}$ and $\varrho_\infty$ satisfy (10.3.1)–(10.3.3), for every $S \subset B'$ the given data satisfy all assumptions of the existence theorem 10.1.1 and Lemma 10.2.1. By Theorem 10.1.1, problem (9.2.4) has a renormalized solution $(\mathbf{u}, \varrho) \in L^\infty(0, T; L^\gamma(B \setminus S)) \times L^2(0, T; W^{1,2}_{S_n}(B))$ which satisfies estimates (10.1.1)–(10.1.2). Moreover, by Lemma 10.2.1 the work functional $W(S, \mathbf{u}, \varrho)$ is well defined. Hence the set $\mathcal{R}(S)$ is not empty and we can define

$$m = \inf_{S \in \mathcal{A}} \Big\{ \inf_{(\mathbf{u}, \varrho) \in \mathcal{R}(S)} W(S, \mathbf{u}, \varrho) \Big\}.$$

Choose sequences $S_n \in \mathcal{A}$ and $(\mathbf{u}_n, \varrho_n) \in \mathcal{R}(S_n)$ such that

$$W(S_n, \mathbf{u}_n, \varrho_n) \to m \quad \text{as } n \to \infty. \tag{10.3.5}$$

Since $S_n \in \mathcal{A} \subset \mathcal{O}(k, \mathcal{T}, B')$, it follows from Theorem 8.4.2 that there is a subsequence, still denoted by $S_n$, such that $S_n \xrightarrow{\mathcal{S}} S$. Moreover, by Definition 10.3.1, we have $S \in \mathcal{A}$. By the existence theorem 10.1.1, $(\mathbf{u}_n, \varrho_n)$ satisfies the energy estimate (10.1.1) and hence we can assume, passing to a subsequence, that

$$\varrho_n \rightharpoonup \varrho \text{ weakly}^\star \text{ in } L^\infty(0,T; L^\gamma(B)), \quad \mathbf{u}_n \rightharpoonup \mathbf{u} \text{ weakly in } L^2(0,T; W^{1,2}(B)).$$

Recall that the functions $\varrho_n$ are extended to $B \times (0,T)$ by setting $\varrho_n = \varrho_\infty(\cdot, 0)$ on $S_n \times (0,T)$, and $\mathbf{u}_n = 0$ a.e. on $S_n \times (0,T)$. Hence the triplets $(S_n, \mathbf{u}_n, \varrho_n)$ meet all requirements of the stability theorem 9.3.1. Consequently, $(\mathbf{u}, \varrho)$ is a renormalized solution to problem (9.2.4) in $(B \setminus S) \times (0,T)$ and so it belongs to $\mathcal{R}(S)$. It remains to note that by the continuity theorem 10.2.5, we have $W(S, \mathbf{u}, \varrho) = \lim_{n\to\infty} W(S_n, \mathbf{u}_n, \varrho_n) = m$. □

# Chapter 11

# Sensitivity analysis. Shape gradient of the drag functional

## 11.1 Introduction

In this chapter we study domain dependence of stationary solutions to compressible Navier-Stokes equations. The main goal is to prove the existence and obtain a robust representation for the derivatives of solutions with respect to smooth deformations of the flow domain. We apply these results to analysis of the drag minimization problem for the flow around a moving body, and derive formulae for the variations of the drag with respect to the variations of the body.

The stationary boundary value problem for compressible Navier-Stokes equations can be formulated as follows.

**Problem 11.1.1.** *Assume that a gas occupies the flow domain $\Omega = B \setminus S$, where $B \subset \mathbb{R}^3$ is a bounded hold-all domain, with an obstacle $S$ in its interior.*

*Find a couple $(\mathbf{u}, \varrho)$ such that*

$$\Delta \mathbf{u} + \lambda \nabla \operatorname{div} \mathbf{u} = \mathsf{Re}\, \varrho \mathbf{u} \cdot \nabla \mathbf{u} + \frac{\mathsf{Re}}{\mathsf{Ma}^2} \nabla p(\varrho) \quad \text{in } \Omega, \tag{11.1.1a}$$

$$\operatorname{div}(\varrho \mathbf{u}) = 0 \quad \text{in } \Omega, \tag{11.1.1b}$$

$$\mathbf{u} = \mathbf{U} \quad \text{on } \Sigma, \quad \mathbf{u} = 0 \quad \text{on } \partial S, \tag{11.1.1c}$$

$$\varrho = \varrho_0 \quad \text{on } \Sigma_{\text{in}}. \tag{11.1.1d}$$

*where $\mathbf{U}$ is a given vector field and $\Sigma = \partial B$. The inlet $\Sigma_{\text{in}}$ and the outlet $\Sigma_{\text{out}}$ are defined by*

$$\Sigma_{\text{in}} = \{x \in \Sigma : \mathbf{U} \cdot \mathbf{n} < 0\}, \quad \Sigma_{\text{out}} = \{x \in \Sigma : \mathbf{U} \cdot \mathbf{n} > 0\},$$

*respectively. Here $\mathbf{n}$ stands for the outward normal to $\partial(B \setminus S) = \Sigma \cup \partial S$. For* *simplicity of calculations we assume that the mass force equals zero, and $\varrho_0$ is a given positive constant.*

It is assumed that the body $S$ moves through the gas with a constant speed $\mathbf{U}_\infty$, and that equations (11.1.1) are written in the moving frame attached to the body. In this framework the drag is the component parallel to $\mathbf{U}_\infty$ of the power of the hydrodynamical force acting on the body. It equals $\mathsf{Re}^{-1}J(\Omega)$, where we denote by $J(\Omega)$ the *drag functional* given by the boundary and volume integrals of the form

$$\begin{aligned} J(\Omega) &= -\mathbf{U}_\infty \cdot \int_{\partial S} \Big(\nabla \mathbf{u} + (\nabla \mathbf{u})^\top + (\lambda - 1)\operatorname{div} \mathbf{u}\,\mathbb{I} - \frac{\mathsf{Re}}{\mathsf{Ma}^2} p\mathbb{I}\Big) \cdot \mathbf{n}\, dS \\ &= -\int_\Omega \mathbf{U}_\infty \cdot \Big[\Big(\nabla \mathbf{u} + (\nabla \mathbf{u})^\top + (\lambda - 1)\operatorname{div} \mathbf{u}\,\mathbb{I} - \frac{\mathsf{Re}}{\mathsf{Ma}^2} p\mathbb{I}\Big) \nabla\eta + \mathsf{Re}\,\eta\varrho\mathbf{u}\nabla\mathbf{u}\Big]\, dx, \end{aligned}$$

where $\eta \equiv 1$ in the vicinity of $S$, and $\eta \equiv 0$ in the vicinity of the boundary $\Sigma = \partial B$ of the hold-all domain $B$. Notice that the latter integral is independent of the choice of $\eta$.

For fixed $B$, $\mathbf{U}$, $\mathbf{U}_\infty$, and $\varrho_0$ the drag depends only on the shape of the body $S$. The minimization of the drag is an important applied problem. If $S_\varepsilon$, $\varepsilon \in \mathbb{R}$, is a one-parameter family of deformations of $S$ with $S_0 = S$, then the functional $J(\Omega_\varepsilon)$, $\Omega_\varepsilon = B \setminus S_\varepsilon$ becomes a function of $\varepsilon$. The study of this function for all possible deformations of $S$ and the calculation of the first order derivative $dJ(\Omega_\varepsilon)/d\varepsilon$ is the main task of *shape sensitivity analysis* performed in this chapter.

### 11.1.1 Change of variables in Navier-Stokes equations

In order to perform the shape sensitivity analysis in three space dimensions we choose a vector field $\mathbf{T} \in C^2(\mathbb{R}^3)^3$ vanishing in a neighborhood of $\Sigma$, and introduce the mapping $y = x + \varepsilon\mathbf{T}(x)$, which defines the perturbation $S_\varepsilon = y(S)$ of the obstacle $S$. For small $\varepsilon$, the mapping $x \mapsto y$ takes diffeomorphically the flow region $\Omega = B \setminus S$ onto $\Omega_\varepsilon = B \setminus S_\varepsilon$. In the perturbed domain Problem 11.1.1 reads

**Problem 11.1.2.** *Find a solution $(\bar{\mathbf{u}}_\varepsilon, \bar\varrho_\varepsilon)$, to the following boundary value problem posed in the variable domain $\Omega_\varepsilon = B \setminus S_\varepsilon$, for the shape parameter $\varepsilon \in (-\delta, \delta)$ with $\delta > 0$:*

$$\Delta\bar{\mathbf{u}}_\varepsilon + \lambda\nabla \operatorname{div} \bar{\mathbf{u}}_\varepsilon = \mathsf{Re}\,\bar\varrho_\varepsilon\bar{\mathbf{u}}_\varepsilon \cdot \nabla\bar{\mathbf{u}}_\varepsilon + \frac{\mathsf{Re}}{\mathsf{Ma}^2}\nabla p(\bar\varrho_\varepsilon) \quad \text{in } B \setminus S_\varepsilon, \tag{11.1.2a}$$

$$\operatorname{div}(\bar\varrho_\varepsilon\bar{\mathbf{u}}_\varepsilon) = 0 \quad \text{in } B \setminus S_\varepsilon, \tag{11.1.2b}$$

$$\bar{\mathbf{u}}_\varepsilon = \mathbf{U} \quad \text{on } \Sigma, \qquad \bar{\mathbf{u}}_\varepsilon = 0 \quad \text{on } \partial S_\varepsilon, \tag{11.1.2c}$$

$$\bar\varrho_\varepsilon = \varrho_0 \quad \text{on } \Sigma_{\text{in}}. \tag{11.1.2d}$$

The shape functional in the variable domain $B \setminus S_\varepsilon$ is written in the form

$$J(\Omega_\varepsilon) = -\mathsf{Re}\,\mathbf{U}_\infty \cdot \int_{B\setminus S_\varepsilon} \eta \bar{\varrho}_\varepsilon \bar{\mathbf{u}}_\varepsilon \nabla \bar{\mathbf{u}}_\varepsilon \, dx$$
$$- \mathbf{U}_\infty \cdot \int_{B\setminus S_\varepsilon} \Big( \nabla \bar{\mathbf{u}}_\varepsilon + (\nabla \bar{\mathbf{u}}_\varepsilon)^\top + (\lambda - 1) \operatorname{div} \bar{\mathbf{u}}_\varepsilon \, \mathbb{I} - \frac{\mathsf{Re}}{\mathsf{Ma}^2} p(\bar{\varrho}_\varepsilon)\mathbb{I} \Big) \cdot \nabla \eta \, dx. \tag{11.1.3}$$

Now, we perform a change of variables in equations (11.1.2) as well as in the drag functional (11.1.3) in order to reduce Problem 11.1.2 in the variable domain $\Omega_\varepsilon$, depending on a small parameter $\varepsilon$, to a problem in the fixed domain $\Omega$, named the *reference domain.* Denote by $\mathbb{M}$ the Jacobi matrix of the mapping $x \mapsto x + \varepsilon \mathbf{T}(x)$ and by $\mathfrak{g}$ the determinant of $\mathbb{M}$, i.e.,

$$\mathbb{M}(x) = \mathbb{I} + \varepsilon D\,\mathbf{T}(x), \quad \mathfrak{g}(x) = \det(\mathbb{I} + \varepsilon D\,\mathbf{T}(x)). \tag{11.1.4}$$

Further the notation $\mathbf{N}$ stands for the adjugate matrix

$$\mathbf{N} = (\det \mathbb{M})\, \mathbb{M}^{-1} \quad \text{with} \quad \det \mathbf{N} = \mathfrak{g}^2. \tag{11.1.5}$$

**Definition 11.1.3.** • In shape sensitivity analysis the *material derivatives* of solutions to stationary Navier-Stokes equations are obtained by the differentiation of the mapping

$$\varepsilon \mapsto (\bar{\mathbf{u}}_\varepsilon(x + \varepsilon \mathbf{T}(x)), \bar{\varrho}_\varepsilon(x + \varepsilon \mathbf{T}(x))), \quad x \in \Omega.$$

- The *shape derivatives* of solutions to stationary Navier-Stokes equations are defined by derivatives of the mapping

$$\varepsilon \mapsto (\bar{\mathbf{u}}_\varepsilon(y), \bar{\varrho}_\varepsilon(y)), \quad y \in \Omega_\varepsilon.$$

- The *shape gradient* of the drag functional for stationary Navier-Stokes equations is given by the derivative of the function

$$\varepsilon \mapsto J(\Omega_\varepsilon).$$

Introduce the functions

$$\mathbf{u}_\varepsilon(x) = \mathbf{N}(x)\, \bar{\mathbf{u}}_\varepsilon(x + \varepsilon \mathbf{T}(x)), \quad \varrho_\varepsilon(x) = \bar{\varrho}_\varepsilon(x + \varepsilon \mathbf{T}(x)), \quad x \in \Omega. \tag{11.1.6}$$

The function $\mathbf{u}_\varepsilon$ is called the *Piola transformation* of $\bar{\mathbf{u}}_\varepsilon$.

**Lemma 11.1.4.** *Let $(\bar{\mathbf{u}}_\varepsilon(y), \bar{\varrho}_\varepsilon(y))$ be a solution to Problem 11.1.2. Then the couple $(\mathbf{u}_\varepsilon(x), \varrho_\varepsilon(x))$ defined by (11.1.6) satisfies*

$$\Delta \mathbf{u}_\varepsilon + \nabla\Big(\lambda \mathfrak{g}^{-1} \operatorname{div} \mathbf{u}_\varepsilon - \frac{\mathsf{Re}}{\mathsf{Ma}^2} p(\varrho_\varepsilon)\Big) = \mathscr{A}(\mathbf{u}_\varepsilon) + \mathsf{Re}\, \mathscr{B}(\varrho_\varepsilon, \mathbf{u}_\varepsilon, \mathbf{u}_\varepsilon) \quad \text{in } \Omega, \tag{11.1.7a}$$

$$\operatorname{div}(\varrho_\varepsilon \mathbf{u}_\varepsilon) = 0 \quad \text{in } \Omega, \tag{11.1.7b}$$

$$\mathbf{u}_\varepsilon = \mathbf{U} \quad \text{on } \Sigma, \quad \mathbf{u}_\varepsilon = 0 \quad \text{on } \partial S, \tag{11.1.7c}$$

$$\varrho_\varepsilon = \varrho_0 \quad \text{on } \Sigma_{\text{in}}. \tag{11.1.7d}$$

*Here, the linear operator $\mathscr{A}$ and the trilinear form $\mathscr{B}$ are defined by*

$$\begin{aligned} \mathscr{A}(\mathbf{u}) &= \Delta \mathbf{u} - \mathbf{N}^{-\top} \operatorname{div}\big(\mathfrak{g}^{-1} \mathbf{N}\mathbf{N}^\top \nabla(\mathbf{N}^{-1}\mathbf{u})\big), \\ \mathscr{B}(\varrho, \mathbf{u}, \mathbf{w}) &= \varrho \mathbf{N}^{-\top}(\mathbf{u}\, \nabla(\mathbf{N}^{-1}\mathbf{w})). \end{aligned} \tag{11.1.8}$$

*Proof.* Let $\mathbf{a} \in C^1(\Omega_\varepsilon)$ be a vector field and $\tilde{\mathbf{a}}(x) = \mathbf{a}(y(x))$. For any $\phi \in C^1(\Omega_\varepsilon)$ we have $(\nabla_y \phi)(y(x)) = (\mathbb{M}^{-\top} \nabla_x \widetilde{\phi})(x)$, where $\widetilde{\phi}(x) = \phi(y(x))$. It follows that the identities

$$\begin{aligned} \int_\Omega (\operatorname{div}_y \mathbf{a})(y(x))\, \tilde{\phi}(x) \det \mathbb{M}\, dx &= \int_{\Omega_\varepsilon} \operatorname{div}_y \mathbf{a}\phi(y)\, dy = -\int_{\Omega_\varepsilon} \mathbf{a} \cdot \nabla_y \phi\, dy \\ &= -\int_\Omega \tilde{\mathbf{a}} \cdot \mathbb{M}^{-\top} \nabla_x \tilde{\phi}(x) \det \mathbb{M}\, dx = \int_\Omega \operatorname{div}_x((\det \mathbb{M})\, \mathbb{M}^{-1}\tilde{\mathbf{a}})\, \tilde{\phi}(x)\, dx \\ &= \int_\Omega \operatorname{div}_x(\mathbf{N}\tilde{\mathbf{a}})\, \tilde{\phi}(x)\, dx \end{aligned}$$

hold true for all $\phi \in C_0^\infty(\Omega_\varepsilon)$. This leads to

$$(\operatorname{div}_y \mathbf{a})(y(x)) = \mathfrak{g}^{-1} \operatorname{div}_x \big(\mathbf{N}(x)\, \mathbf{a}(y(x))\big). \tag{11.1.9}$$

Setting $\mathbf{a}(y) = \bar{\varrho}_\varepsilon(y)\bar{\mathbf{u}}_\varepsilon(y)$ and noting that, by (11.1.6), $\mathbf{N}(x)\mathbf{a}(y(x)) = \varrho_\varepsilon(x)\mathbf{u}_\varepsilon(x)$ we obtain

$$0 = \operatorname{div}_y(\bar{\varrho}_\varepsilon(y)\bar{\mathbf{u}}_\varepsilon(y)) \equiv \mathfrak{g}^{-1} \operatorname{div}(\varrho_\varepsilon(x)\mathbf{u}_\varepsilon(x)),$$

which implies the modified mass balance equation (11.1.7b). Repeating these arguments gives

$$(\operatorname{div}_y(\bar{\mathbf{u}}_\varepsilon))(y(x)) = \mathfrak{g}^{-1}(x) \operatorname{div}_x \mathbf{u}_\varepsilon(x).$$

From this and the identity $\mathbb{M}^{-\top} = \mathfrak{g}^{-1}\mathbf{N}^\top$ we obtain

$$\nabla_y \Big(\lambda \operatorname{div}_y \bar{\mathbf{u}}_\varepsilon - \frac{\mathsf{Re}}{\mathsf{Ma}^2} p(\bar{\varrho}_\varepsilon)\Big) = \mathfrak{g}^{-1}\mathbf{N}^\top \nabla_x \Big(\lambda \mathfrak{g}^{-1} \operatorname{div}_x \mathbf{u}_\varepsilon - \frac{\mathsf{Re}}{\mathsf{Ma}^2} p(\varrho_\varepsilon)\Big). \tag{11.1.10}$$

Next set $\mathbf{a} = \nabla_y \bar{\mathbf{u}}_\varepsilon$. We have

$$\begin{aligned}\tilde{\mathbf{a}} &= (\nabla_y \bar{\mathbf{u}}_\varepsilon)(y(x)) = \mathbb{M}^{-\top}\nabla_x\big(\bar{\mathbf{u}}_\varepsilon(y(x))\big) = \mathbb{M}^{-\top}\nabla_x(\mathbf{N}^{-1}\mathbf{u}_\varepsilon(x)) \\ &= \mathfrak{g}^{-1}\mathbf{N}^\top\nabla_x(\mathbf{N}^{-1}\mathbf{u}_\varepsilon(x)),\end{aligned}$$

which along with (11.1.9) yields

$$\operatorname{div}_y(\nabla_y\bar{\mathbf{u}}_\varepsilon)(y(x)) = \mathfrak{g}^{-1}\operatorname{div}_x\big(\mathfrak{g}^{-1}\mathbf{N}\mathbf{N}^\top\nabla_x(\mathbf{N}^{-1}\mathbf{u}_\varepsilon)\big)(x).$$

In view of the identity $\Delta = \operatorname{div}\nabla$ we have

$$\begin{aligned}\big(\Delta\bar{\mathbf{u}}_\varepsilon\big)(y(x)) &= \mathfrak{g}^{-1}(x)\operatorname{div}\big(\mathfrak{g}^{-1}\mathbf{N}\mathbf{N}^\top\nabla(\mathbf{N}^{-1}\mathbf{u}_\varepsilon)\big)(x) \\ &= \mathfrak{g}^{-1}(x)\mathbf{N}^\top(\Delta\mathbf{u}_\varepsilon - \mathscr{A}(\mathbf{u}_\varepsilon))(x).\end{aligned} \tag{11.1.11}$$

Next note that the components $(\bar{\mathbf{u}}_\varepsilon\nabla\bar{\mathbf{u}}_\varepsilon)_i$ of the vector $\bar{\mathbf{u}}_\varepsilon\nabla\bar{\mathbf{u}}_\varepsilon$ satisfy

$$\begin{aligned}(\bar{\mathbf{u}}_\varepsilon\nabla\bar{\mathbf{u}}_\varepsilon)_i(y(x)) &= \bar{\mathbf{u}}_\varepsilon(y(x))\cdot\nabla_y\bar{u}_{\varepsilon,i}(y(x)) = \bar{\mathbf{u}}_\varepsilon(y(x))\cdot\Big(\mathbb{M}^{-\top}(x)\nabla_x\big(\bar{u}_{\varepsilon,i}(y(x))\big)\Big) \\ &= \bar{\mathbf{u}}_\varepsilon(y(x))\cdot\Big(\mathfrak{g}^{-1}(x)\mathbf{N}^\top(x)\nabla_x\big(\bar{u}_{\varepsilon,i}(y(x))\big)\Big) \\ &= \mathfrak{g}^{-1}(x)\mathbf{N}(x)\bar{\mathbf{u}}_\varepsilon(y(x))\cdot\nabla_x\big(\bar{u}_{\varepsilon,i}(y(x))\big) = \mathfrak{g}^{-1}\mathbf{u}_\varepsilon(x)\cdot\nabla_x(\mathbf{N}^{-1}(x)\mathbf{u}_\varepsilon(x))_i.\end{aligned}$$

This gives

$$(\bar{\varrho}_\varepsilon\bar{\mathbf{u}}_\varepsilon\nabla\bar{\mathbf{u}}_\varepsilon)(y(x)) = \mathfrak{g}^{-1}(x)\mathbf{N}^\top(x)\mathscr{B}(\varrho_\varepsilon(x),\mathbf{u}_\varepsilon(x),\mathbf{u}_\varepsilon(x)). \tag{11.1.12}$$

Substituting (11.1.10)–(11.1.12) into (11.1.2a) and multiplying both sides of the resulting equality by $\mathfrak{g}\mathbf{N}^{-\top}$, we obtain (11.1.7a). □

**Corollary 11.1.5.** *The expression for the drag functional $J(\Omega_\varepsilon)$ in the reference domain reads*

$$\begin{aligned}J(\Omega_\varepsilon) = &-\mathsf{Re}\,\mathbf{U}_\infty\cdot\int_\Omega\varrho_\varepsilon\mathbf{u}_\varepsilon\nabla(\mathbf{N}^{-1}\mathbf{u}_\varepsilon)\,\eta\,dx \\ &-\mathbf{U}_\infty\cdot\int_\Omega\big[\mathfrak{g}^{-1}(\mathbf{N}^\top\nabla(\mathbf{N}^{-1}\mathbf{u}_\varepsilon)+\nabla(\mathbf{N}^{-1}\mathbf{u}_\varepsilon)^\top\mathbf{N}-\operatorname{div}\mathbf{u}_\varepsilon\mathbb{I}\big)-q_\varepsilon\mathbb{I}\big]\mathbf{N}^\top\nabla\eta\,dx\end{aligned} \tag{11.1.13}$$

*where $\eta$ is an arbitrary smooth function such that $\eta\equiv 1$ in a neighborhood of $S$ and $\eta\equiv 0$ in a neighborhood of $\Sigma$, and the* effective viscous pressure *$q_\varepsilon$ is given by*

$$q_\varepsilon = \frac{\mathsf{Re}}{\mathsf{Ma}^2}\,p(\varrho_\varepsilon) - \lambda\mathfrak{g}^{-1}\operatorname{div}\mathbf{u}_\varepsilon, \tag{11.1.14}$$

By abuse of notation we will omit the subscript $\varepsilon$ and write $(\mathbf{u},\varrho)$ instead of $(\mathbf{u}_\varepsilon,\varrho_\varepsilon)$. Thus we come to the following problem:

**Problem 11.1.6.** *Find* $(\mathbf{u}, \varrho)$ *such that*

$$\Delta \mathbf{u} + \nabla\Big(\lambda \mathfrak{g}^{-1} \operatorname{div} \mathbf{u} - \frac{\mathsf{Re}}{\mathsf{Ma}^2} p(\varrho)\Big) = \mathscr{A}(\mathbf{u}) + \mathsf{Re}\, \mathscr{B}(\varrho, \mathbf{u}, \mathbf{u}) \quad \text{in } \Omega, \tag{11.1.15a}$$

$$\operatorname{div}(\varrho \mathbf{u}) = 0 \quad \text{in } \Omega, \tag{11.1.15b}$$

$$\mathbf{u} = \mathbf{U} \quad \text{on } \Sigma, \quad \mathbf{u} = 0 \quad \text{on } \partial S, \tag{11.1.15c}$$

$$\varrho = \varrho_0 \quad \text{on } \Sigma_{\text{in}}. \tag{11.1.15d}$$

*Here* $\mathfrak{g} = \sqrt{\det \mathbf{N}}$, *and the operators* $\mathscr{A}$ *and* $\mathscr{B}$ *are defined by* (11.1.8).

Observe that equations (11.1.15) and the functional (11.1.13) depend only on the matrix $\mathbf{N}$ and do not depend on $\varepsilon$ directly. In this framework, the fact that $\mathbf{N}$ is the adjugate matrix to the Jacobi matrix $\mathbb{I} + \varepsilon D\,\mathbf{T}$ is not so important. Hence our task is to show that Problem 11.1.6 is well-posed and its solution is differentiable with respect to $\mathbf{N}$. We begin by analysing the well-posedness.

### 11.1.2 Extended problem. Perturbation theory

There are numerous papers dealing with the homogeneous boundary value problem for stationary compressible Navier-Stokes equations with small data. We just recall that there are three different approaches to this problem proposed in [11], [103], and [96], respectively. The basic results on the local existence and uniqueness of strong solutions are assembled in [101]. For an interesting overview see [104].

The in/out flow boundary value problems were studied in [69]–[70], where the local existence and uniqueness results were obtained in the two-dimensional case under the assumption that the velocity $\mathbf{u}$ is close to a given constant vector. The case of small Reynolds and Mach numbers was treated in [108]. The difficulties that appear include:

- The problem of total mass control. It is important to notice that, in contrast to the case of zero velocity boundary conditions when the total mass of the gas is prescribed, for in/out flow the problem of controlling the total mass of the gas remains essentially unsolved.
- The problem of singularities developed by solutions at the interface between $\Sigma_{\text{in}}$ and $\Sigma \setminus \Sigma_{\text{in}}$.
- The formation of a boundary layer near the inlet for small Mach numbers.

The main idea of the approach developed in [11], [96] and [108] is to express $\operatorname{div} \mathbf{u}$ in terms of $\varrho$, next to substitute this expression in the mass balance equations, thus reducing the original problem to a boundary value problem for a transport equation. Unfortunately, the resulting transport equation contains nonlocal operators. However this scheme gives the existence and uniqueness of strong solutions for $\mathsf{Re}, \mathsf{Ma}^2 \ll 1$.

In this book we use the simple algebraic scheme proposed in [103]. The basic idea is to introduce the *effective viscous pressure* $q$ and a large parameter $\sigma_0$,

$$q := \frac{\mathsf{Re}}{\mathsf{Ma}^2} p(\varrho) - \lambda \mathfrak{g}^{-1} \operatorname{div} \mathbf{u}, \quad \sigma_0 = \frac{\mathsf{Re}}{\lambda \mathsf{Ma}^2}, \tag{11.1.16}$$

and write equations (11.1.15) in the form

$$\Delta \mathbf{u} - \nabla q = \mathscr{A}(\mathbf{u}) + \mathsf{Re}\, \mathscr{B}(\varrho, \mathbf{u}, \mathbf{u}) \quad \text{in } \Omega, \tag{11.1.17a}$$

$$\operatorname{div} \mathbf{u} = \mathfrak{g}\sigma_0 p(\varrho) - \frac{\mathfrak{g} q}{\lambda} \quad \text{in } \Omega, \tag{11.1.17b}$$

$$\mathbf{u} \cdot \nabla \varrho + \mathfrak{g}\sigma_0 p(\varrho)\varrho = \frac{\mathfrak{g} q}{\lambda} \varrho \quad \text{in } \Omega, \tag{11.1.17c}$$

$$\mathbf{u} = \mathbf{U} \quad \text{on } \Sigma, \quad \mathbf{u} = 0 \quad \text{on } \partial S, \tag{11.1.17d}$$

$$\varrho = \varrho_0 \quad \text{on } \Sigma_{\text{in}}, \tag{11.1.17e}$$

System (11.1.17) consists of perturbed Stokes equations (11.1.17a)–(11.1.17b) for $(\mathbf{u}, q)$ and a perturbed transport equation (11.1.17c) for $\varrho$. It is easy to see that equations (11.1.17b) and (11.1.17c) are equivalent to (11.1.16) and the mass balance equation $\operatorname{div}(\varrho \mathbf{u}) = 0$. Therefore, equations (11.1.17) are equivalent (11.1.15). It was shown in [103] and [107] that problem (11.1.17) has a strong solution if the parameters satisfy the conditions

$$\lambda \gg 1, \quad \mathsf{Re} \ll 1, \quad \sigma_0 \gg 1,$$

which are more restrictive than the standard conditions $\mathsf{Re} \ll 1$, $\mathsf{Ma}^2 \ll 1$ adopted in [11], [96], and [108].

**Perturbations.** We will look for a local solution to problem (11.1.15) which is close to an approximate solution $(\mathbf{u}_\star, q_\star, \varrho_\star)$. In order to define such a solution notice that for small Mach and Reynolds numbers equations (11.1.17a)–(11.1.17b) are close to Stokes equations and the density $\varrho$ is close to $\varrho_0$. Thus we take

$$\varrho_\star = \varrho_0 \tag{11.1.18}$$

and choose $(\mathbf{u}_\star, q_\star)$ as the solution to the following boundary problem for the Stokes equation:

$$\begin{gathered} \Delta \mathbf{u}_\star - \nabla q_\star = 0, \quad \operatorname{div} \mathbf{u}_\star = 0 \quad \text{in } \Omega, \\ \mathbf{u}_\star = \mathbf{U} \quad \text{on } \Sigma, \quad \mathbf{u}_\star = 0 \quad \text{on } \partial S, \quad \Pi q_\star = q_\star, \end{gathered} \tag{11.1.19}$$

where

$$\Pi q := q - \frac{1}{\operatorname{meas} \Omega} \int_\Omega q \, dx. \tag{11.1.20}$$

The solution to problem (11.1.15) is decomposed into the approximate solution $(\mathbf{u}_\star, q_\star, \varrho_\star)$ and small perturbations $(\mathbf{v}, \pi, \varphi, m)$,

$$\mathbf{u} = \mathbf{u}_\star + \mathbf{v}, \quad \varrho = \varrho_\star + \varphi, \quad q = q_\star + \lambda\sigma_0 p(\varrho_\star) + \pi + \lambda m, \tag{11.1.21}$$

with some unknown functions $\vartheta = (\mathbf{v}, \pi, \varphi)$ and an unknown constant $m$.

**Remark 11.1.7.** We point out that our decomposition (11.1.21) holds exclusively in the reference domain $\Omega = B \setminus S$. The unknown functions $\vartheta$ and the unknown constant $m$ contain the necessary and sufficient information on the shape dependence of solutions to the equations (11.1.15) upon the shape parameter $\varepsilon \to 0$. This information is used in order to determine the material derivatives of the solutions to the model (11.1.2) with respect to the shape perturbation. However, the Stokes boundary value problem (11.1.19) defined in the reference domain $\Omega$ is meaningless for the model (11.1.2) defined in the variable domain $\Omega_\varepsilon$.

Substituting decomposition (11.1.21) into (11.1.17) we obtain the following boundary value problem for $\vartheta = (\mathbf{v}, \pi, \varphi)$:

$$\Delta \mathbf{v} - \nabla \pi = \mathscr{A}(\mathbf{u}) + \mathsf{Re}\,\mathscr{B}(\varrho, \mathbf{u}, \mathbf{u}) \quad \text{in } \Omega, \tag{11.1.22a}$$

$$\operatorname{div} \mathbf{v} = \mathfrak{g}\Big(\frac{\sigma}{\varrho_\star}\varphi - \Psi[\vartheta] - m\Big) \quad \text{in } \Omega, \tag{11.1.22b}$$

$$\mathbf{u} \cdot \nabla \varphi + \sigma \varphi = \Psi_1[\vartheta] + m\mathfrak{g}\varrho \quad \text{in } \Omega, \tag{11.1.22c}$$

$$\mathbf{v} = 0 \quad \text{on } \partial\Omega, \quad \varphi = 0 \quad \text{on } \Sigma_{\text{in}}, \quad \Pi\pi = \pi, \tag{11.1.22d}$$

where

$$\Psi[\vartheta] = \frac{q_\star + \pi}{\lambda} - \frac{\sigma}{p'(\varrho_\star)\varrho_\star} H(\varphi), \quad \Psi_1[\vartheta] = \mathfrak{g}\Big(\varrho\Psi[\vartheta] - \frac{\sigma}{\varrho_\star}\varphi^2\Big) + \sigma\varphi(1 - \mathfrak{g}), \tag{11.1.22e}$$

$$\sigma = \sigma_0 p'(\varrho_\star)\varrho_\star, \quad H(\varphi) = p(\varrho_\star + \varphi) - p(\varrho_\star) - p'(\varrho_\star)\varphi, \tag{11.1.22f}$$

the vector field $\mathbf{u}$ and the function $\varrho$ are given by (11.1.21).

**Determination of $m$. Extended system.** Finally, we specify the constant $m$. In our framework, in contrast to the case of homogeneous boundary value problem, the answer to the question how $m$ can be determined is not trivial. Note that since $\operatorname{div} \mathbf{u}$ has null mean value, the right hand side of equation (11.1.22c) must satisfy the compatibility condition

$$m \int_\Omega \mathfrak{g}\, dx = \int_\Omega \mathfrak{g}\Big(\frac{\sigma}{\varrho_\star}\varphi - \Psi[\vartheta]\Big)\, dx,$$

which formally determines $m$. However, this choice of $m$ leads to essential mathematical difficulties. To make this issue clear note that in the simplest case $\mathfrak{g} = 1$ we have $m = \varrho_\star^{-1}\sigma(\mathbb{I} - \Pi)\varphi + O(|\vartheta|^2) + O(\lambda^{-1})$, and the principal part of equations (11.1.22) becomes

$$\begin{pmatrix} \Delta & -\nabla & 0 \\ \operatorname{div} & 0 & -\dfrac{\sigma}{\varrho_\star} \\ 0 & 0 & \mathbf{u}\nabla + \sigma \end{pmatrix} \begin{pmatrix} \mathbf{v} \\ \pi \\ \varphi \end{pmatrix} + \begin{pmatrix} 0 \\ m \\ -m\varrho_\star \end{pmatrix} \sim \begin{pmatrix} \Delta \mathbf{v} - \nabla \pi \\ \operatorname{div} \mathbf{v} - \dfrac{\sigma}{\varrho_\star}\Pi\varphi \\ \mathbf{u}\nabla\varphi + \sigma\Pi\varphi \end{pmatrix}.$$

Hence, the question of invertibility of the principal part of equations (11.1.22) can be reduced to the question of solvability of the boundary value problem for the nonlocal transport equation

$$\mathbf{u} \cdot \nabla\varphi + \sigma\Pi\varphi = f \quad \text{in } \Omega, \quad \varphi = 0 \quad \text{on } \Sigma_{\text{in}}. \tag{11.1.23}$$

The following simple example shows that there is a drastic difference between properties of solutions to the nonlocal problem (11.1.23) and solutions to the local boundary value problem

$$\mathbf{u} \cdot \nabla\varphi + \sigma\varphi = f \quad \text{in } \Omega, \quad \varphi = 0 \quad \text{on } \Sigma_{\text{in}}.$$

**Example 11.1.8.** Let us consider the one-dimensional boundary value problems

$$\partial_{x_1}\bar{\varphi} + \sigma\bar{\varphi} = 1, \quad 0 \leq x_1 \leq 1, \quad \bar{\varphi}(0) = 0,$$

and

$$\partial_{x_1}\varphi + \sigma\Big(\varphi - \int_0^1 \varphi(x_1)\, dx_1\Big) = 1, \quad 0 \leq x_1 \leq 1, \quad \varphi(0) = 0.$$

Simple calculations give

$$\bar{\varphi}(x_1) = \frac{1}{\sigma}(1 - e^{-\sigma x_1}), \quad \varphi(x_1) = \frac{1 - e^{-\sigma x_1}}{1 - e^{-\sigma}}.$$

Hence $\bar{\varphi} = O(\sigma^{-1})$ but $\varphi = O(1)$ as $\sigma \to \infty$. Thus for the nonlocal transport equation, the advantage of having a large parameter $\sigma$ is lost.

We point out that this difficulty disappears for the homogeneous boundary value problem, i.e., for homogeneous Dirichlet conditions for the velocity, since in this case the mean value of $\varrho$ over $\Omega$ is prescribed and $\varphi = \Pi\varphi$. The difficulty of determining $m$ is only present in the in/out flow problem. In order to cope with it we introduce an extra unknown function $\zeta$ satisfying the adjoint transport equation and add an extra equation for $m$. By doing so we split the problem into the problem of determining $\vartheta = (\mathbf{v}, \pi, \varphi)$ and the problem of finding $m$. The augmented problem has the following form:

**Problem 11.1.9.** *Given $(\mathbf{u}_\star, q_\star, \varrho_\star)$, find $(\mathbf{v}, \pi, \varphi)$ and a constant $m$, depending on $\varepsilon$, such that*

$$\begin{aligned}
&\Delta\mathbf{v} - \nabla\pi = \mathscr{A}(\mathbf{u}) + \mathsf{Re}\mathscr{B}(\varrho, \mathbf{u}, \mathbf{u}) \quad \text{in } \Omega,\\
&\operatorname{div}\mathbf{v} = \mathfrak{g}\Big(\frac{\sigma}{\varrho_\star}\varphi - \Psi[\vartheta] - m\Big) \quad \text{in } \Omega,\\
&\mathbf{u} \cdot \nabla\varphi + \sigma\varphi = \Psi_1[\vartheta] + m\mathfrak{g}\varrho \quad \text{in } \Omega,\\
&\mathbf{v} = 0 \quad \text{on } \partial\Omega, \quad \varphi = 0 \quad \text{on } \Sigma_{\text{in}}, \quad \Pi\pi = \pi.
\end{aligned} \tag{11.1.24a}$$

*The constant $m$ is determined from the following relations:*

$$m = \varkappa \int_\Omega (\varrho_\star^{-1}\Psi_1[\vartheta]\,\zeta - \mathfrak{g}\Psi[\vartheta])\, dx, \quad \varkappa = \Big(\int_\Omega \mathfrak{g}(1 - \zeta - \varrho_\star^{-1}\zeta\varphi)\, dx\Big)^{-1}, \tag{11.1.24b}$$

*where the auxiliary function* $\zeta$ *is a solution to the adjoint boundary value problem*

$$-\operatorname{div}(\mathbf{u}\zeta) + \sigma\zeta = \sigma\mathfrak{g} \quad \text{in } \Omega, \quad \zeta = 0 \quad \text{on } \Sigma_{\text{out}}. \tag{11.1.24c}$$

*Here we denote*

$$\begin{aligned}
\Psi[\vartheta] &= \frac{q_\star + \pi}{\lambda} - \frac{\sigma}{p'(\varrho_\star)\varrho_\star} H(\varphi),\\
H(\varphi) &= p(\varrho_\star + \varphi) - p(\varrho_\star) - p'(\varrho_\star)\varphi, \qquad (11.1.24\text{d})\\
\Psi_1[\vartheta] &= \mathfrak{g}\Big(\varrho\Psi[\vartheta] - \frac{\sigma}{\varrho_\star}\varphi^2\Big) + \sigma\varphi(1-\mathfrak{g}),
\end{aligned}$$

$$\begin{aligned}
\mathscr{A}(\mathbf{u}) &= \Delta\mathbf{u} - \mathbf{N}^{-\top}\operatorname{div}\big(\mathfrak{g}^{-1}\mathbf{N}\mathbf{N}^\top\nabla(\mathbf{N}^{-1}\mathbf{u})\big),\\
\mathscr{B}(\varrho,\mathbf{u},\mathbf{u}) &= \varrho\mathbf{N}^{-\top}(\mathbf{u}\nabla(\mathbf{N}^{-1}\mathbf{u})),
\end{aligned} \tag{11.1.24e}$$

*with the velocity field* $\mathbf{u} := \mathbf{u}_\star + \mathbf{v}$ *and the density* $\varrho := \varrho_\star + \varphi$.

Equations (11.1.24) define a complete system of differential equations and boundary conditions for the functions $\vartheta, \zeta$ and the constant $m$. It is easy to check that every solution to (11.1.24) also satisfies equations (11.1.22). The important feature of equations (11.1.24) is that the right hand side in the expression (11.1.24b) for $m$ is small for large $\sigma$, $\lambda$ and sufficiently small $\varphi$, provided that $\varkappa$ is bounded from above. We employ this fact in the next section devoted to the existence theory.

# 11.2 Existence of solutions

## 11.2.1 Function spaces. Results

**Function spaces.** To establish existence results for problem (11.1.22), we begin with the definition of an admissible class of solutions to this problem.

Notice that we are looking for strong solutions, which means that in equations (11.1.22) the unknown functions and their derivatives are integrable, i.e., these equations are satisfied a.e. in $\Omega$. To this end it suffices to assume that $\mathbf{v} \in W^{2,2}(\Omega)$ and $(\pi, \varphi, \zeta) \in W^{1,2}(\Omega)$. Next, we are looking for solutions such that $(\pi, \varphi, \zeta)$ are continuous and $\mathbf{v}$ is continuously differentiable in $\Omega$. In order to satisfy this demand it suffices to take $\mathbf{v} \in W^{s+1,r}(\Omega)$ and $(\pi, \varphi, \zeta) \in W^{s,r}(\Omega)$ with $sr > 3$. To address both issues we introduce the Banach spaces

$$\begin{aligned}
X^{s,r} &= W^{s,r}(\Omega) \cap W^{1,2}(\Omega),\\
Y^{s,r} &= W^{s+1,r}(\Omega) \cap W^{2,2}(\Omega), \quad 0 < s < 1 < r < \infty,
\end{aligned} \tag{11.2.1}$$

equipped with the norms

$$\|u\|_{X^{s,r}} = \|u\|_{W^{s,r}(\Omega)} + \|u\|_{W^{1,2}(\Omega)}, \quad \|u\|_{Y^{s,r}} = \|u\|_{W^{s+1,r}(\Omega)} + \|u\|_{W^{2,2}(\Omega)}.$$

Throughout of the rest of this chapter we intensively exploit the fact that the spaces $X^{s,r}$ and $Y^{s,r}$ are reflexive. The proof of this fact is nontrivial and is given below.

It is well known (see [2, Thm. 3.5]) that for any bounded domain $\Omega$ with $C^1$ boundary, any integer $m \geq 0$ and $1 < r < \infty$, the Sobolev space $W^{m,r}(\Omega)$ is reflexive. The space $W_0^{1,r}(\Omega)$ is obviously reflexive as a closed subspace of a reflexive Banach space (see [2, Thm. 1.21]).

**Lemma 11.2.1.** *Let $\Omega$ be a bounded domain with $C^1$ boundary and $0 < s < 1 < r < \infty$. Then the spaces $W^{s,r}(\Omega)$ and $W^{s+1,r}(\Omega)$ are reflexive.*

*Proof.* First observe that $W^{1,r}(\Omega)$ and $W_0^{1,r}(\Omega)$ are dense in $L^r(\Omega)$ since $C_0^\infty(\Omega)$ is dense in $L^r(\Omega)$. By Lemma 1.1.12, recalling that $W^{s,r}(\Omega) = [L^r(\Omega), W^{1,r}(\Omega)]_{s,r}$ we obtain

$$(W^{s,r}(\Omega))' = [L^r(\Omega)', (W^{1,r}(\Omega))']_{s,r}. \tag{11.2.2}$$

Next notice that $W_0^{1,r}(\Omega) \subset W^{1,r}(\Omega) \subset L^r(\Omega)$. Hence

$$L^r(\Omega)' \subset (W^{1,r}(\Omega))' \subset (W_0^{1,r}(\Omega))'.$$

In other words $L^r(\Omega)'$ and $(W^{1,r}(\Omega))'$ are linear subspaces of $(W_0^{1,r}(\Omega))'$. Recall (see Section 1.5) that $(W_0^{1,r}(\Omega))' = W^{-1,r'}(\Omega)$ is the completion of $L^{r'}(\Omega) = L^r(\Omega)'$ in the $(W_0^{1,r}(\Omega))'$-norm. Hence the intersection $L^r(\Omega)' \cap (W^{1,r}(\Omega))' = L^r(\Omega)'$ is dense in the ambient space $(W_0^{1,r}(\Omega))'$. Therefore we may apply Lemma 1.1.12 to obtain

$$\begin{aligned}((W^{s,r}(\Omega))')' &\equiv ([L^r(\Omega)', (W^{1,r}(\Omega))']_{s,r})' \\ &= [(L^r(\Omega)')', ((W^{1,r}(\Omega))')']_{s,r} = [L^r(\Omega), W^{1,r}(\Omega)]_{s,r} = W^{s,r}(\Omega),\end{aligned}$$

since $(L^r(\Omega)')' = L^r(\Omega)$ and $((W^{1,r}(\Omega))')' = W^{1,r}(\Omega)$. Thus $W^{s,r}(\Omega)$ is reflexive. Since $W^{s+1,r}(\Omega)$ is the interpolation space $[W^{1,r}(\Omega), W^{2,r}(\Omega)]_{s,r}$ and the spaces $W^{1,r}(\Omega)$, $W^{2,r}(\Omega)$ are reflexive, we can apply the same arguments to conclude that $W^{s+1,r}(\Omega)$ is reflexive. □

**Lemma 11.2.2.** *For any functional $f \in (X^{s,r})'$ there are continuous functionals $g \in (W^{s,r}(\Omega))'$ and $h \in (W^{1,2}(\Omega))'$ such that*

$$\langle f, \varphi \rangle = \langle g, \varphi \rangle + \langle h, \varphi \rangle \quad \textit{for all } \varphi \in X^{s,r}.$$

*Proof.* Applying Lemma 1.1.11 with $A_0 = L^r(\Omega)$ and $A_1 = W^{1,r}(\Omega)$ we find that $W^{1,r}(\Omega)$ is dense in the interpolation space $W^{s,r}(\Omega)$. In particular, $C^\infty(\Omega)$ is dense in $W^{s,r}(\Omega)$, since it is dense in $W^{1,r}(\Omega)$. Next, $C^\infty(\Omega)$ is dense in $W^{1,2}(\Omega)$. As $C^\infty(\Omega) \subset W^{s,r}(\Omega) \cap W^{1,2}(\Omega)$, we conclude that the latter intersection is dense in both $W^{s,r}(\Omega)$ and $W^{1,2}(\Omega)$. Moreover both these spaces contain $W^{1,m}(\Omega)$ for $m > 2+r$, and hence their duals are embedded in $(W^{1,m}(\Omega))'$. By Theorem 1.1.1, for any functional $f \in (W^{s,r}(\Omega) \cap W^{1,2}(\Omega))'$ there are continuous functionals $g \in (W^{s,r}(\Omega))'$ and $h \in (W^{1,2}(\Omega))'$ such that

$$\langle f, \varphi \rangle = \langle g, \varphi \rangle + \langle h, \varphi \rangle \quad \text{for all } \varphi \in W^{s,r}(\Omega) \cap W^{1,2}(\Omega).$$

It remains to recall that $X^{s,r} = W^{s,r}(\Omega) \cap W^{1,2}(\Omega)$. □

**Lemma 11.2.3.** *Let $\Omega$ be a bounded domain with $C^1$ boundary and $0 < s < 1 < r < \infty$. Then the spaces $X^{s,r}$ and $Y^{s,r}$ are reflexive.*

*Proof.* We give the proof only for $X^{s,r}$. The same proof works for $Y^{s,r}$. In view of Theorem 1.1.3, it suffices to show that $X^{s,r}(\Omega)$ is weakly complete. Choose a sequence $\varphi_n \in X^{s,r}$, $n \geq 1$, such that

$$\langle f, \varphi_n \rangle \quad \text{converges in } \mathbb{R} \quad \text{for all } f \in (X^{s,r})'.$$

We need to show that $\varphi_n$ converges weakly to some $\varphi \in X^{s,r}$. Since $X^{s,r}$ is compactly embedded in $C(\Omega)$, $\varphi_n$ converges in $C(\Omega)$, there is $\varphi \in C(\Omega)$ such that $\varphi_n$ converges uniformly in $\Omega$ to $\varphi$. Next, since $\{\varphi_n\}$ is bounded in $W^{s,r}(\Omega)$ and $W^{1,2}(\Omega)$, there is a subsequence $\{\varphi_m\} \subset \{\varphi_n\}$ such that $\varphi_m \rightharpoonup \varphi^*$ weakly in $W^{s,r}(\Omega)$ and $\varphi_m \rightharpoonup \varphi^{**}$ weakly in $W^{1,2}(\Omega)$ as $m \to \infty$. Obviously $\varphi^* = \varphi^{**} = \varphi$ and $\varphi \in X^{s,r}$. Hence $\varphi_m \rightharpoonup \varphi$ weakly in $W^{s,r}(\Omega)$ and $W^{1,2}(\Omega)$. Repeating this argument we conclude that every subsequence of $\{\varphi_n\}$ contains a subsequence which converges weakly to $\varphi$ in $W^{s,r}(\Omega)$ and $W^{1,2}(\Omega)$. Hence the whole sequence $\{\varphi_n\}$ converges weakly to $\varphi$ in both $W^{s,r}(\Omega)$ and $W^{1,2}(\Omega)$. It remains to show that $\varphi_n \rightharpoonup \varphi$ weakly in $X^{s,r}$. To this end, choose $f \in (X^{s,r})'$. In view of Lemma 11.2.2, there are $g \in (W^{s,r}(\Omega))'$ and $h \in (W^{1,2}(\Omega))'$ such that $f = g + h$. Hence

$$\langle f, \varphi_n \rangle = \langle g, \varphi_n \rangle + \langle h, \varphi_n \rangle \to \langle g, \varphi \rangle + \langle h, \varphi \rangle = \langle f, \varphi \rangle.$$

Therefore, $\varphi_n$ converges weakly to $\varphi$ in $X^{s,r}$. □

While $W^{s,r}(\Omega)$ and $W^{1,2}(\Omega)$ are separable, we cannot conclude from this that $X^{s,r}$ is separable. However, the dual spaces $(X^{s,r})'$ and $(Y^{s,r})'$ are separable:

**Lemma 11.2.4.** *Let $\Omega \subset \mathbb{R}^d$ be a bounded domain with $C^1$ boundary and $0 < s < 1 < r < \infty$. Then $(X^{s,r})'$ and $(Y^{s,r})'$ are separable.*

*Proof.* We prove the separability of $(X^{s,r})'$ . The same proof works for $(Y^{s,r})'$. Notice that $(W^{s,r}(\Omega))'$ is the interpolation space $[L^r(\Omega)', (W^{1,r}(\Omega))']_{s,r}$. The space $L^r(\Omega)' = L^{r'}(\Omega)$ is separable, so contains a countable dense set. By Lemma 1.1.11, $L^r(\Omega)'$ is dense in the interpolation space $(W^{s,r}(\Omega))'$. Hence $(W^{s,r}(\Omega))'$ contains a countable dense subset and consequently is separable. Also, by [2, Thm. 3.8], the space $(W^{1,2}(\Omega))'$ is separable. Next, by Lemma 11.2.2 we have $(X^{s,r})' = (W^{s,r}(\Omega))' + (W^{1,2}(\Omega))'$. Obviously $(X^{s,r})'$ is separable as a sum of separable spaces. □

**Existence theory.** Define the Banach space

$$E = (Y^{s,r})^3 \times (X^{s,r})^2, \tag{11.2.3}$$

and denote by $\mathcal{B}_\tau \subset E$ the closed ball of radius $\tau$ centered at 0. Next, note that for $sr > 3$, elements of the ball $\mathcal{B}_\tau$ satisfy the inequality

$$\|\mathbf{v}\|_{C^1(\Omega)} + \|\pi\|_{C(\Omega)} + \|\varphi\|_{C(\Omega)} \leq c_e(r, s, \Omega)\|\vartheta\|_E \leq c_e\tau, \tag{11.2.4}$$

where the norm in $E$ is defined by

$$\|\vartheta\|_E = \|\mathbf{v}\|_{Y^{s,r}} + \|\pi\|_{X^{s,r}} + \|\varphi\|_{X^{s,r}}.$$

Further, $c$ denotes generic constants, which are different in different places and depend only on $\Omega$, $\mathbf{U}$, $\sigma$ and $r, s$. We assume that the flow domain and the given data satisfy the following conditions.

**Condition 11.2.5.** • $\partial\Omega$ is a closed surface of class $C^3$ and the set $\Gamma = \mathrm{cl}\,\Sigma_{\mathrm{in}} \cap \Sigma \setminus \Sigma_{\mathrm{in}}$ is a closed $C^3$ one-dimensional manifold such that $\Sigma = \Sigma_{\mathrm{in}} \cup \Gamma \cup \Sigma_{\mathrm{out}}$.

- The vector field $\mathbf{U} \in C^3(\partial\Omega)$ satisfies

$$\int_{\partial\Omega} \mathbf{U} \cdot \mathbf{n}\, d\Sigma = 0.$$

  Moreover, we can assume that it is extended to $\mathbb{R}^3$ in such a way that the extension $\mathbf{U} \in C^3(\mathbb{R}^3)$ vanishes in a neighborhood of $S$.

- There is a positive constant $c$ such that

$$\mathbf{U} \cdot \nabla(\mathbf{U} \cdot \mathbf{n}) > c > 0 \quad \text{on } \Gamma.$$

  Since the vector field $\mathbf{U}$ is tangent to $\partial\Omega$ on $\Gamma$, the left hand side is well defined.

The following existence theorem is the main result of this section.

**Theorem 11.2.6.** *Assume that the surface $\Sigma$ and the given vector field $\mathbf{U}$ satisfy Condition 11.2.5. Furthermore, let numbers $r$ and $s$ satisfy*

$$0 < s < 1, \quad 1 < r < \infty, \quad 2s - 3r^{-1} < 1, \quad sr > 3. \tag{11.2.5}$$

*Then there exists $\sigma^* > 1$, depending only on $\mathbf{U}$, $\Omega$ and $s$, $r$, with the following property. For every $\sigma > \sigma^*$ there are positive $\tau^*$ and $c$, depending only on $\mathbf{U}$, $\Omega$, $r$, $s$, and $\sigma$, such that whenever*

$$\tau \in (0, \tau^*], \quad \lambda^{-1}, \mathsf{Re} \in (0, \tau^2], \quad \|\mathbf{N} - \mathbb{I}\|_{C^2(\Omega)} \leq \tau^2, \tag{11.2.6}$$

*Problem 11.1.9 has a solution $\vartheta \in \mathcal{B}_\tau$, $\zeta \in X^{s,r}$, $m \in \mathbb{R}$. The auxiliary function $\zeta$ and the constants $\varkappa, m$ admit the estimates*

$$\|\zeta\|_{X^{s,r}} + |\varkappa| \leq c, \quad |m| \leq c\tau < 1. \tag{11.2.7}$$

*Let* $\Theta_{\mathbf{N}} \subset \mathcal{B}_\tau \times W^{s,r}(\Omega) \times \mathbb{R}$ *be the set of all solutions* $(\vartheta, \zeta, m) = (\mathbf{v}, \pi, \varphi, \zeta, m)$ *corresponding to a matrix-valued function* $\mathbf{N}$. *Then for every* $\mathbf{N}$ *in the ball* $B(\tau^2) = \{\mathbf{N} : \|\mathbb{I} - \mathbf{N}\|_{C^2(\Omega)} \leq \tau^2\}$ *there is a nonempty subset* $\Theta'_{\mathbf{N}} \subset \Theta_{\mathbf{N}}$ *such that* $\bigcup_{\mathbf{N}\in B(\tau^2)} \Theta'_{\mathbf{N}}$ *is relatively compact in* $W^{s+1,r}(\Omega) \times W^{s,r}(\Omega)^3 \times \mathbb{R}$.

The remaining part of this section is devoted to the proof of Theorem 11.2.6. The proof falls into five steps.

### 11.2.2 Step 1. Estimates of composite functions

We begin by proving the following lemmas which furnish regularity properties of composite functions.

**Lemma 11.2.7.** *Let* $\Omega \subset \mathbb{R}^3$ *be a bounded domain with* $C^1$ *boundary and* $0 < s < 1 < r < \infty$. *Then there is* $c$, *depending on* $\Omega$, $s$, $r$, *such that*

$$\|uv\|_{W^{s,r}(\Omega)} \leq c\|u\|_{L^\infty(\Omega)}\|v\|_{W^{s,r}(\Omega)} + c\|v\|_{L^\infty(\Omega)}\|u\|_{W^{s,r}(\Omega)}, \tag{11.2.8}$$

$$\|uv\|_{X^{s,r}} \leq c\|u\|_{L^\infty(\Omega)}\|v\|_{X^{s,r}} + c\|v\|_{L^\infty(\Omega)}\|u\|_{X^{s,r}}. \tag{11.2.9}$$

*If, in addition,* $sr > 3$, *then*

$$\|uv\|_{X^{s,r}} \leq c\|u\|_{X^{s,r}}\|v\|_{X^{s,r}}. \tag{11.2.10}$$

*Proof.* Obviously

$$\|uv\|_{L^r(\Omega)} \leq \|u\|_{L^\infty(\Omega)}\|v\|_{L^r(\Omega)} + \|v\|_{L^\infty(\Omega)}\|u\|_{L^r(\Omega)}. \tag{11.2.11}$$

Next we have

$$\begin{aligned}
|uv|^r_{s,r,\Omega} &\equiv \int_\Omega\int_\Omega |u(x)v(x) - u(y)v(y)|^r\,|x-y|^{-3-sr}\,dxdy \\
&\leq c\int_\Omega\int_\Omega \{|u(x)|^r|v(x)-v(y)|^r + |v(y)|^r|u(x)-u(y)|^r\}|x-y|^{-3-sr}\,dxdy \\
&\leq c\|u\|^r_{L^\infty(\Omega)}\int_\Omega\int_\Omega |v(x)-v(y)|^r|x-y|^{-3-sr}dxdy \\
&\quad + c\|v\|^r_{L^\infty(\Omega)}\int_\Omega\int_\Omega |u(x)-u(y)|^r|x-y|^{-3-sr}\,dxdy \\
&= c\|u\|^r_{L^\infty(\Omega)}|v|^r_{s,r,\Omega} + c\|v\|^r_{L^\infty(\Omega)}|u|^r_{s,r,\Omega}.
\end{aligned} \tag{11.2.12}$$

Combining (11.2.2) and (11.2.11), and recalling the equality $\|u\|_{W^{s,r}(\Omega)} = \|u\|_{L^r(\Omega)} + |u|_{s,r,\Omega}$, we obtain

$$\|uv\|_{W^{s,r}(\Omega)} \leq c\|u\|_{L^\infty(\Omega)}\|v\|_{W^{s,r}(\Omega)} + c\|v\|_{L^\infty(\Omega)}\|u\|_{W^{s,r}(\Omega)}, \tag{11.2.13}$$

which is (11.2.8). The identity $\nabla(uv) = u\nabla v + v\nabla u$ implies

$$\|\nabla(uv)\|_{L^2(\Omega)} \leq \|u\|_{L^\infty(\Omega)}\|\nabla v\|_{L^2(\Omega)} + \|v\|_{L^\infty(\Omega)}\|\nabla u\|_{L^2(\Omega)},$$

and

$$\|uv\|_{W^{1,2}(\Omega)} \le c\|u\|_{L^\infty(\Omega)}\|v\|_{W^{1,2}(\Omega)} + c\|v\|_{L^\infty(\Omega)}\|u\|_{W^{1,2}(\Omega)},$$

which along with (11.2.13) leads to (11.2.9). Since for $sr > 3$ the space $W^{s,r}(\Omega)$ is continuously embedded in $L^\infty(\Omega)$, we have $\|u\|_{L^\infty(\Omega)} \le c\|u\|_{W^{s,r}(\Omega)} \le c\|u\|_{X^{s,r}}$ for all $u \in X^{s,r}$. Combining this with (11.2.9) we obtain (11.2.10). □

Let us consider functions $u, v : \Omega \to B_K$, where $B_K = \{x : |x| \le K\} \subset \mathbb{R}^3$ is the ball of radius $K$ centered at 0.

**Lemma 11.2.8.** *Assume that $u, v \in X^{s,r}$, $s \in (0,1)$, $sr > 3$, and $\mathfrak{f} \in C^2(\Omega \times B_K)$. Then*

$$\|\mathfrak{f}(\cdot, u)\|_{X^{s,r}} \le c(r,s)\|\mathfrak{f}\|_{C^1(\Omega\times B_K)}(1 + \|u\|_{X^{s,r}}), \tag{11.2.14}$$

$$\|\mathfrak{f}(\cdot, v) - \mathfrak{f}(\cdot, u)\|_{X^{s,r}} \le c(r,s)\|\mathfrak{f}\|_{C^2(\Omega\times B_K)}(1 + \|u\|_{X^{s,r}} + \|v\|_{X^{s,r}})\|u - v\|_{X^{s,r}}. \tag{11.2.15}$$

*Proof.* We begin by proving (11.2.14). Notice that for every function $f$ we have

$$\|f\|_{X^{s,r}} \le \|f\|_{L^\infty(\Omega)} + |f|_{s,r,\Omega} + \|\nabla f\|_{L^2(\Omega)}.$$

Since the function $|\mathfrak{f}(x, u(x))|$ is bounded by $\|\mathfrak{f}\|_{C(\Omega\times B_K)}$, it suffices to estimate $|\mathfrak{f}|_{s,r,\Omega}$ and $\|\nabla \mathfrak{f}\|_{L^2(\Omega)}$. We have

$$|\mathfrak{f}(x, u(x)) - \mathfrak{f}(y, u(y))|^r \le c(r)\|\mathfrak{f}\|^r_{C^1(\Omega\times B_K)}(|x-y|^r + |u(x) - u(y)|^r),$$

which along with $-3 - rs + r > -3$ gives

$$\begin{aligned}|\mathfrak{f}(\cdot, u)|^r_{s,r,\Omega} &= \int_\Omega\int_\Omega |\mathfrak{f}(x, u(x)) - \mathfrak{f}(y, u(y))|^r\, |x-y|^{-3-rs}\, dxdy \\ &\le c\|\mathfrak{f}\|^r_{C^1(\Omega\times B_K)} \\ &\quad\times\Big\{\int_\Omega\int_\Omega |x-y|^{-3-rs+r} dxdy + \int_\Omega\int_\Omega |u(x)-u(y)|^r|x-y|^{-3-rs}\, dxdy\Big\} \\ &\le c\|\mathfrak{f}\|^r_{C^1(\Omega\times B_K)}(1 + |u|^r_{s,r,\Omega}) \le c|\mathfrak{f}\|^r_{C^1(\Omega\times B_K)}(1 + \|u\|^r_{X^{s,r}}).\end{aligned} \tag{11.2.16}$$

Next we have $|\nabla \mathfrak{f}(x,u)| \le c\|\mathfrak{f}\|_{C^1(\Omega\times B_K)}(1 + |\nabla u|)$, which yields

$$\|\nabla \mathfrak{f}(\cdot, u)\|_{L^2(\Omega)} \le c\|\mathfrak{f}\|_{C^1(\Omega\times B_K)}(1 + \|u\|_{W^{1,2}(\Omega)}) \le c\|\mathfrak{f}\|_{C^1(\Omega\times B_K)}(1 + \|u\|_{X^{s,r}}).$$

Combining this with (11.2.16) we obtain (11.2.14).

Let us turn to the proof of (11.2.15). We have

$$\mathfrak{f}(x,u) - \mathfrak{f}(x,v) = (u - v)\cdot \int_0^1 \nabla_u \mathfrak{f}(x, tu + (1-t)v)\, dt.$$

It follows from (11.2.14) that

$$\|\nabla_u \mathfrak{f}(\cdot, tu + (1-t)v)\|_{X^{s,r}} \le c\|\mathfrak{f}\|_{C^2(\Omega\times B_K)}(1 + t\|u\|_{X^{s,r}} + (1-t)\|v\|_{X^{s,r}}). \tag{11.2.17}$$

Thus we get

$$\begin{aligned}\Big\|\int_0^1 \nabla_u \mathfrak{f}(\cdot, tu+(1-t)v)\,dt\Big\|_{X^{s,r}} &\le \int_0^1 \|\nabla_u \mathfrak{f}(\cdot, tu+(1-t)v)\|_{X^{s,r}}\,dt\\ &\le c\|\mathfrak{f}\|_{C^2(\Omega\times B_K)}\int_0^1 (1+t\|u\|_{X^{s,r}}+(1-t)\|v\|_{X^{s,r}})\,dt\\ &\le c\|\mathfrak{f}\|_{C^2(\Omega\times B_K)}(1+\|u\|_{X^{s,r}}+\|v\|_{X^{s,r}}),\end{aligned} \tag{11.2.18}$$

From this and estimate (11.2.10) we finally obtain

$$\begin{aligned}\|\mathfrak{f}(\cdot,v)-\mathfrak{f}(\cdot,u)\|_{X^{s,r}} &\le c\|v-u\|_{X^{s,r}}\Big\|\int_0^1 \nabla_u\mathfrak{f}(\cdot,tu+(1-t)v)\,dt\Big\|_{X^{s,r}}\\ &\le c(r,s)\|\mathfrak{f}\|_{C^2(\Omega\times B_K)}\|v-u\|_{X^{s,r}}(1+\|u\|_{X^{s,r}}+\|v\|_{X^{s,r}}).\end{aligned}$$

□

### 11.2.3 Step 2. Stokes equations

In this section we establish the existence and uniqueness results in the Banach space $Y^{s,r}\times X^{s,r}$ for the boundary value problem for Stokes equations

$$\begin{aligned}&\Delta\mathbf{v}-\nabla\pi = F, \quad \operatorname{div}\mathbf{v} = \Pi G \quad \text{in } \Omega,\\ &\mathbf{v}=0 \quad \text{on } \partial\Omega, \quad \Pi\pi=\pi.\end{aligned} \tag{11.2.19}$$

Recall notation (11.2.1) for the Banach spaces $X^{s,r}$ and $Y^{s,r}$ and introduce the Banach space

$$Z^{s,r} = \mathcal{W}^{s-1,r}(\Omega)\cap L^2(\Omega), \quad 0<s<1<r<\infty. \tag{11.2.20}$$

While the Banach spaces $X^{s,r}$ and $Y^{s,r}$ consist of functions of real variables and their structure is clear, the structure of the space $Z^{s,r}$ is more complicated and extra explanations are in order. By Lemma 1.5.1 the space $\mathcal{W}^{s-1,r}(\Omega)$ is the dual space to $\mathcal{W}_0^{1-s,r'}(\Omega)$, where $r'=r/(r-1)$. Moreover, $\mathcal{W}^{s-1,r}(\Omega)$ is the completion of $L^r(\Omega)$ in the norm

$$\|f\|_{\mathcal{W}^{s-1,r}(\Omega)} = \sup_{0\neq u\in\mathcal{W}_0^{1-s,r'}(\Omega)} \frac{1}{\|u\|_{\mathcal{W}_0^{1-s,r'}(\Omega)}}\Big|\int_\Omega fu\,dx\Big|.$$

We can identify $f\in L^r(\Omega)$ with the continuous functional

$$L^{r'}(\Omega)\ni u\mapsto \int_\Omega fu\,dx\in\mathbb{R}$$

and, by doing so, identify $L^r(\Omega)$ with the dual space $L^{r'}(\Omega)'$. Thus $\mathcal{W}^{s-1,r}(\Omega)$ can be regarded as the completion of $L^{r'}(\Omega)'$ in the corresponding norm. Hence $L^r(\Omega)$ is dense in $\mathcal{W}^{s-1,r}(\Omega)$. We identify $L^2(\Omega)$ and $L^2(\Omega)'$ since $L^2(\Omega)$ is a Hilbert space. Thus we get $Z^{s,r} = (\mathcal{W}_0^{1-s,r'}(\Omega))' \cap L^2(\Omega)'$. Next, for $t' > \max\{r', 2\}$ we have $W_0^{1,t'}(\Omega) \subset \mathcal{W}_0^{1-s,r'}(\Omega)$ and $W_0^{1,t'}(\Omega) \subset L^2(\Omega)$. It follows that

$$\mathcal{W}^{s-1,r}(\Omega) \subset W^{-1,t}(\Omega) = (W_0^{1,t'}(\Omega))', \quad t' = t/(t-1),$$

and $L^2(\Omega)' \subset W^{-1,t}(\Omega)$. Hence the intersection $Z^{s,r} = (\mathcal{W}_0^{1-s,r'}(\Omega))' \cap L^2(\Omega)'$ makes sense as a linear subspace of the Banach space $W^{-1,t}(\Omega)$. Equipped with the standard norm of the intersection (see [49, Ch. 1]),

$$\|f\|_{Z^{s,r}} = \|f\|_{\mathcal{W}^{s-1,r}(\Omega)} + \|f\|_{L^2(\Omega)},$$

$Z^{s,r}$ becomes a Banach space. Notice that it is not empty since both $\mathcal{W}^{s-1,r}(\Omega)$ and $L^2(\Omega)$ contain $C_0^\infty(\Omega)$. Now we can formulate the main result of this section.

**Lemma 11.2.9.** *Let $\Omega \subset \mathbb{R}^3$ be a bounded domain with $\partial\Omega \in C^2$ and $(F, G) \in Z^{s,r} \times X^{s,r}$ $(0 < s < 1 < r < \infty)$. Then the boundary value problem (11.2.19) has a unique solution $(\mathbf{v}, \pi) \in Y^{s+1,r}(\Omega) \times X^{s,r}(\Omega)$ such that*

$$\|\mathbf{v}\|_{Y^{s,r}} + \|\pi\|_{X^{s,r}} \le c(\Omega, r, s)(\|F\|_{Z^{s,r}} + \|G\|_{X^{s,r}}). \tag{11.2.21}$$

*Proof.* This is an obvious consequence of Lemma 1.7.9. □

### 11.2.4 Step 3. Transport equations

In this section we collect the auxiliary results concerning existence and uniqueness of solutions to the following boundary value problems for transport equations:

$$\mathscr{L}\varphi := \mathbf{u}\nabla\varphi + \sigma\varphi = f \quad \text{in } \Omega, \quad \varphi = 0 \quad \text{on } \Sigma_{\text{in}}, \tag{11.2.22}$$

$$\mathscr{L}^*\varphi^* := -\operatorname{div}(\varphi^*\mathbf{u}) + \sigma\varphi^* = f \quad \text{in } \Omega, \quad \varphi^* = 0 \quad \text{on } \Sigma_{\text{out}}. \tag{11.2.23}$$

Bounded functions $\varphi$, $\varphi^*$ are called weak solutions to problems (11.2.22), (11.2.23), respectively, if the integral identities

$$\int_\Omega (\varphi\mathscr{L}^*\phi^* - f\phi^*)\,dx = 0, \quad \int_\Omega (\varphi^*\mathscr{L}\phi - f\phi)\,dx = 0, \tag{11.2.24}$$

hold true for all test functions $\phi^*, \phi \in C(\Omega) \cap W^{1,1}(\Omega)$, respectively, such that $\phi^* = 0$ on $\Sigma_{\text{out}}$ and $\phi = 0$ on $\Sigma_{\text{in}}$. Our considerations are based on the following two lemmas, which follow from the general theory of transport equations developed in Chapter 12.

**Lemma 11.2.10.** *Suppose that a bounded domain $\Omega = B \setminus S \subset \mathbb{R}^d$ and a vector field $\mathbf{U} : \partial\Omega \to \mathbb{R}^d$ satisfy Condition 11.2.5. Furthermore assume that a vector field $\mathbf{u} \in C^1(\Omega)$ satisfies the boundary condition*

$$\mathbf{u} = \mathbf{U} \quad \text{on } \partial\Omega.$$

*Then for every $\sigma > 0$ and $f \in L^\infty(\Omega)$ problem (11.2.22) has a unique weak solution $\varphi \in L^\infty(\Omega)$ satisfying*

$$\|\varphi\|_{L^\infty(\Omega)} \le \sigma^{-1}\|f\|_{L^\infty(\Omega)}.$$

*Proof.* It follows from Condition 11.2.5 that $\Gamma = \mathrm{cl}\,\Sigma_{\mathrm{in}} \cap (\Sigma \setminus \Sigma_{\mathrm{in}})$ is a one-dimensional manifold of class $C^3$. Hence $\mathbf{u}$ and $\Omega$ meets all requirements of Theorem 12.1.2 with $c = \sigma$ and $r = \infty$, and the lemma follows directly from that theorem. □

**Lemma 11.2.11.** *Assume that $\partial\Omega$ and $\mathbf{U}$ satisfy Condition 11.2.5, and the vector field $\mathbf{u}$ belongs to $C^1(\Omega)$ and satisfies*

$$\mathbf{u} = \mathbf{U} \quad \text{on } \partial\Omega. \tag{11.2.25}$$

*Furthermore, let $s$ and $r$ be constants satisfying*

$$0 < s \le 1, \quad 1 < r < \infty, \quad 2s - 3r^{-1} < 1. \tag{11.2.26}$$

*Then there are positive constants $\sigma^* > 1$ and $C$, which depend on $\partial\Omega$, $\mathbf{U}$, $s$, $r$, $\|\mathbf{u}\|_{C^1(\Omega)}$ and are independent of $\sigma$, such that for any $\sigma > \sigma^*$ and $f \in W^{s,r}(\Omega) \cap L^\infty(\Omega)$, problem (11.2.22) has a unique solution $\varphi \in W^{s,r}(\Omega) \cap L^\infty(\Omega)$ satisfying*

$$\begin{aligned} &\|\varphi\|_{W^{s,r}(\Omega)} \le C\sigma^{-1}\|f\|_{W^{s,r}(\Omega)} + C\sigma^{-1+\alpha}\|f\|_{L^\infty(\Omega)} \quad \text{for } sr \ne 1,2,\\ &\|\varphi\|_{W^{s,r}(\Omega)} \le C\sigma^{-1}\|f\|_{W^{s,r}(\Omega)} + C\sigma^{-1+\alpha}(1+\log\sigma)^{1/r}\|f\|_{L^\infty(\Omega)} \quad \text{for } sr = 1,2. \end{aligned} \tag{11.2.27}$$

*Here, the* accretivity defect *$\alpha$ is defined by*

$$\alpha(s,r) = \max\{0,\ s - r^{-1},\ 2s - 3r^{-1}\}. \tag{11.2.28}$$

*If, in addition, $sr > 3$ then for any $\sigma > \sigma^*$ and $f \in W^{s,r}(\Omega)$, problem (11.2.22) has a unique solution $\varphi \in W^{s,r}(\Omega) \cap L^\infty(\Omega)$ satisfying*

$$\begin{aligned} &\|\varphi\|_{W^{s,r}(\Omega)} \le C\sigma^{-1+\alpha}\|f\|_{W^{s,r}(\Omega)} \quad \text{for } sr \ne 1,2,\\ &\|\varphi\|_{W^{s,r}(\Omega)} \le C\sigma^{-1}C\sigma^{-1+\alpha}(1+\log\sigma)^{1/r}\|f\|_{W^{s,r}(\Omega)} \quad \text{for } sr = 1,2. \end{aligned} \tag{11.2.29}$$

*Proof.* It follows from Condition 11.2.5 that $\partial\Omega$ and $\mathbf{U}$ also satisfy Condition 12.2.2 and hence meet all requirements of Theorem 12.2.3. Therefore, the lemma is a straightforward consequence of that theorem. Notice only that for $sr > 3$ the space $W^{s,r}(\Omega)$ is continuously embedded is $C(\Omega)$ and hence estimate (11.2.29) follows from (11.2.27). □

The following proposition on solvability of problem (11.2.22) is the main result of this section.

**Proposition 11.2.12.** *Assume that $\Sigma$ and $\mathbf{U}$ satisfy Condition 11.2.5, the exponents $s, r$ satisfy the inequalities*

$$0 < s \le 1, \quad 1 < r < \infty, \quad 2s - 3r^{-1} < 1, \quad sr > 3, \tag{11.2.30}$$

*and the vector field* $\mathbf{u} \in Y^{s,r}(\Omega)$ *satisfies the boundary condition*

$$\mathbf{u} = \mathbf{U} \quad on\ \Sigma, \quad \mathbf{u} = 0 \quad on\ \partial S. \tag{11.2.31}$$

*Then there are constants* $\sigma_0 > 1$ *and* $C > 0$, *depending only on* $\Omega$, $\mathbf{U}$, $s$, $r$, *and* $\|\mathbf{u}\|_{Y^{s,r}}$, *such that: For any* $\sigma > \sigma_0$ *and* $f \in X^{s,r}$, *problem* (11.2.22) *has a unique solution* $\varphi \in X^{s,r}$ *satisfying*

$$\|\varphi\|_{X^{s,r}} \le C\|f\|_{X^{s,r}}, \quad \|\varphi\|_{L^\infty(\Omega)} \le \sigma^{-1}\|f\|_{L^\infty(\Omega)}, \tag{11.2.32}$$

*and problem* (11.2.23) *has a unique solution* $\varphi^* \in X^{s,r}$ *satisfying*

$$\|\varphi^*\|_{X^{s,r}} \le C\|f\|_{X^{s,r}}, \quad \|\varphi^*\|_{L^\infty(\Omega)} \le (\sigma - \|\operatorname{div}\mathbf{u}\|_{C(\Omega)})^{-1}\|f\|_{L^\infty(\Omega)}. \tag{11.2.33}$$

*Proof.* We begin by proving (11.2.32). Since $sr > 3$, the space $Y^{s,r} = W^{s+1,r}(\Omega) \cap W^{2,2}(\Omega)$ is continuously embedded in $C^1(\Omega)$, which yields

$$\|\mathbf{u}\|_{C^1(\Omega)} \le c(\Omega, s, r)\|\mathbf{u}\|_{Y^{s,r}}.$$

Hence $\Omega$, $s$, $r$, and $\mathbf{u}$ meet all requirements of Lemma 11.2.11. It follows that there are $\sigma^* > 1$ and $c_0 > 0$, depending only on $\Omega$, $\mathbf{U}$, $s$, $r$, and $\|\mathbf{u}\|_{Y^{s,r}}$, such that for all $\sigma > \sigma^*$ and $f \in W^{s,r}(\Omega)$, problem (11.2.22) has a unique solution $\varphi \in W^{s,r}(\Omega)$ satisfying

$$\|\varphi\|_{W^{s,r}(\Omega)} \le c_0\sigma^{-1}\|f\|_{W^{s,r}(\Omega)} + c_0\sigma^{-1+\alpha}\|f\|_{L^\infty(\Omega)}, \tag{11.2.34}$$

where $\alpha \in (0,1)$ is defined by (11.2.28).

On the other hand, $\Omega$, $\mathbf{U}$, $\mathbf{u}$, and the exponents $s = 1$, $r = 2$ also satisfy all conditions of Lemma 11.2.11. Therefore, there are $\overline{\sigma}^* > 1$ and $c^* > 0$, depending only on $\Omega$, $\mathbf{U}$, and $\|\mathbf{u}\|_{Y^{s,r}}$, such that for all $\sigma > \overline{\sigma}^*$ and $f \in W^{1,2}(\Omega)$, problem (11.2.22) has a unique solution $\varphi \in W^{1,2}(\Omega)$ satisfying

$$\|\varphi\|_{W^{1,2}(\Omega)} \le c^*\sigma^{-1}\|f\|_{W^{1,2}(\Omega)} + c^*\sigma^{-1/2}(1 + \log\sigma)^{1/2}\|f\|_{L^\infty(\Omega)}. \tag{11.2.35}$$

Next fix $\sigma_0 > \max\{\sigma^*, \overline{\sigma}^*\}$ and set

$$K(\sigma) = \min\{\sigma^{-1+\alpha}, \sigma^{-1/2}(1 + \log\sigma)^{1/2}\} > \sigma^{-1}.$$

Recall that $X^{s,r} = W^{s,r}(\Omega) \cap W^{1,2}(\Omega)$, which along with $rs > 3$ implies $\|f\|_{L^\infty(\Omega)} \le c(\Omega, s, r)\|f\|_{X^{s,r}}$. Combining (11.2.34) and (11.2.35), we conclude that for all $\sigma > \sigma_0$ and $f \in X^{s,r}$, problem (11.2.22) has a unique solution $\varphi \in X^{s,r}$ such that

$$\|\varphi\|_{X^{s,r}} \le 2(c^* + c_0)K(\sigma)\,\|f\|_{X^{s,r}}. \tag{11.2.36}$$

Since $K$ decreases in $\sigma$, we may take $C = 2(c_0 + c^*)K(\sigma_0)$ to obtain the first estimate in (11.2.32). The second follows from Lemma 11.2.10 since $\varphi \in X^{s,r} \subset L^\infty(\Omega)$ is a strong and weak solution to problem (11.2.22).

Let us turn to the proof of (11.2.33). Consider the auxiliary boundary value problem

$$-\mathbf{u}\nabla\varphi^* + \sigma\varphi^* = f \quad \text{in } \Omega, \quad \varphi^* = 0 \quad \text{on } \Sigma_{\text{out}}. \tag{11.2.37}$$

It follows from Condition 11.2.5 that $\Gamma = \operatorname{cl}\Sigma_{\text{out}} \cap (\Sigma \setminus \Sigma_{\text{out}})$. Hence $\Omega$ and the vector field $-\mathbf{U}$ satisfy Condition 11.2.5 with $\Sigma_{\text{in}}$ replaced by $\Sigma_{\text{out}}$. It follows that $\Omega$, $-\mathbf{U}$, $-\mathbf{u}$, and the exponents $s$, $r$ meet all requirements of Lemma 11.2.11. Arguing as in the proof of (11.2.32) we conclude that there are constants $c$ and $\sigma_0 > 1$, depending only on $\Omega$, $\mathbf{U}$, and $\|\mathbf{u}\|_{Y^{s,r}}$, such that for all $\sigma > \sigma_0$ and $f \in X^{s,r}$, problem (11.2.37) has a unique solution $\varphi^* \in X^{s,r}$ such that

$$\|\varphi^*\|_{X^{s,r}} \le cK(\sigma)\|f\|_{X^{s,r}}, \quad \|\varphi^*\|_{L^\infty(\Omega)} \le \sigma^{-1}\|f\|_{L^\infty(\Omega)}. \tag{11.2.38}$$

Introduce the linear operator $\mathcal{A} : X^{s,r} \to X^{s,r}$ which assigns to every $f \in X^{s,r}$ the solution $\varphi^*$ to problem (11.2.37). It follows from (11.2.38) that

$$\|\mathcal{A}f\|_{X^{s,r}} \le cK(\sigma)\|f\|_{X^{s,r}}. \tag{11.2.39}$$

Now we can rewrite equations (11.2.23) in the form

$$\varphi^* = \mathcal{A}(\varphi^* \operatorname{div}\mathbf{u}) + \mathcal{A}f. \tag{11.2.40}$$

Since $\|\operatorname{div}\mathbf{u}\|_{X^{s,r}} \le c\|\mathbf{u}\|_{Y^{s,r}}$, it follows from (11.2.39) and Lemma 11.2.8 that

$$\|\mathcal{A}(\varphi^* \operatorname{div}\mathbf{u})\|_{X^{s,r}} \le cK(\sigma)\,\|\varphi^* \operatorname{div}\mathbf{u}\|_{X^{s,r}} \le cK(\sigma)\|\mathbf{u}\|_{Y^{s,r}}\,\|\varphi^*\|_{X^{s,r}}$$

Notice that $K(\sigma) \to 0$ as $\sigma \to \infty$. Hence we can choose $\sigma_0 > 0$, depending only on $\Omega$, $\mathbf{U}$, $s$, $r$, and $\|\mathbf{u}\|_{Y^{s,r}}$, so large that $cK(\sigma)\|\mathbf{u}\|_{Y^{s,r}} < 1/2$ for all $\sigma > \sigma_0$. For such $\sigma$, the mapping $\varphi^* \mapsto \mathcal{A}(\varphi^* \operatorname{div}\mathbf{u})$ is a contraction in $X^{s,r}$ and its norm is less than $1/2$. By the Banach fixed point theorem (the contraction mapping principle), equation (11.2.40) has a unique solution $\varphi^* \in X^{s,r}$ such that

$$\|\varphi^*\|_{X^{s,r}} \le 2\|\mathcal{A}f\|_{X^{s,r}} \le 2cK(\sigma)\|f\|_{X^{s,r}}.$$

Since $K$ decreases in $\sigma$, these inequalities lead to the first estimate in (11.2.33) with the constant $C = 2K(\sigma_0)$. It remains to note that equation (11.2.23) can be written in the form

$$-\mathbf{u}\cdot\nabla\varphi^* + \sigma\varphi^* = \varphi^* \operatorname{div}\mathbf{u} + f$$

and estimates (11.2.38) for solutions to problem (11.2.37) lead to the inequalities

$$\|\varphi^*\|_{L^\infty(\Omega)} \le \sigma^{-1}\|\operatorname{div}\mathbf{u}\|_{C(\Omega)}\|\varphi^*\|_{L^\infty(\Omega)} + \sigma^{-1}\|f\|_{L^\infty(\Omega)},$$

which imply the second estimate in (11.2.33). □

### 11.2.5 Step 4. Fixed point scheme. Estimates

#### Fixed point scheme

We solve Problem 11.1.9 by an application of the Tikhonov fixed point theorem in the following framework:

Fix $\tau \in (0,1)$ and choose a matrix-valued function $\mathbf{N} \in C^1(\Omega)$ such that $\|\mathbb{I} - \mathbf{N}\|_{C^1(\Omega)} \le \tau$. Next choose $\vartheta = (\mathbf{v}, \pi, \varphi) \in \mathcal{B}_\tau$, and denote

$$\mathbf{u} := \mathbf{u}_\star + \mathbf{v}, \quad q := q_\star + \pi, \quad \varrho := \varrho_\star + \varphi. \tag{11.2.41a}$$

Consider the following problem for new unknowns $\vartheta_1 = (\mathbf{v}_1, \pi_1, \varphi_1)$:

$$\mathbf{u} \cdot \nabla \varphi_1 + \sigma \varphi_1 = \Psi_1[\vartheta] + m\mathfrak{g}\varrho \quad \text{in } \Omega, \quad \varphi_1 = 0 \quad \text{on } \Sigma_{\text{in}}, \tag{11.2.41b}$$

$$\Delta \mathbf{v}_1 - \nabla \pi_1 = \mathscr{A}(\mathbf{u}) + \mathsf{Re}\,\mathscr{B}(\varrho, \mathbf{u}, \mathbf{u}) \equiv F[\vartheta] \quad \text{in } \Omega, \tag{11.2.41c}$$

$$\operatorname{div} \mathbf{v}_1 = \varrho_\star^{-1} \Pi(\mathfrak{g}\sigma\varphi_1 - \mathfrak{g}\varrho_\star \Psi[\vartheta] - \mathfrak{g}m\varrho_\star) \quad \text{in } \Omega, \tag{11.2.41d}$$

$$\mathbf{v}_1 = 0 \quad \text{on } \partial\Omega, \quad \pi_1 - \Pi\pi_1 = 0. \tag{11.2.41e}$$

Here $m$ is given by

$$m = \varkappa \int_\Omega (\varrho_\star^{-1} \Psi_1[\vartheta]\,\zeta - \mathfrak{g}\Psi[\vartheta])\, dx, \quad \varkappa = \Big( \int_\Omega \mathfrak{g}(1 - \zeta - \varrho_\star^{-1}\zeta\varphi)\, dx \Big)^{-1}, \tag{11.2.41f}$$

where the auxiliary function $\zeta$ is a solution to the adjoint boundary value problem

$$-\operatorname{div}(\mathbf{u}\zeta) + \sigma\zeta = \sigma\mathfrak{g} \quad \text{in } \Omega, \quad \zeta = 0 \quad \text{on } \Sigma_{\text{out}}. \tag{11.2.41g}$$

Notice that $m$ and $\zeta$ are completely defined by $\vartheta$. Relations (11.2.41) form a complete system of equations and boundary conditions for $\vartheta_1$ and define a mapping $\Xi : \vartheta \mapsto \vartheta_1$. We claim that for a suitable choice of the constant $\tau$, $\Xi$ is an automorphism of the ball $\mathcal{B}_\tau$. Our strategy is the following. First we estimate the right hand sides of equations (11.2.41) in $Z^{s,r}$ and $X^{s,r}$ norms. Then we apply Lemma 11.2.9 and Proposition 11.2.12 to estimate $\vartheta_1 = \Xi(\vartheta)$.

#### Auxiliary lemmas

**Lemma 11.2.13.** *Let $\Omega$ be a bounded domain with $C^1$ boundary and $0 < s < 1 < r < \infty$. Then the operators* $\operatorname{div} : W^{s,r}(\Omega) \to \mathcal{W}^{s-1,r}(\Omega)$ *and* $\nabla : W^{s,r}(\Omega) \to \mathcal{W}^{s-1,r}(\Omega)$ *are continuous. In particular, there is $c$, depending only on $\Omega$, $s$, $r$, such that*

$$\|\operatorname{div} \mathbf{u}\|_{\mathcal{W}^{s-1,r}(\Omega)} \le c\|\mathbf{u}\|_{W^{s,r}(\Omega)}, \quad \|\nabla \mathbf{u}\|_{\mathcal{W}^{s-1,r}(\Omega)} \le c\|\mathbf{u}\|_{W^{s,r}(\Omega)}.$$

*Proof.* It follows from the definitions (see Section 1.5) that the mappings $\operatorname{div} : L^r(\Omega) \to W^{-1,r}(\Omega)$ and $\operatorname{div} : W^{1,r}(\Omega) \to L^r(\Omega)$ are continuous. Since $W^{s,r}(\Omega) = [L^r(\Omega), W^{1,r}(\Omega)]_{s,r}$ and

$$\mathcal{W}^{s-1,r}(\Omega) = [L^r(\Omega), \mathcal{W}^{-1,r}(\Omega)]_{1-s,r} \equiv [\mathcal{W}^{-1,r}(\Omega), L^r(\Omega)]_{s,r}$$

are interpolation spaces, the continuity of div : $W^{s,r}(\Omega) \to \mathcal{W}^{s-1,r}(\Omega)$ follows from the interpolation lemma 1.1.13. The same arguments imply the continuity of $\nabla : W^{s,r}(\Omega) \to \mathcal{W}^{s-1,r}(\Omega)$. □

**Lemma 11.2.14.** *Let $\Omega$ be a bounded domain with $C^1$ boundary and $0 < s < 1 < r < \infty$. Then for every matrix-valued function $\mathbf{M} \in C^1(\Omega)$ and $\mathbf{w} \in \mathcal{W}^{s-1,r}(\Omega)$,*

$$\|\mathbf{M}\mathbf{w}\|_{\mathcal{W}^{s-1,r}(\Omega)} \le c(\Omega, s, r)\|\mathbf{M}\|_{C^1(\Omega)}\|\mathbf{w}\|_{\mathcal{W}^{s-1,r}(\Omega)}.$$

*Proof.* Lemma 1.5.1 implies that $\mathcal{W}^{s-1,r}(\Omega)$ is the dual space $(\mathcal{W}_0^{1-s,r'}(\Omega))'$. By definition, we have

$$\langle \mathbf{M}\mathbf{w}, \varphi\rangle := \langle \mathbf{w}, \mathbf{M}^\top \varphi\rangle \quad \text{for all } \varphi \in \mathcal{W}_0^{1-s,r'}(\Omega),$$

where $\langle \cdot, \cdot\rangle$ is the duality pairing between $(\mathcal{W}_0^{1-s,r'}(\Omega))'$ and $\mathcal{W}_0^{1-s,r'}(\Omega)$. Obviously the mapping $\varphi \mapsto \mathbf{M}^\top\varphi$ is continuous from $L^{r'}(\Omega)$ to $L^{r'}(\Omega)$ and from $W_0^{1,r'}(\Omega)$ to $W_0^{1,r'}(\Omega)$. Moreover, we have

$$\|\mathbf{M}^\top\varphi\|_{L^{r'}(\Omega)} \le c\|\mathbf{M}^\top\|_{C^1(\Omega)}\|\varphi\|_{L^{r'}(\Omega)},$$
$$\|\mathbf{M}^\top\varphi\|_{W_0^{1,r'}(\Omega)} \le c\|\mathbf{M}^\top\|_{C^1(\Omega)}\|\varphi\|_{W_0^{1,r'}(\Omega)}.$$

Since $\mathcal{W}_0^{1-s,r'}(\Omega) = [L^r(\Omega), W_0^{1,r'}(\Omega)]_{1-s,r'}$ is an interpolation space, it follows from Lemma 1.1.13 that

$$\|\mathbf{M}^\top\varphi\|_{\mathcal{W}_0^{1-s,r'}(\Omega)} \le c\|\mathbf{M}^\top\|_{C^1(\Omega)}\|\varphi\|_{\mathcal{W}_0^{1-s,r'}(\Omega)}.$$

Thus we get

$$|\langle \mathbf{M}\mathbf{w}, \varphi\rangle| = |\langle \mathbf{w}, \mathbf{M}^\top\varphi\rangle| \le \|\mathbf{w}\|_{\mathcal{W}^{s-1,r}(\Omega)}\|\mathbf{M}^\top\varphi\|_{\mathcal{W}_0^{1-s,r'}(\Omega)}$$
$$\le c\|\mathbf{M}^\top\|_{C^1(\Omega)}\|\mathbf{w}\|_{\mathcal{W}^{s-1,r}(\Omega)}\|\varphi\|_{\mathcal{W}_0^{1-s,r'}(\Omega)}.$$

It remains to note that

$$\|\mathbf{M}\mathbf{w}\|_{\mathcal{W}^{s-1,r}(\Omega)} = \sup_{\|\varphi\|_{\mathcal{W}_0^{1-s,r'}(\Omega)}=1} |\langle \mathbf{M}\mathbf{w}, \varphi\rangle|.$$ □

**Lemma 11.2.15.** *Let $\mathbf{N} \in C^2(\Omega)$ satisfy $\|\mathbb{I}-\mathbf{N}\|_{C^2(\Omega)} \le \tau^2$. Then there are absolute constants $\tau_0 > 0$ and $c > 0$ such that for all $\tau \in (0, \tau_0)$,*

$$\|\mathbb{I} - \mathbf{N}^{-1}\|_{C^2(\Omega)} \le c\tau^2, \quad \|1 - \mathfrak{g}^{\pm 1}\|_{C^2(\Omega)} \le c\tau^2. \tag{11.2.42}$$

*Proof.* Observe that $\|\mathbf{A}\mathbf{B}\|_{C^2(\Omega)} \le c\|\mathbf{A}\|_{C^2(\Omega)}\|\mathbf{B}\|_{C^2(\Omega)}$ for all matrix-valued functions $\mathbf{A}$ and $\mathbf{B}$. Here $c$ depends only on the dimension of $\mathbb{R}^3$, i.e., it is an absolute constant. From this and the identity

$$\mathbf{N}^{-1} - \mathbb{I} = \sum_{k\ge 1}(\mathbb{I} - \mathbf{N})^k$$

we obtain

$$\|\mathbf{N}^{-1} - \mathbb{I}\|_{C^2(\Omega)} \le \sum_{k\ge 1} c^{k-1}\|\mathbb{I} - \mathbf{N}\|^k_{C^2(\Omega)} \le \sum_{k\ge 1} c^{k-1}\tau^{2k} = \tau^2/(1 - c\tau^2).$$

Choosing $\tau_0$ sufficiently small we obtain the first estimate in (11.2.42). In order to prove the second it suffices to note that $\mathfrak{g} = \sqrt{\det \mathbf{N}}$ is an analytic function of the entries of $\mathbf{N}$, and $\mathfrak{g} = 1$ for $\mathbf{N} = \mathbb{I}$. □

Let us turn to equations (11.2.41b)–(11.2.41e). Our task is to estimate their right hand sides and, by doing so, estimate their solutions.

**Estimates of right hand sides in (11.2.41b)–(11.2.41e)**

**Lemma 11.2.16.** *Let*

$$\sigma > 1, \quad \vartheta \in \mathcal{B}_\tau, \quad 0 < \tau \le \tau_0 < 1, \tag{11.2.43}$$

*where $\tau_0$ is given by Lemma 11.2.15. Then there is $c > 0$, depending only on $\Omega$, $\mathbf{U}$ and $s$, $r$, such that the functions $\Psi$, $\Psi_1$, $\mathscr{A}$, and $\mathscr{B}$, given by* (11.1.24d) *and* (11.1.24e) *satisfy*

$$\|\mathscr{A}(\mathbf{u}) + \operatorname{Re}\mathscr{B}(\varrho, \mathbf{u}, \mathbf{u})\|_{Z^{s,r}} \le c\tau^2, \tag{11.2.44}$$

$$\|\Psi[\vartheta]\|_{X^{s,r}} \le c\sigma\tau^2, \tag{11.2.45}$$

$$\|\Psi_1[\vartheta]\|_{X^{s,r}} \le c\sigma\tau^2. \tag{11.2.46}$$

*Proof.* We begin by proving (11.2.44). It follows from (11.1.8) that

$$\mathscr{A}(\mathbf{u}) = (\mathbb{I} - \mathbf{N}^{-\top})\operatorname{div}(\nabla\mathbf{u}) + \mathbf{N}^{-\top}\operatorname{div}\big((\mathbb{I} - \mathbf{A})\nabla u + \mathbf{A}\nabla(\mathbf{B}\mathbf{u})\big),$$

where

$$\mathbf{A} = \mathfrak{g}^{-1}\mathbf{N}\mathbf{N}^\top, \quad \mathbf{B} = \mathbb{I} - \mathbf{N}^{-1}.$$

Applying Lemmas 11.2.13 and 11.2.14 we obtain

$$\|\mathscr{A}(\mathbf{u})\|_{W^{s-1,r}(\Omega)} \le c\tau^2\|\nabla\mathbf{u}\|_{W^{s,r}(\Omega)} + c\|(\mathbb{I} - \mathbf{A})\nabla\mathbf{u}\|_{W^{s,r}(\Omega)} + \|\mathbf{A}\nabla(\mathbf{B}\mathbf{u})\|_{W^{s,r}(\Omega)}.$$

Since $sr > 3$, estimate (11.2.8) implies

$$\begin{aligned}
&\|(\mathbb{I} - \mathbf{A})\nabla\mathbf{u}\|_{W^{s,r}(\Omega)} + \|\mathbf{A}\nabla(\mathbf{B}\mathbf{u})\|_{W^{s,r}(\Omega)}\\
&\qquad \le c\|\mathbb{I} - \mathbf{A}\|_{W^{s,r}(\Omega)}\|\nabla\mathbf{u}\|_{W^{s,r}(\Omega)} + c\|\mathbf{A}\|_{W^{s,r}(\Omega)}\|\nabla(\mathbf{B}\mathbf{u})\|_{W^{s,r}(\Omega)}\\
&\qquad \le c\|\mathbb{I} - \mathbf{A}\|_{C^2(\Omega)}\|\nabla\mathbf{u}\|_{W^{s,r}(\Omega)} + c\|\mathbf{A}\|_{C^2(\Omega)}\|\nabla(\mathbf{B}\mathbf{u})\|_{W^{s,r}(\Omega)}.
\end{aligned} \tag{11.2.47}$$

Notice that

$$\mathbb{I} - \mathbf{A} = (\mathbb{I} - \mathbf{N})^\top - (\mathbf{N} - \mathbb{I})\mathbf{N}^\top - (\mathfrak{g}^{-1} - 1)\mathbf{N}\mathbf{N}^\top.$$

Lemma 11.2.15 implies

$$\|\mathbb{I}-\mathbf{A}\|_{C^2(\Omega)} + \|\mathbf{B}\|_{C^2(\Omega)} \le c\tau^2, \quad \|\mathbf{A}\|_{C^2(\Omega)} \le c.$$

Combining the estimates obtained we arrive at

$$\|\mathscr{A}(\mathbf{u})\|_{\mathcal{W}^{s-1,r}(\Omega)} \le c\tau^2\|\nabla\mathbf{u}\|_{W^{s,r}(\Omega)} + c\|\nabla(\mathbf{B}\mathbf{u})\|_{W^{s,r}(\Omega)}.$$

On the other hand, the identity

$$\nabla(\mathbf{B}\mathbf{u}) = \nabla\mathbf{B}\mathbf{u} + \nabla\mathbf{u}\mathbf{B}^\top, \text{ where } \nabla(\mathbf{B}\mathbf{u})_{ij} = \partial_{x_i}(B_{jk}u_k),\ (\nabla\mathbf{B}\mathbf{u})_{ij} = \partial_{x_i}(B_{jk})u_k,$$

and estimate (11.2.42) imply

$$\begin{aligned}\|\nabla(\mathbf{B}\mathbf{u})\|_{W^{s,r}(\Omega)} &\le \|\nabla\mathbf{B}\|_{W^{s,r}(\Omega)}\|\mathbf{u}\|_{W^{s,r}(\Omega)} + c\|\mathbf{B}\|_{W^{s,r}(\Omega)}\|\nabla\mathbf{u}\|_{W^{s,r}(\Omega)}\\ &\le c\|\mathbf{B}\|_{C^2(\Omega)}(\|\nabla\mathbf{u}\|_{W^{s,r}(\Omega)} + \|\mathbf{u}\|_{W^{s,r}(\Omega)})\\ &\le c\tau^2(\|\nabla\mathbf{u}\|_{W^{s,r}(\Omega)} + \|\mathbf{u}\|_{W^{s,r}(\Omega)}).\end{aligned} \tag{11.2.48}$$

Thus we get

$$\|\mathscr{A}(\mathbf{u})\|_{\mathcal{W}^{s-1,r}(\Omega)} \le c\tau^2(\|\nabla\mathbf{u}\|_{W^{s,r}(\Omega)} + \|\mathbf{u}\|_{W^{s,r}(\Omega)}) \le c\tau^2\|\mathbf{u}\|_{W^{s+1,r}(\Omega)}.$$

We also have the obvious estimate

$$\|\mathscr{A}(\mathbf{u})\|_{L^2(\Omega)} \le \|\mathbb{I}-\mathbf{N}\|_{C^2(\Omega)}\|\mathbf{u}\|_{W^{2,2}(\Omega)} \le c\tau^2\|\mathbf{u}\|_{W^{2,2}(\Omega)}.$$

On the other hand, since $\varrho_\star = \text{const}$ and $\mathbf{u}_\star \in C^2(\Omega)$, equalities (11.2.41a) imply

$$\begin{aligned}\|\varrho\|_{X^{s,r}} + \|\mathbf{u}\|_{X^{s,r}} + \|\nabla\mathbf{u}\|_{X^{s,r}} &\le c + \|\varphi\|_{X^{s,r}} + \|\mathbf{v}\|_{X^{s,r}} + \|\mathbf{v}\|_{Y^{s,r}}\\ &\le c + c\tau \le 2c.\end{aligned} \tag{11.2.49}$$

Recalling the identity

$$\|\mathbf{u}\|_{Y^{s+1,r}} = \|\mathbf{u}\|_{W^{s+1,r}(\Omega)} + \|\mathbf{u}\|_{W^{2,2}(\Omega)}$$

we obtain

$$\begin{aligned}\|\mathscr{A}(\mathbf{u})\|_{Z^{s,r}} = \|\mathscr{A}(\mathbf{u})\|_{\mathcal{W}^{s-1,r}(\Omega)} + \|\mathscr{A}(\mathbf{u})\|_{L^2(\Omega)} &\le c\tau^2\|\mathbf{u}\|_{Y^{s+1,r}}\\ &\le c\tau^2.\end{aligned} \tag{11.2.50}$$

Next, since $X^{s,r} \hookrightarrow Z^{s,r}$ is continuous, the product estimate (11.2.8) implies

$$\mathsf{Re}\,\|\mathscr{B}(\varrho,\mathbf{u},\mathbf{u})\|_{Z^{s,r}} \le \mathsf{Re}\,\|\mathscr{B}(\varrho,\mathbf{u},\mathbf{u})\|_{X^{s,r}} \le c\,\mathsf{Re}\,\|\varrho\|_{X^{s,r}}\|\mathbf{u}\|_{X^{s,r}}\|\nabla\mathbf{u}\|_{X^{s,r}}.$$

From this, (11.2.49), and the inequality $\mathsf{Re} \le \tau^2$ we obtain

$$\mathsf{Re}\,\|\mathscr{B}(\varrho,\mathbf{u},\mathbf{u})\|_{Z^{s,r}} \le c\tau^2.$$

Combining this estimate with (11.2.50) we arrive at (11.2.44).

Our next task is to estimate the function

$$\Psi[\vartheta] = \frac{q_\star + \pi}{\lambda} - \frac{\sigma}{p'(\varrho_\star)\varrho_\star} H(\varphi), \quad H(\varphi) = p(\varrho_\star + \varphi) - p(\varrho_\star) - p'(\varrho_\star)\varphi.$$

Since $p \in C^3(\Omega)$, it follows by Taylor expansion that $H$ admits a representation $H = \mathfrak{f}(\varphi)\varphi^2$, where $\mathfrak{f}$ is a $C^1$ function on the real line. It follows from the product estimates in Lemma 11.2.7 and from the estimates for composite functions in Lemma 11.2.8 that

$$\|H\|_{X^{s,r}} \leq c\|\mathfrak{f}(\varphi)\|_{X^{s,r}}\|\varphi\|^2_{X^{s,r}} \leq c(1 + \|\varphi\|_{X^{s,r}})\|\varphi\|^2_{X^{s,r}}.$$

Since $\vartheta = (\mathbf{v}, \pi, \varphi) \in \mathcal{B}_\tau$, we have $\|\varphi\|_{X^{s,r}} \leq \tau$ and $\|\pi\|_{X^{s,r}} \leq c\tau$. Recalling the inequality $\lambda^{-1} \leq \tau^2 < 1$ we obtain the estimate

$$\|\Psi[\vartheta]\|_{X^{s,r}} \leq \frac{1}{\lambda}(c + \|\pi\|_{X^{s,r}}) + c\sigma\|H\|_{X^{s,r}} \leq \frac{c}{\lambda} + c\sigma\tau^2 \leq c\tau^2 + c\sigma\tau^2,$$

which implies (11.2.45) since $\sigma > 1$. Finally, let us estimate the function

$$\Psi_1[\vartheta] = \mathfrak{g}\left(\varrho\Psi[\vartheta] - \frac{\sigma}{\varrho_\star}\varphi^2\right) + \sigma\varphi(1 - \mathfrak{g}).$$

By Lemma 11.2.15, for $\tau \in (0, \tau_0]$ we have

$$\begin{aligned} &\|\mathfrak{g}\|_{C^2(\Omega)} \leq c\|\mathbf{N}\|^3_{C^2(\Omega)} \leq c, \\ &\|1 - \mathfrak{g}\|_{C^2(\Omega)} \leq c\|\mathbf{N}\|^2_{C^2(\Omega)}\|\mathbb{I} - \mathbf{N}\|_{C^2(\Omega)} \leq c\tau^2, \end{aligned} \tag{11.2.51}$$

so that

$$\begin{aligned} \|\Psi_1[\vartheta]\|_{X^{s,r}} &\leq c\|\varrho\|_{X^{s,r}}\|\Psi[\vartheta]\|_{X^{s,r}} + c\sigma\|\varphi\|^2_{X^{s,r}} + c\sigma\tau^2\|\varphi\|_{X^{s,r}} \\ &\leq c\tau^2 + \tau^2 + c\sigma\tau^2. \end{aligned}$$

Since $\sigma > 1$ this obviously yields (11.2.46). □

### Estimates of solutions to problem (11.2.41)

**Lemma 11.2.17.** *Let $\tau \in (0, \tau_0]$, where $\tau_0$ is given by Lemma 11.2.15, and $m \in \mathbb{R}$ be given by (11.2.41f). Furthermore, assume that $s$, $r$ satisfy (11.2.5). Then there are $\sigma_1 > 1$ and $c > 0$, depending only on $\Omega$, $\mathbf{U}$ and $s$, $r$, with the following property. For every $\sigma > \sigma_1$ and $\vartheta \in \mathcal{B}_\tau$, problem (11.2.41) has a unique solution $\vartheta_1 = (\mathbf{v}_1, \pi_1, \varphi_1) \in E$, where $E$ is given by (11.2.3), such that*

$$\|\varphi_1\|_{X^{s,r}} + \|\mathbf{v}_1\|_{Y^{s,r}} + \|\pi_1\|_{X^{s,r}} \leq c\sigma\tau^2 + c|m|. \tag{11.2.52}$$

*Proof.* First we estimate the solution $\varphi_1$ to problem (11.2.41b)–(11.2.41e). By (11.2.41a),

$$\|\mathbf{u}\|_{Y^{s,r}} \leq \|\mathbf{u}_\star\|_{C^2(\Omega)} + \|\mathbf{v}\|_{Y^{s,r}} \leq c + c\|\mathbf{v}\|_{Y^{s,r}} \leq c + c\tau \leq 2c.$$

From this and the assumptions of Theorem 11.2.6 we conclude that $\mathbf{u}$ and $s$, $r$ meet all requirements of Proposition 11.2.12. Hence there exist $C > 0$ and $\sigma_1 > 1$, depending only on $\Omega$, $\mathbf{U}$ and $s$, $r$, such that for all $\sigma > \sigma_1$ problem (11.2.41b)–(11.2.41e) has a unique solution $\varphi_1 \in X^{s,r}$ satisfying

$$\|\varphi_1\|_{X^{s,r}} \leq C\|\Psi_1[\vartheta] + m\mathfrak{g}\varrho\|_{X^{s,r}}.$$

It now follows from (11.2.46) that

$$\|\varphi_1\|_{X^{s,r}} \leq c\tau^2 + c\sigma\tau^2 + c|m| \leq c\sigma\tau^2 + c|m| \tag{11.2.53}$$

since $\sigma > 1$. Next we estimate the solution $\mathbf{v}_1$, $\pi_1$ to problem (11.2.41c)–(11.2.41e). Notice that the right hand side of the equation for $\operatorname{div} \mathbf{v}_1$ in (11.2.41d) admits the estimate

$$\|\varrho_\star^{-1}\Pi(\mathfrak{g}\sigma\varphi_1 - \mathfrak{g}\varrho_\star\Psi[\vartheta] - \mathfrak{g}m\varrho_\star)\|_{X^{s,r}} \leq c\sigma\|\varphi_1\|_{X^{s,r}} + c\|\Psi[\vartheta]\|_{X^{s,r}} + c|m|.$$

From this and (11.2.45), (11.2.53) we obtain

$$\|\varrho_\star^{-1}\Pi(\mathfrak{g}\sigma\varphi_1 - \mathfrak{g}\varrho_\star\Psi[\vartheta] - \mathfrak{g}m\varrho_\star)\|_{X^{s,r}} \leq c\tau^2 + c\sigma\tau^2 + c|m|.$$

Combining this with (11.2.44) and applying Lemma 11.2.9 we arrive at

$$\|\mathbf{v}_1\|_{Y^{s,r}} + \|\pi_1\|_{X^{s,r}} \leq c\sigma\tau^2 + c|m|. \tag{11.2.54}$$

Estimates (11.2.54) and (11.2.53) imply that for all $\sigma > \sigma_1 > 1$ and $\tau \in (0, \tau_0]$, problem (11.2.41) has a unique solution satisfying (11.2.52). □

**Estimate of $m$**

**Proposition 11.2.18.** *Let all assumptions of Theorem 11.2.6 be satisfied and $\sigma_1$ be given by Lemma 11.2.17. Then there is $\sigma_2 \geq \sigma_1$, depending only on $\Omega$, $\mathbf{U}$, and $s$, $r$, with the following property. For every $\sigma > \sigma_2$, there is $\tau_1 \in (0, \tau_0]$, depending only on $\Omega$, $\mathbf{U}$, $s$, $r$, and $\sigma$, such that for all $\vartheta \in \mathcal{B}_\tau$ with $\tau \in (0, \tau_1)$ and $\alpha \in (0, s - 3/r)$,*

$$\|\zeta\|_{X^{s,r}} = \|\zeta\|_{W^{s,r}(\Omega)} + \|\zeta\|_{W^{1,2}(\Omega)} \leq c\sigma, \tag{11.2.55}$$

$$\varkappa \leq c\sigma^{1/\alpha}, \tag{11.2.56}$$

$$|m| \leq c\sigma^{1/\alpha+1}\tau^2, \tag{11.2.57}$$

*where $c$ depends on $\sigma$, $\Omega$, $\mathbf{U}$, $s$, $r$, and $\alpha$.*

*Proof.* Observe that $\Omega$, $\mathbf{U}$, $\mathbf{u}$, and the exponents $s$, $r$ meet all requirements of Proposition 11.2.12. On the other hand, equations (11.2.41g) are equivalent to (11.2.23) with $\varphi^*$ and $f$ replaced by $\zeta$ and $\sigma$. In view of Proposition 11.2.12, there exist $c > 0$ and $\sigma_2 > 1$, depending only on $\Omega$, $\mathbf{U}$, and $s$, $r$, such that for all $\sigma > \sigma_2$ problem (11.2.41g) has a unique solution $\zeta \in X^{s,r}$ satisfying (11.2.55). We can assume that $\sigma_2 \geq \sigma_1$.

Since $sr > 3$, the embedding $W^{s,r}(\Omega) \hookrightarrow C^\alpha(\Omega)$ is bounded for $0 < \alpha < s - 3/r$, which yields

$$\|\zeta\|_{C^\alpha(\Omega)} \leq c\sigma \quad \text{for all } \sigma > \sigma_2. \tag{11.2.58}$$

Next, since $\operatorname{div} \mathbf{u} = \operatorname{div} \mathbf{v}$ satisfies

$$\|\operatorname{div} \mathbf{v}\|_{C(\Omega)} \leq \|\mathbf{v}\|_{C^1(\Omega)} \leq c\|\mathbf{v}\|_{Y^{s,r}} \leq c\tau,$$

Proposition 11.2.12 implies

$$\|\zeta\|_{L^\infty(\Omega)} \leq \sigma/(\sigma - \|\operatorname{div} \mathbf{v}\|_{C(\Omega)}) \leq c\sigma(\sigma - c\tau)^{-1} \leq (1 - c\tau)^{-1}.$$

Choose $\tau_1 > 0$, depending only on $\Omega$, $\mathbf{U}$ and $s$, $r$, so small that $c\tau_1 < 1/2$. We have

$$\|\zeta\|_{L^\infty(\Omega)} \leq 2 \quad \text{for all } \tau \in (0, \tau_1]. \tag{11.2.59}$$

Next, formula (11.2.41f) for $m$ and estimates (11.2.45), (11.2.46), (11.2.59) imply

$$|m| \leq |\varkappa| c(\|\Psi[\vartheta]\|_{X^{s,r}} + \|\Psi_1[\vartheta]\|_{X^{s,r}}) \leq |\varkappa|(c\tau^2 + c\sigma\tau^2). \tag{11.2.60}$$

It remains to estimate the constant $\varkappa$. By (11.2.41f),

$$\varkappa^{-1} = \int_\Omega (1 - \zeta)\, dx + \int_\Omega (\mathfrak{g} - 1)\zeta\, dx - \int_\Omega \mathfrak{g}\varrho_\star^{-1}\zeta\varphi\, dx.$$

Since the embedding $X^{s,r} \hookrightarrow C(\Omega)$ is bounded we have

$$\|\varphi\|_{C(\Omega)} \leq c\|\varphi\|_{X^{s,r}} \leq c\tau.$$

On the other hand, inequalities (11.2.51) and (11.2.59) imply

$$|\zeta| \leq 2, \quad |1 - \mathfrak{g}| \leq c\tau^2.$$

Thus we get

$$\varkappa^{-1} \geq \int_\Omega (1 - \zeta)\, dx - c\tau^2 - c\tau. \tag{11.2.61}$$

Now represent $1 - \zeta$ in the form $1 - \zeta = (1 - \zeta)^+ + (1 - \zeta)^-$, where

$$(1 - \zeta)^+ = \max\{1 - \zeta, 0\}, \quad (1 - \zeta)^- = \min\{1 - \zeta, 0\}.$$

Our further considerations are based on the following two lemmas.

**Lemma 11.2.19.** *There is a constant c, depending only on $\Omega$, $\mathbf{U}$ and s, r, such that for all $\tau \in (0, \tau_0)$ and $\sigma > \sigma_1$, where $\tau_0$ is given by Lemma 11.2.15 and $\sigma_1$ is given by Lemma 11.2.17, we have*

$$\|(1-\zeta)^-\|_{C(\Omega)} \le c\sigma^{-1}\tau + c\tau^2. \tag{11.2.62}$$

*Proof.* Since the mapping $\zeta \mapsto (1-\zeta)^-$ is Lipschitz and $\zeta \in W^{1,2}(\Omega)$ is a continuous function, $(1-\zeta)^-$ is a continuous function in $W^{1,2}(\Omega)$. Moreover (see [133, Corollary 2.1.8]), we have

$$\nabla(1-\zeta)^- = \begin{cases} -\nabla\zeta & \text{a.e. on } \{x : 1-\zeta(x) < 0\}, \\ 0 & \text{a.e. on } \{x : 1-\zeta(x) \ge 0\}. \end{cases}$$

On the other hand, $\zeta$ is a strong solution to equation (11.2.41g), which along with the equality $\operatorname{div}\mathbf{u} = \operatorname{div}\mathbf{v}$ yields

$$-\mathbf{u}\cdot\nabla\zeta = \zeta\operatorname{div}\mathbf{v} + \zeta(\mathfrak{g}-1) + (1-\zeta)\sigma$$

Since $(1-\zeta)^- = 1-\zeta$ on $\{x : 1-\zeta < 0\}$ and $(1-\zeta)^- = 0$ on $\{x : 1-\zeta \ge 0\}$, we have

$$\begin{aligned} &-\mathbf{u}\cdot\nabla(1-\zeta)^- + \sigma(1-\zeta)^- = f \quad \text{a.e. in } \Omega, \\ &f = -\zeta\operatorname{div}\mathbf{v} + \sigma(\mathfrak{g}-1) \quad \text{if } 1-\zeta < 0 \quad \text{and} \quad f = 0 \quad \text{if } 1-\zeta < 0. \end{aligned} \tag{11.2.63}$$

Moreover, since $\zeta = 0$ on $\Sigma_{\text{out}}$, the function $(1-\zeta)^-$ vanishes on $\Sigma_{\text{out}}$. Now choose $\phi \in C^1(\Omega)$ vanishing in a neighborhood of $\Sigma \setminus \Sigma_{\text{out}}$. Multiplying (11.2.63) by $\phi$ and integrating the result over $\Omega$ we obtain

$$\int_\Omega (1-\zeta)^-(\operatorname{div}(\phi\mathbf{u}) + \sigma\phi - \phi f)\,dx = 0.$$

Hence $(1-\zeta)^- \in C(\Omega)$ is a bounded weak solution to problem (11.2.22) with $\mathbf{u}$, $\mathbf{U}$, and $\Sigma_{\text{in}}$ replaced by $-\mathbf{u}$, $-\mathbf{U}$, and $\Sigma_{\text{out}}$, respectively. Since $-\mathbf{u}$, $-\mathbf{U}$, and $\Omega$ meet all requirements of Lemma 11.2.10, we obtain

$$\begin{aligned} \|(1-\zeta)^-\|_{L^\infty(\Omega)} &\le \sigma^{-1}\|f\|_{L^\infty(\Omega)} \le 2\sigma^{-1}\|\operatorname{div}\mathbf{v}\|_{L^\infty(\Omega)} + \|\mathfrak{g}-1\|_{C^2(\Omega)} \\ &\le 2\sigma^{-1}\|\mathbf{v}\|_{C^1(\Omega)} + \tau^2 \le c\sigma^{-1}\|\mathbf{v}\|_{Y^{s,r}} + \tau^2 \le c\sigma^{-1}\tau + \tau^2. \end{aligned} \qquad \square$$

**Lemma 11.2.20.** *Let $\sigma_2 > 1$, $\alpha$, and $\tau_1$ be given by Proposition 11.2.18. Then there is $c > 0$, depending only on $\Omega$, $\mathbf{U}$, s, r, and $\alpha$, such that for all $\sigma > \sigma_2$,*

$$\int_\Omega (1-\zeta)^+\,dx \ge c^{-1}\sigma^{-1/\alpha}. \tag{11.2.64}$$

*Proof.* Since $\zeta$ vanishes at $\Sigma_{\text{out}}$, inequality (11.2.58) implies

$$|\zeta(x)| \le c\sigma\rho^\alpha \quad \text{for all } x \text{ with } \operatorname{dist}(x, \Sigma_{\text{out}}) \le \rho.$$

Choosing $\rho = (c\sigma)^{-1/\alpha}$, where $c$ is the constant from (11.2.58), we obtain

$$|\zeta(x)| \le 1/2 \quad \text{if} \quad \operatorname{dist}(x, \Sigma_{\text{out}}) \le \rho.$$

Denote by $O(\rho)$ the $\rho$-neighborhood of $\Sigma_{\text{out}}$ in $\mathbb{R}^3$. Since $\Sigma_{\text{out}}$ is a $C^1$ nonempty manifold, there is a constant $c_1$, depending only on $\Omega$, such that $\operatorname{meas}(O(\rho) \cap \Omega) \ge c^{-1}\varrho$. Thus we get

$$\int_{O(\rho)\cap\Omega} (1-\zeta)\, dx \ge 2c_1^{-1}(c\sigma)^{-1/\alpha}.$$

Next, as the nonnegative function $(1-\zeta)^+$ coincides with $1-\zeta$ in $O(\rho)\cap\Omega$, we get

$$\int_\Omega (1-\zeta)^+\, dx \ge \int_{O(\rho)\cap\Omega} (1-\zeta)\, dx \ge 2c_1^{-1}(c\sigma)^{-1/\alpha}.$$ □

We are now in a position to estimate the constant $\varkappa$. Substituting (11.2.62) and (11.2.64) into (11.2.61) we obtain

$$\begin{aligned}\varkappa^{-1} &\ge \int_\Omega (1-\zeta)^+\, dx - c\tau^2 - c\tau \ge \int_\Omega (1-\zeta)^+\, dx - \int_\Omega |(1-\zeta)^-|\, dx - c\tau^2 - c\tau \\ &\ge c^{-1}\sigma^{-1/\alpha} - c\sigma^{-1}\tau - c\tau^2 - c\tau.\end{aligned}$$

Since $\sigma > 1$ and $\tau < 1$ we can rewrite this inequality in the form

$$\varkappa^{-1} \ge c^{-1}\sigma^{-1/\alpha}(1 - c\tau\sigma^{1/\alpha}). \tag{11.2.65}$$

Now fix $\sigma > \sigma_1$ and choose $\tau^* > 0$ so small that

$$c\tau\sigma^{1/\alpha} \le 1/2 \quad \text{for all } \tau \in (0, \tau^*).$$

For such $\tau$ we have $0 \le \varkappa \le 2c_1\sigma^{1/\alpha}$, which gives (11.2.56). From this and (11.2.60) we obtain

$$|m| \le c\sigma^{1/\alpha}(\tau^2 + \sigma\tau^2),$$

which leads to (11.2.57) since $\sigma > 1$. □

We are now in a position to prove that for small $\tau$ the mapping $\Xi$ takes the ball $\mathcal{B}_\tau$ into itself. The corresponding result is given by

**Lemma 11.2.21.** *Under the assumptions of Theorem 11.2.6, there exists $\sigma^* > 1$, depending only on $\mathbf{U}$, $\Omega$ and $s$, $r$, with the following property. For every $\sigma > \sigma^*$ there are positive $\tau^*$ and $c$, depending only on $\mathbf{U}$, $\Omega$, $r$, $s$, and $\sigma$, such that whenever*

$$\tau \in (0, \tau^*], \quad \lambda^{-1}, \mathsf{Re} \in (0, \tau^2], \quad \|\mathbf{N} - \mathbb{I}\|_{C^2(\Omega)} \le \tau^2, \tag{11.2.66}$$

*the mapping $\Xi$ takes the ball $\mathcal{B}_\tau$ into itself.*

*Proof.* Set $\sigma^* = \max\{\sigma_1, \sigma_2\}$, where $\sigma_i$ are given by Lemma 11.2.17 and Proposition 11.2.18. Next, take $\tau_2 = \min\{\tau_0, \tau_1\}$, where $\tau_i$ are defined by Lemma 11.2.15 and Proposition 11.2.18. It follows from (11.2.52) and (11.2.57) that for $\sigma > \sigma^*$ and $\tau \in (0, \tau_2)$, the solution $\vartheta_1 := \Xi(\vartheta)$ to problem (11.2.41) satisfies

$$\|\varphi_1\|^2_{X^{s,r}} + \|\mathbf{v}_1\|_{Y^{s,r}} + \|\pi_1\|_{X^{s,r}} \le c\sigma\tau^2 + c\sigma^{1/\alpha+1}\tau^2.$$

Since $\sigma$ is fixed, we can choose $\tau^* \in (0, \tau_2]$ so small that

$$c\sigma\tau + c\sigma^{1/\alpha+1}\tau < 1 \quad \text{for all } \tau \in (0, \tau^*].$$

It remains to note that for such $\tau$,

$$\|\vartheta_1\|_E = \|\varphi_1\|_{X^{s,r}} + \|\mathbf{v}_1\|_{Y^{s,r}} + \|\pi_1\|_{X^{s,r}} < \tau.$$ □

### 11.2.6 Step 5. Proof of Theorem 11.2.6

**Sequential weak continuity of $\Xi$.** It follows from Lemma 11.2.21 that for a fixed $\sigma > \sigma^*$ and all $\tau \in (0, \tau^*)$ the mapping $\Xi$ takes the ball $\mathcal{B}_\tau \subset E$ into itself. In order to apply the fixed point theory we show that $\Xi$ is sequentially weakly continuous. To do so, assume that all assumptions of Lemma 11.2.21 are satisfied and $\mathbf{N}$, $\lambda$, $\mathsf{Re}$, $\tau$ satisfy condition (11.2.66). Choose a sequence $\vartheta_n \in \mathcal{B}_\tau$ such that $\vartheta_n = (\mathbf{u}_n, \pi_n, \varphi_n)$ converges weakly in $E = (Y^{s,r})^3 \times X^{s,r} \times X^{s,r}$. It follows from Lemma 11.2.3 that $E$ is reflexive as a product of reflexive spaces. Hence there is $\vartheta \in E$ such that $\vartheta_n \rightharpoonup \vartheta$ weakly in $E$. Since the ball $\mathcal{B}_\tau$ is closed and convex, $\vartheta$ belongs to $\mathcal{B}_\tau$. By Lemma 11.2.21, the corresponding elements $\vartheta_{1,n} = \Xi(\vartheta_n)$ belong to the ball $\mathcal{B}_\tau$, and the sequence of functions $\zeta_n$ is bounded in $X^{s,r}$. There are subsequences $\{\vartheta_{1,j}\} \subset \{\vartheta_{1,n}\}$ and $\{\zeta_j\} \subset \{\zeta_n\}$ such that $\vartheta_{1,j}$ converges weakly in $E$ to some $\vartheta_1 \in \mathcal{B}_\tau$ and $\zeta_j$ converges weakly in $X^{s,r}$ to some $\zeta \in X^{s,r}$. Since the embedding $E \hookrightarrow C(\Omega)$ is compact, we have $\vartheta_n \to \vartheta$, $\vartheta_{1,j} \to \vartheta_1$ in $C(\Omega)$, and

$$\nabla\zeta_j \rightharpoonup \nabla\zeta \quad \text{weakly in } L^2(\Omega), \quad \zeta_j \to \zeta \quad \text{in } C(\Omega).$$

Substituting $\vartheta_j$ and $\vartheta_{1,j}$ into equations (11.2.41) and letting $j \to \infty$ we find that the limits $\vartheta$ and $\vartheta_1$ also satisfy these equations. Thus, $\vartheta_1 = \Xi(\vartheta)$. In view of Lemma 11.2.21, for given $\vartheta \in \mathcal{B}_\tau$, problem (11.2.41) has a unique solution $\vartheta_1 \in E$. Hence all weakly convergent subsequences of $\vartheta_{1,n}$ have the common limit $\vartheta_1$. Therefore, the whole sequence $\vartheta_{1,n} = \Xi(\vartheta_n)$ converges weakly to $\Xi(\vartheta)$. Hence the mapping $\Xi : \mathcal{B}_\tau \to \mathcal{B}_\tau$ is sequentially weakly continuous. Notice that, by Lemma 11.2.4, $E' = (Y^{s,r})^{3'} \times (X^{s,r})^{2'}$ is separable as a Cartesian product of separable spaces. Therefore we can apply the Tikhonov fixed point theorem 1.1.18 to conclude that there is $\vartheta \in \mathcal{B}(\tau)$ such that $\vartheta = \Xi(\vartheta)$.

**Proof of Theorem 11.2.6.** It remains to prove that the fixed point $\vartheta = \Xi(\vartheta) \in E$ satisfies the original equations (11.1.24). It follows from the definition of the mapping $\Xi$ and equations (11.2.41) that $\vartheta$ and the corresponding function $\zeta$ serve as a strong solution to the boundary value problem

$$\mathbf{u} := \mathbf{u}_\star + \mathbf{v}, \quad q := q_\star + \pi, \quad \varrho := \varrho_\star + \varphi, \tag{11.2.67}$$

$$\mathbf{u} \cdot \nabla\varphi + \sigma\varphi = \Psi_1[\vartheta] + m\mathfrak{g}\varrho \quad \text{in } \Omega, \quad \varphi = 0 \quad \text{on } \Sigma_{\text{in}}, \tag{11.2.68}$$

$$\begin{aligned}
&\Delta\mathbf{v} - \nabla\pi = \mathscr{A}(\mathbf{u}) + \mathsf{Re}\,\mathscr{B}(\varrho, \mathbf{u}, \mathbf{u}) \equiv F[\vartheta] \quad \text{in } \Omega,\\
&\operatorname{div}\mathbf{v} = \varrho_\star^{-1}\Pi(\mathfrak{g}\sigma\varphi - \mathfrak{g}\varrho_\star\Psi[\vartheta] - \mathfrak{g}m\varrho_\star) \quad \text{in } \Omega, \qquad\qquad (11.2.69)\\
&\mathbf{v} = 0 \quad \text{on } \partial\Omega, \quad \pi - \Pi\pi = 0,
\end{aligned}$$

$$m = \varkappa \int_\Omega (\varrho_\star^{-1}\Psi_1[\vartheta]\,\zeta - \mathfrak{g}\Psi[\vartheta])\,dx, \quad \varkappa = \Big(\int_\Omega \mathfrak{g}(1 - \zeta - \varrho_\star^{-1}\zeta\varphi)\,dx\Big)^{-1}, \tag{11.2.70}$$

$$-\operatorname{div}(\mathbf{u}\zeta) + \sigma\zeta = \sigma\mathfrak{g} \quad \text{in } \Omega, \quad \zeta = 0 \quad \text{on } \Sigma_{\text{out}}. \tag{11.2.71}$$

The only difference between problems (11.2.67)–(11.2.71) and (11.1.24) is the presence of the projection $\Pi$ on the right hand side of (11.2.69). Hence, it suffices to show that

$$\Pi(\varrho_\star^{-1}\mathfrak{g}\sigma\varphi - \mathfrak{g}\Psi[\vartheta] - \mathfrak{g}m) = \varrho_\star^{-1}\mathfrak{g}\sigma\varphi - \mathfrak{g}\Psi[\vartheta] - \mathfrak{g}m.$$

By expression (1.7.6) for $\Pi$, this means that the function

$$\Upsilon = \varrho_\star^{-1}\mathfrak{g}\sigma\varphi - \mathfrak{g}\Psi[\vartheta] - \mathfrak{g}m$$

has zero mean over $\Omega$, i.e.,

$$\int_\Omega \Upsilon(x)\,dx = 0. \tag{11.2.72}$$

In order to prove (11.2.72), we note that $\varphi$ is a strong solution to problem (11.2.68). Next, both $\zeta$ and $\varphi$ are continuous and belong to $W^{1,2}(\Omega)$. Moreover, $\zeta$ vanishes on $\Sigma_{\text{out}}$. Multiplying (11.2.68) by $\zeta$, integrating the result by parts, and recalling (11.2.71) we obtain

$$\sigma \int_\Omega \varphi\mathfrak{g}\,dx = \int_\Omega \zeta(\Psi_1[\vartheta] + m\mathfrak{g}\varrho)\,dx. \tag{11.2.73}$$

Next, multiplying the formula for $m$ in (11.2.70) by $\varkappa$ and noting that $1 + \varrho_\star^{-1}\varphi = \varrho/\varrho_\star$ we obtain

$$m \int_\Omega \mathfrak{g}(1 - \varrho_\star^{-1}\zeta\varrho)\,dx = \int_\Omega \varrho_\star^{-1}\Psi_1[\vartheta]\zeta\,dx - \int_\Omega \mathfrak{g}\Psi[\vartheta]\,dx. \tag{11.2.74}$$

It follows from the expression for $\Upsilon$ that

$$-\mathfrak{g}\Psi[\vartheta] = \Upsilon + \mathfrak{g}m - \varrho_\star^{-1}\sigma\mathfrak{g}\varphi,$$

which leads to

$$-\int_\Omega \mathfrak{g}\Psi[\vartheta]\,dx = \int_\Omega \Upsilon\,dx + m\int_\Omega \mathfrak{g}\,dx - \int_\Omega \varrho_\star^{-1}\sigma\mathfrak{g}\varphi\,dx.$$

Substituting this into (11.2.74) we obtain

$$0 = \int_\Omega \varrho_\star^{-1}(\Psi_1[\vartheta] + m\mathfrak{g}\varrho)\zeta\,dx - \int_\Omega \varrho_\star^{-1}\sigma\mathfrak{g}\varphi\,dx + \int_\Omega \Upsilon\,dx.$$

Since $\varrho_\star = \text{const}$, it now follows from (11.2.73) that

$$0 = \int_\Omega \varrho_\star^{-1}\sigma\,\varphi\mathfrak{g}\,dx - \int_\Omega \varrho_\star^{-1}\sigma\mathfrak{g}\varphi\,dx + \int_\Omega \Upsilon\,dx,$$

which implies (11.2.72). Hence the fixed point $\vartheta \in \mathcal{B}_\tau$ is a solution to problem (11.1.24). Next, estimates (11.2.7) for $\zeta$, $m$ and $\varkappa$ follow from Proposition 11.2.18.

Let us prove that the set of solutions $(\vartheta, \zeta, m) \in \mathcal{B}_\tau \times X^{s,r} \times \mathbb{R}$ to Problem 11.1.9 contains a subset which is relatively compact in $W^{s+1,r}(\Omega) \times W^{s,r}(\Omega)^3 \times \mathbb{R}$. To this end choose $s_0 > s$ such that $s_0$ and $r$ satisfy inequalities (11.2.5) with $s$ replaced by $s_0$. Now we can choose $\sigma^* > 1$ and $\tau^*(\sigma)$ such that whenever $\sigma > \sigma^*$, $\tau \in (0, \tau^*(\sigma)]$ and

$$\lambda^{-1}, \mathsf{Re} \in (0, \tau^2], \quad \|\mathbf{N} - \mathbb{I}\|_{C^2(\Omega)} \le \tau^2,$$

Problem 11.1.9 has a solution $(\vartheta, \zeta, m) \in \mathcal{B}^0_\tau \times W^{s_0,r}(\Omega) \times \mathbb{R}$, where

$$\mathcal{B}^0_\tau = \{\vartheta = (\mathbf{v}, \pi, \varphi) : \|\mathbf{v}\|_{Y^{s_0,r}} + \|\pi\|_{X^{s_0,r}} + \|\varphi\|_{X^{s_0,r}} \le \tau \le 1\}.$$

Notice that $\mathcal{B}^0_\tau \subset \mathcal{B}_\tau$. Since the embeddings

$$Y^{s_0,r} \hookrightarrow W^{s_0+1,r}(\Omega), \quad X^{s_0,r} \hookrightarrow W^{s_0,r}(\Omega)$$

are continuous, $\mathcal{B}^0_\tau$ is a bounded subset of $W^{s_0+1,r}(\Omega) \times W^{s_0,r}(\Omega)^2$. On the other hand, estimate (11.2.7) with $s$ replaced by $s_0$ implies that the functions $\zeta$ and constants $m$ are bounded in $X^{s_0,r} \times \mathbb{R}$. Hence the constructed set of solutions $(\vartheta, \zeta, m)$ to Problem 11.1.9 is bounded in $W^{s_0+1,r}(\Omega) \times W^{s_0,r}(\Omega)^3 \times \mathbb{R}$. It remains to note that as $s_0 > s$, this set is relatively compact in $W^{s+1,r}(\Omega) \times W^{s,r}(\Omega)^3 \times \mathbb{R}$.

## 11.3 Dependence of solutions on N

### 11.3.1 Problem formulation. Basic equations. Transposed problem

Theorem 11.2.6 yields the existence of solutions (may be not unique) $\vartheta \in \mathcal{B}_\tau$ to Problem 11.1.9 for all sufficiently small $\mathsf{Re}$, $\lambda$, $\sigma^{-1}$ and all matrices $\mathbf{N}$ close to $\mathbb{I}$. Recall that $\mathbf{N}$ is completely defined by a deformation $x + \varepsilon\mathbf{T}(x)$ and is given by (11.1.5). The main goal of the sensitivity analysis is to prove differentiability of solutions to Problem 11.1.9 with respect to $\varepsilon$. The first step in this direction is to investigate the dependence of these solutions on the matrix $\mathbf{N}$. At this stage the special structure of $\mathbf{N}$ is not important and the results will be valid for an arbitrary smooth matrix-valued function $\mathbf{N}(x)$, $x \in \Omega$. In this section we prove that, under the assumptions of Theorem 11.2.6, the solution to Problem 11.1.9 is unique and the mapping $\mathbf{N} \mapsto \vartheta$ is Lipschitz in a weak norm. The simplest way to do this is to take two different matrices $\mathbf{N}_i$, $i = 0, 1$, and to compare the corresponding solutions $(\vartheta_i, \zeta_i, m_i) \in E \times X^{s,r} \times \mathbb{R}$ (with $\vartheta_i = (\mathbf{v}_i, \pi_i, \varphi_i)$) to Problem 11.1.9. Denote by

$$\begin{aligned} &\mathbf{w} = \mathbf{v}_0 - \mathbf{v}_1, \quad \omega = \pi_0 - \pi_1, \quad \psi = \varphi_0 - \varphi_1, \\ &\xi = \zeta_0 - \zeta_1, \quad n = m_0 - m_1, \end{aligned} \tag{11.3.1}$$

the difference between these solutions. It follows from (11.1.24) that the functions $(\mathbf{w}, \omega, \psi, \xi)$ and the constant $n$ satisfy

$$\mathbf{u}_0 \nabla \psi + \sigma \psi = -\mathbf{w} \cdot \nabla \varphi_1 + b_{11} \psi + b_{12} \omega + b_{13} n + b_{10} \mathfrak{d} \quad \text{in } \Omega, \tag{11.3.2a}$$

$$\begin{aligned} &\Delta \mathbf{w} - \nabla \omega = \mathscr{A}_0(\mathbf{w}) + \mathrm{Re}\, \mathscr{C}(\psi, \mathbf{w}) + \mathscr{D} \quad \text{in } \Omega, \\ &\operatorname{div} \mathbf{w} = b_{21} \psi + b_{22} \omega + b_{23} n + b_{20} \mathfrak{d} \quad \text{in } \Omega, \end{aligned} \tag{11.3.2b}$$

$$-\operatorname{div}(\mathbf{u}_0 \xi) + \sigma \xi = \operatorname{div}(\zeta_1 \mathbf{w}) + \sigma \mathfrak{d} \quad \text{in } \Omega, \tag{11.3.2c}$$

$$\begin{aligned} &\mathbf{w} = 0 \quad \text{on } \partial\Omega, \quad \psi = 0 \quad \text{on } \Sigma_{\mathrm{in}}, \qquad \xi = 0 \quad \text{on } \Sigma_{\mathrm{out}}, \\ &\omega - \Pi \omega = 0, \quad n = \varkappa_0 \int_\Omega (b_{31} \psi + b_{32} \omega + b_{34} \xi + b_{30} \mathfrak{d})\, dx. \end{aligned} \tag{11.3.2d}$$

Here the coefficients are given by

$$\begin{aligned}
b_{11} &= \sigma(1 - \mathfrak{g}_0) + \mathfrak{g}_0 \Psi[\vartheta_0] - \mathfrak{g}_0 \varrho_1 \Phi(\varphi_0, \varphi_1) - \frac{\sigma \mathfrak{g}_0}{\varrho_\star}(\varphi_0 + \varphi_1) + \mathfrak{g}_0 m_1, \\
b_{12} &= \lambda^{-1} \varrho_1 \mathfrak{g}_0, \quad b_{13} = \mathfrak{g}_0 \varrho_0, \quad b_{10} = \varrho_1 \Psi[\vartheta_1] - \frac{\sigma}{\varrho_\star} \varphi_1^2 - \sigma \varphi_1 + m_1 \varrho_1, \\
b_{21} &= \mathfrak{g}_0 \Big( \frac{\sigma}{\varrho_\star} + \Phi(\varphi_0, \varphi_1) \Big), \quad b_{22} = -\mathfrak{g}_0 / \lambda, \\
b_{23} &= -\mathfrak{g}_0, \quad b_{20} = \sigma \varphi_1 \varrho_\star^{-1} - \Psi[\vartheta_1] - m_1, \\
b_{31} &= \varrho_\star^{-1} \zeta_1 \Big( \sigma(1 - \mathfrak{g}_0) + \mathfrak{g}_0 \Psi[\vartheta_0] - \mathfrak{g}_0 \varrho_1 \Phi(\varphi_0, \varphi_1) - \frac{\mathfrak{g}_0 \sigma}{\varrho_\star}(\varphi_0 + \varphi_1) \Big) \\
&\quad + \mathfrak{g}_0 \Phi(\varphi_0, \varphi_1) + m_1 \mathfrak{g}_0 \varrho_\star^{-1} \zeta_1, \\
b_{32} &= \mathfrak{g}_0 \lambda^{-1} (\varrho_\star^{-1} \zeta_1 \varrho_1 - 1), \quad b_{34} = \varrho_\star^{-1} \Psi_1[\vartheta_0] + m_1 \mathfrak{g}_0 (1 + \varrho_\star^{-1} \varphi_0), \\
b_{30} &= \varrho_\star^{-1} \zeta_1 (b_{10} - m_1 \varrho_1) - \Psi[\vartheta_1] - m_1 (1 - \zeta_1 - \varrho_\star^{-1} \zeta_1 \varphi_1), \\
&\Phi(\varphi_0, \varphi_1) = (p'(\varrho_\star) \varrho_\star)^{-1} \sigma \int_0^1 H'(\varphi_0 s + \varphi_1 (1 - s))\, ds, \\
&\varkappa_0 = \Big( \int_\Omega \mathfrak{g}_0 (1 - \zeta_0 - \varrho_\star^{-1} \zeta_0 \varphi_0)\, dx \Big)^{-1},
\end{aligned} \tag{11.3.3}$$

the linear operator $\mathscr{C}$ is given by

$$\mathscr{C}(\psi, \mathbf{w}) = \mathscr{B}_0(\psi, \mathbf{u}_0, \mathbf{u}_0) + \mathscr{B}_0(\varrho_1, \mathbf{w}, \mathbf{u}_0) + \mathscr{B}_0(\varrho_1, \mathbf{u}_1, \mathbf{w}),$$

the linear operators $\mathscr{A}_i$ and trilinear forms $\mathscr{B}_i$ are given by

$$\begin{aligned}\mathscr{A}_i(\mathbf{u}) &= \Delta\mathbf{u} - \mathbf{N}_i^{-\top}\,\operatorname{div}\big(\mathfrak{g}_i^{-1}\mathbf{N}_i\mathbf{N}_i^\top\nabla(\mathbf{N}_i^{-1}\mathbf{u})\big),\\ \mathscr{B}_i(\varrho,\mathbf{u},\mathbf{w}) &= \varrho\mathbf{N}_i^{-\top}(\mathbf{u}\,\nabla(\mathbf{N}_i^{-1}\mathbf{w})),\end{aligned} \tag{11.3.4}$$

the vector fields $\mathbf{u}_i$ and functions $\varrho_i$ are given by

$$\mathbf{u}_i = \mathbf{u}_\star + \mathbf{v}_i, \quad \varrho_i = \varrho_\star + \varphi_i, \quad i = 0, 1,$$

the functions $\mathscr{D}$ and $\mathfrak{d}$ are given by

$$\begin{aligned}\mathscr{D} &= \mathscr{A}_0(\mathbf{u}_1) - \mathscr{A}_1(\mathbf{u}_1) + \mathsf{Re}(\mathscr{B}_0(\varrho_1,\mathbf{u}_1,\mathbf{u}_1) - \mathscr{B}_1(\varrho_1,\mathbf{u}_1,\mathbf{u}_1)),\\ \mathfrak{d} &= \mathfrak{g}_0 - \mathfrak{g}_1 \equiv \sqrt{\det\mathbf{N}_0} - \sqrt{\det\mathbf{N}_1}.\end{aligned} \tag{11.3.5}$$

We consider $\mathscr{D}$ and $\mathfrak{d}$ as *given functions*, and equalities (11.3.2) as a linear system of equations and boundary conditions for unknowns $\mathbf{w}$, $\omega$, $\psi$, $\xi$, and $n$. Our goal is to establish the well-posedness of problem (11.3.2) and to estimate its solutions in terms of $\mathscr{D}$ and $\mathfrak{d}$.

The main difficulty here is that we cannot apply Lemmas 11.2.10 and 11.2.11 to the transport equation (11.3.2a) since the term $\mathbf{w}\cdot\nabla\varphi_1$ in (11.3.2a) is neither smooth nor bounded. However, as shown in Lemma 11.3.7 below, this term belongs to some dual space. Thus we come to the idea to consider $(\mathbf{w},\omega,\psi,\xi)$ as a *very weak solution* to problem (11.3.2) (see [50, Ch. IV.9] for the theory of very weak solutions of Stokes equations). The construction of very weak solutions to a boundary value problem is built upon the analysis of the transposed problem, which in our case is formulated as follows. For a given vector field $\mathbf{H}$, functions $G$, $F$, $M$ and a constant $e$ find a vector field $\mathbf{h}$, functions $g$, $\varsigma$, $\upsilon$, and a constant $l$ satisfying

$$\begin{aligned}&\Delta\mathbf{h} - \nabla g = \mathscr{F}(\mathbf{h}) + \mathsf{Re}\,\mathscr{H}(\mathbf{h}) - \varsigma\nabla\varphi_1 - \zeta_1\nabla\upsilon + \mathbf{H} \quad \text{in } \Omega,\\ &\operatorname{div}\mathbf{h} = \Pi(b_{22}g + b_{12}\varsigma + b_{32}l) + \Pi G \quad \text{in } \Omega,\end{aligned} \tag{11.3.6a}$$

$$\begin{aligned}&-\operatorname{div}(\mathbf{u}_0\varsigma) + \sigma\varsigma = \mathsf{Re}\,\mathscr{M}(\mathbf{h}) + b_{11}\varsigma + b_{21}g + b_{31}l + F \quad \text{in } \Omega,\\ &\mathbf{u}_0\nabla\upsilon + \sigma\upsilon = b_{34}l + M \quad \text{in } \Omega,\end{aligned} \tag{11.3.6b}$$

$$\begin{aligned}&\mathbf{h} = 0 \quad \text{on } \partial\Omega, \quad \varsigma = 0 \quad \text{on } \Sigma_{\text{out}}, \quad \upsilon = 0 \quad \text{on } \Sigma_{\text{in}},\\ &g - \Pi g = 0, \quad l = \varkappa_0\int_\Omega(b_{13}\varsigma + b_{23}g)\,dx + e,\end{aligned} \tag{11.3.6c}$$

where the coefficients $b_{ij}$ and the constant $\varkappa_0$ are defined by (11.3.3), and the linear operators $\mathscr{F}$, $\mathscr{H}$, $\mathscr{M}$ are given by

$$\begin{aligned}\mathscr{F}(\mathbf{h}) &= \Delta\mathbf{h} - \mathbf{N}_0^{-\top}\operatorname{div}\big(\mathfrak{g}_0\mathbf{N}_0\mathbf{N}_0^\top\nabla(\mathbf{N}_0^{-1}\mathbf{h})\big),\\ \mathscr{H}(\mathbf{h}) &= \varrho_1\nabla(\mathbf{N}_0^{-1}\mathbf{u}_0)\mathbf{N}_0^{-1}\mathbf{h} - \mathbf{N}_0^{-\top}\operatorname{div}(\varrho_1\mathbf{u}_1\otimes(\mathbf{N}_0^{-1}\mathbf{h})),\\ \mathscr{M}(\mathbf{h}) &= (\mathbf{u}_0\nabla(\mathbf{N}_0^{-1}\mathbf{u}_0))\cdot\mathbf{N}_0^{-1}\mathbf{h}.\end{aligned} \tag{11.3.7}$$

**Remark 11.3.1.** Here we use the following notation. If $\mathbf{a}$ is a vector with components $a_j$ and $\mathbb{A}$ is a matrix with entries $A_{ij}$, then $\nabla\mathbf{a}$ denotes the matrix with entries $(\nabla\mathbf{a})_{ij} = \partial_{x_i}a_j$, and $\operatorname{div}\mathbb{A}$ denotes the vector with components $(\operatorname{div}\mathbb{A})_j = \partial_{x_i}A_{ij}$.

Our strategy is the following. First we establish the existence and uniqueness of weak and strong solutions to problem (11.3.6). Using this we show that the differences $(\mathbf{w}, \omega, \psi, \xi)$ given by (11.3.1) are a very weak solution to problem (11.3.2). Next we prove that the existence result for the transposed problem implies the desired estimates for a very weak solution to problem (11.3.2). We start by proving the well-posedness of problem (11.3.6).

### 11.3.2 Transposed problem

In this section we prove the existence and uniqueness of solutions to problem (11.3.6). We assume that the following conditions are satisfied.

**Condition 11.3.2.**

- The $C^3$ surface $\partial\Omega$ and the vector field $\mathbf{U} \in C^3(\Omega)$ satisfy Condition 11.2.5.
- The exponents $s$ and $r$ satisfy

$$1/2 < s < 1, \quad r < \infty, \quad 2s - 3r^{-1} < 1, \quad sr > 3, \quad (1-s)r > 3. \tag{11.3.8}$$

- $\sigma^* > 1$ and $\tau^*(\sigma) \le 1$ are given by the requirements of Theorem 11.2.6.
- $\sigma > \sigma^*$ and $\tau \in (0, \tau^*(\sigma)]$.
- We have

$$\lambda^{-1}, \mathsf{Re} \in (0, \tau^2], \quad \|\mathbf{N}_i - \mathbb{I}\|_{C^2(\Omega)} \le \tau^2. \tag{11.3.9}$$

- $\tau \in (0, \tau_0]$, where $\tau_0$ is the absolute constant given by Lemma 11.2.15, so

$$\|\mathbf{N}_i^{\pm 1} - \mathbb{I}\|_{C^2(\Omega)} + \|\mathfrak{g}_i^{\pm 1} - 1\|_{C^2(\Omega)} \le c\tau^2. \tag{11.3.10}$$

- $\vartheta_i = (\mathbf{v}_i, \pi_i, \varphi_i) \in \mathcal{B}_\tau$, $\zeta_i \in W^{s,r}(\Omega)$, $m_i \in \mathbb{R}$, $i = 0, 1$, are solutions to Problem 11.1.9 and satisfy inequalities (11.2.7).
- $\mathbf{u}_i = \mathbf{u}_\star + \mathbf{v}_i$ and $\varrho_i = \varrho_\star + \varphi_i$.

Throughout this section we will use the following notation for products of Sobolev spaces.

**Definition 11.3.3.** For every $s \in (0,1)$ and $r \in (1,\infty)$, we set

$$\begin{aligned}
\mathcal{U}^{s,r} &= (\mathcal{W}^{s-1,r}(\Omega))^3 \times (W^{s,r}(\Omega))^3 \times \mathbb{R},\\
\mathcal{V}^{s,r} &= (W^{s+1,r}(\Omega))^3 \times (W^{s,r}(\Omega))^3 \times \mathbb{R},
\end{aligned}$$

and

$$\mathcal{E}^{s,r} = (Z^{s,r})^3 \times (X^{s,r})^3 \times \mathbb{R},$$
$$\mathcal{F}^{s,r} = (Y^{s,r})^3 \times (X^{s,r})^3 \times \mathbb{R}.$$

We are now in a position to formulate the main result of this section on existence and uniqueness of solutions to the transposed problem.

**Theorem 11.3.4.** *Let Condition 11.3.2 be satisfied. Then there are $c$ and $\sigma_c \geq \sigma^*$, depending only on $\Omega$, $\mathbf{U}$, $s$, $r$, and $\tau_c(\sigma) \in (0, \min\{\tau_0, \tau^*(\sigma)\}]$, depending on $\Omega$, $\mathbf{U}$, $s$, $r$, and $\sigma$, with the following property. If*

$$\sigma > \sigma_c \quad \textit{and} \quad 0 < \tau \leq \tau_c(\sigma),$$

*then for every $\mathfrak{f} = (\mathbf{H}, G, F, M, e) \in \mathcal{U}^{s,r}$, problem (11.3.6) has a unique solution $\mathfrak{h} = (\mathbf{h}, g, \varsigma, v, l) \in \mathcal{V}^{s,r}$ which satisfies*

$$\|\mathfrak{h}\|_{\mathcal{V}^{s,r}} \leq c\|\mathfrak{f}\|_{\mathcal{U}^{s,r}}. \tag{11.3.11}$$

*If, in addition, $\mathfrak{f} \in \mathcal{E}^{s,r}$, then*

$$\|\mathfrak{h}\|_{\mathcal{F}^{s,r}} \leq c\|\mathfrak{f}\|_{\mathcal{E}^{s,r}}. \tag{11.3.12}$$

**Remark 11.3.5.** In the case $\mathfrak{f} \in \mathcal{E}^{s,r}$ the solution $\mathfrak{h}$ to the transposed problem (11.3.6) is understood in the strong sense, which creates no difficulties. The case $\mathfrak{f} \in \mathbf{U}^{s,r}$ needs clarification. In this case we assume that the right hand side of the Stokes equations (11.3.6a) belongs to $\mathcal{W}^{s-1,r}(\Omega) \times W^{s,r}(\Omega)$ and $(\mathbf{h}, g) \in W^{s+1,r}(\Omega) \times W^{s,r}(\Omega)$ satisfies (11.3.6a) in the sense of Lemma 1.7.9. As for the transport equations (11.3.6b), we assume that they are satisfied in the weak sense (see Section 11.2.4).

The rest of this section is devoted to the proof of Theorem 11.3.4. We split it into four steps.

**Step 1. Multiplicators.** We first show that for $\varphi, \varsigma \in W^{s,r}(\Omega)$, the quantity $\varsigma\nabla\varphi$ belongs to the dual space $\mathcal{W}^{s-1,r}(\Omega)$. This is based on the following lemma on multiplicators in fractional Sobolev spaces.

**Lemma 11.3.6.** *Let the exponents $s$ and $r$ satisfy*

$$1/2 < s < 1, \quad r < \infty, \quad rs > 3, \quad (1-s)r > 3, \tag{11.3.13}$$

*and suppose functions $\overline{\varsigma} \in W^{s,r}(\mathbb{R}^3)$, $\overline{\mathbf{w}} \in W^{1-s,r'}(\mathbb{R}^3)$, $r' = r/(r-1)$, are compactly supported in a ball $B \subset \mathbb{R}^3$. Then*

$$\|\overline{\varsigma}\,\overline{\mathbf{w}}\|_{W^{1-s,r'}(\mathbb{R}^3)} \leq c\|\overline{\mathbf{w}}\|_{W^{1-s,r'}(\mathbb{R}^3)}\|\overline{\varsigma}\|_{W^{s,r}(\mathbb{R}^3)}, \tag{11.3.14}$$

*where $c$ depends on $s$, $r$, and the diameter of $B$.*

*Proof.* Since $\bar\varsigma$ and $\overline{\mathbf{w}}$ are supported in $B$, we have

$$\|\bar\varsigma\,\overline{\mathbf{w}}\|_{W^{1-s,r'}(\mathbb{R}^3)} = \|\bar\varsigma\,\overline{\mathbf{w}}\|_{L^{r'}(B)} + \Big(\int_{B\times B} |x-y|^{-3-(1-s)r'} |\bar\varsigma\,\overline{\mathbf{w}}(x) - \bar\varsigma\,\overline{\mathbf{w}}(y)|^{r'}\, dxdy\Big)^{1/r'}.$$

The identity $\bar\varsigma\,\overline{\mathbf{w}}(x) - \bar\varsigma\,\overline{\mathbf{w}}(y) = \bar\varsigma(x)(\overline{\mathbf{w}}(x) - \overline{\mathbf{w}}(y)) + \overline{\mathbf{w}}(y)(\bar\varsigma(x) - \bar\varsigma(y))$ implies

$$\begin{aligned}\|\bar\varsigma\,\overline{\mathbf{w}}\|_{W^{1-s,r'}(\mathbb{R}^3)} \le \|\bar\varsigma\,\overline{\mathbf{w}}\|_{L^{r'}(B)} &+ \Big(\int_{B\times B} |x-y|^{-3-(1-s)r'} |\bar\varsigma(x)|^{r'} |\overline{\mathbf{w}}(x) - \overline{\mathbf{w}}(y)|^{r'}\, dxdy\Big)^{1/r'}\\ &+ \Big(\int_{B\times B} |x-y|^{-3-(1-s)r'} |\overline{\mathbf{w}}(y)|^{r'} |\bar\varsigma(x) - \bar\varsigma(y)|^{r'}\, dxdy\Big)^{1/r'}\end{aligned}$$

By an embedding theorem (see Section 1.5), we have

$$|\bar\varsigma(x)| \le c\|\bar\varsigma\|_{W^{s,r}(\mathbb{R}^3)} \quad \text{in } \mathbb{R}^3 \text{ for } sr > 3,$$

which gives

$$\begin{aligned}&\|\bar\varsigma\,\overline{\mathbf{w}}\|_{L^{r'}(B)} + \Big(\int_{B\times B} |x-y|^{-3-(1-s)r'} |\bar\varsigma(x)|^{r'} |\overline{\mathbf{w}}(x) - \overline{\mathbf{w}}(y)|^{r'}\, dxdy\Big)^{1/r'}\\ &\le c\|\bar\varsigma\|_{W^{s,r}(\mathbb{R}^3)}\Big\{\|\overline{\mathbf{w}}\|_{L^{r'}(B)} + \Big(\int_{B\times B} |x-y|^{-3-(1-s)r'} |\overline{\mathbf{w}}(x) - \overline{\mathbf{w}}(y)|^{r'}\, dxdy\Big)^{1/r'}\Big\}\\ &\le c\|\overline{\mathbf{w}}\|_{W^{1-s,r'}(\mathbb{R}^3)}\, \|\bar\varsigma\|_{W^{s,r}(\mathbb{R}^3)}.\end{aligned}$$

Thus we get

$$\begin{aligned}\|\bar\varsigma\,\overline{\mathbf{w}}\|_{W^{1-s,r'}(\mathbb{R}^3)} \le c\|\overline{\mathbf{w}}\|_{W^{1-s,r'}(\mathbb{R}^3)}\, \|\bar\varsigma\|_{W^{s,r}(\mathbb{R}^3)}&\\ + \Big(\int_{B\times B} |x-y|^{-3-(1-s)r'} |\overline{\mathbf{w}}(y)|^{r'} |\bar\varsigma(x) - \bar\varsigma(y)|^{r'}\, dxdy\Big)^{1/r'}.&\end{aligned} \tag{11.3.15}$$

In order to estimate the integral on the right hand side, we use the identity

$$\begin{aligned}&|x-y|^{-3-(1-s)r'} |\overline{\mathbf{w}}(y)|^{r'} |\bar\varsigma(x) - \bar\varsigma(y)|^{r'}\\ &\qquad = \big(|x-y|^{-\alpha}|\overline{\mathbf{w}}(y)|^{r'}\big)\big((|x-y|^{-3-sr})^{r'/r}\, |\bar\varsigma(x) - \bar\varsigma(y)|^{r'}\big),\end{aligned}$$

where

$$\alpha = 3 + (1-s)r' - (3+sr)r'/r = 3\frac{r-2}{r-1} + (1-2s)\frac{r}{r-1}.$$

Applying the Hölder inequality we obtain

$$\begin{aligned}\Big(\int_{B\times B}&|x-y|^{-3-(1-s)r'}|\overline{\mathbf{w}}(y)|^{r'}|\overline{\varsigma}(x)-\overline{\varsigma}(y)|^{r'}\,dxdy\Big)^{1/r'}\\&\le\Big(\int_{B\times B}|x-y|^{-\alpha p}|\overline{\mathbf{w}}(y)|^{pr'}|\,dxdy\Big)^{1/(pr')}\\&\quad\times\Big(\int_{B\times B}|x-y|^{-3-sr}|\overline{\varsigma}(x)-\overline{\varsigma}(y)|^{r}\,dxdy\Big)^{1/r}\\&\le c\Big(\int_{B\times B}|x-y|^{-\alpha p}|\overline{\mathbf{w}}(y)|^{pr'}\,dxdy\Big)^{1/(pr')}\|\overline{\varsigma}\|_{W^{s,r}(\mathbb{R}^3)},\end{aligned}\tag{11.3.16}$$

where

$$1/p=1-r'/r,\quad p=(r-1)/(r-2),\quad pr'=r/(r-2).$$

Notice that $r>2$ since $sr>3$. We have

$$p\alpha=3+(1-2s)r/(r-2)<3\quad\text{since}\quad s>1/2,$$

which gives

$$\int_B|x-y|^{-\alpha p}dx\le c.$$

Inserting this in (11.3.16) we obtain

$$\begin{aligned}\Big(\int_{B\times B}&|x-y|^{-3-(1-s)r'}|\overline{\mathbf{w}}(y)|^{r'}|\overline{\varsigma}(x)-\overline{\varsigma}(y)|^{r'}\,dxdy\Big)^{1/r'}\\&\le c\Big(\int_{B\times B}|\overline{\mathbf{w}}(y)|^{r/(r-2)}\,dxdy\Big)^{(r-2)/r}\|\overline{\varsigma}\|_{W^{s,r}(\mathbb{R}^3)}.\end{aligned}\tag{11.3.17}$$

Next an embedding theorem (see Section 1.5) implies

$$\|\overline{\mathbf{w}}\|_{L^t(\mathbb{R}^3)}\le c\|\overline{\mathbf{w}}\|_{W^{1-s,r'}(\mathbb{R}^3)}$$

for

$$1/t=1/r'-(1-s)/3,\quad t=\frac{r}{r-1-\frac{(1-s)r}{3}}.$$

Since $(1-s)r>3$, we have $t>r/(r-2)$, which leads to

$$\|\overline{\mathbf{w}}\|_{L^{r/(r-2)}(B)}\le c\|\overline{\mathbf{w}}\|_{L^t(\mathbb{R}^3)}\le c\|\overline{\mathbf{w}}\|_{W^{1-s,r'}(\mathbb{R}^3)}.$$

Inserting this into (11.3.17) we obtain

$$\begin{aligned}\Big(\int_{B\times B}&|x-y|^{-3-(1-s)r'}|\overline{\mathbf{w}}(y)|^{r'}|\overline{\varsigma}(x)-\overline{\varsigma}(y)|^{r'}\,dxdy\Big)^{1/r'}\\&\le c\|\overline{\mathbf{w}}\|_{W^{1-s,r'}(\mathbb{R}^3)}\|\overline{\varsigma}\|_{W^{s,r}(\mathbb{R}^3)}.\end{aligned}$$

Combining this estimate with (11.3.15) we arrive at (11.3.14). □

**Lemma 11.3.7.** *Let $\Omega \subset \mathbb{R}^3$ be a bounded domain with $C^1$ boundary. Let the exponents $s \in (0,1)$ and $r \in (1,\infty)$ satisfy the inequalities*

$$1/2 < s < 1, \quad r < \infty, \quad (1-s)r > 3, \quad sr > 3, \tag{11.3.18}$$

*and let $\varphi, \varsigma \in W^{s,r}(\Omega) \cap W^{1,2}(\Omega)$ and $\mathbf{w} \in \mathcal{W}_0^{1-s,r'}(\Omega)$. Then there is a constant $c$, depending only on $s$, $r$, and $\Omega$, such that the trilinear form*

$$\mathfrak{B}(\mathbf{w}, \varphi, \varsigma) = \int_\Omega \varsigma \mathbf{w} \cdot \nabla \varphi \, dx$$

*satisfies*

$$|\mathfrak{B}(\mathbf{w}, \varphi, \varsigma)| \leq c \|\mathbf{w}\|_{\mathcal{W}_0^{1-s,r'}(\Omega)} \|\varphi\|_{W^{s,r}(\Omega)} \|\varsigma\|_{W^{s,r}(\Omega)}, \tag{11.3.19}$$

*and can be continuously extended to $\mathfrak{B} : \mathcal{W}_0^{1-s,r'}(\Omega)^3 \times W^{s,r}(\Omega)^2 \to \mathbb{R}$. In particular,*

$$\varsigma \nabla \varphi \in \mathcal{W}^{s-1,r}(\Omega) \quad \textit{and} \quad \|\varsigma \nabla \varphi\|_{\mathcal{W}^{s-1,r}(\Omega)} \leq c \|\varphi\|_{W^{s,r}(\Omega)} \|\varsigma\|_{W^{s,r}(\Omega)}. \tag{11.3.20}$$

*Proof.* Let $B$ be a ball in $\mathbb{R}^3$ such that $\Omega \Subset B$. Since $\partial\Omega$ is $C^1$, the functions $\varphi$ and $\varsigma$ have extensions $\overline{\varphi}, \overline{\varsigma} \in W^{s,r}(\mathbb{R}^3) \cap W^{1,2}(\mathbb{R}^3)$ such that $\overline{\varphi}$ and $\overline{\varsigma}$ are compactly supported in $B$ and

$$\|\overline{\varphi}\|_{W^{s,r}(\mathbb{R}^3)} \leq c \|\varphi\|_{W^{s,r}(\Omega)}, \quad \|\overline{\varsigma}\|_{W^{s,r}(\mathbb{R}^3)} \leq c \|\varsigma\|_{W^{s,r}(\Omega)},$$

where $c$ depends on $s$, $r$ and $B$.

By the definition of $W_0^{1,r'}(\Omega)$ (see Section 1.5), the extension operator $T : W_0^{1,r'}(\Omega) \to W^{1,r'}(\mathbb{R}^3)$ defined by

$$T\mathbf{w} = \mathbf{w} \quad \text{in } \Omega \quad \text{and} \quad T\mathbf{w} = 0 \quad \text{in } \mathbb{R}^3 \setminus \Omega$$

is continuous. Obviously it maps continuously $L^{r'}(\Omega)$ to $L^{r'}(\mathbb{R}^3)$. Since $\mathcal{W}_0^{1-s,r'}(\Omega) = [L^{r'}(\Omega), W_0^{1,r'}(\Omega)]_{1-s,r'}$ and $W^{1-s,r'}(\mathbb{R}^3) = [L^{r'}(\mathbb{R}^3), W^{1,r'}(\mathbb{R}^3)]_{1-s,r'}$ are interpolation spaces, Lemma 1.1.13 shows that $T : \mathcal{W}_0^{1-s,r'}(\Omega) \to W^{1-s,r'}(\mathbb{R}^3)$ is also continuous. Therefore, for every $\mathbf{w} \in \mathcal{W}_0^{1-s,r'}(\Omega)$, the extension $\overline{\mathbf{w}} := T\mathbf{w}$ satisfies

$$\|\overline{\mathbf{w}}\|_{W^{1-s,r'}(\mathbb{R}^3)} \leq c(s,r') \|\mathbf{w}\|_{\mathcal{W}_0^{1-s,r'}(\Omega)}. \tag{11.3.21}$$

Obviously we have

$$\mathfrak{B}(\mathbf{w}, \varphi, \varsigma) = \int_{\mathbb{R}^3} \overline{\mathbf{w}} \cdot \nabla \overline{\varphi} \, \overline{\varsigma} \, dx.$$

It follows that

$$\begin{aligned} |\mathfrak{B}(\mathbf{w}, \varphi, \varsigma)| &\leq \|\overline{\mathbf{w}}\,\overline{\varsigma}\|_{W^{1-s,r'}(\mathbb{R}^3)} \|\nabla \overline{\varphi}\|_{W^{s-1,r}(\mathbb{R}^3)} \\ &\leq c \|\overline{\mathbf{w}}\,\overline{\varsigma}\|_{W^{1-s,r'}(\mathbb{R}^3)} \|\overline{\varphi}\|_{W^{s,r}(\mathbb{R}^3)} \leq c \|\overline{\mathbf{w}}\,\overline{\varsigma}\|_{W^{1-s,r'}(\mathbb{R}^3)} \|\varphi\|_{W^{s,r}(\Omega)}. \end{aligned}$$

Application of Lemma 11.3.6 completes the proof. □

**Step 2. Estimates of the operators $\mathscr{F}$, $\mathscr{H}$, $\mathscr{M}$ and the coefficients $b_{ij}$.** Our next task is to estimate the norms of the differential operators $\mathscr{F}$, $\mathscr{H}$, $\mathscr{M}$ appearing in equations (11.3.6) and next to estimate the coefficients of these equations. Recall that $\mathscr{F}$, $\mathscr{H}$, $\mathscr{M}$ are defined by (11.3.7). We begin by estimating the operator $\mathscr{F}$.

**Lemma 11.3.8.** *Let Condition 11.3.2 be satisfied. Then there is a constant $c > 0$, depending only on $s$, $r$, and $\Omega$, such that*

$$\|\mathscr{F}(\mathbf{h})\|_{\mathcal{W}^{s-1,r}(\Omega)} \le c\tau^2 \|\mathbf{h}\|_{W^{s+1,r}(\Omega)} \quad \textit{for all } \mathbf{h} \in W^{s+1,r}(\Omega), \tag{11.3.22}$$

$$\|\mathscr{F}(\mathbf{h})\|_{Z^{s,r}} \le c\tau^2 \|\mathbf{h}\|_{Y^{s,r}} \quad \textit{for all } \mathbf{h} \in Y^{s,r}. \tag{11.3.23}$$

*Proof.* We have

$$\mathscr{F}(\mathbf{h}) = \operatorname{div}\big((\mathbb{I} - \mathbb{S})\nabla \mathbf{h} + \mathbb{S}\nabla((\mathbb{I} - \mathbf{N}_0^{-1})\mathbf{h})\big) + \nabla(\mathbf{N}_0^{-\top}) \cdot \big(\mathbf{N}_0^{\top}\mathbb{S}\nabla(\mathbf{N}_0^{-1}\mathbf{h})\big),$$

where

$$\mathbb{S} = \mathfrak{g}_0 \mathbf{N}_0^{-\top}\mathbf{N}_0\mathbf{N}_0^{\top},$$

and the notation $\nabla(\mathbb{A}) \cdot \mathbb{B}$ stands for the column vector with components

$$(\nabla(\mathbb{A}) \cdot \mathbb{B})_i = \partial_{x_j}(\mathbb{A})_{im}(\mathbb{B})_{mj}.$$

Estimates (11.3.10) yield

$$\|\mathbb{I} - \mathbb{S}\|_{C^2(\Omega)} \le c\tau^2, \quad \|\mathbb{I} - \mathbb{N}_0^{-1}\|_{C^2(\Omega)} \le c\tau^2. \tag{11.3.24}$$

On the other hand, Lemma 11.2.13 implies

$$\begin{aligned} &\big\| \operatorname{div}\big((\mathbb{I} - \mathbb{S})\nabla \mathbf{h} + \mathbb{S}\nabla((\mathbb{I} - \mathbf{N}_0^{-1})\mathbf{h})\big)\big\|_{\mathcal{W}^{s-1,r}(\Omega)} \\ &\qquad\qquad \le c\|(\mathbb{I} - \mathbb{S})\nabla \mathbf{h} + \mathbb{S}\nabla((\mathbb{I} - \mathbf{N}_0^{-1})\mathbf{h})\|_{W^{s,r}(\Omega)}. \end{aligned}$$

By (11.3.24), we have

$$\begin{aligned} &\|(\mathbb{I} - \mathbb{S})\nabla \mathbf{h} + \mathbb{S}\nabla((\mathbb{I} - \mathbf{N}_0^{-1})\mathbf{h})\|_{W^{s,r}(\Omega)} \\ &\quad \le c\|\mathbb{I} - \mathbb{S}\|_{C^2(\Omega)}\|\nabla \mathbf{h}\|_{W^{s,r}(\Omega)} + c\|S\|_{C^2(\Omega)}\|\nabla((\mathbb{I} - \mathbf{N}_0^{-1})\mathbf{h})\|_{W^{s,r}(\Omega)} \\ &\quad \le c\tau^2\|\nabla \mathbf{h}\|_{W^{s,r}(\Omega)} + c\|\nabla((\mathbb{I} - \mathbf{N}_0^{-1})\mathbf{h})\|_{W^{s,r}(\Omega)} \\ &\quad \le c\tau^2\|\mathbf{h}\|_{W^{s+1,r}(\Omega)} + c\|(\mathbb{I} - \mathbf{N}_0^{-1})\mathbf{h}\|_{W^{s+1,r}(\Omega)} \\ &\quad \le c(\tau^2 + \|\mathbb{I} - \mathbf{N}_0^{-1}\|_{C^2(\Omega)})\|\mathbf{h}\|_{W^{s+1,r}(\Omega)} \le c\tau^2\|\mathbf{h}\|_{W^{s+1,r}(\Omega)}. \end{aligned}$$

Thus we get

$$\big\| \operatorname{div}\big((\mathbb{I} - \mathbb{S})\nabla \mathbf{h} + \mathbb{S}\nabla((\mathbb{I} - \mathbf{N}_0^{-1})\mathbf{h})\big)\big\|_{\mathcal{W}^{s-1,r}(\Omega)} \le c\tau^2\|\mathbf{h}\|_{W^{s+1,r}(\Omega)}. \tag{11.3.25}$$

Next, it follows from (11.3.24) that

$$\|\nabla(\mathbf{N}_0^{-\top})\|_{C^1(\Omega)} = \|\nabla(\mathbf{N}_0^{-\top} - \mathbb{I})\|_{C^1(\Omega)} \le \|\mathbf{N}_0^{-\top} - \mathbb{I}\|_{C^2(\Omega)} \le c\tau^2.$$

Applying Lemmas 11.2.13 and 11.2.14 we obtain

$$\begin{aligned}
&\|\nabla(\mathbf{N}_0^{-\top})\cdot(\mathbf{N}_0^{\top}\mathbb{S}\nabla(\mathbf{N}_0^{-1}\mathbf{h}))\|_{\mathcal{W}^{s-1,r}(\Omega)}\\
&\quad\le \|\nabla(\mathbf{N}_0^{-\top})\|_{C^1(\Omega)}\|\mathbf{N}_0^{\top}\mathbb{S}\nabla(\mathbf{N}_0^{-1}\mathbf{h})\|_{\mathcal{W}^{s-1,r}(\Omega)}\\
&\quad\le c\tau^2\|\mathbf{N}_0^{\top}\mathbb{S}\|_{C^1(\Omega)}\|\nabla(\mathbf{N}_0^{-1}\mathbf{h})\|_{\mathcal{W}^{s-1,r}(\Omega)}\\
&\quad\le c\tau^2\|\mathbf{N}_0^{-1}\mathbf{h}\|_{W^{s,r}(\Omega)}\le c\tau^2\|\mathbf{h}\|_{W^{s,r}(\Omega)}\le c\tau^2\|\mathbf{h}\|_{W^{s+1,r}(\Omega)}. \qquad (11.3.26)
\end{aligned}$$

Combining (11.3.25) and (11.3.26) we arrive at (11.3.22). Next, estimates (11.3.24) imply

$$\|\mathscr{F}(\mathbf{h})\|_{L^2(\Omega)}\le c\tau^2\|\mathbf{h}\|_{W^{2,2}(\Omega)} \quad \text{for all } \mathbf{h}\in W^{2,2}(\Omega).$$

Combining this with (11.3.22) and recalling

$$\begin{aligned}
\|\mathscr{F}(\mathbf{h})\|_{Z^{s,r}}&=\|\mathscr{F}(\mathbf{h})\|_{\mathcal{W}^{s-1,r}(\Omega)}+\|\mathscr{F}(\mathbf{h})\|_{L^2(\Omega)},\\
\|\mathbf{h}\|_{Y^{s,r}}&=\|\mathbf{h}\|_{W^{s+1,r}(\Omega)}+\|\mathbf{h}\|_{W^{2,2}(\Omega)}
\end{aligned}$$

we arrive at (11.3.23). □

**Lemma 11.3.9.** *Let Condition 11.3.2 be satisfied. Then there is a constant $c>0$, depending only on $s$, $r$, $\Omega$, and $\mathbf{U}$, such that*

$$\|\mathscr{H}(\mathbf{h})\|_{\mathcal{W}^{s-1,r}(\Omega)}\le c\|\mathbf{h}\|_{W^{s+1,r}(\Omega)} \quad \textit{for all } \mathbf{h}\in W^{s+1,r}(\Omega), \qquad (11.3.27)$$

$$\|\mathscr{H}(\mathbf{h})\|_{Z^{s,r}}\le c\|\mathbf{h}\|_{Y^{s,r}} \quad \textit{for all } \mathbf{h}\in Y^{s,r}. \qquad (11.3.28)$$

*Proof.* Since $W^{s,r}(\Omega)$ is a Banach algebra, we have

$$\begin{aligned}
&\|\varrho_1\nabla(\mathbf{N}_0^{-1}\mathbf{u}_0)\mathbf{N}_0^{-1}\mathbf{h}\|_{\mathcal{W}^{s-1,r}(\Omega)}\le\|\varrho_1\nabla(\mathbf{N}_0^{-1}\mathbf{u}_0)\mathbf{N}_0^{-1}\mathbf{h}\|_{W^{s,r}(\Omega)}\\
&\quad\le c\|\varrho_1\|_{W^{s,r}(\Omega)}\|\nabla(\mathbf{N}_0^{-1}\mathbf{u}_0)\|_{W^{s,r}(\Omega)}\,\|\mathbf{N}_0^{-1}\mathbf{h}\|_{W^{s,r}(\Omega)}\\
&\quad\le c\|\mathbf{N}_0^{-1}\mathbf{u}_0\|_{W^{s+1,r}(\Omega)}\|\mathbf{h}\|_{W^{s,r}(\Omega)}\le c\|\mathbf{u}_0\|_{W^{s+1,r}(\Omega)}\|\mathbf{h}\|_{W^{s,r}(\Omega)}\\
&\quad\le c\|\mathbf{h}\|_{W^{s,r}(\Omega)}. \qquad (11.3.29)
\end{aligned}$$

Next applying Lemmas 11.2.13 and 11.2.14 we obtain

$$\begin{aligned}
&\|\mathbf{N}_0^{-\top}\operatorname{div}(\varrho_1\mathbf{u}_1\otimes(\mathbf{N}_0^{-1}\mathbf{h}))\|_{\mathcal{W}^{s-1,r}(\Omega)}\le c\|\operatorname{div}(\varrho_1\mathbf{u}_1\otimes(\mathbf{N}_0^{-1}\mathbf{h}))\|_{\mathcal{W}^{s-1,r}(\Omega)}\\
&\quad\le c\|\varrho_1\mathbf{u}_1\otimes(\mathbf{N}_0^{-1}\mathbf{h})\|_{W^{s,r}(\Omega)}\le c\|\varrho_1\mathbf{u}_1\|_{W^{s,r}(\Omega)}\|\mathbf{N}_0^{-1}\mathbf{h}\|_{W^{s,r}(\Omega)}\\
&\quad\le c\|\mathbf{N}_0^{-1}\mathbf{h}\|_{W^{s,r}(\Omega)}\le c\|\mathbf{h}\|_{W^{s,r}(\Omega)}. \qquad (11.3.30)
\end{aligned}$$

Combining (11.3.29) and (11.3.30) we arrive at (11.3.27). It remains to prove inequality (11.3.28). Since

$$\begin{aligned}
\|\mathscr{H}(\mathbf{h})\|_{Z^{s,r}}&=\|\mathscr{H}(\mathbf{h})\|_{\mathcal{W}^{s-1,r}(\Omega)}+\|\mathscr{H}(\mathbf{h})\|_{L^2(\Omega)},\\
\|\mathbf{h}\|_{Y^{s,r}}&=\|\mathbf{h}\|_{W^{s+1,r}(\Omega)}+\|\mathbf{h}\|_{W^{2,2}(\Omega)},
\end{aligned}$$

it suffices to show that

$$\|\mathscr{H}(\mathbf{h})\|_{L^2(\Omega)} \le c\|\mathbf{h}\|_{Y^{s,r}}. \tag{11.3.31}$$

Since $X^{s,r}$ is continuously embedded in $C(\Omega)$ and $Y^{s,r}$ is continuously embedded in $C^1(\Omega)$, we have

$$\|\varrho_1\|_{C(\Omega)} \le c\|\varrho_1\|_{X^{s,r}} \le c, \quad \|\mathbf{u}_i\|_{C^1(\Omega)} \le c\|\mathbf{u}_i\|_{Y^{s,r}} \le c. \tag{11.3.32}$$

Thus we get

$$\begin{aligned}
\|\varrho_1\nabla(\mathbf{N}_0^{-1}\mathbf{u}_0)\mathbf{N}_0^{-1}\mathbf{h}\|_{L^2(\Omega)} &\le \|\varrho_1\|_{C(\Omega)}\|\nabla(\mathbf{N}_0^{-1}\mathbf{u}_0)\|_{C(\Omega)}\|\mathbf{N}_0^{-1}\mathbf{h}\|_{L^2(\Omega))}\\
&\le c\|\varrho_1\|_{W^{s,r}(\Omega)}\|\nabla(\mathbf{N}_0^{-1}\mathbf{u}_0)\|_{W^{s,r}(\Omega)}\,\|\mathbf{N}_0^{-1}\mathbf{h}\|_{W^{s,r}(\Omega)}\\
&\le c\|\mathbf{N}_0^{-1}\mathbf{u}_0\|_{W^{s+1,r}(\Omega)}\|\mathbf{h}\|_{W^{s,r}(\Omega)} \le c\|\mathbf{u}_0\|_{W^{s+1,r}(\Omega)}\|\mathbf{h}\|_{W^{s,r}(\Omega)}\\
&\le c\|\mathbf{h}\|_{W^{s,r}(\Omega)} \le c\|\mathbf{h}\|_{Y^{s,r}}.
\end{aligned} \tag{11.3.33}$$

Next we have

$$\begin{aligned}
\|\mathbf{N}_0^{-\top}\operatorname{div}(\varrho_1\mathbf{u}_1\otimes(\mathbf{N}_0^{-1}\mathbf{h}))\|_{L^2(\Omega)} &\le c\|\varrho_1\mathbf{u}_1\otimes(\mathbf{N}_0^{-1}\mathbf{h})\|_{W^{1,2}(\Omega)}\\
&\le c\|\varrho_1\mathbf{u}_1\|_{W^{1,2}(\Omega)}\|\mathbf{h}\|_{L^\infty(\Omega)} + c\|\mathbf{h}\|_{W^{1,2}(\Omega)}.
\end{aligned} \tag{11.3.34}$$

Since the embedding $W^{1,2}(\Omega) \hookrightarrow X^{s,r}$ is continuous, inequalities (11.3.32) imply

$$\|\varrho_1\mathbf{u}_1\|_{W^{1,2}(\Omega)} \le c\|\varrho_1\|_{W^{1,2}(\Omega)} \le c.$$

From this and (11.3.34) we obtain

$$\|\mathbf{N}_0^{-\top}\operatorname{div}(\varrho_1\mathbf{u}_1\otimes(\mathbf{N}_0^{-1}\mathbf{h}))\|_{L^2(\Omega)} \le c\|\mathbf{h}\|_{L^\infty(\Omega)} + c\|\mathbf{h}\|_{W^{1,2}(\Omega)} \le c\|\mathbf{h}\|_{W^{2,2}(\Omega)}.$$

Combining this with (11.3.33) we arrive at

$$\|\mathscr{H}(\mathbf{h})\|_{L^2(\Omega)} \le c\|\mathbf{h}\|_{Y^{s,r}},$$

which along with (11.3.27) yields (11.3.31). □

**Lemma 11.3.10.** *Let Condition 11.3.2 be satisfied. Then there is $c > 0$, depending only on $s$, $r$, and $\Omega$, such that*

$$\|\mathscr{M}(\mathbf{h})\|_{W^{s,r}(\Omega)} \le c\|\mathbf{h}\|_{W^{s,r}(\Omega)} \quad \textit{for all } \mathbf{h}\in W^{s+1,r}(\Omega), \tag{11.3.35}$$

$$\|\mathscr{M}(\mathbf{h})\|_{X^{s,r}} \le c\|\mathbf{h}\|_{Y^{s,r}} \quad \textit{for all } \mathbf{h}\in Y^{s,r}. \tag{11.3.36}$$

*Proof.* The proof imitates the proofs of Lemmas 11.3.8 and 11.3.9. Recall from (11.3.7) that

$$\mathscr{M}(\mathbf{h}) = \mathbf{u}_0\nabla(\mathbf{N}_0^{-1}\mathbf{u}_0)\cdot\mathbf{N}_0^{-1}\mathbf{h}.$$

We have

$$
\begin{aligned}
\|\mathbf{u}_0\nabla(\mathbf{N}_0^{-1}\mathbf{u}_0)\cdot\mathbf{N}_0^{-1}\mathbf{h}\|_{W^{s,r}(\Omega)} &\le \|\mathbf{u}_0\nabla(\mathbf{N}_0^{-1}\mathbf{u}_0)\|_{W^{s,r}(\Omega)}\|\mathbf{N}_0^{-1}\mathbf{h}\|_{W^{s,r}(\Omega)}\\
&\le c\|\mathbf{u}_0\|_{W^{s,r}(\Omega)}\|\nabla(\mathbf{N}_0^{-1}\mathbf{u}_0)\|_{W^{s,r}(\Omega)}\ \|\mathbf{N}_0^{-1}\mathbf{h}\|_{W^{s,r}(\Omega)}\\
&\le c\|\mathbf{N}_0^{-1}\mathbf{u}_0\|_{W^{s+1,r}(\Omega)}\|\mathbf{h}\|_{W^{s,r}(\Omega)} \le c\|\mathbf{u}_0\|_{W^{s+1,r}(\Omega)}\|\mathbf{h}\|_{W^{s,r}(\Omega)}\\
&\le c\|\mathbf{h}\|_{W^{s,r}(\Omega)}, \qquad (11.3.37)
\end{aligned}
$$

which leads to (11.3.35). Next we have

$$
\begin{aligned}
&\|\mathbf{u}_0\nabla(\mathbf{N}_0^{-1}\mathbf{u}_0)\mathbf{N}_0^{-1}\mathbf{h}\|_{W^{1,2}(\Omega)}\\
&\quad\le c\|\mathbf{u}_0\nabla(\mathbf{N}_0^{-1}\mathbf{u}_0)\|_{L^\infty(\Omega)}\|\mathbf{h}\|_{W^{1,2}(\Omega)} + \|\mathbf{u}_0\nabla(\mathbf{N}_0^{-1}\mathbf{u}_0)\|_{W^{1,2}(\Omega)}\|\mathbf{h}\|_{L^\infty(\Omega)}\\
&\quad\le c\|\mathbf{h}\|_{W^{1,2}(\Omega)} + \|\mathbf{u}_0\nabla(\mathbf{N}_0^{-1}\mathbf{u}_0)\|_{W^{1,2}(\Omega)}\|\mathbf{h}\|_{L^\infty(\Omega)}.
\end{aligned}
$$

On the other hand, inequalities (11.3.32) imply

$$
\begin{aligned}
\|\mathbf{u}_0\nabla(\mathbf{N}_0^{-1}\mathbf{u}_0)\|_{W^{1,2}(\Omega)} &\le c\|\nabla(\mathbf{N}_0^{-1}\mathbf{u}_0)\|_{W^{1,2}(\Omega)} \le c\|\mathbf{N}_0^{-1}\mathbf{u}_0\|_{W^{2,2}(\Omega)} \le c\|\mathbf{u}_0\|_{Y^{s,r}}\\
&\le c.
\end{aligned}
$$

Thus we get

$$
\|\mathbf{u}_0\nabla(\mathbf{N}_0^{-1}\mathbf{u}_0)\mathbf{N}_0^{-1}\mathbf{h}\|_{W^{1,2}(\Omega)} \le c\|\mathbf{h}\|_{W^{1,2}(\Omega))} + \|\mathbf{h}\|_{L^\infty(\Omega))} \le c\|\mathbf{h}\|_{Y^{s,r}}.
$$

Combining this with (11.3.35) we obtain (11.3.36). □

The next lemma gives estimates for the coefficients of equations (11.3.6).

**Lemma 11.3.11.** *Let Condition 11.3.2 be satisfied and $b_{ij}$ be given by (11.3.3). Then*

$$
\begin{aligned}
&\|b_{11}\|_{X^{s,r}} + \|b_{10}\|_{X^{s,r}} + \|b_{20}\|_{X^{s,r}} + \|b_{31}\|_{X^{s,r}} + \|b_{34}\|_{X^{s,r}}\\
&\qquad\qquad + \|b_{30}\|_{X^{s,r}} + \|b_{22}\|_{X^{s,r}} + \|b_{32}\|_{X^{s,r}} \le c\tau, \qquad (11.3.38)\\
&\|b_{13}\|_{X^{s,r}} + \|b_{23}\|_{X^{s,r}} + \|b_{21}\|_{X^{s,r}} \le c,
\end{aligned}
$$

*where $c$ depends on $\Omega$, $\mathbf{U}$, $s$, $r$, and $\sigma$.*

*Proof.* It follows from (11.3.2) that

$$
\|\mathfrak{g}_0 - 1\|_{C^2(\Omega)} \le c\tau^2. \qquad (11.3.39)
$$

Next, in view of Condition 11.3.2, the functions $\vartheta_i \in \mathcal{B}_\tau$, and $\sigma > \sigma^*$, $\tau \in (0, \tau^*(\sigma)]$ meet all requirements of Lemma 11.2.16. Hence we obtain the following estimates for the functions $\Psi$ and $\Psi_1$ given by (11.1.24d):

$$
\|\Psi[\vartheta]\|_{X^{s,r}} \le c\sigma\tau^2, \quad \|\Psi_1[\vartheta]\|_{X^{s,r}} \le c\sigma\tau^2. \qquad (11.3.40)
$$

Next notice that the function $H$ defined by (11.1.24d) has the derivative $H'(\varphi) = p'(\varrho_\star + \varphi) - p'(\varrho_\star)$ which belongs to $C^2(\mathbb{R})$ and vanishes at 0. Hence $\Phi(\varphi_0, \varphi_1)$, defined in (11.3.3), also belongs to $C^2(\mathbb{R}^2)$ and vanishes at 0. Thus we get

$$\Phi(\varphi_0, \varphi_1) = \varphi_0 \Phi_0(\varphi_0, \varphi_1) + \varphi_1 \Phi_1(\varphi_0, \varphi_1),$$

where $\Phi_i \in C^1(\mathbb{R}^2)$, $i = 0, 1$. Since $\vartheta_i = (\mathbf{v}_i, \pi_i, \varphi_i)$, $i = 0, 1$, belong to the ball $\mathcal{B}_\tau$, we have

$$\|\varphi_i\|_{X^{s,r}} \le \tau. \tag{11.3.41}$$

It follows from this and Lemmas 11.2.7 and 11.2.8 that

$$\begin{aligned} \|\Phi(\varphi_0, \varphi_1)\|_{X^{s,r}} \le c\|\varphi_0\|_{X^{s,r}} \|\Phi_0(\varphi_0, \varphi_1)\|_{X^{s,r}} \\ + c\|\varphi_1\|_{X^{s,r}} \|\Phi_1(\varphi_0, \varphi_1)\|_{X^{s,r}} \le c\tau. \end{aligned} \tag{11.3.42}$$

Finally note that by Condition 11.3.2, $m_i$ and $\zeta_i$, $i = 0, 1$, satisfy all requirements of Theorem 11.2.6 and hence satisfy

$$\|\zeta_i\|_{X^{s,r}} + |\varkappa_0| \le c, \quad |m_i| \le c\tau. \tag{11.3.43}$$

Applying Lemma 11.2.7 to formulae (11.3.3) for the coefficients $b_{kl}$ and using the estimates (11.3.39)–(11.3.43) we obtain

$$\|b_{11}\|_{X^{s,r}} + \|b_{10}\|_{X^{s,r}} + \|b_{20}\|_{X^{s,r}} + \|b_{31}\|_{X^{s,r}} + \|b_{34}\|_{X^{s,r}} + \|b_{30}\|_{X^{s,r}} \le c\tau. \tag{11.3.44}$$

Since $\lambda^{-1} \in (0, \tau^2]$, we have

$$\|b_{22}\|_{X^{s,r}} + \|b_{32}\|_{X^{s,r}} \le c\lambda^{-1} \le c\tau. \tag{11.3.45}$$

On the other hand, it follows from (11.3.3) that the coefficients $b_{13}$, $b_{21}$, $b_{23}$ are not small and admit the estimate

$$\|b_{13}\|_{X^{s,r}} + \|b_{21}\|_{X^{s,r}} + \|b_{23}\|_{X^{s,r}} \le c.$$

Combining this with (11.3.44) and (11.3.45) we arrive at (11.3.38). □

**Step 3. Truncated problem.** It follows from Lemmas 11.3.7–11.3.11 that some terms in equations (11.3.6) are small while $\tau$ is small. If we neglect these terms, we come to the following truncated problem:

$$\begin{aligned} &\Delta \mathbf{h} - \nabla g = \mathbf{H} - \zeta_1 \nabla \upsilon \quad \text{in } \Omega, \\ &\operatorname{div} \mathbf{h} = \Pi G \quad \text{in } \Omega, \end{aligned} \tag{11.3.46a}$$

$$\begin{aligned} &-\mathbf{u}_0 \nabla \varsigma + \sigma \varsigma = b_{21} g + F \quad \text{in } \Omega, \\ &\mathbf{u}_0 \nabla \upsilon + \sigma \upsilon = M \quad \text{in } \Omega, \end{aligned} \tag{11.3.46b}$$

$$\begin{aligned} &\mathbf{h} = 0 \quad \text{on } \partial\Omega, \quad \varsigma = 0 \quad \text{on } \Sigma_{\text{out}}, \quad \upsilon = 0 \quad \text{on } \Sigma_{\text{in}}, \\ &g - \Pi g = 0, \quad l = \varkappa_0 \int_\Omega (b_{13} \varsigma + b_{23} g)\, dx + e, \end{aligned} \tag{11.3.46c}$$

where the coefficients $b_{ij}$ and the constant $\varkappa_0$ are defined by (11.3.3).

**Lemma 11.3.12.** *Let Condition 11.3.2 be satisfied. Then there are constants $\sigma_3 \geq \sigma^*$ and $c$, depending only on $\Omega$, $\mathbf{U}$, $s$, $r$, with the following property. If $\sigma > \sigma_3$, then for every $\mathfrak{f} = (\mathbf{H}, G, F, M, e) \in \mathcal{U}^{s,r}$, problem (11.3.6) has a solution $\mathfrak{h} = (\mathbf{h}, g, \varsigma, \upsilon, l) \in \mathcal{V}^{s,r}$ which satisfies*

$$\|\mathfrak{h}\|_{\mathcal{V}^{s,r}} \leq c\|\mathfrak{f}\|_{\mathcal{U}^{s,r}}. \tag{11.3.47}$$

*If, in addition, $\mathfrak{f} \in \mathcal{E}^{s,r}$, then*

$$\|\mathfrak{h}\|_{\mathcal{F}^{s,r}} \leq c\|\mathfrak{f}\|_{\mathcal{E}^{s,r}}. \tag{11.3.48}$$

*Proof.* Observe that the truncated system (11.3.12) is in triangular form. Hence we can solve it step by step starting with the boundary value problem for $\upsilon$. By (11.3.32), we have

$$\|\mathbf{u}_0\|_{C^1(\Omega)} \leq c\|\mathbf{u}_0\|_{Y^{s,r}} \leq c.$$

It follows from this and Condition 11.3.2 that $\Omega$, $\mathbf{u}_0$ and $s$, $r$ meet all requirements of Lemma 11.2.11 and Proposition 11.2.12. Applying Lemma 11.2.11 we find that there is $\sigma_3 > 1$, depending only on $\Omega$, $\mathbf{U}$, and $s$, $r$, with the following property. For every $\sigma > \sigma_3$ and $M \in W^{s,r}(\Omega)$ the boundary value problem

$$\mathbf{u}_0 \nabla \upsilon + \sigma \upsilon = M \quad \text{in } \Omega, \qquad \upsilon = 0 \quad \text{on } \Sigma_{\text{in}}$$

has a solution satisfying

$$\|\upsilon\|_{W^{s,r}(\Omega)} \leq c\|M\|_{W^{s,r}(\Omega)}. \tag{11.3.49}$$

If, in addition, $M \in X^{s,r}$, then by Proposition 11.2.12,

$$\|\upsilon\|_{X^{s,r}} \leq c\|M\|_{X^{s,r}}. \tag{11.3.50}$$

Here $c$ depends only on $\Omega$, $\mathbf{U}$, and $s$, $r$. Obviously we can take $\sigma_3 \geq \sigma^*$.

Let us turn to the Stokes equations (11.3.46a). Notice that in view of Condition (11.3.2), the functions $(\vartheta_i, \zeta_i)$ solve Problem 11.1.9 and meet all requirements of Theorem 11.2.6. In particular, we have $\|\zeta_1\|_{X^{s,r}} \leq c$. Since $X^{s,r} = W^{s,r}(\Omega) \cap W^{1,2}(\Omega)$, we can apply Lemma 11.3.7 to obtain

$$\begin{aligned} \|\zeta_1 \nabla \upsilon\|_{\mathcal{W}^{s-1,r}(\Omega)} &\leq c\|\zeta_1\|_{W^{s,r}(\Omega)}\|\upsilon\|_{W^{s,r}(\Omega)} \leq c\|\zeta_1\|_{X^{s,r}}\|\upsilon\|_{W^{s,r}(\Omega)} \\ &\leq c\|\upsilon\|_{W^{s,r}(\Omega)}. \end{aligned} \tag{11.3.51}$$

Next, since $\|\zeta\|_{C(\Omega)} \leq c\|\zeta_1\|_{X^{s,r}} \leq c$, we have

$$\|\zeta_1 \nabla \upsilon\|_{L^2(\Omega)} \leq c\|\zeta_1\|_{C(\Omega)}\|\nabla \upsilon\|_{L^2(\Omega)} \leq c\|\upsilon\|_{W^{1,2}(\Omega)} \leq c\|\upsilon\|_{X^{s,r}}.$$

Combining this with (11.3.51) and noting that $Z^{s,r} = \mathcal{W}^{1-s,r}(\Omega) \cap L^2(\Omega)$ we arrive at

$$\|\zeta_1 \nabla \upsilon\|_{Z^{s,r}} \leq c\|\upsilon\|_{X^{s,r}}. \tag{11.3.52}$$

Inequalities (11.3.49)–(11.3.52) imply

$$\|\zeta_1 \nabla v\|_{\mathcal{W}^{s-1,r}(\Omega)} \le c\|M\|_{W^{s,r}(\Omega)}, \quad \|\zeta_1 \nabla v\|_{Z^{s,r}} \le c\|M\|_{X^{s,r}}. \tag{11.3.53}$$

In view of Lemma 1.7.9, for $\mathbf{H} - \zeta_1 \nabla v \in \mathcal{W}^{s-1,r}(\Omega)$ and $G \in W^{s,r}(\Omega)$, equations (11.3.46a) have a solution $(\mathbf{h}, g) \in W^{s+1,r}(\Omega) \times W^{s,r}(\Omega)$ satisfying conditions (11.3.46c). This solution satisfies

$$\|\mathbf{h}\|_{W^{s+1,r}(\Omega)} + \|g\|_{W^{s,r}(\Omega)} \le c(\|\mathbf{H} - \zeta_1 \nabla v\|_{\mathcal{W}^{s-1,r}(\Omega)} + \|G\|_{W^{s,r}(\Omega)}).$$

If, in addition, $\mathbf{H} \in Z^{s,r}$ and $G \in X^{s,r}$, then by Lemma 11.2.9,

$$\|\mathbf{h}\|_{Y^{s,r}} + \|g\|_{X^{s,r}} \le c(\Omega, r, s)(\|\mathbf{H} - \zeta_1 \nabla v\|_{Z^{s,r}} + \|G\|_{X^{s,r}(\Omega)}).$$

From this and (11.3.53) we obtain

$$\|\mathbf{h}\|_{W^{s+1,r}(\Omega)} + \|g\|_{W^{s,r}(\Omega)} \le c(\|\mathbf{H}\|_{\mathcal{W}^{s-1,r}(\Omega)} + \|G\|_{W^{s,r}(\Omega)} + \|M\|_{W^{s,r}(\Omega)}), \tag{11.3.54}$$

$$\|\mathbf{h}\|_{Y^{s,r}} + \|g\|_{X^{s,r}} \le c(\|\mathbf{H}\|_{Z^{s,r}} + \|G\|_{X^{s,r}} + \|M\|_{X^{s,r}}). \tag{11.3.55}$$

Now the functions $\mathbf{h}$, $g$, $v$ are defined and we can solve the transport equations (11.3.46b) for $\varsigma$. Notice that, by Condition 11.3.2, $\mathbf{U}$ and $\Omega$ satisfy Condition 11.2.5. In particular, $-\mathbf{U}$ and $\Omega$ also satisfy Condition 11.2.5. Hence $-\mathbf{u}_0$ and $\Omega$ meet all requirements of Lemma 11.2.11 and Proposition 11.2.12 with $\Sigma_{\text{in}}$ replaced by $\Sigma_{\text{out}}$. Hence we can choose $\sigma_3$, depending only on $\Omega$, $\mathbf{U}$, and $s$, $r$, such that for $\sigma > \sigma_3$ and $F, g \in W^{s,r}(\Omega)$ (respectively $F, g \in X^{s,r}$), the boundary value problem

$$-\mathbf{u}_0 \nabla \varsigma + \sigma \varsigma = b_{21} g + F \quad \text{in } \Omega, \quad \varsigma = 0 \quad \text{on } \Sigma_{\text{out}}$$

has a solution satisfying

$$\begin{aligned}\|\varsigma\|_{W^{s,r}(\Omega)} &\le c\|F\|_{W^{s,r}(\Omega)} + c\|b_{21} g\|_{W^{s,r}(\Omega)} \le c\|F\|_{W^{s,r}(\Omega)} + c\|g\|_{W^{s,r}(\Omega)} \\ &\le c(\|F\|_{W^{s,r}(\Omega)} + \|\mathbf{H}\|_{\mathcal{W}^{s-1,r}(\Omega)} + \|G\|_{W^{s,r}(\Omega)} + \|M\|_{W^{s,r}(\Omega)}),\end{aligned} \tag{11.3.56}$$

$$\begin{aligned}\|\varsigma\|_{X^{s,r}} &\le c\|F\|_{X^{s,r}} + c\|b_{21} g\|_{X^{s,r}} \le c\|F\|_{X^{s,r}} + c\|g\|_{X^{s,r}} \\ &\le c(\|\mathbf{H}\|_{Y^{s,r}} + \|F\|_{X^{s,r}} + \|G\|_{X^{s,r}} + \|M\|_{X^{s,r}}).\end{aligned} \tag{11.3.57}$$

Thus we have defined the functions $\mathbf{h}$, $g$, $\varsigma$ and $v$. It remains to note that the constant $l$ is completely determined by $\varsigma$ and $g$. Moreover, relation (11.3.46c) along with estimates (11.3.38) and (11.3.43) yields

$$|l| \le c(\|g\|_{L^1(\Omega)} + \|\varsigma\|_{L^1(\Omega)}).$$

Combining this with estimates (11.3.49), (11.3.50) and (11.3.54)–(11.3.57) we obtain (11.3.47) and (11.3.48). □

**Step 4. Proof of Theorem 11.3.4.** We prove Theorem 11.3.4 by applying the contraction mapping principle. To this end we write the equations and boundary conditions (11.3.6) in the form of an operator equation. Introduce the linear differential operator

$$\mathfrak{V}: \begin{pmatrix} \mathbf{h} \\ g \\ \varsigma \\ \upsilon \\ l \end{pmatrix} \mapsto \begin{pmatrix} \mathscr{F}(\mathbf{h}) + \mathsf{Re}\,\mathscr{H}(\mathbf{h}) - \varsigma\nabla\varphi \\ \Pi(b_{22}g + b_{12}\varsigma + b_{32}l) \\ \mathsf{Re}\,\mathscr{M}(\mathbf{h}) + b_{11}\varsigma + b_{31}l \\ b_{34}\,l \\ 0 \end{pmatrix}.$$

**Lemma 11.3.13.** *Let Condition 11.3.2 be satisfied. Then there is a constant $c$, depending only on $\Omega$, $\mathbf{U}$, and $s$, $r$, such that for $\mathfrak{h} = (\mathbf{h}, g, \varsigma, \upsilon, l)$,*

$$\|\mathfrak{V}\,\mathfrak{h}\|_{\mathcal{U}^{s,r}} \leq c\tau\|\mathbf{h}\|_{\mathcal{V}^{s,r}(\Omega)}, \tag{11.3.58}$$

$$\|\mathfrak{V}\,\mathfrak{h}\|_{\mathcal{E}^{s,r}} \leq c\tau\|\mathfrak{h}\|_{\mathcal{F}^{s,r}}. \tag{11.3.59}$$

*Proof.* Let us prove (11.3.58). Set

$$\overline{\mathfrak{h}} = (\overline{\mathbf{h}}, \overline{g}, \overline{\varsigma}, \overline{\upsilon}, \overline{l}) = \mathfrak{V}\mathfrak{h}.$$

Our first task is to estimate $\overline{\mathbf{h}}$. By Condition 11.3.2, we have $\vartheta_i \in \mathcal{B}_\tau$ and hence $\|\varphi_1\|_{X^{s,r}} \leq \tau$. Since the embeddings $X^{s,r} \hookrightarrow W^{s,r}(\Omega) \hookrightarrow C(\Omega)$ are continuous,

$$\|\varphi_1\|_{W^{s,r}(\Omega)} \leq \|\varphi_1\|_{X^{s,r}} \leq c\tau \quad \text{and} \quad \|\varsigma\|_{C(\Omega)} \leq \|\varsigma\|_{X^{s,r}}. \tag{11.3.60}$$

Application of Lemma 11.3.7 gives

$$\|\varsigma\nabla\varphi_1\|_{\mathcal{W}^{s-1,r}(\Omega)} \leq c\|\varsigma\|_{W^{s,r}(\Omega)}\|\varphi_1\|_{W^{s,r}(\Omega)} \leq c\tau\|\varsigma\|_{W^{s,r}(\Omega)}. \tag{11.3.61}$$

Since $X^{s,r} = W^{s,r}(\Omega) \cap W^{1,2}(\Omega)$, inequality (11.3.60) yields

$$\|\varsigma\nabla\varphi_1\|_{L^2(\Omega)} \leq c\|\varsigma\|_{C(\Omega)}\|\nabla\varphi_1\|_{L^2(\Omega)} \leq c\|\varphi_1\|_{X^{s,r}}\|\varsigma\|_{X^{s,r}} \leq c\tau\|\varsigma\|_{X^{s,r}}.$$

Combining this with (11.3.61) and noting that $Z^{s,r} = \mathcal{W}^{1-s,r}(\Omega) \cap L^2(\Omega)$ we arrive at

$$\|\varsigma\nabla\varphi_1\|_{Z^{s,r}} \leq c\|\varsigma\|_{X^{s,r}}. \tag{11.3.62}$$

Since $\mathsf{Re} \in (0, \tau^2]$ and $\tau \leq 1$ estimates (11.3.22), (11.3.27) and (11.3.61) imply

$$\begin{aligned} \|\overline{\mathbf{h}}\|_{\mathcal{W}^{s-1,r}(\Omega)} &\leq \|\mathscr{F}(\mathbf{h})\|_{\mathcal{W}^{s-1,r}(\Omega)} + \mathsf{Re}\,\|\mathscr{H}(\mathbf{h})\|_{\mathcal{W}^{s-1,r}(\Omega)} + \|\varsigma\nabla\varphi_1\|_{\mathcal{W}^{s-1,r}(\Omega)} \\ &\leq c\tau\|\mathbf{h}\|_{W^{s+1,r}(\Omega)} + c\tau\|\varsigma\|_{W^{s,r}(\Omega)}. \end{aligned} \tag{11.3.63}$$

Finally, estimates (11.3.23), (11.3.28) and (11.3.62) yield

$$\begin{aligned} \|\overline{\mathbf{h}}\|_{Z^{s,r}} &\leq \|\mathscr{F}(\mathbf{h})\|_{Z^{s,r}} + \mathsf{Re}\,\|\mathscr{H}(\mathbf{h})\|_{Z^{s,r}} + \|\varsigma\nabla\varphi_1\|_{Z^{s,r}} \\ &\leq c\tau\|\mathbf{h}\|_{Y^{s,r}} + c\tau\|\varsigma\|_{X^{s,r}}. \end{aligned} \tag{11.3.64}$$

Let us estimate $\overline{g}$. Recalling that $W^{s,r}(\Omega)$ and $X^{s,r}$ are Banach algebras and applying Lemma 11.3.11 we obtain

$$\begin{aligned}\|\overline{g}\|_{W^{s,r}(\Omega)} &\leq c(\|b_{22}\|_{W^{s,r}(\Omega)}\|g\|_{W^{s,r}(\Omega)} + \|b_{12}\|_{W^{s,r}(\Omega)}\|\varsigma\|_{W^{s,r}(\Omega)} + |l|\|b_{22}\|_{W^{s,r}(\Omega)}) \\ &\leq c\tau(\|g\|_{W^{s,r}(\Omega)} + \|\varsigma\|_{W^{s,r}(\Omega)} + |l|)\end{aligned} \tag{11.3.65}$$

and

$$\begin{aligned}\|\overline{g}\|_{X^{s,r}} &\leq c(\|b_{22}\|_{X^{s,r}}\|g\|_{X^{s,r}} + \|b_{12}\|_{X^{s,r}}\|\varsigma\|_{X^{s,r}} + |l|\,\|b_{22}\|_{X^{s,r}}) \\ &\leq c\tau(\|g\|_{X^{s,r}} + \|\varsigma\|_{X^{s,r}} + |l|).\end{aligned} \tag{11.3.66}$$

Next applying Lemmas 11.3.10 and 11.3.11 we get

$$\begin{aligned}\|\overline{\varsigma}\|_{W^{s,r}(\Omega)} &\leq c(\operatorname{Re}\|\mathscr{M}(\mathbf{h})\|_{W^{s,r}(\Omega)} + \|b_{11}\|_{W^{s,r}(\Omega)}\|\varsigma\|_{W^{s,r}(\Omega)} + |l|\,\|b_{31}\|_{W^{s,r}(\Omega)}) \\ &\leq c\tau(\|\mathbf{h}\|_{W^{s+1,r}(\Omega)} + \|\varsigma\|_{W^{s,r}(\Omega)} + |l|)\end{aligned} \tag{11.3.67}$$

and

$$\begin{aligned}\|\overline{\varsigma}\|_{X^{s,r}} &\leq c(\operatorname{Re}\|\mathscr{M}(\mathbf{h})\|_{X^{s,r}} + \|b_{11}\|_{X^{s,r}}\|\varsigma\|_{X^{s,r}} + |l|\,\|b_{31}\|_{X^{s,r}}) \\ &\leq c\tau(\|\mathbf{h}\|_{Y^{s,r}} + \|\varsigma\|_{X^{s,r}} + |l|).\end{aligned} \tag{11.3.68}$$

Finally, from Lemma 11.3.11 we obtain

$$\begin{aligned}\|\overline{v}\|_{W^{s,r}(\Omega)} &\leq c|l|\,\|b_{34}\|_{W^{s,r}(\Omega)} \leq c\tau|l|, \\ \|\overline{v}\|_{X^{s,r}} &\leq c|l|\,\|b_{34}\|_{X^{s,r}} \leq c\tau|l|.\end{aligned} \tag{11.3.69}$$

Combining estimates (11.3.63)–(11.3.69) we obtain (11.3.58) and (11.3.59). □

We are now in a position to complete the proof of Theorem 11.3.4. Set $\sigma_c = \min\{\sigma^*, \sigma_3\}$, where $\sigma^*$ is a constant from Condition 11.3.2 and $\sigma_3$ is given by Lemma 11.3.12. Let $\sigma$ satisfy Condition 11.3.2 and $\sigma > \sigma_c$. Denote by $\mathfrak{U}$ the mapping which assigns to every vector $(\mathbf{H}, G, F, M, l)$ the solution $(\mathbf{h}, g, \varsigma, v, l)$ of the truncated problem (11.3.46). Thus the transposed problem (11.3.6) can be written in the form of the operator equation

$$(\mathrm{Id} - \mathfrak{U}\mathfrak{V})\,\mathfrak{h} = \mathfrak{U}\,\mathfrak{f}. \tag{11.3.70}$$

In view of Lemmas 11.3.12 and 11.3.13, the operators

$$\mathfrak{U} : \mathcal{U}^{s,r} \to \mathcal{V}^{s,r}, \quad \mathfrak{U}\mathfrak{V} : \mathcal{V}^{s,r} \to \mathcal{V}^{s,r}$$

are continuous. Moreover, inequalities (11.3.47) and (11.3.58) imply

$$\|\mathfrak{U}\|_{\mathscr{L}(\mathcal{U}^{s,r}\to\mathcal{V}^{s,r})} \leq c_U, \quad \|\mathfrak{U}\mathfrak{V}\|_{\mathscr{L}(\mathcal{V}^{s,r}\to\mathcal{V}^{s,r})} \leq c_V\tau. \tag{11.3.71}$$

Next, it also follows from Lemmas 11.3.12 and 11.3.13 that the operators

$$\mathfrak{U} : \mathcal{E}^{s,r} \to \mathcal{F}^{s,r}, \quad \mathfrak{U}\mathfrak{V} : \mathcal{F}^{s,r} \to \mathcal{F}^{s,r}$$

are continuous. Inequalities (11.3.48) and (11.3.59) imply

$$\|\mathfrak{U}\|_{\mathscr{L}(\mathcal{E}^{s,r}\to\mathcal{F}^{s,r})}\leq c_E,\quad \|\mathfrak{U}\mathfrak{V}\|_{\mathscr{L}(\mathcal{F}^{s,r}\to\mathcal{F}^{s,r})}\leq c_F\tau. \tag{11.3.72}$$

The constants $c_U$, $c_V$, $c_E$, and $c_F$ are completely determined by the constant $c$ in inequalities (11.3.47), (11.3.58), (11.3.59) and depend only on $\Omega$, $U$, $s$, $r$, and $\sigma$. Notice that, by Condition 11.3.2, the parameter $\tau$ satisfies $0<\tau\leq\tau^*(\sigma)$ and $\tau\leq\tau_0$, where $\tau^*(\sigma)$ depends on $\Omega$, $U$, $s$, $r$, and $\sigma$, while $\tau_0$ is an absolute constant. Now choose a constant $\tau_c$ such that

$$0<\tau_c\leq\min\{\tau_0,\tau^*(\sigma)\},\quad \tau_c c_V+\tau_c c_F<1/2.$$

It follows from (11.3.71), (11.3.72) that for $\tau\in(0,\tau_c]$ the norms $\|\mathfrak{U}\mathfrak{V}\|_{\mathscr{L}(\mathcal{V}^{s,r}\to\mathcal{V}^{s,r})}$ and$\|\mathfrak{U}\mathfrak{V}\|_{\mathscr{L}(\mathcal{F}^{s,r}\to\mathcal{F}^{s,r})}$ are less than 1/2. By the contraction mapping principle, for such $\tau$ and any $\mathfrak{f}\in\mathcal{U}^{s,r}$ (resp. $\mathfrak{f}\in\mathcal{E}^{s,r}$), the operator equation (11.3.70) has a unique solution $\mathfrak{h}\in\mathcal{V}^{s,r}$ (resp. $\mathfrak{h}\in\mathcal{F}^{s,r}$). This solution satisfies the estimates (11.3.58), (11.3.59). It remains to note that the operator equation (11.3.70) is equivalent to the transposed problem (11.3.6).

### 11.3.3 Special dual space

Recall that our goal is to establish well-posedness for the linear boundary value problem (11.3.2) in dual spaces. In this section we consider in detail the properties of the Banach space dual to $W^{s,r}(\Omega)$.

For every $1<r<\infty$ and $v\in L^r(\Omega)$ define the functional

$$\mathcal{L}_v(u)=\langle v,u\rangle=\int_\Omega v(x)u(x)\,dx,\quad u\in L^{r'}(\Omega),\quad r'=r/(r-1).$$

By the Hölder inequality, $\mathcal{L}_v\in L^{r'}(\Omega)'$. Moreover, in view of Theorem 1.2.27, $\|\mathcal{L}_v\|_{L^{r'}(\Omega)'}=\|v\|_{L^r(\Omega)}$, and the mapping $v\mapsto\mathcal{L}_v$ defines an isometry of $L^r(\Omega)$ onto $L^{r'}(\Omega)'$. We will identify $L^r(\Omega)$ and $L^{r'}(\Omega)'$. Since $W^{s,r'}(\Omega)$, $s>0$, is continuously embedded into $L^{r'}(\Omega)$, $\mathcal{L}_v$ is a continuous functional on $W^{s,r'}(\Omega)$ with the finite norm

$$\|\mathcal{L}_v\|_{(W^{s,r'}(\Omega))'}:=\sup_{u\in W^{s,r'}(\Omega)\setminus\{0\}}\frac{|\mathcal{L}_v(u)|}{\|u\|_{W^{s,r'}(\Omega)}}.$$

**Definition 11.3.14.** For every $s\in(0,1]$ and $r\in(1,\infty)$, the Banach space $\mathbb{W}^{-s,r}(\Omega)$ is the completion of $L^r(\Omega)$ in the norm

$$\|v\|_{\mathbb{W}^{-s,r}(\Omega)}=\|\mathcal{L}_v\|_{(W^{s,r'}(\Omega))'}.$$

**Lemma 11.3.15.** *Let $\Omega$ be a bounded domain in $\mathbb{R}^d$ with $C^1$ boundary and $s\in(0,1]$, $r\in(1,\infty)$. Then $\mathbb{W}^{-s,r}(\Omega)$ is algebraically and topologically isomorphic to the* *dual space* $(W^{s,r}(\Omega))'$ *and can be identified with it. Moreover,* $\mathbb{W}^{-s,r}(\Omega)$ *is the interpolation space*

$$\mathbb{W}^{-s,r}(\Omega) = [L^r(\Omega), \mathbb{W}^{-1,r}(\Omega)]_{s,r} = ([L^{r'}(\Omega), W^{1,r'}(\Omega)]_{s,r'})'. \tag{11.3.73}$$

*Proof.* By definition, $\mathbb{W}^{-s,r}(\Omega)$ is the completion of the linear space $\mathfrak{L} = L^{r'}(\Omega)'$ of functionals $\mathcal{L}_v$, $v \in L^r(\Omega)$, in the $(W^{s,r'}(\Omega))'$-norm. Hence $\mathbb{W}^{-s,r}(\Omega)$ is a closed subspace of $(W^{s,r'}(\Omega))'$. To prove that $\mathbb{W}^{-s,r}(\Omega)$ coincides with $(W^{s,r'}(\Omega))'$, it suffices to show that $\mathfrak{L}$ is dense in $(W^{s,r'}(\Omega))'$.

We start with the case $s = 1$. If $\mathfrak{L}$ is not dense in $(W^{1,r'}(\Omega))'$, then there exists $\mathcal{T} \in (W^{1,r'}(\Omega))''$ such that $\mathcal{T} \neq 0$ and $\mathcal{T}(\mathcal{L}_v) = 0$ for all $v \in L^r(\Omega)$. Since the space $W^{1,r}(\Omega)$ is reflexive (see [2, Thm. 3.5]), there is $\varphi \in W^{1,r}(\Omega)$ such that

$$\|\varphi\|_{W^{1,r}(\Omega)} = \|\mathcal{T}\|_{(W^{1,r'}(\Omega))''} \neq 0 \quad \text{and} \quad \mathcal{T}(f) = f(\varphi) \quad \text{for all } f \in (W^{1,r'}(\Omega))'. \tag{11.3.74}$$

Thus we get

$$\mathcal{T}(\mathcal{L}_v) = \mathcal{L}_v(\varphi) = \int_\Omega v\varphi\, dx = 0 \quad \text{for all } v \in L^r(\Omega).$$

Hence $\varphi = 0$, which contradicts (11.3.74). It follows that $\mathfrak{L}$ is dense in $(W^{1,r'}(\Omega))'$. Hence $\mathbb{W}^{-1,r}(\Omega) = (W^{1,r'}(\Omega))'$.

Let us turn to the case $s \in (0,1)$. Notice that $W^{1,r'}(\Omega)$ is dense in $L^{r'}(\Omega)$. Therefore, we can apply Lemma 1.1.12 to obtain

$$([L^{r'}(\Omega), W^{1,r'}(\Omega)]_{s,r'})' = [L^{r'}(\Omega)', (W^{1,r'}(\Omega))']_{s,r}. \tag{11.3.75}$$

Since, $[L^{r'}(\Omega), W^{1,r'}(\Omega)]_{s,r'} = W^{s,r'}(\Omega)$, we have

$$(W^{s,r'}(\Omega))' = [L^{r'}(\Omega)', (W^{1,r'}(\Omega))']_{s,r}. \tag{11.3.76}$$

Next, the inclusion $W^{1,r'}(\Omega) \subset L^{r'}(\Omega)$ implies $L^{r'}(\Omega)' \subset (W^{1,r'}(\Omega))'$. From this and Lemma 1.1.11 we find that $\mathfrak{L} = L^{r'}(\Omega)'$ is dense in $[L^{r'}(\Omega)', (W^{1,r'}(\Omega))']_{s,r}$ and hence in $(W^{s,r'}(\Omega))'$. Thus $(W^{s,r'}(\Omega))' = \mathbb{W}^{-s,r}(\Omega)$. It remains to note that relations (11.3.73) follow from (11.3.75) and (11.3.76). □

### 11.3.4 Very weak solutions

This step is crucial for further analysis. We replace equations (11.3.2) by an integral identity, which leads to the notion of a very weak solution to (11.3.2). For $s \in (0,1)$, $r \in (1,\infty)$, denote by $\langle \cdot, \cdot \rangle_1$ the duality pairing between $\mathcal{W}^{s-1,r}(\Omega)$ and $\mathcal{W}_0^{1-s,r'}(\Omega)$, and by $\langle \cdot, \cdot \rangle_0$ the duality pairing between $\mathbb{W}^{-s,r'}(\Omega)$ and $W^{s,r}(\Omega)$.

**Definition 11.3.16.** Let Condition 11.3.2 be satisfied and $\sigma > \sigma_c$, $\tau \in (0, \tau_c(\sigma)]$, where $\sigma_c$ and $\tau_c(\sigma)$ are given by Theorem 11.3.4. The vector field $\mathbf{w} \in \mathcal{W}_0^{1-s,r'}(\Omega)$, functionals $(\omega, \psi, \xi) \in \mathbb{W}^{-s,r'}(\Omega)^3$ and constant $n$ are said to be a *very weak solution* to problem (11.3.2) if whenever $(\mathbf{H}, G, F, M, e) \in \mathcal{W}^{s-1,r}(\Omega) \times (W^{s,r}(\Omega))^3 \times \mathbb{R}$,

$$\langle \mathbf{H}, \mathbf{w}\rangle_1 + \langle \omega, G\rangle_0 + \langle \psi, F\rangle_0 + \langle \xi, \mathcal{M}\rangle_0 + \varkappa_0^{-1} ne \\ = \int_\Omega \big(\mathbf{h} \cdot \mathscr{D} + g\, b_{20}\mathfrak{d} + \varsigma\, b_{10}\mathfrak{d} + \upsilon\, \sigma \mathfrak{d} + l\, b_{30}\mathfrak{d}\big)\, dx, \quad (11.3.77)$$

where $(\mathbf{h}, g, \varsigma, \upsilon, l)$ is the solution to the transposed problem (11.3.6).

**Condition 11.3.17.**

- The $C^3$ surface $\partial\Omega$ and the vector field $\mathbf{U} \in C^3(\Omega)$ satisfy Condition 11.2.5.
- The exponents $s$ and $r$ satisfy

$$1/2 < s < 1, \quad r < \infty, \quad 2s - 3r^{-1} < 1, \quad sr > 3, \quad (1-s)r > 3.$$

- $\sigma_c > 1$ and $\tau_c(\sigma) \in (0, 1]$ satisfy the hypotheses of of Theorem 11.3.4. In particular $\tau_c(\sigma) \le \tau_0$, where $\tau_0$ is given by Lemma 11.2.15.
- $\sigma > \sigma_c$ and $\tau \in (0, \tau_c(\sigma)]$.
- We have

$$\lambda^{-1}, \mathsf{Re} \in (0, \tau^2], \quad \|\mathbf{N}_i - \mathbb{I}\|_{C^2(\Omega)} \le \tau^2.$$

- $\vartheta_i = (\mathbf{v}_i, \pi_i, \varphi_i) \in \mathcal{B}_\tau$, $\zeta_i \in W^{s,r}(\Omega)$, $m_i \in \mathbb{R}$, $i = 0, 1$, are solutions to Problem 11.1.9 and satisfy inequalities (11.2.7).
- $\mathbf{u}_i = \mathbf{u}_\star + \mathbf{v}_i$ and $\varrho_i = \varrho_\star + \varphi_i$.

We set

$$\mathbf{w} = \mathbf{v}_0 - \mathbf{v}_1, \quad \omega = \pi_0 - \pi_1,$$
$$\psi = \varphi_0 - \varphi_1, \quad \xi = \zeta_0 - \zeta_1, \quad n = m_0 - m_1.$$

The following theorem is the main result of this section.

**Theorem 11.3.18.** *Assume that Condition 11.3.17 is satisfied. Then* $(\mathbf{w}, \omega, \psi, \xi, n) \in \mathcal{W}_0^{1-s,r'}(\Omega) \times \mathbb{W}^{-s,r'}(\Omega)^3 \times \mathbb{R}$ *is a very weak solution to problem* (11.3.2) *and*

$$\|\mathbf{w}\|_{\mathcal{W}_0^{1-s,r'}(\Omega)} + \|\omega\|_{\mathbb{W}^{-s,r'}(\Omega)} + \|\psi\|_{\mathbb{W}^{-s,r'}(\Omega)} + \|\xi\|_{\mathbb{W}^{-s,r'}(\Omega)} + |n| \\ \le c(\|\mathscr{D}\|_{L^1(\Omega)} + \|\mathfrak{d}\|_{L^1(\Omega)}), \quad (11.3.78)$$

*where $c$ depends on $\Omega$, $\mathbf{U}$, $s$, $r$, and $\sigma$.*

*Proof.* The proof falls into two steps.

**Step 1.** First we deduce the Green formula for problem (11.3.2). Observe that under the assumptions of Theorem 11.3.18 the embeddings

$$Y^{s,r} \hookrightarrow W^{s+1,r}(\Omega) \hookrightarrow C^1(\Omega), \quad X^{s,r} \hookrightarrow W^{s,r}(\Omega) \hookrightarrow C(\Omega) \tag{11.3.79}$$

are continuous. Hence $\mathbf{u}_i, \mathbf{v}_i, \mathbf{w} \in Y^{s,r} = W^{s+1,r}(\Omega) \cap W^{2,2}(\Omega)$ are continuously differentiable and belong to $W^{2,2}(\Omega)$. In its turn, the functions $(\varrho_i, \pi_i, \varphi_i, \zeta_i)$ and $(\omega, \psi, \xi)$ belong to $X^{s,r}$. Hence they are continuous and belong to $W^{1,2}(\Omega)$. It follows that equations (11.3.2) are satisfied in the strong sense. Next choose $\mathbf{h} \in Y^{s,r}$, $(g, \varsigma, \upsilon) \in X^{s,r}$, and $l \in \mathbb{R}$ such that

$$\mathbf{h} = 0 \quad \text{on } \partial\Omega, \quad \varsigma = 0 \quad \text{on } \Sigma_{\text{out}}, \quad \upsilon = 0 \quad \text{on } \Sigma_{\text{in}}.$$

Notice that $\mathbf{h}$ is continuously differentiable and belongs to $W^{2,2}(\Omega)$, while $g$, $\varsigma$, and $\upsilon$ are continuous and belong to $W^{1,2}(\Omega)$. Multiplying the Stokes equations (11.3.2b) by $\mathbf{h}$ and $g$ respectively, integrating the result by parts over $\Omega$, and combining the integral identities obtained we arrive at

$$\begin{aligned}\int_\Omega \mathbf{w} \cdot (\Delta \mathbf{h} - \nabla g)\, dx + \int_\Omega \omega \operatorname{div} \mathbf{h} = \int_\Omega \mathbf{h} \cdot \big(\mathscr{A}_0(\mathbf{w}) + \mathsf{Re}\, \mathscr{C}(\psi, \mathbf{w}) + \mathscr{D}\big)\, dx \\ + \int_\Omega g(b_{21}\psi + b_{22}\omega + b_{23}n + b_{20}\mathfrak{d})\, dx.\end{aligned} \tag{11.3.80}$$

It follows from (11.3.4) and (11.3.7) that

$$\int_\Omega \mathbf{h} \cdot (\mathscr{A}_0(\mathbf{w}) + \mathsf{Re}\, \mathscr{C}(\psi, \mathbf{w}))\, dx = \int_\Omega \mathbf{w} \cdot (\mathscr{F}(h) + \mathsf{Re}\, \mathscr{H}(\mathbf{h}))\, dx + \int_\Omega \psi\, \mathsf{Re}\, \mathscr{M}(\mathbf{h})\, dx.$$

Inserting this into (11.3.80) we obtain

$$\begin{aligned}\int_\Omega \mathbf{w} \cdot \big(\Delta \mathbf{h} - \nabla g - \mathscr{F}(h) - \mathsf{Re}\, \mathscr{H}(\mathbf{h})\big)\, dx + \int_\Omega \omega(\operatorname{div} \mathbf{h} - b_{22} g)\, dx \\ - \int_\Omega \psi(\mathsf{Re}\, \mathscr{M}(\mathbf{h}) + b_{21} g)\, dx - \int_\Omega (n b_{23} g + b_{20}\mathfrak{d}\, g)\, dx = \int_\Omega \mathbf{h} \cdot \mathscr{D}.\end{aligned} \tag{11.3.81}$$

Next, multiplying the transport equation (11.3.2a) by $\varsigma$, integrating the result by parts over $\Omega$, and noting that $\psi\varsigma$ vanishes at $\partial\Omega$ we arrive at

$$\begin{aligned}\int_\Omega \psi\big(-\operatorname{div}(\mathbf{u}_0 \varsigma) + \sigma\varsigma - b_{11}\varsigma\big)\, dx + \int_\Omega \mathbf{w} \cdot (\varsigma \nabla \varphi_1)\, dx \\ - \int_\Omega (\omega b_{12}\varsigma + n b_{13}\varsigma + b_{10}\mathfrak{d}\varsigma)\, dx = 0.\end{aligned} \tag{11.3.82}$$

Multiplying the transport equation (11.3.2c) by $\upsilon$, integrating the result by parts over $\Omega$, and noting that $\xi\upsilon$ vanishes at $\partial\Omega$ we obtain

$$\int_\Omega \xi\big(\mathbf{u}_0 \nabla \upsilon + \sigma \upsilon\big)\, dx + \int_\Omega \mathbf{w} \cdot (\zeta_1 \nabla \upsilon)\, dx - \int_\Omega \sigma \mathfrak{d} \upsilon\, dx = 0. \tag{11.3.83}$$

Finally, multiplying the formula for $n$ in (11.3.2d) by $\varkappa_0^{-1}(\operatorname{meas}\Omega)^{-1}l$ and integrating over $\Omega$, we arrive at

$$\varkappa_0^{-1}nl - \int_\Omega (\psi b_{31}l + \omega b_{32}l + \xi b_{34}l + b_{30}\mathfrak{d}l)\,dx = 0. \tag{11.3.84}$$

Adding (11.3.81)–(11.3.84) we arrive at the Green formula for problem (11.3.2):

$$\begin{aligned}
&\int_\Omega \mathbf{w}\cdot\big(\Delta\mathbf{h} - \nabla g - \mathscr{F}(h) - \operatorname{Re}\mathscr{H}(\mathbf{h}) + \varsigma\nabla\varphi_1 + \zeta_1\nabla\upsilon\big)\,dx \\
&\qquad + \int_\Omega \omega(\operatorname{div}\mathbf{h} - b_{22}g - b_{12}\varsigma - b_{32}l)\,dx \\
&\qquad + \int_\Omega \psi\big(-\operatorname{div}(\mathbf{u}_0\varsigma) + \sigma\varsigma - \operatorname{Re}\mathscr{M}(\mathbf{h}) - b_{21}g - b_{11}\varsigma - b_{31}l\big)\,dx \\
&\qquad + \int_\Omega \xi\big(\mathbf{u}_0\nabla\upsilon + \sigma\upsilon - b_{34}l\big)\,dx + n\Big(\varkappa_0^{-1}l - \int_\Omega (b_{23}g + b_{13}\varsigma)\,dx\Big) \\
&\qquad\qquad = \int_\Omega (\mathbf{h}\cdot\mathscr{D} + gb_{20}\mathfrak{d} + \varsigma b_{10}\mathfrak{d} + \upsilon\sigma\mathfrak{d} + lb_{30}\mathfrak{d})\,dx.
\end{aligned} \tag{11.3.85}$$

Notice that $\mathbf{h}$ is continuously differentiable and vanishes on $\partial\Omega$, hence $\Pi\operatorname{div}\mathbf{h} = \operatorname{div}\mathbf{h}$. On the other hand, in view of (11.3.2d), we have $\Pi\omega = \omega$. From this and the symmetry of the operator $\Pi$ given by (1.7.6) we conclude that

$$\begin{aligned}
&\int_\Omega \omega(\operatorname{div}\mathbf{h} - b_{22}g - b_{12}\varsigma - b_{32}l)\,dx \\
&\qquad\qquad = \int_\Omega \omega\big(\operatorname{div}\mathbf{h} - \Pi(b_{22}g + b_{12}\varsigma + b_{32}l)\big)\,dx.
\end{aligned} \tag{11.3.86}$$

**Step 2.** Recall the notation for the Banach spaces

$$\mathcal{U}^{s,r} = \mathcal{W}^{s-1,r}(\Omega)\times(W^{s,r}(\Omega))^3\times\mathbb{R}, \quad \mathcal{V}^{s,r} = W^{s+1,r}(\Omega)\times(W^{s,r}(\Omega))^3\times\mathbb{R},$$
$$\mathcal{E}^{s,r} = Z^{s,r}\times(X^{s,r})^3\times\mathbb{R}, \quad \mathcal{F}^{s,r} = Y^{s,r}\times(X^{s,r})^3\times\mathbb{R}.$$

Introduce also the temporary notation

$$\mathfrak{f} = (\mathbf{H}, G, F, M, e), \quad \mathfrak{h} = (\mathbf{h}, g, \varsigma, \upsilon, l).$$

Now our task is to prove that equality (11.3.77) holds for every $\mathfrak{f}\in\mathcal{U}^{s,r}$. The proof is based on the following

**Lemma 11.3.19.** *Under the assumptions of Theorem 11.3.18,*

$$L^r(\Omega)\times(W^{1,r}(\Omega))^3\times\mathbb{R} \quad \textit{is contained in } \mathcal{E}^{s,r} \textit{ and is dense in } \mathcal{U}^{s,r}.$$

*Proof.* Since $r > 2$, we have the inclusions

$$L^r(\Omega) \subset \mathcal{W}^{s-1,r}(\Omega) \cap L^2(\Omega) = Z^{s,r} \quad \text{and} \quad W^{1,r}(\Omega) \subset W^{s,r}(\Omega) \cap W^{1,2}(\Omega) = X^{s,r}.$$

Therefore, $L^r(\Omega) \times (W^{1,r}(\Omega))^3 \times \mathbb{R}$ is contained in $\mathcal{E}^{s,r}$.

Let us prove that it is dense in $\mathcal{U}^{s,r}$. Notice that $\mathcal{W}^{s-1,r}(\Omega)$ is obtained by completing $L^r(\Omega)$ in the dual norm defined by (1.5.5). Hence $L^r(\Omega)$ is dense in $\mathcal{W}^{s-1,r}(\Omega)$. It remains to prove that $W^{1,r}(\Omega)$ is dense in $W^{s,r}(\Omega)$. By definition, $W^{s,r}(\Omega)$ is the interpolation space $[L^r(\Omega), W^{1,r}(\Omega)]_{s,r}$. Moreover, we have $W^{1,r}(\Omega) \subset L^r(\Omega)$. Hence Lemma 1.1.11 shows that $W^{1,r}(\Omega)$ is dense in $W^{s,r}(\Omega)$. Therefore, $L^r(\Omega) \times (W^{1,r}(\Omega))^3 \times \mathbb{R}$ is dense in $\mathcal{U}^{s,r}$. □

Now choose $\mathfrak{f} = (\mathbf{H}, G, F, M, e) \in \mathcal{U}^{s,r}$. In view of Lemma 11.3.19, there is a sequence

$$\mathfrak{f}_n = (\mathbf{H}_n, G_n, F_n, M_n, e_n) \in L^r(\Omega) \times (W^{1,r}(\Omega))^3 \times \mathbb{R}, \quad n \geq 1,$$

such that

$$\mathfrak{f}_n \to \mathfrak{f} \quad \text{in } \mathcal{U}^{s,r} \quad \text{as } n \to \infty. \tag{11.3.87}$$

It follows from Condition 11.3.17 that the matrices $\mathbf{N}_i$, parameters $\mathsf{Re}$, $\lambda$ and $\sigma$, $\tau$, and functions $\vartheta_i$ satisfy all hypotheses of Theorem 11.3.4. Hence problem (11.3.6) with the given $\mathfrak{f}$ and $\mathfrak{f}_n$ has solutions

$$\mathfrak{h} = (\mathbf{h}, g, \varsigma, v, l) \in \mathcal{V}^{s,r}, \quad \mathfrak{h}_n = (\mathbf{h}_n, g_n, \varsigma_n, v_n, l_n) \in \mathcal{F}^{s,r} \subset \mathcal{V}^{s,r},$$

satisfying

$$\|\mathfrak{h}\|_{\mathcal{V}^{s,r}} \leq c\|\mathfrak{f}\|_{\mathcal{U}^{s,r}}, \quad \|\mathfrak{h} - \mathfrak{h}_n\|_{\mathcal{V}^{s,r}} \leq c\|\mathfrak{f} - \mathfrak{f}_n\|_{\mathcal{U}^{s,r}} \to 0 \quad \text{as } n \to \infty, \tag{11.3.88}$$

where $c$ depends only on $\Omega$, $\mathbf{U}$, $s$, $r$, and $\sigma$. It follows from equations (11.3.6) that the Green formula (11.3.85) holds for $\mathfrak{h}_n$. Substituting $\mathfrak{h}_n$ into (11.3.85), using identity (11.3.86), and recalling equations (11.3.6) we obtain

$$\int_\Omega (\mathbf{w} \cdot \mathbf{H}_n + \omega \Pi G_n + \psi F_n + \xi M_n)\, dx + \varkappa_0^{-1} n e_n$$
$$= \int_\Omega (\mathbf{h}_n \cdot \mathscr{D} + g_n b_{20}\mathfrak{d} + \varsigma_n b_{10}\mathfrak{d} + v_n \sigma \mathfrak{d} + l_n b_{30}\mathfrak{d})\, dx.$$

Since the operator $\Pi$ is symmetric and $\Pi\omega = \omega$ we can rewrite this identity in the equivalent form

$$\int_\Omega (\mathbf{w} \cdot \mathbf{H}_n + \omega G_n + \psi F_n + \xi M_n)\, dx + \varkappa_0^{-1} n e_n$$
$$= \int_\Omega (\mathbf{h}_n \cdot \mathscr{D} + g_n b_{20}\mathfrak{d} + \varsigma_n b_{10}\mathfrak{d} + v_n \sigma \mathfrak{d} + l_n b_{30}\mathfrak{d})\, dx. \tag{11.3.89}$$

Next we show that the integral on the left hand side of this equality may be replaced by a duality pairing. To do this we have to prove that $(\mathbf{H}_n, \omega, \psi, \xi)$ and $(\mathbf{w}, G_n, F_n, M_n)$ belong to dual spaces. Observe that Condition 11.3.17 implies $r > 3$ and hence $1 < r' = r/(r-1) < r$. It follows that

$$\omega, \psi, \xi \in W^{s,r}(\Omega) \subset L^{r'}(\Omega) \subset \mathbb{W}^{-s,r'}(\Omega) \quad \text{and} \quad \mathbf{H}_n \in L^r(\Omega) \subset \mathcal{W}^{s-1,r}(\Omega),$$

which yields

$$(\mathbf{H}_n, \omega, \psi, \xi) \in \mathcal{W}^{s-1,r}(\Omega) \times (\mathbb{W}^{-s,r'}(\Omega))^3. \tag{11.3.90}$$

Next, embedding (11.3.79) implies that $\mathbf{w} \in Y^{s,r}$ belongs to $C^1(\Omega)$. Moreover, equations (11.3.2) imply that $\mathbf{w}$ vanishes on $\partial\Omega$ and hence $\mathbf{w} \in W_0^{1,r'}(\Omega)$. Recall that $\mathcal{W}_0^{1-s,r'}(\Omega)$ is the interpolation space $[L^{r'}(\Omega), W_0^{1,r'}(\Omega)]_{1-s,r'}$, and that $W_0^{1,r'}(\Omega) \subset L^{r'}(\Omega)$. Lemma 1.1.11 shows that $W_0^{1,r'}(\Omega) \subset \mathcal{W}_0^{1-s,r'}(\Omega)$. It follows that $\mathbf{w} \in \mathcal{W}_0^{1-s,r'}(\Omega)$. On the other hand, $G_n, F_n, M_n \in W^{s,r}(\Omega)$. Thus

$$(\mathbf{w}, G_n, F_n, M_n) \in \mathcal{W}_0^{1-s,r'}(\Omega) \times (W^{s,r}(\Omega))^3. \tag{11.3.91}$$

Obviously, $\mathcal{W}^{s-1,r}(\Omega) \times (\mathbb{W}^{-s,r'}(\Omega))^3 = (\mathcal{W}_0^{1-s,r'}(\Omega) \times (W^{s,r}(\Omega))^3)'$. From this and (11.3.90), (11.3.91) we conclude that the integral on the left hand side of (11.3.89) can be replaced by a duality pairing. Thus we get

$$\langle \mathbf{H}_n, \mathbf{w}\rangle_1 + \langle \omega, G_n\rangle_0 + \langle \psi, F_n\rangle_0 + \langle \xi, M_n\rangle_0 + \varkappa_0^{-1} n e_n$$
$$= \int_\Omega \big(\mathbf{h}_n \cdot \mathscr{D} + g_n\, b_{20}\mathfrak{d} + \varsigma_n\, b_{10}\mathfrak{d} + \upsilon_n\, \sigma\mathfrak{d} + l_n\, b_{30}\mathfrak{d}\big)\, dx. \tag{11.3.92}$$

Now our task is to pass to the limit as $n \to \infty$ in (11.3.92) and, by doing so, to derive identity (11.3.77). Relation (11.3.87) implies

$$\mathbf{H}_n \to \mathbf{H} \quad \text{in } \mathcal{W}^{s-1,r}(\Omega) \quad \text{and} \quad (G_n, F_n, M_n) \to (G, F, M) \quad \text{in } W^{s,r}(\Omega),$$

which leads to

$$\langle \mathbf{H}_n, \mathbf{w}\rangle_1 + \langle \omega, G_n\rangle_0 + \langle \psi, F_n\rangle_0 + \langle \xi, M_n\rangle_0 + \varkappa_0^{-1} n e_n$$
$$\to \langle \mathbf{H}, \mathbf{w}\rangle_1 + \langle \omega, G\rangle_0 + \langle \psi, F\rangle_0 + \langle \xi, M\rangle_0 + \varkappa_0^{-1} n e.$$

In view of (11.3.79), the embedding $\mathcal{V}^{s,r} \hookrightarrow C^1(\Omega) \times (C(\Omega))^3 \times \mathbb{R}$ is continuous. It follows from this and (11.3.88) that $\mathbf{h}_n$, $g_n$, $\varsigma_n$ $\upsilon_n$ converge to $\mathbf{h}$, $g$, $\varsigma$, $\upsilon$ uniformly in $\Omega$. Since $\mathscr{D}, \mathfrak{d} \in L^2(\Omega)$, the Lebesgue dominated convergence theorem yields

$$\int_\Omega \big(\mathbf{h}_n \cdot \mathscr{D} + g_n\, b_{20}\mathfrak{d} + \varsigma_n\, b_{10}\mathfrak{d} + \upsilon_n\, \sigma\mathfrak{d} + l_n\, b_{30}\mathfrak{d}\big)\, dx$$
$$\to \int_\Omega \big(\mathbf{h} \cdot \mathscr{D} + g\, b_{20}\mathfrak{d} + \varsigma\, b_{10}\mathfrak{d} + \upsilon\, \sigma\mathfrak{d} + l\, b_{30}\mathfrak{d}\big)\, dx.$$

Letting $n \to \infty$ in (11.3.92) we arrive at (11.3.77). Hence $(\mathbf{w}, \omega, \psi, \xi, n)$ is a very weak solution to problem (11.3.2). It remains to prove estimate (11.3.78). It suffices to show that for all $\mathfrak{f} = (\mathbf{H}, G, F, M, e) \in \mathcal{U}^{s,r}$,

$$\big|\langle \mathbf{H}, \mathbf{w}\rangle_1 + \langle \omega, G\rangle_0 + \langle \psi, F\rangle_0 + \langle \xi, M\rangle_0 + \varkappa_0^{-1} n e_n\big| \\ \leq c(\|\mathscr{D}\|_{L^1(\Omega)} + \|\mathfrak{d}\|_{L^1(\Omega)})\,\|\mathfrak{f}\|_{\mathcal{U}^{s,r}}. \quad (11.3.93)$$

Notice that in view of Lemma 11.3.11 and embedding (11.3.79), the coefficients $b_{23}$ and $b_{13}$ are bounded by a constant depending only on $\Omega$, $\mathbf{U}$, $s$, $r$, and $\sigma$. From this and (11.3.77) we obtain

$$\big|\langle \mathbf{H}, \mathbf{w}\rangle_1 + \langle \omega, G\rangle_0 + \langle \psi, F\rangle_0 + \langle \xi, M\rangle_0 + \varkappa_0^{-1} n e\big| \\ \leq c(\|\mathscr{D}\|_{L^1(\Omega)} + \|\mathfrak{d}\|_{L^1(\Omega)})\,\|\mathfrak{h}\|_{C(\Omega)^6\times\mathbb{R}}, \quad (11.3.94)$$

where $\mathfrak{h} = (\mathbf{h}, g, \varsigma, \upsilon, l)$ is the solution to problem (11.3.6) corresponding to $\mathfrak{f} = (\mathbf{H}, G, F, M, e)$. Since the embedding $\mathcal{V}^{s,r} \hookrightarrow C^1(\Omega) \times (C(\Omega))^3 \times \mathbb{R}$ is continuous, inequality (11.3.88) implies

$$\|\mathfrak{h}\|_{C(\Omega)^6\times\mathbb{R}} \leq c\|\mathfrak{h}\|_{\mathcal{V}^{s,r}} \leq c\|\mathfrak{f}\|_{\mathcal{U}^{s,r}}.$$

Inserting this into (11.3.94) we arrive at (11.3.93). □

### 11.3.5 Uniqueness and existence. Main theorem

In this section we assemble the obtained results on existence, uniqueness, and compactness properties of solutions to Problem 11.1.9 in the form of the following general theorem.

**Theorem 11.3.20.** *Assume that the surface $\Sigma$ and the given vector field $\mathbf{U}$ satisfy Condition 11.2.5. Furthermore, let numbers $r$ and $s$ satisfy*

$$1/2 < s < 1, \quad r < \infty, \quad 2s - 3r^{-1} < 1, \quad sr > 3, \quad (1-s)r > 3. \quad (11.3.95)$$

*Then there exists $\sigma_\star > 1$, depending only on $\mathbf{U}$, $\Omega$ and $s$, $r$, with the following property. For every $\sigma > \sigma_\star$ there are positive $\tau_\star(\sigma)$ and $c$, depending only on $\mathbf{U}$, $\Omega$, $r$, $s$, and $\sigma$, such that whenever*

$$\tau \in (0, \tau_\star], \quad \lambda^{-1}, \mathsf{Re} \in (0, \tau^2], \quad \|\mathbf{N} - \mathbb{I}\|_{C^2(\Omega)} \leq \tau^2, \quad (11.3.96)$$

*Problem 11.1.9 has a unique solution $\vartheta \in \mathcal{B}_\tau$, $\zeta \in X^{s,r}$, $m \in \mathbb{R}$. The auxiliary function $\zeta$ and the constants $\varkappa, m$ satisfy*

$$\|\zeta\|_{X^{s,r}} + |\varkappa| \leq c, \quad |m| \leq c\tau < 1. \quad (11.3.97)$$

*The totality of solutions $\{\vartheta, \zeta, m\}$ to Problem 11.1.9 corresponding to all matrix-valued functions $\mathbf{N}$ in the ball $B(\tau^2) = \{\mathbf{N} : \|\mathbb{I} - \mathbf{N}\|_{C^2(\Omega)} \leq \tau^2\}$ is a relatively compact subset of $W^{s+1,r}(\Omega) \times W^{s,r}(\Omega)^3 \times \mathbb{R}$.*

*Moreover, if $\vartheta_i \in \mathcal{B}_\tau$, $\zeta_i$, $m_i$, $i = 0, 1$, are solutions to Problem 11.1.9 with* $\mathbf{N}_i \in B(\tau^2)$, *then the transposed problem* (11.3.6) *has a unique solution satisfying inequality* (11.3.11).

*Proof.* Set $\sigma_\star = \max\{\sigma^*, \sigma_c\}$ and $\tau_\star(\sigma) = \min\{\tau^*(\sigma), \tau_c(\sigma), \tau_0\}$, where the constants $\sigma^*$ and $\tau^*(\sigma)$ are given by Theorem 11.2.6, $\sigma_c$ and $\tau_c(\sigma)$ are given by Theorem 11.3.4, and $\tau_0$ is given by Lemma 11.2.15. For such a choice, the hypotheses of Theorem 11.3.20 are stronger than those of Theorems 11.2.6, 11.3.4 and 11.3.18. Applying Theorem 11.2.6 we conclude that if $\sigma > \sigma_\star$ and if $\tau$, $\lambda$, $\mathsf{Re}$, $\mathbf{N}$ satisfy condition (11.3.96) and $\tau \in [0, \tau_\star)$, then Problem 11.1.9 has a solution $\vartheta \in \mathcal{B}_\tau$, $\zeta \in X^{s,r}$, $m \in \mathbb{R}$, which satisfies estimate (11.3.97).

Let us prove that this solution is unique. If for some $\mathbf{N}$, the problem has two solutions $(\vartheta_i, \zeta_i, m_i)$, $i = 0, 1$, with $\vartheta_i \in \mathcal{B}_\tau$, then the corresponding differences $\mathbf{w}$, $\psi$, $\omega$, $\xi$ satisfy all hypotheses of Theorem 11.3.78 with $\mathfrak{d} = 0$ and $\mathscr{D} = 0$. Therefore, the elements $\mathbf{w}$, $\psi$, $\omega$, $\xi$ are all 0. Hence the solutions $(\vartheta_i, \zeta_i, m_i)$ coincide.

It follows that the set $\Theta_{\mathbf{N}} \subset \mathcal{B}_\tau \times X^{s,r} \times \mathbb{R}$ of solutions $(\vartheta, \zeta, m)$ corresponding to a matrix $\mathbf{N}$ has only one element. Applying Theorem 11.2.6 we conclude that the totality $\bigcup_{\mathbf{N} \in B(\tau^2)} \Theta_{\mathbf{N}}$ of all such solutions is relatively compact in the product $W^{s+1,r}(\Omega) \times (W^{s,r}(\Omega))^3 \times \mathbb{R}$. □

## 11.4 Shape derivative

### 11.4.1 Preliminaries. Results

In the previous sections we developed the theory of the perturbation problem 11.1.9. In our analysis the structure of the matrix $\mathbf{N}$ did not play any role, and the results obtained hold true for an arbitrary smooth $\mathbf{N}$ close to the unit matrix. But in the original formulation of the domain dependence problem given in Section 11.1, the matrix $\mathbf{N}$ is determined by the family of deformations $x \mapsto x + \varepsilon \mathbf{T}(x)$ of the unperturbed flow domain $\Omega$. By (11.1.4), $\mathbf{N}$ is the adjugate matrix of the Jacobi matrix of the mapping $\mathrm{Id} + \varepsilon \mathbf{T}$,

$$\mathbf{N}(\varepsilon) = \mathfrak{g}(\varepsilon)(\mathbb{I} + \varepsilon D\,\mathbf{T})^{-1}, \quad \mathfrak{g}(\varepsilon) = \det(\mathbb{I} + \varepsilon D\,\mathbf{T}). \tag{11.4.1}$$

In this framework, the solution to Problem 11.1.9 becomes a function of the parameter $\varepsilon$. Investigating the differentiability of this function is the main goal of sensitivity analysis. In this section we consider this question in detail and show that the relevant derivative exists in a weak sense. In order to give a rigorous formulation of this fact we need some preliminary results which are given in the following auxiliary lemmas.

**Lemma 11.4.1.** *There are $\varepsilon_1 > 0$ and $c, c_1$, depending only on $\|\mathbf{T}\|_{C^2(\Omega)}$, such that for every $\varepsilon \in [0, \varepsilon_1]$,*

$$\mathbf{N}(\varepsilon)^{\pm 1} = \mathbb{I} \pm \varepsilon \mathbb{D} + \varepsilon^2 \mathbb{D}^{\pm}, \quad \mathfrak{g}(\varepsilon)^{\pm 1} = 1 \pm \varepsilon \operatorname{div} \mathbf{T} + \varepsilon^2 \mathfrak{g}^{\pm}, \tag{11.4.2}$$

*where* $\mathbb{D} = \operatorname{div} \mathbf{T}\,\mathbb{I} - D\mathbf{T}$ *and*

$$\|\mathbb{D}^{\pm}\|_{C^2(\Omega)} + \|\mathfrak{g}^{\pm}\|_{C^2(\Omega)} \le c. \tag{11.4.3}$$

*The deviations* $\mathbf{N}(\varepsilon)^{\pm 1} - \mathbb{I}$ *and* $\mathfrak{g}(\varepsilon)^{\pm 1} - 1$ *satisfy*

$$\|\mathbf{N}(\varepsilon)^{\pm 1} - \mathbb{I}\|_{C^2(\Omega)} + \|\mathfrak{g}(\varepsilon)^{\pm 1} - 1\|_{C^2(\Omega)} \le c_1\varepsilon. \tag{11.4.4}$$

*Proof.* Direct calculations show that

$$\mathbf{N}(\varepsilon) = \mathbb{I} + \varepsilon\mathbb{D} + \varepsilon^2\mathbb{D}^+ \text{ where } \mathbb{D}^+ = [\partial_{x_2}\mathbf{T} \times \partial_{x_3}\mathbf{T}, \partial_{x_3}\mathbf{T} \times \partial_{x_1}\mathbf{T}, \partial_{x_1}\mathbf{T} \times \partial_{x_2}\mathbf{T}]^\top.$$

It follows that $\mathbb{D}^+$ satisfies (11.4.3). Next we have

$$\mathbf{N}(\varepsilon)^{-1} = 1 - \varepsilon\mathbb{D} + \varepsilon^2\mathbb{D}^- \quad \text{where} \quad \mathbb{D}^- = -\mathbb{D}^+ + \sum_{k\ge 0}(-1)^k\varepsilon^k(\mathbb{D} + \varepsilon\mathbb{D}^+)^{k+2}.$$

Obviously, there is $\varepsilon_0$, depending on $\|\mathbf{T}\|_{C^2(\Omega)}$, such that the series above converges in $C^2(\Omega)$ for all $\varepsilon \in [0, \varepsilon_0]$. This leads to estimate (11.4.3) for $\mathbb{D}^-$. Next, the determinant $\mathfrak{g}(\varepsilon)$ has a representation $\mathfrak{g}(\varepsilon) = 1 + \varepsilon \operatorname{div} \mathbf{T} + \varepsilon^2\mathfrak{g}^+$, where $\mathfrak{g}^+$ is a cubic polynomial of the entries of the Jacobi matrix $D\mathbf{T}$. Arguing as before we obtain the estimates (11.4.3) for the functions $\mathfrak{g}^{\pm}$. It remains to note that (11.4.4) is a straightforward consequence of (11.4.3). □

Lemma 11.4.1 implies that for sufficiently small $\varepsilon$ the matrices $\mathbf{N}(\varepsilon)$ satisfy the assumption of Theorem 11.3.20. Hence we can apply that theorem to analyze solutions to Problem 11.1.9. To be more precise we formulate explicit conditions on the flow domain and given data, which provide the existence and uniqueness of solutions to Problem 11.1.9 with the matrix $\mathbf{N} = \mathbf{N}(\varepsilon)$. We assume that the following condition will be satisfied throughout the section.

**Condition 11.4.2.**

- The $C^3$ surface $\partial\Omega$ and the vector field $\mathbf{U} \in C^3(\Omega)$ satisfy Condition 11.2.5.
- The exponents $s$ and $r$ satisfy

$$1/2 < s < 1, \quad r < \infty, \quad 2s - 3r^{-1} < 1, \quad sr > 3, \quad (1-s)r > 3.$$

- $\sigma_\star > 1$ and $\tau_\star(\sigma) \in (0, 1]$ satisfy the hypotheses of of Theorem 11.3.20.
- $\sigma > \sigma_\star$ and $\tau \in (0, \tau_c(\sigma)]$, $\tau_c(\sigma)$ is given by Theorem 11.3.4, and

$$\lambda^{-1}, \mathsf{Re} \in (0, \tau^2],$$

- $\varepsilon \in [0, \varepsilon_\star]$, where $\varepsilon_\star = \min\{c_1^{-1}\tau^2, \varepsilon_1\}$ with the constants $c_1$ and $\varepsilon_1$ given by Lemma 11.4.1.

**Lemma 11.4.3.** *Let Condition 11.4.2 be satisfied. Then for every* $\varepsilon \in [0, \varepsilon_\star]$ *and* $\mathbf{N} = \mathbf{N}(\varepsilon)$ *Problem 11.1.9 has a unique solution* $(\vartheta(\varepsilon), \zeta(\varepsilon), m(\varepsilon)) \in \mathcal{B}_\tau \times X^{s,r} \times \mathbb{R}$. *The totality of all such solutions is a relatively compact subset of* $\mathcal{B}_\tau \times X^{s,r} \times \mathbb{R}$.

*Proof.* By Lemma 11.4.1 and the choice of $\varepsilon_\star$, the inequality $\|\mathbf{N}(\varepsilon) - \mathbb{I}\|_{C^2(\Omega)} \leq \tau^2$ holds for all $\varepsilon \in [0, \varepsilon_\star]$. It follows from Condition 11.4.2 that $\Omega$, $\mathbf{U}$, $s$, $r$, $\lambda$, Re and $\mathbf{N} = \mathbf{N}(\varepsilon)$ satisfy all hypotheses of Theorem 11.3.20. Application of this theorem completes the proof. □

By abuse of notation we will write

$$\mathbf{v} = \mathbf{v}(0), \quad \pi = \pi(0), \quad \varphi = \varphi(0), \quad \zeta = \zeta(0), \quad m = m(0). \tag{11.4.5}$$

By Lemma 11.4.3, the quantities

$$\begin{aligned} &\mathbf{w}(\varepsilon) = \mathbf{v} - \mathbf{v}(\varepsilon), \quad \omega(\varepsilon) = \pi - \pi(\varepsilon), \quad \psi(\varepsilon) = \varphi - \varphi(\varepsilon), \\ &\xi(\varepsilon) = \zeta - \zeta(\varepsilon), \quad n(\varepsilon) = m - m(\varepsilon) \end{aligned}$$

are well defined and $(\mathbf{w}(\varepsilon), \omega(\varepsilon), \psi(\varepsilon), \xi(\varepsilon), n(\varepsilon))$ belongs to the Banach space $\mathcal{F}^{s,r}$ given by Definition 11.3.3. Now set

$$\begin{aligned} &\mathbf{w}_\varepsilon = -\varepsilon^{-1}\mathbf{w}(\varepsilon), \quad \omega_\varepsilon = -\varepsilon^{-1}\omega(\varepsilon), \quad \psi_\varepsilon = -\varepsilon^{-1}\psi(\varepsilon), \\ &\xi_\varepsilon = -\varepsilon^{-1}\xi(\varepsilon), \quad n_\varepsilon = -\varepsilon^{-1}n(\varepsilon). \end{aligned} \tag{11.4.6}$$

With this notation the *material derivatives* of the solution to Problem 11.1.9 at $\varepsilon = 0$ are given by

$$\mathbf{w} = \lim_{\varepsilon \to 0} \mathbf{w}_\varepsilon, \quad \omega = \lim_{\varepsilon \to 0} \omega_\varepsilon, \quad \psi = \lim_{\varepsilon \to 0} \psi_\varepsilon, \quad \xi = \lim_{\varepsilon \to 0} \xi_\varepsilon, \quad n = \lim_{\varepsilon \to 0} n_\varepsilon,$$

provided that the limits exist in some sense.

**Remark 11.4.4.** Let us point out that according to Definition 11.1.3 the material derivative for the velocity field in the reference domain, defined by

$$\mathbf{u} := \mathbf{N}(\varepsilon)^{-1}(\mathbf{u}_\star + \mathbf{v}(\varepsilon)),$$

obviously takes the form $\dot{\mathbf{u}} := \mathbf{w} - \mathbb{D}(\mathbf{u}_\star + \mathbf{v}(0))$. Thus, the element $\mathbf{w}$ is only a part of the material derivative for the velocity field we are looking for. However, in the following the element $\mathbf{w}$ is called the material derivative.

It is easy to derive formally the equations for the material derivatives by substituting $\vartheta = \vartheta(\varepsilon)$ and $\mathbf{N} = \mathbf{N}(\varepsilon)$ into the basic equations (11.1.24) and differentiating the result with respect to $\varepsilon$ at zero. This formal procedure gives the following system of equations and boundary conditions for the material derivatives

$\mathbf{w}$, $(\omega, \psi, \zeta)$, and $n$:

$$\mathbf{u}\nabla\psi + \sigma\psi = -\mathbf{w}\cdot\nabla\varphi_1 + a_{11}\psi + a_{12}\omega + a_{13}n + a_{10}\mathfrak{d}_0 \quad \text{in } \Omega, \tag{11.4.7a}$$

$$\begin{aligned}&\Delta\mathbf{w} - \nabla\omega = \mathsf{Re}\varrho\mathbf{u}\nabla\mathbf{w} + \mathsf{Re}\,\varrho\mathbf{w}\nabla\mathbf{u} + \mathsf{Re}\,\psi\mathbf{u}\nabla\mathbf{u} + \mathscr{D}_0 \quad \text{in } \Omega,\\ &\operatorname{div}\mathbf{w} = a_{21}\psi + a_{22}\omega + a_{23}n + a_{20}\mathfrak{d}_0 \quad \text{in } \Omega,\end{aligned} \tag{11.4.7b}$$

$$-\operatorname{div}(\mathbf{u}\xi) + \sigma\xi = \operatorname{div}(\zeta\mathbf{w}) + \sigma\mathfrak{d}_0 \quad \text{in } \Omega, \tag{11.4.7c}$$

$$\mathbf{w} = 0 \quad \text{on } \partial\Omega, \quad \psi = 0 \quad \text{on } \Sigma_{\text{in}}, \qquad \xi = 0 \quad \text{on } \Sigma_{\text{out}}, \tag{11.4.7d}$$

$$\omega - \Pi\omega = 0, \quad n = \varkappa_0\int_\Omega (a_{31}\psi + b_{32}\omega + b_{34}\xi + b_{30}\mathfrak{d})\,dx. \tag{11.4.7e}$$

Here the coefficients $a_{ij}$, $\varkappa$ and the functions $\mathscr{D}_0$, $\mathfrak{d}_0$ are given by

$$\begin{aligned}
&a_{11} = \Psi[\vartheta] - \frac{2\sigma}{\varrho_\star}\varphi + m, \quad a_{12} = \lambda^{-1}\varrho, \quad a_{13} = \varrho,\\
&a_{10} = \varrho\Psi[\vartheta] - \frac{\sigma}{\varrho_\star}\varphi^2 - \sigma\varphi + m\varrho, \quad a_{21} = \sigma\varrho_\star{}^{-1}, \quad a_{22} = -\lambda^{-1},\\
&a_{23} = -1, \quad a_{20} = \sigma\varphi\varrho_\star^{-1} - \Psi[\vartheta] - m,\\
&a_{31} = \varrho_\star^{-1}\zeta\Big(\Psi[\vartheta] - \frac{2\sigma}{\varrho_\star}\varphi_0\Big) + m\varrho_\star^{-1}\zeta,\\
&a_{32} = \lambda^{-1}(\varrho_\star^{-1}\zeta\varrho - 1), \quad a_{34} = \varrho_\star^{-1}\Psi_1[\vartheta] + m(1 + \varrho_\star^{-1}\varphi),\\
&a_{30} = \varrho_\star^{-1}\zeta(a_{10} - m\varrho) - \Psi[\vartheta] - m(1 - \zeta - \varrho_\star^{-1}\zeta\varphi),\\
&\varkappa = \Big(\int_\Omega (1 - \zeta - \varrho_\star^{-1}\zeta\varphi)\,dx\Big)^{-1},
\end{aligned} \tag{11.4.8}$$

$$\begin{aligned}&\mathscr{D}_0 = \mathbb{D}^\top\Delta\mathbf{u} + \Delta(\mathbb{D}\mathbf{u}) + \operatorname{div}(\mathbb{T}\nabla\mathbf{u}) - \mathsf{Re}\,\varrho(\mathbb{D}^\top\mathbf{u}\nabla\mathbf{u} + \mathbf{u}\nabla(\mathbb{D}\mathbf{u})),\\ &\mathfrak{d}_0 = \operatorname{div}\mathbf{T}\,\mathbb{I}, \quad \mathbb{T} = \operatorname{div}\mathbf{T}\,\mathbb{I} - D\mathbf{T} - (D\mathbf{T})^\top, \quad \mathbb{D} = \operatorname{div}\mathbf{T}\,\mathbb{I} - D\mathbf{T}.\end{aligned} \tag{11.4.9}$$

In these formulae $\vartheta = (\mathbf{v}, \pi, \varphi)$, $\zeta$ and the constant $m$ are given by (11.4.5), and $\mathbf{u} = \mathbf{u}_\star + \mathbf{v}$, $\varrho = \varrho_\star + \varphi$. It is important to note that the corresponding *transposed problem* is formulated as follows: For given $\mathbf{f} = (\mathbf{H}, G, F, M, e)$, find $\mathfrak{h} = (\mathbf{h}, g, \varsigma, \upsilon, l)$ satisfying

$$\begin{aligned}&\Delta\mathbf{h} - \nabla g = \mathsf{Re}\,\mathscr{H}_0(\mathbf{h}) - \varsigma\nabla\varphi - \zeta\nabla\upsilon + \mathbf{H} \quad \text{in } \Omega,\\ &\operatorname{div}\mathbf{h} = \Pi(a_{22}g + a_{12}\varsigma + a_{32}l) + \Pi G \quad \text{in } \Omega,\end{aligned} \tag{11.4.10a}$$

$$-\operatorname{div}(\mathbf{u}\varsigma) + \sigma\varsigma = \mathsf{Re}\,\mathscr{M}_0(\mathbf{h}) + a_{11}\varsigma + a_{21}g + a_{31}l + F \quad \text{in } \Omega, \tag{11.4.10b}$$

$$\mathbf{u}\nabla\upsilon + \sigma\upsilon = a_{34}l + M \quad \text{in } \Omega, \tag{11.4.10c}$$

$$\mathbf{h} = 0 \quad \text{on } \partial\Omega, \quad \varsigma = 0 \quad \text{on } \Sigma_{\text{out}}, \quad \upsilon = 0 \quad \text{on } \Sigma_{\text{in}}, \tag{11.4.10d}$$

$$g - \Pi g = 0, \quad l = \varkappa\int_\Omega (a_{13}\varsigma + a_{23}g)\,dx + e. \tag{11.4.10e}$$

Here the operators $\mathscr{H}_0$, $\mathscr{M}_0$ are given by

$$\mathscr{H}_0(\mathbf{h}) = \varrho\nabla(\mathbf{u})\mathbf{h} - \operatorname{div}(\varrho\mathbf{u}\otimes\mathbf{h}), \quad \mathscr{M}_0(\mathbf{h}) = (\mathbf{u}\nabla\mathbf{u})\cdot\mathbf{h}. \tag{11.4.11}$$

Notice that the linearized equations (11.4.7) are obtained by formal differentiation. It is not clear whether the limits $\mathbf{w}$, $(\omega, \psi, \zeta)$, and $n$ really exist and satisfy (11.4.7). We claim that they exist in a weak sense and the material derivatives are defined as elements of the space $\mathcal{W}^{1-s,r'}(\Omega) \times (\mathbb{W}^{-s,r'}(\Omega))^3 \times \mathbb{R}$. Moreover, we claim that the material derivatives are given by a very weak solution to the linear problem (11.4.7). Definition 11.3.16 of a very weak solution requires the well-posedness of the transposed problem (11.4.10) associated with the linearized equations (11.4.7). This fact results from the following

**Lemma 11.4.5.** *Let Condition 11.4.2 be satisfied. Then for every* $(\mathbf{H}, G, F, M, e) \in \mathcal{U}^{s,r}$*, problem* (11.4.10) *has a unique solution* $\mathfrak{h} = (\mathbf{h}, g, \varsigma, v, l) \in \mathcal{V}^{s,r}$.

*Proof.* Notice that problem (11.4.10) is a particular case of problem (11.3.6) with $\mathbf{N}_0 = \mathbb{I}$, $\mathfrak{g}_0 = 1$ and

$$\vartheta_i = \vartheta(0), \quad \zeta_i = \zeta(0), \quad m_i = m(0), \quad \mathbf{u}_i = \mathbf{u}_\star + \mathbf{v}(0), \quad \varrho_i = \varrho_\star + \varphi(0)$$

for $i = 0, 1$. Hence the statement is a straightforward consequence of Theorem 11.3.20. $\square$

Now we are in a position to formulate the main result of this section.

**Theorem 11.4.6.** *Let Condition 11.4.2 be satisfied. Then there are* $(\mathbf{w}, \omega, \psi, \xi, n) \in \mathcal{W}_0^{1-s,r'}(\Omega) \times (\mathbb{W}^{-s,r'}(\Omega))^3 \times \mathbb{R}$ *such that*

$$\mathbf{w}_\varepsilon \rightharpoonup \mathbf{w} \text{ weakly in } \mathcal{W}_0^{1-s,r'}(\Omega), \quad (\omega_\varepsilon, \psi_\varepsilon, \xi_\varepsilon) \rightharpoonup (\omega, \psi, \xi) \text{ weakly in } \mathbb{W}^{-s,r'}(\Omega),$$

*and* $n_\varepsilon \to n$ *in* $\mathbb{R}$ *as* $\varepsilon \to 0$. *Moreover, for every* $(\mathbf{H}, G, F, M, e) \in \mathcal{U}^{s,r}$,

$$\begin{aligned}\langle \mathbf{H}, \mathbf{w}\rangle_1 + \langle \omega, G\rangle_0 + \langle \psi, F\rangle_0 + \langle \xi, \mathcal{M}\rangle_0 + \varkappa^{-1} n e \\ = \int_\Omega \big(\mathbf{h} \cdot \mathscr{D}_0 + g\, a_{20}\mathfrak{d}_0 + \varsigma\, a_{10}\mathfrak{d}_0 + v\,\sigma\mathfrak{d}_0 + l\, a_{30}\mathfrak{d}_0\big)\, dx. \quad (11.4.12)\end{aligned}$$

*Here* $\mathfrak{h} = (\mathbf{h}, g, \varsigma, v, l) \in \mathcal{V}^{s,r}$ *is the solution to problem* (11.4.10) *given by Lemma 11.4.6,* $\varkappa$ *is given by* (11.4.8)*, and* $\mathscr{D}_0$ *and* $\mathfrak{d}_0$ *are given by* (11.4.9).

The next section is devoted to the proof of this theorem.

## 11.4.2 Proof of Theorem 11.4.6

We split the proof into a sequence of lemmas. The first is just a particular case of Theorem 11.3.18. In order to formulate it we introduce the following notation. Assume that $\mathbf{N}(\varepsilon)$, $\mathfrak{g}(\varepsilon)$ are defined by (11.4.1), and $\vartheta(\varepsilon)$, $\zeta(\varepsilon)$, $m(\varepsilon)$ are given by Lemma 11.4.3. Next set

$$\mathbf{u} = \mathbf{u}_\star + \mathbf{v}, \quad \mathbf{u}(\varepsilon) = \mathbf{u}_\star + \mathbf{v}(\varepsilon), \quad \varrho = \varrho_\star + \varphi, \quad \varrho(\varepsilon) = \varrho_\star + \varphi(\varepsilon). \quad (11.4.13)$$

Introduce the functions $b_{\varepsilon,ij}$, $\mathscr{D}(\varepsilon)$, $\mathfrak{d}(\varepsilon)$ and operators $\mathscr{H}_\varepsilon$, $\mathscr{M}_0$ by

$$
\begin{aligned}
&b_{\varepsilon,11} = \Psi[\vartheta] - \varrho(\varepsilon)\Phi(\varphi,\varphi(\varepsilon)) - \frac{\sigma}{\varrho_\star}(\varphi+\varphi(\varepsilon)) + m(\varepsilon),\\
&b_{\varepsilon,12} = \lambda^{-1}\varrho(\varepsilon),\ b_{\varepsilon,13} = \varrho,\ b_{\varepsilon,10} = \varrho(\varepsilon)\Psi[\vartheta(\varepsilon)] - \frac{\sigma}{\varrho_\star}\varphi(\varepsilon)^2 - \sigma\varphi(\varepsilon) + m(\varepsilon)\varrho(\varepsilon),\\
&b_{\varepsilon,21} = \frac{\sigma}{\varrho_\star} + \Phi(\varphi,\varphi(\varepsilon)), \quad b_{\varepsilon,22} = -\lambda^{-1},\\
&b_{\varepsilon,23} = -1, \quad b_{\varepsilon,20} = \sigma\varphi(\varepsilon)\varrho_\star^{-1} - \Psi[\vartheta(\varepsilon)] - m(\varepsilon), \qquad (11.4.14)\\
&b_{\varepsilon,31} = \varrho_\star^{-1}\zeta(\varepsilon)\Big(\Psi[\vartheta] - \varrho(\varepsilon)\Phi(\varphi,\varphi(\varepsilon)) - \frac{\sigma}{\varrho_\star}(\varphi+\varphi(\varepsilon))\Big)\\
&\qquad - \Phi(\varphi,\varphi(\varepsilon)) + m(\varepsilon)\varrho_\star^{-1}\zeta(\varepsilon),\\
&b_{\varepsilon,32} = \lambda^{-1}(\varrho_\star^{-1}\zeta(\varepsilon)\varrho(\varepsilon) - 1), \quad b_{\varepsilon,34} = \varrho_\star^{-1}\Psi_1[\vartheta] + m(\varepsilon)(1+\varrho_\star^{-1}\varphi),\\
&b_{\varepsilon,30} = \varrho_\star^{-1}\zeta(\varepsilon)(b_{\varepsilon,10} - m(\varepsilon)\varrho(\varepsilon)) - \Psi[\vartheta(\varepsilon)] - m(\varepsilon)(1-\zeta(\varepsilon) - \varrho_\star^{-1}\zeta(\varepsilon)\varphi(\varepsilon)),\\
&\Phi(\varphi_0,\varphi_1) = (p'(\varrho_\star)\varrho_\star)^{-1}\sigma\int_0^1 H'(\varphi_0 s + \varphi_1(1-s))\,ds,\\
&H'(\varphi) = p'(\varrho_\star+\varphi) - p'(\varrho_\star), \quad \varkappa = \Big(\int_\Omega (1-\zeta-\varrho_\star^{-1}\zeta\varphi)\,dx\Big)^{-1},
\end{aligned}
$$

$$\mathscr{H}_\varepsilon(\mathbf{h}) = \varrho(\varepsilon)\nabla\mathbf{u}\,\mathbf{h} - \operatorname{div}(\varrho(\varepsilon)\mathbf{u}(\varepsilon)\otimes\mathbf{h}), \quad \mathscr{M}_0(\mathbf{h}) = (\mathbf{u}\nabla\mathbf{u})\cdot\mathbf{h}, \tag{11.4.15}$$

$$
\begin{aligned}
\mathscr{D}(\varepsilon) = {}& \mathbf{N}(\varepsilon)^{-\top}\operatorname{div}\big(\mathfrak{g}(\varepsilon)^{-1}\mathbf{N}(\varepsilon)\mathbf{N}(\varepsilon)^\top\nabla(\mathbf{N}(\varepsilon)^{-1}\mathbf{u}(\varepsilon))\big) - \Delta\mathbf{u}(\varepsilon)\\
&+ \mathsf{Re}\,\varrho(\varepsilon)\Big(\mathbf{u}(\varepsilon)\nabla\mathbf{u}(\varepsilon) - \mathbf{N}(\varepsilon)^{-\top}\big(\mathbf{u}(\varepsilon)\nabla(\mathbf{N}(\varepsilon)^{-1}\mathbf{u}(\varepsilon))\big)\Big),
\end{aligned} \tag{11.4.16}
$$

$$\mathfrak{d}(\varepsilon) = 1 - \mathfrak{g}(\varepsilon). \tag{11.4.17}$$

Let us consider the transposed problem

$$
\begin{aligned}
&\Delta\mathbf{h}_\varepsilon - \nabla g_\varepsilon = \mathsf{Re}\,\mathscr{H}_\varepsilon(\mathbf{h}_\varepsilon) - \varsigma_\varepsilon\nabla\varphi(\varepsilon) - \zeta(\varepsilon)\nabla\upsilon_\varepsilon + \mathbf{H} \quad \text{in } \Omega,\\
&\operatorname{div}\mathbf{h}_\varepsilon = \Pi(b_{\varepsilon,22}g_\varepsilon + b_{\varepsilon,12}\varsigma_\varepsilon + b_{\varepsilon,32}l_\varepsilon) + \Pi G \quad \text{in } \Omega,
\end{aligned} \tag{11.4.18a}
$$

$$
\begin{aligned}
&-\operatorname{div}(\mathbf{u}\varsigma_\varepsilon) + \sigma\varsigma_\varepsilon\\
&\qquad = \mathsf{Re}\,\mathscr{M}(\mathbf{h}_\varepsilon) + b_{\varepsilon,11}\varsigma_\varepsilon + b_{\varepsilon,21}g_\varepsilon + b_{\varepsilon,31}l_\varepsilon + F \quad \text{in } \Omega,\\
&\mathbf{u}\nabla\upsilon + \sigma\upsilon = b_{\varepsilon,34}l + M \quad \text{in } \Omega,
\end{aligned} \tag{11.4.18b}
$$

$$
\begin{aligned}
&\mathbf{h}_\varepsilon = 0 \quad \text{on } \partial\Omega, \quad \varsigma_\varepsilon = 0 \quad \text{on } \Sigma_{\text{out}}, \quad \upsilon_\varepsilon = 0 \quad \text{on } \Sigma_{\text{in}},\\
&g_\varepsilon - \Pi g_\varepsilon = 0, \quad l = \varkappa\int_\Omega (b_{\varepsilon,13}\varsigma_\varepsilon + b_{\varepsilon,23}g_\varepsilon)\,dx + e.
\end{aligned} \tag{11.4.18c}
$$

**Lemma 11.4.7.** *Let Condition 11.4.2 be satisfied. Then for each* $\mathfrak{f} = (\mathbf{H}, G, F, M, e) \in \mathcal{U}^{s,r}$, *problem* (11.4.18) *has a unique solution* $\mathfrak{h}_\varepsilon = (\mathbf{h}_\varepsilon, g_\varepsilon, \varsigma_\varepsilon, \upsilon_\varepsilon, l_\varepsilon) \in \mathcal{V}^{s,r}$. *This solution satisfies the estimate*

$$\|\mathfrak{h}_\varepsilon\|_{\mathcal{V}^{s,r}} \le c\|\mathfrak{f}\|_{\mathcal{U}^{s,r}}, \tag{11.4.19}$$

*where c is independent of* $\varepsilon$. *Moreover, the functions* $(\mathbf{w}_\varepsilon, \omega_\varepsilon, \psi_\varepsilon, \xi_\varepsilon, n_\varepsilon)$ *given by* (11.4.6) *belong to the class* $\mathcal{W}_0^{1-s,r'} \times (\mathbb{W}^{-s,r'}(\Omega))^3 \times \mathbb{R}$ *and satisfy the identity*

$$\langle \mathbf{H}, \mathbf{w}_\varepsilon\rangle_1 + \langle \omega_\varepsilon, G\rangle_0 + \langle \psi_\varepsilon, F\rangle_0 + \langle \xi_\varepsilon, \mathcal{M}\rangle_0 + \varkappa^{-1} n_\varepsilon e$$
$$= -\frac{1}{\varepsilon}\int_\Omega \big(\mathbf{h}_\varepsilon \cdot \mathscr{D}(\varepsilon) + g_\varepsilon\, b_{\varepsilon,20}\mathfrak{d}(\varepsilon) + \varsigma_\varepsilon\, b_{\varepsilon,10}\mathfrak{d}(\varepsilon) + v_\varepsilon\, \sigma\mathfrak{d}(\varepsilon) + l_\varepsilon\, b_{\varepsilon,30}\mathfrak{d}(\varepsilon)\big)\, dx, \tag{11.4.20}$$

*where* $(\mathbf{h}_\varepsilon, g_\varepsilon, \varsigma_\varepsilon, v_\varepsilon, l_\varepsilon)$ *is the solution to problem* (11.4.18).

*Proof.* Observe that in view of Lemma 11.4.1 and Condition 11.4.2, the inequality $\|\mathbf{N}(\varepsilon)-\mathbb{I}\|_{C^2(\Omega)} \le \tau^2$ holds for all $\varepsilon \in [0, \varepsilon_\star]$. Hence $\Omega$, $\mathbf{U}$, $s$, $r$, $\lambda$, $\mathsf{Re}$ and $\mathbf{N} = \mathbf{N}(\varepsilon)$ satisfy all hypotheses of Theorem 11.3.20. Moreover, the functions and matrices

$$\begin{aligned}
&\mathbf{N}_0 = \mathbb{I}, \quad \mathbf{N}_1 = \mathbf{N}(\varepsilon), \quad \mathfrak{g}_0 = 1,\\
&(\vartheta_0, \zeta_0, m_0) = (\vartheta, \zeta, m), \quad (\vartheta_1, \zeta_1, m_1) = (\vartheta(\varepsilon), \zeta(\varepsilon), m(\varepsilon)),\\
&\mathbf{u}_0 = \mathbf{u}_\star + \mathbf{v}, \quad \mathbf{u}_1 = \mathbf{u}(\varepsilon), \quad \varrho_0 = \varrho_\star + \varphi = \varrho, \quad \varrho_1 = \varrho(\varepsilon)
\end{aligned} \tag{11.4.21}$$

satisfy all hypotheses of Theorems 11.3.4 and 11.3.18. In particular, formulae (11.4.14)–(11.4.15) imply that problem (11.4.18) is a special case of (11.3.6). Therefore the solvability of problem (11.4.18) and estimates (11.4.19) follow directly from Theorem 11.3.4. Next, applying Theorem 11.3.18 we find that the differences $(\mathbf{w}(\varepsilon), \omega(\varepsilon), \psi(\varepsilon), \xi(\varepsilon), n(\varepsilon))$ satisfy identity (11.3.77) with $\mathscr{D}$ and $\mathfrak{d}$ replaced by $\mathscr{D}(\varepsilon)$ and $\mathfrak{d}(\varepsilon)$, $b_{ij}$ replaced by $b_{\varepsilon,ij}$, and $(\mathbf{h}, g, \varsigma, v, l)$ replaced by $(\mathbf{h}_\varepsilon, g_\varepsilon, \varsigma_\varepsilon, v_\varepsilon, l_\varepsilon)$. Thus we get

$$\langle \mathbf{H}, \mathbf{w}(\varepsilon)\rangle_1 + \langle \omega(\varepsilon), G\rangle_0 + \langle \psi(\varepsilon), F\rangle_0 + \langle \xi(\varepsilon), \mathcal{M}\rangle_0 + \varkappa^{-1} n(\varepsilon) e$$
$$= \int_\Omega \big(\mathbf{h}_\varepsilon \cdot \mathscr{D}(\varepsilon) + g_\varepsilon\, b_{\varepsilon,20}\mathfrak{d}(\varepsilon) + \varsigma_\varepsilon\, b_{\varepsilon,10}\mathfrak{d}(\varepsilon) + v_\varepsilon\, \sigma\mathfrak{d}(\varepsilon) + l_\varepsilon\, b_{\varepsilon,30}\mathfrak{d}(\varepsilon)\big)\, dx,$$

Substituting expressions (11.4.6) into this identity we arrive at (11.4.20). □

In order to prove Theorem 11.4.6 we have to pass to the limit in (11.4.20) as $\varepsilon \to 0$. To this end, we investigate all terms in this identity and begin with the evaluation of $\mathscr{D}(\varepsilon)$ and $\mathfrak{d}(\varepsilon)$ given by (11.4.16).

**Lemma 11.4.8.** *Let Condition 11.4.2 be satisfied. Then*

$$\mathscr{D}(\varepsilon) = -\varepsilon\mathscr{D}_1(\varepsilon) + \varepsilon^2\mathscr{D}_2(\varepsilon), \quad \mathfrak{d}(\varepsilon) = -\varepsilon \operatorname{div} \mathbf{T} + \varepsilon^2\mathfrak{d}_2(\varepsilon), \tag{11.4.22}$$

*where*

$$\begin{aligned}
\mathscr{D}_1(\varepsilon) &= \operatorname{div}(\mathbb{T}\nabla\mathbf{u}(\varepsilon)) + \mathbb{D}^\top\Delta\mathbf{u}(\varepsilon) + \Delta(\mathbb{D}\mathbf{u}(\varepsilon))\\
&\qquad - \mathsf{Re}\,\varrho(\varepsilon)\big(\mathbb{D}^\top\mathbf{u}(\varepsilon)\nabla\mathbf{u}(\varepsilon) + \mathbf{u}(\varepsilon)\nabla(\mathbb{D}\mathbf{u}(\varepsilon))\big),\\
\mathbb{T} &= \operatorname{div}\mathbf{T}\,\mathbb{I} - D\mathbf{T} - (D\mathbf{T})^\top, \quad \mathbb{D} = \operatorname{div}\mathbf{T}\,\mathbb{I} - D\mathbf{T}.
\end{aligned} \tag{11.4.23}$$

*The remainders $\mathscr{D}_2(\varepsilon)$ and $\mathfrak{d}_2(\varepsilon)$ satisfy the estimates*

$$\|\mathscr{D}_2(\varepsilon)\|_{L^2(\Omega)} + \|\mathfrak{d}_2(\varepsilon)\|_{C(\Omega)} \le c, \tag{11.4.24}$$

*where $c$ is independent of $\varepsilon$.*

*Proof.* By Lemma 11.4.1, we have

$$\begin{aligned}\mathbb{S}(\varepsilon) &:= \mathfrak{g}(\varepsilon)^{-1}\mathbf{N}(\varepsilon)\mathbf{N}(\varepsilon)^\top \\ &= (1 - \varepsilon \operatorname{div} \mathbf{T} + \varepsilon^2 \mathfrak{g}^-)(\mathbb{I} + \varepsilon\mathbb{D} + \varepsilon\mathbb{D}^+)(\mathbb{I} + \varepsilon\mathbb{D} + \varepsilon\mathbb{D}^+)^\top = \mathbb{I} - \varepsilon\mathbb{T} + \varepsilon^2\mathbb{S}_2(\varepsilon),\end{aligned}$$

where the matrix $\mathbb{S}_2$ satisfies the estimate

$$\|\mathbb{S}_2(\varepsilon)\|_{C^2(\Omega)} \le c. \tag{11.4.25}$$

Substituting $\mathbb{S}(\varepsilon)$ and decompositions (11.4.2) into (11.4.16) we arrive at expansion (11.4.22) with

$$\begin{aligned}\mathscr{D}_2(\varepsilon) = {} & (\mathbb{D}^-)^\top \operatorname{div}\big(\mathbb{S}(\varepsilon)\nabla(\mathbf{N}(\varepsilon)^{-1}\mathbf{u}(\varepsilon))\big) \\ & + \mathbf{N}(\varepsilon)^{-\top} \operatorname{div}\big(\mathbb{S}_2(\varepsilon)\nabla(\mathbf{N}(\varepsilon)^{-1}\mathbf{u}(\varepsilon))\big) + \mathbf{N}(\varepsilon)^{-\top} \operatorname{div}\big(\mathbb{S}(\varepsilon)\nabla(\mathbb{D}^-(\varepsilon)\mathbf{u}(\varepsilon))\big) \\ & + \mathbb{D}^\top\Delta(\mathbb{D}\mathbf{u}(\varepsilon)) - \mathbb{D}^\top \operatorname{div}(\mathbb{T}\nabla\mathbf{u}(\varepsilon)) - \operatorname{div}\big(\mathbb{T}\nabla(\mathbb{D}\mathbf{u}(\varepsilon))\big) \\ & + \mathsf{Re}\,\varrho(\varepsilon)\Big(\mathbb{D}^\top\mathbf{u}(\varepsilon)\nabla(\mathbb{D}\mathbf{u}(\varepsilon)) - (\mathbb{D}^-)^\top\mathbf{u}(\varepsilon)\nabla(\mathbf{N}(\varepsilon)^{-1}\mathbf{u}(\varepsilon)) \\ & - (\mathbb{D}^-)^\top\mathbf{u}(\varepsilon)\nabla(\mathbf{N}(\varepsilon)^{-1}\mathbf{u}(\varepsilon))\Big) - \varepsilon\mathbb{D}^\top \operatorname{div}\big(\mathbb{T}\nabla(\mathbb{D}\mathbf{u}(\varepsilon))\big).\end{aligned} \tag{11.4.26}$$

It follows from Lemma 11.4.3 that $\vartheta(\varepsilon) = (\mathbf{v}(\varepsilon), \omega(\varepsilon), \varphi(\varepsilon))$ belongs to the ball $\mathcal{B}_\tau$ and hence

$$\|\mathbf{v}(\varepsilon)\|_{Y^{s,r}} + \|\varphi(\varepsilon)\|_{X^{s,r}} \le \tau.$$

Since $Y^{s,r} = W^{1+s,r}(\Omega) \cap W^{2,2}(\Omega)$ and $X^{s,r} = W^{s,r}(\Omega) \cap W^{1,2}(\Omega)$, $sr > 3$, the functions $\mathbf{u}(\varepsilon) = \mathbf{u}_\star + \mathbf{v}(\varepsilon)$ and $\varrho(\varepsilon) = \varrho_\star + \varphi(\varepsilon)$ satisfy

$$\|\mathbf{u}(\varepsilon)\|_{W^{2,2}(\Omega)} + \|\varrho(\varepsilon)\|_{C(\Omega)} \le c, \tag{11.4.27}$$

where $c$ is independent of $\varepsilon$. It follows from this, expression (11.4.26), and estimates (11.4.3), (11.4.25) that $\mathscr{D}_2(\varepsilon)$ satisfies (11.4.24). It remains to note that $\mathfrak{d}_2(\varepsilon) = -\mathfrak{g}^+(\varepsilon)$ and inequality (11.4.24) for $\mathfrak{d}_2$ is a direct consequence of (11.4.3). □

**Lemma 11.4.9.** *Let Condition 11.4.2 be satisfied and $(\mathbf{w}_\varepsilon, \omega_\varepsilon, \psi_\varepsilon, n_\varepsilon)$ be given by (11.4.6). Then there exists $c$ independent of $\varepsilon$ such that*

$$\|\mathbf{w}_\varepsilon\|_{\mathcal{W}_0^{1-s,r'}(\Omega)} + \|\omega_\varepsilon\|_{\mathbb{W}^{-s,r'}(\Omega)} + \|\psi_\varepsilon\|_{\mathbb{W}^{-s,r'}(\Omega)} + \|\xi_\varepsilon\|_{\mathbb{W}^{-s,r'}(\Omega)} + |n_\varepsilon| \le c. \tag{11.4.28}$$

*Proof.* Recall the notation

$$\mathcal{U}^{s,r} = \mathcal{W}^{s-1,r}(\Omega) \times (W^{s,r}(\Omega))^3 \times \mathbb{R}, \quad \mathcal{V}^{s,r} = W^{s+1,r}(\Omega) \times (W^{s,r}(\Omega))^3 \times \mathbb{R},$$

It suffices to show that for all $\mathfrak{f} = (\mathbf{H}, G, F, M, e) \in \mathcal{U}^{s,r}$,

$$\big|\langle \mathbf{H}, \mathbf{w}_\varepsilon\rangle_1 + \langle \omega_\varepsilon, G\rangle_0 + \langle \psi_\varepsilon, F\rangle_0 + \langle \xi_\varepsilon, M_n\rangle_0 + \varkappa_0^{-1} n_\varepsilon e\big| \le c\|\mathfrak{f}\|_{\mathcal{U}^{s,r}}. \tag{11.4.29}$$

Notice that in view of Lemma 11.3.11 and the embedding $X^{s,r} \hookrightarrow C(\Omega)$, the coefficients $b_{\varepsilon,23}$ and $b_{\varepsilon,13}$ are bounded by a constant depending only on $\Omega$, $\mathbf{U}$, $s$, $r$, and $\sigma$. From this and (11.4.20) we obtain

$$\begin{aligned}\big|\langle \mathbf{H}, \mathbf{w}_\varepsilon\rangle_1 + \langle \omega_\varepsilon, G\rangle_0 + \langle \psi_\varepsilon, F\rangle_0 + \langle \xi_\varepsilon, M\rangle_0 + \varkappa_0^{-1} n_\varepsilon e\big| \\ \le \varepsilon^{-1}(\|\mathscr{D}(\varepsilon)\|_{L^1(\Omega)} + \|\mathfrak{d}(\varepsilon)\|_{L^1(\Omega)})\|\mathfrak{h}_\varepsilon\|_{C(\Omega)^6\times\mathbb{R}},\end{aligned} \tag{11.4.30}$$

where $\mathfrak{h}_\varepsilon = (\mathbf{h}_\varepsilon, g_\varepsilon, \varsigma_\varepsilon, \upsilon_\varepsilon, l_\varepsilon) \in \mathcal{V}^{s,r}$ is the solution to problem (11.4.18). Since the embedding $\mathcal{V}^{s,r} \hookrightarrow C^1(\Omega) \times (C(\Omega))^3 \times \mathbb{R}$ is continuous, inequality (11.4.19) implies

$$\|\mathfrak{h}_\varepsilon\|_{C(\Omega)^6\times\mathbb{R}} \le c\|\mathfrak{h}_\varepsilon\|_{\mathcal{V}^{s,r}} \le c\|\mathfrak{f}_\varepsilon\|_{\mathcal{U}^{s,r}}.$$

On the other hand, Lemma 11.4.8 yields

$$\varepsilon^{-1}\|\mathscr{D}(\varepsilon)\|_{L^2(\Omega)} + \varepsilon^{-1}\|\mathfrak{d}(\varepsilon)\|_{C(\Omega)} \le c.$$

Inserting these results into (11.4.30) we arrive at (11.4.29). □

Although Lemma 11.4.3 states the existence and uniqueness of solutions $(\vartheta(\varepsilon), \zeta(\varepsilon), m(\varepsilon))$ to the basic Problem 11.1.9, it says nothing about the dependence of these solutions on $\varepsilon$. Now we are in a position to prove they are continuous at $\varepsilon = 0$.

**Lemma 11.4.10.** *Let Condition 11.4.2 be satisfied. Then*

$$(\vartheta(\varepsilon), \zeta(\varepsilon), m(\varepsilon)) \to (\vartheta, \zeta, m) \quad \textit{in } Y^{s,r} \times (X^{s,r})^3 \times \mathbb{R} \quad \textit{as } \varepsilon \to 0.$$

*Proof.* Recall that $\mathcal{F}^{s,r} = Y^{s,r} \times (X^{s,r})^3 \times \mathbb{R}$ and introduce the temporary notation

$$\begin{aligned}\mathbf{q}(\varepsilon) &= (\vartheta(\varepsilon), \zeta(\varepsilon), m(\varepsilon)) \equiv (\mathbf{v}(\varepsilon), \omega(\varepsilon), \varphi(\varepsilon), \zeta(\varepsilon), m(\varepsilon)),\\ \mathbf{q} &= (\vartheta, \zeta, m) \equiv (\mathbf{v}, \omega, \varphi, \zeta, m),\\ \mathcal{T}^{s,r} &= \mathcal{W}_0^{1-s,r'}(\Omega) \times (\mathbb{W}^{-s,r'}(\Omega))^3 \times \mathbb{R}.\end{aligned}$$

We must prove that $\mathbf{q}(\varepsilon) \to \mathbf{q}$ in $\mathcal{F}^{s,r}$. In view of Lemma 11.4.3, the sequence $\mathbf{q}(\varepsilon)$, $\varepsilon \in (0, \varepsilon_\star]$, is relatively compact in $\mathcal{F}^{s,r}$. Hence it contains a subsequence, still denoted by $\mathbf{q}(\varepsilon)$, which converges in $\mathcal{F}^{s,r}$ to some $\mathbf{q}^*$. Notice that $\mathbf{v}(\varepsilon) \to \mathbf{v}^*$ in $Y^{s,r}$ and hence $\mathbf{v}(\varepsilon) \to \mathbf{v}^*$ in $C^1(\Omega)$. Since $\mathbf{v}(\varepsilon)$ vanishes at $\partial\Omega$, it follows that $\mathbf{v}(\varepsilon) \to \mathbf{v}^*$ in $W_0^{1,r'}(\Omega)$. On the other hand, $W_0^{1,r'}(\Omega)$ is continuously embedded and dense in the interpolation space $\mathcal{W}_0^{1-s,r'}(\Omega) = [L^{r'}(\Omega), W_0^{1,r'}(\Omega)]_{1-s,r'}$. Therefore, we have $\mathbf{v}(\varepsilon) \to \mathbf{v}^*$ in $W^{1-s,r'}(\Omega)$.

Next, $(\omega(\varepsilon), \varphi(\varepsilon), \zeta(\varepsilon)) \to (\omega^*, \varphi^*, \zeta^*)$ in $W^{s,r}(\Omega)$ and hence in $L^r(\Omega)$. Since $r > r'$, the space $L^r(\Omega)$ is continuously embedded in $L^{r'}(\Omega)$. In its turn, $L^{r'}(\Omega)$ is continuously embedded and dense in $\mathbb{W}^{-s,r'}(\Omega) = [L^{r'}(\Omega), W^{-1,r'}(\Omega)]_{s,r'}$. Hence $(\omega(\varepsilon), \varphi(\varepsilon), \zeta(\varepsilon)) \to (\omega^*, \varphi^*, \zeta^*)$ in $\mathbb{W}^{-s,r'}(\Omega)$. Thus $\mathbf{q}(\varepsilon) \to \mathbf{q}^*$ in $\mathcal{T}^{s,r}$. On the other hand, formula (11.4.6) and Lemma 11.4.9 imply

$$\mathbf{q} - \mathbf{q}(\varepsilon) = \varepsilon(\mathbf{w}_\varepsilon, \omega_\varepsilon, \psi_\varepsilon, n_\varepsilon) \to 0 \quad \text{in } \mathcal{T}^{s,r}.$$

It follows that the limit $\mathbf{q}^* = \mathbf{q}$ is independent of the choice of a converging subsequence of $\mathbf{q}(\varepsilon)$. Hence the whole sequence $\mathbf{q}(\varepsilon)$ converges to $\mathbf{q}$ in $\mathcal{T}^{s,r}$. □

Our last task is to pass to the limit as $\varepsilon \to 0$ in the transposed equations (11.4.18). We begin with the study of the coefficients of these equations.

**Lemma 11.4.11.** *Let Condition 11.4.2 be satisfied. Let $b_{\varepsilon,ij}$, $\mathscr{H}_\varepsilon$, and $\mathscr{D}(\varepsilon)$, $\mathfrak{d}(\varepsilon)$ be given by* (11.4.14). *Then*

$$\begin{aligned} &b_{\varepsilon,ij} \to a_{ij} \quad \text{in } X^{s,r}, \\ &\mathscr{H}_\varepsilon \to \mathscr{H}_0 \quad \text{in } \mathcal{L}(W^{s+1,r}(\Omega) \to \mathcal{W}^{s-1,r}(\Omega)), \\ &-\varepsilon^{-1}\mathscr{D}(\varepsilon) \to \mathscr{D}_0 \quad \text{in } L^2(\Omega), \quad -\varepsilon^{-1}\mathfrak{d}(\varepsilon) \to \mathfrak{d}_0 \quad \text{in } C(\Omega), \end{aligned} \tag{11.4.31}$$

*where $a_{ij}$ are given by* (11.4.8), *$\mathscr{D}_0$ and $\mathfrak{d}_0$ are given by* (11.4.9), *and $\mathscr{H}_0$ is given by* (11.4.11).

*Proof.* It follows from (11.4.14) that $b_{\varepsilon,ij}$ are $C^3$ functions of $\vartheta$, $\zeta$ and $\vartheta(\varepsilon)$, $\zeta(\varepsilon)$ defined over the whole space and $a_{ij} = b_{0,ij}$. On the other hand, in view of Lemma 11.4.10, $\vartheta(\varepsilon)$ and $\zeta(\varepsilon)$ converge to $\vartheta$ and $\zeta$ in $X^{s,r}$ as $\varepsilon \to 0$. It follows from Lemma 11.2.8 that the composition operator, which assigns to each element of $X^{s,r}$ the composition of $C^2$ function with this element, is continuous in $X^{s,r}$. Hence the first relation in (11.4.31) is a straightforward consequence of Lemma 11.2.8.

Let us estimate the norm of $\mathscr{H}_\varepsilon - \mathscr{H}_0$. Choose $\mathbf{h} \in W^{s+1,r}(\Omega)$ such that $\|\mathbf{h}\|_{W^{s+1,r}(\Omega)} \le 1$. We have

$$\mathscr{H}_\varepsilon(\mathbf{h}) - \mathscr{H}_0(\mathbf{h}) = (\varphi(\varepsilon) - \varphi)\nabla\mathbf{u}\,\mathbf{h} + \operatorname{div}\big((\varrho(\varepsilon)\mathbf{u}(\varepsilon) - \varrho\mathbf{u}) \otimes \mathbf{h}\big).$$

Here we use identities (11.4.13). Applying Lemma 11.2.13 to the last term on the right hand side we arrive at

$$\begin{aligned} &\|\mathscr{H}_\varepsilon(\mathbf{h}) - \mathscr{H}_0(\mathbf{h})\|_{\mathcal{W}^{s-1,r}(\Omega)} \\ &\qquad \le c\|\varphi(\varepsilon) - \varphi)\nabla\mathbf{u}\,\mathbf{h}\|_{W^{s,r}(\Omega)} + c\|(\varrho(\varepsilon)\mathbf{u}(\varepsilon) - \varrho\mathbf{u}) \otimes \mathbf{h}\|_{W^{s,r}(\Omega)}. \end{aligned} \tag{11.4.32}$$

Since $W^{s,r}(\Omega)$ is a Banach algebra and $\|\nabla\mathbf{u}\|_{W^{s,r}(\Omega)} \le c\|\mathbf{u}\|_{W^{s+1,r}(\Omega)} \le c$, we have

$$\begin{aligned} \|(\varphi(\varepsilon) - \varphi)\nabla\mathbf{u}\,\mathbf{h}\|_{W^{s,r}(\Omega)} &\le c\|\varphi(\varepsilon) - \varphi\|_{W^{s,r}(\Omega)}\|\nabla\mathbf{u}\|_{W^{s,r}(\Omega)}\|\mathbf{h}\|_{W^{s,r}(\Omega)} \\ &\le c\|\varphi(\varepsilon) - \varphi\|_{W^{s,r}(\Omega)}. \end{aligned} \tag{11.4.33}$$

Next we have

$$\begin{aligned}\|(\varrho(\varepsilon)\mathbf{u}(\varepsilon)-\varrho\mathbf{u})\otimes\mathbf{h}\|_{W^{s,r}(\Omega)} &\leq c\|\varrho(\varepsilon)\mathbf{u}(\varepsilon)-\varrho\mathbf{u}\|_{W^{s,r}(\Omega)}\|\mathbf{h}\|_{W^{s,r}(\Omega)}\\ &\leq c\|\varrho(\varepsilon)\mathbf{u}(\varepsilon)-\varrho\mathbf{u}\|_{W^{s,r}(\Omega)}.\end{aligned}$$

Using the identity

$$\varrho(\varepsilon)\mathbf{u}(\varepsilon)-\varrho\mathbf{u}=(\varphi(\varepsilon)-\varphi)\mathbf{u}(\varepsilon)+\varrho(\mathbf{v}(\varepsilon)-\mathbf{v})$$

and noting that $\mathbf{u}(\varepsilon)$, $\varrho(\varepsilon)$ are uniformly bounded in $W^{s,r}(\Omega)$ we obtain

$$\|\varrho(\varepsilon)\mathbf{u}(\varepsilon)-\varrho\mathbf{u}\|_{W^{s,r}(\Omega)}\leq c\|\varphi(\varepsilon)-\varphi\|_{W^{s,r}(\Omega)}+c\|\mathbf{v}(\varepsilon)-\mathbf{v}\|_{W^{s,r}(\Omega)}.$$

Thus we get

$$\|(\varrho(\varepsilon)\mathbf{u}(\varepsilon)-\varrho\mathbf{u})\otimes\mathbf{h}\|_{W^{s,r}(\Omega)}\leq c\|\varphi(\varepsilon)-\varphi\|_{W^{s,r}(\Omega)}+c\|\mathbf{v}(\varepsilon)-\mathbf{v}\|_{W^{s,r}(\Omega)}. \tag{11.4.34}$$

Inserting (11.4.33) and (11.4.34) into (11.4.32) leads to

$$\|\mathscr{H}_\varepsilon(\mathbf{h})-\mathscr{H}_0(\mathbf{h})\|_{\mathcal{W}^{s-1,r}(\Omega)}\leq c\|\varphi(\varepsilon)-\varphi\|_{W^{s,r}(\Omega)}+c\|\mathbf{v}(\varepsilon)-\mathbf{v}\|_{W^{s,r}(\Omega)}\to 0$$

as $\varepsilon\to 0$. It follows that $\mathscr{H}_\varepsilon(\mathbf{h})\to\mathscr{H}_0(\mathbf{h})$ in $\mathcal{W}^{s-1,r}(\Omega)$ uniformly on the unit ball in $W^{s+1,r}(\Omega)$, which leads to the second relation in (11.4.31).

It remains to prove the third. Lemma 11.4.8 and expression (11.4.9) for $\mathscr{D}_0$ imply

$$\mathscr{D}_0+\varepsilon^{-1}\mathscr{D}(\varepsilon)=\mathscr{D}_3(\varepsilon)+\operatorname{Re}\mathscr{D}_4(\varepsilon)+\varepsilon\mathscr{D}_2(\varepsilon), \tag{11.4.35}$$

where

$$\begin{aligned}\mathscr{D}_3(\varepsilon)&=\operatorname{div}\big(\mathbb{T}\nabla(\mathbf{u}-\mathbf{u}(\varepsilon))\big)+\mathbb{D}^\top\Delta(\mathbf{u}-\mathbf{u}(\varepsilon))+\Delta\big(\mathbb{D}(\mathbf{u}-\mathbf{u}(\varepsilon))\big),\\ \mathscr{D}_4(\varepsilon)&=\varrho(\varepsilon)\big(\mathbb{D}^\top\mathbf{u}(\varepsilon)\nabla\mathbf{u}(\varepsilon)+\mathbf{u}(\varepsilon)\nabla(\mathbb{D}\mathbf{u}(\varepsilon))\big)-\varrho\big(\mathbb{D}^\top\mathbf{u}\nabla\mathbf{u}+\mathbf{u}\nabla(\mathbb{D}\mathbf{u})\big),\\ \mathbb{T}&=\operatorname{div}\mathbf{T}\,\mathbb{I}-D\mathbf{T}-(D\mathbf{T})^\top,\quad \mathbb{D}=\operatorname{div}\mathbf{T}\,\mathbb{I}-D\mathbf{T}.\end{aligned}$$

Lemma 11.4.10 and identities (11.4.13) imply that $\mathbf{u}(\varepsilon)\to\mathbf{u}$ in $Y^{s,r}=W^{s+1,r}(\Omega)\cap W^{2,2}(\Omega)$. Since $sr>3$, the embedding $W^{s+1,r}(\Omega)\hookrightarrow C^1(\Omega)$ is continuous. Thus

$$\mathbf{u}(\varepsilon)\to\mathbf{u}\quad\text{in } W^{2,2}(\Omega) \text{ and in } C^1(\Omega)\quad\text{as } \varepsilon\to 0.$$

Next, $\varrho(\varepsilon)\to\varrho$ in $X^{s,r}=W^{s,r}(\Omega)\cap W^{1,2}(\Omega)$. Since the embedding $W^{s,r}(\Omega)\hookrightarrow C(\Omega)$ is continuous, we have

$$\varrho(\varepsilon)\to\varrho\quad\text{in } W^{1,2}(\Omega) \text{ and in } C(\Omega)\quad\text{as } \varepsilon\to 0.$$

It follows that

$$\mathscr{D}_3(\varepsilon)\to 0\quad\text{in } L^2(\Omega)\quad\text{and}\quad\mathscr{D}_4(\varepsilon)\to 0\quad\text{in } C(\Omega)\quad\text{as } \varepsilon\to 0.$$

Estimate (11.4.24) implies that $\varepsilon\|\mathscr{D}_2(\varepsilon)\|_{L^2(\Omega)} \le c\varepsilon \to 0$ as $\varepsilon \to 0$. Letting $\varepsilon \to 0$ in (11.4.35) we obtain

$$\mathscr{D}_0 + \varepsilon^{-1}\mathscr{D}(\varepsilon) \to 0 \quad \text{in } L^2(\Omega) \quad \text{as } \varepsilon \to 0.$$

It remains to note that in view of (11.4.24),

$$\mathfrak{d}_0 + \varepsilon^{-1}\mathfrak{d}(\varepsilon) = \varepsilon\mathfrak{d}_2(\varepsilon) \to 0 \quad \text{in } C(\Omega) \quad \text{as } \varepsilon \to 0. \qquad \square$$

The next lemma shows that solutions to the transposed problem converge to solutions of problem (11.4.10) as $\varepsilon \to 0$.

**Lemma 11.4.12.** *Let Condition 11.4.2 be satisfied. Let* $\mathfrak{h}_\varepsilon = (\mathbf{h}_\varepsilon, g_\varepsilon, \varsigma_\varepsilon, \upsilon_\varepsilon, l_\varepsilon)$, $\varepsilon \in (0, \varepsilon_\star]$, *be the solutions to problem* (11.4.18). *Then*

$$\mathfrak{h}_\varepsilon \rightharpoonup \mathfrak{h} \quad \textit{weakly in } W^{s+1,r}(\Omega) \times (W^{s,r}(\Omega))^3 \times \mathbb{R},$$

*where* $\mathfrak{h} = (\mathbf{h}, g, \varsigma, \upsilon, l)$ *is the solution to problem* (11.4.10).

*Proof.* We have to pass to the limit as $\varepsilon \to 0$ in equations (11.4.18). Recall the notation $\mathcal{V}^{s,r} = W^{s+1,r}(\Omega) \times (W^{s,r}(\Omega))^3 \times \mathbb{R}$. By Lemma 11.2.13 the space $\mathcal{V}^{s,r}$ is reflexive. In view of Lemma 11.4.7 the sequence $\mathfrak{h}_\varepsilon$ is bounded in $\mathcal{V}^{s,r}$. Hence there are $\mathfrak{h} \in \mathcal{V}^{s,r}$ and a subsequence, still denoted by $\mathfrak{h}_\varepsilon$, such that $\mathfrak{h}_\varepsilon$ converges weakly to $\mathfrak{h}$ as $\varepsilon \to 0$. Let us prove that $\mathfrak{h}$ satisfies equations (11.4.10). We begin with the Stokes equations (11.4.18a) which are the integral part of problem (11.4.18). These equations can be written in the form

$$\begin{aligned} &\Delta\mathbf{h}_\varepsilon - \nabla g_\varepsilon = \mathbf{Q}_\varepsilon, \quad \operatorname{div}\mathbf{h}_\varepsilon = \Pi P_\varepsilon \quad \text{in } \Omega,\\ &\mathbf{h}_\varepsilon = 0 \quad \text{on } \partial\Omega, \quad g_\varepsilon = \Pi g_\varepsilon, \end{aligned} \tag{11.4.36}$$

where

$$\begin{aligned} \mathbf{Q}_\varepsilon &= \mathsf{Re}\,\mathscr{H}_\varepsilon(\mathbf{h}_\varepsilon) - \varsigma_\varepsilon\nabla\varphi(\varepsilon) - \zeta(\varepsilon)\nabla\upsilon_\varepsilon + \mathbf{H},\\ P_\varepsilon &= b_{\varepsilon,22}g_\varepsilon + b_{\varepsilon,12}\varsigma_\varepsilon + b_{\varepsilon,32}l_\varepsilon + G. \end{aligned}$$

Set

$$\mathbf{Q} = \mathsf{Re}\,\mathscr{H}_0(\mathbf{h}) - \varsigma\nabla\varphi - \zeta\nabla\upsilon + \mathbf{H}, \quad P = a_{22}g + a_{12}\varsigma + a_{32}l + G.$$

We have

$$\begin{aligned} \mathbf{Q} - \mathbf{Q}_\varepsilon = {}& \mathsf{Re}(\mathscr{H}_0 - \mathscr{H}_\varepsilon)(\mathbf{h}_\varepsilon) - \varsigma_\varepsilon\nabla(\varphi - \varphi(\varepsilon)) - (\zeta - \zeta(\varepsilon))\nabla\upsilon_\varepsilon\\ &+ \mathsf{Re}\,\mathscr{H}_0(\mathbf{h} - \mathbf{h}_\varepsilon) - (\varsigma - \varsigma_\varepsilon)\nabla\varphi - \zeta\nabla(\upsilon - \upsilon_\varepsilon). \end{aligned}$$

Lemma 11.4.11 implies

$$\begin{aligned} \|(\mathscr{H}_0 - \mathscr{H}_\varepsilon)(\mathbf{h}_\varepsilon)\|_{\mathcal{W}^{s-1,r}(\Omega)} &\le \|\mathscr{H}_0 - \mathscr{H}_\varepsilon\|_{\mathcal{L}(W^{s+1,r}(\Omega)\to\mathcal{W}^{s-1,r}(\Omega))}\|\mathbf{h}_\varepsilon\|_{W^{s+1,r}(\Omega)}\\ &\le c\|\mathscr{H}_0 - \mathscr{H}_\varepsilon\|_{\mathcal{L}(W^{s+1,r}(\Omega)\to\mathcal{W}^{s-1,r}(\Omega))} \to 0 \quad \text{as } \varepsilon \to 0. \end{aligned} \tag{11.4.37}$$

In view of Lemma 11.3.7 the bilinear mapping

$$(W^{s,r}(\Omega))^2 \ni (\varphi, \psi) \mapsto \psi\nabla\varphi \in \mathcal{W}^{s-1,r}(\Omega)$$

is continuous. It follows from this and Lemma 11.4.10 that

$$\begin{aligned}
&\|\varsigma_\varepsilon \nabla(\varphi - \varphi(\varepsilon))\|_{\mathcal{W}^{s-1,r}(\Omega)} + \|(\zeta - \zeta(\varepsilon))\nabla \upsilon_\varepsilon\|_{\mathcal{W}^{s-1,r}(\Omega)} \\
&\quad \le c\|\varphi - \varphi(\varepsilon)\|_{W^{s,r}(\Omega)}\|\varsigma_\varepsilon\|_{W^{s,r}(\Omega)} + c\|\zeta - \zeta(\varepsilon)\|_{W^{s,r}(\Omega)}\|\upsilon_\varepsilon\|_{W^{s,r}(\Omega)} \\
&\quad \le c\|\varphi - \varphi(\varepsilon)\|_{W^{s,r}(\Omega)} + c\|\zeta - \zeta(\varepsilon)\|_{W^{s,r}(\Omega)} \to 0 \quad \text{as } \varepsilon \to 0. \qquad (11.4.38)
\end{aligned}$$

The operator $\mathscr{H}_0$ is a special case of the operator $\mathscr{H}$ given by (11.3.7) with $\mathbf{N}_0 = \mathbb{I}$ and $\mathbf{u}_0 = \mathbf{u}$. Lemma 11.3.10 shows that $\mathscr{H}_0 : W^{s+1,r}(\Omega) \to W^{s,r}(\Omega)$ is continuous, and hence is continuous in the weak topology, which leads to

$$\mathscr{H}_0(\mathbf{h} - \mathbf{h}_\varepsilon) \rightharpoonup 0 \quad \text{weakly in } \mathcal{W}^{s-1,r}(\Omega) \quad \text{as } \varepsilon \to 0. \qquad (11.4.39)$$

Next, Lemma 11.3.7 implies that the trilinear form

$$(W^{s,r}(\Omega))^2 \times \mathcal{W}_0^{1-s,r'}(\Omega) \ni (\varphi, \psi, \mathbf{w}) \mapsto \langle \psi\nabla\varphi, \mathbf{w}\rangle_1 \in \mathbb{R},$$

where $\langle \cdot, \cdot \rangle_1$ denotes the duality pairing between $\mathcal{W}^{s-1,r}(\Omega)$ and $\mathcal{W}_0^{1-s,r'}(\Omega)$, is continuous. Hence for fixed $\varphi \in W^{s,r}(\Omega)$ and $\mathbf{w} \in \mathcal{W}_0^{s,r}(\Omega)$ the mapping $\psi \mapsto \langle \psi\nabla\varphi, \mathbf{w}\rangle_1$ defines a continuous linear functional on $W^{s,r}(\Omega)$. Since $\varsigma - \varsigma_\varepsilon \rightharpoonup 0$ weakly in $W^{s,r}(\Omega)$, it follows that

$$\langle (\varsigma - \varsigma_\varepsilon)\nabla\varphi, \mathbf{w}\rangle_1 \to 0 \quad \text{for all } \mathbf{w} \in \mathcal{W}_0^{1-s,r'}(\Omega) \quad \text{as } \varepsilon \to 0.$$

This means that $(\varsigma - \varsigma_\varepsilon)\nabla\varphi \to 0$ weakly$^\star$ in $\mathcal{W}^{s-1,r}(\Omega)$. Notice that $\mathcal{W}^{s-1,r}(\Omega)$ is reflexive as the dual space to the reflexive space $\mathcal{W}_0^{1-s,r'}(\Omega)$. Thus we get

$$(\varsigma - \varsigma_\varepsilon)\nabla\varphi \rightharpoonup 0 \quad \text{weakly in } \mathcal{W}^{s-1,r}(\Omega) \quad \text{as } \varepsilon \to 0. \qquad (11.4.40)$$

The same arguments give

$$\zeta\nabla(\upsilon - \upsilon_\varepsilon) \rightharpoonup 0 \quad \text{weakly in } \mathcal{W}^{s-1,r}(\Omega) \quad \text{as } \varepsilon \to 0.$$

Combining this with (11.4.38)–(11.4.40) we obtain

$$\mathbf{Q}_\varepsilon \rightharpoonup \mathbf{Q} \quad \text{weakly in } \mathcal{W}^{s-1,r}(\Omega) \quad \text{as } \varepsilon \to 0. \qquad (11.4.41)$$

Let us consider the sequence $P_\varepsilon$. In view of Lemma 11.4.11, the coefficients $b_{\varepsilon,22}$, $b_{\varepsilon,12}$ , $b_{\varepsilon,32}$ converge to $a_{22}$, $a_{12}$, $a_{32}$ in $W^{s,r}$. On the other hand $g_\varepsilon$, $\varsigma_\varepsilon$ converge to $g$, $\varsigma$ weakly in $W^{s,r}$. Hence

$$b_{\varepsilon,22}g_\varepsilon + b_{\varepsilon,12}\varsigma_\varepsilon + b_{\varepsilon,32}l_\varepsilon \to a_{22}g + a_{12}\varsigma + a_{32}l \quad \text{weakly in } \mathcal{W}^{s-1,r}(\Omega) \quad \text{as } \varepsilon \to 0,$$

which leads to

$$P_\varepsilon \rightharpoonup P \quad \text{weakly in } \mathcal{W}^{s-1,r}(\Omega) \quad \text{as } \varepsilon \to 0. \tag{11.4.42}$$

Let us turn to the Stokes problem (11.4.36). Lemma 1.7.9 implies that the operator $T : \mathcal{W}^{s-1,r}(\Omega) \times W^{s,r}(\Omega) \to W^{s+1,r}(\Omega) \times W^{s,r}(\Omega)$, which assigns to a couple $(\mathbf{Q}_\varepsilon, \mathbf{P}_\varepsilon)$ the solution $(\mathbf{h}_\varepsilon, g_\varepsilon)$ of problem (11.4.36), is continuous and hence it is continuous in the weak topology. From this and (11.4.41), (11.4.42) we conclude that the weak limit $(\mathbf{h}, g)$ of the solutions to problem (11.4.36) satisfies

$$\begin{aligned} &\Delta \mathbf{h} - \nabla g = \mathbf{Q}, \quad \operatorname{div} \mathbf{h} = \Pi P \quad \text{in } \Omega, \\ &\mathbf{h} = 0 \quad \text{on } \partial\Omega, \quad g = \Pi g, \end{aligned}$$

which coincides with (11.4.10a).

It remains to prove that $\varsigma$ and $\upsilon$ satisfy the transport equations (11.4.10b) and (11.4.10c). Notice that these equations are understood in the weak sense. Let us consider the boundary value problem (11.4.18b), (11.4.18c) for the function $\varsigma_\varepsilon$. Recall that $\varsigma_\varepsilon$ is a weak solution to this problem if and only if the integral identity

$$\int_\Omega \Big(\varsigma_\varepsilon \mathbf{u} \nabla \phi + \phi\big(\sigma \varsigma_\varepsilon - \mathsf{Re}\, \mathscr{M}(\mathbf{h}_\varepsilon) - b_{\varepsilon,11}\varsigma_\varepsilon - b_{\varepsilon,21} g_\varepsilon - b_{\varepsilon,31} l_\varepsilon - F\big)\Big)\, dx = 0 \tag{11.4.43}$$

holds for every $\phi \in C^1(\Omega)$ vanishing in a neighborhood of $\partial\Omega \setminus \Sigma_{\text{out}}$. Since the embedding $W^{s+1,r}(\Omega) \hookrightarrow W^{s,r}(\Omega)$ is compact, the sequence $\mathbf{h}_\varepsilon$ converges to $\mathbf{h}$ strongly in $W^{s,r}(\Omega)$. On the other hand, Lemma 11.3.10 implies that the operator $\mathscr{M}_0 : W^{s,r}(\Omega) \to W^{s,r}(\Omega)$ is continuous. Thus we get

$$\mathscr{M}_0(\mathbf{h}_\varepsilon) \to \mathscr{M}_0(\mathbf{h}) \quad \text{in } W^{s,r}(\Omega) \quad \text{as } \varepsilon \to 0.$$

It follows that $\mathscr{M}_0(\mathbf{h}_\varepsilon) \to \mathscr{M}_0(\mathbf{h})$ in $C(\Omega)$. Since the embedding $W^{s,r}(\Omega) \hookrightarrow C(\Omega)$ is compact the functions $(g_\varepsilon, \varsigma_\varepsilon, \upsilon_\varepsilon)$ converge to $(g, \varsigma, \upsilon)$. In view of Lemma 11.4.11, the functions $b_{\varepsilon,ij}$ converge to $a_{ij}$ in $C(\Omega)$. Letting $\varepsilon \to 0$ in (11.4.43) we arrive at

$$\int_\Omega \Big(\varsigma \mathbf{u} \nabla \phi + \phi\big(\sigma \varsigma - \mathsf{Re}\, \mathscr{M}(\mathbf{h}) - a_{11}\varsigma - a_{21} g - a_{31} l - F\big)\Big)\, dx = 0,$$

which means that $\varsigma$ satisfies the transport equation (11.4.10b) and the boundary condition (11.4.10d). The same arguments show that $\upsilon$ also satisfies (11.4.10c), (11.4.10d). Obviously we have

$$\int_\Omega (b_{\varepsilon,13}\varsigma_\varepsilon + b_{\varepsilon,23} g_\varepsilon)\, dx \to \int_\Omega (a_{13}\varsigma + a_{23} g)\, dx \quad \text{as } \varepsilon \to 0 \quad \text{and} \quad g = \Pi g.$$

Assembling the results obtained we conclude that $\mathfrak{h} = (\mathbf{w}, g, \varsigma, \upsilon, l)$ satisfies equations (11.4.10). Thus we have proved that every weak limit $\mathfrak{h} \in \mathcal{V}^{s,r}$ of the sequence $\mathfrak{h}_\varepsilon = (\mathbf{w}_\varepsilon, g_\varepsilon, \varsigma_\varepsilon, \upsilon_\varepsilon, l_\varepsilon)$ is a solution to the boundary value problem (11.4.10). Since such a solution is unique, it follows that the whole sequence $\mathfrak{h}_\varepsilon$ converges weakly in $\mathcal{V}^{s,r}$ to $\mathfrak{h}$. □

**Proof of Theorem 11.4.6.** We are now in a position to complete the proof of Theorem 11.4.6. Observe that in view of Lemma 11.4.9, the sequence $(\mathbf{w}_\varepsilon, \omega_\varepsilon, \psi_\varepsilon, \xi_\varepsilon, n_\varepsilon)$ given by (11.4.6) is bounded in $\mathcal{W}_0^{1-s,r'}(\Omega) \times (\mathbb{W}^{-s,r'}(\Omega))^3 \times \mathbb{R}$. Since this space is reflexive we may assume, passing to a subsequence if necessary, that there are $(\mathbf{w}, \omega, \psi, \xi, n) \in \mathcal{W}_0^{1-s,r'}(\Omega) \times (\mathbb{W}^{-s,r'}(\Omega))^3 \times \mathbb{R}$ such that

$$\mathbf{w}_\varepsilon \rightharpoonup \mathbf{w} \text{ weakly in } \mathcal{W}_0^{1-s,r'}(\Omega), \quad (\omega_\varepsilon, \psi_\varepsilon, \xi_\varepsilon) \rightharpoonup (\omega, \psi, \xi) \text{ weakly in } \mathbb{W}^{-s,r'}(\Omega).$$

On the other hand, Lemma 11.4.7 implies that for $(\mathbf{H}, G, F, M, e) \in \mathcal{W}^{s-1,r}(\Omega) \times (W^{s,r}(\Omega))^3 \times \mathbb{R}$, we have

$$\begin{aligned} &\langle \mathbf{H}, \mathbf{w}_\varepsilon \rangle_1 + \langle \omega_\varepsilon, G \rangle_0 + \langle \psi_\varepsilon, F \rangle_0 + \langle \xi_\varepsilon, \mathcal{M} \rangle_0 + \varkappa^{-1} n_\varepsilon e \\ &\quad = -\frac{1}{\varepsilon} \int_\Omega \big(\mathbf{h}_\varepsilon \cdot \mathscr{D}(\varepsilon) + g_\varepsilon\, b_{\varepsilon,20}\mathfrak{d}(\varepsilon) + \varsigma_\varepsilon\, b_{\varepsilon,10}\mathfrak{d}(\varepsilon) + v_\varepsilon\, \sigma\mathfrak{d}(\varepsilon) + l_\varepsilon\, b_{\varepsilon,30}\mathfrak{d}(\varepsilon)\big)\, dx, \end{aligned} \tag{11.4.44}$$

where $(\mathbf{h}_\varepsilon, g_\varepsilon, \varsigma_\varepsilon, v, l_\varepsilon)$ is the solution to problem (11.4.18). Since $\mathcal{W}_0^{1-s,r'}(\Omega) \times (\mathbb{W}^{-s,r'}(\Omega))^3 \times \mathbb{R}$ is dual to $\mathcal{W}^{s-1,r}(\Omega) \times (W^{s,r}(\Omega))^3 \times \mathbb{R}$ we can pass to the limit on the left hand side of this identity to obtain

$$\begin{aligned} &\langle \mathbf{H}, \mathbf{w}_\varepsilon \rangle_1 + \langle \omega_\varepsilon, G \rangle_0 + \langle \psi_\varepsilon, F \rangle_0 + \langle \xi_\varepsilon, \mathcal{M} \rangle_0 + \varkappa^{-1} n_\varepsilon e \\ &\qquad\qquad \to \langle \mathbf{H}, \mathbf{w} \rangle_1 + \langle \omega, G \rangle_0 + \langle \psi, F \rangle_0 + \langle \xi, \mathcal{M} \rangle_0 + \varkappa_0^{-1} n e \end{aligned}$$

as $\varepsilon \to 0$. Since the embedding of $\mathcal{V}^{s,r} = W^{s+1,r}(\Omega) \times (W^{s,r}(\Omega))^3 \times \mathbb{R}$ into $C(\Omega)^6 \times \mathbb{R}$ is continuous, it follows from Lemma 11.4.12 that

$$(\mathbf{h}_\varepsilon, g_\varepsilon, \varsigma_\varepsilon, v, l_\varepsilon) \to (\mathbf{h}, g, \varsigma, v, l) \quad \text{in } C(\Omega)^6 \times \mathbb{R},$$

Here $(\mathbf{h}, g, \varsigma, v, l) \in \mathcal{V}^{s,r}$ is the solution to problem (11.4.10) given by Lemma 11.4.6. Next, Lemma 11.4.11 implies

$$b_{\varepsilon,ij} \to a_{ij} \quad \text{in } C(\Omega), \quad -\varepsilon^{-1}\mathscr{D}(\varepsilon) \to \mathscr{D}_0 \quad \text{in } L^2(\Omega), \quad -\varepsilon^{-1}\mathfrak{d}(\varepsilon) \to \mathfrak{d}_0 \quad \text{in } C(\Omega).$$

Hence we can pass to the limit on the right hand side of (11.4.44) to obtain

$$\begin{aligned} &-\frac{1}{\varepsilon} \int_\Omega \big(\mathbf{h}_\varepsilon \cdot \mathscr{D}(\varepsilon) + g_\varepsilon\, b_{\varepsilon,20}\mathfrak{d}(\varepsilon) + \varsigma_\varepsilon\, b_{\varepsilon,10}\mathfrak{d}(\varepsilon) + v_\varepsilon\, \sigma\mathfrak{d}(\varepsilon) + l_\varepsilon\, b_{\varepsilon,30}\mathfrak{d}(\varepsilon)\big)\, dx \\ &\qquad\qquad \to \int_\Omega \big(\mathbf{h} \cdot \mathscr{D}_0 + g\, a_{20}\mathfrak{d}_0 + \varsigma\, a_{10}\mathfrak{d}_0 + v\, \sigma\mathfrak{d}_0 + l\, a_{30}\mathfrak{d}\big)\, dx. \end{aligned}$$

Letting $\varepsilon \to 0$ in (11.4.44) we finally obtain

$$\begin{aligned} &\langle \mathbf{H}, \mathbf{w} \rangle_1 + \langle \omega, G \rangle_0 + \langle \psi, F \rangle_0 + \langle \xi, \mathcal{M} \rangle_0 + \varkappa^{-1} n e \\ &\qquad = \int_\Omega \big(\mathbf{h} \cdot \mathscr{D}_0 + g\, a_{20}\mathfrak{d}_0 + \varsigma\, a_{10}\mathfrak{d}_0 + v\, \sigma\mathfrak{d}_0 + l\, a_{30}\mathfrak{d}_0\big)\, dx, \end{aligned} \tag{11.4.45}$$

where $(\mathbf{h}, g, \varsigma, \upsilon, l)$ is a solution to problem (11.4.10). In view of Lemma 11.4.6 such a solution exists and is unique. Thus we have proved that every weak limit $(\mathbf{w}, \omega, \psi, \xi, n)$ of the sequence $(\mathbf{w}_\varepsilon, \omega_\varepsilon, \psi_\varepsilon, \xi_\varepsilon, n_\varepsilon)$ satisfies identity (11.4.45).

It remains to show that this sequence has a unique weak limit point, and hence converges weakly to this point. Assume that there are $(\mathbf{w}, \omega, \psi, \xi, n)$ and $(\mathbf{w}', \omega', \psi', \xi', n')$ in $\mathcal{W}_0^{1-s,r'}(\Omega) \times (\mathbb{W}^{-s,r'}(\Omega))^3 \times \mathbb{R}$ which satisfy (11.4.45). Then for all $(\mathbf{H}, G, F, M, e) \in \mathcal{W}^{s-1,r}(\Omega) \times (W^{s,r}(\Omega))^3 \times \mathbb{R}$,

$$\langle \mathbf{H}, \mathbf{w} - \mathbf{w}' \rangle_1 + \langle \omega - \omega', G \rangle_0 + \langle \psi - \psi', F \rangle_0 + \langle \xi - \xi', \mathcal{M} \rangle_0 + \varkappa^{-1}(n - n')e = 0.$$

Since $\mathcal{W}^{s-1,r}(\Omega) \times (W^{s,r}(\Omega))^3 \times \mathbb{R}$ is dual to $\mathcal{W}_0^{1-s,r'}(\Omega) \times (\mathbb{W}^{-s,r'}(\Omega))^3 \times \mathbb{R}$, it follows that $(\mathbf{w}', \omega', \psi', \xi', n') = (\mathbf{w}, \omega, \psi, \xi, n)$.

### 11.4.3 Conclusion. Material and shape derivatives

Recall that our main goal was to investigate the domain perturbations influence on solutions to Navier-Stokes equations. In order to specify the domain perturbations we considered the one-parameter family of domains $\Omega_\varepsilon = (\mathrm{Id} + \varepsilon\mathbb{T})(\Omega)$, where $\mathbf{T} : \mathbb{R}^3 \to \mathbb{R}^3$ is an arbitrary compactly supported smooth mapping. Thus we have got two boundary value problems. The first is the boundary value problem (11.1.1) for $(\mathbf{u}, \varrho)$ in the unperturbed domain $\Omega$, and the second is the boundary value problem (11.1.2) for $(\bar{\mathbf{u}}_\varepsilon, \bar{\varrho}_\varepsilon)$ in the domain $\Omega_\varepsilon$. We stress that $\mathbf{u}(x)$ and $\varrho(x)$ are the velocity and density calculated at the physical point $x \in \Omega$, and $\bar{\mathbf{u}}_\varepsilon(y)$ and $\bar{\varrho}_\varepsilon(y)$ are the perturbed velocity and density calculated at the physical point $y \in \Omega_\varepsilon$. A comparison between perturbed and unperturbed quantities is difficult since they are defined in different domains. In order to cope with this difficulty we introduced the Piola transform $\mathbf{u}_\varepsilon$ and the transformed density $\varrho_\varepsilon$ defined in $\Omega$ and given by

$$\mathbf{u}_\varepsilon(x) = \mathbf{N}(\varepsilon)(x)\,\bar{\mathbf{u}}_\varepsilon(x + \varepsilon\mathbf{T}(x)), \quad \varrho_\varepsilon(x) = \bar{\varrho}_\varepsilon(x + \varepsilon\mathbf{T}(x)), \quad x \in \Omega,$$

where

$$\mathbf{N}(\varepsilon) = \det(\mathbb{I} + \varepsilon D\mathbf{T})\,(\mathbb{I} + \varepsilon D\mathbf{T})^{-1}.$$

In this setting we refer to $\Omega$ as the *reference domain* in order to emphasize that for $\varepsilon \neq 0$, points of this domain cannot be regarded as points of the "physical" Euclidean space.

The change of independent variables in (11.1.2) leads to Problem 11.1.6 for the transformed functions and next to Problem 11.1.9 for their perturbations in the reference domain. Theorem 11.3.20 and Lemma 11.4.3 state the existence of a one-parameter family of solutions $\vartheta(\varepsilon) = (\mathbf{v}(\varepsilon), \pi(\varepsilon), \varphi(\varepsilon))$ to Problem 11.1.9. In its turn, the functions

$$\mathbf{u}(\varepsilon) = \mathbf{u}_\star + \mathbf{v}(\varepsilon), \quad \varrho(\varepsilon) = \varrho_\star + \varphi(\varepsilon),$$

where $(\mathbf{u}_\star, \varrho_\star)$ is the approximate solution given by (11.1.18), (11.1.19), are a solution to Problem 11.1.6. Theorem 11.4.6 shows that there are $\mathbf{w} \in \mathcal{W}_0^{1-s,r'}(\Omega)$ and $\psi \in \mathbb{W}^{-s,r'}(\Omega)$ such that

$$\varepsilon^{-1}(\mathbf{u}(\varepsilon)-\mathbf{u}) \rightharpoonup \mathbf{w} \text{ weakly in } \mathcal{W}_0^{1-s,r'}(\Omega),\ \varepsilon^{-1}(\varrho(\varepsilon)-\varrho) \rightharpoonup \psi \text{ weakly in } \mathbb{W}^{-s,r'}(\Omega);$$

$\mathbf{w}$ and $\psi$ are named material derivatives of solutions to Problem 11.1.6 and are denoted by

$$\dot{\mathbf{u}}(0) := \mathbf{w}, \quad \dot{\varrho}(0) := \psi.$$

These artificial quantities have no physical meaning because they express the derivative of the Piola transform of the velocity and the derivative of the density in the nonphysical reference domain. The real shape derivatives of the velocity and density at a fixed point of the real physical space are defined as follows:

$$\mathbf{u}'(0) := \lim_{\varepsilon\to 0} \varepsilon^{-1}(\bar{\mathbf{u}}_\varepsilon - \mathbf{u}), \quad \varrho'(0) := \lim_{\varepsilon\to 0} \varepsilon^{-1}(\bar{\varrho}_\varepsilon - \mathbf{u}).$$

It is easy to establish formally the connection between the material and shape derivatives using the formulae

$$\bar{\mathbf{u}}_\varepsilon = (\mathbf{N}(\varepsilon))^{-1} \circ (\mathrm{Id} + \varepsilon\mathbf{T})^{-1}\mathbf{u}(\varepsilon) \circ (\mathrm{Id} + \varepsilon\mathbf{T})^{-1}, \quad \bar{\varrho}_\varepsilon = \varrho(\varepsilon) \circ (\mathrm{Id} + \varepsilon\mathbf{T})^{-1}.$$

Direct calculations show that the shape derivatives can be recovered from the material derivatives:

$$\mathbf{u}'(0) = \dot{\mathbf{u}}(0) - \nabla\mathbf{u}\mathbf{T} + D\mathbf{T}\,\mathbf{u} - \operatorname{div}\mathbf{T}\,\mathbf{u}, \quad \varrho'(0) = \dot{\varrho}(0) - \nabla\varrho\cdot\mathbf{T}.$$

Since $\mathbf{u} \in W^{2,2}(\Omega)$ and $\varrho \in W^{1,2}(\Omega)$, all terms in the right hand sides are well defined. On the other hand, the justification of the boundary value problem obtained for the shape derivatives is a delicate task for Navier-Stokes equations [115].

## 11.5 Shape derivative of the drag functional. Adjoint state

Theorem 11.3.20 and Lemma 11.4.3 imply the existence and uniqueness of solutions to Problem 11.1.9 associated with the matrices $\mathbf{N}(\varepsilon)$. These solutions generate the solutions

$$\begin{aligned} &\mathbf{u}(\varepsilon) = \mathbf{u}_\star + \mathbf{v}(\varepsilon), \quad \varrho(\varepsilon) = \varrho_\star + \varphi(\varepsilon), \\ &q(\varepsilon) = q_\star + \lambda\sigma_0 p(\varrho_\star) + \pi(\varepsilon) + \lambda m(\varepsilon) \end{aligned} \tag{11.5.1}$$

to Problem 11.1.6. Substituting (11.5.1) into expression (11.1.13) for the drag functional $J(\Omega_\varepsilon)$ we obtain the representation for the drag functional as a function of the parameter $\varepsilon$:

$$J(\Omega_\varepsilon) = \mathbf{U}_\infty\cdot\mathbf{J}_1(\Omega_\varepsilon) + \mathbf{U}_\infty\cdot\mathbf{J}_2(\Omega_\varepsilon), \tag{11.5.2}$$

where

$$\mathbf{J}_1(\Omega_\varepsilon) = -\mathsf{Re}\int_\Omega \varrho(\varepsilon)\mathbf{u}(\varepsilon)\nabla\big(\mathbf{N}^{-1}\mathbf{u}(\varepsilon)\big)\eta\,dx,$$

$$\begin{aligned}\mathbf{J}_2(\Omega_\varepsilon) = -\int_\Omega \Big[\mathfrak{g}(\varepsilon)^{-1}\Big(\mathbf{N}(\varepsilon)^\top\nabla\big(\mathbf{N}(\varepsilon)^{-1}\mathbf{u}(\varepsilon)\big)\\ + \nabla\big(\mathbf{N}(\varepsilon)^{-1}\mathbf{u}(\varepsilon)\big)^\top\mathbf{N} - \operatorname{div}\mathbf{u}(\varepsilon)\mathbb{I}\Big) - q(\varepsilon)\mathbb{I}\Big]\mathbf{N}(\varepsilon)^\top\nabla\eta\,dx.\end{aligned} \tag{11.5.3}$$

Recall that $\Omega_\varepsilon = B\setminus S_\varepsilon$, where $B$ is a fixed hold-all domain and $S_\varepsilon = (\mathrm{Id}+\varepsilon\mathbf{T})(S)$ is the compact body placed in $B$. While the solutions $\mathbf{v}(\varepsilon)$, $\varphi(\varepsilon)$, $\pi(\varepsilon)$, $m(\varepsilon)$ to Problem 11.1.6 have no clear physical meaning, the drag $\mathsf{Re}^{-1}J(\Omega_\varepsilon)$ is a real physical quantity. It equals the power developed by the hydrodynamical force acting on the body $S_\varepsilon$. The calculation of the drag and its variations with respect to perturbations of $S$ is a problem of practical significance. In this section we derive an explicit formula for the derivative

$$dJ(\Omega)[\mathbf{T}] := \lim_{\varepsilon\to 0}\frac{1}{\varepsilon}\big(J(\Omega_\varepsilon) - J(\Omega)\big). \tag{11.5.4}$$

Notice that the perturbed domains $\Omega_\varepsilon = (\mathrm{Id}+\varepsilon\mathbf{T})(\Omega)$ are generated by the deformations $\mathrm{Id}+\varepsilon\mathbf{T}$ of the ambient space $\mathbb{R}^3$. Moreover, $\Omega_\varepsilon$ can be represented as the shift of $\Omega$ along the integral lines of the vector field $\mathbb{T}(x,\varepsilon) := (\mathbf{T}\circ(\mathrm{Id}+\varepsilon\mathbf{T})^{-1})(x)$ (see [126] for details). From this viewpoint the quantity (11.5.4) is nothing but the Gâteaux derivative of the shape functional $J(\Omega)$ in the direction $\mathbf{T}$. It is not too difficult to calculate $dJ(\Omega)[\mathbf{T}]$ in terms of the material derivatives of solutions to Problem 11.1.9. The corresponding result is given in the following theorem, which is a straightforward consequence of Theorem 11.4.6.

**Theorem 11.5.1.** *Let all hypotheses of Theorem 11.4.6 be satisfied. Then $dJ(\Omega)[\mathbf{T}]$ is well defined and has the representation*

$$dJ(\Omega)[\mathbf{T}] = L_u(\mathbf{T}) + L_e(\mathbf{T}). \tag{11.5.5}$$

*Here the form $L_u$ is given by*

$$L_u(\mathbf{T}) = \langle\mathbf{B},\mathbf{w}\rangle_1 + \langle\omega, A\rangle_0 + \langle\psi, C\rangle_0, \tag{11.5.6}$$

*where the material derivatives $\mathbf{w}$, $\omega$, and $\psi$ are given by Theorem 11.4.6 and*

$$\begin{aligned}\mathbf{B} &= \Delta\eta\mathbf{U}_\infty + \mathsf{Re}\,\varrho(\mathbf{u}\cdot\nabla\eta)\mathbf{U}_\infty - \mathsf{Re}\,\eta\,\varrho\nabla\mathbf{u}\mathbf{U}_\infty,\\ A &= \nabla\eta\cdot\mathbf{U}_\infty, \quad C = -\mathsf{Re}\,\eta(\mathbf{u}\nabla\mathbf{u})\cdot\mathbf{U}_\infty,\end{aligned} \tag{11.5.7}$$

*and the form $L_e$ is given by*

$$\begin{aligned}L_e(\mathbf{T}) = \mathbf{U}_\infty\cdot\int_\Omega\Big(\big(\nabla\mathbf{u} + (\nabla\mathbf{u})^\top - \operatorname{div}\mathbf{u}\mathbb{I}\big)D\mathbf{T}^\top - \big(\mathbb{D}^\top\nabla\mathbf{u} - \nabla(\mathbb{D}\mathbf{u})\big)\\ - \big(\mathbb{D}^\top\nabla\mathbf{u} - \nabla(\mathbb{D}\mathbf{u})\big)^\top + q\mathbb{D}^\top\Big)\nabla\eta\,dx - \mathsf{Re}\,\mathbf{U}_\infty\cdot\int_\Omega\varrho(\mathbf{u}\cdot\nabla\eta)\,\mathbb{D}\mathbf{u}\,dx,\end{aligned} \tag{11.5.8}$$

*where $\mathbb{D} = \operatorname{div}\mathbf{T}\,\mathbb{I} - D\mathbf{T}$.*

*Proof.* It follows from (11.4.6) and (11.5.1) that

$$\mathbf{u}(\varepsilon) - \mathbf{u} = \varepsilon \mathbf{w}_\varepsilon, \quad q(\varepsilon) - q = \varepsilon\omega_\varepsilon + \varepsilon\lambda n_\varepsilon, \quad \varrho(\varepsilon) - \varrho = \varepsilon\psi_\varepsilon.$$

Substituting these relations into (11.5.2) we obtain

$$\frac{1}{\varepsilon}(J(\Omega_\varepsilon) - J(\Omega)) = L_{\varepsilon,u} + L_{\varepsilon,e}, \tag{11.5.9}$$

where

$$\begin{aligned} L_{\varepsilon,u} = &-\mathsf{Re}\mathbf{U}_\infty \cdot \int_\Omega (\varrho\mathbf{u}\nabla\mathbf{w}_\varepsilon + \psi_\varepsilon\mathbf{u}\nabla\mathbf{u}_\varepsilon + \varrho_\varepsilon\mathbf{w}_\varepsilon\nabla\mathbf{u}_\varepsilon)\eta\, dx \\ &- \mathbf{U}_\infty \cdot \int_\Omega \big(\nabla\mathbf{w}_\varepsilon + (\nabla\mathbf{w}_\varepsilon)^\top - \operatorname{div}\mathbf{w}_\varepsilon\mathbb{I}\big)\nabla\eta\, dx + \mathbf{U}_\infty \cdot \int_\Omega (\omega_\varepsilon + \lambda n_\varepsilon)\nabla\eta\, dx, \end{aligned} \tag{11.5.10}$$

and

$$\begin{aligned} \varepsilon L_{\varepsilon,e} = &\ \mathsf{Re}\,\mathbf{U}_\infty \cdot \int_\Omega \varrho(\varepsilon)\mathbf{u}(\varepsilon)\nabla\big((\mathbb{I} - \mathbf{N}(\varepsilon)^{-1})\mathbf{u}(\varepsilon)\big)\eta\, dx \\ &+ \mathbf{U}_\infty \cdot \int_\Omega \big[\nabla(\mathbf{u}(\varepsilon)) + \nabla(\mathbf{u}(\varepsilon))^\top - \operatorname{div}\mathbf{u}(\varepsilon)\mathbb{I} - q(\varepsilon)\mathbb{I}\big]\nabla\eta\, dx \\ &- \mathbf{U}_\infty \cdot \int_\Omega \Big[\mathfrak{g}(\varepsilon)^{-1}\Big(\mathbf{N}(\varepsilon)^\top\nabla\big(\mathbf{N}(\varepsilon)^{-1}\mathbf{u}(\varepsilon)\big) \\ &+ \nabla\big(\mathbf{N}(\varepsilon)^{-1}\mathbf{u}(\varepsilon)\big)^\top\mathbf{N} - \operatorname{div}\mathbf{u}(\varepsilon)\mathbb{I}\Big) - q(\varepsilon)\mathbb{I}\Big]\mathbf{N}(\varepsilon)^\top\nabla\eta\, dx. \end{aligned} \tag{11.5.11}$$

Notice that $\mathbf{u}$, $\mathbf{u}(\varepsilon)$, $\mathbf{w}_\varepsilon$ are continuously differentiable and belong to $W^{2,2}(\Omega)$. In its turn, $q$, $q(\varepsilon)$, $\varrho$, $\varrho(\varepsilon)$, $\psi_\varepsilon$ are continuous and belong to $W^{1,2}(\Omega)$. Moreover, $\mathbf{w}_\varepsilon$ vanishes at $\partial\Omega$ and $\operatorname{div}(\varrho\mathbf{u}) = 0$ in $\Omega$. Hence we can integrate by parts to obtain

$$\int_\Omega \eta\varrho\mathbf{u}\nabla\mathbf{w}_\varepsilon\, dx = -\int_\Omega \varrho(\nabla\eta\cdot\mathbf{u})\mathbf{w}_\varepsilon\, dx,$$

$$\int_\Omega \big(\nabla\mathbf{w}_\varepsilon + (\nabla\mathbf{w}_\varepsilon)^\top - \operatorname{div}\mathbf{w}_\varepsilon\mathbb{I}\big)\nabla\eta\, dx = -\int_\Omega \Delta\eta\mathbf{w}_\varepsilon\, dx.$$

Since $\eta$ vanishes in a neighborhood of $\partial B$ and equals 1 in a neighborhood of $\partial S$, we have

$$\int_\Omega \nabla\eta\, dx = \int_{\partial S} \mathbf{n}\, dS = 0.$$

Inserting the results obtained into (11.5.10) we arrive at

$$L_{\varepsilon,u} = \int_\Omega (\mathbf{B}_\varepsilon\cdot\mathbf{w}_\varepsilon + \omega_\varepsilon A + \psi_\varepsilon C_\varepsilon)\, dx,$$

where

$$\mathbf{B}_\varepsilon = \Delta\eta\mathbf{U}_\infty + \mathsf{Re}\,\varrho(\mathbf{u}\cdot\nabla\eta)\mathbf{U}_\infty - \mathsf{Re}\,\eta\,\varrho(\varepsilon)\,\nabla\mathbf{u}(\varepsilon)\mathbf{U}_\infty,$$
$$A = \nabla\eta\cdot\mathbf{U}_\infty, \quad C_\varepsilon = -\mathsf{Re}\,\eta(\mathbf{u}\nabla\mathbf{u}(\varepsilon))\cdot\mathbf{U}_\infty.$$

It follows from (11.5.1), Lemma 11.4.10, and the continuity of the embeddings $Y^{s,r}\hookrightarrow C^1(\Omega)$ and $X^{s,r}\hookrightarrow C(\Omega)$ that

$$\mathbf{B}_\varepsilon \to \mathbf{B} \quad \text{in } C(\Omega) \quad \text{as } \varepsilon\to 0.$$

In particular, $\mathbf{B}_\varepsilon \in C(\Omega)\hookrightarrow L^r(\Omega)\hookrightarrow \mathcal{W}^{s-1,r}(\Omega)$ converges to $\mathbf{B}$ in $\mathcal{W}^{s-1,r}(\Omega)$. On the other hand, in view of Theorem 11.4.6 the functions $\mathbf{w}_\varepsilon$ converge to $\mathbf{w}$ weakly in $\mathcal{W}_0^{1-s,r'}(\Omega)$. Thus we get

$$\int_\Omega \mathbf{B}_\varepsilon\cdot\mathbf{w}_\varepsilon\,dx = \langle\mathbf{B}_\varepsilon,\mathbf{w}_\varepsilon\rangle_1 \to \langle\mathbf{B},\mathbf{w}\rangle_1 \quad \text{as } \varepsilon\to 0. \tag{11.5.12}$$

Next, it follows from Theorem 11.4.6 and (11.5.1) that $\mathbf{u}_\varepsilon\to\mathbf{u}$ in $Y^{s,r}$ and hence $\nabla\mathbf{u}_\varepsilon\to\nabla\mathbf{u}$ in $X^{s,r}\hookrightarrow W^{s,r}(\Omega)$. Since $\varrho,\mathbf{u}\in W^{s,r}(\Omega)$ and $W^{s,r}(\Omega)$ is a Banach algebra, we get

$$C_\varepsilon = -\mathsf{Re}\,\eta(\mathbf{u}\nabla\mathbf{u}(\varepsilon))\cdot\mathbf{U}_\infty \to -\mathsf{Re}\,\eta(\mathbf{u}\nabla\mathbf{u})\cdot\mathbf{U}_\infty = C \quad \text{in } W^{s,r}(\Omega).$$

Notice that by Theorem 11.4.6, $\psi_\varepsilon\to\psi$ in $\mathbb{W}^{-s,r}(\Omega)$, which leads to

$$\int_\Omega \psi_\varepsilon C_\varepsilon\,dx = \langle\psi_\varepsilon, C_\varepsilon\rangle_0 \to \langle\psi, C\rangle_0 \quad \text{as } \varepsilon\to 0. \tag{11.5.13}$$

Finally, notice that $A\in C^1(\Omega)\subset W^{s,r}(\Omega)$ and $\omega_\varepsilon\to\omega$ in $\mathbb{W}^{-s,r}(\Omega)$, which gives

$$\int_\Omega \omega_\varepsilon A\,dx = \langle\omega_\varepsilon, A\rangle_0 \to \langle\omega, A\rangle_0 \quad \text{as } \varepsilon\to 0. \tag{11.5.14}$$

Combining (11.5.12)–(11.5.14) we obtain

$$L_{\varepsilon,u}(\mathbf{T}) \to L_u(\mathbf{T}) \quad \text{as } \varepsilon\to 0. \tag{11.5.15}$$

In order to pass to the limit in the expression for $L_{\varepsilon,e}$ notice that $\mathbf{u}_\varepsilon$ are bounded in $C^1(\Omega)$ and $q(\varepsilon),\varrho(\varepsilon)$ are bounded in $C(\Omega)$. Substituting representations (11.4.2) for $\mathbf{N}(\varepsilon)$ and $\mathfrak{g}(\varepsilon)$ into (11.5.11) we obtain

$$\begin{aligned} L_{\varepsilon,e}(\mathbf{T}) = \mathbf{U}_\infty\cdot\int_\Omega \Big( &(\nabla\mathbf{u}(\varepsilon) + (\nabla\mathbf{u}(\varepsilon))^\top - \operatorname{div}\mathbf{u}(\varepsilon)\mathbb{I})D\mathbf{T}^\top \\ &- (\mathbb{D}^\top\nabla\mathbf{u}(\varepsilon) - \nabla(\mathbb{D}\mathbf{u}(\varepsilon))) - (\mathbb{D}^\top\nabla\mathbf{u}(\varepsilon) - \nabla(\mathbb{D}\mathbf{u}(\varepsilon)))^\top + q(\varepsilon)\mathbb{D}^\top\Big)\nabla\eta\,dx \\ &- \mathsf{Re}\,\mathbf{U}_\infty\cdot\int_\Omega \varrho(\varepsilon)(\mathbf{u}(\varepsilon)\cdot\nabla\eta)\,\mathbb{D}\mathbf{u}(\varepsilon)\,dx + O(\varepsilon). \end{aligned}$$

As mentioned above, $\mathbf{u}(\varepsilon) \to \mathbf{u}$ in $C^1(\Omega)$ and $(q(\varepsilon), \varrho(\varepsilon)) \to (q, \varrho)$ in $C(\Omega)$. Thus we get

$$L_{\varepsilon,e}(\mathbf{T}) \to L_e(\mathbf{T}) \quad \text{as } \varepsilon \to 0. \tag{11.5.16}$$

Letting $\varepsilon \to 0$ in (11.5.9) and using (11.5.15), (11.5.16) we arrive at (11.5.5). □

**Remark 11.5.2.** Recall that $\eta \in C^\infty(\Omega)$ is an arbitrary function such that $\eta = 1$ in a neighborhood of $\partial S$ and $\eta = 0$ in a neighborhood of $\Sigma = \partial B$. In particular, $\nabla \eta = 0$ in a neighborhood of $\partial\Omega$. Notice that the functionals $J(\Omega)$ and $dJ(\Omega)[\mathbf{T}]$ are independent of the choice of $\eta$. On the other hand, $\mathbf{T} \in C_0^\infty(\mathbb{R}^3)$ is an arbitrary vector field vanishing in a neighborhood of $\Sigma$. Hence for a given $\mathbf{T}$ we can always choose $\eta$ such that $L_e(\mathbf{T}) = 0$.

Formulae similar to (11.5.5) are widely used for solving shape optimization problems. The standard simple gradient scheme in shape optimization is the following. We choose an arbitrary compact set $S \subset B$ as a starting point of the iteration process. Next we look for a vector field $\mathbf{T}$ such that $dJ(\Omega)[\mathbf{T}] < 0$. The set $S_\varepsilon = (\mathrm{Id} + \varepsilon\mathbf{T})(S)$ may be considered as the successive iteration step. Then we take $\Omega = \Omega(\varepsilon)$ and repeat the calculations. We may hope that after a number of steps we obtain a configuration close to an optimum. Realization of such a scheme requires efficient methods for calculation of the first order shape derivative $dJ$. We refer the reader to [117, 64] for preliminary numerical results of drag minimization within this framework, with the use of numerical iterations [59] for the fixed point, as well as finite elements [117] or finite volumes [64].

From the viewpoint of numerical methods, however, in general representation (11.5.5) falls far short of being a good basis for calculations, in particular when using the level set method [47].

Further we refer to basic Problem 11.1.9 with $\mathbf{N} = \mathbb{I}$ (unperturbed case) as the *state equations*. Solutions $\vartheta = (\mathbf{v}, \pi, \varphi)$, $\zeta$, $m$ to this problem are called the *state variables*. The state variables completely describe the flow in the unperturbed domain $\Omega$ and are independent of the vector field $\mathbf{T}$. Theorem 11.5.1 shows that the derivative of the functional $J(\Omega_\varepsilon)$ consists of two parts, $L_e$ and $L_u$. The geometric part $L_e$ is simply the integral of a linear form of $\mathbf{T}$ over $\Omega$. The coefficients of this form are defined by the state variables. Hence it is not difficult to calculate $L_e$ for all smooth $\mathbf{T}$. In contrast, the dynamical part $L_u$ depends on $\mathbf{T}$ in a complicated manner. In order to calculate $L_u$ we have to find the very weak solution to the linearized problem and then to substitute this solution into (11.5.6). Notice that the very weak solution is given by Theorem 11.4.6 and depends on $\mathbf{T}$ in some implicit way. If we would like to find $L_u$ for another vector field $\mathbf{T}$, then we have to repeat all calculations. Fortunately, the situation can be drastically improved if we represent $L_u$ via the so-called *adjoint state.*

**Adjoint state.** It is a remarkable fact of optimization theory that we do not need the material or shape derivatives in order to calculate the first order shape derivative $dJ$. Moreover we can express $dJ$ as the integral of a linear form of $\mathbf{T}$

with coefficients depending only on the state variables and the adjoint state which is defined as follows.

**Definition 11.5.3.** The *adjoint state* is a solution $(\mathbf{h}, g, \varsigma, \upsilon, l) \in W^{s+1,r}(\Omega) \times (W^{s,r}(\Omega))^3 \times \mathbb{R}$ to the transposed problem

$$\begin{aligned}
&\Delta \mathbf{h} - \nabla g = \mathsf{Re}\,\mathscr{H}_0(\mathbf{h}) - \varsigma\nabla\varphi - \zeta\nabla\upsilon + \mathbf{B} \quad \text{in } \Omega,\\
&\operatorname{div}\mathbf{h} = \Pi(a_{22}g + a_{12}\varsigma + a_{32}l) + \Pi A \quad \text{in } \Omega,\\
&-\operatorname{div}(\mathbf{u}\varsigma) + \sigma\varsigma = \mathsf{Re}\,\mathscr{M}_0(\mathbf{h}) + a_{11}\varsigma + a_{21}g + a_{31}l + C \quad \text{in } \Omega,\\
&\mathbf{u}\nabla\upsilon + \sigma\upsilon = a_{34}l \quad \text{in } \Omega,\\
&\mathbf{h} = 0 \quad \text{on } \partial\Omega, \quad \varsigma = 0 \quad \text{on } \Sigma_{\text{out}}, \quad \upsilon = 0 \quad \text{on } \Sigma_{\text{in}},\\
&g - \Pi g = 0, \quad l = \varkappa \int_\Omega (a_{13}\varsigma + a_{23}g)\,dx,
\end{aligned}$$

where the coefficients $a_{ij}$ are given by (11.4.8), the functions $\mathbf{B}$, $A$, and $C$ are given by (11.5.7), and

$$\mathscr{H}_0(\mathbf{h}) = \varrho\nabla(\mathbf{u})\mathbf{h} - \operatorname{div}(\varrho\mathbf{u}\otimes\mathbf{h}), \quad \mathscr{M}_0(\mathbf{h}) = (\mathbf{u}\nabla\mathbf{u})\cdot\mathbf{h}.$$

The coefficients and given functions are defined by the state variables. Hence the adjoint state is completely defined by the state variables and is independent of $\mathbf{T}$.

Notice that $\mathbf{B} \in \mathcal{W}^{s-1,r}(\Omega)$ and $A, C \in W^{s,r}(\Omega)$. Hence we may apply Lemma 11.4.5 to conclude that the boundary value problem for the adjoint state has a unique solution in the class $W^{s+1,r}(\Omega) \times (\mathbb{W}^{s,r}(\Omega))^3 \times \mathbb{R}$. Therefore, the adjoint state is well defined and it is unique. The following theorem, which is the main result of this section, shows that $L_u$ can be represented as an integral of a linear form of $\mathbf{T}$.

**Theorem 11.5.4.** *Under the assumptions of Theorem 11.5.1,*

$$L_u(\mathbf{T}) = \int_\Omega \big(\mathbf{h}\cdot\mathscr{D}_0 + g\,a_{20}\mathfrak{d}_0 + \varsigma\,a_{10}\mathfrak{d}_0 + \upsilon\,\sigma\mathfrak{d}_0 + l\,a_{30}\mathfrak{d}_0\big)\,dx,$$

*where* $(\mathbf{h}, g, \varsigma, \upsilon, l)$ *is the adjoint state and*

$$\begin{aligned}
&\mathscr{D}_0 = \mathbb{D}^\top\Delta\mathbf{u} + \Delta(\mathbb{D}\mathbf{u}) + \operatorname{div}(\mathbb{T}\nabla\mathbf{u}) - \mathsf{Re}\,\varrho\big(\mathbb{D}^\top\mathbf{u}\nabla\mathbf{u} + \mathbf{u}\nabla(\mathbb{D}\mathbf{u})\big),\\
&\mathfrak{d}_0 = \operatorname{div}\mathbf{T}\,\mathbb{I}, \quad \mathbb{T} = \operatorname{div}\mathbf{T}\,\mathbb{I} - D\mathbf{T} - (D\mathbf{T})^\top, \quad \mathbb{D} = \operatorname{div}\mathbf{T}\,\mathbb{I} - D\mathbf{T}.
\end{aligned}$$

*Proof.* Choosing $(\mathbf{H}, G, F, M, e) = (\mathbf{B}, A, C, 0, 0)$ and applying Theorem 11.4.6 we arrive at

$$\begin{aligned}
L_u(\mathbf{T}) &= \langle\mathbf{B}, \mathbf{w}\rangle_1 + \langle\omega, A\rangle_0 + \langle\psi, C\rangle_0\\
&= \int_\Omega \big(\mathbf{h}\cdot\mathscr{D}_0 + g\,a_{20}\mathfrak{d}_0 + \varsigma\,a_{10}\mathfrak{d}_0 + \upsilon\,\sigma\mathfrak{d}_0 + l\,a_{30}\mathfrak{d}_0\big)\,dx. \qquad \square
\end{aligned}$$

# Chapter 12

# Boundary value problems for transport equations

## 12.1 Introduction

The first order scalar differential equation

$$\mathbf{u} \cdot \nabla\varphi + b\varphi = f, \tag{12.1.1}$$

which is called the *transport equation*, is one of the basic equations of mathematical physics. It is widely used for mathematical modeling of mass and heat transfer and plays an important role in kinetic theory of such phenomena. In the framework of compressible Navier-Stokes equations the most important examples of transport equations are given by the stationary mass balance equation,

$$\operatorname{div}(\varphi\mathbf{u}) = 0, \tag{12.1.2}$$

and the *relaxed* mass balance equation

$$\operatorname{div}(\varphi\mathbf{u}) + \alpha\varphi = h, \tag{12.1.3}$$

where the given vector field $\mathbf{u}$ belongs to the Sobolev space $W^{1,2}(\Omega)$ and satisfies

$$\mathbf{u} = \mathbf{U} \quad \text{on } \partial\Omega. \tag{12.1.4}$$

A typical boundary value problem for the transport equation can be formulated as follows:

Let $\Omega \subset \mathbb{R}^3$ be a bounded domain. Split the boundary of $\Omega$ into three disjoint parts, the inlet $\Sigma_{\text{in}}$, the outgoing set $\Sigma_{\text{out}}$, and the characteristic set $\Sigma_0$, defined by

$$\Sigma_{\text{in}} = \{x \in \partial\Omega : \mathbf{U}{\cdot}\mathbf{n} < 0\}, \quad \Sigma_{\text{out}} = \{x \in \partial\Omega : \mathbf{U}{\cdot}\mathbf{n} > 0\}, \quad \Sigma_0 = \partial\Omega\backslash(\Sigma_{\text{in}}\cup\Sigma_{\text{out}}). \tag{12.1.5}$$

Let $\Gamma = \big(\partial\Omega \setminus \Sigma_{\rm in}\big) \cap \operatorname{cl}\Sigma_{\rm in} \subset \Sigma_0$ be the interface between the inlet and the rest of $\partial\Omega$.

The problem is to find a solution $\varphi$ to the differential equation (12.1.1) such that $\varphi$ takes a prescribed value $\varphi_\star$ at the inlet $\Sigma_{\rm in}$,

$$\varphi = \varphi_\star \quad \text{on } \Sigma_{\rm in}. \tag{12.1.6}$$

The simplest way to find a solution to problem (12.1.1), (12.1.6) is to apply the *method of characteristics* which can be described as follows. Suppose for a moment that the vector field $\mathbf{u}$ has continuous derivatives and does not vanish in $\Omega$. Suppose in addition that for each $x^* \in \Omega$, there is a unique $C^1$ integral curve $x = x(s)$, $0 \le s \le s^*$, of the vector field $\mathbf{u}$, i.e., a solution of the ODE

$$\frac{dx}{ds} = \mathbf{u}(x),$$

such that $x(s^*) = x^*$ and $x(0) \in \Sigma_{\rm in}$. Then the solution to problem (12.1.1), (12.1.6) is given by $\varphi(x^*) = \phi(s^*)$ where $\phi$ is a solution to the Cauchy problem

$$\frac{d\phi(s)}{ds} + b(x(s))\phi(s) = f(x(s)), \quad s \in [0, s^*], \quad \phi(0) = \varphi_b(x(0)).$$

The method of characteristics does not work if the totality of integral curves has a complicated structure, for example if $\mathbf{u}$ has rest points within $\Omega$, and when $\mathbf{u}$ is not smooth and therefore the integral curves are not well defined. To address the first issue note that the theory of linear transport equations is part of a general theory of elliptic-parabolic equations also known as second order equations with nonnegative quadratic form. The theory deals with general second order partial differential equations

$$-\sum_{i,j=1}^{n} a_{ij}\partial_{x_i}\partial_{x_j}\varrho + \mathbf{u}\cdot\nabla\varrho + b\varrho = h \tag{12.1.7}$$

under the assumption

$$\sum_{i,j=1}^{n} a_{ij}(x)\xi_i\xi_j \ge 0 \quad \text{for all } x \in \Omega \text{ and } \xi \in \mathbb{R}^3.$$

Boundary value problems for second order PDE's with nonnegative quadratic form were studied by many authors starting from the pioneering papers of Fichera [41], Kohn and Nirenberg [66], and Oleĭnik and Radkevich [102]. In this section we give a short review of available results.

**Weak solutions.** We start with a discussion of the theory of weak solutions to the boundary value problem for the transport equation. Without loss of generality we may assume that $\varphi_b = 0$ and so the boundary condition takes the form

$$\varphi = 0 \quad \text{on } \Sigma_{\rm in}. \tag{12.1.8}$$

**Definition 12.1.1.** A function $\varphi \in L^1(\Omega)$ is a *weak solution* to problem (12.1.1), (12.1.8) if the integral identity

$$\int_\Omega (\varphi \mathscr{L}^* \zeta - f\zeta)\, dx = 0 \tag{12.1.9}$$

holds for all $\zeta \in C^1(\Omega)$ vanishing on $\Sigma_{\text{out}}$. Here the adjoint operator $\mathscr{L}^*$ is defined by

$$\mathscr{L}^* \zeta := -\operatorname{div}(\mathbf{u}\zeta) + b\zeta. \tag{12.1.10}$$

The first result on the existence of weak solutions to the general equation (12.1.7) is due to Fichera. A complete theory of integrable weak solutions to elliptic-parabolic equations was developed by Oleĭnik and Radkevich. The following theorem on existence and uniqueness of weak solutions to (12.1.1), (12.1.8) is a particular case of Oleĭnik's results (see Theorems 1.5.1 and 1.6.2 in [102]).

**Theorem 12.1.2.** *Let $\Omega \subset \mathbb{R}^3$ be a bounded domain with $C^2$ boundary and let $1 < r \leq \infty$. Assume that the vector field $\mathbf{u} \in C^1(\Omega)$ and the function $b \in C(\Omega)$ satisfy*

$$\delta = \inf_{x \in \Omega} (b(x) - r^{-1} \operatorname{div} \mathbf{u}(x)) > 0.$$

*Then problem* (12.1.1), (12.1.8) *has a weak solution $\varphi \in L^r(\Omega)$ satisfying*

$$\|\varphi\|_{L^r(\Omega)} \leq \delta^{-1} \|f\|_{L^r(\Omega)}. \tag{12.1.11}$$

*If, in addition, $r > 3$ and $\Gamma = \operatorname{cl}(\Sigma_{\text{out}} \cup \Sigma_0) \cap \operatorname{cl} \Sigma_{\text{in}}$ is a one-dimensional $C^1$ manifold, then there is a unique weak solution $\varphi \in L^r(\Omega)$.*

Moreover, in [102] it is shown that weak solutions are continuous at interior points of $\Sigma_{\text{in}}$ and take the boundary value in the classical sense.

**Remark 12.1.3.** The existence of solutions satisfying (12.1.11) was proved in [102] by using the vanishing viscosity method. In view of (12.1.11) the solution obtained by this method is unique. However problem (12.1.1), (12.1.8) may have other solutions which do not satisfy (12.1.11). Hence the problem of uniqueness is not trivial.

**Strong solutions.** The question of regularity of solutions to boundary value problems for transport equations is difficult. All known results [66], [102] relate to the case of multi-connected domains with isolated inlet. We illustrate the theory by two theorems. The first is a consequence of a general result of Kohn and Nirenberg (see [66]) on solvability of boundary value problems for elliptic-parabolic equations.

**Theorem 12.1.4.** *Let $\Omega \subset \mathbb{R}^d$ be a bounded domain with $C^\infty$ boundary, $\mathbf{u}, b \in C^\infty(\Omega)$, and $k \geq 1$ an integer. Furthermore, assume that*

$$\Gamma = \operatorname{cl} \Sigma_{\text{in}} \cap \operatorname{cl}(\Sigma_{\text{out}} \cup \Sigma_0) = \emptyset \tag{12.1.12}$$

*and* $b > c_0 > 0$, *where* $c_0$ *is a large constant depending only on* $\Omega$, $\|\mathbf{u}\|_{C^3(\Omega)}$, $\|b\|_{C^3(\Omega)}$, *and* $k$. *Then for any* $f \in W^{2,k}(\Omega)$ *problem* (12.1.1), (12.1.8) *has a unique solution satisfying*

$$\|\varphi\|_{W^{2,k}(\Omega)} \leq C(k, \Omega, \mathbf{u}, b)\|f\|_{W^{2,k}(\Omega)}.$$

The second result is a consequence of [102, Thm. 1.8.1].

**Theorem 12.1.5.** *Assume that* $\Omega$ *is a bounded domain with* $C^2$ *boundary and* $\mathbf{u}, b, f \in C^1(\mathbb{R}^3)$. *Furthermore, suppose that:*

- *The vector field* $\mathbf{U} = \mathbf{u}|_{\partial\Omega}$ *and the manifold* $\Gamma$ *satisfy condition* (12.1.12).
- *There is* $\Omega' \Supset \Omega$ *such that*

$$b(x) - \sup_{\Omega'}\left\{|\operatorname{div}\mathbf{u}| + \frac{1}{2}\sup_i \sum_{j\neq i}\left|\frac{\partial u_i}{\partial x_j}\right| + \frac{1}{2}\sup_j \sum_{j\neq i}\left|\frac{\partial u_i}{\partial x_j}\right|\right\} > 0.$$

*Then every weak solution* $\varphi \in L^\infty(\Omega)$ *to problem* (12.1.1), (12.1.8) *satisfies the Lipschitz condition in* $\operatorname{cl}\Omega$.

If the vector field $\mathbf{u}$ satisfies the nonpermeability condition

$$\mathbf{u}\cdot\mathbf{n} = 0 \quad \text{on } \partial\Omega,$$

then $\Sigma_{\text{in}} = \Sigma_{\text{out}} = \emptyset$ and we do not need boundary conditions. This particular case was investigated in detail by Beirão da Veiga [13] and Novotný [96], [97]. The case $\Gamma \neq \emptyset$ is still poorly investigated. In the next sections we develop an existence theorem for problem (12.1.1), (12.1.8) in fractional Sobolev space under the assumption that $\Gamma$ satisfies some natural geometric conditions.

## 12.2 Existence theory in Sobolev spaces

The main difficulty of the theory of the boundary value problem for the transport equation in the case of nonempty interface $\Gamma$ is that a solution may develop singularities at $\Gamma$. To make things clear consider the following simple example.

**Example 12.2.1.** Consider the boundary value problem

$$\partial_{x_1}\varphi(x) = f(x) \quad \text{in } \Omega, \quad \varphi = 0 \quad \text{on } \Sigma_{\text{in}}, \tag{12.2.1}$$

in the plane domain

$$\Omega = \{x = (x_1, x_2) : x_2 > (x_1^2 - 1)^2,\, x_1 \in \mathbb{R}\} \subset \mathbb{R}^2.$$

Assume that $f \in C^1(\Omega)$, i.e., $f$ is $C^1$ on $\operatorname{cl}\Omega$ in the notation of Section 1.2.1. In this case the inlet is the union of two arcs,

$$\Sigma_{\text{in}} = \{x : x_2 = (x_1^2 - 1)^2,\, x_1 \in (-\infty, -1)\} \cup \{x : x_2 = (x_1^2 - 1)^2,\, x_1 \in (0, 1)\},$$

and the characteristic set consists of three points

$$\Sigma_0 = \Gamma = \{(-1,0)\} \cup \{(0,1)\} \cup \{(1,0)\}.$$

For $\alpha > 1$ the intersection of the line $\{x_2 = \alpha\}$ and $\Omega$ is the interval

$$I = \{-(1 + x_2^{1/2})^{1/2} < x_1 < (1 + x_2^{1/2})^{1/2}\}$$

Hence for $x_2 > 1$ the solution to problem (12.2.1) is given by

$$\varphi(x) = \int_{-\sqrt{1+\sqrt{x_2}}}^{x_1} f(s, x_2)\, ds \quad \text{for } x_1 \in I.$$

For $0 < \alpha \le 1$ the intersection of the line $\{x_2 = \alpha\}$ and $\Omega$ consists of the disjoint intervals

$$I^- : -(1+x_2^{1/2})^{1/2} < x_1 < -(1-x_2^{1/2})^{1/2},\ I^+ : (1-x_2^{1/2})^{1/2} < x_1 < (1+x_2^{1/2})^{1/2}.$$

Hence for $0 < x_2 \le 1$ the solution to problem (12.2.1) is given by

$$\varphi(x) = \int_{-\sqrt{1+\sqrt{x_2}}}^{x_1} f(s, x_2)\, ds \quad \text{for } x_1 \in I^-$$

and

$$\varphi(x) = \int_{\sqrt{1-\sqrt{x_2}}}^{x_1} f(s, x_2)\, ds \quad \text{for } x_1 \in I^+.$$

It follows from these formulae that $\varphi$ has a jump across the segment $(0, \sqrt{2}) \times \{1\}$ emanating from the point $(0,1) \in \Gamma$. On the other hand, $\varphi$ is continuous at the points $(\pm 1, 0) \in \Gamma$, but the gradient of $\varphi$ develops a singularity, $\nabla\varphi \sim x_2^{-1/2}$, at these points. Hence there is essential difference between the behavior of the solution at $(0,1)$ and $(\pm 1, 0)$. Calculations show that in our case, the vector field $\mathbf{U} = (1,0)$ satisfies

$$\mathbf{U} \cdot \nabla(\mathbf{U} \cdot \mathbf{n}) < 0 \quad \text{at } (0,1) \quad \text{and} \quad \mathbf{U} \cdot \nabla(\mathbf{U} \cdot \mathbf{n}) > 0 \quad \text{at } (\pm 1, 0).$$

Hence imposing the condition $\mathbf{U} \cdot \nabla(\mathbf{U} \cdot \mathbf{n}) > 0$ on $\Gamma$ prevents the formation of discontinuities of solutions to the boundary value problem (12.2.1), but the derivatives of solutions have singularities at $\Gamma$ in any case.

In view of the above discussion we assume that a characteristic set $\Gamma \subset \partial\Omega$ and a given vector field $\mathbf{U}$ satisfy the following condition, referred to as the *emergent vector field condition.*

**Condition 12.2.2.** Assume that $\partial\Omega$ is a closed surface of class $C^3$ and the set $\Gamma$ is a closed $C^3$ one-dimensional manifold. Assume also that $\mathbf{U} \in C^3(\partial\Omega)$. Moreover, there is a positive constant $c$ such that

$$\mathbf{U} \cdot \nabla(\mathbf{U} \cdot \mathbf{n}) > c > 0 \quad \text{on } \Gamma. \tag{12.2.2}$$

Since the vector field $\mathbf{U}$ is tangent to $\partial\Omega$ on $\Gamma$, the left hand side of (12.2.2) is well defined.

This condition is obviously fulfilled for all strictly convex domains and constant vector fields. It has a simple geometric interpretation: $\mathbf{U} \cdot \mathbf{n}$ only vanishes up to the first order at $\Gamma$, and for each point $P \in \Gamma$, the vector $\mathbf{U}(P)$ points to the part of $\partial\Omega$ where $\mathbf{U}$ is an exterior vector field. Note that the emergent vector field condition plays an important role in theory of the oblique derivative problem for elliptic equations (see [58]). The following theorem is the first main result of this chapter. Let us consider the following boundary value problems for linear transport equations:

$$\mathscr{L}\varphi := \mathbf{u}\nabla\varphi + \sigma\varphi = f \quad \text{in } \Omega, \quad \varphi = 0 \quad \text{on } \Sigma_{\text{in}}, \tag{12.2.3}$$

$$\mathscr{L}^*\varphi^* := -\operatorname{div}(\varphi^*\mathbf{u}) + \sigma\varphi^* = f \quad \text{in } \Omega, \quad \varphi^* = 0 \quad \text{on } \Sigma_{\text{out}}. \tag{12.2.4}$$

Bounded functions $\varphi$, $\varphi^*$ are called *weak solutions* to problems (12.2.3), (12.2.4), respectively, if the integral identities

$$\int_\Omega (\varphi\mathcal{L}^*\zeta^* - f\zeta^*)\,dx = 0, \quad \int_\Omega (\varphi^*\mathcal{L}\zeta - f\zeta)\,dx = 0 \tag{12.2.5}$$

hold for all $\zeta^*, \zeta \in C(\Omega) \cap W^{1,1}(\Omega)$, respectively, such that $\zeta^* = 0$ on $\Sigma_{\text{out}}$ and $\zeta = 0$ on $\Sigma_{\text{in}}$.

**Theorem 12.2.3.** *Let $\Omega \subset \mathbb{R}^3$ be a bounded domain. Assume that $\partial\Omega$ and $\mathbf{U}$ satisfy Condition 12.2.2, and the vector field $\mathbf{u}$ belongs to $C^1(\Omega)$ and satisfies*

$$\mathbf{u} = \mathbf{U} \quad on\ \partial\Omega. \tag{12.2.6}$$

*Furthermore, let $s$ and $r$ be constants satisfying*

$$0 < s \le 1, \quad 1 < r < \infty, \quad \kappa := 2s - 3r^{-1} < 1. \tag{12.2.7}$$

*Then there are positive constants $\sigma^* > 1$ and $C$, which depend on $\partial\Omega$, $\mathbf{U}$, $s$, $r$, $\|\mathbf{u}\|_{C^1(\Omega)}$ and are independent of $\sigma$, such that for any $\sigma > \sigma^*$ and $f \in W^{s,r}(\Omega) \cap L^\infty(\Omega)$, problem (12.2.3) has a unique solution $\varphi \in W^{s,r}(\Omega) \cap L^\infty(\Omega)$ satisfying*

$$\begin{aligned} &\|\varphi\|_{W^{s,r}(\Omega)} \le C\sigma^{-1}\|f\|_{W^{s,r}(\Omega)} + C\sigma^{-1+\alpha}\|f\|_{L^\infty(\Omega)} \quad \textit{for } sr \neq 1, 2, \\ &\|\varphi\|_{W^{s,r}(\Omega)} \le C\sigma^{-1}\|f\|_{W^{s,r}(\Omega)} + C\sigma^{-1+\alpha}(1 + \log\sigma)^{1/r}\|f\|_{L^\infty(\Omega)} \quad \textit{for } sr = 1, 2. \end{aligned} \tag{12.2.8}$$

*Here, the* accretivity defect *$\alpha$ is defined by*

$$\alpha(s, r) = \max\{0,\ s - r^{-1},\ 2s - 3r^{-1}\}. \tag{12.2.9}$$

## 12.3 Proof of Theorem 12.2.3

Our strategy is as follows. First, we show that in the vicinity of each $P \in \Sigma_{\text{in}}$ there exist normal coordinates $(y_1, y_2, y_3)$ such that $\mathbf{u} \cdot \nabla_x = \mathbf{e}_1\nabla_y$. Hence, existence of solutions to the transport equation in the neighborhood of $\Sigma_{\text{in}}$ reduces to a boundary value problem for the model equation $\partial_{y_1}\varphi + \sigma\varphi = f$. Next, we prove that the boundary value problem for the model equation has a unique solution in a fractional Sobolev space, which leads to the existence and uniqueness of solutions in the neighborhood of the inlet set. Using the existence of local solutions we reduce problem (12.2.3) to the problem for a modified equation, which does not require boundary data. Application of well known results on solvability of elliptic-hyperbolic equations in the case $\Gamma = \emptyset$ finally gives the existence and uniqueness of solutions to problem (12.2.3).

## 12.3.1 Normal coordinates

First, we introduce some notation which is used throughout this section. For any $a > 0$ we denote by $Q_a$ the cube

$$Q_a := [-a, a]^3 \tag{12.3.1}$$

and by $Q_a^+$ the slab

$$Q_a^+ := [-a, a]^2 \times [0, a] \tag{12.3.2}$$

in the space of points $y = (y_1, y_2, y_3) \in \mathbb{R}^3$. We write $Y$ for $(y_2, y_3)$ so that $y = (y_1, Y)$.

**Definition 12.3.1.** A *standard parabolic neighborhood* associated with the constant $c_0$ is a compact subset of a slab $Q_a^+$, defined by

$$\mathcal{P}_a = \{y = (y_1, Y) \in Q_a^+ : a^-(Y) \le y_1 \le a^+(Y)\}, \tag{12.3.3}$$

where $a^\pm : [-a, a] \times [0, a] \to \mathbb{R}$ are continuous, piecewise $C^1$ functions satisfying

$$\begin{aligned} &-a \le a^-(Y) \le 0 \le a^+(Y) \le a, \\ &-c_0\sqrt{y_3} \le a^-(Y) \le a^+(Y) \le c_0\sqrt{y_3}, \\ &|\partial_{y_2} a^\pm(Y)| \le c_0, \quad |\partial_{y_3} a^\pm(Y)| \le c_0/\sqrt{y_3}. \end{aligned} \tag{12.3.4}$$

Denote by $\Sigma_{\text{in}}^y$ and $\Sigma_{\text{out}}^y$ the surfaces

$$\begin{aligned} \Sigma_{\text{in}}^y &= \{y : Y \in Q_{\text{in}},\ y_1 = a^-(Y)\}, \\ \Sigma_{\text{out}}^y &= \{y : Y \in Q_{\text{out}},\ y_1 = a^+(Y)\}, \end{aligned}$$

where $Q_{\text{in}} = \{Y : a^-(Y) > -a\}$ and $Q_{\text{out}} = \{Y : a^+(Y) < a\}$. It is clear that $\partial\mathcal{P}_a = (\partial Q_a \cap \partial\mathcal{P}_a) \cup \Sigma_{\text{in}}^y \cup \Sigma_{\text{out}}^y$.

**Lemma 12.3.2.** *Suppose that* $\Gamma = \operatorname{cl}\Sigma_{\text{in}} \cap \operatorname{cl}(\Sigma_{\text{out}} \cup \Sigma_0)$ *is a* $C^3$ *manifold, and* $\Gamma$ *and the vector field* $\mathbf{U} \in C^3(\partial\Omega)$ *satisfy Condition 12.2.2. Let* $\mathbf{u} \in C^1(\mathbb{R}^3)$ *be a compactly supported vector field such that* $\mathbf{u} = \mathbf{U}$ *on* $\partial\Omega$. *Denote* $M = \|\mathbf{u}\|_{C^1(\mathbb{R}^3)}$. *Then there are positive constants* $a$, $c$, $C$, $\rho$, *and* $R$, *depending only on* $M$, $\partial\Omega$, *and* $\mathbf{U}$, *with the properties:*

**(P1)** *For any point $P \in \Gamma$, there exists a diffeomorphism $y \mapsto \mathbf{x}(y)$ of the cube $Q_a$ onto a neighborhood $\mathcal{O}_P$ of $P$ which satisfies*

$$\partial_{y_1}\mathbf{x}(y) = \mathbf{u}(\mathbf{x}(y)) \quad \text{in } Q_a, \tag{12.3.5}$$

*and*

$$\|\mathbf{x}\|_{C^1(Q_a)} + \|\mathbf{x}^{-1}\|_{C^1(\mathcal{O}_P)} \le C, \quad |\mathbf{x}(y)| \le C|y|. \tag{12.3.6}$$

**(P2)** *There is a standard parabolic neighborhood $\mathcal{P}_a$ associated with the constant $c$ such that*

$$\mathbf{x}(\mathcal{P}_a) = \mathcal{O}_P \cap \Omega, \quad \mathbf{x}(\Sigma_{\text{in}}^y) = \Sigma_{\text{in}} \cap \mathcal{O}_P, \quad \mathbf{x}(\Sigma_{\text{out}}^y) = \Sigma_{\text{out}} \cap \mathcal{O}_P. \tag{12.3.7}$$

**(P3)** *Denote by $G_a \subset \mathcal{P}_a$ the domain*

$$G_a = \{y = (y_1, Y) \in \mathcal{P}_a : Y \in Q_{\text{in}}\}, \tag{12.3.8}$$

*and by $B_P(\rho)$ the ball $|x - P| \le \rho$. Then*

$$B_P(\rho) \cap \Omega \subset \mathbf{x}(G_a) \subset \mathcal{O}_P \cap \Omega \subset B_P(R) \cap \Omega. \tag{12.3.9}$$

The following lemma shows the existence of normal coordinates in the vicinity of points of the inlet $\Sigma_{\text{in}}$.

**Lemma 12.3.3.** *Let $\mathbf{u}$ and $\mathbf{U}$ meet all requirements of Lemma 12.3.2 and suppose $-\mathbf{U}(P) \cdot \mathbf{n} > N > 0$. Then there are $b > 0$ and $C > 0$, depending only on $N$, $\Omega$ and $M = \|\mathbf{u}\|_{C^1(\Omega)}$, with the following properties. There exists a diffeomorphism $y \mapsto \mathbf{x}(y)$ of the cube $Q_b = [-b, b]^3$ onto a neighborhood $\mathcal{O}_P$ of $P$ which satisfies*

$$\partial_{y_3}\mathbf{x}(y) = \mathbf{u}(\mathbf{x}(y)) \quad \text{in } Q_b, \quad \mathbf{x}(y_1, y_2, 0) \in \partial\Omega \cap \mathcal{O}_P \quad \text{for } |y_2| \le a, \tag{12.3.10}$$

*and*

$$\|\mathbf{x}\|_{C^1(Q_b)} + \|\mathbf{x}^{-1}\|_{C^1(\mathcal{O}_P)} \le C, \quad \mathbf{x}(0) = P. \tag{12.3.11}$$

*Moreover,*

$$B_P(\rho_i) \cap \Omega \subset \mathbf{x}(Q_b \cap \{y_3 > 0\}) \subset B_P(R_i) \cap \Omega \tag{12.3.12}$$

*for $\rho_i = C^{-1}b$ and $R_i = Cb$.*

Lemmas 12.3.2 and 12.3.3 are proved in Section 12.3.6.

### 12.3.2 Model equation

Let $\mathcal{P}_a$ be a standard parabolic neighborhood associated with the constant $c_0$ (see Definition 12.3.1). Consider the boundary value problem

$$\partial_{y_1}\varphi(y) + \sigma\varphi(y) = f(y) \quad \text{in } \mathcal{P}_a, \quad \varphi(y) = 0 \quad \text{for } y_1 = a^-(Y). \tag{12.3.13}$$

**Lemma 12.3.4.** *Suppose that the exponents $r$, $s$ and the accretivity defect $\alpha$ meet all requirements of Theorem 12.2.3, $\sigma > 1$, and $f \in W^{s,r}(\mathcal{P}_a) \cap L^\infty(\mathcal{P}_a)$. Then there is a constant $c$, depending only on $a$, $c_0$, $r$, $s$, such that problem* (12.3.13) *has a solution that satisfies*

$$\begin{aligned} \|\varphi\|_{W^{s,r}(\mathcal{P}_a)} &\leq c\big(\sigma^{-1}\|f\|_{W^{s,r}(\mathcal{P}_a)} + \wp(s,r,\sigma)\|f\|_{L^\infty(\mathcal{P}_a)}\big), \\ \|\varphi\|_{L^\infty(\mathcal{P}_a)} &\leq \sigma^{-1}\|f\|_{L^\infty(\mathcal{P}_a)}, \end{aligned} \tag{12.3.14}$$

*where*

$$\wp(s,r,\sigma) = \begin{cases} \sigma^{-1+\alpha} & \text{for } rs \neq 1,2, \\ \sigma^{-1+\alpha}(1+\log\sigma)^{1/r} & \text{for } rs = 1,2. \end{cases}$$

The proof is given in Appendix 12.3.7. Next, consider the boundary value problem

$$\partial_{y_3}\varphi + \sigma\varphi = f \quad \text{in } Q_a^+ = [-a,a]^2 \times [0,a], \quad \varphi(y) = 0 \quad \text{for } y_3 = 0. \tag{12.3.15}$$

**Lemma 12.3.5.** *Suppose that the exponents $r, s$ and the accretivity defect $\alpha$ meet all requirements of Theorem 12.2.3 and $\sigma > 1$. Then for any $f \in W^{s,r}(\mathcal{P}_a) \cap L^\infty(\mathcal{P}_a)$, problem* (12.3.15) *has a unique solution satisfying*

$$\|\varphi\|_{W^{s,r}(Q_a^+)} \leq c(r,s,a)\big(\sigma^{-1}\|f\|_{W^{s,r}(Q_a^+)} + \wp(s,r,\sigma)\|f\|_{L^\infty(Q_a^+)}\big). \tag{12.3.16}$$

*Proof.* The proof of Lemma 12.3.4 can be used also in this simple case. □

### 12.3.3 Local estimates

It follows from the assumptions of Theorem 12.2.3 that $\mathbf{u}$ and $\Omega$ meet all requirements of Theorem 12.1.2. In view of this theorem for every $\sigma > \|\operatorname{div}\mathbf{u}\|_{C(\Omega)}$ and $f \in L^\infty(\Omega)$ problem (12.2.3) has a unique weak solution $\varphi \in L^\infty(\Omega)$. In this section we estimate the $W^{s,r}$-norm of $\varphi$ in a neighborhood of an arbitrary point of $\Sigma_{\text{in}}$.

It follows from the assumptions of Theorem 12.2.3 that the vector field $\mathbf{u}$ and the manifold $\Gamma$ satisfy all assumptions of Lemma 12.3.2. Therefore, there exist positive numbers $a$, $\rho$ and $R$, depending only on $\Omega$ and $\|\mathbf{u}\|_{C^1(\Omega)}$, such that for all $P \in \Gamma$, the canonical diffeomorphism $\mathbf{x} : Q_a \to \mathcal{O}_P$ is well defined and meets all requirements of Lemma 12.3.2. The following lemma is the main result of this section.

**Lemma 12.3.6.** *Assume that $r$, $s$, $\alpha$ and $\mathbf{U}$ are as in Theorem 12.2.3 and $\|\mathbf{u}\|_{C^1(\Omega)} \leq M$. Then there exists $\sigma^* > 1$, depending on $\Omega$, $s$, $r$, and $M$, such that for any $f \in W^{s,r}(\Omega) \cap L^\infty(\Omega)$ and $\sigma > \sigma^*$, the weak solution $\varphi \in L^\infty(\Omega)$ to problem* (12.2.3) *defined by Theorem 12.1.2 satisfies*

$$\begin{aligned} \|\varphi\|_{W^{s,r}(B_P(\rho)\cap\Omega)} &\leq c(\sigma^{-1}\|f\|_{W^{s,r}(B_P(R)\cap\Omega)} + \wp(s,r,\sigma)\|f\|_{L^\infty(B_P(R)\cap\Omega)}), \\ \|\varphi\|_{L^\infty(B_P(\rho)\cap\Omega)} &\leq \sigma^{-1}\|f\|_{L^\infty(B_P(R)\cap\Omega)}, \end{aligned} \tag{12.3.17}$$

*where the constant c depends only on $\partial\Omega$, $\mathbf{U}$, $M$, $s$, and where $r$, $\rho$, $R$ are defined by Lemma 12.3.2.*

*Proof.* The proof falls into three steps.

**Step 1.** Let $G_a$ be the domain defined in Lemma 12.3.2. Denote by $\mathcal{D}_a \subset \Omega$ the domain $\mathbf{x}(G_a)$ and set

$$\mathbb{M}(y) = \mathbf{x}'(y), \quad |\mathbb{M}|(y) = \det \mathbf{x}'(y), \quad y \in G_a.$$

In view of Lemma 12.3.2 we have $\mathbb{M}, \partial_{y_1}\mathbb{M} \in C^1(G_a)$. Next choose $\eta \in C^1(\mathcal{D}_a)$ and set $\zeta(y) = \eta(\mathbf{x}(y))$. Our first task is to show that for all $h \in L^1(\mathcal{D}_a)$,

$$\int_{\mathcal{D}_a} h(x) \operatorname{div}(\eta(x)\mathbf{u}(x))\, dx = \int_{G_a} h(x(y))\partial_{y_1}\big(\zeta(y)|\mathbb{M}|(y)\big)\, dy. \tag{12.3.18}$$

Since the functions $\operatorname{div}(\eta\mathbf{u})$ and $\partial_{y_1}(|\mathbb{M}|\,\zeta)$ are continuous in $\operatorname{cl}\mathcal{D}_a$ and $\operatorname{cl}G_a$ respectively, it suffices to prove (12.3.18) for $h \in C_0^\infty(\mathcal{D}_a)$. It is easy to check that $h \circ \mathbf{x} \in C^1(G_a)$ is compactly supported in $G_a$ and

$$(\nabla_x h)(\mathbf{x}(y)) = (\mathbb{M}^{-1})^\top \nabla_y h(\mathbf{x}(y)).$$

Thus we get

$$\begin{aligned}
\int_{\mathcal{D}_a} h(x) \operatorname{div}(\eta(x)\mathbf{u}(x))\, dx &= -\int_{\mathcal{D}_a} \eta(x)\nabla_x h(x) \cdot \mathbf{u}(x)\, dx \\
&= -\int_{G_a} \eta(\mathbf{x}(y))\nabla_y\, h(\mathbf{x}(y)) \cdot \mathbb{M}^{-1}\mathbf{u}(\mathbf{x}(y))\, |\mathbb{M}|\, dy \\
&= -\int_{G_a} \zeta(y)\nabla_y\, h(\mathbf{x}(y)) \cdot \big(|\mathbb{M}|\mathbb{M}^{-1}\mathbf{u}(\mathbf{x}(y))\big)\, dy.
\end{aligned} \tag{12.3.19}$$

Notice that the matrix $\mathbb{M}$ has columns $\partial_{y_i}\mathbf{x}(y)$, $i = 1, 2, 3$, and hence

$$|\mathbb{M}|\mathbb{M}^{-1} = \big[\partial_{y_2}\mathbf{x}\times\partial_{y_3}\mathbf{x},\ \partial_{y_3}\mathbf{x}\times\partial_{y_1}\mathbf{x},\ \partial_{y_1}\mathbf{x}\times\partial_{y_2}\mathbf{x}\big]^\top, \quad |\mathbb{M}| = (\partial_{y_2}\mathbf{x}\times\partial_{y_3}\mathbf{x})\cdot\partial_{y_1}\mathbf{x}.$$

On the other hand, Lemma 12.3.2 yields $\mathbf{u}(\mathbf{x}(y)) = \partial_{y_1}\mathbf{x}(y)$. It follows that

$$|\mathbb{M}|\mathbb{M}^{-1}\mathbf{u}(\mathbf{x}(y)) = (|\mathbb{M}|, 0, 0) = |\mathbb{M}|\mathbf{e}_1.$$

Lemma 12.3.2 also implies that $\partial_{y_1}|\mathbb{M}|$ belongs to $C^1(G_a)$. Thus we get

$$\begin{aligned}
-\int_{G_a} &\zeta(y)\nabla_y h(\mathbf{x}(y)) \cdot \big(|\mathbb{M}|\mathbb{M}^{-1}\mathbf{u}(\mathbf{x}(y))\big)\, dy \\
&= -\int_{G_a} \zeta(y)\partial_{y_1} h(\mathbf{x}(y))\, |\mathbb{M}|\, dy = \int_{G_a} h(\mathbf{x}(y))\partial_{y_1}(|\mathbb{M}|\zeta(y))\, dy.
\end{aligned}$$

Combining this with (12.3.19) we arrive at (12.3.18).

**Step 2.** Let $\varphi$ be a bounded weak solution to problem (12.2.3). Let us consider the restriction of $\varphi$ to $\mathcal{D}_a$ and set

$$\overline{\varphi}(y) = \varphi(\mathbf{x}(y)), \quad \overline{f}(y) = f(\mathbf{x}(y)), \quad y \in G_a.$$

Now our task is to show that $\partial_{y_1}\overline{\varphi}$ belongs to $L^\infty(G_a)$ and satisfies the equations

$$\partial_{y_1}\overline{\varphi} + \sigma\overline{\varphi} = \overline{f} \quad \text{in } G_a, \quad \overline{\varphi} = 0 \quad \text{on } \Sigma_{\text{in}}^y. \tag{12.3.20}$$

To this end choose $\zeta \in C^1(G_a)$ vanishing in a neighborhood of $\partial G_a \setminus \Sigma_{\text{in}}^y$. In view of Lemma 12.3.2 (equalities (12.3.7), (12.3.8)) we have $\Sigma_{\text{in}} \cap \partial\mathcal{D}_a = \Sigma_{\text{in}}^y \cap G_a$. Hence the function $\eta = \zeta \circ \mathbf{x}^{-1}$ belongs to $C^1(\mathcal{D}_a)$ and vanishes in a neighborhood of $\partial\mathcal{D}_a \setminus \Sigma_{\text{in}}$. Extend $\eta$ by zero to $\Omega$. The extension belongs to $C^1(\Omega)$ and vanishes in a neighborhood of $\partial\Omega \setminus \Sigma_{\text{in}}$. Since $\varphi$ is a bounded weak solution to problem (12.2.3), we have

$$\int_{\mathcal{D}_a} (\varphi \operatorname{div}(\eta\mathbf{u}) - \sigma\varphi\eta + f\eta)\, dx = \int_\Omega (\varphi \operatorname{div}(\eta\mathbf{u}) - \sigma\varphi\eta + f\eta)\, dx = 0.$$

From this and identity (12.3.18) with $h$ replaced by $\varphi$ we obtain

$$\int_{G_a} (\overline{\varphi}\partial_{y_1}(\zeta|\mathbb{M}|) - \sigma\overline{\varphi}\zeta|\mathbb{M}| + \overline{f}\zeta|\mathbb{M}|)\, dy = 0. \tag{12.3.21}$$

Next, the open set $G_a$ can be covered by a countable collection of cubes

$$\mathcal{C}_n = \mathcal{I}_n^3, \quad \mathcal{I}_n = [\alpha_n, \beta_n], \quad n \geq 1.$$

Denote by $C_0^1(\mathcal{I}_n) \subset C^1(\mathcal{I}_n)$ the Banach space which consists of functions $\upsilon \in C^1(\mathcal{I}_n)$ vanishing at the boundary of $\mathcal{I}_n$. Since $C_0^1(\mathcal{I}_n)$ is a separable Banach space, it contains a countable dense subset $\{\upsilon_{m,n}\}_{m\geq 1}$. Now choose $\varsigma \in C_0^\infty(\mathcal{I}_n^2)$ and set $\zeta(y_1, Y) = \varsigma(Y)\upsilon_{m,n}(y_1)$. Assume that $\zeta$ is extended by zero to $G_a$. It is clear that the extension belongs to $C^1(G_a)$ and is compactly supported in $G_a$. Substituting $\zeta$ in (12.3.21) we obtain

$$\int_{\mathcal{I}_n^2} \varsigma(Y)\Xi_{m,n}(Y)\, dY = 0,$$

where

$$\Xi_{m,n}(Y) = \int_{\mathcal{I}_n} \Big(\overline{\varphi}(y_1, Y)\partial_{y_1}\big(\upsilon_{m,n}(y_1)|\mathbb{M}(y_1, Y)|\big) \\ + \big(\overline{f} - \sigma\overline{\varphi}\big)(y_1, Y)\, \upsilon_{m,n}(y_1)|\mathbb{M}(y_1, Y)|\Big)\, dy_1.$$

Since $\varsigma(Y)$ is an arbitrary function of class $C_0^\infty(\mathcal{I}_n^2)$, we conclude that there is a set $\mathcal{E}_{m,n} \subset \mathbb{R}^2$ of zero measure such that $\Xi_{m,n}(Y) = 0$ for all $Y \in \mathcal{I}_n^2 \setminus \mathcal{E}_{m,n}$. It is clear that the set $\mathcal{E}=\bigcup_{m,n}\mathcal{E}_{m,n}$ has zero measure and $\Xi_{m,n}(Y)=0$ for every $Y\in\mathcal{I}_n^2\setminus\mathcal{E}$ and all $m\geq 1$. Since the $v_{m,n}$ are dense in $C_0^1(\mathcal{I}_n)$, the identity

$$\int_{\mathcal{I}_n}(\overline{\varphi}\partial_{y_1}(v|\mathbb{M}|)-\sigma\overline{\varphi}v|\mathbb{M}|+\overline{f}v|\mathbb{M}|)(y_1,Y)\,dy_1=0 \tag{12.3.22}$$

holds for all $Y\in\mathcal{I}_n^2\setminus\mathcal{E}$ and all $v\in C_0^1(\mathcal{I}_n)$. Since $|\mathbb{M}|^{-1}(\cdot,Y)$ belongs to $C^1(\mathcal{I}_n)$ the function $v/|\mathbb{M}|(\cdot,Y)$ belongs to $C_0^1(\mathcal{I}_n)$. Hence we can replace $v$ by $v/|\mathbb{M}|$ in (12.3.22). Therefore,

$$\int_{\mathcal{I}_n}(\overline{\varphi}\partial_{y_1}v-\sigma\overline{\varphi}v+\overline{f}v)(Y,y_1)\,dy_1=0$$

for all $Y\in\mathcal{I}_n^2\setminus\mathcal{E}$ and all $v\in C_0^1(\mathcal{I}_n)$. This means that for every $Y\in\mathcal{I}_n^2\setminus\mathcal{E}$ the function $\overline{\varphi}(\cdot,Y)$ has a bounded derivative satisfying

$$\partial_{y_1}\overline{\varphi}=-\sigma\overline{\varphi}+\overline{f}\quad\text{in } \mathcal{C}_n.$$

Recalling that the sets $\mathcal{C}_n$ cover $G_a$ we obtain

$$\partial_{y_1}\overline{\varphi}=-\sigma\overline{\varphi}+\overline{f}\quad\text{a.e. in } G_a. \tag{12.3.23}$$

Since $\overline{\varphi}$ and $\overline{f}$ are bounded in $G_a$, it follows that $\partial_{y_1}\overline{\varphi}$ is bounded in $G_a$. In view of Lemma 12.3.2 the open connected set $G_a$ is a union of segments parallel to the $y_1$ axis,

$$G_a=\bigcup_{Y\in Q_{\text{in}}}(a^-(Y),a^+(Y))\times\{Y\}. \tag{12.3.24}$$

Since $\partial_{y_1}\overline{\varphi}$ is bounded there exists a bounded limit

$$\varphi^-(Y)=\lim_{y_1\to a^-(Y)}\overline{\varphi}(y_1,Y),\quad Y\in Q_{\text{in}}.$$

Now we turn to the integral identity (12.3.21). Integrating by parts and using (12.3.23) we obtain

$$\int_{Q_{\text{in}}}\zeta(a^-(Y),Y)\varphi^-(Y)\,dY=0$$

for all $\zeta\in C^1(G_a)$ vanishing in a neighborhood of $\partial G_a\setminus\Sigma_{\text{in}}^y=G_a\setminus\{(a^-(Y),Y):Y\in Q_{\text{in}}\}$. Hence $\varphi^-=0$, which along with (12.3.23) yields (12.3.20). Finally notice that by (12.3.20),

$$\overline{\varphi}(y_1,Y)=\int_{a^-(Y)}^{y_1}e^{\sigma(t-y_1)}\overline{f}(t,Y)\,dt,\quad Y\in Q_{\text{in}},\ a^-(Y)\leq y_1\leq a^+(Y). \tag{12.3.25}$$

**Step 3.** Consider the boundary value problem

$$\partial_{y_1}\tilde{\varphi} + \sigma\tilde{\varphi} = \overline{f} \quad \text{in } \mathcal{P}_a, \quad \tilde{\varphi}(y) = 0 \quad \text{for } y_1 = a^-(Y), \tag{12.3.26}$$

in the parabolic neighborhood $\mathcal{P}_a \supset G_a$ given by Lemma 12.3.2. Obviously,

$$\tilde{\varphi}(y_1, Y) = \int_{a^-(Y)}^{y_1} e^{\sigma(t-y_1)}\overline{f}(t, Y)\,dt, \quad Y \in [-a, a]^2,\ a^-(Y) \le y_1 \le a^+(Y).$$

Since $Q_{\text{in}} \subset [-a, a]^2$, it follows from (12.3.24) and (12.3.25) that $\overline{\varphi} = \tilde{\varphi}$ in $G_a$. Next Lemma 12.3.2 implies $B_P(\rho) \cap \Omega \subset \mathcal{D}_a = \mathbf{x}(G_a)$. Since $\mathbf{x}$ is a diffeomorphism, we conclude that

$$\|\varphi\|_{W^{s,r}(B_P(\rho)\cap\Omega)} \le \|\varphi\|_{W^{s,r}(\mathcal{D}_a)} \le c\|\overline{\varphi}\|_{W^{s,r}(G_a)} \le \|\tilde{\varphi}\|_{W^{s,r}(\mathcal{P}_a)}. \tag{12.3.27}$$

On the other hand, it follows from Lemma 12.3.4 and the inclusion $\mathbf{X}(\mathcal{P}_a) \subset \Omega \cap B_P(\rho)$ that

$$\begin{aligned}\|\tilde{\varphi}\|_{W^{s,r}(\mathcal{P}_a)} &\le c(\sigma^{-1}\|\overline{f}\|_{W^{s,r}(\mathcal{P}_a)} + \wp(s, r, \sigma)\|\overline{f}\|_{L^\infty(\mathcal{P}_a)}) \\ &\le c(\sigma^{-1}\|f\|_{W^{s,r}(\Omega\cap B_P(\rho))} + \wp(s, r, \sigma)\|f\|_{L^\infty(\Omega\cap B_P(\rho))}).\end{aligned}$$

Here, the constant $c$ depends only on $M$, $r$, $s$, $\mathbf{U}$ and $\partial\Omega$. Combining these estimates with (12.3.27) we obtain the first inequality in (12.3.17). The second obviously follows from (12.3.25) and the estimates

$$\|\overline{f}\|_{L^\infty(G_a)} = \|f\|_{L^\infty(\mathcal{D}_a)} \le \|f\|_{L^\infty(\Omega\cap B_P(\rho))}. \qquad \square$$

To formulate a similar result for the interior points of the inlet we define

$$\Sigma'_{\text{in}} = \{x \in \Sigma_{\text{in}} : \text{dist}(x, \Gamma) \ge \rho/3\}, \tag{12.3.28}$$

where the constant $\rho$ is given by Lemma 12.3.2. It is clear that

$$\inf_{P\in\Sigma'_{\text{in}}} -\mathbf{U}(P)\cdot\mathbf{n}(P) \ge N > 0,$$

where the constant $N$ depends only on $M$, $\mathbf{U}$, and $\partial\Omega$. It follows from Lemma 12.3.3 that there are positive numbers $b$, $\rho_i$ and $R_i$ such that for each $P \in \Sigma'_{\text{in}}$, the canonical diffeomorphism $\mathbf{x} : Q_b \to \mathcal{O}_P \subset B_P(R_i)$ is well defined and satisfies the hypotheses of Lemma 12.3.3.

**Lemma 12.3.7.** *Suppose that the exponents $s$, $r$ satisfy condition* (12.2.7) *and that $\|\mathbf{u}\|_{C^1(\Omega)} \le M$. Then there exists $\sigma^* > 1$, depending on $\Omega$, $s$, $r$, and $M$, such that for any $f \in W^{s,r}(\Omega) \cap L^\infty(\Omega)$ and $\sigma > \sigma^*$, the weak solution $\varphi \in L^\infty(\Omega)$ to problem* (12.2.3) *defined by Theorem 12.1.2 satisfies*

$$\begin{aligned}&\|\varphi\|_{W^{s,r}(B_P(\rho_i)\cap\Omega)} \le c\big(\sigma^{-1}\|f\|_{W^{s,r}(B_P(R_i)\cap\Omega)} + \wp(s, r, \sigma)\|f\|_{L^\infty(B_P(R_i)\cap\Omega)}\big), \\ &\|\varphi\|_{L^\infty(B_P(\rho_i)\cap\Omega)} \le \sigma^{-1}\|f\|_{L^\infty(B_P(R_i)\cap\Omega)}.\end{aligned} \tag{12.3.29}$$

*where $c$ depends on $\Sigma$, $M$, $\mathbf{U}$ and the exponents $s$, $r$.*

*Proof.* Using the normal coordinates given by Lemma 12.3.3 and repeating the proof of Lemma 12.3.6 we obtain

$$\partial_{y_3}\overline{\varphi} + \sigma\overline{\varphi} = \overline{f} \quad \text{in } Q_b, \quad \overline{\varphi} = 0 \quad \text{for } y_3 = 0,$$

where $\overline{\varphi} = \varphi \circ \mathbf{x}$ and $\overline{f} = f \circ \mathbf{x}$. Applying Lemma 12.3.4 and arguing as in the proof of Lemma 12.3.6 we obtain (12.3.29). □

### 12.3.4 Estimates near the inlet

We are now in a position to prove local estimates of the solution for the boundary value problem (12.2.3) near the inlet. Let $\Omega_t$ be the tubular neighborhood of the set $\Sigma_{\text{in}}$,

$$\Omega_t = \{x \in \Omega : \text{dist}(x, \Sigma_{\text{in}}) < t\}.$$

**Lemma 12.3.8.** *Let $t = \min\{\rho/2, \rho_i/2\}$ and $T = \max\{R, R_i\}$, where the constants $\rho$, $\rho_i$ and $R$, $R_i$ are defined by Lemmas 12.3.2 and 12.3.3, respectively. Then there exists $\sigma^* > 1$, depending on $\Omega$, $s$, $r$, and $\|\mathbf{u}\|_{C^1(\Omega)}$, such that for any $f \in W^{s,r}(\Omega) \cap L^\infty(\Omega)$ and $\sigma > \sigma^*$, the weak solution $\varphi \in L^\infty(\Omega)$ to problem (12.2.3) satisfies*

$$\begin{aligned} \|\varphi\|_{W^{s,r}(\Omega_t)} &\le C\big(\sigma^{-1}\|f\|_{W^{s,r}(\Omega_T)} + \wp(s,r,\sigma)\|f\|_{L^\infty(\Omega_T)}\big), \\ \|\varphi\|_{L^\infty(\Omega_t)} &\le \sigma^{-1}\|f\|_{L^\infty(\Omega_T)}. \end{aligned} \tag{12.3.30}$$

*where the constant $C$ depends only on $M$, $\partial\Omega$, $\mathbf{U}$ and the exponents $s$, $r$.*

*Proof.* There exists a covering of the characteristic manifold $\Gamma$ by a finite collection of balls $B_{P_i}(\rho/4)$, $1 \le i \le m$, $P_i \in \Gamma$. Obviously, the balls $B_{P_i}(\rho)$ cover the set

$$\mathcal{V}_\Gamma = \{x \in \Omega : \text{dist}(x, \Gamma) < \rho/2\}.$$

By Lemma 12.3.6 for any $P \in \Gamma$, the solution to problem (12.2.3) is uniquely determined in some neighborhood of $P$ containing the ball $B_P(\rho)$. Hence, it suffices to prove estimates (12.3.30). To this end recall that by definition (see (1.5.1))

$$\|u\|_{W^{s,r}(\Omega)} = \|u\|_{L^r(\Omega)} + |u|_{s,r,\Omega},$$

where

$$|u|^r_{s,r,\Omega} = \int_{\Omega\times\Omega} |x-y|^{-3-rs}|u(x) - u(y)|^r\,dxdy. \tag{12.3.31}$$

We have

$$\begin{aligned} |\varphi|^r_{s,r,\mathcal{V}_\Gamma} &= \int_{\mathcal{V}_\Gamma^2} |x-y|^{-3-rs}|\varphi(x) - \varphi(y)|^r\,dxdy \\ &\le \int_{(\mathcal{V}_\Gamma)^2 \cap \{|x-y|<\rho/2\}} |x-y|^{-3-rs}|\varphi(x) - \varphi(y)|^r\,dxdy \\ &\quad + c\rho^{-3-rs}\|\varphi\|^r_{L^\infty(\mathcal{V}_\Gamma)}\,\text{meas}\,(\mathcal{V}_\Gamma)^2. \end{aligned}$$

Since any pair of points $x, y \in \mathcal{V}_\Gamma$ with $|x-y| < \rho/2$ belongs to some ball $B_{P_i}(\rho)$, the first term on the right hand side above does not exceed

$$\sum_i \int_{(B_{P_i}(\rho)\cap\Omega)^2} |x-y|^{-3-rs} |\varphi(x)-\varphi(y)|^r \, dxdy = \sum_i |\varphi|^r_{s,r,B_{P_i}(\rho)\cap\Omega},$$

which leads to

$$\begin{aligned} \|\varphi\|_{W^{s,r}(\mathcal{V}_\Gamma)} &= |\varphi|_{s,r,\mathcal{V}_\Gamma} + \|\varphi\|_{L^r(\mathcal{V}_\Gamma)} \\ &\le c\sum_i \|\varphi\|_{W^{s,r}(B_{P_i}(\rho)\cap\Omega)} + c\|\varphi\|_{L^\infty(\mathcal{V}_\Gamma)}, \end{aligned} \tag{12.3.32}$$

where $c$ depends on $s$, $r$ and $\rho$, i.e., on $s$, $r$, $\mathbf{U}$, $\partial\Omega$ and $M$. By Lemma 12.3.6 in each of these balls, the solution to problem (12.2.3) satisfies (12.3.17), so

$$\begin{aligned} \|\varphi\|_{W^{s,r}(\mathcal{V}_\Gamma)} \le{}& c\wp(s,r,\sigma)\sum_i \|f\|_{L^\infty(B_{P_i}(R)\cap\Omega)} \\ &+ c\sigma^{-1}\sum_i \|f\|_{W^{s,r}(B_{P_i}(R)\cap\Omega)} + c\|\varphi\|_{L^\infty(\mathcal{V}_\Gamma)}. \end{aligned} \tag{12.3.33}$$

On the other hand, $\|\varphi\|_{L^\infty(B_{P_i}(\rho)\cap\Omega)} \le \sigma^{-1}\|f\|_{L^\infty(B_{P_i}(R)\cap\Omega)}$. Moreover, since $B_{P_i}(R)\cap\Omega \subset \Omega_T$ we have

$$\begin{aligned} \wp(s,r,\sigma)\sum_i \|f\|^r_{L^\infty(B_{P_i}(R)\cap\Omega)} + \sigma^{-1}\sum_i \|f\|_{W^{s,r}(B_{P_i}(R)\cap\Omega)} \\ \le m\wp(s,r,\sigma)\|f\|_{L^\infty(\Omega_T)} + m\sigma^{-1}\|f\|_{W^{s,r}(\Omega_T)}. \end{aligned}$$

From this and (12.3.33), we finally obtain the estimates for the solution to problem (12.2.3) in the neighborhood of the characteristic manifold $\Gamma$,

$$\|\varphi\|_{W^{s,r}(\mathcal{V}_\Gamma)} \le c\wp(s,r,\sigma)\|f\|_{L^\infty(\Omega_T)} + c\sigma^{-1}\|f\|_{W^{s,r}(\Omega_T)}, \tag{12.3.34}$$

where $c$ depends only on $M$, $\partial\Omega$, $\mathbf{U}$ and $s$, $r$.

Our next task is to obtain a similar estimate in the neighborhood of the compact set $\Sigma'_{\text{in}} \subset \Sigma_{\text{in}}$ given by (12.3.28). To this end, we define

$$\mathcal{V}_{\text{in}} = \{x \in \Omega : \operatorname{dist}(x, \Sigma'_{\text{in}}) < \rho_i/2\}.$$

Let $B_{P_k}(\rho_i/4)$, $1 \le k \le m$, be a minimal collection of balls of radius $\rho_i/4$ covering $\Sigma'_{\text{in}}$. It is clear that the sets $B_{P_k}(\rho_i)\cap\Omega$ cover $\mathcal{V}_{\text{in}}$. Arguing as in the proof of (12.3.32) we obtain

$$\|\varphi\|_{W^{s,r}(\mathcal{V}_{\text{in}})} \le \sum_k \|\varphi\|_{W^{s,r}(B_{P_k}(\rho_i)\cap\Omega)} + c\|\varphi\|_{L^\infty(\mathcal{V}_{\text{in}})}.$$

From this and Lemma 12.3.7 we obtain

$$\|\varphi\|_{W^{s,r}(\mathcal{V}_{\text{in}})} \le c\sigma^{-1}\sum_k \|f\|_{W^{s,r}(B_{P_k}(R_i)\cap\Omega)} + c\wp(s,r,\sigma)\sum_k \|f\|_{L^\infty(B_{P_k}(R_i)\cap\Omega)}.$$

Thus, we get

$$\|\varphi\|_{W^{s,r}(\mathcal{V}_{\text{in}})} \le c\sigma^{-1}\|f\|_{W^{s,r}(\Omega_T)} + c\wp(r,s,\sigma)\|f\|_{L^\infty(\Omega_T)}.$$

Since $\mathcal{V}_\Gamma$ and $\mathcal{V}_{\text{in}}$ cover $\Omega_t$, this inequality along with (12.3.34) yields (12.3.30). □

### 12.3.5 Partition of unity

Let us turn to the analysis of the general problem

$$\mathscr{L}\varphi := \mathbf{u}\cdot\nabla\varphi + \sigma\varphi = f \quad \text{in } \Omega, \quad \varphi = 0 \quad \text{on } \Sigma_{\text{in}}. \tag{12.3.35}$$

We split the weak solution $\varphi \in L^\infty(\Omega)$ to problem (12.3.35) into two parts, namely the local solution, defined by Lemma 12.3.8, and the remainder vanishing near the inlet. To this end fix $\eta \in C^\infty(\mathbb{R})$ such that

$$0 \le \eta' \le 3, \quad \eta(u) = 0 \quad \text{for } u \le 1 \quad \text{and} \quad \eta(u) = 1 \quad \text{for } u \ge 3/2, \tag{12.3.36}$$

and introduce the one-parameter family of smooth functions

$$\chi_t(x) = \frac{1}{t^3}\int_{\mathbb{R}^3} \varpi\Big(\frac{2(x-y)}{t}\Big)\eta\Big(\frac{\operatorname{dist}(y,\Sigma_{\text{in}})}{t}\Big)\,dy, \tag{12.3.37}$$

where $\varpi \in C^\infty(\mathbb{R}^3)$ is a standard mollifying kernel with $\operatorname{supp}\varpi \subset \{|y| \le 1\}$. It follows that $\chi_t$ is a $C^\infty$ function with $|\nabla\chi_t| \le c/t$ and

$$\chi_t(x) = 0 \quad \text{for } \operatorname{dist}(x,\Sigma_{\text{in}}) \le t/2, \quad \chi_t(x) = 1 \quad \text{for } \operatorname{dist}(x,\Sigma_{\text{in}}) \ge 2t. \tag{12.3.38}$$

The functions $\chi_{t/2}$ and $1-\chi_{t/2}$ form a partition of unity in $\Omega$. Now, let $t$ be as in Lemma 12.3.8 and write

$$\varphi(x) = (1-\chi_{t/2}(x))\varphi(x) + \phi(x). \tag{12.3.39}$$

By (12.3.38) and Lemma 12.3.8, $\phi \in L^\infty(\Omega)$ vanishes in $\Omega_{t/4}$ and satisfies in a weak sense the equations

$$\mathbf{u}\cdot\nabla\phi + \sigma\phi = \chi_{t/2}f + \varphi\mathbf{u}\cdot\nabla\chi_{t/2} =: F \quad \text{in } \Omega, \quad \phi = 0 \quad \text{on } \Sigma_{\text{in}}.$$

Next, introduce a new vector field $\tilde{\mathbf{u}}(x) = \chi_{t/8}(x)\mathbf{u}(x)$. It is easy to see that $\chi_{t/8} = 1$ on the support of $\phi$, and hence $\phi$ is also a weak solution to the modified transport equation

$$\tilde{\mathscr{L}}\phi := \tilde{\mathbf{u}}\cdot\nabla\phi + \sigma\phi = F \quad \text{in } \Omega. \tag{12.3.40}$$

The advantage gained here is that the topology of integral curves of the modified vector field $\tilde{\mathbf{u}}$ drastically differs from that of $\mathbf{u}$. The corresponding outgoing and characteristic sets have a different structure and $\tilde{\Sigma}_{\text{in}} = \emptyset$. In particular, equation (12.3.40) does not require boundary conditions. Finally, the $C^1$-norm of the modified vector field can be bounded by

$$\|\tilde{\mathbf{u}}\|_{C^1(\Omega)} \le M(1 + c_1 t^{-1}), \tag{12.3.41}$$

In the following lemma the existence and uniqueness of solutions to the modified equation is established.

**Lemma 12.3.9.** *Suppose that*

$$\sigma > \sigma^* + 1, \quad \sigma^* = 16M(1 + c_1 t^{-1}) + 16, \quad M = \|\mathbf{u}\|_{C^1(\Omega)}, \tag{12.3.42}$$

*and furthermore* $0 \le s \le 1$, $r > 1$. *Then for any* $F \in W^{s,r}(\Omega) \cap L^\infty(\Omega)$, *equation* (12.3.40) *has a unique weak solution* $\phi \in W^{s,r}(\Omega) \cap L^\infty(\Omega)$ *such that*

$$\|\phi\|_{W^{s,r}(\Omega)} \le c\sigma^{-1}\|F\|_{W^{s,r}(\Omega)}, \quad \|\phi\|_{L^\infty(\Omega)} \le \sigma^{-1}\|F\|_{L^\infty(\Omega)}, \tag{12.3.43}$$

*where* $c$ *depends only on* $r$.

*Proof.* We can assume that $F \in C^1(\Omega)$. By (12.3.41) and (12.3.42), the vector fields $\tilde{\mathbf{u}}$ and $\sigma$ meet all requirements of Theorem 12.1.5. Hence, equation (12.3.40) has a unique solution $\phi \in W^{1,\infty}(\Omega)$. For $i = 1, 2, 3$ and $\tau > 0$, define the finite difference operator

$$\delta_{i\tau}\phi = \frac{1}{\tau}(\phi(x + \tau\mathbf{e}_i) - \phi(x)).$$

It can be easily seen that

$$\tilde{\mathbf{u}} \cdot \nabla\delta_{i\tau}\phi + \sigma\delta_{i\tau}\phi = \delta_{i\tau}F - \delta_{i\tau}\tilde{\mathbf{u}} \cdot \nabla\phi(x + \tau\mathbf{e}_i) \quad \text{in } \Omega \cap (\Omega - \tau\mathbf{e}_i). \tag{12.3.44}$$

Next, set $\eta_h(x) = \eta(\operatorname{dist}(x, \partial\Omega)/h)$, where $\eta$ is defined by (12.3.36). Since $\tilde{\Sigma}_{\text{in}} = \emptyset$, the inequality

$$\limsup_{h\to 0} \int_\Omega g\tilde{\mathbf{u}} \cdot \nabla\eta_h(x)\, dx \le 0 \tag{12.3.45}$$

holds for all nonnegative functions $g \in L^\infty(\Omega)$. Choosing $h > \tau$, multiplying both sides of equation (12.3.44) by $\eta_h|\delta_{i\tau}\phi|^{r-2}\delta_{i\tau}\phi$ and integrating the result over $\Omega \cap (\Omega - \tau\mathbf{e}_i)$ we obtain

$$\int_{\Omega\cap(\Omega-\tau\mathbf{e}_i)} \eta_h|\delta_{i\tau}\phi|^r\Big(\sigma - \frac{1}{r}\operatorname{div}\tilde{\mathbf{u}}\Big)\, dx - \int_{\Omega\cap(\Omega-\tau\mathbf{e}_i)} |\delta_{i\tau}\phi|^r\tilde{\mathbf{u}} \cdot \nabla\eta_h\, dx$$
$$= \int_{\Omega\cap(\Omega-\tau\mathbf{e}_i)} (\delta_{i\tau}F - \delta_{i\tau}\tilde{\mathbf{u}} \cdot \nabla\phi(x + \tau\mathbf{e}_i))\eta_h|\delta_{i\tau}\phi|^{r-2}\delta_{i\tau}\phi\, dx.$$

Letting $\tau \to 0$ and then $h \to 0$ and using inequality (12.3.45) we obtain

$$\int_\Omega |\partial_{x_i}\phi|^r\Big(\sigma - \frac{1}{r}\operatorname{div}\tilde{\mathbf{u}}\Big)\, dx \le \int_\Omega (\partial_{x_i}F - \partial_{x_i}\tilde{\mathbf{u}} \cdot \nabla\phi)|\partial_{x_i}\phi|^{r-2}\partial_{x_i}\phi\, dx. \tag{12.3.46}$$

Next, note that

$$\sum_i \partial_{x_i}\tilde{\mathbf{u}} \cdot \nabla\phi|\partial_{x_i}\phi|^{r-2}\partial_{x_i}\phi \le 3\|\tilde{\mathbf{u}}\|_{C^1(\Omega)} \sum_i |\partial_{x_i}\phi|^r.$$

On the other hand, since $1/r + 3 \le 4$, inequalities (12.3.41) and (12.3.42) imply

$$\sigma - (1/r + 3)\|\tilde{\mathbf{u}}\|_{C^1(\Omega)} \ge \sigma - \sigma^* \ge 1.$$

From this we conclude that

$$(\sigma-\sigma^*)\sum_i\int_\Omega|\partial_{x_i}\phi|^r\,dx\le\sum_i\int_\Omega|\partial_{x_i}\phi|^{r-1}|\partial_{x_i}F|\,dx\le c\|\nabla\phi\|^{r-1}_{L^r(\Omega)}\|\nabla F\|_{L^{r'}(\Omega)},$$

which leads to

$$\|\nabla\phi\|_{L^r(\Omega)}\le c(r)\sigma^{-1}\|\nabla F\|_{L^r(\Omega)}\quad\text{for }\sigma>\sigma^*(M,r).\tag{12.3.47}$$

Next, multiplying both sides of (12.3.40) by $|\phi|^{r-2}\eta_h$ and integrating the result over $\Omega$ we get

$$\int_\Omega(\sigma-r^{-1}\operatorname{div}\tilde{\mathbf{u}})\eta_h|\phi|^r\,dx-\int_\Omega|\phi|^r\tilde{\mathbf{u}}\cdot\nabla\eta_h\,dx=\int_\Omega F\eta_h|\phi|^{r-2}\phi\,dx.$$

The passage $h\to 0$ gives the inequality

$$\int_\Omega(\sigma-r^{-1}\operatorname{div}\tilde{\mathbf{u}})|\phi|^r\,dx\le\int_\Omega|F|\,|\phi|^{r-1}\,dx.$$

Recalling that $\sigma-r^{-1}\operatorname{div}\tilde{\mathbf{u}}\ge\sigma-\sigma^*$, we finally obtain

$$\|\phi\|_{L^r(\Omega)}\le c(r)\sigma^{-1}\|F\|_{L^r(\Omega)}.\tag{12.3.48}$$

Inequalities (12.3.47) and (12.3.48) imply (12.3.43) for $s=0,1$. Consequently, for $\sigma>\sigma^*$, the linear operator $\tilde{\mathscr{L}}^{-1}:F\mapsto\phi$ is continuous between the Banach spaces $L^r(\Omega)$ and $W^{1,r}(\Omega)$ and its norm does not exceed $c(r)\sigma^{-1}$. Recall that $W^{s,r}(\Omega)$ is the interpolation space $[L^r(\Omega),W^{1,r}(\Omega)]_{s,r}$. From interpolation theory we thus conclude that (12.3.43) holds for all $s\in[0,1]$. It remains to note that the $L^\infty$ estimate follows directly from Theorem 12.1.2. □

We are now in a position to complete the proof of Theorem 12.2.3. Fix $\sigma>\sigma^*+1$, where the constant $\sigma^*$ depends only on $\Sigma$, $\mathbf{U}$ and $\|\mathbf{u}\|_{C^1(\Omega)}$, and it is defined by (12.3.42). We can assume that $f\in C^1(\Omega)$. The existence and uniqueness of a weak bounded solution for $\sigma>\sigma^*$ follows from Theorem 12.1.2. Therefore, it suffices to prove estimate (12.2.8) for $\|\varphi\|_{W^{s,r}(\Omega)}$. Since $W^{s,r}(\Omega)\cap L^\infty(\Omega)$ is a Banach algebra, representation (12.3.39) together with inequality (12.3.38) implies

$$\|\varphi\|_{W^{s,r}(\Omega)}\le c(\|\varphi\|_{W^{s,r}(\Omega_t)}+\|\varphi\|_{L^\infty(\Omega_t)})+c\|\phi\|_{W^{s,r}(\Omega)}.\tag{12.3.49}$$

On the other hand, Lemma 12.3.9 along with (12.3.40) yields

$$\begin{aligned}\|\phi\|_{W^{s,r}(\Omega)}&\le c\sigma^{-1}\|F\|_{W^{s,r}(\Omega)}\le c\sigma^{-1}\|\chi_{t/2}f\|_{W^{s,r}(\Omega)}+\sigma^{-1}\|\varphi\mathbf{u}\cdot\nabla\chi_{t/2}\|_{W^{s,r}(\Omega)}\\&\le c\sigma^{-1}(\|f\|_{W^{s,r}(\Omega)}+\|\varphi\|_{W^{s,r}(\Omega_t)})\end{aligned}$$

since $\nabla\chi_{t/2}\in C_0^\infty(\Omega_t)$. Substituting these estimates into (12.3.49) we arrive at

$$\|\varphi\|_{W^{s,r}(\Omega)}\le c\big(\sigma^{-1}\|\varphi\|_{W^{s,r}(\Omega_t)}+\|\varphi\|_{L^\infty(\Omega_t)}+\sigma^{-1}\|f\|_{W^{s,r}(\Omega_t)}+\sigma^{-1}\|f\|_{L^\infty(\Omega)}\big),$$

which along with (12.3.30) leads to estimate (12.2.8), and the theorem follows.

### 12.3.6 Proofs of Lemmas 12.3.2 and 12.3.3

**Proof of Lemma 12.3.2.** We start with the proof of (**P**1). Choose an arbitrary $P \in \Gamma$ and recall the emergent field Condition 12.2.2. Without loss of generality we can assume that Cartesian coordinates $(x_1, x_2, x_3)$ have origin at $P$ and $\mathbf{U}(P) = (U, 0, 0)$ with $U = |\mathbf{U}(P)|$ and $\mathbf{n}(P) = (0, 0, -1)$. It follows from this and Condition 12.2.2 that there is a neighborhood $\mathcal{O} = [-k, k]^2 \times [-t, t]$ of $P$ such that the intersections $\partial\Omega \cap \mathcal{O}$ and $\Gamma \cap \mathcal{O}$ are defined by

$$F_0(x) \equiv x_3 - F(x_1, x_2) = 0, \quad \nabla F_0(x) \cdot \mathbf{U}(x) = 0,$$

and $\Omega \cap \mathcal{O}$ is the epigraph $\{F_0 > 0\} \cap \mathcal{O}$. The function $F$ satisfies

$$\|F\|_{C^2([-k,k]^2)} \le K, \quad F(0,0) = 0, \quad \nabla F(0,0) = 0, \tag{12.3.50}$$

where the constants $k, t < 1$ and $K > 1$ depend only on the curvature of $\partial\Omega$ and are independent of the point $P$. Since the vector field $\mathbf{U}$ is transversal to $\Gamma$, the manifold $\Gamma \cap \mathcal{O}$ admits a parameterization

$$x = \big(\mathfrak{g}(y_2),\, y_2,\, F(\mathfrak{g}(y_2), y_2)\big) \tag{12.3.51}$$

such that $\mathfrak{g}(0) = 0$ and $\|\mathfrak{g}\|_{C^2([-k,k])} \le C$, where the constant $C > 1$ depends only on $\Omega$ and $\mathbf{U}$.

With this notation, the inequality (12.2.2) implies the existence of positive constants $N^\pm$ independent of $P$ such that for $x \in \partial\Omega$ given by the condition $F_0(x_1, x_2, x_3) = x_3 - F(x_1, x_2) = 0$, we have

$$\begin{aligned} N^-\big(x_1 - \mathfrak{g}(x_2)\big) \le -\nabla F_0(x) \cdot \mathbf{U}(x) \le N^+\big(x_1 - \mathfrak{g}(x_2)\big) \quad &\text{for } x_1 > \mathfrak{g}(x_2), \\ -N^-\big(x_1 - \mathfrak{g}(x_2)\big) \le \nabla F_0(x) \cdot \mathbf{U}(x) \le -N^+\big(x_1 - \mathfrak{g}(x_2)\big) \quad &\text{for } x_1 < \mathfrak{g}(x_2). \end{aligned} \tag{12.3.52}$$

Consider the Cauchy problem

$$\begin{aligned} &\partial_{y_1}\mathbf{x} = \mathbf{u}(\mathbf{x}(y)) \quad \text{in } Q_a, \\ &x_1(y) = \mathfrak{g}(y_2), \quad x_2(y) = y_2 \quad \text{for } y_1 = 0, \\ &x_3 = F(\mathfrak{g}(y_2), y_2) + y_3 \quad \text{for } y_1 = 0. \end{aligned} \tag{12.3.53}$$

We can assume that $0 < a < k < 1$. For any such $a$, problem (12.3.53) has a unique solution in $C^1(Q_a)$. Denote $\mathfrak{F}(y) = D_y\mathbf{x}(y)$. Calculations show that

$$\mathfrak{F}_0 := \mathfrak{F}(y)|_{y_1=0} = \begin{pmatrix} u_1 & \mathfrak{g}'(y_2) & 0 \\ u_2 & 1 & 0 \\ u_3 & \partial_{y_2}F(\mathfrak{g}(y_2), y_2) & 1 \end{pmatrix}, \quad \mathfrak{F}(0) = \begin{pmatrix} U & \mathfrak{g}'(0) & 0 \\ 0 & 1 & 0 \\ 0 & 0 & 1 \end{pmatrix},$$

which implies

$$\|\mathfrak{F}(0)^{\pm 1}\| \le C/3, \quad \|\mathfrak{F}_0(y) - \mathfrak{F}(0)\| \le ca. \tag{12.3.54}$$

Differentiation of (12.3.53) leads to an ordinary differential equation for $\mathfrak{F}$,

$$\partial_{y_1}\mathfrak{F} = D_y\mathbf{u}(\mathbf{x})\mathfrak{F}, \quad \mathfrak{F}|_{y_1=0} = \mathfrak{F}_0.$$

From this we get

$$\partial_{y_1}\|\mathfrak{F}-\mathfrak{F}_0\| \le M(\|\mathfrak{F}-\mathfrak{F}_0\| + \|\mathfrak{F}_0\|),$$

and hence $\|\mathfrak{F}-\mathfrak{F}_0\| \le c(M)\|\mathfrak{F}_0\|a$. Combining this with (12.3.54), we arrive at

$$\|\mathfrak{F}(y)-\mathfrak{F}(0)\| \le ca. \tag{12.3.55}$$

This inequality along with the implicit function theorem implies the existence of $a > 0$, depending only on $M$ and $\Omega$, such that the mapping $x = \mathbf{x}(y)$ takes diffeomorphically the cube $Q_a$ onto some neighborhood of the point $P$ and satisfies inequalities (12.3.6).

Let us turn to the proof of (**P2**). Observe that the manifold $\mathbf{x}^{-1}(\partial\Omega\cap\mathcal{O})$ is defined by

$$\Phi_0(y) := x_3(y) - F(x_1(y), x_2(y)) = 0, \quad y \in Q_a.$$

Let us show that $\Phi_0$ is strictly monotone in $y_3$ and has opposite signs on the faces $y_3 = \pm a$. To this end note that the formula for $\mathfrak{F}(0)$ along with (12.3.55) implies

$$|\partial_{y_3}x_3(y) - 1| + |\partial_{y_3}x_1(y)| + |\partial_{y_3}x_2(y)| \le ca \quad \text{in } Q_a.$$

Thus, we get

$$1 - ca \le \partial_{y_3}\Phi_0(y) = \partial_{y_3}x_3(y) - \partial_{x_i}F(x_1,x_2)\partial_{y_3}x_i(y) \le 1 + ca.$$

It follows from (12.3.55) that for $y_3 = 0$, we have $|x_3(y)| \le ca|y|$, which along with (12.3.6) yields the estimate

$$|\Phi_0(y)| \le |x_3(y)| + |F(\mathbf{x}(y))| \le ca|y| + KC|y|^2 \le ca^2 \quad \text{for } y_3 = 0.$$

Hence, there is a positive $a$, depending only on $M$ and $\Omega$, such that

$$1/2 \le \partial_{y_3}\Phi_0(y) \le 2, \quad \pm\Phi_0(y_1,y_2,\pm a) > 0,$$

for all $y \in Q_a$. Therefore, the equation $\Phi_0(y) = 0$ has a unique solution $y_3 = \Phi(y_1, y_2)$ in the cube $Q_a$. Moreover, $\Phi \in C^1([-a,a]^2)$ vanishes for $y_1 = y_3 = 0$. Thus, we get

$$\mathcal{P}_a := \mathbf{x}^{-1}(\mathcal{O}\cap\Omega) = \{\Phi(y_1,y_2) < y_3 < a, \ |y_1|, |y_2| \le a\}.$$

Note that $|\mathbf{u}(\mathbf{x}(y)) - U\mathbf{e}_1| \le M|\mathbf{x}(y)| \le Ca$. Therefore, we can choose $a = a(M,\Omega)$ such that $2U/3 \le u_1 \le 4U/3$ and $C|u_2| \le U/3$ in $Q_a$. Recall that $x_1(y) - \mathfrak{g}(x_2(y))$ vanishes at the plane $y_1 = 0$ and

$$\partial_{y_1}[x_1(y) - \mathfrak{g}(x_2(y))] = u_1(y) - \mathfrak{g}'(x_2(y))u_2(y).$$

We deduce that for a suitable choice of $a$,

$$|y_1|U/3 \le |x_1(y) - \mathfrak{g}(x_2(y))| \le |y_1|5U/3 \quad \text{for } y \in Q_a. \tag{12.3.56}$$

Equations (12.3.53) imply the identity

$$\partial_{y_1}\Phi_0(y) \equiv \nabla F_0(\mathbf{x}(y)) \cdot \mathbf{u}(\mathbf{x}(y)) = \nabla F_0(\mathbf{x}(y)) \cdot \mathbf{U}(\mathbf{x}(y)) \quad \text{for } \Phi_0(y) = 0.$$

Combining this with (12.3.52) and (12.3.56), we obtain the estimates

$$|y_1|N^-U/3 \le |\partial_{y_1}\Phi_0(y)| \le |y_1|N^+U5/3,$$

which, along with the identity,

$$\partial_{y_1}\Phi = -\partial_{y_1}\Phi_0(\partial_{y_3}\Phi_0)^{-1}$$

yields the inequalities

$$\begin{aligned} &-c < \partial_{y_1}\Phi(y_1, y_2) \le cy_1 \quad \text{for } -a < y_1 < 0,\\ &cy_1 < \partial_{y_1}\Phi(y_1, y_2) \le c \quad \text{for } 0 < y_1 < a,\\ &|\partial_{y_2}\Phi(y_1, y_2)| \le c, \quad 0 \le \Phi(y_1, y_2) \le cy_1^2. \end{aligned} \tag{12.3.57}$$

It is clear that for $a$ sufficiently small, depending only on $\mathbf{U}$ and $\Omega$, the functions $\Phi^\pm(y_2) = \Phi(\pm a, y_2)$ satisfy $ca^2 \le \Phi^\pm(y_2) < a$. Set

$$\begin{aligned} Q_{\text{in}} &= \{Y = (y_2, y_3) \in [-a, a] \times [0, a] : 0 < y_3 < \Phi^-(y_2)\},\\ Q_{\text{out}} &= \{Y = (y_2, y_3) \in [-a, a] \times [0, a] : 0 < y_3 < \Phi^+(y_2)\}. \end{aligned}$$

It follows from (12.3.57) that for every $Y \in Q_{\text{in}}$ (resp. $Y \in Q_{\text{out}}$) the equation $y_3 = \Phi(y_1, y_2)$ has a unique solution $y_1 = a^-(Y) < 0$ (resp. $y_1 = a^+(Y) > 0$). We adopt the convention that $a^\pm(Y) = \pm a$ for $y_3 > \Phi^\pm(y_2)$. It remains to note that, by (12.3.57), the functions $a^\pm$ meet all requirements of Lemma 12.3.2.

**Proof of Lemma 12.3.3.** The proof is similar to that of Lemma 12.3.2. We may assume that Cartesian coordinates $(x_1, x_2, x_3)$ are centered at $P$ and $\mathbf{n} = \mathbf{e}_3$. By the smoothness of $\partial\Omega$, there is a neighborhood $\mathcal{O} = [-k, k]^2 \times [-t, t]$ such that the manifold $\partial\Omega \cap \mathcal{O}$ is defined by

$$x_3 = F(x_1, x_2), \quad F(0,0) = 0, \quad |\nabla F(x_1, x_2)| \le K(|x_1| + |x_2|).$$

The constants $k$, $t$ and $K$ depend only on $\Omega$. Consider the initial value problem

$$\partial_{y_3}\mathbf{x} = \mathbf{u}(\mathbf{x}(y)) \quad \text{in } Q_a, \quad \mathbf{x}|_{y_3=0} = (y_1, y_2, F(y_1, y_2)). \tag{12.3.58}$$

We can assume that $0 < b < k < 1$. For any such $b$, problem (12.3.58) has a unique solution in $C^1(Q_b)$. Next, note that for $y_3 = 0$ we have

$$|\mathbf{x}(y)| \le (K+1)|y|, \quad |\mathbf{u}(\mathbf{x}(y)) - \mathbf{u}(0)| \le M(K+1)|y|. \tag{12.3.59}$$

Denote $\mathfrak{F}(y) = D_y\mathbf{x}(y)$. Calculations show that

$$\mathfrak{F}_0 := \mathfrak{F}(y)|_{y_3=0} = \begin{pmatrix} 1 & 0 & u_1 \\ 0 & 1 & u_2 \\ 0 & 0 & u_3 \end{pmatrix}, \quad \mathfrak{F}(0) = \begin{pmatrix} 1 & 0 & u_1(P) \\ 0 & 1 & u_2(P) \\ 0 & 0 & U_n \end{pmatrix},$$

where $U_n = -\mathbf{U}(P)\cdot\mathbf{n}$, which along with (12.3.59) implies

$$\|\mathfrak{F}(0)^{\pm 1}\| \le C/3, \quad \|\mathfrak{F}_0(y) - \mathfrak{F}(0)\| \le cb. \tag{12.3.60}$$

Next, differentiation of (12.3.58) with respect to $y$ leads to

$$\partial_{y_1}\mathfrak{F} = D_y\mathbf{u}(\mathbf{x})\mathfrak{F}, \quad \mathfrak{F}|_{y_3=0} = \mathfrak{F}_0.$$

Arguing as in the proof of Lemma 12.3.2 we obtain $\|\mathfrak{F} - \mathfrak{F}_0\| \le c(M)\|\mathfrak{F}_0\|b$. Combining this with (12.3.60), we arrive at $\|\mathfrak{F}(y) - \mathfrak{F}(0)\| \le cb$. From this and the implicit function theorem we conclude that there is $b > 0$, depending only on $M$ and $\Omega$, such that the mapping $x = \mathbf{x}(y)$ takes diffeomorphically the cube $Q_b$ onto some neighborhood of $P$, and satisfies inequalities (12.3.11). Inclusions (12.3.12) easily follow from (12.3.11).

### 12.3.7 Proof of Lemma 12.3.4

Throughout the section, $c, C$ stand for various constants depending only on the domain $\mathcal{P}_a$ and the exponents $s$, $r$. Furthermore, for any $y = (y_1, y_2, y_3)$ and $z = (z_1, z_2, z_3)$, $Y$ and $Z$ stand for $(y_2, y_3)$ and $(z_2, z_3)$, respectively.

The existence and uniqueness of solutions to problem (12.3.13) is obvious. Multiplying (12.3.13) by $|\varphi|^{r-1}\varphi$ and integrating the result over $\mathcal{P}_a$ we obtain

$$\|\varphi\|_{L^r(\mathcal{P}_a)} \le \sigma^{-1}\|f\|_{L^r(\mathcal{P}_a)} \quad \text{for } r < \infty. \tag{12.3.61}$$

Letting $r \to \infty$ we conclude that (12.3.61) holds for $r = \infty$. Let us turn to the proof of inequality (12.3.14), beginning with the case $s = 1$. First, we derive the estimates for $\partial_{y_k}\varphi$. We have the representation $\partial_{y_k}\varphi = \varphi' + \varphi''$, where

$$\varphi'(y) = \begin{cases} -e^{\sigma(a^-(Y)-y_1)}\partial_{y_k}a^-(Y)f(a^-(Y),Y) & \text{for } k = 2,3, \\ e^{\sigma(a^-(Y)-y_1)}(Y)f(a^-(Y),Y) & \text{for } k = 1, \end{cases}$$

and $\varphi''$ is a solution to the boundary value problem

$$\partial_{y_1}\varphi'' + \sigma\varphi'' = \partial_{y_k}f \quad \text{in } \mathcal{P}_a, \quad \varphi''(y) = 0 \quad \text{for } y_1 = a^-(Y).$$

It follows from (12.3.61) that $\|\varphi''\|_{L^r(\mathcal{P}_a)} \le \sigma^{-1}\|f\|_{W^{1,r}(\mathcal{P}_a)}$. On the other hand, inequalities (12.3.4) yield the estimate

$$\int_{a^-(Y)}^{a^+(Y)} |\varphi'(y_1, Y)|^r\, dy_1 \le c(r)\sigma^{-1}\|f\|^r_{L^\infty(\mathcal{P}_a)} y_3^{-r/2}(1 - e^{r\sigma(a^-(Y)-a^+(Y))}).$$

Since $0 \le a^+ - a^- \le c y_3^{1/2}$ we conclude that

$$\|\varphi'\|^r_{L^r(\mathcal{P}_a)} \le c\sigma^{-1}\|f\|^r_{L^\infty(\mathcal{P}_a)} \int_0^a y_3^{-r/2}(1 - e^{-c\sigma\sqrt{y_3}})\, dy_3. \qquad (12.3.62)$$

We have

$$\sigma^{-1}\int_0^a y_3^{-r/2}(1 - e^{-c\sigma\sqrt{y_3}})\, dy_3$$

$$= \sigma^{r-3}\int_0^{a\sigma^2} t^{-r/2}(1 - e^{-c\sqrt{t}})\, dt \le \begin{cases} c\sigma^{r-3} & \text{for } 2 < r < 3, \\ c\sigma^{-1}(1 + \log\sigma) & \text{for } r = 2, \\ c\sigma^{-1} & \text{for } 1 < r < 2. \end{cases}$$

Thus, we get

$$\|\varphi'\|_{L^r(\mathcal{P}_a)} \le c\|f\|_{L^\infty(\mathcal{P}_a)} \begin{cases} c\sigma^{-1+\alpha} & \text{for } r \in (1,2) \cup (2,3), \\ c\sigma^{-1+\alpha}(1 + \log\sigma)^{1/r} & \text{for } r = 2, \end{cases}$$

where $\alpha = \max\{0, 1 - 1/r, 2 - 3/r\}$. Combining the estimates for $\varphi'$ and $\varphi''$ we obtain (12.3.14).

The proof of (12.3.14) for $0 < s < 1$ is more complicated. By (12.3.61), it suffices to estimate the seminorm

$$|\varphi|_{s,r,\mathcal{P}_a} = \Big\{ \int_{\mathcal{P}_a^2} |\varphi(z) - \varphi(y)|^r |z - y|^{-3-rs}\, dydz \Big\}^{1/r}.$$

Since this expression is invariant with respect to the change $(y, z) \mapsto (z, y)$, we have

$$|\varphi|_{s,r,\mathcal{P}_a} \le (2I)^{1/r}, \quad I = \int_{D_a} |\varphi(z) - \varphi(y)|^r |z - y|^{-3-rs}\, dydz, \qquad (12.3.63)$$

where $D_a = \{(y, z) \in (\mathcal{P}_a)^2 : a^-(Z) \le a^-(Y)\}$. It is easy to see that

$$\varphi(z) - \varphi(y) = \varphi(z_1, Z) - \varphi(y_1, Z) + \int_{a^-(Z)}^{a^-(Y)} e^{\sigma(x_1 - y_1)} f(x_1, Z)\, dx_1$$

$$+ \int_{a^-(Y)}^{y_1} e^{\sigma(x_1 - y_1)}(f(x_1, Z) - f(x_1, Y))\, dx_1 = I_1 + I_2 + I_3. \qquad (12.3.64)$$

Hence, our task is to estimate the integrals

$$J_k = \int_{D_a} |I_k|^r |z - y|^{-3-rs}\, dydz, \quad k = 1, 2, 3. \qquad (12.3.65)$$

The evaluation falls naturally into three steps and it is based on the following proposition.

**Proposition 12.3.10.** *If $r, s > 0$ and $i \neq j \neq k$, $i \neq k$, then*

$$\int_{[-a,a]^2} |z-y|^{-3-rs}\, dy_i dy_j \le c(r,s)|z_k - y_k|^{-1-rs}.$$

*Proof.* Note that the left hand side is equal to

$$|z_k - y_k|^{-1-rs} \int_{[-a,a]^2} \left(1 + \frac{|z_i - y_i|^2 + |z_j - y_j|^2}{|z_k - y_k|^2}\right)^{(-3-rs)/2} \frac{dy_i dy_j}{|z_k - y_k|^2}$$
$$\le c(r,s)|z_k - y_k|^{-1-rs} \int_{\mathbb{R}^2} (1 + |y_i|^2 + |y_j|^2)^{-(3+rs)/2}\, dy_i dy_j. \qquad \square$$

**Step 1.** By the extension principle, the right hand side $f$ has an extension over $\mathbb{R}^3$ which vanishes outside the cube $Q_{3a}$ and satisfies

$$\|f\|_{W^{s,r}(\mathbb{R}^3)} \le c\|f\|_{W^{s,r}(Q_a)}, \quad \|f\|_{L^\infty(\mathbb{R}^3)} \le c\|f\|_{L^\infty(Q_a)}. \tag{12.3.66}$$

Next, recall that $a^-(Z) \le y_1, z_1 \le a$ for all $(y,z) \in D_a$. From this and Proposition 12.3.10 we obtain

$$J_1 \le \int_{[-a,a]^2} \Big\{ \int_{[a^-(Z),a]^2} |\varphi(z_1, Z) - \varphi(y_1, Z)|^r \mathcal{M}_1(z, y_1)\, dy_1 dz_1 \Big\}\, dZ$$
$$\le c \int_{[-a,a]^2} \Big\{ \int_{[a^-(Z),a]^2} |\varphi(z_1, Z) - \varphi(y_1, Z)|^r |z_1 - y_1|^{-1-rs}\, dy_1 dz_1 \Big\}\, dZ$$

where we denote

$$\mathcal{M}_1(z, y_1) = \int_{[-a,a]^2} |z-y|^{-3-rs}\, dy_2 dy_3 \le c|z_1 - y_1|^{-1-rs}.$$

As the right hand side of the last inequality is invariant with respect to $(y_1, z_1) \mapsto (z_1, y_1)$, we have

$$J_1 \le c(r,s) \int_{[-a,a]^2} \Big\{ \int_{D(Z)} |\varphi(z_1, Z) - \varphi(y_1, Z)|^r |y_1 - z_1|^{-1-rs}\, dy_1 dz_1 \Big\}\, dZ, \tag{12.3.67}$$

where $D(Z) = \{(y_1, z_1) : a^-(Z) \le z_1 \le y_1 \le a\}$. Next, for all $(y_1, z_1) \in D(Z)$,

$$\varphi(y_1, Z) = \int_{a^-(Z)}^{y_1} e^{\sigma(t-y_1)} f(t, Z)\, dt = \int_{a^-(Z)-\xi}^{z_1} e^{\sigma(t-z_1)} f(t+\xi, Z)\, dt,$$

where $\xi = y_1 - z_1 > 0$. Thus, we get

$$\varphi(y_1, Z) - \varphi(z_1, Z) = \int_{a^-(Z)}^{z_1} e^{\sigma(t-z_1)} (f(t+\xi, Z) - f(t, Z))\, dt$$
$$+ \int_{a^-(Z)-\xi}^{a^-(Z)} e^{\sigma(t-z_1)} f(t+\xi, Z)\, dt = I_{11} + I_{12}. \tag{12.3.68}$$

Since $f$ is extended to $\mathbb{R}^3$ and vanishes outside $[-a,a]^3$, we have the estimate

$$\int_{D(Z)} |I_{11}|^r |y_1 - z_1|^{-1-rs}\, dy_1 dz_1 \le \int_{a^-(Z)}^{a} \int_0^{2a} |M(z_1,\xi,Z)|^r\, dz_1 d\xi, \tag{12.3.69}$$

where

$$M(z_1,\xi,Z) = \xi^{-s-1/r} \int_{a^-(Z)}^{z_1} e^{\sigma(t-z_1)}(f(t+\xi,Z) - f(t,Z))\, dt.$$

It is easy to see that $M$ satisfies

$$\partial_{z_1} M + \sigma M = K \quad \text{for } z_1 \in (a^-(Z), a), \qquad M = 0 \quad \text{for } z_1 = a^-(Z),$$

where $K(z_1,\xi,Z) = \xi^{-s-1/r}(f(z_1+\xi,Z) - f(z_1,Z))$. Multiplying this equation by $|M|^{r-2}M$ and integrating the result over $(a^-(Z),a)$ we arrive at

$$\begin{aligned} \sigma \int_{a^-(Z)}^{a} |M|^r\, dz_1 &\le \int_{a^-(Z)}^{a} |M|^{r-1}|K|\, dz_1 \\ &\le \Big(\int_{a^-(Z)}^{a} |M|^r\, dz_1\Big)^{1-1/r} \Big(\int_{a^-(Z)}^{a} |K|^r\, dz_1\Big)^{1/r}, \end{aligned}$$

which gives

$$\int_{a^-(Z)}^{a} |M(z_1,\xi,Z)|^r\, dz_1 \le \sigma^{-r}\xi^{-1-rs} \int_{a^-(Z)}^{a} |f(z_1+\xi,Z) - f(z_1,Z)|^r\, dz_1.$$

Combining this inequality with (12.3.69) we obtain the following estimate for the term $I_{11}$ on the right hand side of (12.3.68):

$$\begin{aligned} &\int_{[-a,a]^2} \int_{D(Z)} |I_{11}|^r |y_1 - z_1|^{-1-rs}\, dy_1 dz_1 dZ \\ &\le \sigma^{-r} \int_{[-a,a]^2} \int_{a^-(Z)}^{a} \int_0^{2a} \xi^{-1-rs} |f(z_1+\xi,Z) - f(z_1,Z)|^r\, dz_1 d\xi dZ \\ &\le \sigma^{-r} \int_{\mathbb{R}^4} |f(y_1,Z) - f(z_1,Z)|^r |y_1 - z_1|^{-1-rs}\, dy_1 dz_1 dZ \\ &\le c\sigma^{-r} \|f\|^r_{L^r(\mathbb{R}^2; W^{s,r}(\mathbb{R}))} \le c\sigma^{-r}\|f\|^r_{W^{s,r}(\mathbb{R}^3)}. \end{aligned} \tag{12.3.70}$$

To estimate $I_{12}$ note that

$$|I_{12}| = \Big| \int_{a^-(Z)-\xi}^{a^-(Z)} e^{\sigma(t-z_1)} f(t+\xi,Z)\, dt \Big| \le \|f\|_{L^\infty(Q_{3a})} e^{\sigma(a^-(Z)-z_1)} \sigma^{-1}(1 - e^{-\sigma\xi}),$$

which gives

$$\int_{D(Z)} |I_{12}|^r |y_1 - z_1|^{-1-rs}\, dy_1 dz_1$$
$$\le c\sigma^{-r}\|f\|^r_{L^\infty(Q_{3a})} \int_{a^-(Z)}^{a} e^{r\sigma(a^-(Z)-z_1)}\, dz_1 \int_0^{3a} \xi^{-1-rs}(1-e^{-\sigma\xi})^r d\xi$$
$$\le c\sigma^{-1-r+rs}\|f\|^r_{L^\infty(Q_{3a})} \int_0^\infty \xi^{-1-rs}(1-e^{-\xi})^r d\xi \le c\sigma^{-1-r+rs}\|f\|^r_{L^\infty(Q_{3a})}.$$

We conclude that

$$\int_{[-a,a]^2}\int_{D(Z)} |I_{12}|^r |y_1 - z_1|^{-1-rs}\, dy_1 dz_1 dZ \le c\|f\|^r_{L^\infty(Q_{3a})}\sigma^{-1-r+rs}.$$

Inserting this together with (12.3.70) in (12.3.67), and recalling (12.3.66), we finally obtain

$$J_1^{1/r} \le c\big(\sigma^{-1}\|f\|_{W^{r,s}(Q_a)} + \sigma^{-1+(s-1/r)}\|f\|_{L^\infty(Q_a)}\big). \tag{12.3.71}$$

**Step 2.** Our next task is to estimate $J_2$. It follows from (12.3.64) that

$$|I_2| \le c\|f\|_{L^\infty(\mathcal{P}_a)}\sigma^{-1} e^{\sigma(a^-(Y)-y_1)}\big(1 - e^{\sigma(a^-(Z)-a^-(Y))}\big).$$

Next, inequalities (12.3.4) imply, for all $(Y, Z) \subset D_a$,

$$0 \le a^-(Y) - a^-(Z) = \int_{z_2}^{y_2} \partial_t a^-(t, z_3)\, dt + \int_{z_3}^{y_3} \partial_t a^-(y_2, t)\, dt$$
$$\le c|y_2 - z_2| + c\Big|\int_{z_3}^{y_3} t^{-1/2}\, dt\Big| \le c|z_2 - y_2| + c|\sqrt{z_3} - \sqrt{y_3}|.$$

Thus we get

$$1 - e^{\sigma(a^-(Z)-a^-(Y))} \le 1 - e^{-c|z_2-y_2| - c|\sqrt{z_3}-\sqrt{y_3}|}$$
$$\le 1 - e^{-c|z_2-y_2|} + 1 - e^{-c|\sqrt{z_3}-\sqrt{y_3}|}.$$

It follows that

$$J_2 = \int_{D_a} |I_2|^r |z-y|^{-3-rs} \le c\sigma^{-r}\|f\|^r_{L^\infty(Q_a)}(J_{23} + J_{22}), \quad \text{where} \tag{12.3.72}$$
$$J_{23} = \int_{D_a} e^{r\sigma(a^-(Y)-y_1)}(1 - e^{-c\sigma|\sqrt{y_3}-\sqrt{z_3}|})^r |z-y|^{-3-rs}\, dydz,$$
$$J_{22} = \int_{D_a} e^{r\sigma(a^-(Y)-y_1)}(1 - e^{-c\sigma|y_2-z_2|})^r |z-y|^{-3-rs}\, dydz.$$

It follows from the obvious inclusion

$$D_a \subset \{(y,z) : y_2, z_1, z_2 \in (-a,a),\ y_3, z_3 \in (0,a),\ a^-(Y) < y_1 < a^+(Y)\}$$

that

$$J_{23} \le$$
$$\int_{[0,a]^2} \Big\{ \int_a^a \Big[ \int_{a^-(Y)}^{a^+(Y)} \mathcal{M}_2(y, z_3) e^{r\sigma(a^-(Y)-y_1)}\, dy_1 \Big] dy_2 \Big\} (1 - e^{-c\sigma|\sqrt{y_3}-\sqrt{z_3}|})^r\, dy_3 dz_3,$$

where

$$\mathcal{M}_2(y, z_3) = \int_{[0,a]^2} |z - y|^{-3-rs}\, dz_1 dz_2.$$

Next, Proposition 12.3.10 implies $\mathcal{M}_2(y, z_3) \le c|z_3 - y_3|^{-1-rs}$. From this and the inequality $|a^-(Y) - a^+(Y)| \le c\sqrt{y_3}$ we obtain

$$\int_{a^-(Y)}^{a^+(Y)} \mathcal{M}_2(y, z_3) e^{r\sigma(a^-(Y)-y_1)}\, dy_1 \le c|y_3 - z_3|^{-1-rs} \int_{a^-(Y)}^{a^+(Y)} e^{r\sigma(a^-(Y)-y_1)}\, dy_1$$
$$\le c\sigma^{-1}|y_3 - z_3|^{-1-rs}(1 - e^{r\sigma(a^-(Y)-a^+(Y))}) \le c\sigma^{-1}|y_3 - z_3|^{-1-rs}(1 - e^{-c\sigma\sqrt{y_3}}),$$

which leads to

$$J_{23} \le c\sigma^{-1} \int_0^a S(z_3)\, dz_3, \tag{12.3.73}$$

where

$$S(z_3) = \int_0^a (1 - e^{-c\sigma\sqrt{y_3}})(1 - e^{-c\sigma|\sqrt{y_3}-\sqrt{z_3}|})^r |y_3 - z_3|^{-1-rs}\, dy_3.$$

The change of variable $t = \sqrt{y_3/z_3} - 1$ gives

$$\begin{aligned}
S(z_3) &\le c z_3^{-rs} \int_{-1}^{\infty} (1 - e^{-c\mu(t+1)})(1 - e^{-c\mu|t|})^r |t(t+2)|^{-1-rs}(t+1)\, dt \\
&\le c z_3^{-rs} \int_{-1}^{1} (1 - e^{-c\mu|t|-c\mu})(1 - e^{-c\mu|t|})^r |t|^{-1-rs}\, dt \\
&\quad + c z_3^{-rs} \int_1^{\infty} (1 - e^{-c\mu t})^{r+1} t^{-1-2rs}\, dt \\
&\le c z_3^{-rs} \int_0^1 (1 - e^{-c\mu t-c\mu})(1 - e^{-c\mu t})^r t^{-1-rs}\, dt \\
&\quad + c z_3^{-rs} \int_1^{\infty} (1 - e^{-c\mu t})^{r+1} t^{-1-2rs}\, dt,
\end{aligned}$$

where $\mu = \sigma\sqrt{z_3}$. Another change of variable $\tau = \mu t$ along with the identity $z_3^{-rs} = \sigma^{2rs}\mu^{-2rs}$ yields

$$S(z_3) \le c\sigma^{2rs}(S_1(\mu) + S_\infty(\mu)),$$

where

$$S_1(\mu) = \mu^{-rs} \int_0^\mu (1 - e^{-c\tau - c\mu})(1 - e^{-c\tau})^r \tau^{-1-rs}\, d\tau,$$

$$S_\infty(\mu) = \int_\mu^\infty (1 - e^{-c\tau})^{r+1} \tau^{-1-2rs}\, d\tau.$$

Note that the inequality $0 < s < 1$ guarantees the convergence of these integrals. Inserting this estimate into (12.3.73), we obtain

$$J_{23} \le c\sigma^{2rs-1} \int_0^a (S_1(\mu) + S_\infty(\mu))\, dz_3 = c\sigma^{2rs-3} \int_0^{\sigma\sqrt{a}} (S_1(\mu) + S_\infty(\mu))\mu\, d\mu. \tag{12.3.74}$$

Since $(1 - e^{-c\tau - c\mu})(1 - e^{-c\tau})^r \le c\tau^{r+1} + c\mu\tau^r$ and $(1 - e^{-c\tau})^{r+1} \le c\tau^{r+1}$, we have

$$\mu S_1(\mu) \le c\mu^{1-rs} \int_0^\mu (\tau^{r-rs} + \mu\tau^{r-rs-1})\, d\tau \le c\mu^{r-2rs+2},$$

$$\mu S_\infty(\mu) \le c\mu \int_\mu^1 \tau^{r-2rs}\, d\tau + c\mu \int_1^\infty \tau^{-1-2rs}\, d\tau \le c\mu + c\mu^{r-rs+2}$$

for all $\mu \in (0, \sqrt{a})$. Next we have

$$\mu S_1(\mu) \le c\mu^{1-rs} \int_0^{\sqrt{a}} \tau^{r-1-rs}\, d\tau + c\mu^{1-rs} \int_{\sqrt{a}}^\mu \tau^{-1-rs}\, d\tau \le c\mu^{1-rs},$$

$$\mu S_\infty(\mu) \le c\mu \int_\mu^\infty \tau^{-1-2rs}\, d\tau \le c\mu^{1-2rs}$$

for all $\mu \in (\sqrt{a}, \infty)$. From the inequality $r - 2rs + 2 = -1 + r(1 - 2s + 3r^{-1}) > -1$ we thus conclude that the integrals in (12.3.74) converge at 0, and are finite for each finite $\sigma$. Hence, for all $\sigma > 1$,

$$J_{23} \le c\sigma^{2rs-3} \int_0^{\sqrt{a}} (\mu + \mu^{r-rs+2})\, d\tau + c\sigma^{2rs-3} \int_{\sqrt{a}}^{\sigma\sqrt{a}} (\mu^{1-rs} + \mu^{1-2rs})\, d\mu$$

$$\le c \begin{cases} \sigma^{2rs-3} + \sigma^{rs-1} & \text{for } sr \ne 1, 2, \\ (\sigma^{2rs-3} + \sigma^{rs-1})(1 + \log\sigma) & \text{for } sr = 1, 2. \end{cases}$$

Since $2rs - 3, rs - 1 \le r\alpha$, we conclude that for all $\sigma > 1$,

$$J_{23}^{1/r} \le c\sigma^\alpha \text{ for } sr \ne 1, 2, \quad J_{23}^{1/r} \le c\sigma^\alpha(1 + \log\sigma)^{1/r} \text{ for } sr = 1, 2, \tag{12.3.75}$$

Let us estimate the quantity $J_{22}$. We have

$$J_{22} \le \int_{[-a,a]^2 \times [0,a]} \mathcal{M}_3(Y, z_2)(1 - e^{-c\sigma|y_2 - z_2|})^r\, dy_3 dy_2 dz_2,$$

where

$$\mathcal{M}_3(Y, z_2) = \int_{[-a,a]^2\times[0,a]} e^{\sigma r(a^-(Y)-y_1)}|z-y|^{-3-rs}\,dz_1dy_1dz_3.$$

Since

$$\int_{[-a,a]\times[0,a]} |z-y|^{-3-rs}dz_1dz_3 \le c|y_2-z_2|^{-1-rs},$$

we have

$$\mathcal{M}_3(Y, z_2) \le c\sigma^{-1}|y_2-z_2|^{-1-rs},$$

which yields

$$\begin{aligned}J_{22} &\le c\sigma^{-1}\int_{-\infty}^{\infty}(1-e^{-c\sigma|y_2-z_2|})^r|y_2-z_2|^{-rs-1}dy_2\\ &= c\sigma^{rs-1}\int_{-\infty}^{\infty}(1-e^{-c|t|})^r|t|^{-1-rs}dt \le c\sigma^{rs-1}.\end{aligned}$$

Since $\sigma^{rs-1} \le c\sigma^{r\alpha}$ for all $\sigma \ge 1$, from (12.3.75) and (12.3.72) we conclude that

$$J_2^{1/r} \le \begin{cases} c\|f\|_{L^\infty(\mathcal{P}_a)}\sigma^{-1+\alpha} & \text{for } sr \ne 1,2,\\ c\|f\|_{L^\infty(\mathcal{P}_a)}\sigma^{-1+\alpha}(1+\log\sigma)^{1/r} & \text{for } sr = 1,2.\end{cases} \tag{12.3.76}$$

**Step 3.** Observe that $I_3(y_1, Y, Z)$ defined by (12.3.64) satisfies

$$\partial_{y_1}I_3 + \sigma I_3 = K_3 \quad \text{for } a^-(Y) < y_1 < a, \quad I_3(a^-(Y), Y, Z) = 0,$$

where $K_3(y_1, Y, Z) = f(y_1, Z) - f(y_1, Y)$. Multiplying this equation by $|I_3|^{r-2}I_3$ and integrating the result over $(a^-(Y), a)$ we arrive at

$$\begin{aligned}\sigma\int_{a^-(Y)}^{a}|I_3|^r\,dy_1 &\le \int_{a^-(Y)}^{a}|I_3|^{r-1}|K_3|\,dy_1\\ &\le \Big(\int_{a^-(Y)}^{a}|I_3|^r\,dy_1\Big)^{1-1/r}\Big(\int_{a^-(Y)}^{a}|K_3|^r\,dy_1\Big)^{1/r},\end{aligned}$$

which leads to

$$\int_{a^-(Y)}^{a}|I_3|^r\,dy_1 \le \sigma^{-r}\int_{[-a,a]}|f(y_1,Z)-f(y_1,Y)|^r\,dy_1.$$

Since $a^-(Y) \le y_1$ for all $(y, z) \in D_a$, taking into account the inequality

$$\int_{[-a,a]}|z-y|^{-3-rs}\,dz_1 \le c|Y-Z|^{-2-rs}$$

we conclude that

$$\begin{aligned} J_3 &= \int_{D_a} |I_3|^r |z-y|^{-3-rs}\, dydz \\ &\le c\sigma^{-r} \int_{[-a,a]} \Big\{ \int_{[-a,a]^4} |f(y_1,Z) - f(y_1,Y)|^r |Y-Z|^{-2-rs}\, dY dZ \Big\} dy_1 \\ &\le c\sigma^{-r} \|f\|^r_{L^r(-a,a;W^{s,r}([-a,a]^2))} \le c\sigma^{-r} \|f\|^r_{W^{s,r}(Q_a)}. \end{aligned} \tag{12.3.77}$$

Inserting estimates (12.3.71), (12.3.76), and (12.3.77) into the inequality

$$|\varphi|_{s,r,\mathcal{P}_a} \le J_1^{1/r} + J_2^{1/r} + J_3^{1/r}$$

we conclude that under the assumptions of the lemma,

$$|\varphi|_{s,r,\mathcal{P}_a} \le \begin{cases} c\big(\sigma^{-1}\|f\|_{W^{s,r}(\mathcal{P}_a)} + \sigma^{-1+\alpha}\|f\|_{L^\infty(\mathcal{P}_a)}\big) & \text{for } sr \ne 1,2, \\ c\big(\sigma^{-1}\|f\|_{W^{s,r}(\mathcal{P}_a)} + \sigma^{-1+\alpha}(1+\log\sigma)^{1/r}\|f\|_{L^\infty(\mathcal{P}_a)}\big) & \text{for } sr = 1,2, \end{cases}$$

which completes the proof.

# Chapter 13

# Appendix

## 13.1 Proof of Lemma 2.4.1

Choose an arbitrary vector field $\boldsymbol{\varphi}$ such that

$$\boldsymbol{\varphi} \in C_0^\infty(\mathbb{R}^d \times \mathbb{R}), \quad \operatorname{supp} \boldsymbol{\varphi} \subset D = \bigcup_t \Omega_t \times \{t\}. \tag{13.1.1}$$

Multiplying (2.1.6a) by $\boldsymbol{\varphi}$ and integrating by parts we arrive at

$$\int_D \Big\{ \varrho\, u_i \frac{\partial \varphi_i}{\partial t} + \varrho\, u_i u_j \frac{\partial \varphi_i}{\partial x_j} - \frac{1}{\mathsf{Re}} S_{ij}(\mathbf{u}) \frac{\partial \varphi_i}{\partial x_j} + \frac{1}{\mathsf{Ma}^2} p \frac{\partial \varphi_i}{\partial x_i} + \frac{1}{\mathsf{Fr}_m^2} \varrho\, f_i \varphi_i \Big\} \, dx dt = 0. \tag{13.1.2}$$

Noting that $\partial_{x_i} = U_{ip} \partial_{y_p}$ we have

$$S_{ij}(\mathbf{u}) = \Big[ U_{jk} \frac{\partial u_i}{\partial y_k} + U_{ik} \frac{\partial u_j}{\partial y_k} + (\lambda - 1) U_{pk} \frac{\partial u_p}{\partial y_k} \delta_{ij} \Big] U_{jn} \frac{\partial \varphi_i}{\partial y_n}, \tag{13.1.3}$$

$$\varrho\, u_i u_j \frac{\partial \varphi_i}{\partial x_j} + p \frac{\partial \varphi_i}{\partial x_i} = \varrho\, u_i u_j\, U_{jk} \frac{\partial \varphi_i}{\partial y_k} + p\, U_{ik} \frac{\partial \varphi_i}{\partial y_k}. \tag{13.1.4}$$

Next, set

$$\mathbf{u} = \mathbb{U}(t)\, \mathbf{w}, \quad \boldsymbol{\varphi} = \mathbb{U}(t)\, \boldsymbol{\psi}, \quad \frac{1}{\mathsf{Fr}_m^2} \mathbf{f} = \mathbb{U}\, \mathbf{F}.$$

Since

$$U_{\alpha\beta} U_{\gamma\beta} = U_{\beta\alpha} U_{\beta\gamma} = \delta_{\alpha\gamma},$$

we have

$$U_{jk} U_{jn} \frac{\partial u_i}{\partial y_k} \frac{\partial \varphi_i}{\partial y_n} = U_{jk} U_{jn} U_{ip} U_{iq} \frac{\partial w_q}{\partial y_k} \frac{\partial \psi_p}{\partial y_n} = \frac{\partial w_q}{\partial y_k} \frac{\partial \psi_q}{\partial y_k}.$$

The same arguments give

$$U_{ik}U_{jn}\frac{\partial u_j}{\partial y_k}\frac{\partial \varphi_i}{\partial y_n} = \frac{\partial w_j}{\partial y_i}\frac{\partial \psi_i}{\partial y_j}$$

and

$$U_{pk}U_{in}\frac{\partial u_p}{\partial y_k}\frac{\partial \varphi_i}{\partial y_n} = U_{pk}U_{in}U_{pl}U_{im}\frac{\partial w_l}{\partial y_k}\frac{\partial \psi_m}{\partial y_n} = \frac{\partial w_j}{\partial y_j}\frac{\partial \psi_i}{\partial y_i}.$$

Substituting these into (13.1.3) gives

$$S_{ij}(\mathbf{u})\frac{\partial \varphi_i}{\partial x_j} = S_{ij}(\mathbf{w})\frac{\partial \psi_i}{\partial y_j}. \tag{13.1.5}$$

Arguing as before we obtain

$$u_i u_j U_{jk}\frac{\partial \varphi_i}{\partial y_k} = U_{ip}U_{jq}U_{jk}U_{im}\frac{\partial \psi_m}{\partial y_k}w_p w_q = w_p w_q\frac{\partial \psi_p}{\partial y_q} = w_i w_j\frac{\partial \psi_i}{\partial y_j},$$

$$U_{ik}\frac{\partial \varphi_i}{\partial y_k} = U_{ik}U_{\mathrm{in}}\frac{\partial \psi_n}{\partial y_k} = \frac{\partial \psi_i}{\partial y_i}.$$

Substituting these along with (13.1.5) into (13.1.3)–(13.1.4) and recalling the notation for $\rho$ we obtain

$$\int_D \Big\{\varrho\, u_i u_j\,\frac{\partial \varphi_i}{\partial x_j} - \frac{1}{\mathsf{Re}}S_{ij}(\mathbf{u})\,\frac{\partial \varphi_i}{\partial x_j} + \frac{1}{\mathsf{Ma}^2}p\,\frac{\partial \varphi_i}{\partial x_i} + \frac{1}{\mathsf{Fr}_m^2}\varrho\, f_i\varphi_i\Big\}\,dx\,dt$$
$$= \int_{\Omega_0\times\mathbb{R}} \Big\{\rho\, w_i w_j\,\frac{\partial \psi_i}{\partial y_j} - \frac{1}{\mathsf{Re}}S_{ij}(\mathbf{w})\,\frac{\partial \psi_i}{\partial y_j} + \frac{1}{\mathsf{Ma}^2}p\,\frac{\partial \psi_i}{\partial y_i} + \rho\, F_i\psi_i\Big\}\,dy\,dt. \tag{13.1.6}$$

Next, note that

$$\frac{\partial \varphi_i}{\partial t}\Big|_{x=\mathrm{const}} = \frac{\partial \varphi_i}{\partial t}\Big|_{y=\mathrm{const}} + \frac{\partial y_k}{\partial t}\Big|_{x=\mathrm{const}}\frac{\partial \varphi_i}{\partial y_k}.$$

We have $y = \mathbb{U}^\top(x-\mathbf{a})$. From this and the identity

$$\dot{\mathbb{U}}^\top\,\mathbb{U} + \dot{\mathbb{U}}\,\mathbb{U}^\top = 0$$

we obtain

$$\frac{\partial y_k}{\partial t}\Big|_{x=\mathrm{const}} = \dot{\mathbb{U}}^\top(x-\mathbf{a}) - \mathbb{U}^\top\dot{\mathbf{a}} = \dot{\mathbb{U}}^\top\mathbb{U}y - \mathbb{U}^\top\dot{\mathbf{a}} = -\mathbb{U}^\top(\dot{\mathbb{U}}y + \dot{\mathbf{a}}) = -\mathbf{W}.$$

Thus we get

$$\frac{\partial \varphi_i}{\partial t}\Big|_{x=\mathrm{const}} = \frac{\partial \varphi_i}{\partial t}\Big|_{y=\mathrm{const}} - W_k\frac{\partial \varphi_i}{\partial y_k},$$

and

$$\int_D \varrho u_i \frac{\partial \varphi_i}{\partial t} dx dt = \int_{\Omega_0 \times \mathbb{R}} \rho u_i \left( \frac{\partial \varphi_i}{\partial t} - W_k \frac{\partial \varphi_i}{\partial y_k} \right) dy dt. \tag{13.1.7}$$

Next, we have the equality

$$\begin{aligned} u_i \left( \frac{\partial \varphi_i}{\partial t} - W_k \frac{\partial \varphi_i}{\partial y_k} \right) &= U_{in} w_n \left( \frac{\partial}{\partial t} (U_{ip} \psi_p) - W_k \frac{\partial}{\partial y_k} (U_{ip} \psi_p) \right) \\ &= w_i \frac{\partial \psi_i}{\partial t} - w_i W_k \frac{\partial \psi_i}{\partial y_k} + w_n \psi_p U_{in} \frac{\partial U_{ip}}{\partial t}, \end{aligned}$$

which along with

$$U_{in} \frac{\partial U_{ip}}{\partial t} = \frac{\partial W_n}{\partial y_p}$$

gives

$$\rho u_i \left( \frac{\partial \varphi_i}{\partial t} - W_j \frac{\partial \varphi_i}{\partial y_j} \right) = \rho w_i \left( \frac{\partial \psi_i}{\partial t} - W_j \frac{\partial \psi_i}{\partial y_j} \right) + \rho w_j \psi_i \frac{\partial W_j}{\partial y_i}.$$

Substituting this identity into (13.1.7) and combining the result with (13.1.6) and (13.1.2) we finally obtain

$$\begin{aligned} &\int_D \left\{ \varrho u_i \frac{\partial \varphi_i}{\partial t} + \varrho\, u_i u_j \frac{\partial \varphi_i}{\partial x_j} - \frac{1}{\mathsf{Re}} S_{ij}(\mathbf{u}) \frac{\partial \varphi_i}{\partial x_j} + \frac{1}{\mathsf{Ma}^2} p \frac{\partial \varphi_i}{\partial x_i} + \frac{1}{\mathsf{Fr}_m^2} \varrho\, f_i \varphi_i \right\} dx\, dt \\ &\quad = \int_{\Omega_0 \times \mathbb{R}} \left\{ \rho w_i \left( \frac{\partial \psi_i}{\partial t} - W_j \frac{\partial \psi_i}{\partial y_j} \right) + \rho\, w_i w_j \frac{\partial \psi_i}{\partial y_j} - \frac{1}{\mathsf{Re}} S_{ij}(\mathbf{w}) \frac{\partial \psi_i}{\partial y_j} \right. \\ &\qquad \left. + \frac{1}{\mathsf{Ma}^2} p \frac{\partial \psi_i}{\partial y_i} + \rho\, F_i \psi_i + \rho w_j\, \psi_i \frac{\partial W_j}{\partial y_i} \right\} dy\, dt = 0, \end{aligned} \tag{13.1.8}$$

where $D$ is given by (13.1.1). Our next task is to obtain an equation for the deviation $\mathbf{v} = \mathbf{w} - \mathbf{W}$. Substituting this equality into (13.1.8) and noting that $\mathbb{S}(\mathbf{W}) = 0$ we obtain

$$\begin{aligned} &\int_{\Omega_0 \times \mathbb{R}} \left\{ \rho v_i \frac{\partial \psi_i}{\partial t} + \rho\, v_i v_j \frac{\partial \psi_i}{\partial y_j} - \frac{1}{\mathsf{Re}} S_{ij}(\mathbf{v}) \frac{\partial \psi_i}{\partial y_j} + \frac{1}{\mathsf{Ma}^2} p \frac{\partial \psi_i}{\partial y_i} \right. \\ &\qquad\qquad \left. + \rho\, F_i \psi_i + \rho v_j\, \psi_i \frac{\partial W_j}{\partial y_i} \right\} dy\, dt \\ &\quad + \int_{\Omega_0 \times \mathbb{R}} \left\{ \rho W_i \frac{\partial \psi_i}{\partial t} + \rho\, W_i v_j \frac{\partial \psi_i}{\partial y_j} + \rho\, \psi_i W_j \frac{\partial W_j}{\partial y_i} \right\} dy\, dt = 0. \end{aligned} \tag{13.1.9}$$

Let us turn to the mass balance equation. Multiplying (2.1.6b) by an arbitrary function $\eta \in C_0^\infty(\Omega)$ and integrating by parts we obtain

$$\int_D \varrho \left( \frac{\partial \eta}{\partial t} + u_i \frac{\partial \eta}{\partial x_i} \right) dx dt = 0.$$

Noting that

$$\frac{\partial \eta}{\partial t}\bigg|_{x=\text{const}} = \frac{\partial \eta}{\partial t}\bigg|_{y=\text{const}} - W_i \frac{\partial \eta}{\partial y_i}, \quad u_i \frac{\partial \eta}{\partial x_i} = u_i U_{ip} \frac{\partial \eta}{\partial y_p} = w_p \frac{\partial \eta}{\partial y_p},$$

we obtain for $\zeta(y,t) = \eta(x(y,t),t)$,

$$\int_{\Omega_0 \times \mathbb{R}} \rho \left( \frac{\partial \zeta}{\partial t} + (w_i - W_i) \frac{\partial \zeta}{\partial y_i} \right) dydt = 0.$$

It now follows from the equality $\mathbf{v} = \mathbf{w} - \mathbf{W}$ that for any $\zeta \in C_0^\infty(\Omega_0 \times \mathbb{R})$,

$$\int_{\Omega_0 \times \mathbb{R}} \rho \left( \frac{\partial \zeta}{\partial t} + v_i \frac{\partial \zeta}{\partial y_i} \right) dydt = 0, \tag{13.1.10}$$

which is equivalent to the equation

$$\partial_t \rho + \operatorname{div}(\rho \mathbf{v}) = 0. \tag{13.1.11}$$

In particular, we have

$$\int_{\Omega_0 \times \mathbb{R}} \rho W_i \left( \frac{\partial \psi_i}{\partial t} + v_j \frac{\partial \psi_i}{\partial y_j} \right) dydt = - \int_{\Omega_0 \times \mathbb{R}} \rho \psi_i \left( \frac{\partial W_i}{\partial t} + v_j \frac{\partial W_i}{\partial y_j} \right) dydt.$$

From this and (13.1.9) we obtain the integral identity

$$\begin{aligned} \int_{\Omega_0 \times \mathbb{R}} \bigg\{ \rho v_i \frac{\partial \psi_i}{\partial t} + \rho v_i v_j \frac{\partial \psi_i}{\partial y_j} - \frac{1}{\mathsf{Re}} \mathbb{S}_{ij}(\mathbf{v}) \frac{\partial \psi_i}{\partial y_j} + \frac{1}{\mathsf{Ma}^2} p \frac{\partial \psi_i}{\partial y_i} \\ + \rho F_i \psi_i + \rho v_j \psi_i \left( \frac{\partial W_j}{\partial y_i} - \frac{\partial W_i}{\partial y_j} \right) \bigg\} dydt \\ + \int_{\Omega_0 \times \mathbb{R}} \rho \psi_i \left\{ \frac{1}{2} \frac{\partial |\mathbf{W}|^2}{\partial y_i} - \frac{\partial W_i}{\partial t} \right\} dydt = 0, \end{aligned}$$

which is equivalent to the equation

$$\begin{aligned} \partial_t(\rho \mathbf{v}) + \operatorname{div}(\rho \mathbf{v} \otimes \mathbf{v}) + \frac{1}{\mathsf{Ma}^2} \nabla p - \frac{1}{\mathsf{Re}} \operatorname{div} \mathbb{S}(\mathbf{v}) + \rho\, \mathbb{C} \mathbf{v} \\ = \rho \mathbf{F} + \frac{\rho}{2} \nabla |\mathbf{W}|^2 - \rho\, \partial_t \mathbf{W}. \end{aligned} \tag{13.1.12}$$

Combining this with (13.1.11) we obtain (2.4.5).

Let us calculate the expression for the hydrodynamic force, the power developed by this force and the work of these forces in the moving frame. Let $\tilde{\mathbf{n}}(x,t)$ be the outward normal vector to $\partial\Omega_t$ at $x$ and $\mathbf{n}(y)$ be the outward normal vector to $\partial\Omega_0$ at $y$. It is easy to see that

$$\tilde{\mathbf{n}}|_{x=x(y,t)} = \mathbb{U}(t)\, \mathbf{n}(y).$$

Thus we get the following expression for the components $R_{fi}$ of the hydrodynamic force $\mathbf{R}_f$ at $x = x(y,t)$:

$$\mathsf{Re}\, R_{fi} = \sigma p U_{i\alpha} n_\alpha - S_{ij}(\mathbf{u}) U_{j\alpha} n_\alpha.$$

Next, note that

$$\begin{aligned} S_{ij}(\mathbf{u}) &= U_{jk} U_{in} \frac{\partial w_n}{\partial y_k} + U_{ik} U_{jn} \frac{\partial w_n}{\partial y_k} + (\lambda - 1) U_{pk} U_{pn} \frac{\partial w_n}{\partial y_k} \\ &= U_{jk} U_{in} \frac{\partial w_n}{\partial y_k} + U_{ik} U_{jn} \frac{\partial w_n}{\partial y_k} + (\lambda - 1) \operatorname{div} \mathbf{w}\, \delta_{ij}. \end{aligned}$$

Thus we get

$$S_{ij}(\mathbf{u}) U_{j\alpha} n_\alpha = U_{in} \left( \frac{\partial w_n}{\partial y_\alpha} + \frac{\partial w_\alpha}{\partial y_n} \right) n_\alpha + (\lambda - 1) \operatorname{div} \mathbf{w}\, U_{i\alpha} n_\alpha,$$

which gives an expression for $\mathbf{R}_f$ in the new variables,

$$\mathsf{Re}\, \mathbf{R}_f = \mathbb{U}(-\mathbb{S}(\mathbf{w}) + \sigma p \mathbb{I}) \mathbf{n}. \tag{13.1.13}$$

Noting that the dimensionless velocity of $\partial\Omega$ is $\mathbf{V} = \mathbb{U}\mathbf{W}$ we obtain

$$\mathsf{Re}\, \mathbf{V} \cdot \mathbf{R}_f = \mathbf{W} \cdot (-\mathbb{S}(\mathbf{w}) + \sigma p \mathbb{I}) \mathbf{n}, \tag{13.1.14}$$

which leads to equalities (2.4.6) and (2.4.7).

## 13.2 Normal coordinates

Let $\Sigma$ be a 2-dimensional closed $C^2$ surface in the Euclidean space $\mathbb{R}^3$ of points $x = (x_1, x_2, x_3)$ such that $\Sigma = \partial\Omega \subset \mathbb{R}^3$ for a domain $\Omega$. For any point $\omega \in \partial\Sigma$ denote by $\mathbf{n}(\omega)$ the unit outward normal vector to $\partial\Sigma$ at a point $\omega$ and by $\boldsymbol{\nu}(\omega) = -\mathbf{n}(\omega)$ the inward normal vector to $\Sigma$. It is well known that there exist a neighborhood $\mathcal{O}$ of $\Sigma$ and a number $\mu > 0$ so that the mapping $(\omega, y_3) \mapsto \omega + y_3 \boldsymbol{\nu}(\omega)$ takes diffeomorphically the set $\Sigma \times [-\mu, \mu]$ onto $\mathcal{O}$.

Let us evaluate the derivatives and the Jacobian of this mapping. To this end, we fix $\omega_0 \in \Sigma$. In some neighborhood $\mathcal{O}_0$ of $\omega_0$, the surface $\Sigma$ admits a parametric representation

$$\Sigma \cap \mathcal{O}_0 : \quad \omega = \mathbf{r}(y_1, y_2), \quad |y_i| \le \mu, \quad \mathbf{r}(0,0) = \omega_0, \tag{13.2.1}$$

where the vector-valued function $\mathbf{r}(y_1, y_2)$ is $C^2$ in the rectangle

$$Q = \{(y_1, y_2) : |y_i| \le \mu,\ i = 1, 2\},$$

for an appropriate choice of $\mu$ independent of $\omega_0$. We write

$$y = (\bar{y}, y_3), \quad \bar{y} = (y_1, y_2),$$

and introduce the moving frame

$$\mathbf{e}_i(\bar{y}) = \partial_{y_i}\mathbf{r}(\bar{y}), \quad i = 1, 2, \quad \mathbf{e}_3(\bar{y}) = \boldsymbol{\nu}(\mathbf{r}(\bar{y})) = \frac{1}{|\mathbf{e}_1 \times \mathbf{e}_2|}\mathbf{e}_1 \times \mathbf{e}_2. \tag{13.2.2}$$

This means that the orientation of $\Sigma$ is compatible with the direction of the inward normal vector. Thus, the mapping

$$X : y \mapsto \mathbf{r}(\bar{y}) + y_3\mathbf{e}_3(\bar{y}) \equiv \omega + y_3\boldsymbol{\nu}(\omega)$$

maps diffeomorphically

$$Q \times [-\mu, \mu] \to \mathcal{O}_0 \quad \text{and} \quad Q \times [0, \mu] \to \mathcal{O}_0 \cap \Omega.$$

Recall the notation

$$g_{ij} = \mathbf{e}_i(\bar{y}) \cdot \mathbf{e}_j(\bar{y}), \quad b_{ij} = \mathbf{e}_3(\bar{y}) \cdot \partial^2_{y_i\, y_j}\mathbf{r}(\bar{y}),$$

for the coefficients of the first and second fundamental form of the surface $\Sigma$ and set

$$g^2 = \det(g_{ij}), \quad (g^{\alpha\beta}) = (g_{\alpha\beta})^{-1}.$$

Note that $g dy_1 dy_2 = d\Sigma$ is the surface element of $\Sigma$. In this notation, the mean and Gauss curvatures of $\Sigma$ are defined by

$$\mathcal{H} = \frac{1}{2}(b_1^1 + b_2^2), \quad \mathcal{K} = b_1^1 b_2^2 - b_2^1 b_1^2, \quad \text{where} \quad b_j^i = g^{i\alpha} b_{\alpha j}.$$

Let us consider the structure of the *Jacobi matrix* $X'(y)$ defined by

$$\mathbb{M}(y) \equiv X'(y) := \big[\partial_{y_1}X,\, \partial_{y_2}X,\, \partial_{y_3}X\big], \tag{13.2.3}$$

where the notation

$$[\mathbf{a},\, \mathbf{b},\, \mathbf{c}]$$

stands for the $(3 \times 3)$-matrix with columns $\mathbf{a}$, $\mathbf{b}$, and $\mathbf{c}$. Using the Weingarten equations

$$\partial_{y_j}\mathbf{e}_3 = -b_j^i\,\mathbf{e}_i \quad \text{with } i, j = 1, 2,$$

we obtain the columns of the matrix function $\mathbb{M}$:

$$\partial_{y_j}X = \mathbf{e}_j - y_3(b_j^1\mathbf{e}_1 + b_j^2\mathbf{e}_2), \quad j = 1, 2, \quad \partial_{y_3}X = \mathbf{e}_3. \tag{13.2.4}$$

In order to derive an expression for the inverse $\mathbb{M}^{-1}$ we use the identity

$$\begin{aligned}\mathbb{M}^{-\top} &\equiv \big[\nabla_x y_1(X(y)), \nabla_x y_2(X(y)), \nabla_x y_3(X(y))\big] \\ &= (\det X')^{-1}\big[\partial_{y_2}X \times \partial_{y_3}X,\, \partial_{y_3}X \times \partial_{y_1}X,\, \partial_{y_1}X \times \partial_{y_2}X\big].\end{aligned} \tag{13.2.5}$$

Next, note that for $j = 1, 2$,

$$\mathbf{e}_j \times \mathbf{e}_3 = g^{-1}\mathbf{e}_j \times (\mathbf{e}_1 \times \mathbf{e}_2) = g^{-1}\big((\mathbf{e}_j \cdot \mathbf{e}_2)\mathbf{e}_1 - (\mathbf{e}_j \cdot \mathbf{e}_1)\mathbf{e}_2\big) = g^{-1}(g_{j2}\mathbf{e}_1 - g_{j1}\mathbf{e}_2)$$

and $\mathbf{e}_1 \times \mathbf{e}_2 = g\mathbf{e}_3$. From this and (13.2.4), (13.2.5) we obtain

$$\partial_{y_j} X \times \partial_{y_3} X = g^{-1}(g_{j2}\mathbf{e}_1 - g_{j1}\mathbf{e}_2) - y_3 g^{-1}(b_{j2}\mathbf{e}_1 - b_{j1}\mathbf{e}_2),$$

and

$$\partial_{y_1} X \times \partial_{y_2} X = g(1 - 2\mathcal{H}y_3 + \mathcal{K}y_3^2)\,\mathbf{e}_3.$$

In particular we have the following expression for the Jacobian:

$$|\mathbb{M}| := \det \mathbb{M} \equiv (\partial_{y_1} X \times \partial_{y_2} X) \cdot \partial_{y_3} X = g(1 - 2\mathcal{H}y_3 + \mathcal{K}y_3^2). \tag{13.2.6}$$

We also get the following expression for the rows of $\mathbb{M}^{-1}$ (columns of $\mathbb{M}^{-\top}$):

$$\begin{aligned}
(\nabla_x y_1)(X(y)) &= g^{-1}(1 - 2\mathcal{H}y_3 + \mathcal{K}y_3^2)^{-1}\big\{g_{22}\mathbf{e}_1 - g_{21}\mathbf{e}_2 - y_3(b_{22}\mathbf{e}_1 - b_{21}\mathbf{e}_2)\big\},\\
(\nabla_x y_2)(X(y)) &= -g^{-1}(1 - 2\mathcal{H}y_3 + \mathcal{K}y_3^2)^{-1}\big\{(g_{12}\mathbf{e}_1 - g_{11}\mathbf{e}_2) - y_3(b_{12}\mathbf{e}_1 - b_{11}\mathbf{e}_2)\big\},\\
(\nabla_x y_3)(X(y)) &= \mathbf{e}_3.
\end{aligned}$$

**Normal conformal coordinates.** The formulae can be essentially simplified if the parameterization $x = \mathbf{r}(\bar{y})$ is conformal. This means that

$$g_{12} = 0, \quad g_{11} = g_{22} = g.$$

In this case the expressions for the rows of the inverse Jacobi matrix become

$$\begin{aligned}
(\nabla_x y_1)(X(y)) &= g^{-1}(1 - 2\mathcal{H}y_3 + \mathcal{K}y_3^2)^{-1}\big\{g\mathbf{e}_1 - y_3(b_{22}\mathbf{e}_1 - b_{21}\mathbf{e}_2)\big\},\\
(\nabla_x y_2)(X(y)) &= g^{-1}(1 - 2\mathcal{H}y_3 + \mathcal{K}y_3^2)^{-1}\big\{g\mathbf{e}_2 - y_3(b_{11}\mathbf{e}_2 - b_{12}\mathbf{e}_1)\big\},\\
(\nabla_x y_3)(X(y)) &= \mathbf{e}_3.
\end{aligned}$$

In particular, we have the following expression for the Jacobi matrix on $\Sigma \cap \mathcal{O}_0$:

$$\mathbb{M} \equiv X'(y) = \begin{pmatrix} g^{1/2} & 0 & 0 \\ 0 & g^{1/2} & 0 \\ 0 & 0 & 1 \end{pmatrix}, \quad |\mathbb{M}| = g \quad \text{for } y_3 = 0. \tag{13.2.7}$$

## 13.3 Geometric results. Approximation of unity

In this section we prove an important technical result which shows that in a cylinder $Q = \Omega \times (0, T)$ the constant function 1 can be approximated in a special way by a sequence of Lipschitz functions whose traces on the lateral side $S_T = \partial\Omega \times (0, T)$ vanish outside of the inlet. It is assumed that the flow domain $\Omega$ and a given vector field $\mathbf{U} : \mathbb{R}^d \times [0, T] \to \mathbb{R}^d$ satisfy the following structural conditions:

Assume that $\Omega = B \setminus S$, where $B$ is a hold-all domain and a compact set $S$ is an obstacle placed in $B$. Hence $\partial\Omega$ is a union of two disjoint compact sets: a surface $\partial B$ and a compact set $\partial S$. Denote by $\Sigma$ the cylindrical surface $\partial B \times (0, T)$. By $\Sigma_{\text{in}}$ we denote, as usual, the inlet, that is, the open subset of $\Sigma$ which consists of all points $(x, t) \in \Sigma$ at which $\mathbf{U} \cdot \mathbf{n} < 0$. Recall that $\mathbf{n}$ stands for the outward unit normal vector to $\partial\Omega$.

Furthermore, we assume that $\partial B$ and $\mathbf{U}$ satisfy the following conditions:

**Condition 13.3.1.**
- The vector field $\mathbf{U} \in C^1(\mathbb{R}^d \times [0, T])$ vanishes in the vicinity of the compact set $S \times [0, T]$.
- The surface $\partial B$ is of class $C^2$. Let

$$\Gamma = \Sigma \setminus \Sigma_{\text{in}} \cap \operatorname{cl} \Sigma_{\text{in}} \subset \Sigma$$

be the interface between the inlet set and the outgoing subset of $\Sigma$. Then

$$\limsup_{\sigma \to 0} \frac{1}{\sigma^d} \operatorname{meas} \mathcal{O}_\sigma < \infty, \tag{13.3.1}$$

where $\mathcal{O}_\sigma$ is a tubular neighborhood of $\Gamma$,

$$\mathcal{O}_\sigma = \{(x, t) \in \mathbb{R}^{d+1} : \operatorname{dist}((x, t), \Gamma) \leq \sigma\}.$$

**Remark 13.3.2.** We do not impose any restrictions on the obstacle $S$.

**Theorem 13.3.3.** *Let $\partial\Omega$ and $\mathbf{U}$ satisfy Condition 13.3.1 and suppose $\mathbf{u} - \mathbf{U} \in L^2(0, T; W^{1,2}(\Omega))$. Then there is a sequence of Lipschitz functions $\psi_n : Q \to [0, 1]$, $n \geq 1$, with the following properties: $\psi_n$ vanishes in some neighborhood of $\Sigma \setminus \Sigma_{\text{in}}$, $\psi_n(x, t) \nearrow 1$ as $n \to \infty$ everywhere in $Q$, and*

$$\liminf_{n \to \infty} \int_Q \Phi(\partial_t \psi_n + \nabla \psi_n \cdot \mathbf{u})\, dxdt \leq 0 \tag{13.3.2}$$

*for any nonnegative $\Phi \in L^q(Q)$ with $2 < q \leq \infty$.*

*Proof.* Define

$$M(x) := \operatorname{dist}(x, \partial B), \quad N(x, t) := \operatorname{dist}((x, t), \Sigma \setminus \Sigma_{\text{in}}), \quad (x, t) \in \mathbb{R}^{d+1}, \tag{13.3.3}$$

and choose $\eta \in C^\infty(\mathbb{R})$ such that

$$\eta(s) = 0 \quad \text{for } s \leq 1/2 \quad \text{and} \quad \eta(s) = 1 \quad \text{for } s \geq 1.$$

Next, set

$$\psi_n(x, t) = \eta(nN(x, t)), \quad (x, t) \in Q. \tag{13.3.4}$$

Obviously $M$ and $N$ satisfy the Lipschitz condition. Moreover,

$$|\nabla M| \leq 1, \quad |\nabla N| \leq 1.$$

Since $M(x) \le N(x,t)$, it follows that

$$|\nabla\psi_n| \le cn \quad \text{in } B_{1/n} \times (0,T), \quad |\nabla\psi_n| = 0 \quad \text{in } (B \setminus B_{1/n}) \times (0,T), \tag{13.3.5}$$

where $B_{1/n} \subset B$ is the set of all points $x \in B$ with $\operatorname{dist}(x, \partial B) < 1/n$, in other words $M(x) < 1/n$. Hence the functions $\psi_n$, $n \ge 1$, also satisfy the Lipschitz condition. Obviously they vanish in the $1/(2n)$-neighborhood of $\Sigma \setminus \Sigma_{\text{in}}$ and are equal to 1 outside the $1/n$-neighborhood of $\Sigma \setminus \Sigma_{\text{in}}$. Hence $\psi_n \nearrow 1$ in $Q$ as $n \to \infty$. Let us rewrite the integral in (13.3.2) as

$$\int_Q \Phi(\partial_t\psi_n + \nabla\psi_n \cdot \mathbf{u})\, dxdt = \int_Q \Phi(\partial_t\psi_n + \nabla\psi_n \cdot \mathbf{U})\, dxdt + \int_Q \Phi\, \nabla\psi_n \cdot \mathbf{v}\, dxdt, \tag{13.3.6}$$

where $\mathbf{v} = \mathbf{u} - \mathbf{U}$. Since $\mathbf{v} \in L^2(0,T; W_0^{1,2}(\Omega))$, in particular we can extend it by zero onto $S \times [0,T]$ and assume that $\mathbf{v} \in L^2(0,T; W_0^{1,2}(B))$. By the general property of the space $W_0^{1,2}(B)$, we have

$$\int_B \frac{|\mathbf{v}|^2}{(\operatorname{dist}(x,\partial B))^2}\, dx = \int_B \frac{|\mathbf{v}|^2}{M^2}\, dx \le c(B)\|\mathbf{v}\|^2_{W_0^{1,2}(B)},$$

which gives

$$\int_{B\times[0,T]} \frac{|\mathbf{v}|^2}{M^2}\, dxdt \le c\|\mathbf{v}\|^2_{L^2(0,T;W_0^{1,2}(B))} \le C, \tag{13.3.7}$$

where $C$ is independent of $n$. It now follows from (13.3.5) that

$$\begin{aligned}
\Big|\int_Q \Phi\nabla\psi_n \cdot \mathbf{v}\, dxdt\Big| &= \Big|\int_{B_{1/n}\times[0,T]} \Phi\nabla\psi_n \cdot \mathbf{v}\, dxdt\Big| \\
&\le cn \int_0^T \int_{B_{1/n}} \Phi|\mathbf{v}|\, dxdt = cn \int_0^T \int_{B_{1/n}} M\Phi M^{-1}|\mathbf{v}|\, dxdt \\
&\le c \int_0^T \int_{B_{1/n}} \Phi M^{-1}|\mathbf{v}|\, dxdt \\
&\le c\Big(\int_0^T \int_{B_{1/n}} (M^{-1}|\mathbf{v}|)^{q/(q-1)}\, dxdt\Big)^{(q-1)/q} \Big(\int_0^T \int_{B_{1/n}} \Phi^q\, dxdt\Big)^{1/q} \\
&\le c\Big(\int_0^T \int_{B_{1/n}} (M^{-1}|\mathbf{v}|)^{q/(q-1)}\, dxdt\Big)^{(q-1)/q} \\
&\le c\Big\{\Big(\int_0^T \int_{B_{1/n}} dxdt\Big)^{(q-2)/(2(q-1))} \Big(\int_0^T \int_{B_{1/n}} M^{-2}|\mathbf{v}|^2\, dxdt\Big)^{q/2(q-1)}\Big\}^{(q-1)/q} \\
&\le cn^{-(q-2)/(2(q-1))} \Big(\int_0^T \int_{B\setminus B_{1/n}} (M^{-2}|\mathbf{v}|)^2\, dxdt\Big)^{1/2}.
\end{aligned}$$

Combining this with (13.3.7) we obtain the following estimate for the second integral on the right hand side of (13.3.6):

$$\Big|\int_Q \Phi\nabla\psi_n\cdot\mathbf{v}\,dxdt\Big| \le Cn^{-(q-2)/(2(q-1))} \to 0 \quad \text{as } n\to\infty. \tag{13.3.8}$$

To estimate the first integral on the right of (13.3.6) we split the cylinder $Q$ into three disjoint parts,

$$Q = Z_n \cup G_n \cup O_n,$$

where the closed set $Z_n$ and the open set $G_n$ are defined by

$$\begin{aligned} Z_n &= \big\{(x,t)\in Q : \operatorname{dist}((x,t),\Gamma)\le 2/n\big\},\\ G_n &= \big\{(x,t)\in Q : \operatorname{dist}((x,t),\Gamma) > 2/n \text{ and } \operatorname{dist}((x,t),\Sigma\setminus\Sigma_{\rm in}) < 1/n\big\}. \end{aligned}$$

Notice that $\operatorname{dist}((x,t),\Sigma\setminus\Sigma_{\rm in})\ge 1/n$, hence $\psi_n(x,t)=1$ on $O_n = Q\setminus(G_n\cup Z_n)$. It follows that the derivatives of $\psi_n$ vanish a.e. on $O_n$, which yields

$$\begin{aligned} \int_Q \Phi(\partial_t\psi_n + \nabla\psi_n\cdot\mathbf{U})\,dxdt &= \int_{Z_n} \Phi(\partial_t\psi_n + \nabla\psi_n\cdot\mathbf{U})\,dxdt\\ &\quad + \int_{G_n} \Phi(\partial_t\psi_n + \nabla\psi_n\cdot\mathbf{U})\,dxdt. \end{aligned} \tag{13.3.9}$$

The remaining part of the proof is based on the following lemma.

**Lemma 13.3.4.** *Under the assumptions of Theorem 13.3.4, there is an $n_0$, depending only on $B$, with the following property. If $n > n_0$ and $(x,t)\in G_n$ then*

$$\nabla\psi_n(x,t)\cdot\mathbf{U}(x,t)\le c \quad \text{and} \quad \partial_t\psi_n(x,t)=0, \tag{13.3.10}$$

*where $c$ is independent of $(x,t)$ and $n$.*

*Proof.* Since $\partial B$ is a $C^2$ surface, there is an $n_0$ such that the normal coordinate $x\leftrightarrow \omega+\tau\mathbf{n}(\omega)$, $\omega\in\partial B$, $|\tau|\le 1/n_0$, is well defined and takes diffeomorphically the $1/n_0$-neighborhood of $\partial B$ in $\mathbb{R}^d$ onto the cylinder $\partial B\times(-1/n_0, 1/n_0)$. By construction we have

$$|\tau| = \operatorname{dist}(x,\partial B) \quad \text{for all } x\in B\setminus B_{1/n_0}.$$

Let $(x,t)\in G_n$ with $n>n_0$. Then

$$\operatorname{dist}(x,\partial B) = \operatorname{dist}((x,t),\Sigma) \le \operatorname{dist}((x,t),\Sigma\setminus\Sigma_{\rm in}) \le 1/n < 1/n_0.$$

Hence $x\in B_{1/n_0}$ and

$$x = \omega + \tau\mathbf{n}(\omega), \quad \omega\in\partial B, \quad -1/n < \tau \le 0.$$

Let $(\omega^*, t^*) \in \Sigma \setminus \Sigma_{\text{in}}$ be at minimal distance to $(x, t)$, i.e.

$$|(\omega^*, t^*) - (x, t)| = \operatorname{dist}((x, t), \Sigma \setminus \Sigma_{\text{in}}).$$

For any $(\omega', t') \in \Gamma$ we have

$$|(\omega^*, t^*) - (\omega', t')| \geq |(\omega', t') - (x, t)| - |(\omega^*, t^*) - (x, t)| \geq 2/n - 1/n = 1/n > 0.$$

Hence $(\omega^*, t^*)$ does not belong to $\Gamma$, i.e., $(\omega^*, t^*) \in \Sigma \setminus (\Sigma_{\text{in}} \cup \Gamma)$. Since $\Sigma \setminus (\Sigma_{\text{in}} \cup \Gamma)$ is an open subset of $\partial B \times (0, T)$ it is a classical result that the vector $(\omega^*, t^*) - (x, t)$ is parallel to $(\mathbf{n}^*, 0)$, a normal vector to $\partial B \times (0, T)$ at $(\omega^*, t^*)$. It follows that $(x, t)$ admits the representation

$$x = \omega^* + \tau^* \mathbf{n}(\omega^*) \in B_{1/n_0}, \quad \omega^* \in \partial B, \quad t^* = t,$$

hence

$$-\tau^* = \operatorname{dist}((x, t), \Sigma \setminus \Sigma_{\text{in}}).$$

Because of uniqueness of the normal coordinates we have $\omega = \omega^*$, $\tau = \tau^*$ and, as noted earlier, $t = t^*$. Thus for every $(x, t) \in G_n$ with normal coordinates $(\omega, \tau)$ we have

$$N(x, t) := \operatorname{dist}((x, t), \Sigma \setminus \Sigma_{\text{in}}) = -\tau \quad \text{in } G_n.$$

Notice that $\tau$ is a $C^1$ function of the variable $x$ in $B \setminus B_{n_0}$. Hence $N(x, t)$ is independent of $t$ on the open set $G_n$, which yields $\partial_t \psi_n = 0$ on $G_n$. Next, recalling the formula

$$\nabla \tau(x) = \mathbf{n}(\omega) \quad \text{for } x = \omega + \tau \mathbf{n}(\omega)$$

we obtain for $x = \omega + \tau \mathbf{n}(\omega)$, and $(x, t) \in G_n$,

$$\mathbf{U}(x, t) \cdot \nabla N(x, t) = -\mathbf{U}(x, t) \cdot \nabla \tau(x) = -\mathbf{U}(\omega + \tau \mathbf{n}(\omega), t) \cdot \mathbf{n}(\omega).$$

Since $\mathbf{U}(\omega, t)\mathbf{n}(\omega) \geq 0$ for $(\omega, t) \in \Sigma \setminus \Sigma_{\text{in}}$ we conclude that

$$\mathbf{U}(x, t)\nabla N(x, t) \leq (\mathbf{U}(\omega, t) - \mathbf{U}(\omega + \tau, t))\mathbf{n}(\omega) \leq c|\tau| \leq c/n \quad \text{in } G_n.$$

This obviously yields the inequality

$$\mathbf{U}(x, t) \cdot \nabla \psi_n(x, t) \leq c \quad \text{in } G_n. \qquad \square$$

Continuing with the proof of Theorem 13.3.3, it follows from Lemma 13.3.4 that

$$\Phi(\partial_t \psi_n + \mathbf{U} \cdot \psi_n) \leq c\Phi.$$

Since $G_n \subset B_{1/n} \times (0, T)$ we conclude that

$$\begin{aligned}
&\int_{G_n} \Phi(\partial_t \psi_n + \mathbf{U} \cdot \psi_n)\, dxdt \leq c \int_0^T \int_{B_{1/n}} \Phi\, dxdt \\
&\leq c\Big(\int_0^T \int_{B_{1/n}} dxdt\Big)^{1-1/q} \Big(\int_0^T \int_{B_{1/n}} \Phi^q\, dxdt\Big)^{1/q} \leq c\Big(\int_0^T \int_{B_{1/n}} dxdt\Big)^{1-1/q} \\
&\leq cn^{-(1-1/q)}.
\end{aligned}$$

It follows that

$$\limsup_{n\to\infty} \int_{G_n} \Phi(\partial_t \psi_n + \mathbf{U}\cdot\psi_n)\, dxdt \le 0. \tag{13.3.11}$$

Next, since $|\partial_t \psi_n| + |\nabla \psi_n| \le cn$ we have

$$\begin{aligned}\int_{Z_n} \Phi(\partial_t \psi_n + \mathbf{U}\cdot\psi_n)\, dxdt &\le cn \int_{Z_n} \Phi\, dxdt \\ &\le cn\Big(\int_{Z_n} dxdt\Big)^{1-1/q}\Big(\int_{Z_n} \Phi^q\, dxdt\Big)^{1/q} \\ &\le cn\, \mathrm{meas}(Z_n)^{1-1/q}.\end{aligned}$$

Since $Z_n \subset \mathcal{O}_{2/n}$, it follows from Condition 13.3.1 that $\mathrm{meas}(Z_n) \le cn^{-2}$. Thus we get

$$\limsup_{n\to\infty} \int_{Z_n} \Phi(\partial_t \psi_n + \mathbf{U}\cdot\psi_n)\, dxdt \le \lim_{n\to\infty} cn^{2/q-1} = 0. \tag{13.3.12}$$

Combining (13.3.11) and (13.3.12) with (13.3.9) we obtain (13.3.2), which completes the proof of Theorem 13.3.3. □

## 13.4 Singular limits for normal derivatives of solutions to diffusion equations

This section is devoted to the proof of Theorem 5.3.13. We derive estimates near the boundary for solutions to a linear parabolic problem depending on a small parameter $\varepsilon > 0$. Namely, we consider the first boundary value problem for the second order parabolic equation

$$\partial_t \varrho + \mathrm{div}(\varrho \mathbf{u}) - \varepsilon \Delta \varrho = 0 \quad \text{in } Q, \tag{13.4.1a}$$

$$\varrho = \varrho_\infty \quad \text{on } S_T, \tag{13.4.1b}$$

$$\varrho(x,0) = \varrho_0(x) \quad \text{in } \Omega. \tag{13.4.1c}$$

Here,

$$Q = \Omega \times (0,T), \quad S_T = \partial\Omega \times (0,T),$$

$\Omega \subset \mathbb{R}^d$, $d = 2,3$, is a bounded domain with $C^3$ boundary, and $\mathbf{u}$ is a given vector field. In this framework, (13.4.1a) can be regarded as a singular perturbation of the linear transport equation. Our main goal is to estimate the normal derivative of a solution to problem (13.4.1). The following arguments show that this question is important for the theory of zero viscosity limits of solutions to the diffusion equation. First of all, recall the definition of a weak solution to the first boundary value problem for the nonperturbed transport equation

$$\partial_t \varrho + \mathrm{div}(\varrho \mathbf{u}) = 0 \quad \text{in } Q,$$

$$\varrho = \varrho_\infty \quad \text{on } \Sigma_{\text{in}}, \quad \varrho(x,0) = \varrho_0(x) \quad \text{in } \Omega,$$

where the inlet is defined by

$$\Sigma_{\text{in}} = \{x \in \partial\Omega : \mathbf{U}_\infty(x) \cdot \mathbf{n}(x) < 0\}.$$

Assume for a moment that $\mathbf{u}$, $\varrho_\infty$ and $\varrho_0$ are continuous. Then an integrable function $\varrho$ is a weak solution to this problem if the integral identity

$$\int_Q (\partial_t \psi + \mathbf{u}\nabla\psi)\varrho \, dxdt + \int_\Omega \psi(x,0)\varrho_0 \, dx = \int_{S_T} \psi\varrho_\infty \mathbf{u}\cdot\mathbf{n} \, dSdt$$

holds for all smooth functions $\psi$ vanishing at $S_T \setminus \Sigma_{\text{in}}$ and for $t = T$. On the other hand, multiplying both sides of equation (13.4.1a) by $\psi$ and integrating by parts we obtain, at least formally,

$$\begin{aligned} \int_Q (\partial_t \psi + \mathbf{u}\cdot\nabla\psi)\varrho \, dxdt + \int_\Omega \psi(x,0)\varrho_0 \, dx - \int_{S_T} \psi\varrho_\infty \mathbf{u}\cdot\mathbf{n} \, dSdt \\ = \varepsilon \int_{S_T} \varrho_\infty \partial_n \psi \, dSdt - \varepsilon \int_Q \varrho\Delta\psi \, dxdt - \varepsilon \int_{S_T} \psi \, \partial_n \varrho \, dSdt. \end{aligned}$$

If solutions to problem (13.4.1) are uniformly integrable with respect to $\varepsilon$, the integrals on the right hand side, except the last one, tend to 0 as $\varepsilon \to 0$. Convergence to zero of $\varepsilon\partial_n\varrho$ is not obvious. What is more, in the general case such a statement is false. Indeed, if $\varepsilon\partial_n\varrho$ converges to zero even in the sense of distributions on the whole surface $S_T = \partial\Omega \times (0,T)$, then the limit transport equation has a solution which takes the prescribed boundary data on the whole surface $S_T$, which is impossible. Hence we can expect that $\varepsilon\partial_n\varrho$ converges to zero only on the inlet. For an elliptic regularization, this result was proved by Oleĭnik and Radkevich under the assumption that the set of solutions to the regularized problem is bounded in the $L^\infty$-norm, i.e., that they are uniformly bounded. In fact, Oleĭnik and Radkevich proved that $\partial_n\varrho \sim 1$ in the interior of the inlet. With applications to Navier-Stokes equations in mind, we extend this result to the case of parabolic equations and unbounded solutions. The result is purely local and it is independent of initial data provided that solutions are uniformly bounded in $L^\gamma(Q)$. Throughout this section we assume that $\varrho_\infty$ belongs to $C^2(Q)$, and the vector field $\mathbf{U}$ satisfies the following smoothness and positivity conditions.

**Condition 13.4.1.** For any $(z, t^*) \in \Sigma_{\text{in}}$ there exist positive constants $c_0$, $a$, $b$ with the following properties.

Denote by $Q_a$ the cylinder $B(z,a) \times (t^* - a, t^* + a)$, where $B(z,a) \subset \mathbb{R}^d$ is the ball of radius $a$ centered at $z$.

- In $Q_a \cap Q$ the velocity field $\mathbf{u}$ admits the representation

$$\begin{aligned} &\mathbf{u} = \nabla H + \mathbf{v} \quad \text{in } Q \cap Q_a, \\ &H = 0, \quad \mathbf{v}\cdot\mathbf{n} = 0 \quad \text{on } S_T \cap Q_a. \end{aligned} \tag{13.4.2}$$

Moreover, the function $H$ and the vector field $\mathbf{v}$ satisfy the inequalities

$$\begin{aligned} &\|H\|_{C^2(Q\cap Q_a)} + \|\mathbf{v}\|_{C^1(Q\cap Q_a)} \le c_0, \\ &\sup_{t\in(t^*-a,t^*+a)} \|\mathbf{v}(\cdot,t)\|_{C^1(B(z,a)\cap\Omega)} \le c_0. \end{aligned} \tag{13.4.3}$$

- The potential $H$ satisfies

$$\begin{aligned} &\frac{1}{2}\partial_t H(x,t) + \frac{1}{4}|\nabla H(x,t)|^2 > b > 0 \quad \text{in } Q_a \cap Q, \\ &-\nabla H(x,t)\cdot \mathbf{n}(x) > b > 0 \quad \text{on } Q_a \cap S_T. \end{aligned} \tag{13.4.4}$$

It follows from these conditions that diminishing $a$ if necessary, we can assume that $\varrho \in C^{1,2}(Q_a \cap Q)$, but smoothness properties of $\varrho$ depend on $\varepsilon$ and global estimates of $\varrho$ in the cylinder $Q$. Our task is to prove the following theorem which is equivalent to Theorem 5.3.13.

**Theorem 13.4.2.** *Assume that $\varrho$ is a solution to problem* (13.4.1) *that belongs to $L^2(0,T;W^{2,2}(\Omega))\cap C(0,T;W^{1,2}(\Omega))\cap W^{1,2}(0,T;L^2(\Omega))$ and that $\gamma > d+2$. Then there are positive constants $C_0$, $a_0$, $\varepsilon_0$, depending only on $d$, $\gamma$, $\Omega$, and on the constants in Condition 13.4.1, such that for all $\varepsilon \in (0,\varepsilon_0)$,*

$$\varepsilon\left|\frac{\partial\varrho}{\partial n}\right| \le C_0\varepsilon^\sigma\big(\|\varrho\|_{L^\gamma(Q_a)} + \|\varrho_\infty\|_{C^2(Q_a)}\big) \quad \textit{on } S_T\cap Q_{a_0}, \textit{where } \sigma = 1/2 - d/(2\gamma).$$

We split the proof into six steps. Furthermore, we denote by $c$, $c_i$ generic constants depending only on $c_0$, $a$, $b$, $z$, and $\partial\Omega$. To avoid repetitions the proof is given for $d=3$ only.

**Step 1. Normal coordinates.** The first step is standard: we make a change of the independent variables in order to rectify the boundary of $\Omega$ and reduce an inhomogeneous boundary problem for a homogeneous parabolic equation to a homogeneous boundary problem for an inhomogeneous equation.

Fix $(z,t^*) \in \Sigma_{\text{in}}$. As mentioned in Section 13.2, there exist constants $c_2 > 0$ and $\mu > 0$ with the following properties. There is a mapping $y \mapsto X(y)$ with $X(0) = z$ which takes diffeomorphically the cube

$$D_\mu = \{y \in \mathbb{R}^d : |y_i| < \mu,\ i = 1,2,3\} \tag{13.4.5}$$

onto some neighborhood $G$ of $z$, the flat square $D_\mu \cap \{y_3 = 0\}$ onto the surface $\partial\Omega \cap G$, and the slab

$$D_\mu^+ = \{y \in D_\mu : y_3 > 0\} \tag{13.4.6}$$

onto $G \cap \Omega$. Moreover, this mapping and its inverse satisfy the estimate

$$\|X\|_{C^2(D_\mu)} + \|X^{-1}\|_{C^2(G)} \le c(z,\Omega) \tag{13.4.7}$$

and we have the representation

$$X(Y) = \mathbf{r}(y_1, y_2) + y_3 \boldsymbol{\nu}(y_1, y_2) \tag{13.4.8}$$

where $x = \mathbf{r}(y_1, y_2)$ is a conformal parameterization of $\partial\Omega$ in the vicinity of $z$ and $\boldsymbol{\nu} = -\mathbf{n}(\mathbf{r}(y_1, y_2))$ is the inward normal vector to $\partial\Omega$. Notice that $y_3(x) = \operatorname{dist}(x, \partial\Omega)$ in $G$.

Further, consider the cylinders

$$\mathcal{Q}_\mu = D_\mu \times (t^* - \mu, t^* + \mu), \quad \mathcal{Q}_\mu^+ = D_\mu^+ \times (t^* - \mu, t^* + \mu). \tag{13.4.9}$$

Notice that the mapping $(X, \mathbb{I}) : (y, t) \mapsto (X(y), t)$ takes diffeomorphically the cylinder $\mathcal{Q}_\mu$ onto some neighborhood of $(z, t^*)$. By (13.4.7) we have

$$Q_{\mu/c} \subset (X, \mathbb{I})(\mathcal{Q}_\mu) \subset Q_{c\mu}, \tag{13.4.10}$$

where $c = c(z, \Omega)$ is the constant in (13.4.7). Without loss of generality we can assume that

$$0 < \mu < c(z, \Omega)a, \quad 0 < \mu < a, \tag{13.4.11}$$

and the mapping $(X, \mathbb{I})$ takes the cylinder $\mathcal{Q}_\mu$ into $Q_a$.

Let us recalculate the normal derivative of $\varrho$ in the normal coordinates. To this end notice that (13.2.2) and (13.2.5) imply

$$\frac{\mathbf{e}_1 \times \mathbf{e}_2}{|\mathbf{e}_1 \times \mathbf{e}_2|} = \frac{1}{g} \partial_{y_1} X \times \partial_{y_2} X = \frac{1}{g} |\mathbb{M}| \mathbb{M}^{-\top} \mathbf{k}, \quad \mathbf{k} = (0, 0, 1), \quad |\mathbb{M}| = \det \mathbb{M}.$$

Since by (13.2.7), $\mathbb{M}^{-\top} \nabla_y \varrho = \nabla_x \varrho$ and $|\mathbb{M}| = g$ for $y_3 = 0$, we obtain the following expression for the normal derivatives of $\varrho$:

$$\frac{\partial \varrho}{\partial n} = -\mathbb{M}^{-1} \mathbb{M}^{-\top} \mathbf{k} \cdot \nabla_y \varrho \quad \text{for } y_3 = 0.$$

Again, using (13.2.7) for the restriction of $\mathbb{M}$ to $\{y_3 = 0\}$ we finally obtain

$$\frac{\partial \varrho}{\partial n} = -\frac{\partial \varrho}{\partial y_3} \quad \text{for } y_3 = 0.$$

For technical reasons it is convenient to introduce a new unknown function $\rho(y, t)$ defined by

$$\varrho(X(y), t) = |\mathbb{M}(y)|^{-1} \rho(y, t) + \varrho_\infty(X(y), t). \tag{13.4.12}$$

It is clear that $\rho$ vanishes at $y_3 = 0$. Thus we get

$$\frac{\partial \varrho}{\partial n} = -|\mathbb{M}|^{-1} \frac{\partial \rho}{\partial y_3} - \frac{\partial \varrho_\infty}{\partial y_3}.$$

Consequently, the proposition will be proved if we show that

$$\varepsilon \left| \frac{\partial \rho}{\partial y_3} \right| \leq c \varepsilon^\sigma \big( \|\varrho\|_{L^\gamma(\mathcal{Q}_\mu^+)} + \|\varrho_\infty\|_{C^2(\mathcal{Q}_\mu^+)} \big) \quad \text{for } (y, t) \in \{y_3 = 0\} \cap \mathcal{Q}_{\mu/4}, \tag{13.4.13}$$

where $c$ depends only on $\gamma$, $d$, $z$, $\Omega$, and the constants in Condition 13.4.1; $\varrho$ and $\varrho_\infty$ are considered as functions of $y, t$, e.g. $\varrho = \varrho(X(y), t)$.

**Step 2. Change of independent variables.** Choose $\zeta \in C_0^\infty(G \times (t^* - \mu, t^* + \mu)))$. Multiplying (13.4.1) by $\zeta$ and integrating by parts we obtain

$$\int_{t^*-\mu}^{t^*+\mu} \int_G (\varrho \partial_t \zeta + \varrho \mathbf{u} \cdot \nabla \zeta - \varepsilon \nabla \varrho \cdot \nabla \zeta)\, dx dt = 0.$$

Recall the notation

$$\mathbb{M} := X'(y), \quad |\mathbb{M}| := \det \mathbb{M},$$

and notice that $\nabla_x \zeta = \mathbb{M}^{-\top} \nabla_y \zeta$. This gives

$$\int_{\mathcal{Q}_\mu^+} (\varrho \partial_t \zeta + \mathbb{M}^{-1} \varrho \mathbf{u} \cdot \nabla_y \zeta - \varepsilon \mathbb{M}^{-1} \mathbb{M}^{-\top} \nabla_y \varrho \cdot \nabla_y \zeta) |\mathbb{M}|\, dy dt = 0,$$

which yields the following equation for $\varrho(X(y), t)$ in the cylinder $\mathcal{Q}_\mu^+$:

$$|\mathbb{M}| \partial_t \varrho + \operatorname{div}(|\mathbb{M}| \mathbb{M}^{-1} \varrho \mathbf{u}) - \varepsilon \operatorname{div}(|\mathbb{M}| \mathbb{M}^{-1} \mathbb{M}^{-\top} \nabla \varrho) = 0. \tag{13.4.14}$$

Next, let $\rho(y, t)$ defined by (13.4.12) be the new unknown function. Substituting (13.4.12) into (13.4.14) we obtain the following equation and boundary conditions for $\rho$:

$$\partial_t \rho + \operatorname{div}(\rho \mathbf{W}) - \varepsilon \operatorname{div}(\mathbb{A} \nabla \rho) = f \quad \text{in } \mathcal{Q}_\mu^+, \tag{13.4.15a}$$

$$\rho(y, t) = 0 \quad \text{for } y_3 = 0, \tag{13.4.15b}$$

where

$$\begin{aligned} &\mathbb{A} = \mathbb{M}^{-1} \mathbb{M}^{-\top}, \quad \mathbf{W} = \mathbb{M}^{-1} \mathbf{u} + \frac{\varepsilon}{|\mathbb{M}|} \mathbb{A} \nabla |\mathbb{M}|, \\ &f = -|\mathbb{M}| \partial_t \varrho_\infty - \operatorname{div}(\varrho_\infty |\mathbb{M}| \mathbb{M}^{-1} \mathbf{u}) + \varepsilon \operatorname{div}(|\mathbb{M}| \mathbb{M}^{-1} \mathbb{M}^{-\top} \nabla \varrho_\infty). \end{aligned} \tag{13.4.16}$$

Here, $\varrho_\infty$ is regarded as a function of $y, t$, i.e., $\varrho_\infty = \varrho_\infty(X(y), t)$, and similarly for $\mathbf{u}$. It follows from Condition 13.4.1 that

$$\begin{aligned} &\sup_{t \in (t^* - \mu, t^* + \mu)} \|\mathbf{W}\|_{C^1(D_\mu^+)} \le c_1, \\ &\|\mathbb{A}\|_{C^1(D_\mu^+)} \le c_1, \\ &c_1^{-1} |\boldsymbol{\xi}|^2 \le \mathbb{A} \boldsymbol{\xi} \cdot \boldsymbol{\xi} \le c_1 |\boldsymbol{\xi}|^2, \end{aligned} \tag{13.4.17}$$

where the positive constant $c_1$ depends only on the constant $c_0$ in Condition 13.4.1 and the constant $c$ in (13.4.7). Using representation (13.4.2) and the identity $\nabla_x H = \mathbb{M}^{-\top} \nabla_y H$ we rewrite the expression for $\mathbf{W}$ in the form

$$\mathbf{W} = \mathbb{A} \nabla_y H + \mathbf{V},$$

where

$$\mathbf{V}(y, t) = \mathbb{M}^{-1} \mathbf{v}(X(y), t) + \frac{\varepsilon}{|\mathbb{M}|} \mathbb{A} \nabla |\mathbb{M}|. \tag{13.4.18}$$

It follows from Condition 13.4.1 that

$$\sup_{t\in(t^*-\mu,t^*+\mu)} \|\mathbf{V}\|_{C^1(D_\mu^+)} \le c_2, \quad \|H\|_{C^2(\mathcal{Q}_\mu^+)} \le c_2/2, \tag{13.4.19}$$

where $c_2$ depends only on the constant $c_0$ in Condition 13.4.1 and the constant $c$ in (13.4.7). Moreover, since the vector $\mathbb{M}^{-\top}\mathbf{k}$, $\mathbf{k} = (0,0,1)$, is collinear with the normal vector $\boldsymbol{\nu}$, we have

$$\mathbb{M}^{-1}\mathbf{v}(X(y),t)\cdot\mathbf{k} = 0 \quad \text{for } y_3 = 0. \tag{13.4.20}$$

On the other hand, using formula (13.2.6),

$$|\mathbb{M}| = g(1 - 2\mathcal{H}(y_1,y_2)y_3 + \mathcal{K}(y_1,y_2)y_3^2),$$

where $\mathcal{H}$ and $\mathcal{K}$ are the mean and Gaussian curvatures of $\partial\Omega$, respectively, we obtain

$$\frac{\varepsilon}{|\mathbb{M}|}\mathbb{A}\nabla|\mathbb{M}|\cdot\mathbf{k} = -2\varepsilon\mathcal{H}(y_1,y_2) \quad \text{for } y_3 = 0.$$

Combining this equality with (13.4.20) and using (13.4.19) we conclude that

$$|\mathbf{V}(y,t)\cdot\mathbf{k}| \le c_2(\varepsilon + y_3) \quad \text{in } \mathcal{Q}_\mu^+. \tag{13.4.21}$$

It follows from (13.4.11) and Condition 13.4.1 that

$$\nabla_y H\cdot\mathbf{k} > b \quad \text{for } (y,t)\in\mathcal{Q}_\mu\cap\{y_3 = 0\}.$$

From this and (13.4.19) it follows that for a suitable choice of $\mu$, depending only on the constants $c_2$ and $b$,

$$\nabla_y H\cdot\mathbf{k} > b/2 \quad \text{in } \mathcal{Q}_\mu^+, \tag{13.4.22}$$

which along with (13.4.21) yields for a suitable choice of $\mu$ and $\varepsilon_0$, depending only on $c_2$ and $b$,

$$\mathbf{W}\cdot\mathbf{k} > b/2 - c_2(\varepsilon + y_3) > b/3 \quad \text{in } \mathcal{Q}_\mu^+ \text{ for all } \varepsilon\in(0,\varepsilon_0). \tag{13.4.23}$$

**Step 3. Change of unknown function.** This step is crucial. We eliminate the potential part of the transport term in equation (13.4.15) using an appropriate change of unknown function. To this end, set

$$\rho(y,t) = \exp\left\{\frac{H(y,t)}{2\varepsilon}\right\} w(y,t). \tag{13.4.24}$$

Calculations show that $w$ satisfies

$$\begin{aligned} &\partial_t w - \varepsilon\,\mathrm{div}(\mathbb{A}\nabla w) + \mathbf{V}\cdot\nabla w + \frac{C}{\varepsilon}w = F \quad \text{in } \mathcal{Q}_\mu^+, \\ &w = 0 \quad \text{for } y_3 = 0. \end{aligned} \tag{13.4.25}$$

Here, the coefficient $C$ and the function $F$ are defined by

$$\begin{aligned} C &= \frac{1}{2}\partial_t H + \frac{1}{4}\mathbb{A}\nabla H\cdot\nabla H + \frac{1}{2}\mathbf{V}\cdot\nabla H + \varepsilon\,\mathrm{div}\,\mathbf{V} + \frac{\varepsilon}{2}\,\mathrm{div}(\mathbb{A}\nabla H), \\ F &= \exp\Big\{-\frac{H(y,t)}{2\varepsilon}\Big\} f. \end{aligned} \tag{13.4.26}$$

Let us show that $C$ is bounded from below and above by constants independent of $\varepsilon$. First we observe that

$$\frac{1}{2}\frac{\partial H}{\partial t}\Big|_{x=\mathrm{const}} + \frac{1}{4}|\nabla_x H|^2 = \frac{1}{2}\frac{\partial H}{\partial t}\Big|_{y=\mathrm{const}} + \frac{1}{4}|\mathbb{A}\nabla_y H\cdot\nabla_y H|^2.$$

It now follows from Condition 13.4.1 that

$$\frac{1}{2}\partial_t H + \frac{1}{4}\mathbb{A}\nabla H\cdot\nabla H \geq b \quad \text{in } \mathcal{Q}^+_\mu.$$

On the other hand, since $\nabla_y H$ is collinear with $\mathbf{k}$, it follows from (13.4.21) that

$$\nabla H\cdot\mathbf{V} = (\nabla H\cdot\mathbf{k})(\mathbf{V}\cdot\mathbf{k}) \leq c_2^2\varepsilon \quad \text{for } y_3 = 0.$$

From this and estimates (13.4.19) we conclude that

$$|\nabla H\cdot\mathbf{V}| \leq c_2^2(\varepsilon + y_3) \leq c_2^2(\varepsilon+\mu),$$

hence

$$C \geq b - c_2^2(y_3+\mu) - 2\varepsilon c_2^2.$$

Therefore, we can choose $\mu$ and $\varepsilon_0$, depending only on the constants in Condition 13.4.1 and the constant $c$ in (13.4.7), such that

$$b/2 \leq C \leq c_2^2 \quad \text{in } \mathcal{Q}^+_\mu \text{ for all } \varepsilon\in(0,\varepsilon_0). \tag{13.4.27}$$

Note that for such a choice of $\mu$ we have

$$H>0, \quad |w|\leq|\rho|\leq c(|\varrho|+|\varrho_\infty|), \quad |F|\leq|f|\leq c\|\varrho_\infty\|_{C^2(\mathcal{Q}^+_\mu)} \quad \text{in } \mathcal{Q}^+_\mu. \tag{13.4.28}$$

In particular,

$$\int_{\mathcal{Q}^+_\mu}|w|^\gamma\,dydt \leq c\int_{\mathcal{Q}^+_\mu}(|\varrho|^\gamma + |\varrho_\infty|^\gamma)\,dydt. \tag{13.4.29}$$

**Estimates for truncated functions.** Now choose $\eta_1\in C_0^\infty(\mathbb{R})$ with

$$\eta_1(s) = 1 \quad \text{for } s\in[-\mu/2,\mu/2] \quad \text{and} \quad \eta_1(s)=0 \quad \text{for } |s|\geq\mu$$

and set

$$\eta = \eta_1(y_1)\eta_1(y_2)\eta_1(y_3)\eta(t-t^*), \quad \varphi(y,t) = \eta w(y,t).$$

Multiplying (13.4.25) by $\eta$ we arrive at

$$\begin{aligned}&\partial_t\varphi-\varepsilon\,\mathrm{div}(\mathbb{A}\nabla\varphi)+\mathbf{V}\cdot\nabla\varphi+\frac{C}{\varepsilon}\varphi\\&\qquad\qquad=\mathrm{div}\,\boldsymbol{\Phi}_0+\Phi_1+\eta F\quad\text{in } D_\mu^+\times(0,T),\\&\varphi=0\quad\text{for } y\in\partial D_\mu^+,\\&\varphi(y,0)=0,\end{aligned}\tag{13.4.30}$$

where

$$\begin{aligned}&\boldsymbol{\Phi}_0=-2\varepsilon w\mathbb{A}\nabla\eta,\\&\Phi_1=\big(\mathbf{V}\cdot\nabla\eta+\varepsilon\,\mathrm{div}(\mathbb{A}\nabla\eta)+\partial_t\eta\big)w\quad\text{in } \mathcal{Q}_\mu^+,\end{aligned}\tag{13.4.31}$$

and $\boldsymbol{\Phi}_0=0$, $\Phi_1=0$ otherwise.

**Energy estimates.** Multiplying (13.4.30) by $|\varphi|^{\gamma-2}\varphi$ and integrating by parts we arrive at

$$\begin{aligned}\frac{1}{\gamma}\int_{D_\mu^+\times\{T\}}|\varphi|^\gamma\,dy+\varepsilon(\gamma-1)\int_{D_\mu^+\times(0,T)}&|\varphi|^{\gamma-2}\mathbb{A}\nabla\varphi\cdot\nabla\varphi\,dydt\\&+\frac{1}{\varepsilon}\int_{D_\mu^+\times(0,T)}|\varphi|^\gamma\Big(C-\frac{\varepsilon}{\gamma}\,\mathrm{div}\,\mathbf{V}\Big)\,dydt\\=\int_{D_\mu^+\times(0,T)}\big(|\varphi|^{\gamma-2}\varphi(\Phi_1+\eta F)&-(\gamma-1)|\varphi|^{\gamma-2}\nabla\varphi\cdot\boldsymbol{\Phi}_0\big)\,dydt.\end{aligned}$$

Here, we use the identity

$$\nabla(|\varphi|^{\gamma-2}\varphi)=(\gamma-1)|\varphi|^{\gamma-1}\nabla\varphi.$$

Noting that for sufficiently small $\varepsilon$ we have

$$C-\frac{\varepsilon}{\gamma}\,\mathrm{div}\,\mathbf{V}\geq\frac{b}{2}-\frac{\varepsilon}{\gamma}c_2\geq\frac{b}{3}$$

we obtain, for sufficiently small $\varepsilon$,

$$\begin{aligned}&\frac{b}{3\varepsilon}\int_{D_\mu^+\times(0,T)}|\varphi|^\gamma\,dydt\\&\qquad\leq\int_{D_\mu^+\times(0,T)}\big(|\varphi|^{\gamma-2}\varphi(\Phi_1+\eta F)-(\gamma-1)|\varphi|^{\gamma-2}\nabla\varphi\cdot\boldsymbol{\Phi}_0\big)\,dydt.\end{aligned}$$

Next, notice that (with all integrals over $D_\mu^+ \times (0,T)$)

$$
\begin{aligned}
-\int |\varphi|^{\gamma-2}\nabla\varphi\cdot\mathbf{\Phi}_0\,dydt &= 2\varepsilon\int \eta^{\gamma-2}|w|^{\gamma-2}w\mathbb{A}\nabla(\eta w)\cdot\nabla\eta\,dydt\\
&= 2\varepsilon\int \eta^{\gamma-2}|w|^{\gamma}\mathbb{A}\nabla\eta\cdot\nabla\eta\,dydt + 2\varepsilon\int \eta^{\gamma-1}|w|^{\gamma-2}w\mathbb{A}\nabla\eta\cdot\nabla w\,dydt\\
&= 2\varepsilon\int \eta^{\gamma-2}|w|^{\gamma}\mathbb{A}\nabla\eta\cdot\nabla\eta\,dydt + \frac{2\varepsilon}{\gamma}\int \eta^{\gamma-1}\nabla(|w|^{\gamma})\cdot\mathbb{A}\nabla\eta\,dydt\\
&= 2\varepsilon\int \eta^{\gamma-2}|w|^{\gamma}\mathbb{A}\nabla\eta\cdot\nabla\eta\,dydt - \frac{2\varepsilon}{\gamma}(\gamma-1)\int \eta^{\gamma-2}|w|^{\gamma}\mathbb{A}\nabla\eta\cdot\nabla\eta\,dydt\\
&\quad -\frac{2\varepsilon}{\gamma}\int \eta^{\gamma-1}|w|^{\gamma}\operatorname{div}(\mathbb{A}\nabla\eta)\,dydt.
\end{aligned}
$$

Thus we get

$$
\begin{aligned}
&-\int_{D_\mu^+\times(0,T)} |\varphi|^{\gamma-2}\nabla\varphi\cdot\mathbf{\Phi}_0\,dydt\\
&\qquad = \frac{2\varepsilon}{\gamma}\int_{D_\mu^+\times(0,T)} \left\{\eta^{\gamma-2}|w|^{\gamma}\mathbb{A}\nabla\eta\cdot\nabla\eta - \eta^{\gamma-1}|w|^{\gamma}\operatorname{div}(\mathbb{A}\nabla\eta)\right\}dydt\\
&\qquad = \frac{2\varepsilon}{\gamma}\int_{D_\mu^+\times(0,T)} \left\{|\varphi|^{\gamma-2}|w|^{2}\mathbb{A}\nabla\eta\cdot\nabla\eta - |\varphi|^{\gamma-1}|w|\operatorname{div}(\mathbb{A}\nabla\eta)\right\}dydt.
\end{aligned}
$$

It follows that

$$
(\gamma-1)\left|\int_{D_\mu^+\times(0,T)} |\varphi|^{\gamma-2}\nabla\varphi\cdot\mathbf{\Phi}_0\,dydt\right| \le c\varepsilon\int_{D_\mu^+\times(0,T)} (|\varphi|^{\gamma-2}|w|^2+|\varphi|^{\gamma-1}|w|)\,dydt,
$$

where $c$ depends only on $\gamma$, $\mu$, and the constant $c_1$ in (13.4.17). Using the representation for $\Phi_1$ and inequalities (13.4.19) we obtain

$$
\left|\int_{D_\mu^+\times(0,T)} |\varphi|^{\gamma-2}\varphi(\Phi_1+\eta F)\,dydt\right| \le c\int_{D_\mu^+\times(0,T)} |\varphi|^{\gamma-1}(|w|+|F|)\,dydt, \tag{13.4.32}
$$

where $c$ depends only on the constants $c_1$ and $c_2$ in (13.4.17) and (13.4.19). Combining the results obtained and noting that $\varphi$ is supported in $\mathcal{Q}_\mu^+$, we arrive at

$$
\begin{aligned}
\frac{b}{3\varepsilon}\int_{D_\mu^+\times(0,T)} |\varphi|^{\gamma}\,dydt &= \frac{b}{3\varepsilon}\int_{\mathcal{Q}_\mu^+} |\varphi|^{\gamma}\,dydt\\
&\le c\int_{\mathcal{Q}_\mu^+} \left(|\varphi|^{\gamma-1}(|w|+|F|) + \varepsilon|\varphi|^{\gamma-2}|w|^2\right)dydt.
\end{aligned} \tag{13.4.33}
$$

It follows from the Young inequality that for any positive $\sigma$ and $\kappa<1$,

$$
\varepsilon|\varphi|^{\gamma-2}|w|^2 \le c\varepsilon|\varphi|^{\gamma}\left(\frac{\kappa}{\sigma}\right)^{\gamma/(\gamma-2)} + \varepsilon c|w|^{\gamma}\left(\frac{\sigma}{\kappa}\right)^{\gamma/2}.
$$

Choosing $\sigma = \varepsilon^{2(\gamma-2)/\gamma}$ we obtain

$$\varepsilon|\varphi|^{\gamma-2}|w|^2 \le c\kappa^{\gamma/(\gamma-2)}\frac{1}{\varepsilon}|\varphi|^\gamma + c\kappa^{-\gamma/2}\varepsilon^{\gamma-1}|w|^\gamma.$$

Next, we have

$$|\varphi|^{\gamma-1}|w| \le c|\varphi|^\gamma\left(\frac{\kappa}{\sigma}\right)^{\gamma/(\gamma-1)} + c|w|^\gamma\left(\frac{\sigma}{\kappa}\right)^\gamma.$$

With $\sigma = \varepsilon^{(\gamma-1)/\gamma}$ we obtain

$$|\varphi|^{\gamma-1}|w| \le c\kappa^{\gamma/(\gamma-1)}\frac{1}{\varepsilon}|\varphi|^\gamma + c\kappa^{-\gamma}\varepsilon^{\gamma-1}|w|^\gamma.$$

Similar arguments give

$$|\varphi|^{\gamma-1}|F| \le c\kappa^{\gamma/(\gamma-1)}\frac{1}{\varepsilon}|\varphi|^\gamma + c\kappa^{-\gamma}\varepsilon^{\gamma-1}|F|^\gamma.$$

Inserting these inequalities into (13.4.33) and noting that $\kappa < 1$ we obtain

$$\left(\frac{b}{3} - c\kappa^{\gamma/(\gamma-1)}\right)\int_{D_\mu^+\times(0,T)}|\varphi|^\gamma\,dydt \le c\varepsilon^\gamma\int_{Q_\mu^+}\kappa^{-\gamma}(|w|^\gamma + |F|^\gamma)\,dydt.$$

Choosing $\kappa$ depending only on $\gamma$ and $c$ sufficiently small and recalling that $|w| \le c|\varrho| + |\varrho_\infty|$ we obtain

$$\|\varphi\|_{L^\gamma(D_\mu^+\times(0,T))} \le c\varepsilon(\|\varrho\|_{L^\gamma(Q_\mu^+)} + \|F\|_{L^\gamma(Q_\mu^+)}).$$

Recalling that $|F| \le c\|\varrho_\infty\|_{C^2(Q_\mu^+)}$ we conclude that

$$\|\varphi\|_{L^\gamma(D_\mu^+\times(0,T))} \le c\varepsilon(\|\varrho\|_{L^\gamma(Q_\mu^+)} + \|\varrho_\infty\|_{C^2(Q_\mu^+)}). \tag{13.4.34}$$

**Step 4. Extension.** Since the solution $\varphi$ to problem (13.4.30) and the function $w$ vanish for $y_3 = 0$, we can extend them over the cylinder $D_\mu \times (t^* - \mu, t^* + \mu)$ by setting

$$\begin{aligned}&\varphi(y,t) = -\varphi(\bar{y},t), \quad w(y,t) = -w(\bar{y},t) \quad \text{for } y \in D_\mu^-, \quad \text{where}\\ &\bar{y} = (y_1, y_2, -y_3) \in D_\mu^+.\end{aligned} \tag{13.4.35}$$

The extended $\varphi$ is Lipschitz and compactly supported in $D_\mu \times (t^* - \mu, t^* + \mu)$. Next, define extensions of the matrix $\mathbb{A}$, the vector fields $\mathbf{V}$, $\mathbf{\Phi}_0$, and the functions $C$, $\Phi_1$, $\eta F$ on $D_\mu$ by the formulae

$$\mathbb{A}(y) = \begin{pmatrix} A_{11} & A_{12} & -A_{13}\\ A_{21} & A_{22} & -A_{23}\\ -A_{31} & -A_{32} & A_{33}\end{pmatrix}(\bar{y}), \quad \mathbf{V}(y,t) = \begin{pmatrix}1 & 0 & 0\\ 0 & 1 & 0\\ 0 & 0 & -1\end{pmatrix}\mathbf{V}(\bar{y},t),$$

$$\mathbf{\Phi}_0(y,t) = \begin{pmatrix}-1 & 0 & 0\\ 0 & -1 & 0\\ 0 & 0 & 1\end{pmatrix}\mathbf{\Phi}_0(\bar{y},t), \qquad \begin{aligned}\Phi_1(y,t) &= -\Phi_1(\bar{y},t),\\ C(y) &= C(\bar{y}),\\ \eta F(y) &= -\eta F(\bar{y},t).\end{aligned}$$

We keep the same notation for the extended quantities. By (13.2.7) we have

$$\mathbb{A} = \begin{pmatrix} g^{-1} & 0 & 0 \\ 0 & g^{-1} & 0 \\ 0 & 0 & 1 \end{pmatrix} \quad \text{for } y_3 = 0. \tag{13.4.36}$$

Hence the extended matrix $\mathbb{A}$ and the function $C$ are Lipschitz in $D_\mu$. It follows from (13.4.17) that

$$\begin{aligned} &\lambda \le C \le c_1^2, \quad \text{where } \lambda = b/2 > 0, \\ &|\mathbb{A}(y') - \mathbb{A}(y'')| \le c_1 |y' - y''| \quad \text{for all } y', y'' \in \mathbb{R}^d, \\ &c_1^{-1}|\boldsymbol{\xi}|^2 \le \mathbb{A}\boldsymbol{\xi}\cdot\boldsymbol{\xi} \le c_1|\boldsymbol{\xi}|^2. \end{aligned} \tag{13.4.37}$$

Since $w$ vanishes for $y_3 = 0$, the extended vector field $\mathbf{\Phi}_0$ is Lipschitz in $D_\mu$. In contrast, the extensions of $\mathbf{V}$, $\Phi_1$ and $F$ are discontinuous and have jumps at $y_3 = 0$. Nevertheless, since $\varphi$ vanishes for $y_3 = 0$, we have the representation

$$\mathbf{V}\cdot\nabla\varphi = \operatorname{div}\mathbf{\Psi}_0 + \Psi_1,$$

where

$$\mathbf{\Psi}_0 = \varphi\mathbf{V}, \quad \Psi_1 = -\operatorname{div}\mathbf{V}\,\varphi \quad \text{in } D_\mu^+ \cup D_\mu^-. \tag{13.4.38}$$

It follows from (13.4.19) and (13.4.34) that

$$\begin{aligned} \|\mathbf{\Psi}_0\|_{L^\gamma(\mathcal{Q}_\mu)} + \|\Psi_1\|_{L^\gamma(\mathcal{Q}_\mu)} &\le c\|\varphi\|_{L^\gamma(\mathcal{Q}_\mu)} \\ &\le c\varepsilon(\|\varrho\|_{L^\gamma(\mathcal{Q}_\mu^+)} + \|\varrho_\infty\|_{C^2(\mathcal{Q}_\mu^+)}). \end{aligned} \tag{13.4.39}$$

Notice that the functions $\varphi$, $\Phi_1$, $\Psi_1$, $\eta F$ and the vector fields $\mathbf{\Psi}_0$, $\mathbf{\Phi}_0$ are compactly supported in the cylinder $\mathcal{Q}_\mu$ defined by (13.4.9). We assume that all these functions and vector fields are extended by 0 over $\mathbb{R}^3 \times (0,\infty)$. We also extend $\mathbf{A}(y)$ over $\mathbb{R}^3$. By the sharp form of Whitney's extension theorem (see [31]), we can assume that the extended matrix satisfies (13.4.37) in the whole $\mathbb{R}^3$. Finally, we extend $C$ over $\mathbb{R}^3 \times (0,\infty)$, simply setting $C = \lambda$ outside of $\mathcal{Q}_\mu$.

Thus, we arrive at the Cauchy problem for the extended function $\varphi$ in the whole space $\Omega$:

$$\begin{aligned} &\partial_t\varphi - \varepsilon\operatorname{div}(\mathbb{A}\nabla\varphi) + \frac{C}{\varepsilon}\varphi \\ &\qquad = \operatorname{div}(\mathbf{\Phi}_0 - \mathbf{\Psi}_0) + \Phi_1 - \Psi_1 + \eta F \quad \text{in } \mathbb{R}^d \times (0,\infty), \\ &\varphi = 0 \quad \text{for } t = 0. \end{aligned} \tag{13.4.40}$$

**Step 5. Heat kernel.** Our next task is to estimate the $L^\infty$-norm of the solution $\varphi$ to problem (13.4.40) by using the representation of $\varphi$ via the heat kernel associated with the parabolic differential operator in (13.4.40). We start by estimating this kernel. Assume that a matrix-valued function $\widetilde{\mathbb{A}} : \mathbb{R}^d \to \mathbb{R}^{d^2}$ with $\widetilde{\mathbb{A}}^\top = \widetilde{\mathbb{A}}$ and a bounded function $\tilde{C} : \mathbb{R}^d \to \mathbb{R}$ satisfy the inequalities (13.4.37) with $D_\mu$ replaced by the whole space $\mathbb{R}^d$. For simplicity, we omit the tildes. Let us consider the Cauchy problem

$$\partial_t \varphi - \varepsilon \operatorname{div}(\mathbb{A}\nabla\varphi) + \frac{C}{\varepsilon}\varphi = f \quad \text{in } \mathbb{R}^d \times (0,\infty), \quad \varphi = 0 \quad \text{for } t = 0, \tag{13.4.41}$$

where $f$ is a bounded function. It is well known that the unique strong solution to this problem admits a representation via the *heat kernel* $G_\varepsilon$,

$$\varphi(y,t) = \int_0^t \int_{\mathbb{R}^d} G_\varepsilon(y,t;z,s) f(z,s)\, dz ds. \tag{13.4.42}$$

The next proposition gives Aronson-type estimates for $G_\varepsilon$.

**Proposition 13.4.3.** *Suppose the matrix $\mathbb{A}$ and the coefficient $C$ satisfy* (13.4.37). *Then there are positive constants $\alpha$ and $\beta$, depending only on d and the constants $c_1$, $\lambda$ in* (13.4.37), *such that*

$$\begin{aligned} 0 \le G_\varepsilon(y,t;z,s) &\le \frac{\alpha}{(\varepsilon(t-s))^{d/2}} \exp\Big(-\beta\frac{|y-z|^2}{\varepsilon(t-s)} - \frac{\lambda(t-s)}{\varepsilon}\Big), \\ |\nabla_z G_\varepsilon(y,t;z,s)| &\le \frac{\alpha}{(\varepsilon(t-s))^{(d+1)/2}} \exp\Big(-\beta\frac{|y-z|^2}{\varepsilon(t-s)}\Big). \end{aligned} \tag{13.4.43}$$

The proof is given at the end of this section. The following lemma is the main result of this subsection.

**Lemma 13.4.4.** *Under the assumption of Theorem 13.4.2, there exists $c > 0$, depending only on the constants in Condition 13.4.1, $\partial\Omega$ and $\mu$, such that the solution $\varphi$ to problem* (13.4.30) *satisfies*

$$\sup_{(y,t)\in\mathcal{Q}^+_{\mu/3}} |\varphi(y,t)| \le c\varepsilon^\sigma\Big(\frac{1}{\varepsilon}e^{-\kappa/\varepsilon} + \varepsilon^\sigma\Big)(\|\varrho\|_{L^\gamma(\mathcal{Q}^+_\mu)} + \|\varrho_\infty\|_{C^2(\mathcal{Q}^+_\mu)}). \tag{13.4.44}$$

*Here,*

$$\kappa = \min\Big\{\frac{\beta\mu^2}{144T}, \frac{\mu\lambda}{12}\Big\}, \quad \sigma = \frac{1}{2} - \frac{d}{2\gamma},$$

*and the constant c depends only on d, $\gamma > d+2$, $\partial\Omega$ and the constants in Condition 13.4.1.*

*Proof.* Choose $(x,t) \in \mathcal{Q}^+_{\mu/3}$ and write the solution to problem (13.4.40) for the extended function $\varphi$ in the form

$$\varphi(x,t) = I_1(x,t) + I_2(x,t) + I_3(x,t),$$

where

$$I_1 = -\int_0^t \int_{\mathbb{R}^d} G_\varepsilon(x,t;y,s)(\operatorname{div}\mathbf{\Psi}_0 + \Psi_1)(y,s)\,dyds,$$
$$I_2 = \int_0^t \int_{\mathbb{R}^d} G_\varepsilon(x,t;y,s)(\operatorname{div}\mathbf{\Phi}_0 + \Phi_1)(y,s)\,dyds,$$
$$I_3 = \int_0^t \int_{\mathbb{R}^d} G_\varepsilon(x,t;y,s)(\eta F)(y,s)\,dyds.$$

To estimate $I_1$, since $\mathbf{\Psi}_0, \Psi_1$ are compactly supported we can write

$$I_1(x,t) = \int_0^t \int_{\mathbb{R}^d} \nabla_y G_\varepsilon(x,t;y,s)\cdot\mathbf{\Psi}_0(y,s)dyds - \int_0^t \int_{\mathbb{R}^d} G_\varepsilon(x,t;y,s)\Psi_1(y,s)dyds.$$

Recalling that, by (13.4.38), $|\mathbf{\Psi}_0|, |\Psi_1| \le c\varphi$ and applying Proposition 13.4.3 we obtain

$$|I_1(x,t)| \le c\alpha \int_0^t \int_{\mathbb{R}^d} \frac{1}{(\varepsilon(t-s))^{(d+1)/2}} \exp\Big(-\beta\frac{|y-x|^2}{\varepsilon(t-s)}\Big)|\varphi(y,s)|\,dyds.$$

The Hölder inequality yields

$$|I_1(x,t)|$$
$$\le c\Big\{\int_0^t \int_{\mathbb{R}^d} \frac{1}{(\varepsilon(t-s))^{\frac{(d+1)\gamma}{2(\gamma-1)}}} \exp\Big(-\frac{\gamma\beta|y-x|^2}{(\gamma-1)\varepsilon(t-s)}\Big)\,dyds\Big\}^{\frac{\gamma-1}{\gamma}} \|\varphi\|_{L^\gamma(\mathbb{R}^d\times(0,T))}.$$

Note that

$$\int_{\mathbb{R}^d} \exp\Big(-\frac{\gamma\beta|y-x|^2}{(\gamma-1)\varepsilon(t-s)}\Big)\,dy = c\varepsilon^{d/2}(t-s)^{d/2},$$

where $c$ depends only on $\gamma$, $\beta$, $d$, i.e., only on $\gamma$, $\Omega$ and the constants in Condition 13.4.1. Thus we get

$$|I_1(x,t)| \le c\varepsilon^{-\frac{1}{2}-\frac{d}{2\gamma}}\Big\{\int_0^t (t-s)^{-\frac{d+\gamma}{2(\gamma-1)}}\,ds\Big\}^{\frac{\gamma-1}{\gamma}} \|\varphi\|_{L^\gamma(\mathbb{R}^d\times(0,T))}.$$

Since $\gamma > 2+d$, we have $(d+\gamma)/(2(\gamma-1)) < 1$, which along with (13.4.39) yields

$$|I_1(x,t)| \le c\varepsilon^{-\frac{1}{2}-\frac{d}{2\gamma}}\|\varphi\|_{L^\gamma(\mathbb{R}^d\times(0,T))} \le c\varepsilon^\sigma\big(\|\varrho\|_{L^\gamma(Q_\mu^+)} + \|varrho_\infty\|_{C^2(Q_\mu^+)}\big). \tag{13.4.45}$$

Estimating $I_2$ is similar to those for $I_1$. Arguing as before we obtain

$$\begin{aligned}|I_2(x,t)| &\le c\alpha \int_0^t \int_{\mathbb{R}^d} \frac{1}{(\varepsilon(t-s))^{(d+1)/2}} \exp\Big(-\beta\frac{|y-x|^2}{\varepsilon(t-s)}\Big)|\mathbf{\Phi}_0(y,s)|\,dyds \\ &\quad + c\alpha\int_0^t \int_{\mathbb{R}^d} \frac{1}{(\varepsilon(t-s))^{d/2}} \exp\Big(-\beta\frac{|y-z|^2}{\varepsilon(t-s)} - \frac{\lambda(t-s)}{\varepsilon}\Big)|\Phi_1|\,dyds \\ &= I_{20} + I_{21}.\end{aligned}$$

To estimate the first integral notice that, by (13.4.31), $\mathbf{\Phi}_0(y,t)$ is compactly supported in $\mathcal{Q}_\mu$, where it satisfies

$$|\mathbf{\Phi}_0(y,t)| \le c|w(y,t)|.$$

Moreover, it vanishes for $|y_i| \le \mu/2$, $i=1,2$. Recall that $w$ is given by (13.4.24). It follows that

$$\int_0^t \int_{\mathbb{R}^d} \frac{1}{(\varepsilon(t-s))^{(d+1)/2}} \exp\Big(-\beta\frac{|y-z|^2}{\varepsilon(t-s)}\Big)|\mathbf{\Phi}_0(y,s)|\,dyds$$
$$\le c\int_{\mathcal{E}_t} \frac{1}{(\varepsilon(t-s))^{(d+1)/2}} \exp\Big(-\beta\frac{|y-z|^2}{\varepsilon(t-s)}\Big)|w|\,dyds,$$

where

$$\mathcal{E}_t = \{(y,s) \in \mathcal{Q}_\mu : 0<s<t,\ |y_i| \ge \mu/2,\ i=1,2\}.$$

For every $(x,t) \in \mathcal{Q}_{\mu/3}$ and $(y,s) \in \mathcal{E}_t$, we have $|y-x| \ge \mu/6$, which yields

$$\exp\Big(-\beta\frac{|y-x|^2}{\varepsilon(t-s)}\Big) \le e^{-\kappa/\varepsilon}\exp\Big(-\beta\frac{|y-x|^2}{2\varepsilon(t-s)}\Big).$$

Thus we get

$$I_{20} \le ce^{-\kappa/\varepsilon}\int_{\mathcal{E}_t} \frac{1}{(\varepsilon(t-s))^{(d+1)/2}} \exp\Big(-\beta\frac{|y-x|^2}{2\varepsilon(t-s)}\Big)|w(y,s)|\,dyds.$$

Applying the Hölder inequality we obtain

$$I_{20} \le ce^{-\kappa/\varepsilon}\Big\{\int_{\mathcal{E}_t} \frac{1}{(\varepsilon(t-s))^{\frac{(d+1)\gamma}{2(\gamma-1)}}} \exp\Big(-\frac{\gamma\beta|y-x|^2}{2(\gamma-1)\varepsilon(t-s)}\Big)\,dyds\Big\}^{\frac{\gamma-1}{\gamma}} \|w\|_{L^\gamma(\mathcal{Q}_\mu)}$$
$$\le ce^{-\kappa/\varepsilon}\Big\{\int_0^t\int_{\mathbb{R}^d} \frac{1}{(\varepsilon(t-s))^{\frac{(d+1)\gamma}{2(\gamma-1)}}} \exp\Big(-\frac{\gamma\beta|y-x|^2}{2(\gamma-1)\varepsilon(t-s)}\Big)\,dyds\Big\}^{\frac{\gamma-1}{\gamma}} \|w\|_{L^\gamma(\mathcal{Q}_\mu)}.$$

Recalling that

$$\int_{\mathbb{R}^d} \exp\Big(-\frac{\gamma\beta|y-x|^2}{2(\gamma-1)\varepsilon(t-s)}\Big)\,dy = c\varepsilon^{d/2}(t-s)^{d/2} \tag{13.4.46}$$

we obtain

$$I_{20} \le ce^{-\kappa/\varepsilon}\,\varepsilon^{-\frac12-\frac{d}{2\gamma}}\Big\{\int_0^t (t-s)^{-\frac{d+\gamma}{2(\gamma-1)}}\,ds\Big\}^{\frac{\gamma-1}{\gamma}} \|w\|_{L^\gamma(D_\mu\times(0,T)}$$
$$\le ce^{-\kappa/\varepsilon}\,\varepsilon^{-\frac12-\frac{d}{2\gamma}}\|w\|_{L^\gamma(\mathcal{Q}_\mu)}.$$

Next, note that

$$\|w\|_{L^\gamma(\mathcal{Q}_\mu)} = 2^{1/\gamma}\|w\|_{L^\gamma(\mathcal{Q}_\mu^+)},$$

hence from the equality $\frac{1}{2} + \frac{d}{2\gamma} = 1 - \sigma$ we obtain

$$I_{20} \le c\varepsilon^{\sigma-1} e^{-\kappa/\varepsilon} \|w\|_{L^\gamma(\mathcal{Q}_\mu^+)}.$$

It follows from the expression (13.4.24) for $w$ that

$$|w| = |e^{-H/\varepsilon}\rho| \le |\rho| \le c|\varrho| + |\varrho_\infty| \quad \text{in } \mathcal{Q}_\mu^+, \tag{13.4.47}$$

which leads to

$$I_{20} \le c\varepsilon^{\sigma-1} e^{-\kappa/\varepsilon} (\|\varrho\|_{L^\gamma(\mathcal{Q}_\mu^+)} + \|\varrho_\infty\|_{C^2(\mathcal{Q}_\mu^+)}). \tag{13.4.48}$$

Next, we estimate $I_{21}$. Again note that, by (13.4.31), $\Phi_1$ is compactly supported in $\mathcal{Q}_\mu$, where it does not exceed $c|w|$. Moreover, it vanishes in $\mathcal{Q}_{\mu/2}$. Thus we get

$$I_{21} \le c \int_{\mathcal{F}_t} \frac{1}{(\varepsilon(t-s))^{d/2}} \exp\Big(-\beta \frac{|y-x|^2}{\varepsilon(t-s)} - \frac{\lambda(t-s)}{\varepsilon}\Big) |w| \, dyds,$$

where

$$\mathcal{F}_t = \big\{(y,s) \in \mathcal{Q}_\mu \setminus \mathcal{Q}_{\mu/2} : 0 < s < t\big\}.$$

Notice that for $(y,s) \in \mathcal{F}_t$ and $(x,t) \in \mathcal{Q}_{\mu/3}$ we have either $|x-y| \ge \mu/6$ or $|t-s| \ge \mu/6$, which gives

$$\exp\Big(-\beta \frac{|y-x|^2}{\varepsilon(t-s)} - \frac{\lambda(t-s)}{\varepsilon}\Big) \le e^{-\kappa/\varepsilon} \exp\Big(-\beta \frac{|y-x|^2}{2\varepsilon(t-s)} - \frac{\lambda(t-s)}{2\varepsilon}\Big).$$

Using this result we arrive at

$$I_{21} \le ce^{-\kappa/\varepsilon} \int_{\mathcal{F}_t} \frac{1}{(\varepsilon(t-s))^{d/2}} \exp\Big(-\beta \frac{|y-x|^2}{2\varepsilon(t-s)} - \frac{\lambda(t-s)}{2\varepsilon}\Big) |w| \, dyds.$$

Combining this with the Hölder inequality leads to

$$I_{21} \le ce^{-\kappa/\varepsilon} \|w\|_{L^\gamma(\mathcal{Q}_\mu)}$$
$$\times \Big\{ \int_0^t \int_{\mathbb{R}^d} \frac{1}{(\varepsilon(t-s))^{\frac{\gamma d}{2(\gamma-1)}}} \exp\Big[-\frac{\gamma}{\gamma-1}\Big(\beta \frac{|y-x|^2}{2\varepsilon(t-s)} + \frac{\lambda(t-s)}{2\varepsilon}\Big)\Big] dyds \Big\}^{\frac{\gamma-1}{\gamma}}.$$

Recalling again (13.4.46) we get

$$I_{21} \le ce^{-\kappa/\varepsilon} \|w\|_{L^\gamma(\mathcal{Q}_\mu)} \Big\{ \int_0^t \frac{1}{(\varepsilon(t-s))^{\frac{d}{2(\gamma-1)}}} \exp\Big(-\frac{\gamma\lambda(t-s)}{2(\gamma-1)\varepsilon}\Big) ds \Big\}^{\frac{\gamma-1}{\gamma}}.$$

Since for $\gamma > d+2$,

$$\Big\{ \int_0^t \frac{1}{(\varepsilon(t-s))^{\frac{d}{2(\gamma-1)}}} \exp\Big(-\frac{\gamma\lambda(t-s)}{2(\gamma-1)\varepsilon}\Big) ds \Big\}^{\frac{\gamma-1}{\gamma}} \le c\varepsilon^{\frac{\gamma-1-d}{\gamma}} \le c,$$

we finally obtain from (13.4.47)

$$I_{21} \le ce^{-\kappa/\varepsilon}\|w\|_{L^\gamma(\mathcal{Q}_\mu)} \le ce^{-\kappa/\varepsilon}(\|\varrho\|_{L^\gamma(\mathcal{Q}_\mu^+)} + \|\varrho_\infty\|_{C^2(\mathcal{Q}_\mu^+)}).$$

Combining this estimate with (13.4.48) yields

$$|I_2(x,t)| \le c\varepsilon^{\sigma-1}e^{-\kappa/\varepsilon}(\|\varrho\|_{L^\gamma(\mathcal{Q}_\mu^+)} + \|\varrho_\infty\|_{C^2(\mathcal{Q}_\mu^+)}). \tag{13.4.49}$$

It remains to estimate $I_3$. Recall that $F$ is given by (13.4.26). Since the extended function $\eta F$ is compactly supported in $\mathcal{Q}_\mu$ we have

$$|I_3(x,t)| \le \int_0^t\int_{D_\mu} G_\varepsilon(x,t;y,s)|\eta F(y,s)|\,dyds \le \|\eta F\|_{C(\mathcal{Q}_\mu)}\int_0^t\int_{\mathbb{R}^d} G_\varepsilon(x,t;y,s)\,dyds.$$

It follows from (13.4.43) and the equality

$$\int_{\mathbb{R}^d}\frac{1}{(\varepsilon(t-s))^{d/2}}\exp\Big(-\beta\frac{|y-x|^2}{\varepsilon(t-s)}\Big)\,dy = \int_{\mathbb{R}^d}\exp(-\beta|y|^2)\,dy = c$$

that

$$\begin{aligned}\int_0^t\int_{\mathbb{R}^d} &G_\varepsilon(x,t;y,s)\,dyds\\ &\le c\int_0^t\int_{\mathbb{R}^d}\frac{1}{(\varepsilon(t-s))^{d/2}}\exp\Big(-\beta\frac{|y-x|^2}{\varepsilon(t-s)} - \frac{\lambda}{\varepsilon}(t-s)\Big)\,dyds\\ &= c\int_0^t \exp\Big(-\frac{\lambda}{\varepsilon}(t-s)\Big)\,ds \le c\varepsilon.\end{aligned}$$

Since, by (13.4.28),

$$\|\eta F\|_{C(\mathcal{Q}_\mu^+)} \le c\|\varrho_\infty\|_{C^2(\mathcal{Q}_\mu^+)},$$

we obtain

$$|I_3(x,t)| \le c\varepsilon\|\varrho_\infty\|_{C^2(\mathcal{Q}_\mu^+)}.$$

Combining this with (13.4.45)–(13.4.48) and noting that $0 < \sigma \le 1/2$ we obtain (13.4.44). □

**Corollary 13.4.5.** *There are constants $c_3 \ge 1$ and $\varepsilon_0 > 0$, depending only on $\partial\Omega$, $T$, $d$, $\gamma$, and the constants in Condition 13.4.1, such that the solution $\varphi$ to problem* (13.4.30) *satisfies*

$$\sup_{(y,t)\in\mathcal{Q}_{\mu/3}^+}|\varphi(y,t)| \le c_2\varepsilon^\sigma(\|\varrho\|_{L^\gamma(\mathcal{Q}_\mu^+)} + \|\varrho_\infty\|_{C^2(\mathcal{Q}_\mu^+)}). \tag{13.4.50}$$

*Proof.* It suffices to note that for a suitable choice of $\varepsilon_0$,

$$\frac{1}{\varepsilon}e^{-\kappa/\varepsilon} \le 1 \quad \text{for all } \varepsilon \in (0,\varepsilon_0).$$

□

**Step 6. Majorants.** We are now in a position to complete the proof of Theorem 13.4.2. First of all we use Lemma 13.4.4 to obtain the $L^\infty$ estimate for $\rho$ connected with $w$ via (13.4.24). To this end, note that $w$ coincides with $\varphi$ in the cube $\mathcal{Q}_{\mu/2}$. From this and (13.4.24) we obtain

$$|\rho| \leq e^{H/\varepsilon}|\varphi| \quad \text{in } \mathcal{Q}^+_{\mu/3} = [-\mu/3, \mu/3]^2 \times [0, \mu/3] \times [t^* - \mu/3, t^* + \mu/3].$$

On the other hand, as $H$ vanishes at $y_3 = 0$, estimates (13.4.19) yield $H \leq c_2 y_3$ in $\mathcal{Q}^+_\mu$, which leads to

$$|\rho| \leq e^{c_2 y_3/\varepsilon}|\varphi| \quad \text{in } \mathcal{Q}^+_{\mu/3}, \tag{13.4.51}$$

where $c_2$ is the constant in (13.4.19). Denote

$$\mathcal{Q} = \{(y,t) \in \mathcal{Q}_{\mu/3} : 0 \leq y_3 \leq \varepsilon\} = [-\mu/3, \mu/3]^2 \times [0, \varepsilon] \times [t^* - \mu/3, t^* + \mu/3].$$

It follows from (13.4.51) that

$$|\rho| \leq e^{c_2}|\varphi| \quad \text{in } \mathcal{Q}.$$

Combining this inequality with (13.4.50) we obtain

$$|\rho| \leq \psi(\varepsilon) \quad \text{in } \mathcal{Q}, \tag{13.4.52}$$

where

$$\psi(\varepsilon) = c_3 e^{c_2} \varepsilon^\sigma (\|\varrho\|_{L^\gamma(\mathcal{Q}^+_\mu)} + \|\varrho_\infty\|_{C^2(\mathcal{Q}^+_\mu)}).$$

Next, fix a nonnegative $\zeta \in C^\infty(\mathbb{R}^3)$ such that

$$\zeta(y_1, y_2, t) = 0 \quad \text{for } |y_1|, |y_2|, |t - t^*| \leq \mu/4,$$
$$\zeta(y_1, y_2, t) = 1 \quad \text{for } |y_1|, |y_2|, |t - t^*| \geq \mu/3$$

and

$$|\partial_t \zeta| + |\nabla \zeta| + |\nabla^2 \zeta| \leq 200\mu^{-2}.$$

We set

$$P(y,t) = \psi(\varepsilon)\zeta(y_1, y_2, t) + K y_3.$$

Let us show that for a suitable choice of the positive constant $K$, the function $P$ can be regarded as a majorant for $|\rho|$. First, observe that

$$P(y,t) = \psi(\varepsilon) + K y_3 \quad \text{for } |y_1|, |y_2|, |t - t^*| \geq \mu/3.$$

It follows from this and (13.4.52) the functions $P \pm \rho$ are nonnegative on the bottom and the lateral sides of the slab $\mathcal{Q}$ for any nonnegative $K$. On the top $y_3 = \varepsilon$ we have

$$P(y,t) - |\rho(y,t)| = \psi(\varepsilon)\zeta(y_1, y_2, t) + K\varepsilon - |\rho(y_1, y_2, \varepsilon, t)| \geq K\varepsilon - \psi(\varepsilon).$$

Hence for $K = \psi(\varepsilon)/\varepsilon$, the functions $P \pm \rho$ are nonnegative on $\partial\mathcal{Q}$. To prove that they are nonnegative in the whole cylinder $\mathcal{Q}$ we calculate

$$\partial_t(P \pm \rho) + \operatorname{div}((P \pm \rho)\mathbf{W}) - \varepsilon(\mathbb{A}(P \pm \rho)) =: \mathcal{L}(P \pm \rho).$$

It follows from (13.4.15) that

$$\mathcal{L}(P \pm \rho) = \psi(\varepsilon)\Big\{\frac{1}{\varepsilon}(\mathbf{W}\cdot\mathbf{k} + y_3 \operatorname{div}\mathbf{W} - \varepsilon \operatorname{div}(\mathbb{A}\mathbf{k})) + \mathcal{L}(\zeta)\Big\} \pm f.$$

Here $\mathbf{k} = (0,0,1)$, $f$ is defined by (13.4.16) and satisfies

$$|f| \le 2c_1 \|\varrho_\infty\|_{C^2(\mathcal{Q}_\mu^+)} \quad \text{in } \mathcal{Q}_{\mu+},$$

where $c_1$ is the constant in (13.4.17). It follows from (13.4.17) and (13.4.23) that

$$\mathbf{W}\cdot\mathbf{k} \ge b/3, \quad |\operatorname{div}\mathbf{W}| + |\operatorname{div}(\mathbb{A}\mathbf{k})| \le 2c_1, \quad |\mathcal{L}(\zeta)| \le 400(c_1+1)\mu^{-2}.$$

Thus we get

$$\mathcal{L}(P \pm \rho) \ge \psi(\varepsilon)(b/\varepsilon - 2c_1 - 400(c_1+1)\mu^{-2}) - 2c_1\|\varrho_\infty\|_{C^2(Q_{\mu+})}$$

in $\mathcal{Q}$. Recalling the expression for $\psi(\varepsilon)$ and the inequality $c_3 \ge 1$, we obtain, for all $\varepsilon \in (0, \varepsilon_0)$,

$$\mathcal{L}(P \pm \rho) \ge \{\varepsilon^{\sigma-1}(b/3 - \varepsilon 2c_1 - \varepsilon 400(c_1+1)\mu^{-2}) - 2c_1\}(\|\varrho\|_{L^\gamma(\mathcal{Q}_\mu^+)} + \|\varrho_\infty\|_{C^2(\mathcal{Q}_\mu^+)}).$$

Recall that $\mu$ and $c_1$ depend only on $\partial\Omega$ and the constants in Condition 13.4.1. Hence for a suitable choice of $\varepsilon_0$ we obtain, for all $\varepsilon \in (0, \varepsilon_0)$,

$$\mathcal{L}(P \pm \rho) \ge 0 \quad \text{in } \mathcal{Q}, \qquad P \pm \rho \ge 0 \quad \text{in } \partial\mathcal{Q}.$$

It follows that $P \pm \rho \ge 0$ in $\mathcal{Q}$. Since $\zeta$ vanishes for $|y_1|, |y_2|, |t - t^*| \le \mu/4$, taking into account the expression for $P$, we conclude that

$$|\rho| \le K y_3 = \psi(\varepsilon)\varepsilon^{-1} y_3 \le c_3 e^{c_2}\varepsilon^{\sigma-1}(\|\varrho\|_{L^\gamma(Q_\mu^+)} + \|\varrho_\infty\|_{C^2(Q_\mu^+)})y_3$$

for $|y_1|, |y_2|, |t - t^*| < \mu/4$ and $0 < y_3 < \varepsilon < \varepsilon_0$. Hence the inequality

$$\left|\varepsilon\frac{\partial\rho}{\partial y_3}\right| \le c_3 e^{c_2}\varepsilon^\sigma(\|\varrho\|_{L^\gamma(\mathcal{Q}_\mu^+)} + \|\varrho_\infty\|_{C^2(\mathcal{Q}_\mu^+)}) \tag{13.4.53}$$

holds for all

$$y = (y_1, y_2, 0) \quad \text{and} \quad t \quad \text{with} \quad |y_1|, |y_2|, |t - t^*| \le \mu/4 \quad \text{and} \quad 0 < \varepsilon < \varepsilon_0,$$

which yields (13.4.13), and the theorem follows.

**Proof of Proposition 13.4.3.** The proof is based on the following lemma which is a simplification of the more general Lemma 3.1 in [132]. For any positive $c > 0$ set

$$\mathcal{G}_c(x,t;y,s) = (t-s)^{-d/2} \exp\left\{ -c\frac{|x-y|^2}{t-s} \right\}, \quad x, y \in \mathbb{R}^2,\, t > s.$$

**Lemma 13.4.6.** *For any $0 < a < b$ and $\lambda > 0$,*

$$\int_s^t \int_{\mathbb{R}^d} \frac{e^{-\lambda(\tau-s)}}{(t-\tau)^{1/2}} \mathcal{G}_a(x,t;z,\tau)\mathcal{G}_b(z,\tau;y,s)\,dzd\tau \le \frac{c(a,b,\lambda,d)}{(t-s)^{1/2}} \mathcal{G}_a(x,t;y,s).$$

*Proof.* We can assume that $s = 0$, so the desired inequality reads

$$\int_0^t \int_{\mathbb{R}^d} \frac{e^{-\lambda\tau}}{(t-\tau)^{1/2}} \mathcal{G}_a(x,t;z,\tau)\mathcal{G}_b(z,\tau;y,0)\,dzd\tau \le \frac{c(a,b,\lambda,d)}{t^{1/2}} \mathcal{G}_a(x,t;y,0).$$

Split the integral into

$$\left(\int_0^{\rho t}\int_{\mathbb{R}^d} + \int_{\rho t}^t \int_{\mathbb{R}^d}\right) \frac{e^{-\lambda\tau}}{(t-\tau)^{1/2}} \mathcal{G}_a(x,t;z,\tau)\mathcal{G}_b(z,\tau;y,0)\,dzd\tau = J_1 + J_2,$$

where $\rho \in (0,1)$ is specified below. Let us estimate $J_1$. We have

$$J_1 = \int_0^{\rho t}\int_{\mathbb{R}^d} e^{-\lambda\tau} \frac{\exp\left(-a\frac{|x-z|^2}{t-\tau} - a\frac{|z-y|^2}{\tau}\right)}{(t-\tau)^{(d+1)/2}} \frac{\exp\left(-(b-a)\frac{|z-y|^2}{\tau}\right)}{\tau^{d/2}}\,dzd\tau.$$

Next we have $\tau = \nu t$, $t - \tau = (1-\nu)t$, where $0 < \nu \le \rho$. It follows that

$$\begin{aligned}
\frac{|x-z|^2}{t-\tau} + \frac{|z-y|^2}{\tau} - \frac{|x-y|^2}{t} &= \frac{1}{t}\left(\frac{|x-z|^2}{1-\nu} + \frac{|z-y|^2}{\nu} - |x-y|^2\right)\\
&= \frac{1}{t}\left(\frac{\nu}{1-\nu}|x-z|^2 + \frac{1-\nu}{\nu}|z-y|^2 + 2(x-z)\cdot(y-z)\right)\\
&\ge \frac{1}{t}\left(\frac{\nu}{1-\nu}|x-z|^2 + \frac{1-\nu}{\nu}|z-y|^2 - 2|x-z|\,|y-z|\right) \ge 0,
\end{aligned}$$

and hence

$$\frac{1}{(t-\tau)^{(d+1)/2}} \exp\left(-a\frac{|x-z|^2}{t-\tau} - a\frac{|z-y|^2}{\tau}\right) \le \frac{1}{t^{1/2}(1-\rho)^{(d+1)/2}} \mathcal{G}_a(x,t;y,0).$$

Thus we get

$$\begin{aligned}
J_1 &\le \frac{1}{t^{1/2}(1-\rho)^{(d+1)/2}} \mathcal{G}_a(x,t;y,0) \int_0^{\rho t}\int_{\mathbb{R}^d} e^{-\lambda\tau} \frac{\exp\left(-(b-a)\frac{|z-y|^2}{\tau}\right)}{\tau^{d/2}}\,dzd\tau\\
&\le c(a,b,\rho)\frac{1}{t^{1/2}} \mathcal{G}_a(x,t;y,0) \int_0^\infty e^{-\lambda\tau}\,d\tau \le c(a,b,\lambda,\rho)t^{-1/2}\mathcal{G}_a(x,t;y,0).
\end{aligned}$$

It remains to estimate $J_2$. To this end, split $\mathbb{R}^d$ into

$$\Omega^+ = \{z : |z-y| \geq |x-y|(a/b)^{1/2}\} \quad \text{and} \quad \Omega^- = \{z : |z-y| < |x-y|(a/b)^{1/2}\}.$$

Notice that for all $z \in \Omega^+$ and $\tau \in (\rho t, t)$,

$$\tau^{-d/2} \exp\left(-b\frac{|z-y|^2}{\tau}\right) \leq (\rho t)^{-d/2} \exp\left(-a\frac{|x-y|^2}{t}\right).$$

It follows that

$$\begin{aligned}
&\int_{\rho t}^{t} \int_{\Omega^+} \frac{e^{-\lambda\tau}}{(t-\tau)^{1/2}} \mathcal{G}_a(x,t;z,\tau)\, \mathcal{G}_b(z,\tau;y,0)\, dz d\tau \\
&\quad \leq \frac{\exp\left(-a\frac{|x-y|^2}{t} - \frac{\lambda\rho t}{2}\right)}{(\rho t)^{d/2}} \int_{\rho t}^{t} \int_{\mathbb{R}^d} \frac{e^{-\frac{\lambda}{2}\tau}}{(t-\tau)^{(d+1)/2}} \exp\left(-a\frac{|x-z|^2}{t-\tau}\right) dz d\tau \\
&\quad \leq c e^{-\frac{\lambda\rho t}{2}} \frac{\exp\left(-a\frac{|x-y|^2}{t}\right)}{(\rho t)^{d/2}} \int_0^t \frac{e^{-\frac{\lambda}{2}\tau}}{(t-\tau)^{1/2}}\, d\tau \leq c e^{-\frac{\lambda\rho t}{2}} \frac{\exp\left(-a\frac{|x-y|^2}{t}\right)}{t^{d/2}} \\
&\quad \leq c \frac{\exp\left(-a\frac{|x-y|^2}{t}\right)}{t^{d+1/2}} = c t^{-1/2} \mathcal{G}_a(x,t;y,0). \qquad (13.4.54)
\end{aligned}$$

On the other hand, for $z \in \Omega^-$, we have $|x-z| \geq |x-y|(1-(a/b)^{1/2})$ and $t - \tau \leq (1-\rho)t$, which yields

$$a\frac{|x-z|^2}{t-\tau} \geq \frac{a}{2}\frac{|x-z|^2}{t-\tau} + \frac{a}{2}(1-(a/b)^{1/2})^2 \frac{|x-y|^2}{(1-\rho)t}.$$

Choose $\rho$ so that

$$\frac{1}{2}(1-(a/b)^{1/2})^2 \frac{1}{1-\rho} = 1.$$

Thus we get

$$a\frac{|x-z|^2}{t-\tau} \geq \frac{a}{2}\frac{|x-z|^2}{t-\tau} + a\frac{|x-y|^2}{t}.$$

Hence

$$\begin{aligned}
&\int_{\rho t}^{t} \int_{\Omega^-} \frac{e^{-\lambda\tau}}{(t-\tau)^{1/2}} \mathcal{G}_a(x,t;z,\tau)\, \mathcal{G}_b(z,\tau;y,0)\, dz d\tau \\
&\quad \leq \frac{c}{t^{d/2}} \exp\left(-a\frac{|x-y|^2}{t} - \frac{\lambda\rho t}{2}\right) \int_{\rho t}^{t} \int_{\Omega^-} \frac{e^{-\frac{\lambda}{2}\tau}}{(t-\tau)^{(d+1)/2}} \exp\left(-a\frac{|x-z|^2}{2(t-\tau)}\right) dz d\tau \\
&\quad \leq \frac{c}{t^{d/2}} e^{-\frac{\lambda\rho t}{2}} \exp\left(-a\frac{|x-y|^2}{t}\right) \int_0^t \frac{e^{-\frac{\lambda}{2}\tau}}{(t-\tau)^{1/2}}\, d\tau \leq \frac{c}{t^{(d+1)/2}} \exp\left(-a\frac{|x-y|^2}{t}\right).
\end{aligned}$$

Combining this with (13.4.54) we obtain

$$J_2 \leq \frac{c}{t^{(d+1)/2}} \exp\left(-a\frac{|x-y|^2}{t}\right). \qquad \square$$

**Corollary 13.4.7.** *Under the assumptions of Lemma 13.4.6, for any* $0 < a < b$,

$$\int_s^t \int_{\mathbb{R}^d} \frac{e^{-\lambda(\tau-s)}}{(t-\tau)^{1/2}} \mathcal{G}_b(x,t;z,\tau)\mathcal{G}_b(z,\tau;y,s)\,dzd\tau \leq \frac{c(a,b,\lambda,d)}{(t-s)^{(1/2}} \mathcal{G}_a(x,t;y,s).$$

Let us turn to the proof of Proposition 13.4.3. First of all we use scaling in order to eliminate $\varepsilon$ from the equations. To this end, we set

$$\varphi_\varepsilon(x,t) = \varphi(\varepsilon x, \varepsilon t),\ f_\varepsilon(x,t) = f(\varepsilon x, \varepsilon t),\ \mathbb{A}_\varepsilon(x) = \mathbb{A}(\varepsilon x),\ C_\varepsilon(x,t) = C(\varepsilon x, \varepsilon t).$$

A change of variables in (13.4.41) leads to

$$\partial_t \varphi_\varepsilon - \operatorname{div}(\mathbb{A}_\varepsilon \nabla \varphi_\varepsilon) + C_\varepsilon \varphi_\varepsilon = \varepsilon f_\varepsilon \quad \text{in } \mathbb{R}^d \times (0,\infty), \quad \varphi_\varepsilon = 0 \quad \text{for } t = 0. \tag{13.4.55}$$

It is well known that the solution to this problem has the representation

$$\varphi_\varepsilon(x,t) = \varepsilon \int_0^t \int_{\mathbb{R}^d} G(x,t;y,s) f_\varepsilon(y,s)\,dyds. \tag{13.4.56}$$

Along with equation (13.4.55) we consider the truncated problem

$$\partial_t \varphi_\varepsilon - \operatorname{div}(\mathbb{A}_\varepsilon \nabla \varphi_\varepsilon) = \varepsilon f_\varepsilon \quad \text{in } \mathbb{R}^d \times (0,\infty), \quad \varphi_\varepsilon = 0 \quad \text{for } t = 0. \tag{13.4.57}$$

with heat kernel $G_0(x,t;y,s)$, and the auxiliary problem

$$\partial_t \varphi_\varepsilon - \operatorname{div}(\mathbb{A}_\varepsilon \nabla \varphi_\varepsilon) + \lambda \varphi_\varepsilon = \varepsilon f_\varepsilon \quad \text{in } \mathbb{R}^d \times (0,\infty), \quad \varphi_\varepsilon = 0 \quad \text{for } t = 0, \tag{13.4.58}$$

with heat kernel (fundamental solution) $G_\lambda(x,t;y,s)$. Since $0 < \lambda \leq C$ it follows from the general theory of parabolic PDE's that

$$G(x,t;y,s) \leq G_\lambda(x,t;y,s) \leq G_0(x,t;y,s). \tag{13.4.59}$$

Moreover, since $\mathbb{A}$ is independent of $t$, we have

$$G_\lambda(x,t;y,s) = e^{-\lambda(t-s)} G_0(x,t;y,s). \tag{13.4.60}$$

Since the operator $\operatorname{div}(\mathbb{A}_\varepsilon \nabla \cdot)$ is symmetric, the heat kernels are symmetric with respect to $x$ and $y$. Estimates for Green functions of second order parabolic equations with bounded coefficients were derived in [7] (see also [132] for further discussion). A feature of these estimates is that they are sensitive to lower order terms and in general nonuniform in time. Nevertheless, as shown in [7] and [132], global in time estimates do hold true for the fundamental solution $G_0$:

$$G_0(x,t;y,s) \leq \frac{\alpha_0}{(t-s)^{d/2}} \exp\Big(-\beta_0 \frac{|y-z|^2}{t-s}\Big),$$

$$|\nabla_x G_0(x,t;y,s)| \leq \frac{\alpha_0}{(t-s)^{(d+1)/2}} \exp\Big(-\beta_0 \frac{|y-z|^2}{t-s}\Big).$$

It follows that

$$
\begin{aligned}
G_0(x,t;y,s) &\le \alpha_0 \mathcal{G}_{\beta_0}(x,t;y,s),\\
|\nabla_x G_0(x,t;y,s)| &\le \frac{\alpha_0}{(t-s)^{1/2}} \mathcal{G}_{\beta_0}(x,t;y,s).
\end{aligned}
\tag{13.4.61}
$$

From this we obtain an estimate for $G$:

$$
G(x,t;y,s) \le e^{-\lambda(t-s)} G_0(x,t;y,s) \le \alpha_0 e^{-\lambda(t-s)} \mathcal{G}_{\beta_0}(x,t;y,s). \tag{13.4.62}
$$

We take advantage of strong positivity of $C$ in order to prove that $\nabla G$ admits the same estimate as $\nabla G_0$. To this end, we write (13.4.57) in the form

$$
\partial_t \varphi_\varepsilon - \operatorname{div}(\mathbb{A}_\varepsilon \nabla \varphi_\varepsilon) = -C_\varepsilon \varphi_\varepsilon + \varepsilon f_\varepsilon,
$$

which leads to

$$
\varphi_\varepsilon(x,t) = \int_0^t \int_{\mathbb{R}^d} G_0(x,t;y,s)(-C_\varepsilon \varphi_\varepsilon(y,s) + \varepsilon f_\varepsilon(y,s))\, dyds.
$$

Next, substitute (13.4.56) into the right hand side to obtain

$$
\begin{aligned}
\varphi_\varepsilon(x,t) = &-\varepsilon \int_0^t \int_{\mathbb{R}^d} G_0(x,t;z,\tau) \Big\{ \int_0^\tau \int_{\mathbb{R}^d} G(z,\tau;y,s) C_\varepsilon(z,\tau) f_\varepsilon(y,s)\, dyds \Big\}\, dzd\tau\\
&+ \varepsilon \int_0^t \int_{\mathbb{R}^d} G_0(x,t;y,s) f_\varepsilon(y,s)\, dyds.
\end{aligned}
$$

Recall that $\varphi_\varepsilon$ also admits representation (13.4.56). Since $f_\varepsilon$ is an arbitrary smooth function, after changing the order of integration, we arrive at

$$
G(x,t;y,s) = -\int_s^t \int_{\mathbb{R}^d} G_0(x,t;z,\tau) G(z,\tau;y,s) C_\varepsilon(z,\tau)\, dzd\tau + G_0(x,t;y,s).
$$

Hence

$$
\begin{aligned}
&\nabla_x G(x,t;y,s)\\
&\quad = -\int_s^t \int_{\mathbb{R}^d} \nabla_x G_0(x,t;z,\tau) G(z,\tau;y,s) C_\varepsilon(y,s)\, dzd\tau + \nabla_x G_0(x,t;y,s).
\end{aligned}
$$

It now follows from inequalities (13.4.61)–(13.4.62) that

$$
\begin{aligned}
|\nabla_x G(x,t;y,s)| \le c_1 \alpha_0^2 &\int_s^t \int_{\mathbb{R}^d} \frac{1}{(t-\tau)^{1/2}} \mathcal{G}_{\beta_0}(x,t;z,\tau) e^{-\lambda(\tau-s)} \mathcal{G}_{\beta_0}(z,\tau;y,s)\, dzd\tau\\
&+ \frac{\alpha_0}{(t-s)^{1/2}} \mathcal{G}_{\beta_0}(x,t;y,s).
\end{aligned}
$$

Next, applying Corollary 13.4.7 we obtain, for any $\beta < \beta_0$ and $\alpha = \alpha_0^2 c_1 + \alpha_0$,

$$|\nabla_x G(x,t;y,s)| \leq \alpha \frac{1}{(t-s)^{1/2}} \mathcal{G}_\beta(x,t;y,s).$$

Combining this inequality with (13.4.62) we finally obtain

$$\begin{aligned} G(x,t;y,s) &\leq \frac{\alpha}{(t-s)^{d/2}} \exp\Big(-\beta \frac{|x-y|^2}{t-s} - \lambda(t-s)\Big), \\ |\nabla_x G(x,t;y,s)| &\leq \frac{\alpha}{(t-s)^{(d+1)/2}} \exp\Big(-\beta \frac{|x-y|^2}{t-s}\Big). \end{aligned} \tag{13.4.63}$$

Next, a change of variables in representation (13.4.56) gives

$$\begin{aligned} \varphi(x,t) &= \varepsilon^{-d} \int_0^t \int_{\mathbb{R}^d} G\Big(\frac{x}{\varepsilon}, \frac{t}{\varepsilon}; \frac{y}{\varepsilon}, \frac{s}{\varepsilon}\Big) f(y,s)\, dyds, \\ \nabla\varphi(x,t) &= \varepsilon^{-d-1} \int_0^t \int_{\mathbb{R}^d} [\nabla_x G]\Big(\frac{x}{\varepsilon}, \frac{t}{\varepsilon}; \frac{y}{\varepsilon}, \frac{s}{\varepsilon}\Big) f(y,s)\, dyds. \end{aligned}$$

Thus we get

$$\begin{aligned} G_\varepsilon(x,t;y,s) &= \varepsilon^{-d} G\Big(\frac{x}{\varepsilon}, \frac{t}{\varepsilon}; \frac{y}{\varepsilon}, \frac{s}{\varepsilon}\Big), \\ \nabla_x G_\varepsilon(x,t;y,s) &= \varepsilon^{-d-1} [\nabla_x G]\Big(\frac{x}{\varepsilon}, \frac{t}{\varepsilon}; \frac{y}{\varepsilon}, \frac{s}{\varepsilon}\Big). \end{aligned}$$

Combining these identities with (13.4.63) we obtain the desired estimates for the heat kernel $G_\varepsilon$. This completes the proof of Proposition 13.4.3.

# Bibliography


[1] D. Adams, L. Hedberg, *Function Spaces and Potential Theory*, Springer, Berlin, 1996.

[2] R. A. Adams, *Sobolev Spaces*, Academic Press, New York, 1975.

[3] C. Agmon, A. Douglis, L. Nirenberg, *Estimates near the boundary for solutions of elliptic partial differential equations satisfying general boundary conditions. I.* , Comm. Pure Appl. Math. **12** (1959), 623–727.

[4] H. Amann, *Linear and Quasilinear Parabolic Problems*, Birkhäuser, Boston, MA, 1995.

[5] A. A. Amosov, A. A. Zlotnik, *Properties in the large of quasi-averaged equations of the one dimensional motion of a viscous heat conducting gas*, Dokl. Akad. Nauk **346** (1996), 151–154 (in Russian).

[6] A. A. Amosov, A. A. Zlotnik, *On quasi-averaged equations of the one-dimensional motion of a viscous barotropic medium with rapidly oscillating data*, Comput. Math. Math. Phys. **36** (1996), 203–220.

[7] D. G. Aronson, *Non-negative solutions of linear parabolic equations*, Ann. Scuola Norm. Sup. Pisa **22** (1968), 607–694.

[8] N. S. Bakhvalov, M. E. Eglit, *Processes in periodic media not describable in terms of mean characteristics*, Dokl. Akad. Nauk SSSR **268** (1983), 836–840 (in Russian).

[9] J. M. Ball, *A version of the fundamental theorem for Young measures*, in: PDEs and Continuum Models of Phase Transitions, Lecture Notes in Phys. 344, Springer, Berlin, 1989, 207–215.

[10] V. Barbu, I. Lasiecka, R. Triggiani, *Tangential boundary stabilization of Navier-Stokes equations*, Mem. Amer. Math. Soc. **181** (2006), no. 852.

[11] H. Beirão da Veiga, *Stationary motions and the incompressible limit for compressible viscous fluids*, Houston J. Math. **13** (1987), 527–544.

[12] H. Beirão da Veiga, *An $L^p$-theory for the n-dimensional, stationary, compressible Navier-Stokes equations, and the incompressible limit for compressible fluids. The equilibrium solutions*, Comm. Math. Phys. **109** (1987), 229–248.

[13] H. Beirão da Veiga, *Existence results in Sobolev spaces for a transport equation*, Ricerche Mat. **36** (1987), suppl., 173–184.

[14] J. A. Bello, E. Fernández-Cara, J. Lemoine, J. Simon, *The differentiability of the drag with respect to the variations of a Lipschitz domain in a Navier-Stokes flow*, SIAM J. Control. Optim. **35** (1997), 626–640.

[15] J. A. Bello, E. Fernández-Cara, J. Simon, *Optimal shape design for Navier-Stokes flow*, in: System Modelling and Optimization, P. Kall (ed.), Lecture Notes in Control and Inform. Sci. 180, Springer, Berlin, 1992, 481–489.

[16] J. Bergh, J. Löfström, *Interpolation Spaces. An Introduction*, Springer, Berlin, 1976.

[17] Y. Brenier, *Résolution d'équations quasilinéaires en dimension $N$ d'espace à l'aide d'équations linéaires en dimension $N+1$*, J. Differential Equations **50** (1983), 375–390.

[18] D. Bucur, G. Buttazzo, *Variational Methods in Shape Optimization Problems*, Birkhäuser, Boston, 2005.

[19] X. F. Chen, W. Q. Xie, *Discontinuous solutions of steady state, viscous compressible Navier-Stokes equations*, J. Differential Equations **115** (1995), 99–119.

[20] R. Courant, D. Hilbert, *Methods of Mathematical Physics. Vol. II: Partial Differential Equations*, Interscience Publ., New York, 1962.

[21] M. C. Delfour, J.-P. Zolésio, *Shapes and Geometries*, Adv. Design Control 22, SIAM, Philadelphia, PA, 2001, 2nd augmented ed., 2011.

[22] B. Desjardins, E. Grenier, P.-L. Lions, N. Masmoudi, *Incompressible limit for solutions of the isentropic Navier-Stokes equations with Dirichlet boundary conditions*, J. Math. Pures Appl. **78** (1999), 461–471.

[23] R. J. DiPerna, P.-L. Lions, *Ordinary differential equations, transport theory and Sobolev spaces*, Invent. Math. **48** (1989), 511–547.

[24] N. Dunford, J. T. Schwartz, *Linear Operators. Part 1: General Theory*, Interscience Publ., New York, 1958.

[25] H. Ebert, *Physikalisches Taschenbuch*, Vieweg, Braunschweig, 1957.

[26] R. E. Edwards, *Functional Analysis. Theory and Applications*, Holt, Rinehart, and Winston, New York, 1965.

[27] L. C. Evans, *Partial Differential Equations*, Amer. Math. Soc., Providence, RI, 1998.

[28] L. C. Evans, R. F. Gariepy, *Measure Theory and Fine Properties of Functions*, CRC Press, Boca Raton, FL, 1992.

[29] R. Farwig, *Stationary solutions to compressible Navier-Stokes equations*, Comm. Partial Differential Equations **14** (1989), 1579–1506.

[30] R. Farwig, G. Galdi, H. Sohr, *A new class of weak solutions of the Navier-Stokes equations*, J. Math. Fluid Mech. **8** (2006), 423–443.

[31] C. L. Fefferman, *A sharp form of Whitney's extension theorem*, Ann. of Math. **161** (2006), 509-577.

[32] E. Feireisl, *On compactness of solutions to the compressible isentropic Navier-Stokes equations when the density is not square integrable*, Comment. Math. Univ. Carolin. **42** (2001), 83–98.

[33] E. Feireisl, *Shape optimization in viscous compressible fluids*, Appl. Math. Optim. **47** (2003), 59–78.

[34] E. Feireisl, *Dynamics of Viscous Compressible Fluids*, Oxford Univ. Press, Oxford, 2004.

[35] E. Feireisl, *Asymptotic analysis of the full Navier-Stokes-Fourier system: from compressible to incompressible fluid flows*, Uspekhi Mat. Nauk **62** (2007), no. 3, 169–192 (in Russian).

[36] E. Feireisl, Š. Matuš-Nečasová, H. Petzeltová, I. Straškraba, *On the motion of a viscous compressible fluid driven by a time-periodic external force*, Arch. Ration. Mech. Anal. **149** (1999), 69–96.

[37] E. Feireisl, A. H. Novotný, *Singular Limits in Thermodynamics of Viscous Fluids*, Birkhäuser, Basel, 2009.

[38] E. Feireisl, A. H. Novotný, H. Petzeltová, *On the existence of globally defined weak solutions to the Navier-Stokes equations*, J. Math. Fluid Mech. **3** (2001), 358–392.

[39] E. Feireisl, A. H. Novotný, H. Petzeltová, *On the domain dependence of solutions to the compressible Navier-Stokes equations of a barotropic fluid*, Math. Methods Appl. Sci. **25** (2002), 1045–1073.

[40] W. Feller, *An Introduction to the Probability Theory and its Applications*, 3rd ed., Wiley, New York, 1968.

[41] G. Fichera, *Sulle equazioni differenziali lineari ellittico-paraboliche del secondo ordine*, Atti Accad. Naz. Lincei Mem. Cl. Sci. Fis. Mat. Nat. Sez. I (8) **5** (1956), 1–30.

[42] I. Fonseca, G. Leoni, *Modern Methods in the Calculus of Variations: $L^p$ Spaces*, Springer, New York, 2007.

[43] J. Frehse, S. Goj, M. Steinhauer, *$L^p$-estimates for the Navier-Stokes equations for steady compressible flow*, Manuscripta Math. **116** (2005), 265–275.

[44] J. Frehse, M. Steinhauer, W. Weigant, *The Dirichlet problem for steady viscous compressible flow in three dimensions*, J. Math. Pures Appl. **97** (2012), 85–97.

[45] J. Frehse, M. Steinhauer, W. Weigant, *The Dirichlet problem for viscous compressible isothermal Navier-Stokes equations in two dimensions*, Arch. Ration. Mech. Anal. **198** (2010), 1–12.

[46] G. Frémiot, W. Horn, A. Laurain, M. Rao, J. Sokołowski, *On the analysis of boundary value problems in nonsmooth domains*, Dissertationes Math. **462** (2009), 149 pp.

[47] P. Fulmański, A. Laurain, J.-F. Scheid, J. Sokołowski, *Level set method with topological derivatives in shape optimization*, Int. J. Computer Math. **85** (2008), 1491–1514.

[48] A. Fursikov, *Optimal Control of Distributed Systems. Theory and Applications*, Transl. Math. Monogr. 187, Amer. Math. Soc., 2000.

[49] H. Gajewski, K. Gröger, K. Zakharias, *Nightlineare Operatorgleichungen und Operatordifferentialgleichungen*, Akademie-Verlag, Berlin, 1974.

[50] G. Galdi, *An Introduction to the Mathematical Theory of the Navier-Stokes Equations. Vol. I. Linearized Steady Problems*, Springer, Berlin, 1994; 2nd ed., 2011.

[51] D. Gilbarg, N. S. Trudinger, *Elliptic Partial Differential Equations of Second Order*, Springer, Berlin, 1983.

[52] V. Girinon, *Navier-Stokes equations with nonhomogeneous boundary conditions in a bounded three-dimensional domain*, J. Math. Fluid Mech. **13** (2011), 309–339.

[53] L. Hedberg, *Spectral synthesis in Sobolev spaces, and uniqueness of solutions of the Dirichlet problem*, Acta Math. **147** (1981), 237–264.

[54] A. Henrot, M. Pierre, *Variation et optimisation de formes: une analyse géométrique*, Springer, Paris, 2005.

[55] J. G. Heywood, M. Padula, *On the existence and uniqueness theory for steady compressible viscous flow*, in: Fundamental Directions in Mathematical Fluid Mechanics, Adv. Math. Fluid Mech., Birkhäuser, Basel, 2000, 171–189.

[56] D. Hoff, *Strong convergence to global solutions for multidimensional flows of compressible, viscous fluids with polytropic equations of state and discontinuous initial data*, Arch. Ration. Mech. Anal. **132** (1995), 1–14.

[57] D. Hoff, *Discontinuous solutions of the Navier-Stokes equations for multidimensional flows of heat-conducting fluids*, Arch. Ration. Mech. Anal. **139** (1997), 303–354.

[58] L. Hörmander, *Pseudo-differential operators and non-elliptic boundary problems*, Ann. of Math. **83** (1966), 129–209.

[59] S. Ishikawa, *Fixed points by a new iteration method*, Proc. Amer. Math. Soc. **44** (1974), 147–150.

[60] N. Itaya, *On the initial value problem of the motion of compressible viscous fluid, especially on the problem of uniqueness*, J. Math. Kyoto Univ. **16** (1976), 413–427.

[61] W. Jäger, A. Mikelić, *Couette flows over a rough boundary and drag reduction*, Comm. Math. Phys. **232** (2003), 429–455.

[62] B. Kawohl, O. Pironneau, L. Tartar, J.-P. Zolésio, *Optimal Shape Design*, Lecture Notes in Math. 1740, Springer, Berlin, 2000.

[63] A. V. Kazhikhov, V. A. Solonnikov, *Existence theorems for the equations of motion of a compressible viscous-fluid*, Ann. Rev. Fluid Mechanics **13** (1981), 79–95.

[64] A. Kaźmierczak, P. I. Plotnikov, J. Sokołowski, A. Żochowski, *Numerical method for drag minimization in compressible flows*, in: 15th International Conference on Methods and Models in Automation and Robotics (MMAR), 2010. 97–101; DOI: 10.1109/MMAR.2010.5587258.

[65] R. B. Kellogg, *Discontinuous solutions of the linearized, steady state, compressible, viscous, Navier-Stokes equations*, SIAM J. Math. Anal. **19** (1988), 567–579.

[66] J. J. Kohn, L. Nirenberg, *Degenerate elliptic-parabolic equations of second order*, Comm. Pure Appl. Math. **20** (1967), 797–872.

[67] A. N. Kolmogorov, S. V. Fomin, *Elements of the Theory of Functions and Functional Analysis*, Mir, Moscow, 1989 (in Russian).

[68] M. A. Krasnosel'skii, *Topological Methods in the Theory of Nonlinear Integral Equations*, Macmillan, New York, 1964.

[69] J. R. Kweon, R. B. Kellogg, *Compressible Navier-Stokes equations in a bounded domain with inflow boundary condition*, SIAM J. Math. Anal. **28** (1997), 94–108.

[70] J. R. Kweon, R. B. Kellogg, *Regularity of solutions to the Navier-Stokes equations for compressible barotropic flows on a polygon*, Arch. Ration. Mech. Anal. **163** (2000), 35–64.

[71] J. R. Kweon, R. B. Kellogg, *Singularities in the density gradient*, J. Math. Fluid Mech. **8** (2006), 445–454.

[72] J. R. Kweon, R. B. Kellogg, *Regularity of solutions to the Navier–Stokes system for compressible flows on a polygon*, SIAM J. Math. Anal. **35** (2004), 1451–1485.

[73] J. R. Kweon, R. B. Kellogg, *Compressible Stokes problem on nonconvex polygonal domains*, J. Differential Equations **176** (2001), 290–314.

[74] O. A. Ladyženskaja, V. A. Solonnikov, N. N. Ural'ceva, *Linear and Quasiliniear Equations of Parabolic Type*, Transl. Math. Monogr. 23, Amer. Math. Soc., Providence, RI, 1967.

[75] L. D. Landau, E. M. Lifshitz, *Course of Theoretical Physics. Vol. 6. Fluid Mechanics*, Pergamon Press, Oxford, 1987.

[76] N. S. Landkof, *Foundations of Modern Potential Theory*, Springer, Berlin, 1972.

[77] J.-L. Lions, *Quelques méthodes de résolution des problèmes aux limites non linéaires*, Dunod, Paris, 1969.

[78] J.-L. Lions, E. Magenes, *Problèmes aux limites non homogènes et applications*, Vols. 1, 2, Dunod, Paris, 1968.

[79] P.-L. Lions, *Mathematical Topics in Fluid Dynamics, Vol. 1, Incompressible Models*, Clarendon Press, Oxford, 1996.

[80] P.-L. Lions, *Mathematical Topics in Fluid Dynamics, Vol. 2, Compressible Models*, Clarendon Press, Oxford, 1998.

[81] P.-L. Lions, *Compacité des solutions des équations de Navier-Stokes compressibles isentropiques*, C. R. Acad. Sci. Paris Sér. I Math. **317** (1993), 115–120.

[82] P.-L. Lions, *Bornes sur la densité pour les équations de Navier-Stokes compressibles isentropiques avec conditions aux limites de Dirichlet*, C. R. Acad. Sci. Paris Sér. I Math. **328** (1999), 659–662.

[83] P.-L. Lions, *On some challenging problems in nonlinear partial differential equations*, in: Mathematics: Frontiers and Perspectives, V. Arnold et al. (eds.), Amer. Math. Soc., Providence, RI, 2000, 121–135.

[84] P.-L. Lions, B. Perthame, P. Souganidis, *Existence and stability of entropy solutions for the hyperbolic systems of isentropic gas dynamics in Eulerian and Lagrangian coordinates*, Comm. Pure Appl. Math. **49** (1996), 599–638.

[85] P.-L. Lions, B. Perthame, E. Tadmor, *A kinetic formulation of multidimensional scalar conservation laws and related equations*, J. Amer. Math. Soc. **7** (1994), 169–191.

[86] J. Málek, J. Nečas, M. Rokyta, M. Růžička, *Weak and Measure-Valued Solutions to Evolutionary PDE*, Chapman and Hall, London, 1996.

[87] A. Matsumura, T. Nishida, *The initial value problem for the equations of motion of compressible viscous and heat-conductive gases*, J. Math. Kyoto Univ. **20** (1980), 67–104.

[88] V. G. Maz'ya, *Sobolev Spaces*, Springer, Berlin, 1985.

[89] B. Mohammadi, O. Pironneau, *Shape optimization in fluid mechanics*, Ann. Rev. Fluid Mech. **36** (2004), 255–279.

[90] B. Mohammadi, O. Pironneau, *Applied Shape Optimization for Fluids*, 2nd ed., Oxford Univ. Press, Oxford, 2010.

[91] M. Moubachir, J.-P. Zolésio, *Moving Shape Analysis and Control: Applications to Fluid Structure Interactions*, Chapman & Hall/CRC, Boca Raton, FL, 2006.

[92] J. Nash, *Le problème de Cauchy pour les équations différentielles d'un fluide général*, Bull. Soc. Math. France **90** (1962), 487–497.

[93] S. Novo, *Compressible Navier-Stokes model with inflow-outflow boundary conditions*, J. Math. Fluid Mech. **7** (2005), 485–514.

[94] S. Novo, A. Novotný, *On the existence of weak solutions to the steady compressible Navier-Stokes equations when the density is not square integrable*, J. Math. Kyoto Univ. **42** (2002), 531–550.

[95] S. Novo, A. Novotný, *On the existence of weak solutions to the steady compressible Navier-Stokes equations in domains with conical outlets*, J. Math. Fluid Mech. **8** (2006), 187–210.

[96] A. Novotný, *About steady transport equation. I. $L^p$-approach in domains with smooth boundaries*, Comment. Math. Univ. Carolin. **37** (1996), 43–89.

[97] A. Novotný, *About steady transport equation. II. Schauder estimates in domains with smooth boundaries*, Portugal. Math. **54** (1997), 317–333.

[98] A. Novotný, M. Padula, *Existence and uniqueness of stationary solutions for viscous compressible heat conductive fluid with large potential and small non-potential external forces*, Sibirsk. Mat. Zh. **34** (1993), no. 5, 120–146 (in Russian).

[99] A. Novotný, M. Padula, *$L^p$-approach to steady flows of viscous compressible fluids in exterior domains*, Arch. Ration. Mech. Anal. **126** (1994), 243–297.

[100] A. Novotný, M. Padula, *Physically reasonable solutions to steady compressible Navier-Stokes equations in 3D-exterior domains* ($v_\infty \neq 0$), Math. Ann. **308** (1997), 439–489.

[101] A. Novotný, I. Straškraba, *Introduction to the Mathematical Theory of Compressible Flow*, Oxford Univ. Press, Oxford, 2004.

[102] O. A. Oleĭnik, E. V. Radkevič, *Second Order Equations with Nonnegative Characteristic Form*, Plenum Press, New York, 1973.

[103] M. Padula, *Existence and uniqueness for viscous steady compressible motions*, Arch. Ration. Mech. Anal. **97** (1987), 89–102.

[104] M. Padula, *Steady flows of barotropic viscous fluids*, in: Classical Problems in Mechanics, Quad. Mat. 1, Dept. Math., Seconda Univ. Napoli, Caserta, 1997, 253–345.

[105] P. Pedregal, *Parametrized Measures and Variational Principles*, Progr. Nonlinear Differential Equations Appl. 30, Birkhäuser, Basel, 1997.

[106] B. Perthame, *Kinetic Formulation of Conservation Laws*, Oxford Lecture Ser. Math. Appl. 21, Oxford Univ. Press, Oxford, 2002.

[107] P. I. Plotnikov, E. V. Ruban, J. Sokołowski, *Inhomogeneous boundary value problems for compressible Navier-Stokes equations, well-posedness and sensitivity analysis*, SIAM J. Math. Anal. **40** (2008), 1152–1200.

[108] P. I. Plotnikov, E. V. Ruban, J. Sokołowski, *Inhomogeneous boundary value problems for compressible Navier-Stokes and transport equations*, J. Math. Pures Appl. **92** (2009), 113–162.

[109] P. I. Plotnikov, J. Sokołowski, *Stationary boundary value problems for Navier-Stokes equations with adiabatic exponent* $\gamma < 3/2$, Dokl. Akad. Nauk **397** (2004), 166–169 (in Russian)

[110] P. I. Plotnikov, J. Sokołowski, *On compactness, domain dependence and existence of steady state solutions to compressible isothermal Navier-Stokes equations*, J. Math. Fluid Mech. **7** (2005), 529–573.

[111] P. I. Plotnikov, J. Sokołowski, *Concentrations of solutions to time-discretized compressible Navier -Stokes equations*, Comm. Math. Phys. **258** (2005), 567–608.

[112] P. I. Plotnikov, J. Sokołowski, *Domain dependence of solutions to compressible Navier-Stokes equations*, SIAM J. Control Optim. **45** (2006), 1165–1197.

[113] P. I. Plotnikov, J. Sokołowski, *Stationary solutions of Navier-Stokes equations for diatomic gases*, Uspekhi Mat. Nauk **62** (2007), no. 3, 117–148 (in Russian).

[114] P. I. Plotnikov, J. Sokołowski, *Stationary boundary value problems for compressible Navier-Stokes equations*, in: Handbook of Differential Equations: Stationary Partial Differential Equations, Vol. VI, Elsevier/North-Holland, Amsterdam, 2008, 313–410.

[115] P. I. Plotnikov, J. Sokołowski, *Shape derivative of drag functional,* SIAM J. Control Optim. **48** (2010), 4680–4706.

[116] P. I. Plotnikov, J. Sokołowski, *Inhomogeneous boundary value problem for nonstationary compressible Navier-Stokes equations*, J. Math. Sci. **170** (2010), 34–130.

[117] P. Plotnikov, J. Sokołowski, A. Żochowski, *Numerical experiments in drag minimization for compressible Navier-Stokes flows in bounded domains*, in: Proc. 14th International IEEE/IFAC Conference on Methods and Models in Automation and Robotics MMAR'09, 2009, 4 pp.

[118] A. Sard, *The measure of the critical values of differentiable maps*, Bull. Amer. Math. Soc. **48** (1942), 883–890.

[119] H. Schlichting, *Boundary Layer Theory*, McGraw-Hill, New York, 1955.

[120] L. Schwartz, *Analyse Mathématique I*, Hermann, Paris, 1967.

[121] J. Serrin, *On the uniqueness of compressible fluid motions*, Arch. Ration. Mech. Anal. **3** (1959), 271–288.

[122] J. Serrin, *Mathematical principles of classical fluid mechanics*, in: Handbuch der Physik VIII/I, Springer, Berlin, 1959, 125–263.

[123] J. Simon, *Domain variation for drag in Stokes flow*, in: Control Theory of Distributed Parameter Systems and Applications (Shanghai, 1990), Lecture Notes in Control and Inform. Sci. 159, Springer, Berlin, 1991, 28–42.

[124] T. Slawig, *A formula for the derivative with respect to domain variations in Navier-Stokes flow based on an embedding domain method*, SIAM J. Control Optim. **42** (2003), 495–512.

[125] T. Slawig, *An explicit formula for the derivative of a class of cost functionals with respect to domain variations in Stokes flow*, SIAM J. Control Optim. **39** (2000), 141–158.

[126] J. Sokołowski, J.-P. Zolésio, *Introduction to Shape Optimization. Shape Sensitivity Analysis*, Springer Ser. Comput. Math. 16, Springer, Berlin, 1992.

[127] E. Stein, *Singular Integrals and Differentiability Properties of Functions*, Princeton Univ. Press, Princeton, NJ, 1970.

[128] L. Tartar, *Compensated compactness and applications to partial differential equations*, in: Nonlinear Analysis and Mechanics, Heriot-Watt Symposium, Vol. IV, Pitman Res. Notes in Math. 39, Pitman, Boston, MA, 1979, 136–212.

[129] L. Tartar, *An Introduction to Sobolev Spaces and Interpolation Spaces*, Lecture Notes of the Unione Matematica Italiana 3, Springer, Berlin, and UMI, Bologna, 2007.

[130] A. Valli, W. M. Zajączkowski, *Navier-Stokes equations for compressible fluids: global existence and qualitative properties of the solutions in the general case*, Comm. Math. Phys. **103** (1986), 259–296.

[131] K. Yosida, *Functional Analysis*, Springer, Berlin, 1965.

[132] Q. S. Zhang, *Gaussian bounds for the fundamental solutions of* $\nabla(A\nabla u) + B\nabla u - u_t = 0$, Manuscripta Math. **93** (1997), 381–390.

[133] W. P. Ziemer, *Weakly Differentiable Functions*, Springer, New York, 1989.

# Notation

$C(\Omega)$ The Banach space of bounded continuous functions on $\mathrm{cl}\,\Omega$, p. 9

$C_c(\Omega)$ The linear space of continuous functions compactly supported on $\Omega$, p. 9

$C_0(\Omega)$ The space of bounded continuous functions vanishing at $\partial\Omega$ for bounded or unbounded domains, p. 16

$C^{m+\beta}(\Omega)$ The Banach space of functions on $\mathrm{cl}\,\Omega$ which are continuous and bounded along with all derivatives of order $\leq m$, and whose derivatives of order $m$ are Hölder continuous with exponent $\beta \in [0,1)$, p. 9

$C^{m+\beta}(0,T;X)$ The space of $C^{m+\beta}$ functions with values in a Banach space, p. 11

$\int_I f(s)\,dF(s)$ Lebesgue-Stieltjes integral, p. 21

$L^\infty_w(\Omega;\mathcal{M}(\mathbb{R}))$ The space of Young measures, dual to $L^1(\Omega;C_0(\mathbb{R}))$, p. 29

$[f]_{,k}$, $[f]_{m,}$, $[f]_{m,k}$ Mollifiers of $f$, p. 35

$\Omega$ A bounded domain in $\mathbb{R}^d$

$Q = \Omega \times (0,T)$

$S_T = \partial\Omega \times (0,T)$

$\sqcup_T = S_T \cup (\mathrm{cl}\,\Omega \times \{t=0\})$

$\mathbb{S}(\mathbf{u})$ The viscous stress tensor, p. 47

$\mathfrak{C}[f]$ The nonlinear operator of the kinetic equation, p. 182

$J(\Omega)$ The drag functional for stationary flow, p. 300

$dJ(\Omega)[\mathbf{T}]$ The shape derivative of the drag functional in the direction of the field $\mathbf{T}$, p. 372

$W^{l,r}(\Omega)$ A Sobolev space, p. 32

$W^{s,r}(\Omega)$ A fractional Sobolev space, p. 32

$\mathcal{W}^{s,r}_0(\Omega)$ The interpolation space $[L^r(\Omega), W^{1,r}_0(\Omega)]_{s,r}$, p. 33

$\mathcal{W}^{-s,r}(\Omega)$ The dual of $\mathcal{W}_0^{s,r'}(\Omega)$, p. 34

$\mathbb{W}^{-s,r}(\Omega)$ The dual of $W^{s,r'}(\Omega)$, p. 347

$X^{s,r} = W^{s,r}(\Omega) \cap W^{1,2}(\Omega)$ p. 308

$Y^{s,r} = W^{s+1,r}(\Omega) \cap W^{2,2}(\Omega)$ p. 308

$Z^{s,r} = \mathcal{W}^{s-1,r}(\Omega) \cap L^2(\Omega)$ p. 314

$\mathcal{U}^{s,r} = (\mathcal{W}^{s-1,r}(\Omega))^3 \times (W^{s,r}(\Omega))^3 \times \mathbb{R}$ p. 333

$\mathcal{V}^{s,r} = (W^{s+1,r}(\Omega))^3 \times (W^{s,r}(\Omega))^3 \times \mathbb{R}$ p. 333

$\mathcal{E}^{s,r} = (Z^{s,r})^3 \times (X^{s,r})^3 \times \mathbb{R}$ p. 334

$\mathcal{F}^{s,r} = (Y^{s,r})^3 \times (X^{s,r})^3 \times \mathbb{R}$ p. 334

# Index

# Monografie Matematyczne

[1] S. Banach, *Théorie des opérations linéaires*, 1932
[2] S. Saks, *Théorie de l'integrale*, 1933
[3] C. Kuratowski, *Topologie* I, 1933
[4] W. Sierpiński, *Hypothèse de continu*, 1934
[5] A. Zygmund, *Trigonometrical Series*, 1935
[6] S. Kaczmarz, H. Steinhaus, *Theorie der Orthogonalreihen*, 1935
[7] S. Saks, *Theory of the integral*, 1937
[8] S. Banach, *Mechanika*, T. I, 1947
[9] S. Banach, *Mechanika*, T. II, 1947
[10] S. Saks, A. Zygmund, *Funkcje analityczne*, 1948
[11] W. Sierpiński, *Zasady algebry wyższej*, 1946
[12] K. Borsuk, *Geometria analityczna w n wymiarach*, 1950
[13] W. Sierpiński, *Działania nieskończone*, 1948
[14] W. Sierpiński, *Rachunek różniczkowy poprzedzony badaniem funkcji elementarnych*, 1947
[15] K. Kuratowski, *Wykłady rachunku różniczkowego i całkowego*, T. I, 1948
[16] E. Otto, *Geometria wykreślna*, 1950
[17] S. Banach, *Wstęp do teorii funkcji rzeczywistych*, 1951
[18] A. Mostowski, *Logika matematyczna*, 1948
[19] W. Sierpiński, *Teoria liczb*, 1950
[20] C. Kuratowski, *Topologie* I, 1948
[21] C. Kuratowski, *Topologie* II, 1950
[22] W. Rubinowicz, *Wektory i tensory*, 1950
[23] W. Sierpiński, *Algèbre des ensembles*, 1951
[24] S. Banach, *Mechanics*, 1951
[25] W. Nikliborc, *Równania różniczkowe*, Cz. I, 1951
[26] M. Stark, *Geometria analityczna*, 1951
[27] K. Kuratowski, A. Mostowski, *Teoria mnogości*, 1952
[28] S. Saks, A. Zygmund, *Analytic functions*, 1952
[29] F. Leja, *Funkcje analityczne i harmoniczne*, Cz. I, 1952
[30] J. Mikusiński, *Rachunek operatorów*, 1953
* [31] W. Ślebodziński, *Formes extérieures et leurs applications*, 1954
[32] S. Mazurkiewicz, *Podstawy rachunku prawdopodobieństwa*, 1956
[33] A. Walfisz, *Gitterpunkte in mehrdimensionalen Kugeln*, 1957
[34] W. Sierpiński, *Cardinal and ordinal numbers*, 1965
[35] R. Sikorski, *Funkcje rzeczywiste*, 1958
[36] K. Maurin, *Metody przestrzeni Hilberta*, 1959
[37] R. Sikorski, *Funkcje rzeczywiste*, T. II, 1959
[38] W. Sierpiński, *Teoria liczb* II, 1959

* [39] J. Aczél, S. Gołąb, *Funktionalgleichungen der Theorie der geometrischen Objekte*, 1960
[40] W. Ślebodziński, *Formes extérieures et leurs applications*, II, 1963
[41] H. Rasiowa, R. Sikorski, *The mathematics of metamathematics*, 1963
[42] W. Sierpiński, *Elementary theory of numbers*, 1964
* [43] J. Szarski, *Differential inequalities*, 1965
[44] K. Borsuk, *Theory of retracts*, 1967
[45] K. Maurin, *Methods of Hilbert spaces*, 1967
[46] M. Kuczma, *Functional equations in a single variable*, 1967
[47] D. Przeworska-Rolewicz, S. Rolewicz, *Equations in linear spaces*, 1968
[48] K. Maurin, *General eigenfunction expansions and unitary representations of topological groups*, 1968
[49] A. Alexiewicz, *Analiza funkcjonalna*, 1969
* [50] K. Borsuk, *Multidimensional analytic geometry*, 1969
* [51] R. Sikorski, *Advanced calculus. Functions of several variables*, 1969
[52] W. Ślebodziński, *Exterior forms and their applications*, 1971
[53] M. Krzyżański, *Partial differential equations of second order*, vol. I, 1971
[54] M. Krzyżański, *Partial differential equations of second order*, vol. II, 1971
[55] Z. Semadeni, *Banach spaces of continuous functions*, 1971
[56] S. Rolewicz, *Metric linear spaces*, 1972
[57] W. Narkiewicz, *Elementary and analytic theory of algebraic numbers*, 1974
[58] Cz. Bessaga, A. Pełczyński, *Selected topics in infinite dimensional topology*, 1975
* [59] K. Borsuk, *Theory of shape*, 1975
[60] R. Engelking, *General topology*, 1977
[61] J. Dugundji, A. Granas, *Fixed point theory*, 1982
* [62] W. Narkiewicz, *Classical problems in number theory*, 1986

The volumes marked with * are available at the exchange department of the library of the Institute of Mathematics, Polish Academy of Sciences.

MIX
Papier aus verantwortungsvollen Quellen
Paper from responsible sources
FSC® C105338

If you have any concerns about our products,
you can contact us on
**ProductSafety@springernature.com**

In case Publisher is established outside the EU,
the EU authorized representative is:
**Springer Nature Customer Service Center GmbH**
**Europaplatz 3, 69115 Heidelberg, Germany**

Printed by Libri Plureos GmbH
in Hamburg, Germany